Introductory Algebra and Trigonometry with Applications

Introductory Algebra and Trigonometry with Applications

Paul Calter

Professor of Mathematics Emeritus
Vermont Technical College

Visiting Professor of Mathematics
Dartmouth College

Carol Felsinger Rogers

Professor of Mathematics
Vermont Technical College

JOHN WILEY & SONS, INC.

New York • Chichester • Weinheim • Brisbane • Singapore • Toronto

Illustrations: Diphrent Strokes

This book was set in Times Roman and Helvetica by Bi-Comp, Inc. and was printed and bound by Courier Westford. The cover was printed by Phoenix Color Corp. Inc.

The paper in this book was manufactured by a mill whose forest management programs include sustained yield harvesting of its timberlands. Sustained yield harvesting principles ensure that the number of trees cut each year does not exceed the amount of new growth.

This book is printed on acid-free paper.

Library of Congress Cataloging-in-Publication Data
Calter, Paul.
    Introductory algebra and trigonometry with applications / Paul
Calter, Carol Rogers.
        p.   cm.
    Includes index.
    ISBN 0-471-36876-8
    1. Algebra.   2. Trigonometry.   I. Rogers, Carol (Carol F.)
II. Title.
QA152.2.C35   1997
512--dc20   96-41955
                                                            CIP

Printed in the United States of America

10 9 8 7 6 5 4 3

To Margaret, Amy, Michael, Kim,
Rachel, and Christopher
—P.C.

To Jennifer, Rebecca,
and Marvin
—C.F.R.

Brief Contents

Contents

28

29

Preface

Introductory Algebra and Trigonometry with Applications is intended for students at two-year colleges, such as community colleges, technical schools, and technical colleges. It has been designed to meet the needs of technical students, as well as students who will benefit from an applications approach to learning algebra and trigonometry. Applications have been drawn from many different technical and nontechnical areas, such as the electrical, financial, navigation, business, automation, computer, and allied health fields.

This book can also be used for courses in intermediate or college algebra. There is sufficient material provided here for either a one-semester or a two-semester course, depending upon the pace and depth of presentation. The book can also be used at the high school level for applied mathematics.

Given the wide adoption of the graphing calculator in the mathematics classroom, we have thoroughly integrated this powerful tool into the book, both in the text and in the exercises. However, we have still retained most of the manual methods, such as graphing by computing and plotting a table of point pairs, for those classes not using the graphing calculator.

We have also tried to follow as closely as possible the suggestions given in the American Mathematical Association of Two-Year Colleges (AMATYC) Standards for Introductory College Mathematics before Calculus given in *Crossroads in Mathematics,* as well as the National Council of Teachers of Mathematics (NCTM) Standards and Guidelines for Secondary Schools.

Features of the Book

We realize that mathematics books are not easy to read. Thus, we have been careful to present the material as clearly as possible by using an informal style of writing. Careful page layout and numerous illustrations are intended to make the material as easy to follow as possible. We follow the same intuitive approach as in Calter's *Technical Mathematics, Third Edition*, from which this book is adapted, and give information in very small segments. The book also has the following features:

- Examples: Many fully worked examples form the backbone of the textbook, both for mathematical ideas and for applications. They are particularly chosen to help the student do the exercises. Examples are marked above and below to separate them clearly from the text discussion.
- Exercises: As with other skills, practice is essential for learning mathematics, so we include a large number of exercises. Those given after each section are arranged by difficulty and grouped by type to allow practice on a particular area. Exercises are indicated by title, as well as by exercise number, to further help the students find things in the book.
- Chapter Review Exercises: Every chapter ends with a set of Chapter Review Exercises. In contrast to the Exercises, many are scrambled as to type and difficulty, requiring the student to be able to identify type.

- Answers: Answers to selected exercises are provided. This allows for checking of work and also enables the instructor to assign exercises for which no answer is given.
- Summary of Facts and Formulas: All important formulas are boxed and numbered in the text. We also list these formulas in Appendix A, Summary of Facts and Formulas. This listing can provide a common thread between chapters, and we hope it will help students to see interconnections that might otherwise be overlooked. The formulas are grouped logically in the Summary of Facts and Formulas and are numbered sequentially there. Therefore, the formulas do not necessarily appear in the text in numerical order.
- Common Error Boxes: There are many mistakes that students make over and over, from one class to the next. We have identified many of these and noted them in the text as Common Error boxes.
- Graphing Calculator: As mentioned earlier, the graphing calculator has been fully integrated throughout the text. Calculator instructions are given in the text, and calculator problems are given in the exercises. The book is not designed for use with any particular brand of calculator. Actual keystrokes are shown in the earliest chapters, for operations that are essentially similar from one calculator to another. Thereafter, for operations that vary widely between calculators, written descriptions are given.
- Applications: Applications are provided wherever appropriate as motivation for the students. They are drawn from both technical and nontechnical fields. An Index to Applications makes it easier to find an application in a given field.
- Writing Questions: The "writing-across-the-curriculum" movement aims to use writing as an aid to teaching mathematics, as well as other subjects. For those teachers who would like to integrate some writing into their courses, we have included a writing exercise at the end of each chapter. A complete listing of these exercises is found in the Index to Writing Questions in the Appendix.
- Team Projects: A Team Project is given at the end of most chapters for the instructor who would like to try a cooperative learning approach to some units. All the team projects are listed in the Index to Team Projects.
- Instructor's Solution Manual: An Instructor's Solution Manual, prepared by Susan Porter, contains worked-out solutions to even-numbered problems in the text. This most valuable of supplements is available to any instructor using the textbook.

Despite our best efforts, errors will creep in. If you find any, we hope you will contact the authors at Vermont Technical College, Randolph Center, VT 05061.

Acknowledgments

We are grateful to the following colleagues for their valuable reviews of the manuscript: Dr. Tony Banach, ECPI Computer Institute; Aaron Loggins, Albuquerque Technical-Vocational Institute; Lois McBride, Stark Technical College; Larry Moore, Midlands Technical College; Len Mrachek, Hennepin Technical Center; James Newsom, Tidewater Community College; Charles Oster, Sauk Valley Community College; Dale Thielker, Ranken Technical College; Roy Wilson, Ceritos College.

We would like to thank especially Floyd McPhetres for reviewing most of the manuscript and for providing invaluable suggestions, and Michael Calter for his extensive work on the early chapters. To check for accuracy, all the examples and exercises were solved by Jennifer Rogers, with help from Emily Davis and Marlys

Eddy. Writing Exercises and Team Projects were reviewed by Linda Davis and Ginny Richburg. The *Instructor's Solution Manual* was prepared by Susan Porter. We further appreciate the help given during the production phase by Stephen Helba, Carol Robison, and Steve Robb of Prentice Hall, Cindy Gamo of Diphrent Strokes, and Karen Fortgang and Lisa Garboski of *bookworks*.

Paul Calter
Carol Felsinger Rogers

Arithmetic with Whole Numbers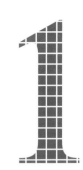

We start our study of mathematics with a review of arithmetic. First we identify the different types of numbers that we will work with in technical mathematics, and then go on to do arithmetic with the simplest kind, the positive whole numbers.

We do addition and subtraction, then multiplication and division, entirely by calculator. We then calculate powers of whole numbers and introduce scientific notation in a simple way. We finally calculate the roots of some whole numbers.

This practice with whole numbers will prepare us for arithmetic with fractions in Chapter 2, and arithmetic with decimal numbers in Chapter 3.

1-1 THE REAL NUMBERS

Integers

The numbers

$$\ldots \ -4, \ -3, \ -2, \ -1, \ 0, \ 1, \ 2, \ 3, \ 4, \ldots$$

are called *integers*. They include zero and negative values as well.

Example 1

Some integers are

$$8, \ 362, \ 3, \ -372, \ 0, \ \text{and} \ 1{,}000{,}000$$

Whole Numbers

The nonnegative integers

$$0, \ 1, \ 2, \ 3, \ldots$$

are called *whole numbers*.

Example 2

Some whole numbers are

$$82, \ 2{,}193, \ 0, \ 5, \ \text{and} \ 1{,}000{,}000$$

Sometimes zero is not included in the definition of the whole numbers, but we will include it in our definition. The whole numbers

$$1, \ 2, \ 3, \ldots$$

are called the *natural numbers*. They do not include zero. The ten whole numbers

$$0, \quad 1, \quad 2, \quad 3, \quad 4, \quad 5, \quad 6, \quad 7, \quad 8, \quad 9$$

are called *digits*.

In this chapter we will do arithmetic with whole numbers only, and we will cover negative numbers in Chapter 6. However, we will show other kinds of numbers now, and tell where in the text we will cover them.

Rational Numbers

The *rational numbers* are all those numbers that can be expressed as the quotient of two integers (except that the denominator may not be zero). The rational numbers include the integers.

Example 3

Some rational numbers are

$$\frac{1}{5}, \quad -\frac{3}{8}, \quad -\frac{27}{13}, \quad \frac{22}{57}, \quad -\frac{3}{5}, \quad 2\frac{1}{3}, \quad \text{and} \quad -4$$

We cover rational numbers in Chapter 2.

Irrational Numbers

Numbers that cannot be expressed as the quotient of two integers are called *irrational*.

Example 4

Some irrational numbers are

$$\sqrt{5}, \quad \sqrt[3]{7}, \quad \text{and} \quad -\sqrt{8}$$

We cover irrational numbers in Chapters 3 and 21.

Real Numbers

The rational and irrational numbers together make up the *real numbers*.

Example 5

Some real numbers are

$$88, \quad -6, \quad \sqrt{72}, \quad 0, \quad -\sqrt{23}, \quad \frac{3}{4}, \quad \text{and} \quad -\frac{55}{57}$$

Imaginary Numbers

Numbers such as $\sqrt{-4}$ do not belong to the real number system. They are called *imaginary numbers*.

Example 6

Some imaginary numbers are

$$\sqrt{-3}, \quad \sqrt{-1}, \quad -\sqrt{-4}, \quad \text{and} \quad \sqrt{-81}$$

Imaginary numbers are discussed in Chapter 29, where we will also cover *complex* numbers. Except when otherwise noted, all the numbers we work with are real numbers.

Decimal Numbers

Most of our computations are with numbers written in the familiar decimal system. We say that the decimal system uses a *base of 10* because it takes 10 units in any place to equal 1 unit in the next higher place.

Example 7

Some decimal numbers are

$$93.2, \quad -29.5, \quad 26, \quad -123, \quad 0.0034, \quad \text{and} \quad -23{,}937.5$$

We will do arithmetic with decimal numbers in Chapter 3.

Place Value

The names of the places relative to the decimal point are shown in Fig. 1-1.

Example 8

What is the place name of the 5 and the 7 in the number 25,471?

Solution: The 5 is in the thousands place, and 7 is in the tens place.

The place names of digits to the right of the decimal point are covered in Chapter 3.

Approximate Numbers

Most of the numbers we deal with in technology are approximate. All numbers that represent measured quantities are approximate.

Example 9

A certain shaft is approximately 2.75 in. in diameter.

Some fractions can be expressed approximately in decimal form.

■ **FIGURE 1-1**
Values of the positions in a decimal number.

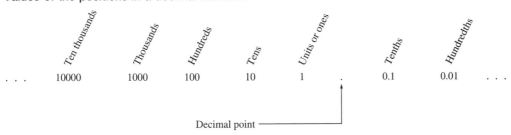

Example 10

$\frac{2}{3}$ is approximately equal to 0.6667.

We will do arithmetic with approximate numbers in Chapter 3.

Exact Numbers

Exact numbers are those that have no uncertainty. Typical exact numbers are those used for counting.

Example 11

A certain classroom contains exactly 32 chairs.

Exact numbers often occur in definitions.

Example 12

There are exactly 24 h in a day, no more, no less.

Exact numbers are usually, but not always, integers.

Example 13

There are exactly 2.54 cm in an inch, by definition.

Not all whole numbers or integers are exact.

Example 14

A certain town has a population of about 23,400 people.

You may assume that all numbers in this chapter are exact.

Number Line

We can graphically represent every real number as a point on a line called the *number line* (Fig. 1-2). It is customary to show the positive numbers to the right of the zero and the negative numbers to the left.

■ **FIGURE 1-2**
The number line.

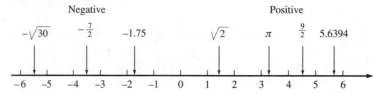

Example 15

Place the following numbers on the number line.

(a) 5   **(b)** -3 **(c)** 2.5   **(d)** $-3\frac{1}{2}$

Solution: The given numbers are shown on the number line of Fig. 1-3.

■ **FIGURE 1-3**

Signs of Equality and Inequality

Several signs are used to show the relative position of two numbers on the number line.

> $a = b$ means that *a equals b,* and that *a* and *b* occupy the same position on the number line.
>
> $a \neq b$ means that *a* and *b* are *not equal* and have different locations on the number line.
>
> $a > b$ means that *a* is *greater than b,* and *a* lies to the right of *b* on the number line.
>
> $a < b$ means that *a* is *less than b,* and *a* lies to the left of *b* on the number line.
>
> $a \geq b$ means that *a* is *equal to or greater than b,* and *a* occupies the same position or lies to the right of *b* on the number line.
>
> $a \leq b$ means that *a* is *equal to or less than b,* and *a* occupies the same position or lies to the left of *b* on the number line.
>
> $a \approx b$ means that *a* is *approximately equal to b* and that *a* and *b* are near each other on the number line.

Example 16

Place the proper equality or inequality sign between $\frac{1}{3}$ and 0.3.

Solution: Since $\frac{1}{3}$ is equal to 0.3333. . . , it is greater than 0.3, so we write

$$\frac{1}{3} > 0.3$$

SECTION 1-1   Exercises   The Real Numbers

Kinds of Numbers

State whether each number is an integer, whole, natural, a digit, rational, irrational, real, imaginary, decimal, positive, or negative. Each number may have more than one name.

1. 7   **2.** -3 **3.** $\dfrac{4}{5}$ **4.** $\sqrt{7}$ **5.** -3.72

6. $\sqrt{-6}$ **7.** $-\dfrac{7}{8}$ **8.** 88.2   **9.** $-\sqrt{82}$ **10.** $2\dfrac{4}{5}$

11. State the place value of the 3 and the 4 in the number 35,841.

12. State the place value of the 1 and the 5 in the number 35,841.

Exact and Approximate Numbers

State whether the number given in each statement is exact or approximate.

13. A person has 10 fingers.

14. A certain table is 7 feet long.

15. The temperature outside is 50 degrees Fahrenheit.

16. A week has seven days.

Number Line

Draw a number line, using a scale to mark divisions from -10 to $+10$. Place each of the following numbers on your number line.

17. 8 **18.** -3 **19.** -6 **20.** 5 **21.** $5\frac{1}{2}$

22. $-\frac{3}{4}$ **23.** 5.4 **24.** 9.7 **25.** -4.6 **26.** -3.2

27. 0 **28.** $-3\frac{1}{4}$

Equality and Inequality Signs

Insert the proper sign of equality or inequality ($=, \approx, <, >$) between each pair of numbers.

29. 8 and 11 **30.** 5 and -3 **31.** -2 and 6

32. -10 and -12 **33.** $\frac{1}{4}$ and 0.25 **34.** $\frac{1}{3}$ and 0.333

1-2

ADDING AND SUBTRACTING WHOLE NUMBERS

Adding Whole Numbers

To add two numbers by calculator, simply enter the first number, press $\boxed{+}$, enter the second number, and press $\boxed{\text{ENTER}}$. (Some calculators use the $\boxed{=}$ key, or $\boxed{\text{EXE}}$, for *execute*, instead of an ENTER key.) The number we get is called the *sum*.

─── **Example 17** ───

Evaluate 2845 + 3273 by calculator.

Solution: There are many types of calculators in use, so the names of the keys and the appearance of the display may be slightly different than shown here. You should, however, be able to use the keystrokes shown here as a guide in operating your own calculator, but be sure to check your operating manual. The keystrokes are

| | 2845 | $\boxed{+}$ | 3273 | $\boxed{\text{ENTER}}$ |
| | | | | 6118 |

Horizontal and Vertical Addition

Numbers to be added may be arranged horizontally or vertically. Of course, when using the calculator it does not make any difference how the numbers are arranged, as they must be keyed in one at a time anyway.

Example 18

The addition of 335, 103, and 224 may be represented horizontally,

$$335 + 103 + 224 = 662$$

or vertically

$$
\begin{array}{r}
335 \\
103 \\
\underline{224} \\
662
\end{array}
$$

Commutative Law

If a and b are two numbers, the commutative law simply states that you can add these numbers in any order.

COMMUTATIVE LAW FOR ADDITION	$a + b = b + a$	1

Example 19

It is no surprise that

$$5 + 2 = 2 + 5$$

Symbols of Grouping

We will often need *symbols of grouping*. They are

( ) parentheses   [ ] brackets   { } braces

We will show the use of these symbols many times as we go along. For now, just remember that the operations *inside* the symbols of grouping must be done before any other operations.

Example 20

In the expression

$$5 - (1 + 3)$$

the 1 and the 3 must be added *before* subtracting from 5.

Associative Law

If a, b, and c are any numbers, the associative law says that you can group numbers to be added in different ways.

ASSOCIATIVE LAW FOR ADDITION	$a + (b + c) = (a + b) + c$	3

* All boxed and numbered formulas are listed in Appendix A, "Summary of Facts and Formulas."

Example 21

Add 1, 3, and 5.

Solution: We might choose to add the 1 and the 3 first,

$$1 + 3 + 5 = (1 + 3) + 5$$
$$= 4 + 5$$
$$= 9$$

Or we might want to add the 3 and the 5 first.

$$1 + 3 + 5 = 1 + (3 + 5)$$
$$= 1 + 8$$
$$= 9$$

The commutative and associative laws combined say that it doesn't matter in what order or in what groupings we add the given numbers. We could even choose to add the 1 and the 5 first.

$$1 + 3 + 5 = (1 + 5) + 3$$
$$= 6 + 3$$
$$= 9$$

We get the same result no matter which way we order and group the numbers.

Subtracting Whole Numbers

The keystrokes for subtraction are the same as for addition, except that the $\boxed{-}$ key is used instead of $\boxed{+}$.

Example 22

Evaluate $8361 - 6582$.

Solution: The keystrokes are

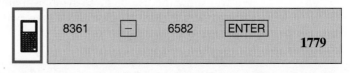

	8361	$\boxed{-}$	6582	$\boxed{\text{ENTER}}$
				1779

> **COMMON ERROR** Many calculators have a $\boxed{(-)}$ key, which looks almost like the $\boxed{-}$ key. The $\boxed{(-)}$ key is used to enter negative numbers. Do not use it for subtraction.

The commutative law *does not apply* to subtraction. In other words, the order in which we subtract terms *is* important.

Example 23

In addition,	$3 + 2$ is equal to	$2 + 3$
In subtraction,	$3 - 2$ is *different* from	$2 - 3$

The associative law also does not apply to subtraction. In other words, the way in which terms are grouped for subtraction *is* important.

Example 24

Show that $7 - (5 - 2)$ is *different* from $(7 - 5) - 2$.

Solution: As we said earlier, we evaluate the quantities inside the parentheses first. So

$$7 - (5 - 2) = 7 - 3 = 4$$

but

$$(7 - 5) - 2 = 2 - 2 = 0$$

Example 25

Evaluate the expression $16 - [(7 - 2) - (6 - 3)]$.

Solution: We start by removing the inner parentheses.

$$16 - [(7 - 2) - (6 - 3)] = 16 - [5 - 3]$$
$$= 16 - 2$$
$$= 14$$

Parentheses on the Calculator

To evaluate an expression containing parentheses on the calculator, we use the $($ key and the $)$ key.

Example 26

Evaluate $82 - (45 - 27)$ by calculator.

Solution: The keystrokes are

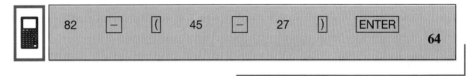

Applications

Problems in mathematics are often stated in verbal form. We discuss word problems in detail in Chapter 11, but we can start now to sharpen our skills by doing simple verbal problems in arithmetic. When doing the exercise, look for words that indicate addition (*sum, increase, gain, accumulation,* etc.), for words that indicate subtraction (*difference, loss, decrease, deduction,* etc.), and for words that stand for equality (*equals, is, are, amounts to,* etc.).

Example 27

A bank account having a balance of $5832 received two deposits of $359 and $38, followed by a withdrawal of $836. Find the new balance.

Solution: Do not hesitate to use a dictionary if you are unfamiliar with the banking terms *balance, deposit,* and *withdrawal.* It should be clear that we must add the deposits to the initial balance, and subtract the withdrawal.

$$\text{New balance} = \$5832 + \$359 + \$38 - \$836$$
$$= \$5393$$

**Adding
Whole Numbers**

Add by calculator.

1.	874	**2.**	364	**3.**	837	**4.**	855

1. 874
        647

2. 364
        542

3. 837
        894

4. 855
        627
        642

5. 6773
        9354
         659

6. 945
        399
        231

7. 7764
        5056
         693
         943
         701
         109
         791

8. 85,326
         8,654
           754
           193
           490
         2,100
           923

9. 63,048
        84,336
         5,487
           902
         1,084
            74
         9,372

10. 76,109
          8,533
            863
         12,959
         10,263
          1,472
         62,923

11. 74,352
          6,475
            496
            379
          5,258
            974

12. 7325
          539
          734
           94
         3749
          741

13. 537 + 829 + 274 + 826

14. 923 + 263 + 724

15. 382 + 730 + 263 + 274

16. 482 + 428 + 284 + 836

**Subtracting
Whole Numbers**

Subtract by calculator.

17.   724
        −384

18. 8263
        −7253

19.   294
        −163

20.   724
        −267

21. 737 − 274

22. 823 − 243

23. 982 − 230 − 263

24. 782 − 328 − 184

**Symbols
of Grouping**

Combine as indicated.

25. 3957 − (284 − 163)

26. 737 − (406 − 264) + 517

27. (634 − 372) − (364 − 164)

28. (847 − 274) − 472

29. (36 − 27) + 83 − (83 − 53)

30. 92 − (34 − 27) + (27 − 18)

31. A house and lot sold together for $50,523, which was $8394 more than its original cost. What was its original cost?

Applications

32. Find the perimeter (the distance around) of a four-sided figure (a quadrilateral) if the lengths of the four sides are 17 ft, 16 ft, 23 ft, and 8 ft.

33. The sum of the sides of a triangle is 734, and two of the sides are 372 and 263. Find the length of the third side.

34. Four cartons, having heights of 34 in., 51 in., 29 in., and 38 in., are placed on top of each other on a skid having a height of 4 in. What is the total height of cartons and skid?

1-3   MULTIPLYING WHOLE NUMBERS

Factors and Product

The numbers we multiply to get a product are called factors.

$$3 \times 5 = 15$$

factors ⟋⟍   ⟍ product

Symbols for Multiplication

The symbols used for multiplication are the times symbol ($\times$), the multiplication dot ($\cdot$), and parentheses.

┌─── **Example 28** ───────────

The product of 3 and 5 can be written

$$3 \times 5, \quad 3 \cdot 5, \quad 3(5), \quad (3)5, \quad \text{or} \quad (3)(5)$$

Notice that the multiplication dot looks almost like a decimal point, and that the times symbol can be mistaken for an x. We avoid these whenever there is a chance of their being confused.

To show the multiplication of two letter quantities (numbers represented by letters) like a and b, or a number and a letter quantity, we simply write the two quantities with no symbol between them.

┌─── **Example 29** ───────────

The expression ab means a times b.
The expression $5a$ means 5 times a.

Multiplying by Calculator

Many calculators use an $\boxed{\times}$ key for multiplication, while some use a $\boxed{*}$ key. The keystrokes used to multiply the factors 13 and 27 are

Commutative Law

If a and b are two numbers, the commutative law for multiplication states that the order of the factors in multiplication is not important.

COMMUTATIVE LAW FOR MULTIPLICATION	$ab = ba$	2

┌─── **Example 30** ───────────

It is no surprise that

$$2 \times 3 = 3 \times 2$$

Associative Law

The associative law for multiplication allows us to group the numbers to be multiplied in different ways.

ASSOCIATIVE LAW FOR MULTIPLICATION	$a(bc) = (ab)c$	4

Example 31

Show that $2(3 \times 4)$ equals $(2 \times 3)4$ and also equals $(2 \times 4)3$.

Solution:
$$2(3 \times 4) = 2(12) = 24$$
$$(2 \times 3)4 = (6)4\ = 24$$
$$(2 \times 4)3 = (8)3\ = 24$$

This last calculation makes use of the commutative law as well as the associative law. We see that the result is the same regardless of which factors we multiply first.

Distributive Law

The distributive law tells us that

DISTRIBUTIVE LAW	$a(b + c) = ab + ac$	5

Example 32

Evaluate $5(3 + 6)$.

Solution: We will do the calculation two different ways.

(a) Let us combine the numbers inside the parentheses first.
$$5(3 + 6) = 5(9) = 45$$

(b) The distributive law enables us to do the same computation in a different way. We can multiply both the 3 and the 6 by 5 and then add.
$$5(3 + 6) = 5(3) + 5(6) = 15 + 30$$
$$= 45 \quad \text{as before}$$

Multiplying by Zero

The product of zero and any number is zero.

Example 33

(a) $56(0) = 0$
(b) $0(385) = 0$
(c) $274(0)(837) = 0$

Applications

Example 34

The power in watts (W) delivered to a device is equal to the product of the current in the device in amperes (A), and the voltage across the device in volts (V). Find the power consumed by a motor that draws 3 A at a voltage of 115 V.

Solution: Multiplying,
$$\text{Power} = \text{current} \times \text{voltage}$$
$$= (3\ \text{A})(115\ \text{V}) = 345\ \text{W}$$

Multiply by calculator.

1. 4853×429
2. 780×347
3. 752×32
4. $863 \times 796 \times 534$
5. 96×4278
6. 322×64
7. 42×862
8. 392×83
9. $49 \times (6 \times 3) \times 8$
10. $(9 \times 89) \times (48 \times 8)$
11. $(836 + 726) \times 14$
12. $74 \times (47 + 53) \times 7$

Applications

13. What is the cost of 23 gallons of paint at $13 per gallon?

14. If 26 tons of gravel are needed for each mile of roadway, how many tons will be required for 125 miles?

15. Two trucks carry 3 tons of asphalt each and a third carries 4 tons. What is the value of the whole, at $33 per ton?

16. How much will 455 brackets weigh if each weighs 13 grams?

17. If there are 360 degrees per revolution, how many degrees are there in 9 revolutions?

1-4

DIVIDING WHOLE NUMBERS

Definitions

The *dividend,* when divided by the *divisor,* gives us the *quotient:*

$$\text{dividend} \div \text{divisor} = \text{quotient}$$

or

$$\frac{\text{dividend}}{\text{divisor}} = \text{quotient}$$

A quantity a/b is also a *fraction.* The fraction line indicates division. The quantity a/b can also be referred to as the *ratio* of a to b. Fractions and ratios are treated in Chapter 2.

Dividing by Calculator

To divide by calculator, enter the dividend, then press $\boxed{\div}$, then enter the divisor and press $\boxed{\text{ENTER}}$, $\boxed{=}$, or $\boxed{\text{EXE}}$, depending on your particular calculator.

┌─── **Example 35** ───
│ To divide 1305 by 145 on a calculator, the keystrokes are

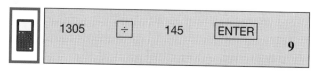

When we added, subtracted, or multiplied two integers, we always got an integer for an answer. This is not always the case when dividing.

Example 36

When we divide 1 by 3, we get 0.33333333333. . . . Here we must choose how many digits we wish to retain. We must also *round* our answer to the chosen number of digits. We cover rounding in Chapter 3.

Zero

Zero divided by any quantity (except zero) is zero.

Example 37

(a) $0 \div 48 = 0$

(b) $0 \div (824 + 293) = 0$

However, division by zero is not defined. It is an illegal operation in mathematics.

Example 38

(a) $48 \div 0$ is undefined

(b) $(284 + 194) \div 0$ is undefined

(c) $0 \div 0$ is undefined

Example 39

Using your calculator, divide 5 by zero.

Solution: The keystrokes are

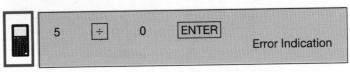

SECTION 1-4   Exercises   Dividing Whole Numbers

Divide by calculator.

1. $2496 \div 832$ 2. $2002 \div 91$ 3. $456 \div 38$

4. $13,893 \div 421$ 5. $2715 \div 543$ 6. $5550 \div 75$

7. $0 \div 7$ 8. $0 \div 2734$ 9. $72 \div 0$

10. $734 \div 0$

Applications

11. How many cubic feet of concrete are needed to make a counterweight weighing 432 lb if the density of concrete is 144 lb/ft³? (Hint: Volume = weight ÷ density)

12. A 2247-acre parcel of land is to be divided into seven equal lots. Find the area of each lot.

13. At the rate of 330 miles in 6 hours, how many miles would a person drive in 9 hours? (Hint: Rate = distance ÷ time)

14. If a factory can assemble 1704 computers in 8 hours, how many are assembled in 3 hours? (Hint: Work rate = amount done ÷ time worked)

15. A voltage of 124 V is measured across a 31-Ω resistor. Find the current through the resistor. (Hint: Current = voltage ÷ resistance)

16. A steel flywheel (density = 450 lb/ft³) weighs 225 lb. Find the volume of the flywheel.

1-5 POWERS AND ROOTS OF WHOLE NUMBERS

Definitions

In the expression

$$2^4$$

The number 2 is called the *base* and the number 4 is called the *exponent*. The expression is read "two to the fourth power." Its value is the product of four 2's.

$$2^4 = (2)(2)(2)(2) = 16$$

Example 40

$$4^3 = (4)(4)(4) = 64$$

Powers by Calculator

To square a number on the calculator, simply enter the number and press the $\boxed{x^2}$ key. Some calculators will give the answer immediately, while on others you will need to press $\boxed{\text{ENTER}}$, $\boxed{=}$, or $\boxed{\text{EXE}}$, as in the following example.

Example 41

To square 5, the keystrokes are

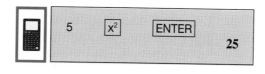

Many calculators have an $\boxed{x^3}$ key for cubing a number. If you do not have one, or to raise a number to other powers, use the *power key*, marked $\boxed{\wedge}$, $\boxed{y^x}$, or $\boxed{x^y}$. We'll use $\boxed{\wedge}$ in the examples. Enter the base, press the power key, then enter the exponent. Finally press the $\boxed{\text{ENTER}}$ or $\boxed{\text{EXE}}$ key to get the result.

Example 42

The keystrokes for finding 5^4 are

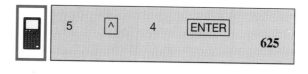

The exponents we have shown were all positive integers. A number may also be raised to a negative exponent like -3, a fractional exponent like $\frac{2}{3}$, or a decimal exponent like 1.53. We explain the meaning of such exponents and show how to calculate with them in later chapters.

Applications

Powers must often be calculated in technical work. Here we do an example from electrical technology.

Example 43

The power (in watts) developed in a resistor is equal to the product of the resistance R (in ohms) and the square of the current I (in amperes). Find the power when a current of 17 amperes flows through a resistance of 3 ohms.

Solution: The power P is equal to

$$P = RI^2$$
$$= (3)(17)^2 = 867 \text{ watts}$$

by calculator.

Root of a Whole Number

If $a^n = b$, then

$$\sqrt[n]{b} = a$$

which is read "the nth root of b equals a." The symbol $\sqrt{}$ is a *radical sign*, b is the *radicand*, and n is the *index* of the radical.

If no index is written, it is assumed to be 2, indicating a *square root*. Thus

$$\sqrt{b} \text{ is the same as } \sqrt[2]{b}$$

Example 44

(a) $\sqrt{9} = 3$ because $3^2 = 9$
(b) $\sqrt[3]{27} = 3$ because $3^3 = 27$
(c) $\sqrt[4]{16} = 2$ because $2^4 = 16$

Roots by Calculator

To find square roots on some calculators, we press the $\boxed{\sqrt{x}}$ (or $\boxed{\sqrt{}}$) key first, enter the number, then press $\boxed{\text{ENTER}}$. On other calculators we enter the number and then press the $\boxed{\sqrt{x}}$ or $\boxed{\sqrt{}}$ key.

Example 45

Find $\sqrt{625}$ by calculator.

Solution: The keystrokes are

or

Many calculators have a $\boxed{\sqrt[3]{\ }}$ key for finding cube roots. If your calculator does not have that key, or to find other roots, we use a *root key,* marked $\boxed{\sqrt[x]{x}}$, $\boxed{\sqrt[y]{y}}$, or something similar. We will use $\boxed{\sqrt[y]{\ }}$ in the examples.

Example 46

Find $\sqrt[4]{81}$, the fourth root of 81.

Solution: The keystrokes are

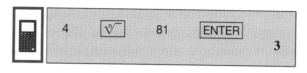

Example 47

A cubical block of concrete has a volume of 6859 cubic inches (in.³). Find the length of one side of the block by taking the cube root of the volume.

Solution: By calculator,

$$\text{Length of one side} = \sqrt[3]{6859} = 19 \text{ in.}$$

SECTION 1-5   Exercises   Powers and Roots of Whole Numbers

Powers by Calculator

Square by calculator.

1. 3^2	**2.** 4^2	**3.** 2^2	**4.** 8^2
5. 11^2	**6.** 14^2	**7.** 23^2	**8.** 21^2

Cube by calculator.

9. 4^3	**10.** 6^3	**11.** 5^3	**12.** 7^3

Evaluate each power by calculator.

13. 5^4	**14.** 3^5	**15.** 2^7	**16.** 4^5
17. 16^2	**18.** 13^3	**19.** 8^3	**20.** 4^7

Roots by Calculator

Evaluate each square root by calculator.

21. $\sqrt{16}$	**22.** $\sqrt{36}$	**23.** $\sqrt{2025}$	**24.** $\sqrt{3721}$

Evaluate each root by calculator.

25. $\sqrt[3]{27}$	**26.** $\sqrt[3]{8}$	**27.** $\sqrt[4]{16}$	**28.** $\sqrt[4]{81}$
29. $\sqrt[5]{32}$	**30.** $\sqrt[5]{1024}$		

Applications

31. Find the power developed by a current of 23 amperes in a resistance of 5 ohms (see Example 43).

32. The number of feet traveled by a falling object in t seconds is approximately equal to $16t^2$. Find the distance traveled in 11 seconds.

33. Find the length of the side of a cube of steel whose volume is 2744 cubic centimeters (cm³). See Example 47.

34. The period (time in seconds for one swing) of a certain pendulum is equal to $\sqrt[2]{289}$. Find this period.

SCIENTIFIC NOTATION

Scientific notation provides a convenient way to represent and to calculate with very large or very small numbers. In this section we give a brief introduction only. We will learn how to convert numbers to and from scientific notation. Our main treatment will come in Chapter 3, where we will do computations in scientific notation.

Powers of Ten

The number 10 raised to a power is called a *power of ten*. Powers of ten are found the same way as powers of any other numbers.

Example 48
(a) $10^3 = 10 \times 10 \times 10 = 1000$
(b) $10^4 = 10 \times 10 \times 10 \times 10 = 10,000$

Summarizing some of the powers of ten in a table:

Powers of Ten
$10^6 = 1,000,000$
$10^5 = 100,000$
$10^4 = 10,000$
$10^3 = 1,000$
$10^2 = 100$
$10^1 = 10$
$10^0 = 1$

Scientific Notation

The numbers needed for technical work are often very large or very small. For example, a certain computer might have a memory of 200,000,000 bits, and might do a certain operation in 0.0000000004 seconds. Numbers such as these, clumsy to express in decimal notation, are written more easily in *scientific notation*. Calculators and computers switch into scientific notation as needed, or when made to do so.

In this section we learn how to write whole numbers in scientific notation. We'll see that the first number above can be written 2×10^8 bits, and in Chapter 6 we show how to write the second number as 4×10^{-10} seconds.

A number is said to be in *scientific notation* when it is written as a number between 1 and 10, multiplied by a power of ten.

Example 49
Some numbers written in scientific notation are

(a) 7×10^6 **(b)** 5×10^{15} **(c)** 6×10^3 **(d)** 3×10^5

Converting Numbers to Scientific Notation

To convert a number to scientific notation:

1. Write the given number as the product of a number between 1 and 10, times another number that is a multiple of ten.
2. Rewrite the multiple of ten as a power of ten.

┌─── **Example 50** ────────

Convert 30,000 to scientific notation.

Solution:

1. We write 30,000 as the product of 3 and 10,000.

$$30{,}000 = 3 \times 10{,}000$$

2. Then we rewrite 10,000 as a power of ten.

$$10{,}000 = 10^4$$

so

$$30{,}000 = 3 \times 10^4$$

┌─── **Example 51** ────────

Convert 1,840,000 to scientific notation.

Solution:

1. We write 1,840,000 as the product of 1.84 and 1,000,000.

$$1{,}840{,}000 = 1.84 \times 1{,}000{,}000$$

2. Then we rewrite 1,000,000 as a power of ten.

$$1{,}000{,}000 = 10^6$$

so

$$1{,}840{,}000 = 1.84 \times 10^6$$

Converting from Scientific Notation

To convert from scientific notation, simply reverse the process.

┌─── **Example 52** ────────

Write 9×10^6 as a whole number.

Solution: Using the fact that

$$10^6 = 1{,}000{,}000$$

we get

$$9 \times 10^6 = 9 \times 1{,}000{,}000$$
$$= 9{,}000{,}000$$

SECTION 1-6   Exercises   Scientific Notation

Powers of Ten

Evaluate each power of ten.

1. 10^5 2. 10^2 3. 10^1 4. 10^0

5. 10^4 6. 10^6 7. 10^7 8. 10^9

Write each number as a power of ten.

9. 1000   10. 100,000   11. 10,000   12. 100

Converting Numbers to Scientific Notation

Convert each number to scientific notation.

13. 30,000   **14.** 900,000   **15.** 2,000,000

16. 800   **17.** 7000   **18.** 30,000,000

Converting from Scientific Notation

Convert each number from scientific notation.

19. 3×10^3 **20.** 8×10^5 **21.** 2×10^2

22. 9×10^4 **23.** 7×10^6 **24.** 5×10^7

1-7

COMBINED OPERATIONS

Order of Operations

If the expression to be evaluated does not contain parentheses, perform the operations in the following order:

1st   powers and roots, in any order
2nd   multiplications and divisions, from left to right
3rd   additions and subtractions, from left to right

We first show a problem containing both addition and multiplication.

Example 53

Evaluate $7 + 3 \times 4$ first without a calculator, and then by calculator.

Solution: The multiplication is done before the addition

$$7 + 3 \times 4 = 7 + 12 = 19$$

A calculator will automatically perform the multiplication before the addition, so we may key the numbers in as written.

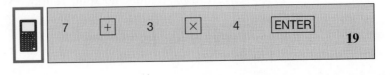

Next we do a calculation having both a power and multiplication.

Example 54

Evaluate 5×3^2.

Solution: We raise to the power before multiplying:

$$5 \times 3^2 = 5 \times 9 = 45$$

A calculator will do the squaring operation on the 3 only, so there is no need to enter the 3^2 operation first. The keystrokes are

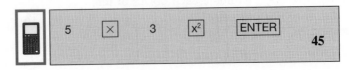

Parentheses

When an expression contains parentheses, first evaluate the expression within the parentheses and then the entire expression.

Example 55

Evaluate $(7 + 3) \times 4$.

Solution:

$$(7 + 3) \times 4 = 10 \times 4 = 40$$

The keystrokes are

Note that the use of parentheses was essential here.

If the sum or difference of more than one number is to be raised to a power, those numbers must be enclosed in parentheses.

Example 56

Evaluate $(5 + 2)^2$.

Solution: We combine the numbers inside the parentheses before squaring.

$$(5 + 2)^2 = 7^2 = 49$$

There are two ways to do this on a calculator. The first way is to use the $\boxed{\text{ENTER}}$, $\boxed{=}$, or $\boxed{\text{EXE}}$ key to combine the 5 and the 2, before squaring. The keystrokes are

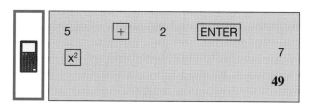

A second way is to use parentheses. The keystrokes are

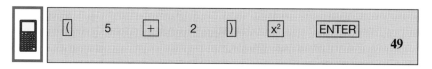

On some calculators the result will be displayed immediately after pressing the $\boxed{x^2}$ key.

Example 57

Evaluate $(2 + 6)(7 + 9)$.

Solution: The two quantities in parentheses are evaluated before multiplying:

$$(2 + 6)(7 + 9) = 8 \times 16 = 128$$

The keystrokes are

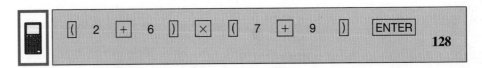

On some calculators you do not need to use the $\times$ key between the two expressions.

Example 58

Evaluate $\dfrac{8 + 4}{9 - 3}$.

Solution: Here the fraction line acts like parentheses, grouping the 8 and the 4, as well as the 9 and the 3. Written on a single line, this problem would be

$$(8 + 4) \div (9 - 3)$$

or $12 \div 6 = 2$.

The calculator keystrokes are almost the same as for the preceding example. Enter the numerator first and use the $\div$ key instead of the $\times$ key.

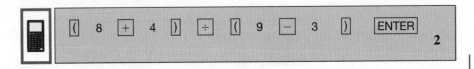

SECTION 1-7 Exercises Combined Operations

Perform each computation by calculator.

1. $(37)(28) + (36)(64)$

2. $(22)(53) - (586)(4) + (47)(59)$

3. $(63 + 36)(97 - 37)$

4. $(89 - 74 + 95)(87 - 49)$

5. $\dfrac{219}{73} + \dfrac{194}{97}$

6. $\dfrac{228}{38} - \dfrac{78}{26} + \dfrac{364}{91}$

7. $\dfrac{648 + 699}{337 + 112}$

8. $\dfrac{809 - 463 + 1858}{958 - 364 + 508}$

9. $(5 + 6)^2$

10. $(422 + 113 - 533)^4$

11. $(423 - 420)^3$

12. $\left(\dfrac{853 - 229}{874 - 562}\right)^2$

13. $\left(\dfrac{141}{47}\right)^3$

14. $\sqrt{434 + 466}$

15. $\sqrt{(8)(72)}$

16. $\sqrt[3]{657 + 553 - 1085}$

17. $\sqrt[4]{(27)(768)}$

18. $\sqrt{\dfrac{2404}{601}}$

19. $\sqrt[4]{\dfrac{1136}{71}}$

20. $\sqrt{961} + \sqrt{121}$

21. $\sqrt[4]{625} + \sqrt{961} - \sqrt[3]{216}$

22. $\sqrt[4]{256} \times \sqrt{49}$

Perform each computation by calculator.

1. 21^2 **2.** $137 + 29 + 474 + 226$

3. $(947 - 324) - 172$ **4.** 52×12 **5.** $237 - 174$

6. 5^3 **7.** $147 \div 21$ **8.** $(13 + 96)(57 - 23)$

9. $0 \div 4$ **10.** 3^4

11. $537 - (206 - 164) + 317$ **12.** $(934 - 242) - (564 - 364)$

13. 2^5 **14.** 4^6 **15.** 25^2

16. $(81)^2$ **17.** $\sqrt{49}$ **18.** $19 \times (5 \times 6) \times 3$

19. $\sqrt[3]{125}$ **20.** $\sqrt[3]{8}$ **21.** $822 - 130 - 133$

22. $5757 - (584 - 233)$ **23.** $36 \times 11 \times 53$ **24.** $(4 \times 29) \times (22 \times 5)$

25. $2552 \div 232$ **26.** $1245 \div 83$ **27.** $44 \div 0$

28. $(7 + 2)^2$ **29.** $(553 - 546)^3$

30. $(22)(74) + (24)(74)$ **31.** $\dfrac{176}{22} + \dfrac{64}{16}$

32. $\sqrt{2209}$ **33.** $\sqrt{(9)(49)}$

34. $\sqrt[4]{625} \times \sqrt{81}$ **35.** $\dfrac{947 + 513}{137 + 228}$

Evaluate each power of ten.

36. 10^3 **37.** 10^4 **38.** 10^5

39. State the place value of the 5 and the 6 in the number 62,571.

Convert each number from scientific notation.

40. 73×10^3 **41.** 4×10^5 **42.** 2×10^4

Place each of the following numbers on a number line.

43. $\dfrac{5}{8}$ **44.** -9 **45.** $\sqrt{7}$

Convert each number to scientific notation.

46. 50,000   **47.** 300,000   **48.** 7,000,000

Insert the proper sign of equality or inequality ($=$, $<$, $>$) between each pair of numbers.

49. 2 and 3   **50.** -7 and -8 **51.** $\dfrac{6}{3}$ and 2

Write each number as a power of ten.

52. 1000   **53.** 100,000   **54.** 1,000,000

55. How many cubic feet of steel are needed to make a casting weighing 2700 lb if the density of steel is 450 lb/ft³? (Hint: Volume = weight ÷ density)

56. A 417-acre parcel of land is to be divided into three equal lots. Find the area of each lot.

57. At the rate of 329 miles in 7 hours, how many miles would a person drive in 4 hours? (Hint: Rate = distance ÷ time)

58. If a person can make 21 lamps in 7 days, how many can be made in 5 days? (Hint: Work rate = amount done ÷ time worked)

59. A voltage of 219 V is measured across a 73-Ω resistor. Find the current through the resistor. (Hint: Current = voltage ÷ resistance)

60. Find the perimeter of a four-sided figure (a quadrilateral) if the lengths of the four sides are 22 cm, 15 cm, 33 cm, and 18 cm.

61. The sum of the sides of a triangle is 224, and two of the sides are 77 and 103. Find the length of the third side.

62. Four boards having lengths of 134 in., 31 in., 229 in., and 308 in. are placed end to end. What is the total length?

63. What is the cost of 47 tons of gravel at $16 per ton?

64. How much will 37 fans weigh if each weighs 4 lb?

65. If there are 360 degrees per revolution, how many degrees are there in 11 revolutions?

66. The number of feet traveled by a falling object in t seconds is approximately equal to $16t^2$. Find the distance traveled in 21 seconds.

67. Find the length of the side of a cubical shipping container whose volume is 216 cubic feet.

WRITING*

68. Suppose someone tells you, "Of all the kinds of numbers, it's most important to understand the difference between exact and approximate numbers." Do you agree? If so, why is this distinction so important? And just what is the difference between exact and approximate numbers? Write a paragraph or so in which you discuss these questions, and file it with your notes on this chapter.

* Many of the writing questions in this book are from P. Calter, *Technical Mathematics, Third Edition* (Upper Saddle River, NJ: Prentice Hall, 1995).

Common Fractions

While the use of calculators and computers, which both work with decimal numbers, have reduced the use of common fractions, it is still important to study them. Fractions are used in many places, such as measurements to fractions of an inch, ratio, and proportion. But more importantly, it will help prepare us for a study of algebraic fractions in Chapter 17, which cannot be handled by most calculators or computers.

2-1   TERMINOLOGY

Numerator and Denominator

Every fraction has a *numerator,* a *denominator,* and a *fraction line.* The numerator is written above the fraction line and the denominator is written below the fraction line.

Example 1

In the fraction $\frac{2}{3}$,

the numerator is 2 and the denominator is 3. You will often see fractions in printed text written with a slanted fraction line, to save space. Thus the fraction in this example may be printed as 2/3. However, we suggest that your handwritten fractions should always be of the *built-up* type, with a horizontal fraction line, for clarity.

■ **FIGURE 2-1**
Three slices of pizza.

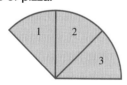

We may think of a fraction as a way to specify a portion of a whole quantity. Thus the fraction $\frac{2}{3}$ says, "*Divide the whole into 3 equal parts, and take 2 of them.*"

Here the denominator 3 *denominates* or *names* the size of each part, here thirds.
The numerator 2 *enumerates* or *counts* the number of thirds to be taken.

Example 2

We may think of $\frac{3}{8}$ of a pizza, Fig. 2-1, as three pieces of the pizza after it has been divided into 8 equal pieces.

A Fraction as Indicating Division

Another way of thinking of a fraction is as the indicated division of two quantities.

Example 3

The fraction $\frac{5}{9}$ may be thought of as 5 divided by 9. It is also called the *quotient* of 5 divided by 9. Thought of in this way, 5 is the *dividend,* 9 is the *divisor,* and the fraction line is a symbol for division.

Fraction Line as a Symbol of Grouping

We have mentioned symbols of grouping such as parentheses ( ), brackets [ ], and braces { }. A *fraction line* is also a symbol of grouping.

Example 4

In the fraction

$$\frac{9}{2+5}$$

the 9 in the numerator cannot be divided by the 2 or by the 5 separately, because they are *grouped* by the fraction line. The 2 and the 5 must be added together before dividing.

Zero Denominator or Numerator

Recall from Chapter 1 that division by zero is not defined. That means that the denominator of a fraction may not be zero.

Example 5

The fraction

$$\frac{5}{0}$$

is not defined.

A zero *numerator,* of course, gives a fraction that is equal to zero.

Example 6

The value of the fraction

$$\frac{0}{5}$$

is zero.

Common Fractions

A *common fraction* is one in which numerator and denominator are both *integers*.

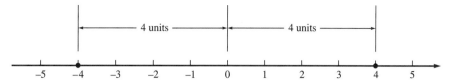

■ FIGURE 2-2
Both +4 and −4 have an absolute value of 4, because both are 4 units from zero.

┌─── **Example 7** ───────
│ The fractions

$$\frac{3}{5}, \quad -\frac{9}{2}, \quad \text{and} \quad \frac{121}{125}$$

are common fractions. The fraction $\dfrac{2.97}{5}$ is *not* a common fraction because numerator and denominator are not both integers. The fractions

$$\frac{x}{2}, \quad \frac{7}{y}, \quad \text{and} \quad \frac{a}{b}$$

are *not* common fractions. They are called *algebraic* fractions. We study them in Chapter 17.

Proper and Improper Fractions

To define proper and improper fractions, we must first say what we mean by the absolute value. The *absolute value* of a number is, geometrically, the distance of the number from zero on the number line, regardless of direction. Thus, +4 and −4 are both 4 units from zero in Fig. 2-2. Thus the absolute value of +4 is 4, and the absolute value of −4 is 4.

A *proper fraction* is one whose numerator is smaller in absolute value than its denominator, while an *improper fraction* is one whose numerator is larger in absolute value than or equal to its denominator.

┌─── **Example 8** ───────
│ Some proper fractions are

$$\frac{2}{3}, \quad \frac{71}{-92}, \quad \text{and} \quad \frac{14}{15}$$

and some improper fractions are

$$\frac{3}{2}, \quad \frac{-32}{21}, \quad \frac{5}{4}, \quad \text{and} \quad \frac{15}{14}$$

Mixed Numbers

A *mixed number* is the sum of a whole number and a proper fraction. Such a sum is written without the + sign. Thus,

$$2 + \frac{3}{4} \text{ is written as } 2\frac{3}{4}.$$

Example 9

Some mixed numbers are

$$3\frac{1}{2}, \quad 72\frac{22}{35}, \quad \text{and} \quad 14\frac{3}{4}$$

Similar Fractions

Similar or *like* fractions are fractions having the same denominator.

Example 10

$$\frac{8}{9} \quad \text{and} \quad \frac{2}{9} \quad \text{are similar fractions}$$

while

$$\frac{3}{5} \quad \text{and} \quad \frac{3}{7} \quad \text{are } not \text{ similar fractions}$$

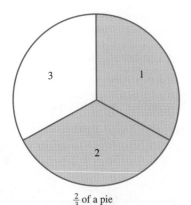

$\frac{2}{3}$ of a pie

Equivalent (Equal) Fractions

Fractions are called *equivalent* (or equal) if they represent the same value.

Example 11

The fractions

$$\frac{2}{3} \quad \text{and} \quad \frac{6}{9}$$

are equivalent fractions. If you ate $\frac{2}{3}$ of a pie or $\frac{6}{9}$ of that pie (Fig. 2-3), it would be the same amount of pie.

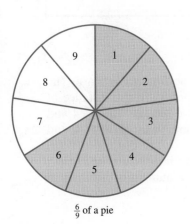

$\frac{6}{9}$ of a pie

■ **FIGURE 2-3**
Two-thirds of a pie.

Complex Fractions

A *complex fraction* is one with more than one fraction line.

Example 12

Some complex fractions are

$$\frac{\frac{3}{5}}{12}, \quad \frac{5}{1+\frac{4}{7}}, \quad \text{and} \quad \frac{\frac{5}{6}-\frac{7}{8}}{\frac{2}{3}+\frac{4}{5}}$$

Label each as a common fraction, algebraic fraction, proper fraction, improper fraction, mixed number, or complex fraction. More than one name may apply.

1. $\dfrac{4}{5}$ **2.** $\dfrac{2}{7}$ **3.** $\dfrac{6}{5}$ **4.** $1\dfrac{3}{4}$

5. $\dfrac{x}{y}$ **6.** $\dfrac{\frac{5}{7}}{3}$ **7.** $\dfrac{12}{7}$ **8.** $\dfrac{z}{2}$

9. $\dfrac{\frac{w}{3}+5}{\frac{x}{4}}$ **10.** $\dfrac{5}{2}$ **11.** $7\dfrac{5}{6}$ **12.** $\dfrac{9}{4}$

State whether the following pairs of fractions are similar, equivalent, or neither.

13. $\dfrac{3}{7}$ and $\dfrac{5}{7}$ **14.** $\dfrac{3}{7}$ and $\dfrac{6}{14}$ **15.** $\dfrac{5}{3}$ and $\dfrac{15}{12}$ **16.** $\dfrac{21}{22}$ and $\dfrac{13}{22}$

2-2 CHANGING THE FORM OF A FRACTION

Fundamental Principle of Fractions

In Sec. 2-1 we said that equivalent fractions $\left(\text{such as } \dfrac{2}{3} \text{ and } \dfrac{6}{9}\right)$ are those that represent the same value. We will now practice writing equivalent fractions. Our tool for doing this is called *the fundamental principle of fractions.*

> If the numerator and the denominator of a fraction are multiplied or divided by the same (nonzero) number, the value of the fraction remains unchanged.

Example 13

If we multiply numerator and denominator of the fraction $\dfrac{5}{8}$, say, by 3,

$$\frac{5}{8} = \frac{5}{8} \cdot \frac{3}{3} = \frac{15}{24}$$

we get the fraction $\dfrac{15}{24}$, which is equal to $\dfrac{5}{8}$. You can verify this by getting a decimal value for each on your calculator.

Reducing to Lowest Terms

One common use for the fundamental principle of fractions is to "simplify" a fraction, or *reduce a fraction to lowest terms.* By this we mean to find an equivalent fraction with the smallest possible denominator.

┌─── **Example 14** ──────────
Reduce the fraction $\frac{26}{65}$ to lowest terms.

Solution: By trial and error, we find that both numerator and denominator are divisible by 13. Dividing,

$$\frac{26}{65} = \frac{26 \div 13}{65 \div 13} = \frac{2}{5}$$

Canceling

As a shortcut, we can reduce a fraction by striking out any factor contained in both numerator and denominator. Recall from Sec. 1-3 that the factors of some number n are those numbers that we multiply to get n. Thus the factors of 12 are 1, 2, 2, and 3.

┌─── **Example 15** ──────────
To reduce the fraction $\frac{26}{65}$, from Example 14, let us first factor numerator and denominator.

$$\frac{26}{65} = \frac{(2)(13)}{(5)(13)}$$

We then strike out the 13 from numerator and denominator, because $\frac{13}{13}$ equals 1.

$$\frac{26}{65} = \frac{(2)\,\cancel{(13)}}{(5)\,\cancel{(13)}} = \frac{2}{5}$$

This is called *canceling*. But be careful when canceling. Any factor to be canceled must be a factor of the *entire* numerator and the *entire* denominator.

┌─── **Example 16** ──────────
In the fraction

$$\frac{5+7}{(7)(8)}$$

there is a 7 in both numerator and denominator. However, 7 is *not a factor* of the numerator so 7 *cannot be canceled*.

Writing a Fraction with a Different Denominator

Another use for the fundamental principle of fractions is to change a fraction to an equivalent fraction with a different denominator. The purpose of this is, of course, to be able to combine (add or subtract) that fraction with another one having the same denominator, as we'll show in Sec. 2-5.

Example 17

Rewrite the fraction $\frac{7}{8}$ with a denominator of 40.

Solution: We must multiply the denominator (8) by 5 to make it 40. But the fundamental principle of fractions says that we must multiply the numerator by the same number to avoid changing the value of the fraction. Multiplying gives

$$\frac{7}{8} = \frac{(7)(5)}{(8)(5)} = \frac{35}{40}$$

We sometimes want to write a whole number as a fraction with a given denominator, so that it may be combined with a given fraction.

Example 18

Rewrite the number 3 as a fraction with a denominator of 9.

Solution: We first write 3 as a fraction, and then multiply by $\frac{9}{9}$.

$$3 = \frac{3}{1} = \frac{(3)(9)}{(1)(9)} = \frac{27}{9}$$

Converting an Improper Fraction to a Mixed Number

Recall from Sec. 2-1 that an *improper fraction* $\left(\text{like } \frac{9}{8}\right)$ is one whose numerator is larger than or equal to its denominator, and that a *mixed number* $\left(\text{such as } 4\frac{1}{3}\right)$ is the sum of a whole number and a proper fraction.

To convert an improper fraction to a mixed number, simply divide the numerator by the denominator. Express any remainder as the numerator of a fraction whose denominator is the same as in the original fraction.

Example 19

Write the improper fraction $\frac{13}{5}$ as a mixed number.

Solution: We divide 13 by 5 and get

$$\frac{13}{5} = 13 \div 5 = 2\frac{3}{5}$$

We show how to reverse this process and convert a mixed number to an improper fraction in Sec. 2-5.

Reduce to lowest terms.

1. $\dfrac{4}{12}$ **2.** $\dfrac{5}{25}$ **3.** $\dfrac{42}{21}$ **4.** $\dfrac{9}{3}$

5. $\dfrac{44}{99}$ **6.** $\dfrac{30}{42}$ **7.** $\dfrac{28}{21}$ **8.** $\dfrac{40}{72}$

Rewrite each fraction with the denominator indicated.

9. $\dfrac{3}{4}$ with denominator 12   **10.** $\dfrac{5}{7}$ with denominator 42

11. $\dfrac{8}{3}$ with denominator 21   **12.** $\dfrac{4}{5}$ with denominator 25

Change each improper fraction to a mixed number.

13. $\dfrac{12}{5}$ **14.** $\dfrac{22}{7}$ **15.** $\dfrac{15}{6}$

16. $\dfrac{103}{25}$ **17.** $\dfrac{82}{7}$ **18.** $\dfrac{19}{3}$

2-3

MULTIPLYING FRACTIONS

To multiply a quantity by a fraction means to take that fractional part *of* the quantity.

■ **FIGURE 2-4**

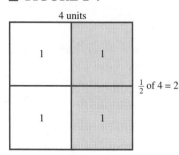

4 units

| 1 | 1 |
| 1 | 1 |

$\frac{1}{2}$ of 4 = 2

■ **FIGURE 2-5**

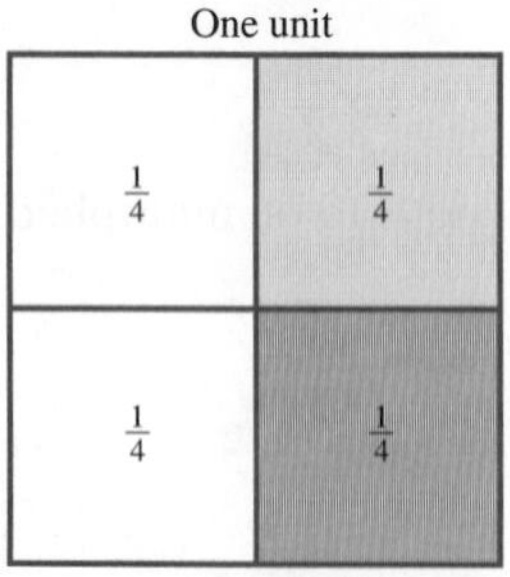

One unit

| $\frac{1}{4}$ | $\frac{1}{4}$ |
| $\frac{1}{4}$ | $\frac{1}{4}$ |

■ **FIGURE 2-6**

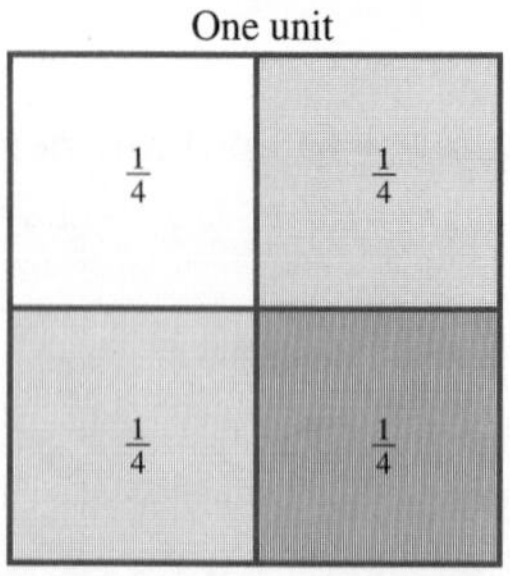

One unit

| $\frac{1}{4}$ | $\frac{1}{4}$ |
| $\frac{1}{4}$ | $\frac{1}{4}$ |

Example 20

(a) To multiply $\dfrac{1}{2}$ and 4,

$$\frac{1}{2} \times 4$$

is the same as *taking one half of four.* We know that half of four is two, so

$$\frac{1}{2} \times 4 = \frac{1}{2} \text{ of } 4 = 2$$

as shown in Fig. 2-4.

(b) $\dfrac{1}{2} \times \dfrac{1}{2}$ is the same as half of one half, or $\dfrac{1}{4}$

as in Fig. 2-5.

(c) $\dfrac{1}{3} \times \dfrac{3}{4}$ means one third of three quarters, or $\dfrac{1}{4}$

as in Fig. 2-6.

In the examples given, we knew the answers from experience. We could also have gotten the same result by using the following rule.

The product of two fractions is a fraction whose numerator is the product of the original numerators and whose denominator is the product of the original denominators.

Example 21

We now repeat Example 20, using the rule. We also reduce our answers to lowest terms.

(a) $\dfrac{1}{2} \times 4 = \dfrac{1}{2} \times \dfrac{4}{1} = \dfrac{(1)(4)}{(2)(1)} = \dfrac{4}{2} = 2$

(b) $\dfrac{1}{2} \times \dfrac{1}{2} = \dfrac{(1)(1)}{(2)(2)} = \dfrac{1}{4}$

(c) $\dfrac{1}{3} \times \dfrac{3}{4} = \dfrac{(1)(3)}{(3)(4)} = \dfrac{3}{12} = \dfrac{1}{4}$

It is usually easier to do the reducing before actually multiplying out.

Example 22

Multiply

$$\frac{5}{7} \times \frac{7}{8}$$

Solution:

$$\frac{5}{7} \times \frac{7}{8} = \frac{(5)(7)}{(7)(8)}$$

Now *before* actually multiplying, let us cancel any factor contained in both numerator and denominator.

$$\frac{5}{7} \times \frac{7}{8} = \frac{(5)(\overset{1}{\cancel{7}})}{(\underset{1}{\cancel{7}})(8)} = \frac{5}{8}$$

The method is the same if there are *more than two* fractions to be multiplied.

Example 23

$$\frac{2}{9} \times \frac{1}{2} \times \frac{9}{11} = \frac{(\overset{1}{\cancel{2}})(1)(\overset{1}{\cancel{9}})}{(\underset{1}{\cancel{9}})(\underset{1}{\cancel{2}})(11)} = \frac{1}{11}$$

Our next example shows canceling when one of the factors to be canceled is a multiple of the other factor to be canceled.

Example 24

Multiply $\dfrac{5}{6}$ and $\dfrac{8}{9}$.

Solution: We see that both the 8 in the numerator and the 6 in the denominator are divisible by 2.

$$\frac{5}{6} \times \frac{8}{9} = \frac{5}{\underset{3}{\cancel{6}}} \times \frac{\overset{4}{\cancel{8}}}{9} = \frac{(5)(4)}{(3)(9)} = \frac{20}{27}$$

Multiplying Mixed Numbers

We may multiply any combination of fractions, mixed numbers, and whole numbers with the same rule for multiplying fractions. However, before multiplying, convert each mixed number to an improper fraction, and write each whole number as a fraction with 1 in the denominator. If you cancel before multiplying, it simplifies the work and reduces the chance for error.

Example 25

$$\frac{1}{5} \times 6 \times \frac{3}{8} \times 5\frac{1}{3} = \frac{1}{5} \times \frac{6}{1} \times \frac{3}{8} \times \frac{16}{3}$$

$$= \frac{(1)(6)(\overset{1}{\cancel{3}})(\overset{2}{\cancel{16}})}{(5)(1)(\underset{1}{\cancel{8}})(\underset{1}{\cancel{3}})} = \frac{12}{5} = 2\frac{2}{5}$$

Example 26

A current of $2\dfrac{4}{5}$ amperes (A) flows in a resistor having a resistance of $21\dfrac{3}{4}$ ohms (Ω). Find the voltage across the resistor by multiplying the current by the resistance.

Solution: Multiplying,

$$\text{Voltage} = \text{current} \times \text{resistance} = 2\frac{4}{5} \times 21\frac{3}{4}$$

$$= \frac{\overset{7}{\cancel{14}}}{5} \times \frac{87}{\underset{2}{\cancel{4}}} = \frac{7}{5} \times \frac{87}{2}$$

$$= \frac{609}{10} = 60\frac{9}{10} \text{ volts}$$

Multiply. Reduce your result to lowest terms.

1. $\dfrac{2}{3} \times \dfrac{3}{7}$ **2.** $\dfrac{1}{7} \times \dfrac{3}{5}$ **3.** $\dfrac{5}{8} \times \dfrac{2}{3}$

4. $\dfrac{5}{12} \times \dfrac{4}{5}$ **5.** $\dfrac{2}{5} \times \dfrac{3}{4}$ **6.** $\dfrac{5}{8} \times \dfrac{16}{17}$

7. $\dfrac{7}{15} \times \dfrac{5}{9}$ **8.** $\dfrac{7}{12} \times \dfrac{8}{11}$ **9.** $\dfrac{1}{2} \times \dfrac{3}{8} \times \dfrac{2}{3}$

10. $\dfrac{7}{12} \times \dfrac{2}{3} \times \dfrac{4}{5}$ **11.** $\dfrac{2}{5} \times \dfrac{3}{4} \times \dfrac{1}{3}$ **12.** $\dfrac{1}{2} \times \dfrac{3}{8} \times \dfrac{4}{5}$

13. $4\dfrac{2}{3} \times \dfrac{3}{4}$ **14.** $\dfrac{5}{8} \times 3\dfrac{1}{2}$ **15.** $2\dfrac{4}{15} \times \dfrac{5}{9}$

16. $\dfrac{5}{12} \times 3\dfrac{8}{11}$ **17.** $3\dfrac{1}{3} \times 2\dfrac{1}{4}$ **18.** $2\dfrac{3}{8} \times 2\dfrac{1}{4}$

19. $3\dfrac{4}{15} \times 4\dfrac{1}{9}$ **20.** $1\dfrac{7}{12} \times 4\dfrac{4}{11}$ **21.** $\dfrac{1}{2} \times 5\dfrac{2}{3} \times \dfrac{3}{4}$

22. $\dfrac{3}{8} \times \dfrac{2}{3} \times 2\dfrac{1}{2}$ **23.** $1\dfrac{7}{15} \times \dfrac{5}{8} \times \dfrac{1}{2}$ **24.** $\dfrac{7}{12} \times \dfrac{3}{5} \times 2\dfrac{6}{11}$

25. $3\dfrac{1}{2} \times 3\dfrac{2}{3} \times \dfrac{1}{4}$ **26.** $\dfrac{7}{8} \times 5\dfrac{2}{3} \times 3\dfrac{1}{3}$ **27.** $2\dfrac{1}{12} \times \dfrac{7}{8} \times 3\dfrac{1}{2}$

28. $\dfrac{5}{12} \times 1\dfrac{2}{5} \times 3\dfrac{2}{11}$

Applications

29. A slider in a certain mechanism moves at the rate of $3\dfrac{5}{8}$ cm/s. How far will it move in $2\dfrac{3}{5}$ seconds?

30. If a certain ore costs $\$55\dfrac{7}{8}$ per ton, how much will $6\dfrac{2}{3}$ tons cost?

31. A particular batch of oak has a density of $32\dfrac{1}{6}$ lb/ft^3. Find the weight of $4\dfrac{3}{5}$ ft^3 of this wood.

32. A certain car can go $33\dfrac{1}{3}$ miles on a gallon of gasoline. How far will it go on $11\dfrac{7}{8}$ gallons?

2-4

DIVIDING FRACTIONS

Reciprocals

To find the *reciprocal* of a fraction, simply *invert* the fraction, so that the numerator and denominator change places.

Example 27

$\frac{2}{3}$ is the reciprocal of $\frac{3}{2}$

$\frac{5}{3}$ is the reciprocal of $\frac{3}{5}$

Dividing Fractions

Let us divide a number, say 3, by 2. We get

$$3 \div 2 = \frac{3}{2}$$

But by our rule for multiplying fractions, we know that

$$\frac{3}{2} = 3 \times \frac{1}{2}$$

In other words,

3 *divided by* 2 is the same as 3 *multiplied by* $\frac{1}{2}$

But 2 and $\frac{1}{2}$ are reciprocals. So we may say that

3 *divided by* 2 is the same as 3 *multiplied by the reciprocal* of 2.

Of course, this does not work only for the numbers 3 and 2. In more general language, we have,

> To divide by a number is the same as to multiply by the reciprocal of that number.

Thus, *to divide by a fraction, invert the fraction and multiply.*

Example 28

Divide $\frac{1}{5} \div \frac{4}{7}$.

Solution:

$$\frac{1}{5} \div \frac{4}{7} = \frac{1}{5} \times \frac{7}{4} = \frac{7}{20}$$

As with multiplication, try to cancel before multiplying. Don't forget to reduce your answer to lowest terms.

Example 29

$$\frac{3}{28} \div \frac{6}{7} = \frac{\overset{1}{\cancel{3}}}{\underset{4}{\cancel{28}}} \times \frac{\overset{1}{\cancel{7}}}{\underset{2}{\cancel{6}}} = \frac{1}{8}$$

Dividing Whole and Mixed Numbers

To find the reciprocal of a mixed number, first rewrite it as an improper fraction, then invert. To find the reciprocal of a whole number, simply write it as a fraction with 1 in the denominator, and invert as before.

Example 30

Find the reciprocals of **(a)** 5 and **(b)** $3\frac{3}{4}$.

Solution:

(a) The reciprocal of 5 is $\frac{1}{5}$.

(b) The mixed number $3\frac{3}{4}$ is equal to the improper fraction $\frac{15}{4}$. The reciprocal of $\frac{15}{4}$ is $\frac{4}{15}$.

If the divisor or the dividend is a mixed number, change it to an improper fraction, and multiply as before. Cancel before multiplying out, and reduce your answer to lowest terms.

Example 31

Divide and simplify.

(a) $6 \div \frac{2}{7} = \frac{6}{1} \div \frac{2}{7} = \frac{\overset{3}{\cancel{6}}}{1} \times \frac{7}{\underset{1}{\cancel{2}}} = \frac{21}{1} = 21$

(b) $\frac{5}{9} \div 3 = \frac{5}{9} \div \frac{3}{1} = \frac{5}{9} \times \frac{1}{3} = \frac{5}{27}$

(c) $2\frac{1}{3} \div \frac{1}{5} = \frac{7}{3} \div \frac{1}{5} = \frac{7}{3} \times \frac{5}{1} = \frac{35}{3} = 11\frac{2}{3}$

(d) $\frac{3}{8} \div 4\frac{1}{2} = \frac{3}{8} \div \frac{9}{2} = \frac{\overset{1}{\cancel{3}}}{\underset{4}{\cancel{8}}} \times \frac{\overset{1}{\cancel{2}}}{\underset{3}{\cancel{9}}} = \frac{1}{12}$

(e) $9 \div 3\frac{1}{2} = \frac{9}{1} \div \frac{7}{2} = \frac{9}{1} \times \frac{2}{7} = \frac{18}{7} = 2\frac{4}{7}$

(f) $1\frac{9}{10} \div 2\frac{4}{5} = \frac{19}{10} \div \frac{14}{5} = \frac{19}{\underset{2}{\cancel{10}}} \times \frac{\overset{1}{\cancel{5}}}{14} = \frac{19}{28}$

Example 32

A block of plastic having a volume of $\frac{4}{5}$ cubic feet weighs $28\frac{3}{8}$ lb. Find the density of the plastic by dividing the weight by the volume.

Solution: Dividing,

$$\text{Density} = \frac{\text{weight}}{\text{volume}} = 28\frac{3}{8}\text{ lb} \div \frac{4}{5}\text{ ft}^3$$

$$= \frac{227}{8} \times \frac{5}{4} = \frac{1135}{32}$$

$$= 35\frac{15}{32}\text{ lb/ft}^3$$

Reciprocals

Find the reciprocal of each number or fraction.

1. 2   **2.** $\dfrac{1}{7}$ **3.** $\dfrac{2}{3}$

4. $\dfrac{4}{5}$ **5.** $\dfrac{12}{13}$ **6.** 16

Dividing Fractions

Divide. Reduce your result to lowest terms.

7. $\dfrac{2}{3} \div \dfrac{5}{7}$ **8.** $\dfrac{2}{7} \div \dfrac{3}{5}$ **9.** $\dfrac{3}{8} \div \dfrac{2}{3}$ **10.** $\dfrac{7}{12} \div \dfrac{4}{5}$

11. $\dfrac{1}{5} \div \dfrac{3}{4}$ **12.** $\dfrac{5}{8} \div \dfrac{13}{17}$ **13.** $\dfrac{1}{15} \div \dfrac{5}{9}$ **14.** $\dfrac{7}{22} \div \dfrac{8}{11}$

15. $14 \div \dfrac{2}{7}$ **16.** $25 \div \dfrac{2}{5}$ **17.** $9 \div \dfrac{2}{3}$ **18.** $15 \div \dfrac{3}{5}$

19. $\dfrac{1}{5} \div 12$ **20.** $\dfrac{7}{8} \div 24$ **21.** $\dfrac{2}{15} \div 21$ **22.** $\dfrac{5}{12} \div 3$

23. $1\dfrac{2}{3} \div \dfrac{3}{4}$ **24.** $\dfrac{3}{8} \div 3\dfrac{1}{2}$ **25.** $3\dfrac{4}{15} \div \dfrac{7}{9}$ **26.** $\dfrac{7}{12} \div 3\dfrac{8}{11}$

27. $2\dfrac{1}{3} \div 5\dfrac{1}{4}$ **28.** $5\dfrac{1}{8} \div 2\dfrac{1}{4}$ **29.** $5\dfrac{4}{15} \div 3\dfrac{1}{9}$ **30.** $2\dfrac{5}{12} \div 4\dfrac{4}{11}$

31. $28 \div 3\dfrac{3}{7}$ **32.** $22 \div 2\dfrac{3}{5}$ **33.** $4 \div 2\dfrac{2}{3}$ **34.** $18 \div 4\dfrac{4}{5}$

35. $3\dfrac{2}{5} \div 16$ **36.** $2\dfrac{5}{8} \div 24$ **37.** $3\dfrac{7}{15} \div 24$ **38.** $3\dfrac{7}{12} \div 3$

39. $34\dfrac{2}{3} \div 6\dfrac{3}{4}$ **40.** $2\dfrac{5}{8} \div 8\dfrac{1}{2}$ **41.** $3\dfrac{4}{15} \div 6\dfrac{5}{9}$ **42.** $2\dfrac{5}{12} \div 6\dfrac{8}{11}$

Applications

43. A parcel of land containing $238\dfrac{2}{3}$ acres is to be divided equally between five heirs. How many acres should each get?

44. A weight of $43\dfrac{3}{5}$ lb is hung from a wire having a cross-sectional area of $\dfrac{1}{64}$ square inches, Fig. 2-7. Find the stress in the wire by dividing the weight by the area.

45. A force of $28\dfrac{3}{4}$ lb will stretch a certain spring a distance of $1\dfrac{5}{16}$ in. from its unstretched or "free" length, Fig. 2-8. Find the spring constant by dividing the force by the distance.

■ **FIGURE 2-7**
A weight hung by a wire.

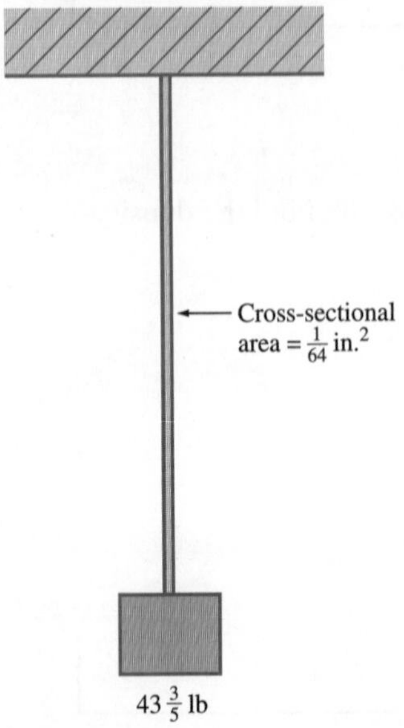

Cross-sectional area $= \dfrac{1}{64}$ in.^{2}

$43\dfrac{3}{5}$ lb

■ **FIGURE 2-8**
A spring stretched by a force.

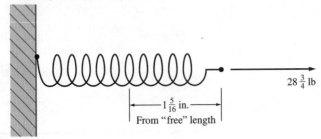

$28\dfrac{3}{4}$ lb

$1\dfrac{5}{16}$ in.
From "free" length

46. The power to a certain resistor is $82\frac{1}{3}$ watts when a current of $3\frac{4}{5}$ amperes flows through the resistor. Find the voltage across the resistor by dividing the power by the current.

2-5

ADDING AND SUBTRACTING FRACTIONS

In this section we show how to add and subtract fractions with the same denominators or with different denominators. We also show how to add and subtract fractions with whole or with mixed numbers. When we add or subtract fractions (or other quantities as well), we say that we *combine* them.

Adding and Subtracting Similar Fractions

Recall that *similar fractions,* or *like* fractions, are those having the same denominator. To add or subtract similar fractions, simply add or subtract the numerators over the common denominator.

—— **Example 33** ——

Combine the fractions

$$\frac{5}{8}+\frac{1}{8}-\frac{3}{8}$$

Solution: The denominators are the same for the three fractions, so we combine the numerators over that denominator.

$$\frac{5}{8}+\frac{1}{8}-\frac{3}{8}=\frac{5+1-3}{8}=\frac{3}{8}$$

Least Common Denominator

A *common denominator* for two or more fractions is a common multiple of the individual denominators.

—— **Example 34** ——

A common denominator for the fractions

$$\frac{3}{5} \quad \text{and} \quad \frac{1}{10}$$

is 50, because 50 is a multiple of both 5 and 10. Other common denominators of these fractions are 10, 20, 30, 40, 60, and so on.

The *least common denominator,* or LCD, of two or more fractions is the smallest of the common denominators of those fractions.

Adding and Subtracting Fractions
with Different Denominators

The first step is to rewrite each given fraction as an equivalent fraction so that both have a common denominator. Any common denominator will work. However, if we use the least common denominator, our result will not have to be reduced later.

┌─ **Example 36** ──────────────────────────────
│ Add $\dfrac{3}{5}$ and $\dfrac{1}{10}$.

Solution: Here the denominators are different, 5 and 10. We saw in Example 35 that the LCD for these fractions is 10. So we rewrite the first fraction as an equivalent fraction having a denominator of 10, and combine as above.

$$\frac{3}{5} + \frac{1}{10} = \frac{(3)(2)}{(5)(2)} + \frac{1}{10} = \frac{6}{10} + \frac{1}{10} = \frac{6+1}{10} = \frac{7}{10}$$

It is often necessary to rewrite *both* given fractions to obtain equivalent fractions with a common denominator.

┌─ **Example 37** ──────────────────────────────
│ Add $\dfrac{3}{4}$ and $\dfrac{1}{3}$.

Solution: The denominators are different here also. This time, though, we cannot change one to equal the other; we must change *both*. The smallest multiple of both 4 and 3 is 12, so 12 is the least common denominator. We thus rewrite each fraction as an equivalent fraction having a denominator of 12.

$$\frac{3}{4} + \frac{1}{3} = \frac{(3)(3)}{(4)(3)} + \frac{(1)(4)}{(3)(4)} = \frac{9}{12} + \frac{4}{12} = \frac{9+4}{12} = \frac{13}{12}$$

or $1\dfrac{1}{12}$, as a mixed number.

When combining fractions with larger denominators, the LCD can more easily be found by separating each denominator into its factors, as shown in the following example.

Example 38

Combine

$$\frac{5}{12} + \frac{1}{18} - \frac{7}{30}$$

Solution: Let us separate each denominator into its factors.

$$\frac{5}{12} + \frac{1}{18} - \frac{7}{30} = \frac{5}{(2)(2)(3)} + \frac{1}{(2)(3)(3)} - \frac{7}{(2)(3)(5)}$$

A common denominator must contain each of the factors from each of the three denominators, or

$$\text{LCD} = (2)(2)(3)(3)(5) = 180$$

Notice that the LCD has *two* 2's. That is because the first denominator, 12, has two 2's as factors. Similarly, our LCD has two 3's, because the denominator 18 has two 3's in its factors. Now we multiply each fraction by the factors needed to make the denominator equal to 180.

$$\frac{5}{(2)(2)(3)} + \frac{1}{(2)(3)(3)} - \frac{7}{(2)(3)(5)}$$

$$= \frac{5}{(2)(2)(3)} \times \frac{(3)(5)}{(3)(5)} + \frac{1}{(2)(3)(3)} \times \frac{(2)(5)}{(2)(5)} - \frac{7}{(2)(3)(5)} \times \frac{(2)(3)}{(2)(3)}$$

$$= \frac{5}{12} \times \frac{15}{15} + \frac{1}{18} \times \frac{10}{10} - \frac{7}{30} \times \frac{6}{6}$$

$$= \frac{75}{180} + \frac{10}{180} - \frac{42}{180} = \frac{43}{180}$$

Combining Fractions and Whole Numbers

To add a fraction and a whole number, simply express them as the parts of a mixed number.

Example 39

$$5 + \frac{1}{8} = 5\frac{1}{8}$$

One way to *subtract* a fraction from a whole number is to *borrow* from the whole number, as shown here.

Example 40

Subtract $\frac{29}{32}$ from 5.

Solution: We borrow 1 from the 5, and call it $\frac{32}{32}$.

$$5 - \frac{29}{32} = 4 + \frac{32}{32} - \frac{29}{32} = 4 + \frac{32 - 29}{32} = 4 + \frac{3}{32} = 4\frac{3}{32}$$

Converting a Mixed Number to an Improper Fraction

To convert a mixed number to an improper fraction, write the integer part as a fraction with the same denominator as the fraction part, and add.

Example 41

Write the mixed number $4\frac{2}{3}$ as an improper fraction.

Solution: We rewrite the integer part (4) as a fraction $\left(\frac{12}{3}\right)$.

$$4\frac{2}{3} = \frac{4}{1} + \frac{2}{3} = \frac{4(3)}{3} + \frac{2}{3}$$

$$= \frac{12}{3} + \frac{2}{3} = \frac{12 + 2}{3} = \frac{14}{3}$$

Adding and Subtracting Fractions and Mixed Numbers

To add or subtract mixed numbers, or mixed numbers and fractions, we combine the whole numbers and fractions *separately*.

Example 42

Combine $6\frac{3}{4} + 2\frac{1}{2} - 3\frac{1}{8}$.

Solution: We group the whole numbers and fractions separately.

$$6\frac{3}{4} + 2\frac{1}{2} - 3\frac{1}{8} = (6 + 2 - 3) + \left(\frac{3}{4} + \frac{1}{2} - \frac{1}{8}\right) = 5 + \left(\frac{3}{4} + \frac{1}{2} - \frac{1}{8}\right)$$

The LCD for the three fractions is 8, so

$$5 + \left(\frac{3}{4} + \frac{1}{2} - \frac{1}{8}\right) = 5 + \left(\frac{6}{8} + \frac{4}{8} - \frac{1}{8}\right) = 5 + \frac{9}{8} = 6\frac{1}{8}$$

When *subtracting* mixed numbers, we must sometimes *borrow* from the whole number part, as we did when subtracting a fraction from a whole number.

Example 43

Subtract $7\frac{1}{5} - 2\frac{4}{5}$.

Solution: We see that we cannot subtract $\frac{4}{5}$ from $\frac{1}{5}$, so we borrow 1 from the 7 and call it $\frac{5}{5}$.

$$7\frac{1}{5} - 2\frac{4}{5} = 6 + \frac{5}{5} + \frac{1}{5} - 2\frac{4}{5}$$

$$= (6 - 2) + \left(\frac{5}{5} + \frac{1}{5} - \frac{4}{5}\right) = 4 + \frac{2}{5} = 4\frac{2}{5}$$

We must sometimes add and subtract fractions, whole numbers, and mixed numbers, as shown here.

Example 44

Combine $9 + 3\frac{5}{6} - 5\frac{2}{5} + \frac{2}{3}$.

Solution: We combine the whole numbers and fractions separately.

$$9 + 3\frac{5}{6} - 5\frac{2}{5} + \frac{2}{3} = (9 + 3 - 5) + \left(\frac{5}{6} - \frac{2}{5} + \frac{2}{3}\right)$$

We combine, using an LCD of 30 for the fractions.

$$(9 + 3 - 5) + \left(\frac{5}{6} - \frac{2}{5} + \frac{2}{3}\right) = 7 + \frac{25}{30} - \frac{12}{30} + \frac{20}{30}$$

$$= 7 + \frac{33}{30} = 7\frac{11}{10} = 8\frac{1}{10}$$

Example 45

A doorway (Fig. 2-9) measures $43\frac{3}{4}$ in. to the outside of the door frame. Each frame has a width of $5\frac{7}{32}$ in. Find the distance between the frames.

Solution: The two frames have a combined width of

$$5\frac{7}{32} + 5\frac{7}{32} = 10\frac{14}{32} = 10\frac{7}{16}$$

■ **FIGURE 2-9**
A doorway.

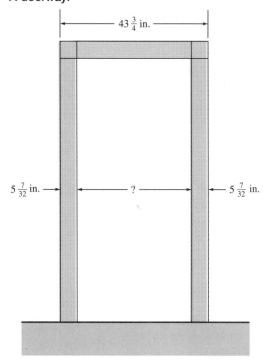

We now subtract this amount from $43\frac{3}{4}$.

$$43\frac{3}{4} - 10\frac{7}{16} = 43\frac{12}{16} - 10\frac{7}{16}$$
$$= 33\frac{5}{16} \text{ in.}$$

Similar Fractions

Combine and reduce to lowest terms.

1. $\dfrac{2}{7} + \dfrac{3}{7}$
2. $\dfrac{5}{11} + \dfrac{3}{11}$

3. $\dfrac{5}{8} + \dfrac{1}{8}$
4. $\dfrac{3}{16} + \dfrac{5}{16}$

5. $\dfrac{9}{13} - \dfrac{6}{13}$
6. $\dfrac{7}{11} - \dfrac{5}{11}$

7. $\dfrac{8}{21} - \dfrac{7}{21}$
8. $\dfrac{7}{15} - \dfrac{4}{15}$

9. $\dfrac{2}{5} + \dfrac{8}{5} - \dfrac{6}{5}$
10. $\dfrac{11}{15} - \dfrac{2}{15} - \dfrac{4}{15}$

11. $\dfrac{4}{13} - \dfrac{1}{13} + \dfrac{5}{13} - \dfrac{3}{13}$
12. $\dfrac{21}{22} - \dfrac{9}{22} - \dfrac{3}{22} + \dfrac{5}{22}$

Finding the LCD

Find the least common denominator for the given fractions.

13. $\dfrac{5}{8}$ and $\dfrac{3}{16}$
14. $\dfrac{2}{5}$ and $\dfrac{4}{15}$

15. $\dfrac{2}{3}$ and $\dfrac{3}{4}$
16. $\dfrac{5}{11}$ and $\dfrac{4}{33}$

17. $\dfrac{3}{5}$ and $\dfrac{4}{7}$
18. $\dfrac{4}{9}$ and $\dfrac{3}{5}$

Fractions with Different Denominators

Combine and simplify.

19. $\dfrac{2}{3} + \dfrac{1}{6}$
20. $\dfrac{5}{11} + \dfrac{3}{22}$

21. $\dfrac{3}{8} + \dfrac{1}{16}$
22. $\dfrac{3}{16} + \dfrac{5}{8}$

23. $\dfrac{9}{11} - \dfrac{5}{22}$
24. $\dfrac{5}{6} - \dfrac{7}{12}$

25. $\dfrac{7}{24} - \dfrac{1}{12}$
26. $\dfrac{7}{15} - \dfrac{11}{30}$

27. $\dfrac{2}{5} + \dfrac{8}{15} - \dfrac{11}{30}$
28. $\dfrac{11}{24} - \dfrac{1}{12} - \dfrac{5}{48}$

29. $\dfrac{4}{13} - \dfrac{1}{26} + \dfrac{5}{39} - \dfrac{3}{52}$
30. $\dfrac{3}{22} + \dfrac{9}{11} - \dfrac{3}{11} + \dfrac{5}{33}$

Fractions and Whole Numbers

Combine the fractions and whole numbers.

31. $\frac{2}{3} + 6$

32. $7 + \frac{3}{22}$

33. $6 + \frac{3}{16}$

34. $\frac{5}{11} + 9$

Fractions and Mixed Numbers

Convert each mixed number to an improper fraction.

35. $3\frac{4}{5}$

36. $5\frac{2}{3}$

37. $1\frac{3}{4}$

38. $7\frac{2}{7}$

Combine the fractions and mixed numbers.

39. $\frac{1}{3} + 5\frac{1}{6}$

40. $3\frac{5}{11} + \frac{5}{22}$

41. $7\frac{3}{8} + \frac{3}{16}$

42. $\frac{5}{16} + 3\frac{3}{8}$

43. $5\frac{7}{11} - \frac{4}{22}$

44. $3\frac{5}{12} - \frac{1}{6}$

45. $2\frac{1}{24} - \frac{7}{12}$

46. $3\frac{1}{15} - \frac{13}{30}$

47. $5\frac{3}{5} + \frac{7}{15} - 1\frac{13}{30}$

48. $2\frac{1}{24} - \frac{11}{12} - 1\frac{1}{48}$

49. $6\frac{2}{13} - \frac{3}{26} + \frac{7}{39} - 1\frac{5}{52}$

50. $8\frac{5}{22} - 2\frac{7}{11} - 1\frac{3}{11} + 2\frac{7}{33}$

Applications

51. A carpenter cuts $35\frac{7}{16}$ in. from a board whose original length was $71\frac{3}{4}$ in. The saw blade makes a cut $\frac{3}{32}$ in. wide. Find the length of the board remaining.

52. A stock whose price was $\$55\frac{1}{2}$ gained $\frac{7}{8}$ and then lost $1\frac{3}{4}$. What was the final price?

53. A book weighing $3\frac{4}{5}$ lb is placed in a box weighing $1\frac{1}{4}$ lb and surrounded by packing material that weighs $\frac{2}{3}$ lb. What is the combined weight?

54. A car trip took $5\frac{3}{4}$ hours, including a rest stop of $\frac{37}{60}$ of an hour. What was the total driving time?

55. A rectangular garden has a length of $32\frac{7}{8}$ ft and a width of $9\frac{3}{5}$ ft. Find the perimeter (the distance around) the garden.

56. A cast part having a length of $4\frac{35}{64}$ in. has $\frac{7}{32}$ ground off one end, Fig. 2-10. That end is then plated with a thickness of $\frac{1}{64}$ in. of nickel. Find the final length of the part.

■ **FIGURE 2-10**
A cast part.

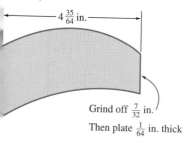

$4\frac{35}{64}$ in.

Grind off $\frac{7}{32}$ in.

Then plate $\frac{1}{64}$ in. thick

COMPLEX FRACTIONS

Recall from Sec. 2-1 that a complex fraction is one that has more than one fraction line.

Example 46

Some complex fractions are

$$\frac{\frac{2}{3}}{17}, \quad \frac{2}{3 + \frac{5}{7}}, \quad \text{and} \quad \frac{\frac{7}{2} - \frac{3}{4}}{\frac{5}{3} + \frac{1}{6}}$$

To *simplify* a complex fraction, we perform the indicated operations to obtain a common fraction or a mixed number.

Example 47

Simplify the complex fraction

$$\frac{\frac{2}{3}}{\frac{5}{7}}$$

Solution: Here we have one fraction $\left(\frac{2}{3}\right)$ divided by another $\left(\frac{5}{7}\right)$. So we perform the indicated division.

$$\frac{\frac{2}{3}}{\frac{5}{7}} = \frac{2}{3} \div \frac{5}{7} = \frac{2}{3} \times \frac{7}{5} = \frac{14}{15}$$

Example 48

Simplify the complex fraction

$$\frac{\frac{3}{5} + \frac{2}{3}}{\frac{1}{3}}$$

Solution: We add the fractions in the numerator, and then divide by the fraction in the denominator.

$$\frac{\frac{3}{5} + \frac{2}{3}}{\frac{1}{3}} = \frac{\frac{9}{15} + \frac{10}{15}}{\frac{1}{3}}$$

$$= \frac{19}{15} \div \frac{1}{3} = \frac{19}{15} \times \frac{3}{1} = \frac{19}{5} = 3\frac{4}{5}$$

Remember that a fraction line is also a symbol of grouping, so all the fractions grouped by a fraction line must be combined first.

Example 49

Simplify the complex fraction

$$\frac{\dfrac{3}{8} - \dfrac{1}{4}}{\dfrac{1}{2} + \dfrac{2}{3}}$$

Solution: We start by combining the fractions in the numerator, and those in the denominator.

$$\frac{\dfrac{3}{8} - \dfrac{1}{4}}{\dfrac{1}{2} + \dfrac{2}{3}} = \frac{\dfrac{3}{8} - \dfrac{2}{8}}{\dfrac{3}{6} + \dfrac{4}{6}} = \frac{\dfrac{1}{8}}{\dfrac{7}{6}} = \frac{1}{8} \div \frac{7}{6} = \frac{1}{\overset{}{\underset{4}{8}}} \times \frac{\overset{3}{\cancel{6}}}{7} = \frac{3}{28}$$

Alternate Solution: Here, we start by multiplying numerator and denominator of the original fraction by the LCD of all individual fractions, 24.

$$\frac{\dfrac{3}{8} - \dfrac{1}{4}}{\dfrac{1}{2} + \dfrac{2}{3}} \cdot \frac{24}{24} = \frac{9 - 6}{12 + 16} = \frac{3}{28}$$

Example 50

A steel plate, Fig. 2-11, is $10\dfrac{7}{8}$ in. long. We want to shorten it by $2\dfrac{11}{16}$ in. and then drill a hole at one quarter of the new length from the end of the plate. The distance d to the hole can be expressed by the complex fraction

$$\frac{10\dfrac{7}{8} - 2\dfrac{11}{16}}{4}$$

Simplify this complex fraction.

■ **FIGURE 2-11**
A steel plate.

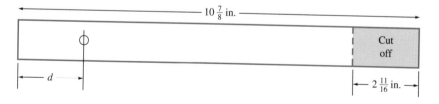

Solution:

$$\frac{10\dfrac{7}{8} - 2\dfrac{11}{16}}{4} = \frac{\dfrac{87}{8} - \dfrac{43}{16}}{4} = \frac{\dfrac{174}{16} - \dfrac{43}{16}}{4}$$

$$= \frac{131}{16} \times \frac{1}{4} = \frac{131}{64} = 2\frac{3}{64} \text{ in.}$$

Simplify.

1. $\dfrac{\frac{1}{4}}{5}$

2. $\dfrac{\frac{2}{5}}{15}$

3. $\dfrac{\frac{4}{5}}{\frac{5}{6}}$

4. $\dfrac{\frac{2}{3}}{4}$

5. $\dfrac{\frac{5}{7}}{\frac{2}{3}}$

6. $\dfrac{\frac{2}{5}}{\frac{1}{7}}$

7. $\dfrac{\frac{1}{3}+\frac{1}{2}}{\frac{1}{4}}$

8. $\dfrac{\frac{2}{3}+\frac{5}{6}}{\frac{1}{3}}$

9. $\dfrac{\frac{2}{5}}{\frac{1}{5}+\frac{3}{5}}$

10. $\dfrac{\frac{3}{4}}{\frac{1}{5}+\frac{3}{10}}$

11. $\dfrac{2}{3+\frac{5}{7}}$

12. $\dfrac{3}{5+\frac{2}{5}}$

13. $\dfrac{\frac{1}{2}-\frac{1}{4}}{\frac{1}{3}+\frac{1}{6}}$

14. $\dfrac{\frac{7}{2}-\frac{3}{4}}{\frac{5}{3}+\frac{1}{6}}$

■ Chapter 2   REVIEW EXERCISES

Perform the indicated operations and reduce to lowest terms.

1. $\dfrac{1}{3} \div \dfrac{5}{8}$

2. $\dfrac{3}{7} \div \dfrac{3}{8}$

3. $\dfrac{5}{8} + \dfrac{7}{16}$

4. $\dfrac{5}{16} + \dfrac{5}{9}$

5. $\dfrac{1}{2} \times \dfrac{5}{8} \times \dfrac{1}{3}$

6. $\dfrac{5}{12} \times \dfrac{2}{5} \times \dfrac{4}{5}$

7. $3\dfrac{7}{15} \times \dfrac{2}{9}$

8. $\dfrac{7}{12} \times 3\dfrac{7}{11}$

9. $\dfrac{5}{12} - \dfrac{1}{24} + \dfrac{3}{24} - \dfrac{3}{48}$

10. $\dfrac{5}{22} + \dfrac{7}{11} - \dfrac{3}{11} + \dfrac{7}{33}$

11. $4 + \dfrac{7}{16}$

12. $\dfrac{7}{11} + 8$

13. $\dfrac{1}{4} \times 3\dfrac{2}{3} \times \dfrac{1}{4}$

14. $\dfrac{5}{8} \times \dfrac{2}{3} \times 5\dfrac{1}{2}$

15. $\dfrac{2}{3} + 7\dfrac{1}{6}$

16. $2\dfrac{5}{11} + \dfrac{7}{22}$

17. $\dfrac{3}{15} \div \dfrac{7}{9}$

18. $\dfrac{5}{22} \div \dfrac{7}{11}$

19. $\dfrac{3}{5} \div 22$

20. $\dfrac{7}{8} \div 44$

21. $1\dfrac{1}{44} - \dfrac{9}{22}$

22. $5\dfrac{2}{15} - \dfrac{11}{30}$

23. $3\dfrac{2}{5} + \dfrac{2}{15} - 3\dfrac{11}{30}$

24. $5\dfrac{1}{8} - \dfrac{7}{12} - 2\dfrac{1}{48}$

25. $7\dfrac{2}{5} \div \dfrac{8}{5}$

26. $\dfrac{5}{22} \div 4\dfrac{8}{11}$

27. $30 \div 5\dfrac{2}{7}$

28. $14 \div 6\dfrac{3}{5}$

29. $8\dfrac{2}{15} \div 14$

30. $5\dfrac{5}{12} \div 4$

31. $\dfrac{7}{8} \times \dfrac{1}{3}$

32. $\dfrac{7}{12} \times \dfrac{3}{5}$

33. $\dfrac{3}{8} + \dfrac{5}{8}$

34. $\dfrac{7}{16} + \dfrac{3}{16}$

35. $\dfrac{\frac{1}{3}}{9}$

36. $\dfrac{\frac{3}{5}}{7}$

37. $\dfrac{\dfrac{2}{3}+\dfrac{3}{5}}{\dfrac{1}{6}}$

38. $\dfrac{\dfrac{2}{7}}{\dfrac{3}{7}+\dfrac{5}{7}}$

39. $\dfrac{6}{3+\dfrac{1}{5}}$

40. $\dfrac{\dfrac{2}{3}-\dfrac{1}{5}}{\dfrac{1}{3}+\dfrac{4}{5}}$

Find the reciprocal.

41. 7

42. $\dfrac{1}{3}$

Convert each mixed number to an improper fraction.

43. $5\dfrac{1}{3}$

44. $2\dfrac{4}{7}$

Convert each improper fraction to a mixed number.

45. $\dfrac{22}{7}$

46. $\dfrac{17}{3}$

47. A farm containing $124\dfrac{3}{5}$ acres is to be divided into 7 equal fields. How many acres should be in each?

48. A force of $14\dfrac{2}{5}$ lb is distributed over an area of $\dfrac{1}{32}$ square inches. Find the stress by dividing the force by the area.

49. A moving sidewalk moves at the rate of $57\dfrac{3}{4}$ ft per minute. How far will it move in $5\dfrac{2}{3}$ minutes?

50. If a certain chemical costs $\$25\dfrac{5}{8}$ per liter, how much will $9\dfrac{2}{3}$ liters cost?

51. A particular plastic has a density of $42\dfrac{5}{6}$ lb/ft³. Find the weight of $8\dfrac{2}{5}$ ft³ of this plastic by multiplying the volume by the density.

52. A certain train can go $53\dfrac{3}{5}$ miles on a gallon of diesel fuel. How far will it go on $6\dfrac{2}{3}$ gallons?

53. A machine weighing $73\dfrac{3}{5}$ lb is placed in a box weighing $11\dfrac{3}{4}$ lb and surrounded by packing material that weighs $4\dfrac{1}{3}$ lb. What is the combined weight?

54. A rectangular courtyard has a length of $7\dfrac{3}{8}$ m and a width of $4\dfrac{4}{5}$ m. Find the perimeter (the distance around) the courtyard.

55. A force of $8\dfrac{1}{4}$ lb. will stretch a certain spring a distance of $\dfrac{3}{16}$ in. Find the spring constant by dividing the force by the distance.

56. The power to a certain resistor is $12\dfrac{2}{3}$ watts when a current of $5\dfrac{1}{5}$ amperes flows through the resistor. Find the voltage across the resistor by dividing the power by the current.

57. Suppose that a vocal member of your local school board says that the study of fractions is no longer important now that we have computers and insists that it be cut from the curriculum to save money.

Write a short letter to the editor of your local paper in which you agree or disagree. Give your reasons for eliminating or retaining the study of fractions.

TEAM PROJECT*

You will find a team project suggested at the end of most chapters. These projects are generally longer and more complicated than the usual exercise, and are meant to be done by several people working as a team over a few days.

If your class decides to do one or more of these, you should divide into teams of about three to six students. Then each team should:

Study the problem.
Estimate the answer(s).
List possible tools and methods for solution. Choose one.
Solve the problem.
Check the answer.
Write a brief report.

Students now have more computational tools at their disposal than anyone has had before. Depending on the problem, you might choose from the following:

Manual computation
Trial and error
Scientific calculator
Graphing calculator
Computer, using BASIC, a spreadsheet like Lotus, mathematical computation software like *MathCad,* or a computer algebra program like *Derive, Maple,* or *Mathematica.*

You might choose *more than one* tool, perhaps doing the computation by calculator and checking the results by computer. For some projects you may want to borrow measuring equipment like a tape measure or a transit, or consult reference books. Sometimes it may be useful to make a model. Be sure to consider *any* aids that may help you.

Your first team project is about the ancient Egyptians' method of decomposing fractions.

* Many of the team projects in this book are from P. Calter, *Technical Mathematics, Third Edition* (Upper Saddle River, NJ: Prentice Hall, 1995).

58. The ancient Egyptians wrote each fraction $\left(\text{except for } \dfrac{2}{3}\right)$ as the sum of unit fractions (a fraction with a numerator of 1). Thus,

$$\frac{3}{5} \text{ would be written as } \frac{1}{2} + \frac{1}{10}$$

and

$$\frac{5}{7} \text{ would be written as } \frac{1}{2} + \frac{1}{7} + \frac{1}{14}$$

This is called the *decomposition* of a fraction. Repetitions were not allowed. Thus $\dfrac{3}{5}$ would not be written as

$$\frac{1}{5} + \frac{1}{5} + \frac{1}{5}$$

Decompose each of the following proper fractions into unit fractions.

$$\frac{3}{4} \quad \frac{2}{5} \quad \frac{4}{5} \quad \frac{5}{6} \quad \frac{2}{7} \quad \frac{3}{7} \quad \frac{4}{7} \quad \frac{6}{7}$$

$$\frac{3}{8} \quad \frac{5}{8} \quad \frac{7}{8} \quad \frac{2}{9} \quad \frac{4}{9} \quad \frac{5}{9} \quad \frac{7}{9} \quad \frac{8}{9}$$

Arithmetic with Decimal Numbers

We did arithmetic with whole numbers in Chapter 1, and arithmetic with fractions in Chapter 2. Here we do arithmetic with decimal numbers. In this chapter we will do arithmetic with *approximate numbers,* the kind of numbers most common in technical work, as well as with exact numbers.

We will learn to distinguish between exact and approximate numbers, to calculate with them, and to round our answers properly. We expand our work with reciprocals, powers, roots, and scientific notation to approximate numbers. Finally, we cover combined operations on the calculator.

We do not yet cover arithmetic with negative numbers; we will do those in Chapter 6.

3-1   DECIMAL NUMBERS

Reading Decimal Numbers

In technical work, numbers are usually written in the familiar *decimal system.*

Example 1

Some decimal numbers are

$$3.50, \quad 137.3, \quad 1.0000, \quad \text{and} \quad 2,746$$

We say that the decimal system uses a *base of 10* because it takes 10 units in any place to equal 1 unit in the next higher place.

The names of the places relative to the decimal point are shown in Fig. 3-1. We use those names when reading a decimal number.

■ **FIGURE 3-1**
Values of the positions in a decimal number.

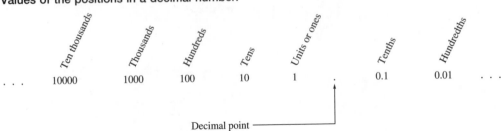

┌─── **Example 2** ───────
Read the numbers
(a) 16,374.2   **(b)** 0.347

Solution: We read
(a) *sixteen thousand, three hundred seventy-four, and two tenths*
(b) *three hundred forty-seven thousandths*

Significant Digits

When we do calculations with approximate numbers, we must be careful about how many digits we keep in our answer. If we keep too few, we may be losing accuracy. If we keep too many, we may be falsely stating the accuracy of our result, and we would certainly be doing extra work by carrying along all those digits.

We will soon see that the number of digits we keep is, of course, dependent on the number of digits in our original numbers. In this section we show how to determine how many digits in our original numbers will have an effect on any calculations we do with those numbers. Such digits are called *significant digits.*

All nonzero digits in a number are, as you might expect, significant digits.

┌─── **Example 3** ───────
The numbers 831.1, 32.62, 8321, and 2.441 each have four significant digits.

Zeros are sometimes significant and sometimes not. In general, zeros not used as placeholders, just to locate the decimal point, *are* significant. This includes zeros located between two nonzero digits. They are *always significant.*

┌─── **Example 4** ───────
The numbers 330.1, 60.02, 8021, and 2.001 each have four significant digits.

Final zeros to the right of the decimal point *are significant.*

┌─── **Example 5** ───────
The numbers 73.30, 2.620, and 3.000 each have four significant digits. The zeros here are not needed to locate the decimal point. Ask yourself why someone bothered to write such a zero in. If it is not needed to locate the decimal point, its only purpose must be to show that that digit is significant and has the value of zero.

Zeros used as placeholders, to locate the decimal point, are *not significant.* These include any zeros to the left of the first nonzero digit.

┌─── **Example 6** ───────
The numbers 0.0274 and 0.000923 each have three significant digits. The zeros in these numbers serve only to locate the decimal point.

Final zeros to the left of the decimal point, such as those in the number 5000, are *not* significant as written. If you want to indicate that some such zeros are significant, place an overscore over the last trailing zero that is significant.

Example 7

The numbers $825,\overline{0}00$, 1,372,000, and $500\overline{0}$ each have four significant digits.

Rounding

We will see, in the next few sections, that the numbers we get from a computation often contain worthless digits that must be thrown away. Whenever we do this, we must *round* our answer.

We'll see later that we sometimes round to a certain number of *decimal places,* and sometimes round to a certain number of *significant digits.* The reasons for doing one or the other will come later; for now we'll just get some practice in rounding.

Round down (do not change the last retained digit) when the first discarded digit is 4 or less.

Example 8

Round to one decimal place (to the nearest tenth): 372.8499.

Solution: The first discarded digit is 4, so we round down and get 372.8.

Example 9

Round to three significant digits: 45.736.

Solution: The first discarded digit is 3, so we round down and get 45.7.

Round up (increase the last retained digit by one) when the first discarded digit is 6 or more, or a 5 followed by a nonzero digit in any of the decimal places to the right.

Example 10

Round to two decimal places (to the nearest hundredth): 372.825001.

Solution: The first discarded digit is a 5, followed eventually by a nonzero digit (1), so we round up and get 372.83.

Example 11

Round to four significant digits: 4.03962.

Solution: The first discarded digit is 6, so we round up and get 4.040.

As we said, we sometimes round to a certain decimal place, and sometimes to a certain number of significant digits. The following example shows some numbers rounded to a given number of decimal places, and to a given number of significant digits.

Example 12

Number	Rounded to three decimal places (nearest thousandth)	Rounded to three significant digits
3.3857	3.386	3.39
0.3656	0.366	0.366
12.2846	12.285	12.3
735.7294	735.729	736

When the first discarded digit is 5 followed by all zeros, it does not usually matter whether you round up or down. However, it is important when adding or subtracting a long column of figures. If, when discarding a 5, you always rounded up, you could bias the result in that direction. To avoid that, you want to round up about as many times as you round down, and a simple way to do that is always to *round to the nearest even digit*. This is just a convention. We could just as easily round to the nearest odd digit. To be consistent in this book, we will always round to the nearest even digit if the first discarded digit is 5 followed by all zeros. This is shown in the following example.

Example 13

Number	Rounded to three decimal places	Rounded to three significant digits
3.3855	3.386	3.39
3.3845	3.384	3.38
12.2855	12.286	12.3
12.2845	12.284	12.3

SECTION 3-1 Exercises Decimal Numbers

State the number of decimal places and of significant digits in each approximate number.

1. 62.5	**2.** 5847	**3.** 9.007	**4.** 6800
5. 700,000	**6.** 9000.0	**7.** 0.85497	**8.** 8.00000

Round each number to two decimal places.

9. 58.309	**10.** 2.997	**11.** 86.03954	**12.** 39.958
13. 689.0495	**14.** 8.678	**15.** 9876.760	**16.** 730.875

Round each number to one decimal place.

17. 45.97	**18.** 364.69	**19.** 9.7503	**20.** 0.845
21. 3984.68	**22.** 84.209	**23.** 4358.38	**24.** 0.385

Round each number to the nearest hundred.

25. 58,029	**26.** 9440	**27.** 8,475,309	**28.** 394,203

Round each number to three significant digits.

29. 7346	**30.** 3745	**31.** 0.04758	**32.** 475,203
33. 230,578	**34.** 0.076543	**35.** 43.555	**36.** 79.650001

ADDING AND SUBTRACTING DECIMAL NUMBERS

Adding and Subtracting Approximate Numbers

We showed how to add and subtract whole numbers in Chapter 1. We assumed them to be exact. But now let us tackle the problem mentioned earlier. How many digits do we keep in our answer when adding or subtracting approximate numbers? We use the following rule:

> **RULE** For each number to be added or subtracted, note the place value of the last significant digit. Identify which of these has the greatest value. Round your answer to that place value.

Example 14

Add 16.8 + 21.4125 + 6.667.

Solution: We use the calculator keystrokes for addition as shown in Chapter 1, and get an unrounded sum of 44.8795. We then note the place value of the last significant digit in each given number; tenths in 16.8, ten-thousandths in 21.4125, and thousandths in 6.667. The greatest of these place values is tenths, so we round our answer to the nearest tenth.

$$
\begin{array}{r|l}
16.8 & \\
21.4 & 125 \\
6.6 & 67 \\
\hline
44.8 & 795
\end{array}
$$

→ discard

So the sum, rounded up to the nearest tenth, is 44.9.

Example 15

Add 354,100 + 2,745.

Solution: Our unrounded sum is 356,845. Lacking any other information, we can only assume that the zeros in 354,100 are not significant, so that this number is "good" only to the nearest hundred. Thus we round our answer to the nearest hundred, getting 356,800.

Adding and Subtracting Exact and Approximate Numbers

Sometimes we must add an exact number and an approximate number. How do we round the answer? Its simple. Treat the exact number as if it had more decimal places than the approximate number.

Example 16

Find the sum, in minutes, of 3 hours and 15.7 minutes.

Solution: Here, the number 3 is exact, and the number 15.7 is approximate. First we convert 3 hours to minutes and get 180 minutes. This is an exact number. To it we must add the approximate number 15.7.

$$\begin{array}{r} 180 \quad \text{min} \\ + \underline{15.7} \quad \text{min} \\ 195.7 \quad \text{min} \end{array}$$

Since 180 is *exact*, we *do not* round our answer to the nearest 10 minutes, but retain as many decimal places as in the approximate number. Our answer is thus 195.7 min.

Applications

Example 17

The sides of a triangular window have lengths of 10.25 ft, 9.1 ft, and 8.375 ft. Find the perimeter (the distance around) of the window.

Solution: We get the perimeter by adding the lengths of the sides.

$$10.25 + 9.1 + 8.375 = 27.725 \text{ ft}$$

which we must round to one decimal place (because 9.1 has just one decimal place), and get a perimeter of 27.7 ft.

SECTION 3-2 Exercises Adding and Subtracting Decimal Numbers

Combine as indicated. Round your answers properly.

1. 638.2
  +723.4

2. 723.6
  +284

3. 11.283
  − 6.26

4. 155.6
  +273.91
  − 18.3

5. 17.3
  +21.93
  − 1.9

6. 9.5
  −9.82
  +2.1

7. 7327 + 7.5

8. 19.69 + 5.8

9. 75.33 − 53.189

10. 936.23 − 362.6

11. 0.005385 + 0.000974 − 0.00527

12. 6322.7 − 13.70 + 39.43 − 7.269

13. 9.273 − 6.3853 − 2.83 + 39.282 − 3.42

14. 54.63 − 23.066 − 23.97 + 133.885 − 7.221

Applications

15. A tractor sold for $30,183, which was $9925 more than its original cost. What was its original cost?

16. Find the perimeter of a four-sided figure (a quadrilateral) if the lengths of the four sides are 22.7 ft, 28.16 ft, 25.27 ft, and 9.556 ft.

17. A company distributed $255,000 to its four partners. One partner got $83,274, a second got $49,222, and a third got $52,000. How much, to the nearest dollar, did the fourth partner get?

18. A circular oil drum has an inside radius of 5.500 ft and a wall thickness of 0.0625 ft. What is the outside diameter of the drum?

19. A batch of concrete is made by mixing 625 lb of aggregate, 264 lb of sand, 153 lb of cement, and 59.25 lb of water. Find the total weight of the mixture.

20. Three resistors, having values of 836 ohms (Ω), 264.75 Ω, and 2844 Ω, are wired in series. Find the total resistance by adding the three individual resistances.

3-3   MULTIPLYING DECIMAL NUMBERS

Multiplying Approximate Numbers

The calculator keystrokes for multiplying approximate numbers are the same as when we multiplied whole numbers in Chapter 1. The difference is that we must now round our product to the proper number of digits. We use the following rule:

> **RULE** When multiplying two or more approximate numbers, round the product to as many significant digits as in the factor having the fewest significant digits.

If both factors have the same number of significant digits, we simply keep that same number of digits in our product.

Example 18

$$23.1 \ \times \ 18.8 \ = \ 434$$

$$\updownarrow \qquad \updownarrow \qquad \updownarrow$$

three   three   three
digits   digits   digits

When the factors have *different* numbers of significant digits, keep the same number of significant digits in your answer as is contained in the factor that has the fewest significant digits.

Example 19

Multiply 23.1 by 18.89 and round the product to the proper number of significant digits.

Solution: Our calculator gives an unrounded product of 436.359. Since the number 23.1 has only three significant digits, we round our answer to three significant digits.

$$23.1 \ \times \ 18.89 \ = \ 436$$

$$\updownarrow \qquad \updownarrow \qquad \updownarrow$$

three   four   keep
digits   digits   three
                           digits

> **COMMON ERROR** Do not confuse significant digits with decimal places. The number 274.56 has five significant digits and two decimal places.
>
> Decimal places determine how we round after adding or subtracting. Significant digits determine how we round after multiplying, and, as we will soon see, after dividing, raising to a power, or taking roots.

Multiplying Exact and Approximate Numbers

When using *exact* numbers in a computation, treat them as if they had *more significant digits* than any of the approximate numbers in that computation.

Example 20

If a certain computer weighs 53.7 lb, how much will 4 such computers weigh?

Solution: Multiplying, we obtain

$$53.7(4) = 214.8 \text{ lb}$$

Since the 4 is an exact number, we retain as many significant digits as contained in 53.7 and round our answer to 215 lb.

SECTION 3-3 Exercises Multiplying Decimal Numbers

Multiplying Approximate Numbers

Multiply. Retain the proper number of digits in your answer.

1. 1.84×2.35	**2.** 8.67×1.26	**3.** 22.35×84.16
4. 1.035×2.943	**5.** 77.283×12.652	**6.** 8.2873×11.847
7. 8.74×7.183	**8.** 30.5×7.265	**9.** 1.4853×14.29
10. 3.800×24.7	**11.** 35.2×0.0062388	**12.** $2.63 \times 8.96 \times 3.34$

13. $6.40 \times 6.78 \times 1.882$ **14.** $1.7322 \times 16.4 \times 0.00335$

15. 14.300×186.92 **16.** 1395×3.958

17. $4.39 \times (6.38 \times 3.58) \times 0.883$ **18.** $(9.28 \times 8.39) \times (4.38 \times 0.853)$

19. $(8.36 + 9.26) \times 14.3$ **20.** $7.43 \times (4.67 + 5.03) \times 7.59$

Multiplying Exact and Approximate Numbers

Assume the integers in these problems to be exact.

21. 34×1.93	**22.** 4.003×6	**23.** 364×2.38
24. $23 \times 0.473 \times 3$	**25.** $2.836 \times 4 \times 3.72$	**26.** $2 \times 8 \times 1.004$

Applications

Work the financial calculations to the nearest penny, regardless of the significant digits in the original numbers.

27. If 30.25 ft of hookup wire is needed for one VCR, how many feet will be required for 145 VCRs, and what will be its cost at $0.0163 per ft?

28. If there are 360 degrees per revolution, how many degrees are there in 0.175 revolutions?

29. The current to a motor is 2.552 A when the line voltage is 115 V. Find the power consumed in the motor by multiplying the current by the voltage.

30. A flywheel rotates 73.5 times per minute. How many revolutions will it make in 2.28 minutes?

31. What is the cost of 25 cans of paint if each costs $17.23?

32. Two joggers start from the same place on a straight road, and run in opposite directions. One runs at the rate of 8.30 mi/h and the other at 11.4 mi/h. How far apart will they be at the end of 0.750 hour?

33. One inch equals exactly 2.54 cm. Convert 15.285 in. to cm.

34. What will be the cost of building a wall 10.4 m long, at $584 per m?

3-4   DIVIDING DECIMAL NUMBERS

Dividing Approximate Numbers

The rule for rounding in division is almost the same as with multiplication:

> **RULE** After dividing approximate numbers, round the quotient to as many significant digits as there are in the original number having the fewest significant digits.

Example 21

Divide 18.63 by 2.15 by calculator.

Solution: The keystrokes are the same as those we used for dividing whole numbers in Chapter 1, and we get a quotient of 8.665116279. Since 2.15 has three significant digits, we round our quotient up to 8.67.

Example 22

A spacecraft is seen to go a distance of 1120 miles in 2.75 hours. Find the speed of the spacecraft.

Solution: The speed equals the distance divided by the time. Dividing,

$$\text{Speed} = \frac{\text{distance}}{\text{time}} = \frac{1120\,\text{mi}}{2.75\,\text{h}} = 407\,\text{mi/h}$$

Reciprocal of a Decimal Number

In Chapter 2 we defined the reciprocal of any number as *1 divided by the number.*

Example 23

The reciprocal of 2 is $\frac{1}{2}$.

To find reciprocals by calculator, simply enter the number and press the $\boxed{x^{-1}}$ key or the $\boxed{1/x}$ key. Keep as many significant digits in your answer as there are significant digits in the original number.

Example 24

Find the reciprocal of 3.74 by calculator.

Solution: The keystrokes are

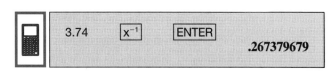

3.74 x^{-1} ENTER

.267379679

Since 3.74 has three significant digits, we round our answer to 0.267.

Example 25

When solving a certain circuit for the equivalent resistance R, we find that R in ohms (Ω) is equal to the reciprocal of 0.285. Find R.

Solution: Taking the reciprocal,

$$R = \frac{1}{0.285} = 3.51 \ \Omega$$

SECTION 3-4   Exercises   Dividing Decimal Numbers

Divide. Round your answer to the proper number of digits. For this set of problems only, assume that the whole numbers are exact.

1. $6.34 \div 1.25$	**2.** $7.47 \div 2.46$	**3.** 2.645×1.463
4. $41.35 \div 4.837$	**5.** $183 \div 2.352$	**6.** $21.53 \div 7.47$
7. $1.574 \div 5.23$	**8.** $2.705 \div 1.21$	**9.** $397.3 \div 1832.2$
10. $9.782 \div 0.07819$	**11.** $162.6 \div 89.3$	**12.** $37.4 \div 6.214$
13. $86.74 \div 5$	**14.** $8 \div 4.495$	**15.** $723.1 \div 9$

Reciprocals of a Decimal Number

Find the reciprocal of each number, retaining the proper number of digits in your answer.

16. 843	**17.** 0.0004580	**18.** 893,926	**19.** 93.78
20. 0.00375	**21.** 937.8	**22.** 6.338	**23.** 9.62
24. 93.2	**25.** 562	**26.** 7.001	**27.** 9.52

28. If 735 bonds are worth $25,836, what is the value of each bond?

Applications

29. A steel casting (density = 455 lb/ft³) weighs 629.3 lb. Find the volume of the casting. (Volume = weight ÷ density)

30. How many cubic feet of water will weigh 8362 lb if the density of water is 62.4 lb/ft³?

31. At the rate of 861.9 km in 1.335 hours, how many kilometers would an aircraft fly in 10.25 hours? (Distance = rate × time)

32. A voltage of 163.4 V is measured across a 1745-Ω resistor. Find the current through the resistor. (Current = voltage ÷ resistance)

33. A steel girder 25.82 ft long is divided into three equal sections. Find the length of each section.

34. How long will it take 9 people to do the same work that 17 people can do in 8.50 days?

35. If two straight lines are perpendicular, the slope of one line is the negative reciprocal of the slope of the other. If the slope of a line is −1.73, find the slope of a perpendicular to that line.

36. When an object is placed 284 cm in front of a certain thin lens having a focal length *f*, the image will be formed 173 cm from the lens. The distances are related by

$$\frac{1}{f} = \frac{1}{284} + \frac{1}{173}$$

Find *f*.

37. The sine of an angle θ (written $\sin \theta$) is equal to the reciprocal of the cosecant of θ (written $\csc \theta$). Find $\sin \theta$ if $\csc \theta = 5.26$.

38. Find the equivalent resistance of a 451-Ω resistor and a 879-Ω resistor, connected in parallel. (Find the sum of the reciprocals of the given resistances. Then take the reciprocal of that sum.)

3-5   POWERS AND ROOTS OF DECIMAL NUMBERS

Powers

As we showed in Chapter 1, to square a number on the calculator, simply enter the number and press the $\boxed{x^2}$ key, and to raise a number to other powers, use the power key, marked $\boxed{\wedge}$, $\boxed{y^x}$, or $\boxed{x^y}$. In our examples we'll use $\boxed{\wedge}$. When finding the power of an approximate number, round your result to the number of significant digits contained in the original number (the base).

Example 26
Find $(2.93)^3$ by calculator.

Solution: Here, the base (2.93) is approximate, and has three significant digits. The keystrokes are

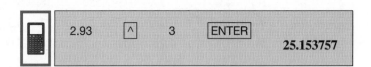

We round the result to 25.2.

Fractional Exponents

We will soon see that a fractional exponent is another way of indicating a radical. For now, we just evaluate fractional exponents on the calculator. We enter the base, press the power key, then enter *the exponent in parentheses,* as shown in the following example.

Example 27

Find $8^{2/3}$ by calculator.

Solution: The keystrokes are

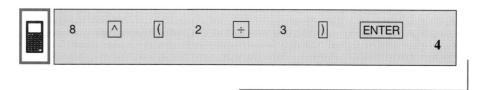

8   ∧   ((2   ÷   3   ))   ENTER

4

To raise an approximate number to a fractional exponent we use the same keys as in the preceding example. However, now we round our answer to as many significant digits as contained in the base.

Example 28

Evaluate $(62.4)^{2/3}$.

Solution: The keystrokes are

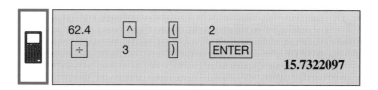

62.4   ∧   ((2
÷   3   ))   ENTER

15.7322097

which we round to 15.7.

Roots

In Sec. 1-5 we took roots of whole numbers. Now we will take roots of approximate numbers. We keep as many significant digits in our answer as contained in the number of which we are taking the root.

To find *square* roots or *cube* roots, enter the number and then press the $\boxed{\sqrt{x}}$ key or the $\boxed{\sqrt[3]{}}$ key. To find other roots, use the root key, marked $\boxed{\sqrt[y]{}}$, $\boxed{\sqrt[x]{y}}$, or $\boxed{\sqrt[y]{x}}$. We'll indicate the root key by $\boxed{\sqrt[y]{}}$ in the examples. (See the following section if you have no root key.) Retain as many significant digits in your answer as there are significant digits in the original number.

Example 29

Find $\sqrt[4]{97.6}$ (the fourth root of 97.6).

Solution: The keystrokes are

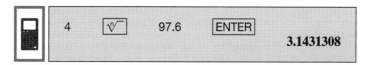

4   $\boxed{\sqrt[y]{}}$ 97.6   ENTER

3.1431308

which we round to 3.14.

Roots and Powers Related

If your calculator does not have a root key, you can find roots by using the power key. For this, you need to know the relationship between roots and exponents.

To take the nth root of a number x is the same as raising x to the $1/n^{th}$ power.

In Chapters 7 and 20 we will explain this idea in much more detail. For now, we will use it only for computing roots.

Example 30

Find $\sqrt[5]{94.7}$ using the power key.

Solution: Taking the 5th root of a number is equal to raising that number to the 1/5th power.

$$\sqrt[5]{94.7} = (94.7)^{1/5}$$

On the calculator:

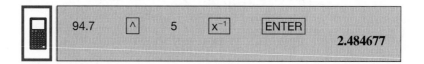

94.7	$\wedge$	5	x^{-1}	ENTER
				2.484677

which we round to 2.48.

Example 31

The volume of a certain cube is 73.5 cubic centimeters. Find the side of the cube by taking the cube root of the volume.

Solution: Taking the cube root,

$$\sqrt[3]{73.5} = 4.188859 \text{ cm}$$

which we round to three significant digits, getting 4.19 cm.

SECTION 3-5   Exercises   Powers and Roots of Decimal Numbers

Powers

Evaluate each expression. Retain the correct number of digits in your answer.

1. $(2.13)^2$	**2.** $(1.34)^3$	**3.** $(3.22)^5$	**4.** $(1.08)^2$
5. $(0.511)^5$	**6.** $(0.291)^4$	**7.** $(1.002)^9$	**8.** $(21.5)^3$
9. $(6.24)^5$	**10.** $(7.003)^3$	**11.** $(4.89)^4$	**12.** $(3.33)^3$
13. $(8.53)^{4/5}$	**14.** $(5.83)^{2/3}$	**15.** $(6.94)^{3/4}$	**16.** $(5.397)^{1/2}$
17. $(76.9)^{5/7}$	**18.** $(7.534)^{3/5}$	**19.** $(6.53)^{3/4}$	**20.** $(2.094)^{2/5}$
21. $(8.365)^{5/7}$	**22.** $(8.65)^{3/2}$	**23.** $(9265)^{1/3}$	**24.** $(7.52)^{2/3}$

Roots

Evaluate each radical by calculator, retaining the proper number of digits in your answer.

25. $\sqrt{84.2}$	**26.** $\sqrt{7.527}$	**27.** $\sqrt[3]{53.0}$
28. $\sqrt{492}$	**29.** $\sqrt{8426}$	**30.** $\sqrt[3]{4973}$
31. $\sqrt[5]{843}$	**32.** $\sqrt[3]{53.8}$	**33.** $\sqrt[3]{87.36}$

34. Evaluate the geometric mean B between 14.3 and 25.6 if

$$B = \sqrt{(14.3)(25.6)}$$

35. Find the volume of a cube of side 8.24 in. by cubing this dimension.

36. The volume of a 2.85-m-radius sphere is $(4/3)\pi(2.85)^3$ m³. Find this volume.

37. The distance traveled by a falling body, starting from rest, is equal to $16t^2$, where t is the elapsed time. The distance fallen in 5.25 seconds is thus $16(5.25)^2$ ft. Evaluate this quantity.

38. An investment of \$8263 at a compound interest rate of 7.5%, left for 5.25 years, will be worth $8263(1.075)^{5.25}$ dollars. Find this amount to the nearest dollar.

39. Find the period T (time for one swing) of a simple pendulum 1.88 ft long, if

$$T = 2\pi \sqrt{\frac{1.88}{32.2}} \quad \text{seconds}$$

40. Find the magnitude of the impedance Z in a circuit having a resistance of 375 Ω and a reactance of 624 Ω if

$$Z = \sqrt{(375)^2 + (6.24)^2} \quad \Omega$$

41. The power dissipated by a current I flowing through a resistance R is equal to I^2R. Therefore, the power in a 1829-Ω resistor carrying a current of 1.263 A is $(1.263)^2(1829)$ W. Evaluate this expression.

3-6   SCIENTIFIC NOTATION

In Sec. 1-6 we introduced scientific notation with whole numbers and showed, for example, that the number 30,000 can be written as 3×10^4. We apply scientific notation to decimal numbers here and do some calculations in scientific notation. We will not do powers of ten with negative exponents yet. We will save those for the chapter on signed numbers.

Converting from Decimal Form to Scientific Notation

Rewrite the given number with just one digit to the left of the decimal point, and discard any nonsignificant zeros. Then multiply by that multiple of ten that makes it equal to the given number. Finally write the multiple of ten as a power of ten.

Example 32

Convert 53,100 to scientific notation.

Solution: We place the decimal point between the 5 and the 3. The two zeros are not significant (see Sec. 3-1) so we discard them, getting 5.31. Finally, the 5.31 must be multiplied by 10,000 so that our new expression is equal to the given number.

$$53{,}100 = 5.31 \times 10{,}000$$

We now express 10,000 as a power of ten (10^4).

$$53{,}100 = 5.31 \times \underline{10{,}000} = 5.31 \times 10^4$$

To convert from scientific notation to decimal form, simply reverse the process.

Example 33

Convert 3.84×10^5 to decimal form.

Solution:

$$3.84 \times 10^5 = 3.84 \times 100{,}000$$
$$= 384{,}000$$

Entering Numbers in Scientific Notation in a Calculator

A calculator display will, as we have seen, switch to scientific notation by itself if the display gets too long for it to show. We can also force a calculator to display in scientific notation by pressing the key that switches the display between fixed notation and scientific notation. Markings for this key vary, but some are $\boxed{\text{FE}}$, $\boxed{\text{FSE}}$, and $\boxed{\text{SCI}}$. Many calculators have a *Mode* menu, from which you can select the kind of display you want.

We can enter numbers directly in scientific notation by means of the ENTER EXPONENT key, usually marked $\boxed{\text{EE}}$, $\boxed{\text{EXP}}$, or $\boxed{\text{EEX}}$. We will use $\boxed{\text{EE}}$ in our examples.

Example 34

The keystrokes to enter the number 8.25×10^8 are

With the calculator in "scientific" mode, the display will show something like 8.25E8 or 8.25 08, depending on the calculator.

We now show how to add, subtract, multiply, and divide numbers in scientific notation.

Adding and Subtracting Numbers in Scientific Notation

If two or more numbers to be added have the *same power of ten,* simply add the numbers and keep the same power of ten, as shown in the following example.

Example 35

Add (3.14×10^4) and (5.27×10^4).

Solution:

$$(3.14 \times 10^4) + (5.27 \times 10^4) = (3.14 + 5.27) \times 10^4$$

When adding 3.14 and 5.27, we follow the rule for adding approximate numbers and retain two decimal places in our answer.

$$(3.14 + 5.27) \times 10^4 = 8.41 \times 10^4$$

Example 36

$$(4.86 \times 10^5) - (2.63 \times 10^5) + (3.63 \times 10^5) = (4.86 - 2.63 + 3.63) \times 10^5$$
$$= 5.86 \times 10^5$$

Example 37

Add (5.83×10^5) and (1.85×10^5) by calculator.

Solution: The keystrokes are

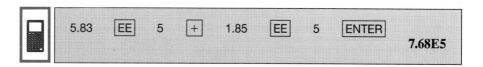

with the calculator in Scientific Notation mode, or 7.68×10^5.

If the powers of ten are different, they must be made equal before the numbers can be combined. A shift of the decimal point of one place to the left will increase the exponent by 1. Conversely, a shift of the decimal point one place to the right will decrease the exponent by 1.

Example 38

$$(3.41 \times 10^5) + (6.5 \times 10^4) = (3.41 \times 10^5) + (0.65 \times 10^5)$$
$$= 4.06 \times 10^5$$

Example 39

$$(2.80 \times 10^4) - (1.6 \times 10^3) + (3 \times 10^2) = (2.80 \times 10^4) - (0.16 \times 10^4) + (0.03 \times 10^4)$$
$$= 2.67 \times 10^4$$

By calculator, the keystrokes are

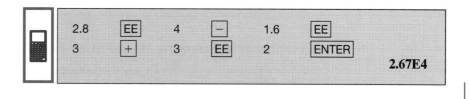

Multiplying Numbers in Scientific Notation

We multiply powers of ten by *adding their exponents*. We are really using one of the laws of exponents that we will study in detail in Chapter 7.

Example 40

$$10^2 \times 10^4 = 10^{2+4} = 10^6$$

To multiply two numbers in scientific notation, multiply the decimal parts and the powers of ten separately.

Example 41

Multiply (4.15×10^4) by (2.63×10^3).

Solution: We multiply the decimal parts by calculator, and round the product to three significant digits $(4.15 \times 2.63 = 10.9)$, and multiply the powers of ten by adding their exponents.

$$(4.15 \times 10^4)(2.63 \times 10^3) = (4.15 \times 2.63)(10^4 \times 10^3)$$
$$= 10.9 \times 10^{4+3} = 10.9 \times 10^7$$
$$= 1.09 \times 10^8$$

Dividing Numbers in Scientific Notation

We divide powers of ten by *subtracting their exponents:*

Example 42:

$$\frac{10^7}{10^2} = 10^{7-2} = 10^5$$

As with multiplication, we divide the decimal parts and the powers of ten *separately.* When dividing the decimal parts, we follow the rules for rounding for division.

Example 43

$$\frac{9.62 \times 10^7}{3.83 \times 10^2} = \frac{9.62}{3.83} \times \frac{10^7}{10^2} = 2.51 \times 10^{7-2} = 2.51 \times 10^5$$

Computing in Scientific Notation by Calculator

Simply follow the same calculator procedures as you would for decimal numbers and let the calculator worry about the powers of ten.

Example 44

Multiply (7.42×10^6) by (5.639×10^3).

Solution: The keystrokes are

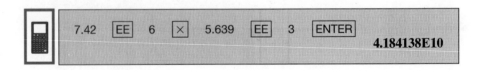

We round to three significant digits, getting 4.18×10^{10}.

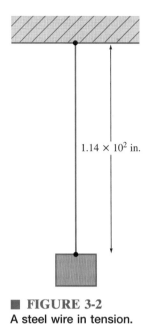

1.14×10^2 in.

■ **FIGURE 3-2**
A steel wire in tension.

Example 45

The elongation e of a steel wire having a length of 1.14×10^2 in. and a cross-sectional area of 1.27×10^{-2} in.^{2}, with a tensile load of 1.57×10^3 lb is given by

$$e = \frac{(1.57 \times 10^3)(1.14 \times 10^2)}{(1.27 \times 10^{-2})(3.11 \times 10^7)} \quad \text{in.}$$

where 3.11×10^7 is the modulus of elasticity of steel, in pounds per square inch, Fig. 3-2. Find e.

Solution: The keystrokes are

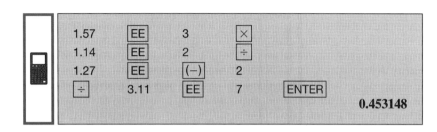

1.57	EE	3	×
1.14	EE	2	÷
1.27	EE	(−)	2
÷	3.11	EE	7	ENTER

0.453148

which we round to 0.453 in.

Converting to and from Scientific Notation

Write each number in scientific notation. Be sure to retain the proper number of digits.

1. 2,373 **2.** 56,632 **3.** 7,260 **4.** 123.9 **5.** 1,920,000

6. 673,900 **7.** 5228 **8.** 64,991 **9.** 7,010 **10.** 325,800,000

Convert each number from scientific notation to decimal notation.

11. 9.023×10^2 **12.** 8.9902×10^4 **13.** 4.01×10^3

14. 7.100×10^4 **15.** 9.520×10^5 **16.** 2.94×10^6

Perform the following computations in scientific notation. Combine the powers of ten by hand or with your calculator.

Adding and Subtracting Numbers in Scientific Notation

17. $(8.40 \times 10^3) + (7.90 \times 10^4)$ **18.** $(42.0 \times 10^3) + 4900$

19. $(4.8862 \times 10^3) + (6.0 \times 10^2)$ **20.** $995 - (7.00 \times 10^2)$

21. $(8.30 \times 10^2) + (5.30 \times 10^2)$ **22.** $(9.50 \times 10^3) - (7.45 \times 10^3)$

Multiplying Numbers in Scientific Notation

23. $10^3 \times 10^4$ **24.** $10^5 \times 10^3$

25. $10^3 \times 10^6$ **26.** $(7.39 \times 10^6)(4.62 \times 10^5)$

27. $(5.03 \times 10^3)(6.17 \times 10^8)$ **28.** $(5.49 \times 10^5)(4.91 \times 10^3)$

29. $(3.86 \times 10^2)(1.28 \times 10^7)$ **30.** $(3.92 \times 10^3)(4.15 \times 10^5)$

31. $(4.23 \times 10^6)(5.94 \times 10^2)$

Dividing Numbers in Scientific Notation

32. $10^{10} \div 10^6$ **33.** $10^8 \div 10^5$

34. $10^9 \div 10^7$ **35.** $(3.25 \times 10^2) \div 0.00416$

36. $45,400 \div (5.25 \times 10^3)$ **37.** $(8 \times 10^7) \div (4 \times 10^3)$

38. $(2 \times 10^6) \div (4 \times 10^4)$ **39.** $(6 \times 10^9) \div 200,000,000$

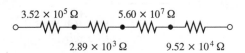

FIGURE 3-3
Resistors in series.

40. $(8.17 \times 10^8) \div 9,020,000$

41. Four resistors, having resistances of 3.52×10^5 Ω, 2.89×10^3 Ω, 5.60×10^7 Ω, and 9.52×10^4 Ω, are wired in series, Fig. 3-3. Find the total resistance by adding the four resistances.

42. Find the equivalent resistance if the four resistors in Exercise 41 are wired in parallel, Fig. 3-4. (Add the reciprocals of the four resistors. Then take the reciprocal of that sum.)

43. Find the current through a resistor if it dissipates 7.82×10^3 watts when the voltage drop across it is 3.74×10^3 V. (Divide the power by the voltage drop.)

44. The voltage across a 3.92×10^4 Ω resistor is 7.39×10^2 V. Find the power dissipated in the resistor. (Divide the square of the voltage by the resistance.)

45. A tensile load of 8.89×10^3 lb is supported by a member having a cross-sectional area of 3.62×10^2 in.^{2}. Find the stress in the member. (Divide the load by the area.)

46. A bar having a cross-sectional area of 0.97 in.^{2} and a length of 3.62×10^2 in. is subjected to a tensile load of 5.27×10^3 lb. The elongation is 0.0446 in. Find the modulus of elasticity E of the material. (Multiply the load by the length, and divide by the area and by the elongation.)

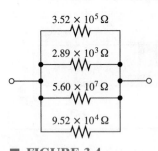

FIGURE 3-4
Resistors in parallel.

3-7 · COMBINED OPERATIONS

In Sec. 1-7 we showed how to do combined operations with whole numbers on the calculator. Combined operations with approximate numbers are done the same way. However, we must round our answer properly using the rules given earlier in this chapter.

Example 46

Evaluate the expression

$$\left(\frac{118.8 + 4.23}{\sqrt{136}} \right)^3$$

Solution: The keystrokes are

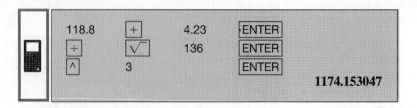

How many digits should we keep? In the numerator, we added a number with one decimal place to another with two decimal places, so we are allowed to keep just one. Thus the numerator, after addition, is good to one decimal place, or, in this case, four significant digits. The denominator, however, has just three significant digits, so we round our answer to three significant digits, getting 1170.

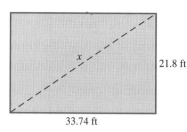

■ FIGURE 3-5
A rectangular courtyard.

Example 47

A rectangular courtyard (Fig. 3-5) having sides of 21.8 ft and 33.74 ft has a diagonal measurement x given by the expression

$$x = \sqrt{(21.8)^2 + (33.74)^2}$$

Evaluate the expression to find x.

Solution: The keystrokes are

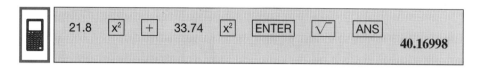

or

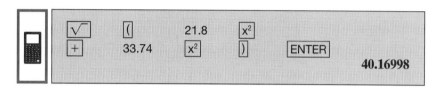

In either case, we round our answer to three significant digits, because 21.8 has only three, and get $x = 40.2$ ft.

SECTION 3-7 Exercises Combined Operations

Perform the computations, keeping the proper number of digits in your answer.

1. $(7.37)(3.28) + (8.36)(9.64)$

2. $(522)(9.53) - (586)(4.70) + (847)(7.59)$

3. $(63.5 + 83.6)(8.37 \times 10^2 - 1.72 \times 10^2)$

4. $(8.93 - 3.74 + 9.05)(68.7 - 64.9)$

5. $\dfrac{583}{473} + \dfrac{946}{907}$

6. $\dfrac{6.73}{8.38} - \dfrac{5.97}{8.06} + \dfrac{8.63}{1.91}$

7. $\dfrac{6.47 + 8.604}{3.37 + 90.8}$

8. $\dfrac{809 - 463 + 744}{758 - 964 + 508}$

9. $(5.37 + 8.36)^2$

10. $(4.25 + 4.36 - 5.24)^4$

11. $(6.423 + 9.05)^3$

12. $\left(\dfrac{45.3 - 8.34}{8.74 - 5.62}\right)^{2.5}$

13. $\left(\dfrac{8.90}{4.70}\right)^2$

14. $\sqrt{4.34 \times 10^3 + 5.66 \times 10^2}$

15. $\sqrt[3]{657 + 553 - 842}$

16. $\sqrt{(28.1)(5.94 \times 10^2)}$

17. $\sqrt[5]{(9.06)(4.86)(7.93)}$

18. $\sqrt{\dfrac{653}{601}}$

19. $4\sqrt{\dfrac{4.50}{7.81}}$

20. $\sqrt{9.74} + \sqrt{12.5}$

21. $\sqrt[4]{528} + \sqrt{94.2} - \sqrt[3]{284}$

22. $\sqrt[4]{653} \times \sqrt{55.3}$

Find the reciprocal of each number, retaining the proper number of digits in your answer.

1. 328 **2.** 21.78 **3.** 2.358

4. 3.82 **5.** 53.2 **6.** 8.62

Write each number in scientific notation.

7. 2300 **8.** 3000 **9.** 83.9×10^4

10. 24,491 **11.** 3020 **12.** 34,000

Convert each number from scientific notation to decimal notation.

13. 3.023×10^2 **14.** 1.9902×10^8 **15.** 8.01×10^3

Evaluate each expression, retaining the correct number of digits in your answer.

16. $(7.33)^3$ **17.** $(2.53)^4$ **18.** $(7.83)^2$

19. $(2.94)^3$ **20.** $(6.327)^2$ **21.** $(4.69)^5$

22. $\sqrt{34.2}$ **23.** $\sqrt{3.527}$ **24.** $\sqrt[3]{23.0}$

25. $\sqrt{892}$ **26.** $(3.2534)^3$ **27.** $(2.53)^4$

28. $(0.0440)^{0.552}$ **29.** $(3.365)^{3.7}$ **30.** $(4.65)^{3/2}$

31. $(3261)^{1/3}$ **32.** $\sqrt{3626}$ **33.** $\sqrt[3]{8373}$

34. $\sqrt[5]{837}$ **35.** $3\sqrt{92.8}$

36. $(322)(7.53) - (186)(7.70) + (347)(259)$

37. $(7.93 - 8.74 + 3.05)(18.7 - 4.9)$ **38.** $\dfrac{383}{173} + \dfrac{746}{307}$

39. $\dfrac{3.47 + 4.604}{7.37 + 30.8}$ **40.** $\dfrac{309 - 263 + 444}{458 - 264 + 808}$

41. $(8.25 + 2.36 - 9.24)^4$ **42.** $\left(\dfrac{75.3 - 2.34}{5.74 - 4.62}\right)^{2.5}$

43. $\sqrt{7.34 + 3.66}$ **44.** $\sqrt{(68.1)(3.94)}$

45. $\sqrt{\dfrac{553}{301}}$ **46.** $\sqrt{6.74} + \sqrt{22.5}$

47. $\sqrt[4]{253} \times \sqrt{75.3}$ **48.** $(4 \times 10^4)(2 \times 10^2)$

49. $(3 \times 10^3)(5 \times 10^2)$ **50.** $(2.39 \times 10^6)(8.60 \times 10^5)$

51. $(1.03 \times 10^3)(3.17 \times 10^8)$ **52.** $(2.95 \times 10^2)^3$

53. $\dfrac{4.39 \times 10^8}{3.22 \times 10^7} - 8.56$ **54.** $(2.52 \times 10^3)^3$

55. $\dfrac{(3.24 \times 10^6) - (1.35 \times 10^3)}{(2.12 \times 10^7) + (1.58 \times 10^2)}$ **56.** $(2.52 \times 10^3)^2$

57. $(8.20 \times 10^3)^4$ **58.** $(2.40 \times 10^3) + (4.90 \times 10^4)$

59. $(62.5 \times 10^3) + 1920$

60. $(3.25 \times 10^5) - (2.30 \times 10^4) + (4.00 \times 10^3)$

61. $(1.35 \times 10^2) + (2.34 \times 10^3) + (3.16 \times 10^4)$

62. $(2.26 \times 10^6) + (9.17 \times 10^4)$

63. A new memory unit containing 5.34×10^5 bits of memory is added to a computer that already has a memory of 8.84×10^7 bits. Find the total number of bits after the new unit is installed.

64. The wind pressure against the side of a certain building is 1.05 lb/ft². If the area of the side of the building is 4.58×10^3 ft², find the total force of the wind.

65. How long will it take a rocket traveling at a rate of 7.8×10^7 meters per hour to travel the 3.8×10^8 m from the earth to the moon?

WRITING

66. Suppose you have submitted a report that contains calculations in which you have rounded the answers according to the rules given in this chapter. Jones, your company rival, has sharply attacked your work, calling it "inaccurate" because you did not keep enough digits, and your boss seems to agree.

Write a memo to your boss defending your rounding practices. Point out why it is misleading to retain too many digits. Do not write more than one page. You may use numerical examples to prove your point.

67. Study your calculator and its manual specifically on the different display formats (normal, scientific notation, and so forth). List the different formats available to you and explain the differences between them. Write a few lines explaining how to switch from one format to another.

68. Starting with the template in Fig. 3-6, a new template is to be made by increasing dimension A by 25.0%, and decreasing dimension B by 30.0%. Find the dimension x (in millimeters) for the new template.

■ **FIGURE 3-6**
A template, not drawn to scale.

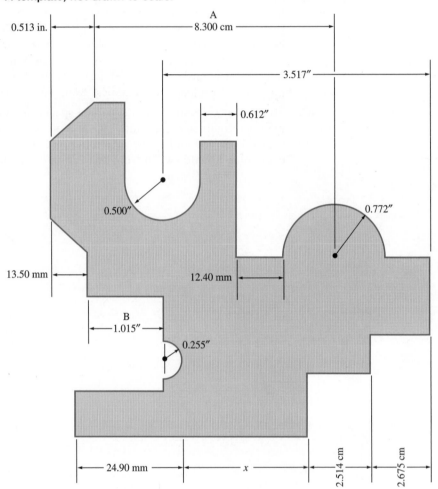

Units of Measurement

When we measure any physical quantity, such as the length of a board, we make use of *units of measure,* such as the foot or meter. In general use today are the *British* system of units (feet, pounds, gallons, etc.) and the SI or *metric* system of units (meters, kilograms, liters, etc.). SI stands for Le Systeme International d'Unites, or the *International System of Units.*

In this chapter we learn about different units of measure and how to convert between them. We also learn how to add, subtract, multiply, and divide *denominate numbers,* those that have units (2.73 cm is a denominate number).

4-1 CONVERTING UNITS

Systems of Units

Each physical quantity can be expressed in any one of a bewildering variety of units. Length, for example, can be measured in meters, inches, nautical miles, angstroms, feet, and so on. In this section we will convert from one British unit to another. In the next section we will convert from one metric unit to another, and also convert between British and metric units.

Conversion Factors

We can convert from any unit of length to any other unit of length by multiplying by a *conversion factor.*

Example 1

Convert 1.532 miles to feet.

Solution: From the Table of Conversion Factors, Appendix B, we find the relationship between miles and feet.

$$5280 \text{ feet} = 1 \text{ mile}$$

Dividing both sides by 1 mile we get the conversion factor:

$$\frac{5280 \text{ feet}}{1 \text{ mile}} = 1$$

Multiplying yields

$$1.532 \text{ miles} = 1.532 \text{ miles} \times \frac{5280 \text{ ft}}{1 \text{ mi}} = 8089 \text{ ft}$$

Note that we have rounded our answer to four significant digits, because the number used in the calculation (1.532) has four significant digits (the 5280 is exact).

Suppose, in the first step of the preceding example, that we had divided both sides by 5280 ft instead of by 1 mile. We would have gotten another conversion factor:

$$\frac{1 \text{ mi}}{5280 \text{ ft}} = 1$$

Thus, each relation between two units of measurement gives us *two* conversion factors. Each of these is sometimes called a *unity ratio;* that is, a fraction that is equal to unity, or one. Since each of these conversion factors is equal to 1, we may multiply any quantity by a conversion factor without changing the value of that quantity. It will, however, cause the units to change. But which of the two conversion factors should we use? It is simple. *Multiply by the conversion factor that will allow you to cancel the units you wish to eliminate.*

Significant Digits

You should try to use a conversion factor that is exact, or one that contains as many significant digits as (or preferably, one more than) your original number. Then you should round your answer to as many significant digits as in the original number. The Table of Conversion Factors shows whether a conversion factor is exact or approximate, but most other tables of conversion factors do not.

Example 2

Convert 934 acres to square miles (mi^2).

Solution: From the Table of Conversion Factors we find the equation

$$1 \text{ mi}^2 = 640 \text{ acres}$$

where 640 is an exact number. We must write our conversion factor so that the unwanted unit (acres) is in the denominator, so that acres will cancel. So our conversion factor is

$$\frac{1 \text{ mi}^2}{640 \text{ acres}} = 1$$

Multiplying, we obtain

$$934 \text{ acres} = 934 \text{ acres} \times \frac{1 \text{ mi}^2}{640 \text{ acres}}$$

$$= 1.46 \text{ mi}^2$$

The conversion factor used here is exact, so we have rounded our answer to three significant digits.

Using More Than One Conversion Factor

Sometimes you may not be able to find a *single* conversion factor linking the units you want to convert. You may have to use *more than one.*

Example 3

Convert 7375 yards to nautical miles.

Solution: In the Table of Conversion Factors we find no conversion factor between nautical miles and yards, but we see that

$$1 \text{ nautical mile} = 6076 \text{ ft} \quad \text{and} \quad 3 \text{ ft} = 1 \text{ yd}$$

So

$$7375 \text{ yd} = 7375 \text{ yd} \times \frac{3 \text{ ft}}{1 \text{ yd}} \times \frac{1 \text{ nau mi}}{6076 \text{ ft}} = 3.641 \text{ nautical miles}$$

Convert.

1. 152 inches to feet
2. 6.53 miles to yards
3. 762.0 feet to inches
4. 627 feet to yards
5. 29 tons to pounds
6. 88.90 pounds to ounces
7. 89,600 pounds to tons
8. 8552 ounces to pounds
9. 1.25 miles to nautical miles
10. 2.63 bushels to cubic feet
11. 375 cubic inches to cubic feet
12. 34.5 board feet to cubic inches
13. 3.75 cubic feet to board feet
14. 88.2 cubic yards to cubic feet
15. 2.88 US gallons to cubic inches
16. 5.82 cubic feet to US gallons
17. 87.2 cubic inches to US gallons
18. 392 acres to square miles
19. 8834 square feet to acres
20. 994 square yards to acres

Applications

21. A certain farm has an area of 93,850 square feet. How many acres is that?
22. Convert all the dimensions for the part in Fig. 4-1 to inches.
23. The jet fuel tank, Fig. 4-2, has a volume of 15.7 ft³. How many gallons of jet fuel will it hold?
24. A certain circuit board weighs 0.176 lb. Find its weight in oz.

■ **FIGURE 4-1**

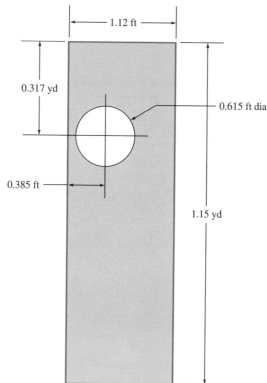

■ **FIGURE 4-2**
A jet fuel tank.

THE METRIC SYSTEM

Metric Units

The *metric system* is a system of weights and measures developed in France in 1793 and has now been adopted by most countries of the world. It is widely used in scientific work in the United States.

The basic unit of length in the metric system is the *meter* (m). The unit of area is the *are,* or 100 square meters. The unit of volume is the *liter* (L), the volume of a cube one tenth of a meter on a side. The unit of weight is the *gram* (g), the theoretical weight of a cube of distilled water measuring $\frac{1}{100}$ of a meter on a side.

Metric Prefixes

Converting between metric units is made easy because larger and smaller metric units are related to the basic units by *factors of ten*. These larger or smaller units are indicated by placing a *prefix* before the basic unit. A prefix is a group of letters placed at the beginning of a word to modify the meaning of that word. For example, the prefix *kilo* means 1000, or 10^3. Thus, a *kilo*gram is 1000 grams.

Other metric prefixes are given in Table 4-1.

■ **TABLE 4-1 Metric Prefixes**

Amount	Multiples and submultiples	Prefixes	Symbols	Pronunciations	Means
1 000 000 000 000	10^{12}	tera	T	ter´á	One trillion times
1 000 000 000	10^9	giga	G	ji´ga	One billion times
1 000 000	10^6	mega	M*	meg´á	One million times
1 000	10^3	kilo	k*	kil´o	One thousand times
100	10^2	hecto	h	hek´to	One hundred times
10	10	deka	da	dek´a	Ten times
0.1	10^{-1}	deci	d	des´i	One tenth of
0.01	10^{-2}	centi	c*	sen´ti	One hundredth of
0.001	10^{-3}	milli	m*	mil´i	One thousandth of
0.000 001	10^{-6}	micro	μ*	mi´kro	One millionth of
0.000 000 001	10^{-9}	nano	n	nan´o	One billionth of
0.000 000 000 001	10^{-12}	pico	p	pē´co	One trillionth of
0.000 000 000 000 001	10^{-15}	femto	f	fem´to	One quadrillionth of
0.000 000 000 000 000 001	10^{-18}	atto	a	at´to	One quintillionth of

* Most commonly used.

Example 4

(a) A *kilo*meter (km) is a thousand meters, because *kilo* means one thousand.

$$1 \text{ km} = 1000 \text{ m}$$

(b) A *centi*meter (cm) is one hundredth of a meter, because *centi* means one hundredth.

$$1 \text{ cm} = \frac{1}{100} \text{ m}$$

(c) A *milli*meter (mm) is one thousandth of a meter, because *milli* means one thousandth.

$$1 \text{ mm} = \frac{1}{1000} \text{ m}$$

Converting from One Metric Unit to Another

Converting from one metric unit to another is usually a matter of multiplying or dividing by a power of ten. Most of the time, the names of the units will tell how they are related, so that we do not even have to look them up.

Example 5

Convert 72,925 meters to kilometers (km).

Solution: A *kilo*meter is a thousand meters,

$$\frac{1 \text{ km}}{1000 \text{ m}} = 1$$

so, as before,

$$72{,}925 \text{ m} = 72{,}925 \text{ m} \times \frac{1 \text{ km}}{1000 \text{ m}} = 72.925 \text{ km}$$

For more unusual metric units, such as Newtons, simply look up the conversion factor in a table and convert as we did for British units.

Example 6

Convert 2.75 Newtons (N) to dynes.

Solution: These two metric units of force do not have any basic units in their names, nor any prefixes. Thus we cannot tell just from their names how they are related to each other. However, from the Table of Conversion Factors we find that

$$1 \text{ Newton} = 10^5 \text{ dynes}$$

Converting in the usual way,

$$2.75 \text{ Newtons} = 2.75 \text{ Newtons} \times \frac{10^5 \text{ dynes}}{1 \text{ Newton}} = 2.75 \times 10^5 \text{ dynes}$$

or 275,000 dynes.

Converting Between British and Metric Units

We convert between British and metric units in the same way that we converted within each system.

Example 7

Convert 2.84 US gallons to liters.

Solution: From the Table of Conversion Factors we find that

$$1 \text{ gallon (US)} = 3.785 \text{ liters}$$

Converting,

$$2.84 \text{ gallons} = 2.84 \text{ gallons} \times \frac{3.785 \text{ liters}}{1 \text{ gallon}} = 10.7 \text{ liters}$$

rounded to three significant digits.

Convert. Write your answer in scientific notation if the numerical value is greater than 1000 or less than 0.1. Be careful to retain the proper number of significant digits.

1. 364,000 m to kilometers
2. 0.000473 V to millivolts
3. 735,900 g to kilograms
4. 7.68×10^{-5} kW to watts
5. 6.2×10^9 ohms to megohms
6. 825×10^4 N to kilonewtons
7. 9348 picofarads to microfarads
8. 84,398 nanoseconds to milliseconds
9. 2.75 kilograms to grams
10. 5.77 kiloohms to ohms
11. 364,000 m to feet
12. 6.83 in. to millimeters
13. 735,900 lb to kilograms
14. 1.27 hp to watts
15. 4.66 gallons to liters
16. 825×10^4 dynes to newtons
17. 3.94 yards to meters
18. 83.4 cm³ to gallons
19. 0.275 lb to grams
20. 2.840 km to yards

Applications

21. A certain circuit has a response time of 2.88 milliseconds. What is that in nanoseconds?

22. A certain laptop computer weighs 6.35 kg. What is its weight in pounds?

23. A generator has an output of 5.34×10^6 millivolts. What is the output in kilovolts?

24. Convert all the dimensions in Fig. 4-3 to centimeters.

■ **FIGURE 4-3**

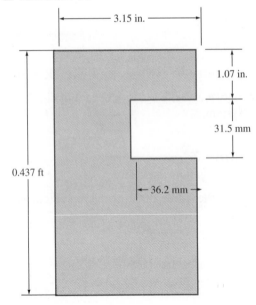

ARITHMETIC WITH NUMBERS HAVING UNITS OF MEASURE

In this section we show how to add, subtract, multiply, and divide *denominate* numbers (numbers having units of measure). We will also find powers and roots of denominate numbers.

Adding and Subtracting Denominate Numbers

If the numbers to be added or subtracted have the same units, simply combine them as usual. The answer will have the same units as the original numbers. Be sure to keep the proper number of decimal places.

If the numbers have *different* units, they must be converted to all have the same units, before combining.

Example 8

Add 3.823 in. and 9.48 cm, and express the result in inches.

Solution: We convert 9.48 cm to in., knowing that 1 in. = 2.54 cm.

$$9.48 \text{ cm} = 9.48 \text{ cm} \times \frac{1 \text{ in.}}{2.54 \text{ cm}} = 3.73 \text{ in.}$$

Adding,

$$3.823 \text{ in.} + 3.73 \text{ in.} = 7.55 \text{ in. rounded}$$

Multiplying Denominate Numbers

Simply multiply the two numbers, keeping the proper number of significant digits. The units in the product will be *the product of the units in the original numbers.*

Example 9

(a) 2.87 inches $\times$ 4.13 pounds = 11.9 inch-pounds
(b) 10.2 cm $\times$ 2.05 cm = 20.9 cm^2

Dividing Denominate Numbers

Divide the given numbers as usual. The units in the quotient will be the quotient of the units in the original numbers.

Example 10

Divide 394 miles by 7.25 hours.

Solution: Dividing,

$$\frac{394 \text{ miles}}{7.25 \text{ hours}} = 54.3 \text{ mi/h}$$

If the units of dividend and divisor *are the same,* we get a number with *no units.* Such *dimensionless quantities* are of great importance in technical work.

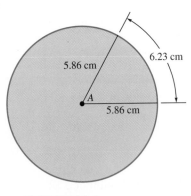

■ FIGURE 4-4
An arc of a circle.

Example 11

An arc 6.23 cm in length is cut off by two radii in a circle with a radius of 5.86 cm, Fig. 4-4. Divide the arc length by the radius.

Solution: Dividing,

$$\frac{6.23 \text{ cm}}{5.86 \text{ cm}} = 1.06$$

The centimeters cancel, leaving the quotient 1.06 without units. (We'll see later that this dimensionless quotient is the *radian measure* of the angle *A* shown in the figure.)

Powers of Denominate Numbers

When raising a denominate number to a power, *be sure to raise the units* to the same power.

Example 12

A certain square board is 5.34 in. on a side. Square that number to get the area of the board.

Solution: Squaring both the number and the units gives

$$(5.34 \text{ in.})^2 = (5.34)^2 (\text{in.})^2$$
$$= 28.5 \text{ in.}^2$$

Roots of Denominate Numbers

When taking the root of a denominate number, we must also take the same root of the units of measure.

Example 13

A certain cube has a volume of 837 cubic centimeters (cm^3). Find the side of that cube by taking the cube root of the given number.

Solution: Taking the cube root of both the number and the units,

$$\sqrt[3]{837 \text{ cm}^3} = \sqrt[3]{837} \, \sqrt[3]{\text{cm}^3} = 9.42 \text{ cm}$$

SECTION 4-3 **Exercises** **Arithmetic with Numbers having Units of Measure**

Be sure to include the proper units with your numerical results.

Adding and Subtracting Denominate Numbers

1. Add 64.8 cm and 34.1 in. Give the result in inches.
2. Add 43.1 yd and 32.9 m. Give the result in meters.
3. Add 33.6 acres and 41.5 ares. Give the result in acres.
4. Add 66.3 ft³ and 21.7 m³. Give the result in cubic meters.

Multiplying Denominate Numbers

5. Multiply 53.4 in. and 5.93 lb.
6. Multiply 34.8 in. and 3.72 cm. Give the result in square inches.
7. Multiply 3.41 cm, 4.93 cm, and 1.82 cm.
8. Multiply 2.73 m, 1.83 yd, and 5.22 ft. Give the result in cubic meters.

Dividing Denominate Numbers

9. Divide 83.2 miles by 2.84 hours.
10. Divide 823 ft by 34.7 seconds.
11. Divide 56.2 m by 34.1 seconds.
12. Divide 88.8 km by 2.44 seconds.
13. Divide 567 cm by 22.5 cm.
14. Divide 83.2 cm by 30.2 in.
15. Divide 294 yd by 162 m.

Powers of Denominate Numbers

16. Find the square of 2.83 in.
17. Find the square of 5.23 cm.
18. Find the square of 3.75 ft.
19. Find the cube of 44.3 mm.
20. Find the cube of 1.66 yd.
21. Find the cube of 4.72 in.

Roots of Denominate Numbers

22. Find the square root of 622 ft^2.
23. Find the square root of 25.8 mm^2.
24. Find the cube root of 73.6 cubic yards.
25. Find the cube root of 255 cubic centimeters.

Applications

26. A building lot, Fig. 4-5, has an area of 2.74 acres. An adjoining lot containing 15,400 ft^2 is obtained. How many acres are in the combined lots?

27. A landing strip, Fig. 4-6, is 355 m long and 48.5 ft wide. Find the area of the strip in square meters by multiplying the length by the width and doing the proper conversion of units.

28. A cylindrical storage tank, Fig. 4-7, has a volume of 3740 ft^3. The area of one end is 185 m^2. Find the height h of the tank, in feet, by dividing the volume by the area of the end and making the proper units conversions.

29. A cube of ceramic, Fig. 4-8, measures 2.45 in. on a side. Find its volume by cubing the length of the side.

30. A square computer chip, Fig. 4-9, has a face area of 1.74 mm^2. Find the length L of a side by taking the square root of the area.

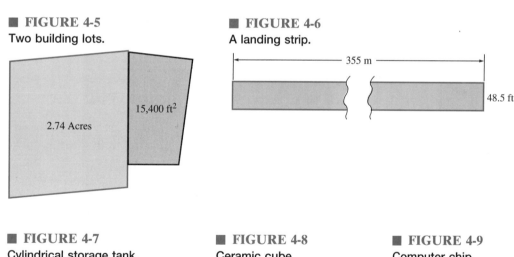

■ **FIGURE 4-5**
Two building lots.

15,400 ft^2

2.74 Acres

■ **FIGURE 4-6**
A landing strip.

355 m

48.5 ft

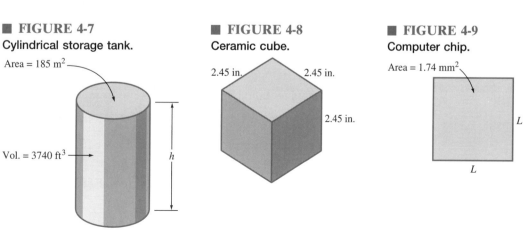

■ **FIGURE 4-7**
Cylindrical storage tank.

Area = 185 m^2

Vol. = 3740 ft^3

h

■ **FIGURE 4-8**
Ceramic cube.

2.45 in.   2.45 in.

2.45 in.

■ **FIGURE 4-9**
Computer chip.

Area = 1.74 mm^2

L

L

CONVERTING AREAS AND VOLUMES

If the conversion factor needed to convert an area or a volume can be found in a table of conversion factors, do the conversion as was shown for other units. If the conversion factor needed is not in the table, it can often be obtained by squaring or cubing a conversion factor for length.

Example 14

Convert 4285 square yards to acres.

Solution: The Table of Conversion Factors has no conversion factor for square yards. However we see

$$1 \text{ yd} = 3 \text{ ft}$$

Squaring yields

$$1 \text{ yd}^2 = (1 \text{ yd})^2 = (3 \text{ ft})^2 = 9 \text{ ft}^2$$

Also from the table,

$$1 \text{ acre} = 43{,}560 \text{ ft}^2$$

So

$$4285 \text{ yd}^2 = 4285 \text{ yd}^2 \times \frac{9 \text{ ft}^2}{1 \text{ yd}^2} \times \frac{1 \text{ acre}}{43{,}560 \text{ ft}^2} = 0.8853 \text{ acres}$$

SECTION 4-4   Exercises   Converting Areas and Volumes

Convert the following areas.

1. 2840 yd^2 to acres
2. 1636 m^2 to ares
3. 24.8 ft^2 to m^2
4. 59.2 in.^{2} to cm^2
5. 557 mm^2 to in.^{2}
6. 372.0 m^2 to ft^2
7. 982 km^2 to acres
8. 5.95 acres to m^2
9. 8845 ft^2 to yd^2
10. 7.66 in^2 to mm^2

Convert the following volumes.

11. 7360 ft^3 to in.^{3}
12. 4.83 m^3 to yd^3
13. 73.8 yd^3 to m^3
14. 19.2 in.^{3} to cm^3
15. 267 mm^3 to in.^{3}
16. 8.22 gal to ft^3
17. 112 liters to gallons
18. 1.43 cm^3 to m^3
19. 2640 ft^3 to yd^3
20. 1.36 in.^{3} to mm^3

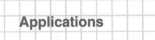

Applications

21. The surface area of a certain lake, Fig. 4-10, is 7360 yd^2. Convert this to square meters.

22. A solar collector, Fig. 4-11, has an area of 8834 in.^{2}. Convert this to square meters.

23. The volume of a balloon, Fig. 4-12, is 8360 ft^3. Convert this to cubic inches.

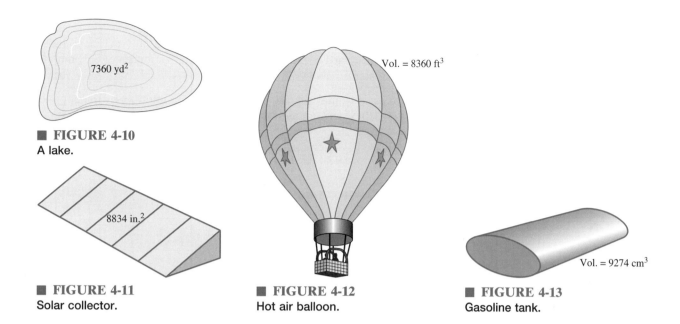

■ **FIGURE 4-10**
A lake.

■ **FIGURE 4-11**
Solar collector.

■ **FIGURE 4-12**
Hot air balloon.

■ **FIGURE 4-13**
Gasoline tank.

24. The volume of a certain gasoline tank, Fig. 4-13, is 9274 cm³. Convert this to gallons.

4-5

CONVERTING RATES

A *rate* is the amount of one quantity expressed per unit of another quantity.

┌─── **Example 15** ───
Some typical rates are

(a) Flow rate (gal/min)
(b) Rate of travel (km/h)
(c) Application rate (lb/h)
(d) Unit price (dollars/lb)

The units given here are, of course, not the only ones that could be used. Flow rate, for example, could be in liters per hour, or any unit of volume per unit of time.

A good clue that you are dealing with a rate is the word *per,* as in dollars *per* pound. Note that each rate contains *two* units of measure.

┌─── **Example 16** ───
Miles per hour has miles in the numerator and hours in the denominator.

It may be necessary to convert either or both of those units to other units. Sometimes a single conversion factor can be found (such as 1 m/h = 1.4667 ft/s), but more often you will have to convert each unit with a separate conversion factor.

Example 17

Chromium is to be plated onto steel automobile parts at the rate of 1.75 pounds of chromium per square yard of steel. Convert this rate to ounces of chromium per square meter of steel.

Solution: We write the original quantity as a fraction, and multiply by the appropriate factors, themselves written as fractions.

$$1.75 \text{ lb/yd}^2 = \frac{1.75 \text{ lb}}{\text{yd}^2} \times \frac{1 \text{ yd}^2}{9 \text{ ft}^2} \times \frac{10.76 \text{ ft}^2}{1 \text{ m}^2} \times \frac{16 \text{ oz}}{1 \text{ lb}} = 33.5 \text{ oz/m}^2$$

SECTION 4-5 | Exercises | Converting Rates

Convert units on the following time rates.

1. 5.86 ft/s to miles per hour

2. 777 gal/min to cubic meters per hour

3. 66.2 mi/h to kilometers per hour

4. 52.0 knots to miles per minute

5. 953 births/year to births per week

6. 8336 Btu/h to foot-pounds per minute

7. 88.9 ft²/min to acres per year

Convert units on the following unit prices

8. $1.25/g to dollars per kilogram

9. $800/acre to cents per square meter

10. $3.54/lb to cents per ounce

11. $4720/ton to cents per pound

12. An airplane is cruising at a speed of 785 mi/h. Convert this speed to km/h.

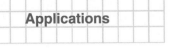

13. Insecticide is applied to a crop at the rate of 5.00 gal/acre. How many liters/are is this?

14. A certain ore costs $25.35 per ton. Find the cost in lire per metric ton, if the exchange rate is 1250 lire to the dollar.

■ Chapter 4 | REVIEW EXERCISES

Convert.

1. 3646 Btu/h to foot-pounds per minute

2. 22.3 miles to kilometers

3. 2.28×10^9 ohms to megohms

4. $7360/ton to cents per pound

5. 7250 picofarads to microfarads

6. 274 acres to square meters

7. 6253 cm² to square inches

8. 2.74 yd³ to cubic meters

9. 9.26 ft²/min to acres per year

10. 385×10^4 N to kilonewtons

11. 228 kg to slugs

12. 72.2 tons to kilograms

13. 83.2 m to kilometers

14. 49.2 mi/h to kilometers per hour

15. 283,200 g to kilograms

16. 34.9×10^{-5} kW to watts

17. $995/acre to cents per square meter

18. 28,300 nanoseconds to milliseconds

19. 635 lb to newtons

20. 88.2 hp to kilowatts

21. 28.2 ft/s to miles per hour

22. $9.85/g to dollars per kilogram

23. A car travels at a speed of 55.5 mi/h. Convert this speed to km/h.

24. A farm has an area of 474 acres. Another parcel containing 0.225 square miles is obtained. How many acres are in the combined parcels?

25. A driveway is 22.5 ft long and 3.10 m wide. Find the area of the driveway in square feet by multiplying the length by the width and doing the proper conversion of units.

26. A cylindrical bar has a volume of 0.372 ft^3. The length of the bar is 1.75 m. Find the area of one end of the bar, in square inches, by dividing the volume by the length and making the proper units conversions.

27. A cubical crate measures 6.45 ft on a side. Find its volume by cubing the length of the side.

28. A square tile has an area of 94.6 in.^{2}. Find the length of a side by taking the square root of the area.

WRITING

29. Your company is agonizing over whether to convert to metric or stay with the English system of measurements. Write a letter to your company president giving specific reasons why you think you should "go metric."

30. Suppose that you work in the same company described in Exercise 29. Write a letter to your company president giving specific reasons why you think you should *not* convert to metric.

TEAM PROJECT

31. The Chinese have a different system of weights than the English or the metric system. These are the *tael*, the *catty* (16 taels), and the *picul* (100 catties). One catty is equal to 0.605 kg. Determine

(a) how many ounces are equal to one tael.

(b) how many piculs are in one ton.

(c) how many catties are needed to make 7.00 kg.

Percentage

Of all the mathematical topics we cover in this text, probably the one most used in everyday life is percentage. The daily news usually has several references to percentage; the inflation rate is now such-and-such percent, the unemployment rate is such-and-such percent, and so on. In this chapter we define *percent,* show how to convert numbers and fractions to and from percents, and work numerous types of percentage problems.

5-1   CONVERTING TO AND FROM PERCENT

Definition

The word *percent* means by the hundred, or per hundred. A percent thus gives the number of parts *in every hundred.*

> **Example 1**
> If we say that a certain alloy is 25% zinc by weight, we mean that 25 out of every 100 lb of alloy will be zinc.

Percent as a Rate

The word *rate* is often used to indicate a percent, or percent rate, as in "rejection rate," "rate of inflation," or "growth rate."

> **Example 2**
> A germination rate for seeds of 94% means that, on the average, 94 seeds out of every 100 would be expected to germinate.

Percent as a Fraction

Percent is another way of expressing a fraction having 100 as a denominator.

> **Example 3**
> If we say that a mason has finished 60% of a chimney, we mean that $\frac{60}{100}\left(\text{or } \frac{3}{5}\right)$ of the chimney is finished.

Converting Decimals to Percent

We will now get some practice in converting decimals and fractions to percent, and vice versa. This will give us the skills needed to do the percentage problems coming later.

To convert decimals to percent, simply move the decimal point two places to the right and write the percent symbol (%) following the number.

Example 4

(a) $0.50 = 50\%$

(b) $2.70 = 270\%$

(c) $0.006 = 0.6\%$

Converting Fractions or Mixed Numbers to Percent

First write the fraction or mixed number as a decimal, and then proceed as above.

Example 5

(a) $\frac{1}{5} = 0.20 = 20\%$

(b) $\frac{7}{4} = 1.75 = 175\%$

(c) $2\frac{1}{2} = 2.5 = 250\%$

With a fraction that yields a repeating decimal, we must either be told how many significant digits to retain or decide how many from the other numbers in the problem.

Example 6

Change $\frac{1}{3}$ to a percent, keeping four significant digits.

Solution:

$$\frac{1}{3} = 0.33333\ldots = 33.33\%$$

Converting Percent to Decimals

Move the decimal point two places to the left and remove the percent sign.

Example 7

(a) $22\% = 0.22$

(b) $2.3\% = 0.023$

(c) $235\% = 2.35$

Converting Percent to a Fraction

Write a fraction with 100 in the denominator and the percent in the numerator. Remove the percent sign and reduce the fraction to lowest terms.

Example 8

(a) $25\% = \dfrac{25}{100} = \dfrac{1}{4}$

(b) $62.5\% = \dfrac{62.5}{100} = \dfrac{625}{1000} = \dfrac{5}{8}$

(c) $120\% = \dfrac{120}{100} = \dfrac{6}{5} = 1\dfrac{1}{5}$

SECTION 5-1 | Exercises | Converting to and from Percent

Convert each decimal to a percent.

1. 6.29	**2.** 0.377	**3.** 0.0046	**4.** 0.748
5. 0.653	**6.** 0.0827	**7.** 3.959	**8.** 0.594

Convert each fraction to a percent. Round to three significant digits.

9. $\dfrac{3}{5}$ **10.** $\dfrac{1}{4}$ **11.** $\dfrac{9}{10}$ **12.** $\dfrac{7}{9}$

13. $\dfrac{2}{7}$ **14.** $\dfrac{7}{11}$ **15.** $\dfrac{1}{25}$ **16.** $\dfrac{4}{9}$

17. $\dfrac{1}{3}$ **18.** $\dfrac{3}{19}$ **19.** $\dfrac{7}{15}$ **20.** $\dfrac{11}{60}$

Convert each percent to a decimal.

21. 45%   **22.** 3.64%   **23.** $365\dfrac{1}{2}\%$ **24.** $18\dfrac{1}{3}\%$

25. 7.4%   **26.** 13%   **27.** 52%   **28.** 305%

Convert each percent to a fraction.

29. 12.5%   **30.** $37\dfrac{1}{2}\%$ **31.** 225%   **32.** 7%

33. 62.5%   **34.** 120%   **35.** 45%

5-2   SOLVING PERCENTAGE PROBLEMS

Amount, Base, and Rate

Percentage problems always involve three quantities:

1. The *percent rate*
2. The *base,* the quantity we are taking the percent of
3. The *amount* we get when we take the percent of the base, also called the *percentage*

The rate, base, and amount are related by the following equation:

PERCENTAGE	Amount = base × rate where the rate is expressed as a decimal.	13

In the percentage problems that follow, you will be given *two* of these three quantities (amount, base, or rate) and be required to find the third.

Finding the Amount When the Base and Rate Are Known

We simply substitute the given base and rate into the percentage equation and solve for the amount.

Example 9 ————————
What is 45 percent of 60?

Solution: In this problem the rate is 45%, so

$$\text{Rate} = 0.45$$

But is 60 the amount or the base?

> **TIP** If you have trouble telling which number is the base and which number is the amount, look for the key phrase "percent of." The quantity following this phrase is always the base.

We look for the key phrase "percent of."

What is 45 *percent of* 60?

The quantity following *percent of* is the base, so the base here is 60.

$$\text{Base} = 60$$

Then,

$$\text{Amount} = \text{base} \times \text{rate} = 60(0.45) = 27$$

COMMON ERROR Don't forget to convert the percent rate to a decimal when using the percentage equation.

Example 10 ————————
Find 0.5% of 7000.

Solution: We substitute with

$$\text{Base} = 7000 \quad \text{and} \quad \text{Rate} = 0.5\% = 0.005 \text{ (not 0.5)}$$

So

$$\text{Amount} = \text{base} \times \text{rate} = 7000(0.005) = 35$$

Finding the Base when a Percent of It Is Known

We see from the percentage equation that the base equals the amount divided by the rate (expressed as a decimal), or

$$\text{Base} = \frac{\text{amount}}{\text{rate}}$$

Example 11

20% of what number is 55?

Solution: Finding the key phrase,

20 *percent of* [what number] is 55?

The quantity following *percent of* is "what number," so it is clear that we are looking for the base. So

Amount = 55   and   Rate = 0.20

Then,

$$\text{Base} = \frac{\text{amount}}{\text{rate}} = \frac{55}{0.20} = 275$$

Example 12

525 is 75% of what number?

Solution: From the percentage equation,

$$\text{Base} = \frac{\text{amount}}{\text{rate}} = \frac{525}{0.75} = 700$$

Finding the Percent One Number Is of Another Number

From the percentage equation, the rate equals the amount divided by the base, or

$$\text{Rate} = \frac{\text{amount}}{\text{base}}$$

Example 13

65 is what percent of 125?

Solution: From the percentage equation, with amount = 65 and base = 125,

$$\text{Rate} = \frac{\text{amount}}{\text{base}} = \frac{65}{125} = 0.52 = 52\%$$

Example 14

What percent of 1.8 is 1.35?

Solution: From the percentage equation,

$$\text{Rate} = \frac{\text{amount}}{\text{base}} = \frac{1.35}{1.8} = 0.75$$
$$= 75\%$$

SECTION 5-2   Exercises   Solving Percentage Problems

Finding the Amount   Find:

1. 15% of 500 tons	**2.** 45% of 300 mi	**3.** 70% of 1400 kg
4. $66\frac{1}{3}$% of 915 gal	**5.** 6% of 25 bushels	**6.** $16\frac{2}{3}$% of $42
7. 25% of 505 km	**8.** 20% of 625 liters	**9.** 55% of 962 acres

Finding the Base

Find the number of which:

10. 58 is $12\frac{1}{2}\%$

11. 75 is 1.5%

12. $\frac{2}{5}$ is $5\frac{1}{2}\%$

13. 100 is $33\frac{1}{3}\%$

14. 54 is 54%

15. 54 is 75%

16. $\frac{1}{4}$ is 25%

17. $\frac{3}{7}$ is $16\frac{2}{3}\%$

18. $\frac{6}{10}$ is $66\frac{2}{3}\%$

Finding the Rate

What percent of:

19. 32 is 16?

20. 64 is 8?

21. 35 is 7?

22. 960 men is 480 men?

23. 650 h are 195 h?

24. 600 tons are 250 tons?

25. 125 qt are 12.5 qt?

26. 75 gal are 15 gal?

27. 60 cm are 4 cm?

Profit and Loss

28. Find the profit or loss:
   (a) on land that cost $15,600 and was sold at a loss of 30%.
   (b) on goods that cost $7,830 and were sold at a gain of 23%.

Applications

29. A resistance, now 6050 Ω, is to be increased by 35%. How much resistance should be added?

30. It is estimated that $\frac{1}{2}\%$ of the earth's surface receives more energy than the total projected needs for the year 2000. Assuming the earth's surface area to be 1.97×10^8 mi², find the required area in acres.

31. As an incentive to install solar equipment, a tax credit of 35% of the first $1000 and 20% of the next $6400 spent on solar equipment is proposed. How much credit would a homeowner get when installing $4500 worth of solar equipment?

32. A woman receives a salary of $32,000 a year. She pays $20\frac{1}{2}\%$ of it for board, 9% of it for clothing, and 14% for incidentals. What are her yearly expenses?

33. A 15.4-m-long girder grew 0.15% when subjected to summer temperatures. Find the growth in cms.

34. It is estimated that the United States has tar sands containing 3.3×10^{10} barrels of oil and that 90% of the tar sands are in Utah. If it is possible to extract 9% of the oil from the sands, how much oil could be extracted from the Utah tar sands?

35. How much metal will be obtained from 590 tons of ore if the metal is $20\frac{1}{4}\%$ of the ore?

36. If a person rents a house for $4300 a year, which is 16% of the value of the property, what is its value?

37. A Department of Energy report on an experimental electric car gives the range of the car as 160 km and states that this is "50% better than on earlier electric vehicles." What was the range of earlier electric vehicles?

38. A man withdrew 40% of his bank deposits and spent 20% of the money drawn in the purchase of a radio worth $325. How much money did he have in the bank?

39. A merchant sold 75% of his goods for $2751. How much were the goods worth?

40. Solar panels provide 65% of the heat for a certain building. If $700/year is now being spent for heating oil, how much would be spent if the solar panels were not used?

41. A fire destroyed $16\frac{2}{3}\%$ of a stock of goods, and the value of the goods destroyed in the fire was $500. What was the value of the entire stock?

42. The distance between two stations on a certain railroad is 243.6 mi, which is $83\frac{1}{3}\%$ of the total length of the road. What is the total length of the road?

43. If the United States imports 8.62 million barrels of oil per day and if this is 47.3% of its needs, how much oil is needed per day?

44. A 45,050-liter-capacity tank contains 7320 liters of water. Express this amount of water in the tank as a percentage of the total capacity.

45. In a journey of 4590 km, a person traveled 1020 km by car and the rest by rail. What percentage of the distance was traveled by rail?

46. A power supply has a dc output 200.0 V with a ripple of 6.0 V peak to peak. Express the ripple as a percentage of the dc output voltage.

47. The construction of a factory cost $380,000 for materials and $490,000 for labor. What percentage of the total was the labor cost?

48. If the United States needs 18.9 million barrels of oil per day and imports 9.14 million barrels per day, what percentage of its oil needs are imported?

49. By insulating, a homeowner's yearly fuel consumption dropped from 894 gallons to 576 gallons. His present oil consumption is what percent of the former?

50. A city lot was sold for $159,000 at a profit of 12%. What did it cost?

51. A merchant gained $45,500 in one year on goods sold at a profit of 25%. What was the cost of the goods?

52. A machine was sold for $187, which was a loss of 15%. What was the cost?

53. An architect charged $\frac{1}{4}\%$ for plans and specifications and $1\frac{3}{8}\%$ for superintending a building that cost $32,000. What was the amount of her fee?

54. Find the selling price of an item having a list price of $348.99, discounted 14%.

55. A person paid $5789.70 for a car, including a 4% sales tax. Find the price of the car without the tax.

5-3   PERCENT CHANGE

Percents are often used to *compare* two quantities.

Example 15

If you wanted to compare the price of oil from last year to this year, you might say

The price of oil this year is 8% greater than last year.

Finding the Percent Change

To get an equation for percent change, we start with the percentage equation

$$\text{Amount} = \text{base} \times \text{rate}$$

When the two numbers being compared involve a change from one to the other, the *original value* is usually taken as the base. We then replace "Amount" with "new value − original value," since this is the amount by which the quantity changes.

$$(\text{New value} - \text{original value}) = (\text{original value}) \times \text{rate}$$

Dividing by "original value" gives the rate, or percent change. Then multiplying by 100 and affixing the percent sign, we get the percent change expressed as a percent.

$$\text{Percent change} = \frac{\text{new value} - \text{original value}}{\text{original value}} \times 100 \qquad 14$$

Example 16

A certain price rose from $1.35 to $1.50. Find the percent change in price.

Solution: We use the original value, $1.35, as the base. From the percent change equation,

$$\text{Percent change} = \frac{1.50 - 1.35}{1.35} \times 100 = 11.1\% \text{ increase}$$

Be sure to show the *direction* of change with a positive or negative sign, or with a word like *increase* or *decrease*.

Finding the New Value

We often want to find the new value when the original value is changed by a given percent. We see from the equation for percent change that

$$\text{New value} = \text{original value} + (\text{original value} \times \text{percent change})$$

Example 17

Find the cost of a $250 radio after the price increases by 5%.

Solution: The original value is $250 and the percent change, expressed as a decimal, is 0.05. So

$$\text{New value} = 250 + 250(0.05) = \$262.50$$

Thus the new value in Example 17 is equal to 100% of the old value, plus 5% of the old value, or 105% of the old value. In general, to increase a quantity by P percent, simply multiply that quantity by $(1 + P)$, where P is in decimal form.

Example 18

The price of the $250 radio of Example 17 after increasing by 5%, is

$$\text{New value} = 250 \times 1.05 = \$262.50$$

as before.

Similarly, to find the new value after a decrease of P percent, multiply the original value by $(1 - P)$, where again, P is in decimal form.

Example 19

What will be the value of a $4750 machine after a decrease of 12% in value?

Solution: The new value will be 12% less than the original, or 88% of the original. Thus,

$$\text{New value} = (\text{original value}) \times 0.88$$
$$= (4750)0.88 = \$4180$$

Finding the Original Value

Finally, we do a problem where we have the new value and the percent change, and we want to find the original value.

Example 20

What number, increased by 15% of itself, is 364?

Solution: The percent change is 15%, so the new value is 1.15 times as great as the original value. If we call the original value x, then

$$364 = 1.15x$$

We get x simply by dividing both sides by 1.15. Thus

$$x = \frac{364}{1.15} = 317 \quad \text{(rounded)}$$

SECTION 5-3 | **Exercises** | **Percent Change**

Finding Percent Change

Find the percent change when a quantity changes:

1. from 43.4 to 85.8
2. from 96 to 34
3. from 556 to 763
4. from 0.00904 to 0.00793

Finding the New Value

Find the new value when

5. 273 is increased by 2.45%
6. 4.83 is increased by 34%
7. 8475 is decreased by 2%
8. 8.36 is decreased by 52%

Finding the Original Amount

What number increased by:

9. 15% of itself = 414?
10. 19% of itself = 595?
11. $5\frac{1}{2}$% of itself = 422?
12. $12\frac{1}{2}$% of itself = 900?

What number decreased by:

13. 10% of itself = 18?
14. 40% of itself = 12?
15. 20% of itself = 620?
16. 90% of itself = 1.84?

Applications

17. The temperature in a building rose from 21°C to 23°C during the day. Find the percent change in temperature.

18. A casting initially weighing 205 lb has 13% of its material shaved off. Find its final weight.

19. The profits of a mill in two years were $66,990, and the profits for the second year were 3% greater than the profits the first year. What were the profits each year?

20. An automobile gas turbine is expected to provide a 30% improvement in fuel economy over a convenient spark-ignition engine. If a car gets 33.1 mi/gal, what mileage would you expect if its conventional engine is replaced by a gas turbine of the same horsepower?

21. A certain common stock rose from a value of 23\frac{1}{2}$ per share to 24\frac{3}{8}$ per share. What was the percent change in value?

22. A house that costs $598 per year to heat has insulation installed in the attic, causing the fuel bill to drop to $501 per year. Find the percent change in the cost.

23. A mechanic who received $84 per day had her salary raised by 7%. What were her daily wages then?

24. A crew paved 1350 m of roadway on one day and 1520 m the following day. What is the percent change in the amount they paved?

25. A certain country's energy consumption of 37 million barrels of oil equivalent per day is expected to climb to 48 million in six years. Find the percent increase in consumption.

26. The population of a certain town is 6024, which is $16\frac{2}{3}$% more than it was 4 years ago. What was the population then?

27. The temperature of a room rose from 65°F to 72°F. Find the percent increase.

28. A train running at 24 mi/h increases its speed by $37\frac{1}{2}$%. How fast does it then go?

29. If the energy consumption next year will be 3.6% higher than the 37 million barrels per day consumed now, how many barrels per day will be needed next year?

5-4

PERCENT ERROR, PERCENT CONCENTRATION, AND PERCENT EFFICIENCY

Percent Error

Percent error is a common way of expressing the accuracy of a measurement. The percent error is the difference between the measured value and the known or "true" value, expressed as a percent of the known value.

$$\text{Percent error} = \frac{\text{measured value} - \text{known value}}{\text{known value}} \times 100 \qquad \boxed{15}$$

Example 21

A gauge block that is certified to be 2.50 in. long is measured with a caliper. The reading is 2.53 in. What is the percent error in the reading?

Solution: From the percent error equation,

$$\text{Percent error} = \frac{2.53 - 2.50}{2.50} \times 100 = +1.2\%$$

As with percent change, be sure to specify the *direction* of the error, either with words or with a plus or minus sign.

Percent Concentration

In a mixture of two or more ingredients, the *percent concentration* of one ingredient is defined as

$$\text{Percent concentration of ingredient } A = \frac{\text{amount of } A}{\text{amount of mixture}} \times 100 \qquad \boxed{16}$$

--- **Example 22** ---

A certain antifreeze mixture contains 22 liters of antifreeze and 112 liters of water. Find the percent of antifreeze in the mixture.

Solution: The total amount of mixture is

$$22 + 112 = 134 \text{ liters}$$

So, by the equation for percent concentration,

$$\text{Percent antifreeze} = \frac{22}{134} \times 100 = 16.4\%$$

COMMON ERROR The denominator in the equation for percent concentration must be the *total amount* of mixture, or the sum of all the ingredients. Do not use just one of the ingredients.

Percent Efficiency

With any device, such as a motor or an amplifier, power is put into the device and power is taken out of the device, often in a different form.

--- **Example 23** ---

A motor takes power in electrical form and delivers power in mechanical form.

It is inevitable that the power output of any device is *always less* than the power input, because of power losses within the device. The *efficiency* of the device is a measure of those losses. The percent efficiency is defined as the output divided by the input, expressed as a percent.

$$\text{Percent efficiency} = \frac{\text{output}}{\text{input}} \times 100 \qquad \boxed{17}$$

--- **Example 24** ---

A certain electric motor consumes 981 W and has an output of 1.21 hp. Find the efficiency of the motor (1 hp = 746 W).

Solution: Since output and input must be in the same units, we must convert either to horsepower or to watts. Converting the output to watts, we obtain

$$\text{Output} = 1.21 \text{ hp} \left(\frac{746 \text{ W}}{\text{hp}} \right) = 903 \text{ W}$$

Then

$$\text{Percent efficiency} = \frac{903}{981} \times 100 = 92.0\%$$

SECTION 5-4 Exercises Percent Error, Percent Concentration, and Percent Efficiency

Percent Error

1. A certain quantity is measured at 237 units but is known actually to be 241 units. What is the percent error in the measurement?

2. A shaft is known to have a diameter of 43.000 cm. It is measured at 42.872 cm. Find the percent error in the measurement.

3. A certain capacitor has a working voltage of 150 V dc −20%, +50%. Within what two voltages would the actual working voltage lie?

4. A resistor is labeled as 42500 Ω with a tolerance of ±6%. Between what two values is the actual resistance expected to lie?

5. A voltmeter is checked by measuring the EMF of a standard cell, known to be 1.5831 V, and the reading obtained is 1.5893 V. What is the percent error in the reading?

Percent Concentration

6. A solution is made by mixing 125 liters of alcohol and 275 liters of water. What is the percent concentration of alcohol?

7. Twelve cubic feet of cement is contained in a concrete mixture that is 16% cement by volume. What is the volume of the total mixture?

8. How many liters of alcohol are contained in 400 liters of a gasohol mixture that is 4% alcohol by volume?

9. How many liters of gasoline are there in 345 gal of a methanol-gasoline blend that is 15% methanol by volume?

10. A scheme for transporting powdered coal from a mine to a power station calls for mixing the powdered coal with oil so that the resulting slurry is 20% oil by weight. While traveling through the pipeline, 70% of the powdered coal becomes liquefied. What percent of the slurry reaching the power plant will be liquid?

Percent Efficiency

11. A certain device consumes 16 hp and delivers 10 hp. What is its efficiency?

12. An electric motor consumes 2350 W. Find the horsepower it can deliver if it is 70% efficient. (1 hp = 746 W)

13. A water pump requires an input of $1\frac{1}{2}$ hp and delivers 20,000 lb of water per hour to a house 36 ft above the pump. Find the percent efficiency. Hint: Multiply the flow rate times the height to get footpounds per hour, and convert that to horsepower.

14. A certain speed reducer delivers 2.3 hp with a power input of 3.1 hp. Find the percent efficiency of the speed reducer.

15. A certain motor consumes 2200 W and has an efficiency of 65%. What is the output in horsepower?

1. Find the percent change when a quantity changes:
   (a) from 23.4 to 37.8   (b) from 66 to 54

2. A certain quantity is measured at 127 units but is known actually to be 131 units. What is the percent error in the measurement?

3. Find the new value when:
   (a) 773 is increased by 8.75%   (b) 24.5 is decreased by 7%

4. A solution is made by mixing 775 liters of alcohol and 365 liters of water. What is the percent concentration of alcohol?

5. What number decreased by:
   (a) 15% of itself is 263   (b) 30% of itself is 452

6. A concrete mixture that is 18% cement by volume contains 23.6 ft^3 of cement. What is the volume of the total mixture?

7. Convert each percent to a fraction.
   (a) 52.5%   (b) $47\frac{1}{2}$%   (c) 325%   (d) 9%

8. How many liters of alcohol are contained in 522 liters of a gasohol mixture that is 5% alcohol by volume?

9. Convert each fraction to a percent. Round to three digits.
   (a) $\frac{1}{7}$ (b) $\frac{7}{15}$ (c) $\frac{3}{25}$ (d) $\frac{5}{7}$

10. How many liters of gasoline are there in 145 gallons of a methanol-gasoline blend that is 14% methanol by volume?

11. Convert each decimal to a percent.
    (a) 8.24   (b) 0.627   (c) 0.0027   (d) 0.358

12. The temperature in a building rose from 25°C to 28°C during the day. Find the percent change in temperature.

13. What number increased by:
    (a) 25% of itself is 214?   (b) 29% of itself is 95?

14. A 23,200-liter-capacity tank contains 17,600 liters of water. Express this amount of water in the tank as a percentage of the total capacity.

15. Find:
    (a) 48% of 2830 kg   (b) $26\frac{1}{3}$% of 362 gallons

    (c) 45% of 736 km   (d) 16% of 837 liters

16. Find the dollar loss on land that cost $65,200 and was sold at a loss of 3%.

17. Find the number of which:
    (a) 937 is $13\frac{1}{3}$%   (b) 93 is 24%   (c) 92 is 25%

18. A merchant gained $82,300 in one year on goods sold at a profit of 12%. What was the cost of the goods?

19. Convert each percent to a decimal.
    (a) 15%   (b) 7.64%   (c) $65\frac{1}{2}$%   (d) $28\frac{1}{3}$%

20. A certain speed reducer delivers 5.3 hp with a power input of 6.1 hp. Find its percent efficiency.

21. What percent of:

   (a) 736 men is 263 men?   **(b)** 826 hours is 315 hours?

   (c) 25 gal is 13 gal?   **(d)** 37 cm is 9 cm?

WRITING

22. This chapter starts by saying, "Of all the mathematical topics we cover in this text, probably the one most used in everyday life is percentage." Do you agree? Write a few paragraphs saying if you agree or not, and back your reasons up with specific examples from personal experience.

TEAM PROJECT

23. Write an hour exam that will test for the most important skills covered in this chapter on percentage. Include at least two word problems. Be sure that it is not too long and hard, and not too short and easy. Then solve the exam and make revisions if necessary. Produce an answer key, to be handed in with the final version of the exam.

6

Arithmetic with Signed Numbers

Up to this point we have done arithmetic with positive numbers and zero only. But positive numbers alone are not enough to enable us to do the calculations we may need to do. Just the simple notion of indicating a loss on a balance sheet requires a *negative* number. Negative temperatures are familiar to everyone. In technical work we encounter negative forces, negative voltages, negative pressures, and so on.

In this chapter we extend our arithmetic to handle negative numbers by learning a few rules of signs. We will follow the same order we used with whole numbers, fractions, decimal numbers, and denominate numbers. After some definitions, we will do addition and subtraction, then multiplication, division, and finally powers and roots.

We learn here how to calculate using positive and negative numbers, but that is not our only goal. We are also laying the groundwork for studying *algebraic* quantities in the following chapters.

6-1   SIGNED NUMBERS

Negative Numbers

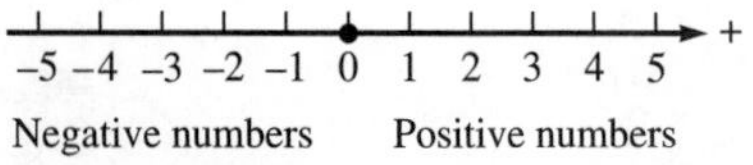

Negative numbers   Positive numbers

■ FIGURE 6-1

The number line.

Recall that a positive number is a number that is greater than zero. Now we define a *negative number* as a number that is *less* than zero. On the number line (Fig. 6-1) they are all the numbers lying to the left of zero. Negative numbers may be integers or fractions, rational or irrational. To distinguish negative numbers from positive numbers we always place a negative sign ($-$) in front of a negative number.

Example 1

Some negative numbers are

$$-5, \quad -6.293, \quad -\frac{2}{3}, \quad -2\frac{7}{8}, \quad \text{and} \quad -\sqrt{5}$$

We usually omit writing the positive sign ($+$) in front of a positive number. Thus a number without a sign is always assumed to be positive. In this chapter we will often write in a ($+$) sign for emphasis.

Example 2

Some positive numbers are

$$5, \quad +5, \quad \frac{2}{3}, \quad +\frac{2}{3}, \quad \text{and} \quad \sqrt{5}$$

The Opposite of a Number

The *opposite* of a number n is that number which, when added to n, gives a sum of zero.

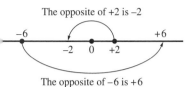

FIGURE 6-2
The opposite of a number.

> **Example 3**
> The opposite of 2 is -2.
> The opposite of -6 is $+6$.

Geometrically, the opposite of a number n lies on the opposite side of the zero point of the number line from n, and at an equal distance from the origin (Fig. 6-2).

Signs of Equality and Inequality

In Chapter 1 we introduced the signs of equality and inequality:

$$a = b \quad a \neq b \quad a > b \quad a < b \quad a \leq b \quad a \geq b \quad a \approx b$$

We used these signs to show the relative position of positive numbers on the number line. Here we use them for negative numbers as well. Notice that the arrow ($>$) always points to the number which is more to the left on the number line.

FIGURE 6-3

> **Example 4**
> Place the proper equality or inequality sign between -2.15 and -2.83.
> **Solution:** Since -2.15 lies to the right of -2.83 on the number line (Fig. 6-3), then -2.15 is *greater than* -2.83. So we write
> $$-2.15 > -2.83$$

Absolute Value

The *absolute value* of a number is, geometrically, the distance of that number from the zero point of the number line, regardless of direction. Thus the absolute value of a number is never negative.

FIGURE 6-4

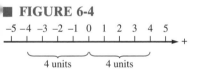

> **Example 5**
> The numbers 4 and -4 both have an absolute value of 4, because both are 4 units from the zero point of the number line (Fig. 6-4).

Absolute value is indicated by enclosing the number between *vertical bars*.

> **Example 6**
> **(a)** $|-4| = 4$ **(b)** $|+4| = 4$ **(c)** $|5 - 8| = |-3| = 3$

Negative Numbers on the Calculator

Note that the ($-$) sign is used for *two different things:*

1. To indicate a negative quantity
2. For the operation of subtraction

This difference is clear on most calculators, which usually have separate keys for these two functions. The $\boxed{-}$ key is used for subtraction but is not used to enter a negative number.

Some calculators have a $\boxed{(-)}$ key (sometimes called a *negation key*) for entering a negative number. Notice that the negative sign is in parentheses on this key, while it is not in parentheses on the subtraction key. Some calculators use a $\boxed{-x}$ key for the same purpose. To enter a negative number on such a calculator, press the $\boxed{(-)}$ key *first,* and then enter the number.

On other calculators, the change-sign key is used to enter a negative number. The change-sign key is marked differently on different calculators, with $\boxed{+/-}$ or $\boxed{CHS}$ being common. Check your manual to see how this key is marked on your calculator. To enter a negative number on a calculator having a change-sign key, simply enter the number first, without sign, and then press the change-sign key.

SECTION 6-1 Exercises Signed Numbers

Negative Numbers

Trace or photocopy the number line of Fig. 6-1, or make your own. On it locate the following numbers.

1. -3　　**2.** 6　　**3.** $-\dfrac{3}{2}$　　**4.** -2.74

Signs of Equality and Inequality

Place the proper equality or inequality sign between each pair of numbers.

5. 5 and 1　　　**6.** 5 and -1　　　　**7.** -5 and 1

8. -5 and -1　　**9.** -12.2 and -12.3　　**10.** -15.4 and -16.8

Absolute Value

Evaluate each expression.

11. $|-6|$　　**12.** $|-24|$　　**13.** $|3-6|$

14. $|6-3|$　　**15.** $|15-16|$　　**16.** $|1-7|$

Negative Numbers on the Calculator

Enter each negative number on your calculator.

17. -23　　**18.** -82　　**19.** -3.72　　**20.** -4.836

6-2 ADDING AND SUBTRACTING SIGNED NUMBERS

Adding Signed Numbers

Let's say that we have a shoe box (Fig. 6-5) into which we toss all our uncashed checks and unpaid bills until we have time to deal with them. Let's further assume that the total checks minus the total bills in the shoebox is $500.

■ **FIGURE 6-5**
A shoe box with bills and checks.

We can think of the amount of a check as a positive number, because it increases our wealth, and the amount of a bill as a negative number because it decreases our wealth. We thus represent a check for $100 as $(+100)$ and a bill for $200 as (-200).

Now, let's *add a check* for $100 to the box. If we had $500 at first, we must now have $600.

$$500 + (+100) = 600$$

or

$$500 + 100 = 600$$

Here we have added a positive number, and our total has increased by the amount. That is easy to understand. But what does it mean to *add a negative number?*

To find out, let us now *add a bill* for $100 to the box. If we had $500 at first, we must now have $100 less, or $400. Representing the bill by (-100) we have

$$500 + (-100) = 400$$

This means that to *add a negative number* is no different than to *subtract the absolute value* of that number.

$$500 + (-100) = 500 - 100$$

This gives us our rule of signs for addition. If a and b are two positive quantities, then

RULE OF SIGNS FOR ADDITION	$a + (-b) = a - b$	6

The operation of adding a negative quantity $(-b)$ to the quantity a is equivalent to the operation of subtracting the positive quantity b from a.

Example 7

(a) $7 + 2 = 9$

(b) $7 + (-2) = 7 - 2 = 5$

(c) $-7 + 2 = -5$

(d) $-7 + (-2) = -7 - 2 = -9$

Subtracting Signed Numbers

Let us return to our shoebox problem. But now instead of adding checks or bills to the box, we will *subtract* (remove) checks or bills from the box.

First we remove (subtract) a check for $100 from the box. If we had $500 at first, we must now have $400.

$$500 - (+100) = 400$$

or

$$500 - 100 = 400$$

Here we have *subtracted a positive number,* and *our total has decreased* by the amount subtracted, as expected.

Now let us see what it means to *subtract a negative number.* We will remove (subtract) a bill for $100 from the box. If we had $500 at first, we must now have $100 more, or $600, since we have removed a bill. Representing the bill by (-100) we have

$$500 - (-100) = 600$$

Thus *to subtract a negative number* is the same as to *add the absolute value* of that number.

$$500 - (-100) = 500 + 100$$

Thus, if *a* and *b* are positive quantities, our rule of signs for subtraction is

RULE OF SIGNS FOR SUBTRACTION	$a - (-b) = a + b$	7

The operation of subtracting a negative quantity (−b) from the quantity a is equivalent to the operation of adding the positive quantity b to a.

--- **Example 8** ---

(a) $7 - 2 = 5$
(b) $7 - (-2) = 7 + 2 = 9$
(c) $-7 - 2 = -9$
(d) $-7 - (-2) = -7 + 2 = -5$

--- **Example 9** ---

A surveyor places a stake at *A*, Fig. 6-6, and calls its elevation 0 ft. She then goes downhill and places stake *B*, which she labels −5.85 ft. She then places stake *C* 3.77 ft below *B*. What is the elevation of *C*?

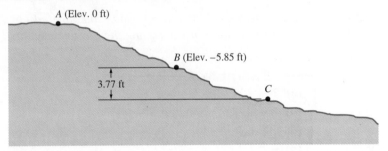

■ **FIGURE 6-6**
Placement of surveyor's stakes.

Solution: We must subtract 3.77 ft from −5.85 ft.

$$\text{Elevation of } C = -5.85 - 3.77 = -9.62 \text{ ft}$$

SECTION 6-2 **Exercises** **Adding and Subtracting Signed Numbers**

Adding Signed Numbers Add.

1. $(+5) + (-3)$ **2.** $(+4) + (-11)$
3. $(+77) + (-24)$ **4.** $(-7) + (+2)$
5. $(-9) + (+11)$ **6.** $(-23) + (+82)$
7. $(-9) + (-3)$ **8.** $(-4) + (-41)$

9. $(-35) + (-54)$ 10. $(+5.52) + (-3.14)$

11. $(+4.26) + (-11.82)$ 12. $(+72.5) + (-27.2)$

13. $(-7.82) + (+2.26)$ 14. $(-9.77) + (+11.26)$

15. $(-23.2) + (+82.7)$ 16. $(-9.33) + (-3.25)$

17. $(-4.94) + (-41.22)$ 18. $(-35.5) + (-54.9)$

Subtracting Signed Numbers

Subtract.

19. $(+7) - (-3)$ 20. $(+9) - (-61)$

21. $(+77) - (-49)$ 22. $(-7) - (+4)$

23. $(-5) - (+66)$ 24. $(-43) - (+84)$

25. $(-9) - (-3)$ 26. $(-2) - (-96)$

27. $(-37) - (-79)$ 28. $(+7.74) - (-3.69)$

29. $(+9.46) - (-16.84)$ 30. $(+74.7) - (-47.2)$

31. $(-7.84) - (+4.26)$ 32. $(-9.77) - (+61.46)$

33. $(-43.2) - (+84.7)$ 34. $(-9.33) - (-3.47)$

35. $(-9.94) - (-96.42)$ 36. $(-37.7) - (-54.9)$

Applications

37. A balance sheet showed a balance of $-\$3482.77$, and another bill came in for $1735.83. Find the new balance.

38. During a certain month the temperature dropped from $+13.8°F$ to $-11.4°F$. Find the difference in the temperatures.

39. A submarine, Fig. 6-7, was at an elevation of -238 ft, and then descended another 177 ft. Find the new elevation.

40. The voltage in a certain circuit was -33.8 V. It was then reduced by 11.3 V. Find the new voltage.

■ **FIGURE 6-7**
A diving submarine.

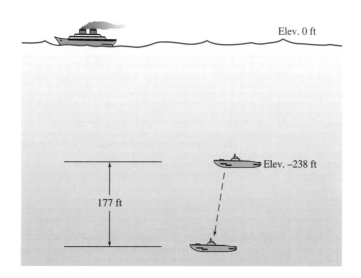

Elev. 0 ft

Elev. –238 ft

177 ft

6-3 MULTIPLYING SIGNED NUMBERS

Rules of Signs

To get our rules of signs for multiplication, we use the idea of *multiplication as repeated addition*. For example, to multiply 3 by 4 means to add four 3's (or three 4's).

$$3 \times 4 = 3 + 3 + 3 + 3$$

or

$$3 \times 4 = 4 + 4 + 4$$

Let us return to our shoebox example. Recall that it contains uncashed checks and unpaid bills. Let's first add 5 checks ($+5$), each worth \$100 ($+100$), to the box. The value of the contents of the box then increases by \$500 ($+500$). Multiplying,

(Number of checks) $\times$ (value of one check) = change in value of contents

$$(+5)(+100) = +500$$

Thus a positive number times a positive number gives a positive product. This is nothing new.

Now let's add 5 *bills* to the box, thus decreasing its value by \$500. To show this multiplication, we use ($+5$) for the number of bills, (-100) for the value of each bill, and (-500) for the change in value of the box contents.

$$(+5)(-100) = -500$$

Here, a positive number times a negative number gives a negative product.

Next we *remove* 5 checks from the box, thus decreasing its value by \$500.

$$(-5)(+100) = -500$$

Here again, the product of a positive number and a negative number is negative. Thus it doesn't matter whether the negative number is the first or the second. This, of course, is what you would expect from the commutative property, which applies to negative numbers as well as to positive numbers.

Finally, we remove 5 bills from the box, causing its value to increase by \$500.

$$(-5)(-100) = +500$$

Here, the product of two negative numbers is positive.

We summarize these findings to get our *rules of signs for multiplication*. If a and b are two positive numbers, then

RULE OF SIGNS FOR MULTIPLICATION	$(+a)(+b) =$ $(-a)(-b) = +ab$	9

The product of two numbers of like sign is positive.

and

RULE OF SIGNS FOR MULTIPLICATION	$(+a)(-b) =$ $(-a)(+b) = -ab$	10

The product of two numbers of opposite sign is negative.

Example 10

(a) $(3)(-4) = -12$

(b) $(-3)(4) = -12$

(c) $(-3)(-4) = 12$

(d) $(-9.641)(-8.31) = 80.1$ (rounded)

Multiplying a String of Numbers

When we multiply two negative numbers we get a positive product. So when multiplying a string of numbers, if an even number of factors are negative, the answer will be positive.

Example 11

(a) $(4)(-3)(-2)(-1) = -24$

(b) $(4)(-3)(-2)(1) = 24$

(c) $(-1.38)(-2.73)(-1.84) = -6.93$ (rounded)

Example 12

The velocity v of an object thrown upward or downward is found by adding the initial velocity v_0 to the product of the acceleration a (32.2 ft/s^2) and the time t, or

$$v = v_0 + at$$

Find the velocity after 1.50 seconds of a ball thrown upward with an initial velocity of 15.6 ft/s.

Solution: If we take the upward direction as positive, Fig. 6-8, then the initial velocity is positive because the ball is thrown upward, and the acceleration is negative because it acts downward. So

$$v_0 = +15.6 \text{ ft/s} \quad \text{and} \quad a = -32.2 \text{ ft/s}^2$$

Substituting gives

$$v = +15.6 + (-32.2)(+1.50) = -32.7 \text{ ft/s}$$

Here the negative sign indicates that the velocity is in the negative direction, or downward.

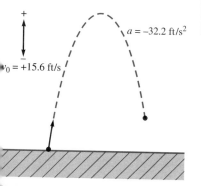

FIGURE 6-8
A ball thrown upward.

$a = -32.2 \text{ ft/s}^2$

$v_0 = +15.6 \text{ ft/s}$

SECTION 6-3 Exercises Multiplying Signed Numbers

Multiply. Treat all integers as exact. Treat all decimal fractions as approximate, and round your answer to the proper number of significant digits.

1. $(+8)(-4)$
2. $(+3)(-2)$
3. $(+7)(-11)$
4. $(-2)(+3)$
5. $(-9)(+3)$
6. $(-5)(-8)$
7. $(-3)(-5)$
8. $(-6)(-8)$
9. $(-2)(-23)$
10. $(+2.18)(-3.24)$
11. $(+82.3)(-15.2)$
12. $(+1.57)(-4.11)$
13. $(-412)(+2.43)$
14. $(-1.29)(+723)$

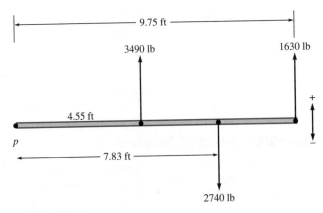

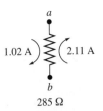

■ FIGURE 6-9
A beam.

■ FIGURE 6-10
Voltage across a resistor.

15. $(-4.85)(-4.28)$

16. $(-173)(-3.25)$

17. $(-5.26)(-618)$

18. $(-722)(-1.23)$

19. $(-2)(+3)(-7)$

20. $(-3)(-9)(+3)$

21. $(+6)(-5)(-8)$

22. $(-6)(-3)(-5)$

23. $(+3)(-6)(-8)$

24. $(-4)(-2)(-23)$

25. $(+1.52)(+3.48)(-1.74)$

26. $(-2.31)(+32.4)(-25.5)$

27. $(-1.72)(+4.67)(-2.81)$

28. $(+2.61)(-602)(+1.63)$

29. $(-1.84)(-2.22)(+523)$

30. $(+2.81)(-7.35)(-1.25)$

Applications

31. Find the velocity after 1.25 seconds of a stone thrown upward with a velocity of 35.8 ft/s. See Example 12.

32. A broker bought 500 shares of stock for $35.25 per share, whose price then rose $1.25 per share and later dropped $0.42 per share. Find the final value of the stock.

33. A beam, Fig. 6-9, has three forces acting on it. If we consider the upward forces to be positive, the three forces are $+3490$ lb, $+1630$ lb, and -2740 lb. The *moment* of each force about p is equal to the force times its distance from p. Thus the sum M of the moments is

$$M = (+3490)(4.55) + (+1630)(9.75) + (-2740)(7.83) \text{ ft-lb}$$

Evaluate M.

34. The total voltage drop across a resistor is equal to the resistance times the current in a resistor. A 285-Ω resistor, Fig. 6-10, has two currents flowing through it. If we take the flow direction from a to b as positive, the currents are $+1.02$A and -2.11A. The sum of the voltage drops across it is then

$$V = (285)(1.02) + (285)(-2.11) \text{ volts}$$

Find V.

6-4 DIVIDING SIGNED NUMBERS

Rules of Signs

We will use the rules of signs for multiplication to get the rules of signs for division. We know that the product of a negative number and a positive number is negative. Thus,

$$(-2)(+3) = -6$$

If we divide both sides of this equation by (+3) we get

$$-2 = \frac{-6}{+3}$$

From this we see that *a negative number divided by a positive number gives a negative quotient.*

Again starting with

$$(-2)(+3) = -6$$

we divide both sides by (−2) and get

$$+3 = \frac{-6}{-2}$$

Here we see that *a negative number divided by a negative number gives a positive product.*

We also know that the product of two negative numbers is positive. Thus

$$(-2)(-3) = +6$$

Dividing both sides by (−3) we get

$$-2 = \frac{+6}{-3}$$

Thus *a positive number divided by a negative number gives a negative quotient.*

We combine these findings with the fact that the quotient of two positive numbers is positive and get our *rules of signs for division.*

RULE OF SIGNS FOR DIVISION	$\dfrac{+a}{+b} = \dfrac{-a}{-b} = +\dfrac{a}{b}$	11

The quotient is positive when dividend and divisor have the same sign.

and

RULE OF SIGNS FOR DIVISION	$\dfrac{+a}{-b} = \dfrac{-a}{+b} = -\dfrac{a}{b}$	12

The quotient is negative when dividend and divisor have opposite signs.

─── **Example 13** ───
(a) $8 \div (-4) = -2$
(b) $(-8) \div 4 = -2$
(c) $(-8) \div (-4) = 2$
(d) $\dfrac{125}{-25} = -5$

Example 14

Divide 85.4 by -2.5386 and round to the proper number of significant digits.

Solution: By calculator,

$$85.4 \div 2.5386 = 33.6405893$$

which we round to three significant digits and determine the sign by inspection, getting

$$85.4 \div (-2.5386) = -33.6.$$

Example 15

An airplane descending for a landing has a change in altitude of -750 ft every 5.0 minutes. Find the rate of change of altitude in feet per minute.

Solution: Dividing,

$$\text{Rate of change} = \frac{-750 \text{ ft}}{5.0 \text{ min}} = -150 \text{ ft/min}$$

SECTION 6-4   Exercises   Dividing Signed Numbers

Divide. Treat all integers as exact and round the quotient to four decimal places. Treat all decimal fractions as approximate, and round your answer to the proper number of significant digits.

1.	$(+16) \div (-4)$	**2.**	$(+6) \div (-2)$
3.	$(+4) \div (-12)$	**4.**	$(-9) \div (+3)$
5.	$(-6) \div (+3)$	**6.**	$(-7) \div (-8)$
7.	$(-8) \div (-5)$	**8.**	$(-3) \div (-8)$
9.	$(-8) \div (-23)$	**10.**	$(+3.16) \div (-7.14)$
11.	$(+22.3) \div (-45.2)$	**12.**	$(+7.57) \div (-3.11)$
13.	$(-515) \div (+8.43)$	**14.**	$(-135) \div (+323)$
15.	$(-1.86) \div (-2.28)$	**16.**	$(-153) \div (-2.25)$
17.	$(-6.26) \div (-118)$	**18.**	$(-612) \div (-6.23)$

Applications

19. A current in a certain circuit is changing at a rate of -173 volts every 15.0 minutes. What is the rate of change in volts per minute?

20. Descending a certain flight of 25 steps, we change in elevation by -18.75 ft. How much do we change in elevation in one step?

6-5

POWERS AND ROOTS OF SIGNED NUMBERS

We did powers and roots of positive whole numbers in Sec. 1-5 and powers and roots of positive approximate numbers in Sec. 3-5. Here we extend those operations to *negative* numbers, both exact and approximate.

Powers of a Negative Number

Let us raise a negative number, say (-5), to an even power, say 2.

$$(-5)^2$$

Recalling that $(-5)^2$ is a shorthand way of writing (-5) times (-5), we get

$$(-5)^2 = (-5)(-5) = +25$$

Since the product of an even number of negative numbers is always positive, we see that *a negative base raised to an even power gives a positive number.*

Similarly, if we raise (-5) to an odd power, say 3, we get

$$(-5)^3 = (-5)(-5)(-5) = -125$$

Since the product of an odd number of negative numbers is negative, we find that *a negative base raised to an odd power gives a negative number.*

Example 16

(a) $(-2)^2 = (-2)(-2) = 4$
(b) $(-2)^3 = (-2)(-2)(-2) = -8$
(c) $(-1)^{24} = 1$
(d) $(-1)^{25} = -1$

COMMON ERROR Be careful how you read and write powers of negative numbers. The exponent does not apply to the negative sign unless that sign is included inside parentheses. Thus,

$$(-3)^2 = (-3)(-3) = +9$$

but

$$-3^2 = -(3)(3) = -9$$

Powers of Negative Numbers by Calculator

Recall that to raise a number to any power, we use a *power key* marked something like $\boxed{\wedge}$, $\boxed{x^y}$, or $\boxed{y^x}$, depending on your calculator. To raise a *negative* number to a power, enclose the number, including the negative sign, in parentheses.

Example 17

To find $(-2.95)^3$ by calculator, the keystrokes are

$$\boxed{(} \quad \boxed{(-)} \quad 2.95 \quad \boxed{)} \quad \boxed{\wedge} \quad 3 \quad \boxed{\text{ENTER}}$$
$$-25.672375$$

which we round to -25.7.

If you try to find the power of a negative number by your calculator, you might get an error indication. Some calculators will not accept a negative base, even though this is a valid operation.

Then how do you do it? Simply enter the base as positive, find the power, and determine the sign by inspection.

Example 18

Find $(-1.45)^5$.

Solution: From the calculator,

$$(+1.45)^5 = 6.41$$

Since we know that a negative number raised to an odd power is negative, we write

$$(-1.45)^5 = -6.41$$

Negative Exponent

A number can be raised to a negative exponent on the calculator using the same key we used for positive exponents.

Example 19

Evaluate $(3.85)^{-3}$.

Solution: On a calculator having a negation key $\boxed{(-)}$, the keystrokes are

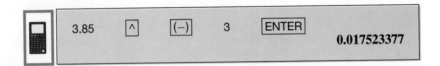

which we round to 0.0175.

Principal Root

The principal root of a positive number is defined as the positive root. Thus, $\sqrt{4} = +2$, not ± 2.

The principal root is negative only when we take an odd root of a negative number.

Example 20

$$\sqrt[3]{-27} = -3$$

Because $(-3)(-3)(-3) = -27$.

Odd Roots of Negative Numbers by Calculator

An even root of a negative number (such as $\sqrt{-4}$) is not a real number. It is what we call an *imaginary number*. We will study these in Chapter 29. But an *odd* root of a negative number is *not* imaginary. It is a real, negative number.

Example 21

$$\sqrt[3]{-8} = -2$$

because

$$(-2)^3 = -8$$

Most calculators have a key for taking the *n*th root of a number. If your calculator does not have such a key, you can take the *n*th root of a number by raising the number to the $1/n$th power.

Example 22

Find the fifth root of -357.

Solution: The keystrokes are

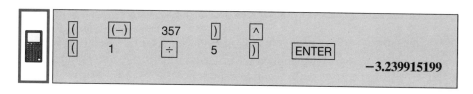

giving -3.24, rounded.

As with powers, some calculators will not accept a negative radicand. Fortunately, we can outsmart these calculators and take odd roots of negative numbers anyway.

Example 23

Find $\sqrt[7]{-1054}$.

Solution: We know that an odd root of a negative number is real and negative. So we take the seventh root of $+1054$, by calculator,

$$\sqrt[7]{+1054} = 2.703 \text{ (rounded)}$$

and we have only to append the minus sign.

$$7\sqrt{-1054} = -2.703$$

SECTION 6-5 Exercises Powers and Roots of Signed Numbers

Treat all integers as exact. Treat all decimal fractions as approximate, and round your answer properly.

Powers of a Negative Number

Raise each number to the power indicated.

1. $(-2)^5$	**2.** $(-3)^3$	**3.** $(-5)^2$	**4.** $(-4)^3$
5. $(-6)^2$	**6.** $(-4)^4$	**7.** $(-2)^6$	**8.** $(-5)^3$
9. $(-1.24)^5$	**10.** $(-2.13)^3$	**11.** $(-2.35)^2$	**12.** $(-3.14)^3$
13. $(-2.36)^2$	**14.** $(-2.24)^4$	**15.** $(-2.11)^6$	**16.** $(-1.25)^3$

Negative Exponent

Raise each number to the power indicated. Round to three significant digits.

17. $(3)^{-4}$	**18.** $(4)^{-3}$	**19.** $(2)^{-3}$	**20.** $(5)^{-3}$
21. $(7)^{-2}$	**22.** $(2)^{-4}$	**23.** $(3)^{-6}$	**24.** $(7)^{-3}$
25. $(3.14)^{-5}$	**26.** $(8.23)^{-3}$	**27.** $(5.55)^{-2}$	**28.** $(4.14)^{-3}$
29. $(7.16)^{-2}$	**30.** $(5.27)^{-4}$	**31.** $(6.21)^{-6}$	**32.** $(8.25)^{-3}$

Evaluate each root.

33. $\sqrt[3]{-8}$ **34.** $\sqrt[3]{-64}$ **35.** $\sqrt[5]{-32}$ **36.** $\sqrt[3]{-125}$

37. $\sqrt[5]{-243}$ **38.** $\sqrt[3]{-27}$ **39.** $\sqrt[3]{-8.4}$ **40.** $\sqrt[3]{-27.6}$

41. $\sqrt[3]{-64.5}$ **42.** $\sqrt[3]{-125.8}$ **43.** $\sqrt[3]{-23.1}$ **44.** $\sqrt[3]{-7.42}$

45. $\sqrt[3]{-9.52}$ **46.** $\sqrt[3]{-16.3}$ **47.** $\sqrt[5]{-53.1}$ **48.** $\sqrt[3]{-9.12}$

49. $\sqrt[5]{-5.12}$ **50.** $\sqrt[3]{-96.3}$

6-6   SCIENTIFIC NOTATION

In Sec. 1-6 we introduced scientific notation with whole numbers, and in Sec. 3-6 we applied scientific notation to decimal numbers. Now we will do scientific notation with negative exponents.

Negative Powers of Ten

For the sake of clarity, let us call a number such as 1000 a *multiple* of ten, and a number such as 10^3 a *power* of ten. In Sec. 1-6 we gave a table of multiples and powers of ten. We repeat that table here, extended to include negative exponents.

Powers of Ten	
10^6	= 1,000,000
10^5	= 100,000
10^4	= 10,000
10^3	= 1,000
10^2	= 100
10^1	= 10
10^0	= 1
10^{-1}	= 0.1
10^{-2}	= 0.01
10^{-3}	= 0.001
10^{-4}	= 0.0001
10^{-5}	= 0.00001
10^{-6}	= 0.000001

This table extends indefinitely, for both positive and negative exponents.

Converting from Decimal Form to Scientific Notation

To convert a number from decimal form to scientific notation, rewrite the given number with just one digit to the left of the decimal point, and discard any nonsignificant zeros. Then multiply by that multiple of ten that makes it equal to the given number. Finally write the multiple of ten as a power of ten.

Example 24

Convert 0.00531 to scientific notation.

Solution: We place the decimal point between the 5 and the 3. The three zeros are not significant (see Sec. 3-1) so we discard them, getting 5.31. Finally, the 5.31 must be multiplied by 0.001 so that our new expression is equal to the given number.

$$0.00531 = 5.31 \times 0.001$$

We now express 0.001 as a power of ten (10^{-3}).

$$0.00531 = 5.31 \times 0.001 = 5.31 \times 10^{-3}$$

To convert from scientific notation to decimal form, simply reverse the process.

Example 25

Convert 3.84×10^{-5} to decimal form.

Solution:

$$3.84 \times 10^{-5} = 3.84 \times 0.00001$$
$$= 0.0000384$$

Entering Numbers in Scientific Notation in a Calculator

We can enter numbers directly in scientific notation by means of the ENTER EXPONENT key, usually marked $\boxed{\text{EE}}$, $\boxed{\text{EXP}}$, or $\boxed{\text{EEX}}$. We will use $\boxed{\text{EE}}$ in our examples.

Example 26

The keystrokes to enter the number 7.25×10^5 are

or 725,000.

The way we enter a *negative* exponent depends on whether our calculator uses a $\boxed{(-)}$ negation key or a change-sign key.

Example 27

The keystrokes to enter the number 8.25×10^{-2} on a calculator that has a $\boxed{(-)}$ negation key are

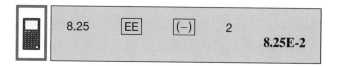

or 0.0825. On a calculator that has a change-sign key, the keystrokes are

Note that we enter the exponent and then change its sign.

We add, subtract, multiply, and divide numbers in scientific notation that have negative exponents exactly the way we did with those having positive exponents. See Sec. 3-6.

Write each number in scientific notation. Be sure to retain the proper number of digits.

1. 0.373   **2.** 0.00632   **3.** 0.0260   **4.** 0.000239

5. 0.0020000   **6.** 0.0003950   **7.** 0.0228   **8.** 0.0000491

Convert each number from scientific notation to decimal notation.

9. 2.323×10^{-2} **10.** 3.2902×10^{-4} **11.** 7.01×10^{-3}

12. 2.950×10^{-4} **13.** 3.370×10^{-5} **14.** 5.14×10^{-6}

Perform the following computations in scientific notation. Combine the powers of ten by hand or with your calculator.

15. $(18.4 \times 10^{-3}) + (17.9 \times 10^{-4})$ **16.** $(13.2 \times 10^{-3}) + 0.0234$

17. $(28.3 \times 10^{-2}) + (15.3 \times 10^{-2})$ **18.** $(26.5 \times 10^{-3}) - (17.4 \times 10^{-3})$

19. $(3.59 \times 10^{-6})(6.23 \times 10^{-5})$ **20.** $(1.03 \times 10^{-3})(8.27 \times 10^{-3})$

21. $(6.19 \times 10^{5})(9.11 \times 10^{-3})$ **22.** $(7.16 \times 10^{-2})(9.28 \times 10^{7})$

23. $(23.2 \times 10^{2}) \div 0.00824$ **24.** $72,400 \div (2.82 \times 10^{-3})$

25. $(3.18 \times 10^{-7}) \div (9.34 \times 10^{-3})$ **26.** $(11.2 \times 10^{3}) \div (7.34 \times 10^{-4})$

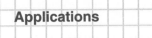

Applications

27. Find the current through a resistor if it dissipates 2.32×10^{-2} watts when the voltage drop across it is 6.72×10^{-3} V. (Divide the power by the voltage drop.)

28. The voltage across a 6.14×10^{4} Ω resistor is 2.27×10^{-2} V. Find the power dissipated in the resistor. (Divide the square of the voltage by the resistance.)

29. A tensile load of 1.39×10^{3} lb is supported by a wire having a cross-sectional area of 6.12×10^{-2} in.². Find the stress in the member. (Divide the load by the area.)

30. A wire 9.19×10^{2} cm long when loaded is seen to stretch 9.66×10^{-2} cm. Find the strain in the wire. (Divide the amount stretched by the length.)

31. A rod having a cross-sectional area of 8.72×10^{-2} in.² and a length of 5.82×10^{2} in. is subjected to a tensile load of 7.14×10^{3} lb. The elongation is 9.26×10^{-2} in. Find the modulus of elasticity E of the material. (Multiply the load by the length, and divide by the area and by the elongation.)

6-7 COMBINED OPERATIONS

In Sec. 1-7 we showed how to do combined operations with whole numbers on the calculator, and in Sec. 3-7 we showed the same for decimal numbers. Now we include negative numbers.

Example 28

Evaluate the expression

$$\left(\frac{-41.8 + 4.23}{-21.3\sqrt[3]{-136}}\right)^{2}$$

Solution: On a calculator having parentheses and a $\boxed{(-)}$ negation key, we can do the computation in a single line. The keystrokes are

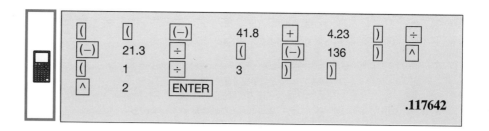

which we round to three significant digits, getting 0.118.

However, many students would prefer to do the computation in stages, using the ENTER key to obtain the intermediate values. Then pressing the ANS key will make that intermediate result available for the next step.

Example 29

Repeating Example 28,

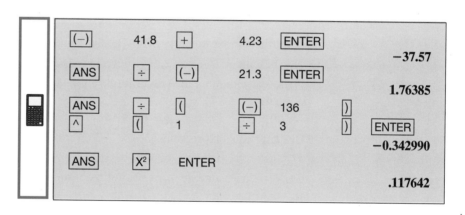

as before.

SECTION 6-7   Exercises   Combined Operations

Perform the computations, keeping the proper number of digits in your answer.

1. $(-2.67)(7.18) + (-3.34)(-2.64)$

2. $(312)(5.53) - (186)(-7.70) + (-247)(-3.39)$

3. $(-13.7 + 33.6)(2.37 \times 10^{-2} - 7.72 \times 10^{-2})$

4. $\dfrac{381}{-623} - \dfrac{-446}{317}$

5. $\dfrac{-2.13}{5.32} - \dfrac{-7.97}{3.06} + \dfrac{3.23}{-1.51}$

6. $\dfrac{3.27 + (-6.204)}{(-7.27) + 68.8}$

7. $\dfrac{463 - (-284) + 274}{274 - 750 + (-284)}$

8. $(3.57 - 7.36)^{-2}$

9. $(7.45 + 9.36 - 2.74)^{-4}$

10. $(2.723 - 3.55)^{-3}$

11. $\sqrt{7.24 \times 10^{-3} + 3.86 \times 10^{-2}}$

12. $\sqrt[3]{-284 - 926 - 264}$

13. $\sqrt{(-88.1)(-3.54 \times 10^{-2})}$

14. $\sqrt[5]{(3.06)(-8.36)(4.93)}$

15. $\sqrt{\dfrac{-385}{-973}}$

16. $\sqrt[4]{\dfrac{-7.50}{-3.61}}$

17. $\sqrt[4]{278} - \sqrt{25.2} - \sqrt[3]{-836}$

■ Chapter 6   REVIEW EXERCISES

Place the proper equality or inequality sign between each pair of numbers.

1. -3 and -1 **2.** -14.4 and -14.5 **3.** -13.2 and -18.6

Write each number in scientific notation. Be sure to retain the proper number of digits.

4. 0.595   **5.** 0.00854   **6.** 0.0005730

Evaluate each expression.

7. $|8 - 5|$ **8.** $|13 - 18|$ **9.** $|1 - 9|$

Convert each number from scientific notation to decimal notation.

10. 4.545×10^{-4} **11.** 5.4704×10^{-2} **12.** 4.730×10^{-2}

Perform each computation, rounding your answer properly.

13. $(16.2 \times 10^{-5}) + (19.7 \times 10^{-2})$ **14.** $(15.4 \times 10^{-5}) + 0.0452$

15. $(48.3 \times 10^{-5}) - (19.2 \times 10^{-5})$ **16.** $(1.05 \times 10^{-5})(6.49 \times 10^{-5})$

17. $(+4.16)(-5.42)$ **18.** $(+1.39)(-2.11)$

19. $(-214)(+4.25)$ **20.** $(-1.42)^3$

21. $(-4.15)^5$ **22.** $(-4.53)^4$

23. $(-5.12)^5$ **24.** $(5.12)^{-3}$

25. $(6.45)^{-5}$ **26.** $(3.33)^{-4}$

27. $(2.12)^{-5}$ **28.** $\sqrt[7]{-6.00}$

29. $\sqrt[7]{-82.0}$ **30.** $\sqrt[3]{-54.0}$

31. $\sqrt[7]{-143.0}$ **32.** $\sqrt[3]{-425.0}$

33. $\sqrt[5]{-55.0}$ **34.** $\sqrt[5]{-6.80}$

35. $\sqrt[5]{-49.0}$ **36.** $(-195)(-5.43)$

37. $(-3.48)(-816)$ **38.** $(-944)(-1.45)$

39. $(+1.34)(+5.26)(-1.92)$ **40.** $(+5.18) \div (-9.12)$

41. $(+9.39) \div (-5.11)$ **42.** $(-313) \div (+6.25)$

43. $(-153) \div (+545)$ **44.** $(-45.4) + (+64.9)$

45. $(-7.55) + (-5.43)$ **46.** $(-2.72) + (-21.44)$

47. $(-4.89)(9.16) + (-5.52)(-4.82)$

48. $(514)(3.35) - (168)(-9.90) + (-429)(-5.57)$

49. $(-15.9 + 55.8)(4.59 \times 10^{-4} - 9.94 \times 10^{-4})$

50. $(5.39 - 9.58)^{-4}$

51. $(9.23 + 7.58 - 4.92)^{-2}$ **52.** $(4.945 - 5.33)^{-5}$

53. $(+9.92) - (-5.87)$ **54.** $(+7.28) - (-18.62)$

55. $(-9.62) - (+2.48)$ **56.** $(-7.99) - (+81.28)$

57. $(-7.55) - (-5.29)$ **58.** $(-7.72) - (-78.24)$

59. $(-135) \div (-4.43)$ **60.** $(-8.48) \div (-116)$

61. $(-1.94)(+2.89)(-4.61)$ **62.** $(+4.81)(-804)(+1.85)$

63. $(-1.62)(-4.44)(+345)$ **64.** $(+4.61)(-9.53)(-1.43)$

65. A wire 7.17×10^4 cm long when loaded is seen to stretch 7.88×10^{-4} cm. Find the strain in the wire. (Divide the amount stretched by the length.)

66. A current in a certain circuit is changing at a rate of -195 volts every 13 minutes. What is the rate of change in volts per minute?

67. A broker bought 300 shares of stock for $84.26 per share, which then rose $1.43 per share and later dropped $0.24 per share. Find the final value of the stock.

68. The total voltage drop across a resistor is equal to the resistance times the current in a resistor. A 463-Ω resistor has two currents flowing through it. The currents are $+1.04$A and -4.11A. The sum of the voltage drops across it is then,

$$V = (463)(1.04) + (463)(-4.11) \text{ volts}$$

Find V.

69. A balance sheet showed a balance of $-$5264.99, and another bill came in for $1953.65. Find the new balance.

70. A tensile load of 1.57×10^5 lb is supported by a wire having a cross-sectional area of 8.14×10^{-4} in.^{2}. Find the stress in the member. (Divide the load by the area.)

71. During a certain month the temperature dropped from $+15.6°$F to $-11.2°$F. Find the difference in the temperatures.

72. A submarine was at an elevation of -456 ft, and then descended another 199 ft. Find the new elevation.

73. The voltage in a certain circuit was -55.6 V. It was then reduced by 11.5 V. Find the new voltage.

74. Find the current through a resistor if it dissipates 4.54×10^{-4} watts when the voltage drop across it is 8.94×10^{-5} V. (Divide the power by the voltage drop.)

75. The voltage across a 8.12×10^2 Ω resistor is 4.49×10^{-4} V. Find the power dissipated in the resistor. (Divide the square of the voltage by the resistance.)

76. A rod having a cross-sectional area of 6.94×10^{-4} in.^{2} and a length of 3.64×10^4 in. is subjected to a tensile load of 9.12×10^5 lb. The elongation is 7.48×10^{-4} in. Find the modulus of elasticity E of the material. (Multiply the load by the length, and divide by the area and by the elongation.)

WRITING

77. Does your calculator have two separate keys marked with a minus sign? Why? Is there any difference between them, and if so, what? When would you use each? Write a paragraph or two explaining these keys and answering these questions.

7 Introduction to Algebra

In Chapter 1 we showed how to raise a number to a power. For example, 3^2, or 3 raised to the power 2, means

$$3^2 = (3)(3)$$

In a similar way, x^2, or x raised to the power 2, means

$$x^2 = (x)(x)$$

where x can stand for any number, not just 3. Going further, we can represent the exponent by a symbol, say n. Thus

$$x^n = (x)(x)(x)(x) \ldots (x)$$

n factors

Thus, 3^2 is an arithmetic expression, while x^n is an *algebraic* expression. In this chapter we will give definitions which apply to algebraic expressions and learn how to add and subtract algebraic expressions, how to remove symbols of grouping, and how to handle expressions with exponents.

7-1 ALGEBRAIC EXPRESSIONS

Mathematical Expressions

A grouping of mathematical symbols, such as signs of operation, numbers, and letters is called a *mathematical expression*.

Example 1

Some mathematical expressions are
(a) $x^2 + 3x - 4$ **(b)** $2 \tan 5x$ **(c)** $3 + 8 \ln x$

Algebraic Expressions

A mathematical expression that contains only numbers, algebraic symbols (such as x and y), and operations (addition, subtraction, multiplication, division, roots, and powers) is called an *algebraic expression*. Example 1(a) above shows an algebraic expression. The others are called transcendental expressions. We will study those in later chapters.

Equations

When one expression is set equal to another expression, we get an *equation*.

Some equations are
(a) $x + 8 = 3x - 1$
(b) $4x^2 - 2x + x = 0$
(c) $y = x^2 + 9$

Constants

A quantity that does not change value in a particular problem is called a *constant*. It can be a number, such as 3, 9.36, or π, or a letter. Letters that represent constants are usually chosen from the beginning of the alphabet (a, b, c, etc.).

─── **Example 3** ───────

The constants in the expression

$$2x^2 + 3x + 4$$

are 2, 3, and 4.

An expression in which the constants are represented by letters is called a *literal expression*.

─── **Example 4** ───────

The constants in the literal expression

$$ax^2 + bx + c$$

are a, b, and c.

Variables

A quantity that may change in a particular problem is called a *variable*. Letters that represent variables are usually chosen from the end of the alphabet (x, y, z, etc.).

─── **Example 5** ───────

In the expression

$$ax + by + cz$$

the letters a, b, and c would usually represent constants, x, y, and z would usually represent variables.

Symbols of Grouping

Parentheses ( ), brackets [ ], and braces { }, are called *symbols of grouping*. As their name implies, these are used to group parts of the expression together, and affect the meaning of the expression.

─── **Example 6** ───────

The expression

$$\{[(2 - x^2) + 4] + (5 - 3x)\}$$

is an example of the use of symbols of grouping.

We will show how to remove symbols of grouping in Sec. 7-2. We will also see in Sec. 8-6 that symbols of grouping may be used to indicate multiplication.

Terms

The plus and minus signs separate an expression into *terms*.

Example 7

The expression $2x^2 + 5x + 3$ has three terms.

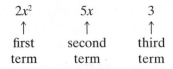

An exception is when the plus or minus sign is within a symbol of grouping.

Example 8

The expression $[(2x + 4x^2) - 5] + (3x + 4)$ has two terms.

$$\underbrace{[(2x + 4x^2) - 5]}_{\text{first term}} \qquad \underbrace{(3x + 4)}_{\text{second term}}$$

Factors

A divisor of a term is called a *factor* of that term.

Example 9

Some factors of ax^2 are a, x, and x^2, because a, x, and x^2 divide evenly into ax^2.

We do not usually state the separate factors of a number as factors of a term.

Example 10

The factors of $6ax$ are usually given as 6, a, and x, *not* as 2, 3, a, and x.

Coefficient

In elementary algebra, the *coefficient* of a term is the numerical part of the term. It is usually written before the variables.

Example 11

The coefficient of xy in the term $4xy$ is 4. It is also correct to say that $4x$ is the coefficient of y, but we will not use the word *coefficient* in that way.

When we say "coefficient," it will always refer to the constant part of the term, also called the *numerical coefficient*. It will include *any constants in literal form*.

Example 12

If b is understood to be a constant in the term $2bxy^2$, the coefficient is then $2b$.

If there is no numerical coefficient written before a term, it is understood to be 1, since

$$x = 1x$$

Example 13

In the expression $-x^3$, the coefficient is -1.

> **COMMON ERROR** Don't forget to include the negative sign with the coefficient.

Degree

The power to which the variable in a term is raised is called the *degree* of that term.

Example 14

(a) $4y$ is a term of degree 1, also called a first-degree term.

(b) $2x^2$ is a term of degree 2, or a second-degree term.

(c) $3z^8$ is a term of degree 8, or an eighth-degree term.

The term *degree* is used *only* when the exponents are positive integers.

Example 15

You would not say that $y^{1/2}$ is a "half-degree" term.
You would not say that x^{-2} is a "negative-two-degree" term.

If a term has more than one variable, *add their powers* to obtain the degree of the term.

Example 16

(a) $5x^3y^3$ is a term of degree 6, since the sum of the exponents 3 and 3 is 6.

(b) $4x^2yz$ is a term of degree 4, since $4x^2yz = 4x^2y^1z^1$, and the sum of the exponents is $2 + 1 + 1 = 4$.

The *degree of an expression* is the same as that of the highest degree term in that expression.

Example 17

$4x^2 - 5x + 1$ is an expression of degree 2.

Mathematical Expressions

Which of the following mathematical expressions are also algebraic expressions?

1. $x + 2y$ **2.** $y - \log x$

3. $3 \sin x$ **4.** $x^2 - z^3$

Algebraic Expressions

Which of the following algebraic expressions are literal expressions?

5. $5xy - 2x$ **6.** $ax + by$

7. $2az - 3bx$ **8.** $4x^2 + 4y^2$

Terms

How many terms are there in each expression?

9. $x^3 - 2x$ **10.** $(5y + y^2) - 5$

11. $(z - 9)(z + 4)$ **12.** $5(1 + x) + 3(4 - 3x)$

Factors

Write the factors of each expression.

13. $3ax$ **14.** $9xyz$

15. $7x^2y^3$ **16.** $6a^2bx$

Coefficient

Write the coefficient of each term. Assume that letters from the beginning of the alphabet $(a, b, c, \ldots)$ are constants.

17. $6x^2$ **18.** x

19. $-x$ **20.** $3cx^3$

21. $\dfrac{ay^2}{2}$ **22.** $(1/3)(x + 1)(b)$

23. $(c - 2)3x^4$ **24.** $\dfrac{a}{y + 1}$

Degree

State the degree of each term.

25. $3x$ **26.** $4y^2$

27. $3xy$ **28.** $5x^2y^3$

State the degree of each expression.

29. $3x + 4$ **30.** $5 - xy$

31. $3x^2 - 2x + 5$ **32.** $2xy^2 + xy - 4$

7-2   ADDING AND SUBTRACTING POLYNOMIALS

Polynomials

An algebraic expression in which the power of every variable is a positive integer is called a *polynomial*.

Example 18

Some polynomials are

$$3x^4, \qquad 4x^2 - 5x - 4, \qquad \text{and} \qquad 2xy + y^3$$

Some expressions that are *not* polynomials are

$$1/x, \qquad 3x^{-4}, \qquad 4x^2 - 5\sqrt{x} - 4, \qquad \text{and} \qquad 2xy + y^{1/3}$$

Combining Like Terms

Terms that differ only in their coefficients are called *like* terms. Their variable parts are the same.

Example 19

$4xz$ and $-5xz$ are like terms.

We add and subtract algebraic expressions by *combining like terms*. Like terms are added by adding their coefficients. We also call this *collecting terms*.

Example 20

(a) $3y + 4y = 7y$

(b) $18z - 9z = 9z$

Don't forget that any term with no numerical coefficient has an unwritten coefficient of 1.

Example 21

(a) $4x + x = 5x$

(b) $-3y + y = -2y$

Commutative Law of Addition

We used the commutative law (Eq. 1) when adding numbers in Chapter 1. We saw that it means that the order in which we add the terms does not affect the sum. Thus $3 + 4 = 4 + 3$. The commutative law applies, of course, when the expression contains letters. Thus $b + c = c + b$. This law enables us to rearrange the terms of an expression to make it easier to simplify.

Example 22

Simplify the expression

$$8y + 5x - 5z + 2z - x - 4y$$

Solution: We use the commutative law to rearrange the expression to get like terms together. We can then easily collect terms.

$$8y + 5x - 5z + 2z - x - 4y = 8y - 4y + 5x - x - 5z + 2z$$
$$= 4x + 4y - 3z$$

With practice, you will soon be able to omit the first step and collect terms by inspecting the original expression.

The procedure is no different when the terms have *decimal coefficients*. Just be sure to round properly.

Example 23

$$3.92x - 4.02y - 2.24x + 1.85y = 3.92x - 2.24x - 4.02y + 1.85y$$
$$= 1.68x - 2.17y$$

Removing Symbols of Grouping

To add or subtract expressions we must first remove any parentheses or other symbols of grouping, such as brackets and braces. If the parentheses are preceded only by a $(+)$ sign (or by no sign), they may simply be removed.

Example 24

(a) $(x - y + z) = x - y + z$

(b) $(a - b) + (c + d) = a - b + c + d$

When the parentheses are preceded only by a $(-)$ sign, change the sign of every term within the parentheses before removing them.

Example 25

$$-(x - y + z) = -x + y - z$$

An expression may contain braces or brackets as well as parentheses. If a group of terms is inside another group, simplify the innermost group first.

Example 26

Simplify the expression

$$3[x - 2(x + y)]$$

Solution: We start removing the parentheses by multiplying $(x + y)$ by (-2),

$$3[x - 2(x + y)] = 3[x - 2x - 2y]$$
$$= 3[-x - 2y]$$
$$= -3x - 6y$$

Example 27

Simplify the expression

$$3a + \{d - [2e + [-3b - (7c + 4f)]]\}$$

Solution: The braces and one set of brackets are preceded by a $(+)$ sign so we may remove them immediately, getting

$$3a + d - [2e - 3b - (7c + 4f)]$$

Next we remove parentheses within the remaining square brackets. Since these are preceded by a $(-)$ sign, we change the signs of the terms.

$$3a + d - [2e - 3b - 7c - 4f]$$

We remove the brackets, changing the signs of the terms as we do.

$$3a + d - 2e + 3b + 7c + 4f$$

Finally, it is common practice to write the terms in alphabetical order. Our simplified expression is then

$$3a + 3b + 7c + d - 2e + 4f$$

Once the parentheses are gone, like terms may be combined.

Example 28

$$-(a - b) - (a + b) = -a + b - a - b = -2a$$

Combining More Than Two Expressions

The procedure is no different when more than two expressions are combined.

Example 29

$$(3a + 5b - 2c) - (2a - b - 4c) - (a + 3b + c)$$
$$= 3a + 5b - 2c - 2a + b + 4c - a - 3b - c$$
$$= 3a - 2a - a + 5b + b - 3b - 2c + 4c - c$$
$$= 3b + c$$

Vertical Addition and Subtraction

It is often easier to arrange the expressions *vertically,* with like terms in the same column.

Example 30

Combine

$$(3x - 2y + 4z) - (7y + 2w - 5x) - (3w + 2z + 4y - x)$$

Solution: We remove parentheses and write the expressions one above the other, with like terms in the same vertical column, and collect terms.

$$
\begin{array}{r}
3x - 2y + 4z \\
5x - 7y - 2w \\
\underline{x - 4y - 2z - 3w} \\
9x - 13y + 2z - 5w
\end{array}
$$

Instructions Given Verbally

In preparation for verbal problems to come later, we give some problems in verbal form.

Example 31

Find the sum of $x - 2y + z$ and $3x + y + z$.

Solution: We have

$$(x - 2y + z) + (3x + y + z)$$
$$= x - 2y + z + 3x + y + z$$
$$= x + 3x - 2y + y + z + z$$
$$= 4x - y + 2z$$

Example 32

Subtract $5a - 2b + 6c$ from the sum of $8a - 4b - 3c$ and $a - b + 6c$.

Solution: We add the last two expressions,

$$(8a - 4b - 3c) + (a - b + 6c) = 9a - 5b + 3c$$

Then we subtract the first expression.

$$9a - 5b + 3c - (5a - 2b + 6c)$$
$$= 9a - 5b + 3c - 5a + 2b - 6c$$
$$= 4a - 3b - 3c$$

SECTION 7-2 | **Exercises** | **Adding and Subtracting Polynomials**

Combining Like Terms

Combine as indicated, and simplify.

1. $8y + 2y$
2. $6x - 8x$
3. $38.2a - 17.2a$
4. $2.94z + 5.37z$
5. $5ac + 9ac - 20ac$
6. $9xyz - 2xyz + 7xyz$
7. $7.39y - 6.62y + 1.94y$
8. $23.9ab + 54.9ab - 65.1ab$
9. $5x + 2x - 8x - x$
10. $9a - 2a + 7a - 3a$
11. $88x + 23y - 17z + 68y - 36x + 39z$
12. $a - 6b + 2c + a + 6b + n - 2c$
13. $1.95x - 4.38z + 2.83a - 5.21z - 9.27x$
14. $33.9ab - 82.4ac + 29.3ad - 84.2ac + 73.2ab$

Removing Symbols of Grouping

Remove symbols of grouping and collect terms.

15. $(4z + 2) + (z - 5)$
16. $(2x + 5) + (x - 2)$
17. $(c - 5) - (6 + 3c)$
18. $(2x + 6a) - (4x - a)$
19. $(1.24z + 2.22) + (9.24z - 1.25)$
20. $(24.2x + 8.25) + (83.6x - 2.92)$
21. $(0.826c - 5.37) - (2.76 + 0.273c)$
22. $(7262x + 1.26a) - (2844x - 8.23a)$
23. $(5 - 2a^3 + 3a^4) + (4a^4 - 6a^3 - 7)$
24. $(y - z - b) - (b + y + z)$
25. $(5bc + 6c - a) - (8c - 2a + 3bc)$
26. $(2z + 5c - 3a) - (6a + 2c - 4z)$

27. $(2.25c - 9.28b + 3.82a) - (-3.92a - 1.72b - 8.33c)$

28. $(23y^2 + 4y^3 - 12) - (11y^3 - 8y^2 + y)$

29. $24ab - (16ab - 3x^2 + 7z - 2y^2)$

30. $(18y^2 - 12xy) - (6y^2 + xy - a)$

31. $(4y + 2y^2) - (3y - 6b + 4y^2 + 5)$

32. $(-6x - z) - \{3y + [7x - (3z + 8y + x)]\}$

33. $-2a + \{d - [6c + [-5b - (4a + 9b)] - (2d - a + 3c)]\}$

34. $-6x - \{9y + [5x - (-2y - 8b)] - [2b + a]\}$

Combining More Than Two Expressions

Combine and simplify. Use either horizontal or vertical addition and subtraction.

35. $(4ab + 6bc + 8cd) + (-6ab - 3ab) + (4cd - 6bc)$

36. $(2a + 3b - 1) + (2c + d - b) - (3a - 4c + 5 - 6d)$

37. $(3b - 7p + 4r) + (3s - 11p - 19r) - (-3p + r - 10s)$
    $- (-5b + 2p - 2r + 4s)$

38. $(2bx + 9x + 4a) - (4bx - a) + (3b + a)$

39. $(-a + 5x) + (a^3 + 3a - 11x^2 + 2a^2 - 3) - (7a^2 + 3a^3 - 6x^3 + 12)$

40. $(2y - x + 3z - 13) - (4x + 2 - 5y) + (6z + 8)$

Instructions Given Verbally

41. Add $a - c + b$ and $b + c - a$.

42. Find the sum of $6bc + n^2 + 3p$ and $-5x + 3n^2$.

43. Subtract $5a + 7d - 4b + 6c$ from $8b - 10c + 3a - d$.

44. Subtract $2xy + 4y - 3x$ from $5x - 2xy + 8y$.

45. What is the sum of $24by^5 - 14bx^4$, $-72bx^5 + 2by^5 - 3bx^4$, and $9bx^4 + 23by^4 - 21by^5$?

46. Add $3b(y^2 - z) - 6nx^2$ to the sum of $2b(z + y^2) + 3nx^2$ and $4b(-z - y^2) - 2nx^2$.

47. What is the sum of $5a^2 - 8c + 5bd$, $4a^2 - 3bd + 3c$, $3bd - 6c + 12a^2$, and $5a^2 - 3c$?

48. Add $3(x + z) - 5(y + z)$, $2(x + y) + 12(z + y)$, and $4(x + y) - 6(x + z)$.

49. Subtract $4(b + c)$ from the sum of $6(b + c)$ and $3(b + c)$.

50. From the sum of $4a - 6b + c$ and $7d - 2a + 8b$, subtract the sum of $3d + 4c - 2a$ and $-5c + 6b - 2a$.

7-3   LAWS OF EXPONENTS

Definitions

We have done some work with powers of numbers in Chapter 1. Now we will expand those ideas to include powers of algebraic expressions. Here we will deal only with expressions that have integers (positive or negative, and zero) as exponents.

Example 33

We study such expressions as

$$x^4, \qquad b^{-3}, \qquad a^0, \qquad a^3b^2d^{-1}, \qquad \text{and} \qquad (x + y)^3$$

Recall from Chapter 1 that a positive exponent shows how many times the base is to be multiplied by itself.

Example 34

In the expression 2^5, the base is 2 and the exponent is 5.

Its meaning is

$$2^5 = (2)(2)(2)(2)(2) = 32$$

In general,

POSITIVE INTEGRAL EXPONENT	x^n means $\underbrace{(x)(x)(x)(x) \cdot \cdot \cdot (x)}_{n \text{ factors}}$	18

> **COMMON ERROR** An exponent applies only to the symbol directly in front of it. Thus,
>
> $$5y^3 = 5(y^3)$$
>
> but
>
> $$5y^3 \neq 5^3 y^3$$

Multiplying Powers

Let us multiply x^3 by x^4, two quantities that have the same base x. From the definition of an exponent we have

$$x^3 = x \cdot x \cdot x$$

and

$$x^4 = x \cdot x \cdot x \cdot x$$

Multiplying gives

$$x^3 \cdot x^4 = (x \cdot x \cdot x)(x \cdot x \cdot x \cdot x)$$
$$= x \cdot x \cdot x \cdot x \cdot x \cdot x \cdot x$$
$$= x^7$$

We see that the exponent in the result is the sum of the two original exponents.

$$x^3 \cdot x^4 = x^{3+4} = x^7$$

This gives our first law of exponents.

PRODUCT	$x^m x^n = x^{m+n}$	19

When multiplying powers having the same base, add the exponents.

Example 35 ─────────
$$x^2(x^5) = x^{2+5} = x^7$$

Every quantity has an exponent of 1, even though it is not usually written. Thus

$$x = x^1$$

Example 36 ─────────

(a) $(x)(x^2) = (x^1)(x^2) = x^{1+2} = x^3$

(b) $(b)(b^3)(b^2) = (b^1)(b^3)(b^2) = b^{1+3+2} = b^6$

Quotients

Let us divide x^4 by x^2. By Eq. 18,

$$\frac{x^4}{x^2} = \frac{(x)(x)(x)(x)}{(x)(x)} = \frac{(\cancel{x})(\cancel{x})}{(\cancel{x})(\cancel{x})}(x)(x)$$
$$= (x)(x) = x^2$$

The same result could have been obtained by subtracting exponents,

$$\frac{x^4}{x^2} = x^{4-2} = x^2$$

We state it as another law of exponents:

QUOTIENT	$\dfrac{x^m}{x^n} = x^{m-n}$ $(x \neq 0)$	20

When dividing powers having the same base, subtract the exponents.

Example 37 ─────────

(a) $\dfrac{x^2 y^5}{xy^3} = x^{2-1}y^{5-3} = xy^2$

(b) $\dfrac{x^{a+b}}{x^{a-b}} = x^{a+b-(a-b)} = x^{2b}$

Power Raised to a Power

Now let us take a quantity which itself is raised to a power and raise that expression to another power. Let us raise y^3 to the power 2.

$$(y^3)^2$$

By Eq. 18,

$$(y^3)^2 = (y^3)(y^3)$$
$$= (y)(y)(y)(y)(y)(y)$$
$$= y^6$$

a result that could have been obtained by multiplying the exponents.

$$(y^3)^2 = y^{3(2)} = y^6$$

In general,

POWERS	$(x^m)^n = x^{mn}$	21

When a power is raised to a power, multiply the exponents.

Example 38

(a) $(n^4)^3 = n^{4(3)} = n^{12}$

(b) $(10^2)^4 = 10^{2(4)} = 10^8$

Product Raised to a Power

Let us now find a rule for raising a product, such as ab, to a power, say 2.

$$(ab)^2$$

By Eq. 18

$$(ab)^2 = (ab)(ab)$$
$$= (a)(b)(a)(b)$$
$$= (a)(a)(b)(b)$$
$$= a^2b^2$$

In general,

PRODUCT RAISED TO A POWER	$(xy)^n = x^n y^n$	22

When a product is raised to a power, each factor may be separately raised to the power.

Example 39

(a) $(abc)^4 = a^4 b^4 c^4$

(b) $(3y)^3 = 3^3 y^3 = 27y^3$

(c) $(4.23 \times 10^2)^3 = (4.23)^3 \times (10^2)^3 = 75.7 \times 10^6$

(d) $(2y^4 z^n)^3 = 2^3 (y^4)^3 (z^n)^3 = 8y^{12} z^{3n}$

(e) $(-x^2 z)^5 = (-1)^5 (x^2)^5 (z)^5 = -x^{10} z^5$

> **COMMON ERROR** There is no similar rule for the sum of two quantities raised to a power.
>
> $$(x + y)^n \neq x^n + y^n$$

Quotient Raised to a Power

We can show, in a similar way to that used for products, that

$$\left(\frac{x}{y}\right)^2 = \left(\frac{x}{y}\right)\left(\frac{x}{y}\right) = \frac{x^2}{y^2}$$

Or, in general,

QUOTIENT RAISED TO A POWER	$\left(\dfrac{x}{y}\right)^n = \dfrac{x^n}{y^n}$ $(y \neq 0)$	23

When a quotient is raised to a power, the numerator and denominator may be separately raised to the power.

Example 40

(a) $\left(\dfrac{2}{a}\right)^3 = \dfrac{2^3}{a^3} = \dfrac{8}{a^3}$

(b) $\left(\dfrac{2x}{4y}\right)^2 = \dfrac{2^2 x^2}{4^2 y^2} = \dfrac{x^2}{4y^2}$

(c) $\left(\dfrac{3c^2}{4d^4}\right)^3 = \dfrac{3^3 (c^2)^3}{4^3 (d^4)^3} = \dfrac{27c^6}{64d^{12}}$

(d) $\left(-\dfrac{x^2}{y}\right)^4 = \dfrac{(-1)^4 (x^2)^4}{y^4} = \dfrac{x^8}{y^4}$

Zero Exponent

If we divide x^n by itself we get, by Eq. 20

$$\frac{x^n}{x^n} = x^{n-n} = x^0$$

But any expression divided by itself equals 1, so

ZERO EXPONENT	$x^0 = 1 \qquad (x \neq 0)$	24

Any expression (except 0) raised to the zero power equals 1.

Example 41

(a) $(9626)^0 = 1$
(b) $(abc)^0 = 1$
(c) $(4x^2 + 9x - 35)^0 = 1$
(d) $7a^0 = 7(1) = 7$

Negative Exponent

We now divide x^0 by x^a. By Eq. 20

$$\frac{x^0}{x^a} = x^{0-a} = x^{-a}$$

Since $x^0 = 1$, we get

NEGATIVE EXPONENT	$x^{-n} = \dfrac{1}{x^n}$ $x \neq 0$	25

Example 42

(a) $6^{-1} = \dfrac{1}{6^1} = \dfrac{1}{6}$

(b) $x^{-2} = \dfrac{1}{x^2}$

(c) $2x^{-3} = 2\left(\dfrac{1}{x^3}\right) = \dfrac{2}{x^3}$

(d) $\dfrac{3}{x^{-2}} = 3x^2$

(e) $4x^{-2} + 2y^{-3} = \dfrac{4}{x^2} + \dfrac{2}{y^3}$

SECTION 7-3   Exercises   Laws of Exponents

Definitions

Evaluate each expression.

1. 3^3
2. $(3)^5$
3. $(-2)^4$
4. $(-2)^5$
5. $(0.001)^3$
6. $(-5)^3$

Multiplying Powers

Multiply.

7. $(x^4)(x^2)$
8. $(y^b)(y^3)$
9. $(a^3)(a^6)$
10. $(10^5)(10^9)$
11. $(10^2)(10^6)$
12. $(z^{11})(z^2)$

Quotients

Divide. Write your answers without negative exponents.

13. $\dfrac{a^6}{a^4}$
14. $\dfrac{2^4}{2^3}$
15. $\dfrac{y^{a+1}}{y^{a-2}}$
16. $\dfrac{10^6}{10^2}$
17. $\dfrac{10^{b-1}}{10^{a-3}}$
18. $\dfrac{10^4}{10^3}$
19. $\dfrac{b^{-4}}{b^{-5}}$
20. $\dfrac{x^{-6}}{y}$

Power Raised to a Power

Simplify.

21. $(a^2)^4$
22. $(2^5)^2$
23. $(z^c)^a$
24. $(y^{-1})^{-3}$
25. $(a^{x-1})^3$

Product or Quotient Raised to a Power

Raise to the power indicated.

26. $(ac)^3$
27. $(3a)^2$
28. $(4a^3c^2)^4$
29. $(2xyz)^5$
30. $\left(\dfrac{2}{3}\right)^3$
31. $\left(-\dfrac{2}{5}\right)^3$
32. $\left(\dfrac{a}{b}\right)^3$
33. $\left(\dfrac{4x^2}{3y^2}\right)^2$
34. $\left(\dfrac{2ab^3}{3c^2d}\right)^3$

Zero Exponent

Evaluate.

35. $(2x^2 - 8x + 32)^0$ **36.** $108a^3c^0$ **37.** $\dfrac{82}{y^0}$

38. $\dfrac{c}{y^0}$ **39.** $\dfrac{(z^{-n})(z^2)}{z^{2-n}}$ **40.** $4\left(\dfrac{a^7}{y^4}\right)^0$

Negative Exponent

Write each expression with positive exponents only.

41. x^{-1} **42.** $(-b)^{-3}$ **43.** $\left(\dfrac{2}{x}\right)^{-4}$

44. $ab^{-5}c^{-2}$ **45.** $\left(\dfrac{4x^3}{3y^2}\right)^{-3}$ **46.** $a^{-5}bc^{-2}$

47. $4w^{-4} - 3z^{-2}$ **48.** $\left(\dfrac{a}{b}\right)^{-4}$

Express without fractions, using negative exponents where needed.

49. $\dfrac{1}{a}$ **50.** $\dfrac{5}{x^3}$ **51.** $\dfrac{c^2}{d^3}$

52. $\dfrac{w^4z^{-2}}{x^{-3}}$ **53.** $\dfrac{y^2}{x^{-4}}$ **54.** $\dfrac{c^{-1}d^{-4}}{b^{-2}c^{-3}}$

■ Chapter 7 REVIEW EXERCISES

Evaluate each expression.

1. $(-3)^4$ **2.** $(0.002)^3$ **3.** $(-4)^3$

4. Subtract $2w + 3z - 4x + 6y$ from $4x - 11y + 3w - z$.

5. What is the sum of $7w^2 - 2y + 3xz$, $3w^2 - 5xz + 3y$, $8xz - 3y + 15w^2$, $3y + 2w^2 - 2xz$, and $8w^2 - 3y$?

Divide. Write your answers without negative exponents.

6. $\dfrac{w^6}{w^4}$ **7.** $\dfrac{b^{w+1}}{b^{w+2}}$ **8.** $\dfrac{10^{a+1}}{10^{w-3}}$ **9.** $\dfrac{x^{-4}}{x^{-5}}$

Simplify.

10. $(w^2)^4$ **11.** $(1^5)^2$ **12.** $(x^y)^w$ **13.** $(b^{-1})^{-3}$

Multiply.

14. $(a^4)(a^2)$ **15.** $(b^x)(b^3)$ **16.** $(z^{w-2})(z^{1-w})$ **17.** $(10^5)(10^9)$

Raise to the given power.

18. $(ay)^3$ **19.** $(4x^3y^2)^4$ **20.** $(2abz)^5$

21. $\left(\dfrac{1}{3}\right)^3$ **22.** $\left(-\dfrac{2}{7}\right)^2$ **23.** $\left(\dfrac{3ax^3}{4y^2z}\right)^3$

Combine as indicated and simplify.

24. $8b + 2b$ **25.** $5a + 2a - 8a - a$ **26.** $5wy + 9wy - 20wy$

27. Add $w - 6x + 2y$ and $w + 6x + y - 2z$.

28. What is the sum of $4ax + 6xy - 8yz$, $-6wx - 3ax$, $4yz - 6xy$, and $5ax - 3xy$?

Write without fractions, using negative exponents where needed.

29. $\dfrac{1}{w}$ **30.** $\dfrac{w^4z^{-2}}{a^{-3}}$ **31.** $\dfrac{y^{-1}z^{-4}}{x^{-2}y^{-3}}$

Evaluate.

32. $(2a^2 - 8a + 32)^0$ **33.** $108w^3y^0$

34. $\dfrac{32}{a^0}$ **35.** $7\left(\dfrac{w^3}{z^2}\right)^0$

How many terms are there in each expression?

36. $a^3 - 2a$ **37.** $(5b + b^2) - 5$ **38.** $(z - 9)(z + 4)$

Write each expression without negative exponents.

39. a^{-1} **40.** $(-x)^{-3}$ **41.** $wx^{-5}y^{-2}$

42. $\left(\dfrac{2}{a}\right)^{-4}$ **43.** $\left(\dfrac{4a^3}{3b^2}\right)^{-3}$ **44.** $\left(\dfrac{w}{x}\right)^{-4}$

Remove symbols of grouping and collect terms.

45. $(4z + 2) + (z - 5)$ **46.** $2a^3 - (4a - w + b)$

47. $13b^2 - 14ab - (2b^2 + ab - w)$

48. $(5xy + 2y - w) - (6w + 3xy) + (-4y - 7w)$

49. $(z + 6w - 5y) - (3x + 4z) - (w - 3y) + (-x - 2z)$

50. $-5a - \{3b + [2a - (-6b - 3x)] - [8x + w]\}$

51. Add $-w + 5a$, $w^3 + 3w - 11a^2 + 2w^2 - 3$, $7w^2 + 3w^3 - 6a^3 + 12$, and $9w^2 - 5w^3 + 2a^2 - 5 + 7a$.

52. Subtract $23b^2 + 4b^3 - 12$ from $11b^3 - 8b^2 + b$.

WRITING

53. Suppose your friend is sick and has missed the introduction to algebra, and is still out of class. Write a note to your friend explaining in your own words what you think algebra is, and how it is related to the arithmetic that you both have just finished studying.

Multiplication of Algebraic Expressions

In Chapter 7 we learned how to add and subtract algebraic expressions. Here we learn how to multiply them, and in Chapter 9 we divide algebraic expressions.

We start by multiplying an expression having just one term (monomial) by another monomial. We then go on to multiply a monomial by an expression having more than one term (called a multinomial). Finally, we multiply a multinomial by another multinomial. In this chapter we will multiply only polynomials, those multinomials whose exponents are all positive integers, but the methods we will learn apply equally well to any multinomials.

8-1 — PRODUCT OF TWO MONOMIALS

Symbols and Definitions

Multiplication is indicated in several ways: by the usual $\times$ symbol, by a dot, or by parentheses, brackets, or braces. Thus, the product of b and d could be written

$$b \times d \quad b \cdot d \quad b(d) \quad (b)d \quad (b)(d)$$

Most common of all is to use no symbol at all. The product of b and d would usually be written bd. Avoid using the $\times$ symbol because it could get confused with the letter x.

We get a product when we multiply two or more factors.

$$(\text{factor})(\text{factor})(\text{factor}) = \text{product}$$

Rules of Signs

When we multiply two factors that have the *same* sign, we get a product that is *positive*. When we multiply two factors that have *opposite* signs, we get a product that is *negative*. If a and b are positive quantities, we have

RULES OF SIGNS	$(+a)(+b) = (-a)(-b) = +ab$	9
	$(+a)(-b) = (-a)(+b) = -(+a)(+b) = -ab$	10

The product of two factors of like signs is positive, of unlike signs is negative.

Example 1

(a) $(+x)(+z) = xz$

(b) $(+x)(-z) = -xz$

(c) $(-x)(+z) = -xz$

139

When we multiply *more* than two factors, every pair of negative factors will give a positive product. Thus, if there is an even number of negative factors, the final result will be positive; if there is an odd number of negative factors, the result will be negative.

Example 2

(a) $(a)(-b)(-c) = abc$

(b) $(-p)(-q)(-r) = -pqr$

(c) $(-w)(-x)(-y)(-z) = wxyz$

Commutative Law

When we multiplied *numbers,* we saw that the order of multiplication did not make any difference, since 2×3 gives the same product as 3×2.

In symbols,

COMMUTATIVE LAW FOR MULTIPLICATION	$ab = ba$	2

Example 3

Multiply $-2y$ by $4x$.

Solution:

$$(-2y)(4x) = -2(y)(4)(x)$$

By Eq. 2,
$$= -2(4)(y)(x)$$
$$= -8yx$$

or
$$= -8xy$$

We usually write the letters in alphabetical order.

Multiplying Monomials

Recall that a monomial is an algebraic expression having one term, such as the expressions $3y^2$ and $(8x)^3$. To multiply monomials, we use the laws of exponents and the rules of signs.

Example 4

$$(-5xy^2)(2x^3y^2) = (-5)(2)(x)(x^3)(y^2)(y^2)$$
$$= -10x^{1+3}y^{2+2}$$
$$= -10x^4y^4$$

It is no harder to multiply three or more monomials than to multiply two monomials.

Example 5

Multiply $-3a$, $-2a^2b$, and ab^2c^3.

Solution: By Eq. 9,

$$-3a(-2a^2b)(ab^2c^3) = -3(-2)(1)(a)(a^2)(a)(b)(b^2)(c^3)$$
$$= 6a^{1+2+1}b^{1+2}c^3$$
$$= 6a^4b^3c^3$$

If there are *literals* in the exponents, simply combine them using the laws of exponents we have already studied.

Example 6

Multiply $5a^nx^2$ by $-3a^3$.

Solution: By Eq. 10,

$$5a^nx^2(-3a^3) = 5(-3)(a^n)(a^3)(x^2)$$
$$= -15a^{n+3}x^2$$

The procedure is no different when the quantities to be multiplied are approximate numbers. We must, however, retain the proper number of digits in our answer. Recall from Sec. 3-3 that *when multiplying approximate numbers we retain as many significant digits in our product as contained in the factor having the fewest significant digits.*

Example 7

$$(3.848x^2)(5.24xy^2) = 20.2x^3y^2$$

Since one of our numerical factors has four significant digits and the other has only three, we round our product to three significant digits.

SECTION 8-1 Exercises   Product of Two Monomials

Multiply the following monomials, and simplify.

1. $(x)(-y)$
2. $(-w)(-z)$
3. $(a)(-b)(c)$
4. $(-x)(-y)(z)$
5. $(2.82x)(3.26y)$
6. $(33.5a)(3.72ab)$
7. $(-8xy)(-2x)$
8. $(-3ab)(-5ac)$
9. $a^2(a^5)$
10. $xy^2(5x^2)$
11. $(3.72x^2a)(-4.26xa^3)$
12. $(-6.72ab^4)(5.27a^2bc)$
13. $(5x^2y)(-3xyz^2)(2x^2y^3z)$
14. $(-3a^2bc)(4abc^3)(-6ab^2)$
15. $8x(xy)(5yz)(-x^2y^3z)$
16. $(-5c^2)(-2bc)(-4a^2c)(-6ab^3)(-3b^2c)$
17. $(3.11a^mx^2)(2.94a^3)$
18. $(6.82b^3x^3)(5.16b^2x)$
19. $(x^my^n)(3x^3y^2)$
20. $(2w^3y^3)(7w^ay^b)$

8-2

PRODUCT OF A MULTINOMIAL AND A MONOMIAL

A *multinomial* is an algebraic expression having more than one term.

Example 8

Some multinomials are

(a) $6x - 8$

(b) $7x^2 + 2x - 1$

(c) $x^{-2} + 5$

Recall that a *polynomial* is a monomial or a multinomial in which the powers to which the variable is raised are all positive integers. The first two expressions in the preceding example are polynomials, but the third is not. The examples in this chapter will show multiplication of polynomials only, but the methods we show are valid for any multinomials. We have not yet covered the rules needed for multiplying other multinomials, such as those containing radicals, negative exponents, logarithms, and so forth. In later chapters we will show multiplication of such expressions.

A *binomial* is a polynomial with two terms, and a *trinomial* is a polynomial having three terms. In Example 8, the first expression is a binomial and the second is a trinomial.

To multiply a multinomial and a monomial, we use the distributive law (Eq. 5):

DISTRIBUTIVE LAW	$a(b + c) = ab + ac$	5

Example 9
$$2x(x + 1) = 2x(x) + 2x(1) = 2x^2 + 2x$$

Be especially careful when multiplying negative quantities.

Example 10
$$-y(y^2 - 3) = -y(y^2) - y(-3) = -y^3 + 3y$$

Example 11
$$5x(-2x^2 - 1) = 5x(-2x^2) + 5x(-1)$$
$$= -10x^3 - 5x$$

We can extend the distributive law, which was written for a monomial times a binomial, to a multinomial having any number of terms. We simply multiply every term in the multinomial by the monomial.

Example 12

Multiply $2y^2 + 5y - 8$ by $-4y^2$.

Solution:

$$-4y^2(2y^2 + 5y - 8) = -4y^2(2y^2) + (-4y^2)(5y) - (-4y^2)(8)$$
$$= -8y^4 - 20y^3 + 32y^2$$

SECTION 8-2   Exercises   Product of a Multinomial and a Monomial

Multiply the multinomial by the monomial, and simplify.

1. $3(-2 - x)$ **2.** $x(b + 2)$ **3.** $2(a + 3b)$

4. $x(x - 5)$ **5.** $3.83b(b^2 + 1.27)$ **6.** $2.03x(1.27x - 2.36)$

7. $3x(-7 - 10x)$ **8.** $b^4(b^2 + 8)$ **9.** $a^2b(2a + b - ab)$

10. $-6x^3y^3(3xy + 5x^2y^3 - 2xy^2)$ **11.** $2ab(9a^2 + 6ab - 3b^2)$

12. $-4.27m^2(2.83m^4 + 6.82m^2 - 3.25m^3 + 2.47)$

13. $-5.16xy(1.23x^2y - 5.83xy^2 + 4.27x^2y^2 - 2.94xy)$

14. $6mn^2(5m^3n + 4mn^2 - 2m^2n + mn)$

8-3

PRODUCT OF TWO BINOMIALS

To multiply any multinomials, we multiply every term in one multinomial by each term in the other multinomial, and combine like terms. Here we apply this rule to binomials and later use it for multinomials having any number of terms.

Example 13
$$(x - 2)(x + 3) = (x)(x) + (x)(3) + (-2)(x) + (-2)(3)$$
$$= x^2 + 3x - 2x - 6$$
$$= x^2 + x - 6$$

We can also multiply two multinomials by using the distributive rule.

Example 14

Repeating Example 13 using the distributive rule gives,

$$(x - 2)(x + 3) = x(x + 3) - 2(x + 3)$$
$$= x^2 + 3x - 2x - 6$$
$$= x^2 + x - 6$$

as before.

FOIL Rule

One way to keep track of the terms when multiplying binomials is by the FOIL rule. Multiply the **F**irst terms, then **O**uter, **I**nner, and **L**ast terms. These products can, of course, be done in any order, but we can avoid getting mixed up if we always follow the same FOIL order.

Example 15
Repeating Example 13,

$$
\begin{array}{cccc}
\textbf{F} & \textbf{O} & \textbf{I} & \textbf{L} \\
\downarrow & \downarrow & \downarrow & \downarrow
\end{array}
$$

$$(x - 2)(x + 3) = x^2 + 3x - 2x - 6$$
$$= x^2 + x - 6$$

You might prefer to arrange the expressions *vertically*.

Example 16
Repeating Example 13, we get,

$$
\begin{array}{r}
x - 2 \\
x + 3 \\
\hline
3x - 6 \\
x^2 - 2x \quad\;\; \\
\hline
x^2 + \; x - 6
\end{array}
$$

Product of the Sum and Difference of Two Terms

The product of two binomials gives us expressions that will occur often enough later for us to take special note of them here. These are called *special products*. We will need them especially when we study factoring.

Let us multiply the binomials $(x - y)$ and $(x + y)$. We get

$$(x - y)(x + y) = x^2 + xy - xy - y^2$$

The sum of the two middle terms is zero, so we get

DIFFERENCE OF TWO SQUARES	$(x - y)(x + y) = x^2 - y^2$	31

Thus, the product is the *difference of two squares:* the squares of the original numbers.

Example 17
$$(x + 4)(x - 4) = x^2 - 4^2 = x^2 - 16$$

Example 18
$$(a + b)(a - b) = a^2 - b^2$$

Example 19
$$(2x + 3y)(2x - 3y) = (2x)^2 - (3y)^2 = 4x^2 - 9y^2$$

Example 20

$$(7a^2x + 3by^2)(7a^2x - 3by^2) = (7a^2x)^2 - (3by^2)^2$$
$$= 49a^4x^2 - 9b^2y^4$$

Product of Two Binomials Having the Same First Term

When we multiply $(x + a)$ and $(x + b)$, we get

$$(x + a)(x + b) = x^2 + ax + bx + ab$$

Combining like terms, we obtain a trinomial in which the coefficient of the first term (called the *leading coefficient*) is equal to 1.

TRINOMIAL WITH LEADING COEFFICIENT = 1	$(x + a)(x + b) = x^2 + (a + b)x + ab$	35

Example 21

$$(x - 3)(x + 5) = x^2 + (-3 + 5)x + (-3)5$$
$$= x^2 + 2x - 15$$

General Quadratic Trinomial

When we multiply two binomials of the form $ax + b$ and $cx + d$, we get

$$(ax + b)(cx + d) = acx^2 + adx + bcx + bd$$

Combining like terms, we obtain

GENERAL QUADRATIC TRINOMIAL	$(ax + b)(cx + d) = acx^2 + (ad + bc)x + bd$	36

Example 22

$$(4x - 2)(5x - 3) = 4(5)x^2 + [(4)(-3) + (-2)(5)]x + (-2)(-3)$$
$$= 20x^2 - 22x + 6$$

SECTION 8-3   Exercises   Product of Two Binomials

Multiply the following binomials, and simplify.

1. $(x + y)(x + z)$
2. $(4a - 3)(a + 2)$
3. $(4m + n)(2m^2 - n)$
4. $(y + 2)(y - 2)$
5. $(2x - y)(x + y)$
6. $(a^2 - 3b)(a^2 + 5b)$
7. $(4xy^2 - 3a^3b)(3xy^2 + 4a^3b)$
8. $(2m^2 - 2n^2)(2m^2 + 2n^2)$
9. $(a - 7x)(2a + 3x)$
10. $(3x - z^2)(4x - 3z^2)$
11. $(ax - 5b)(ax + 5b)$
12. $(5y^2 + 3z)(5y^2 - 3z)$
13. $(2.93x - 1.11y)(x + y)$
14. $(2.84a^2 - 3.82b)(a^2 + 5.11b)$

15. $(4.03y^2 - 3.92a^3b)(3.26y^2 + 4.73a^3b)$

16. $(2.83m^2 - 2.12n^2)(2.83m^2 + 2.12n^2)$

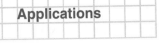

17. A rectangle has its length L increased by 2 units and its width W decreased by 3 units. Write an expression for the area of the rectangle, and multiply out.

18. A car traveling at a rate R for a time T will go a distance equal to RT. If the rate is decreased by 8.50 mi/h and the time is increased by 2.40 h, write an expression for the new distance traveled, and multiply out.

8-4   PRODUCT OF TWO MULTINOMIALS

We multiply multinomials in the same way that we multiplied other expressions; *multiply every term in one multinomial by every term in the other.* Then combine like terms.

Example 23

$$(a + 1)(a^2 + a + 1) = a(a^2) + a(a) + a(1) + 1(a^2) + 1(a) + 1(1)$$
$$= a^3 + a^2 + a + a^2 + a + 1$$
$$= a^3 + 2a^2 + 2a + 1$$

Example 24

$$(x + 2)(x^2 + 4x - 3) = x(x^2) + x(4x) + x(-3) + 2(x^2) + 2(4x) + 2(-3)$$
$$= x^3 + 4x^2 - 3x + 2x^2 + 8x - 6$$
$$= x^3 + 6x^2 + 5x - 6$$

Example 25

$$(w - x^2)(w^3 + aw^2 + bx^2 - x^3)$$
$$= w^4 + aw^3 + bwx^2 - wx^3 - w^3x^2 - aw^2x^2 - bx^4 + x^5$$

Example 26

Multiply $(3 + x)(4 - x)(5 - x)(6 + x)$.

Solution: Let us first multiply one pair of binomials, say, $(5 - x)(6 + x)$.

$$(5 - x)(6 + x) = 30 + 5x - 6x - x^2 = 30 - x - x^2$$

Then let us multiply that product by $(4 - x)$.

$$(4 - x)[(5 - x)(6 + x)] = (4 - x)(30 - x - x^2)$$
$$= 120 - 4x - 4x^2 - 30x + x^2 + x^3$$
$$= 120 - 34x - 3x^2 + x^3$$

Finally we multiply that product by $(3 + x)$.

$$(3 + x)[(4 - x)(5 - x)(6 + x)] = (3 + x)(120 - 34x - 3x^2 + x^3)$$
$$= 360 - 102x - 9x^2 + 3x^3 + 120x - 34x^2 - 3x^3 + x^4$$
$$= 360 + 18x - 43x^2 + x^4$$

Example 27

$$(x^2 + 3x - 1)(x^2 - 3x + 2) = x^4 - 3x^3 + 2x^2 + 3x^3 - 9x^2 + 6x - x^2 + 3x - 2$$
$$= x^4 - 8x^2 + 9x - 2$$

SECTION 8-4 Exercises Product of Two Multinomials

Multiply the following binomials and trinomials, and simplify.

1. $(x - 3)(x + 4 - y)$ **2.** $(a - d)(a - 2d + 5)$

3. $(w^2 + w - 5)(4w - 2)$ **4.** $(a^2 - 5)(3a^2 - 7a - 4)$

5. $(b^7 - 2.82b^5 + 4.27b^3)(b + 2.93)$ **6.** $(x + 3.88)(x^3 - 2.15x - 6.03)$

7. $(1 + c^2)(4c^2 + 7c - 3)$ **8.** $(2x^2 - 6xy + 3y^2)(3x + 3y)$

9. $(b^2 - bx + x^2)(b + x)$ **10.** $(a^2 + 2a - 2)(a + 1)$

11. $(b^2 + bx + x^2)(b - x)$ **12.** $(b^2 + bx - x^2)(b - x)$

Multiply the following binomials and polynomials, and simplify.

13. $(a^2 + 5a - xy)(a + z)$ **14.** $(c^2 - cm + cn + mn)(c - m)$

15. $(y^2 - x^2)(y^3 + ay^2 - abxy + bx^2 - x^3)$

Multiply the following monomials and binomials, and simplify.

16. $4(5b + 3c)(5b + 3c)$ **17.** $(4w - 5z)(4w - 5z)d$

18. $z(22.1a - 3.03b)(2.26a + 38.2b)$ **19.** $(d - 4.11)(d - 4.93)(d + 2.26)$

20. $(m + 3)(m + 3)(m - 5)$ **21.** $(c + m)(c + n)(c - p)$

22. $(z + 5)(z - 2)(z - 5)(z + 2)$ **23.** $(2 + y)(2 - y)(2 - y)(4 + y^2)$

Multiply the following trinomials, and simplify.

24. $(x - y - z)(x + y + z)$ **25.** $(5x^3 + 2xy^2 - 2x)(5x^2 - 2x)$

26. $(5x - y + 2x)(4x - y + 6)$ **27.** $(x + y - z)(x - y - z)$

28. $(a^2 - 5.93a + 31.4)(a^2 - 5.37a + 4.03)$

29. $(m^3 - 4.83 + 32.4m)(m^2 - 3.37m + 2.26)$

30. $(am - ym + yx)(am + ym - yx)$

8-5

POWERS OF MULTINOMIALS

We see from Eq. 18 that raising an expression to a power is the same as multiplying the expression by itself the proper number of times, provided that the power is a positive integer.

Example 28

$$(x + 2)^2 = (x + 2)(x + 2)$$
$$= x^2 + 2x + 2x + 4$$
$$= x^2 + 4x + 4$$

Perfect Square Trinomial

Let us square a binomial, such as $(x + y)$. We get

$$(x + y)^2 = (x + y)(x + y) = x^2 + xy + xy + y^2$$

Collecting terms, we get an expression that is called a *perfect square trinomial.*

PERFECT SQUARE TRINOMIAL	$(x + y)^2 = x^2 + 2xy + y^2$	37

The first and last terms of the perfect square trinomial are the squares of the two terms of the binomial; its middle term is twice the product of the terms of the binomial. Similarly, when we square $(x - y)$, we get

PERFECT SQUARE TRINOMIAL	$(x - y)^2 = x^2 - 2xy + y^2$	38

Example 29

$$(2x - 3)^2 = (2x)^2 - 2(2x)(3) + 3^2$$
$$= 4x^2 - 12x + 9$$

Cube of a Binomial

The following example shows how to cube a binomial. The same method can, of course, be extended to cube any multinomial, or to raise a multinomial to any positive integer power.

Example 30

Cube $(x + 1)$.

Solution: By the definition of an exponent, we get

$$(x + 1)^3 = (x + 1)(x + 1)(x + 1)$$

Let us first multiply out $(x + 1)(x + 1)$.

$$(x + 1)^3 = (x + 1)(x^2 + x + x + 1)$$
$$= (x + 1)(x^2 + 2x + 1)$$

Then by the distributive law,

$$(x + 1)^3 = x(x^2 + 2x + 1) + 1(x^2 + 2x + 1)$$
$$= x^3 + 2x^2 + x + x^2 + 2x + 1$$
$$= x^3 + 3x^2 + 3x + 1$$

Example 31

$$(x + 2)^3 = (x + 2)(x + 2)(x + 2)$$
$$= (x + 2)(x^2 + 4x + 4) \text{ (from Example 28)}$$
$$= x^3 + 4x^2 + 4x + 2x^2 + 8x + 8$$

Combining like terms,

$$(x + 2)^3 = x^3 + 6x^2 + 12x + 8$$

Square each binomial.

1. $(x + y)$	**2.** $(m + n)$	**3.** $(a - d)$
4. $(z - w)$	**5.** $(B + D)$	**6.** $(C - D)$
7. $(4.92y + 3.12z)$	**8.** $(2.45d - 1.93x)$	**9.** $(5n + 6x)$
10. $(d^3 + d^2)$	**11.** $(1 - w)$	**12.** $(cy^2 - c^3y)$
13. $(b^3 - 13)$	**14.** $(a^n + b^3)$	

Square each trinomial.

15. $(x + y + z)$	**16.** $(x - y - z)$	**17.** $(a + b - 1)$
18. $(5a^3 - 3a + 16)$	**19.** $(c^2 - cd + d^2)$	**20.** $(w^2 - 5w + 2)$

Cube each binomial.

21. $(x - y)$	**22.** $(1.93 - 3.24a)$	**23.** $(3.02m + 2.16n)$
24. $(a^2 + 1)$	**25.** $(c + d)$	**26.** $(4p - q)$

Applications

27. A square of side x has each side increased by 2 units. Write an expression for the area of the new square, and multiply out.

28. When a current I flows through a resistance R, the power in the resistance is I^2R. If the current is increased by 2.50 amperes, write an expression for the new power, and multiply out.

29. If the radius r of a sphere is decreased by 2.00 units, write an expression for the new volume of the sphere, and multiply out. (Volume $= \dfrac{4}{3}\pi r^3$)

8-6

REMOVING SYMBOLS OF GROUPING

We learned earlier that symbols of grouping indicate that the enclosed expression is to be taken as a whole. Here we see that they also *indicate multiplication*.

Example 32

(a) $x(y)$ is the product of x and y.
(b) $x(y + 4)$ is the product of x and $(y + 4)$.
(c) $(x - 3)(y + 1)$ is the product of $(x - 3)$ and $(y + 1)$.
(d) $-(y + 1)$ is the product of -1 and $(y + 1)$.

After the multiplication has been performed, we simply remove the parentheses.

Example 33

(a) $x(y - 3) = xy - 3x$
(b) $8(2y + 1) - 3 = 16y + 8 - 3 = 16y + 5$

> **COMMON ERROR** Don't forget to multiply every term within the grouping by the preceding factor.
>
> $$-(3x + 4) \neq -3x + 4$$

> **COMMON ERROR** Multiply the terms within the parentheses only by the factor directly preceding it.
>
> $$a - 3(x + y) \neq (a - 3)(x + y)$$

If there are groupings within groupings, start simplifying with the innermost groupings.

Example 34

(a) $4[2(x + 1) + 3] = 4[2x + 2 + 3]$
$$= 4[2x + 5]$$
$$= 8x + 20$$

(b) $a + 4[3 + 2(b - 3)] = a + 4[3 + 2b - 6]$
$$= a + 4[2b - 3]$$
$$= a + 8b - 12$$

(c) $2\{[(1 - m) + (n + 3)] - 5\} - 4 = 2\{[1 - m + n + 3] - 5\} - 4$
$$= 2\{[4 - m + n] - 5\} - 4$$
$$= 2\{-m + n - 1\} - 4$$
$$= -2m + 2n - 2 - 4$$
$$= -2m + 2n - 6$$

SECTION 8-6 | Exercises | Removing Symbols of Grouping

Remove symbols of grouping, and simplify.

1. $a + (b + a)$ 2. $x + (2 - x)$

3. $x + (x - y)$ 4. $-(a + 3.92) - (a - 4.14)$

5. $x - 2.66[y - 3.02(x + 6.22y) + 4.98y]$

6. $[(5x - 4y) - (2x - 5y)][(-3x - 6y) - (2x + 4y)]$

7. $\{[3 - (x + 7)] - 3x\} - (x + 7)$

8. $[a(a + b) - a^2](a^2 + b^2)(a^3 - 4ab + b^3)$

9. $6p - \{3p + [2q - ((5p + 4q) + p) - (3p + 2)] - 2p\}$

10. $[(5a - 2)(2a + 3) - 6a^2][(3a - 4)(2a - 4) + 5a^2]$

Multiply and simplify.

1. $(3ay^2)(-4a^3y)$
2. $(3x^2y)(-5xyz^2)(4x^2y^3z)$
3. $(a - 3)(a + 4 - b)$
4. $(2 + x)(2 - x)(2 - x)(4 + x^2)$
5. $(x - y)(x - 2.45y + 5.83)$
6. $[z^2 + (x - y)z - xy](x + z)$
7. $(x^2 - 3y)(x^2 + 5y)$
8. $4(5p + 3q)(5p + 3q)$
9. $(4a + 2b)(2a^2 - b)$
10. $(4.45x - 5.92y)(4.23x - 5.86y)z$
11. $(c + 3)(c + 3)(c - 5)$
12. $(a - b - c)(a + b + c)$
13. $(p^2 - px + x^2)(p + x)$
14. $(x^2 + 2x - 2)(x + 1)$
15. $(5a^3 + 2ab^2 - 2a)(5a^2 - 2a)$
16. $(a + b)(a + c)$
17. $(p + q - r)(p - q - r)(p + q + r)$
18. $x^2y(2x + y - xy)$
19. $(3m^3 + m^2 - 5m + 3)(m^2 - 5m + 4)$
20. $(a + 2)(a - 2)$
21. $(a^2 - b^2)(a^2 + b^2 - ab)$
22. $x^4(x^2 + 8)$
23. $(bm - by)(bm + by + nz)$
24. $3a(-2 - 10a)$
25. $(-5z^2)(-2yz)(-4x^2z)(-6xy^3)(-3y^2z)$
26. $ab^2(5a^2)$

Square each binomial.

27. $(2p + 3q)$
28. $(x - 5.03y)$
29. $(x^3 + x^2)$

Square each trinomial.

30. $(x^3 - 3.33x + 1.84)$
31. $(x^2 - xy + y^2)$
32. $(a^2 - 5a + 2)$

Remove symbols of grouping, and simplify.

33. $x - 2[y - 3(x + 2y) + 3y]$
34. $\{[4 - (x + 4)] - 2x\} - (x + 6)$
35. $2x - \{3y + [2x - (4y + 2x) + y - (4x + 2)] - 2y\}$

Cube each binomial.

36. $(a - 2b)$
37. $(4x + 3y)$
38. $(x^2 + 2)$

WRITING

39. Your friend refuses to learn the FOIL rule. "I want to learn math, not memorize a bunch of tricks!" he declares. What do you think? Write a paragraph or so giving your opinion on the value or harm in learning devices like the FOIL rule.

40. Figure 8-1 shows a geometric representation of $(a + b)^2$, the square of the binomial $(a + b)$. Use the figure to evaluate $(a + b)^2$. Then make a similar sketch to evaluate each of the following:

(a) $(2a + b)^2$

(b) $(a + 2)(a + 3)$

(c) $(a + b)(c + d)$

(d) $(a + b + 1)^2$

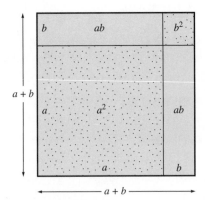

■ **FIGURE 8-1**
A geometric representation of $(a + b)^2$.

Division of Algebraic Expressions

This chapter completes our study of the four basic operations with algebraic expressions; we have already covered addition, subtraction, and multiplication of algebraic expressions, and now we do division. As with multiplication, we start with monomials and gradually work up to longer expressions.

9-1 QUOTIENT OF TWO MONOMIALS

Symbols for Division

The following symbols are commonly used to indicate division.

$$x \div y \qquad \frac{x}{y} \qquad x/y$$

We get a *quotient* when we divide the *dividend* by the *divisor*.

$$\text{quotient} = \frac{\text{dividend}}{\text{divisor}}$$

The quantity x/y $\left(\text{or } \dfrac{x}{y}\right)$ is also called a *fraction*, where x is the *numerator* and y is the *denominator*. It is also spoken of as the *ratio* of x to y.

Division by Zero

Division by zero is not permitted in algebra.

Example 1

In the fraction

$$\frac{x + 8}{x - 3}$$

x cannot equal three, or the illegal operation of division by zero will result. We say $x \neq 3$.

Rules of Signs

If the dividend and the divisor have the same sign, the quotient is positive.

$$\frac{+a}{+b} = \frac{-a}{-b} = \frac{a}{b}$$

If the dividend and divisor have the opposite signs, the quotient is negative.

$$\frac{+a}{-b} = \frac{-a}{+b} = -\frac{a}{b}$$

The fraction itself has an algebraic sign which, when negative, reverses the sign of the quotient. These three ideas are summarized in the following rules:

RULES OF SIGNS FOR DIVISION	$\dfrac{+a}{+b} = \dfrac{-a}{-b} = -\dfrac{-a}{+b} = -\dfrac{+a}{-b} = \dfrac{a}{b}$	11
	$\dfrac{+a}{-b} = \dfrac{-a}{+b} = -\dfrac{-a}{-b} = -\dfrac{a}{b}$	12

These rules show that any pair of negative signs may be removed without changing the value of the fraction.

Example 2

Simplify $-\dfrac{y}{-x}$.

Solution: Removing the pair of negative signs, we obtain

$$-\frac{y}{-x} = \frac{y}{x}$$

Dividing a Monomial by a Monomial

The quotient of any quantity (except zero) and itself equals one.

Example 3

$$\frac{a}{a} = 1 \qquad (a \neq 0)$$

Thus if the same factor appears both in the divisor and the dividend, they may be eliminated because their quotient is 1. This is called *canceling*.

Example 4

Divide $8xy$ by $2x$.

Solution:

$$\frac{8xy}{2x} = \frac{8}{2} \cdot \frac{x}{x} \cdot y = 4y$$

When the quantities to be divided have *exponents,* we make use of the law of exponents for division, Eq. 20.

QUOTIENT	$\dfrac{x^m}{x^n} = x^{m-n}$	20

Example 5

Divide y^7 by y^5.

Solution: By Eq. 20,

$$\frac{y^7}{y^5} = y^{7-5} = y^2$$

Any numerical coefficients are divided separately.

Example 6

Divide $10y^5$ by $2y^2$.

Solution:

$$\frac{10y^5}{2y^2} = \frac{10}{2} \cdot \frac{y^5}{y^2}$$
$$= 5y^{5-2} = 5y^3$$

When there are more than one unknowns in the expressions to be divided, treat each unknown separately.

Example 7

Divide $21x^2y^4z^3$ by $7xy^2z^2$.

Solution:

$$\frac{21x^2y^4z^3}{7xy^2z^2} = \frac{21}{7} \cdot \frac{x^2}{x} \cdot \frac{y^4}{y^2} \cdot \frac{z^3}{z^2}$$
$$= 3xy^2z$$

When dividing expressions it is common to get a quotient containing negative exponents.

Example 8

$$\frac{5y^4}{y^6} = 5y^{4-6} = 5y^{-2}$$

You may leave the answer in this form, or eliminate the negative exponent by using Eq. 25. Thus,

$$5y^{-2} = \frac{5}{y^2}$$

The process of dividing a monomial by a monomial is also referred to as *simplifying* a fraction, or *reducing a fraction to lowest terms*.

Example 9

Simplify the fraction

$$\frac{4xy^4z}{12x^3yz^2}$$

Solution: The procedure is no different than if we had been asked to divide $4xy^4z$ by $12x^3yz^2$.

$$\frac{4xy^4z}{12x^3yz^2} = \frac{4}{12}x^{1-3}y^{4-1}z^{1-2}$$

$$= \frac{1}{3}x^{-2}y^3z^{-1}$$

or,

$$= \frac{y^3}{3x^2z}$$

When the expressions to be divided have negative exponents, we simply apply Eq. 25, as before.

Example 10

Divide $16x^{-3}y^2z^{-3}$ by $4x^2y^4z^{-5}$.

Solution: Proceeding as before, we obtain

$$\frac{16x^{-3}y^2z^{-3}}{4x^2y^4z^{-5}} = \frac{16}{4}x^{-3-2}y^{2-4}z^{-3-(-5)}$$

$$= 4x^{-5}y^{-2}z^2$$

$$= \frac{4z^2}{x^5y^2}$$

Alternate Solution: We may choose to eliminate the negative exponents first.

$$\frac{16x^{-3}y^2z^{-3}}{4x^2y^4z^{-5}} = \frac{16y^2z^5}{4x^2(x^3)(y^4)(z^3)} = \frac{4z^2}{x^5y^2}$$

SECTION 9-1 Exercises Quotient of Two Monomials

Divide the following monomials.

1. x^7 by x^4
2. $21a^4$ by $3a^2$
3. $5xyz$ by xy
4. m^3n by mn
5. $4a^2d$ by $-2ad$
6. $-360x^4y^2$ by $-30x^2y$
7. $31ab^9$ by ab^8
8. $54x^2z$ by $-9x$
9. $-99ad$ by $3a$
10. $49xyz$ by $-7y$
11. $42p^5q^4r^2 \div 7p^3qr$
12. $50x^3y^5z^3 \div -10xy^3z^2$
13. $-32m^2nx \div 4mx$
14. $48cd^2z^3 \div -24cd$
15. $-36a^4b^2c \div 9ab$
16. $45m^2q$ by $-5mq$
17. $-24m^3n^3z$ by $4m^3z$
18. $-27a^5p^2$ by $-9a^3q^2$
19. $25a^4bcxyz$ by $5a^2bcxz$
20. $18d^3f^2$ by $3d^2f$
21. $32a^2bc \div -8ab$
22. $-12m^2n^3 \div 4mn^2$
23. $-36a^2by^2 \div 12a^2y$
24. $44a^2b^3c^4 \div 11a^2bc$
25. $64x^2y^2$ by $8xy$
26. $24pq^2r^3s$ by $8r$

27. $x^2y^6z^2$ by x^2z^2 **28.** $(a - x)^5$ by $(a - x)^2$

29. a^x by a^y **30.** $66c^2dy^3 \div (-22cy)$

31. $-35a^3b^2z \div 7ab^2$ **32.** $19e^2m^2n^2 \div (-em^2n)$

33. $95abc \div 5a^2b^3c$ **34.** $45a^2b^2d^2 \div 15abd$

9-2   DIVIDING A POLYNOMIAL BY A MONOMIAL

To divide a polynomial by a monomial, we simply divide each term of the polynomial by the monomial.

Example 11

Divide $6x^3 - 3x^2$ by $3x$.

Solution:

$$\frac{6x^3 - 3x^2}{3x} = \frac{6x^3}{3x} - \frac{3x^2}{3x} = 2x^2 - x$$

Example 12

Divide $8x^2 - 4x + 4$ by $2x$.

Solution:

$$\frac{8x^2 - 4x + 4}{2x} = \frac{8x^2}{2x} - \frac{4x}{2x} + \frac{4}{2x}$$

This is really a consequence of the distributive law, Eq. 5.

$$\frac{8x^2 - 4x + 4}{2x} = \frac{1}{2x}(8x^2 - 4x + 4) = \left(\frac{1}{2x}\right)(8x^2) - \left(\frac{1}{2x}\right)(4x) + \left(\frac{1}{2x}\right)(4)$$

Each of these terms is now simplified as in the preceding section.

$$\frac{8x^2}{2x} - \frac{4x}{2x} + \frac{4}{2x} = 4x - 2 + \frac{2}{x}$$

COMMON ERROR Do not to forget to divide *every* term of the polynomial by the monomial.

$$\frac{4x^2 - 2x + 5}{2x} \neq 2x - 1 + 5$$

COMMON ERROR There is no similar rule for dividing a monomial by a polynomial.

$$\frac{a}{b + c} \neq \frac{a}{b} + \frac{a}{c}$$

Example 13

Divide $4xy - 8x^2y + 3xy^2 - 2x^2y^2$ by $16xy$.

Solution:

$$\frac{4xy - 8x^2y + 3xy^2 - 2x^2y^2}{16xy} = \frac{4xy}{16xy} - \frac{8x^2y}{16xy} + \frac{3xy^2}{16xy} - \frac{2x^2y^2}{16xy}$$

$$= \frac{1}{4} - \frac{x}{2} + \frac{3y}{16} - \frac{xy}{8}$$

We may also combine these terms over the common denominator 16, getting

$$\frac{4 - 8x + 3y - 2xy}{16}$$

Alternate Solution: We can get the same result by canceling, noting that every term in both numerator and denominator contains an x and a y.

$$\frac{4\cancel{x}\cancel{y} - 8x^2\cancel{y} + 3\cancel{x}y^2 - 2x^2y^2}{16\cancel{x}\cancel{y}} = \frac{4 - 8x + 3y - 2xy}{16}$$

SECTION 9-2 | **Exercises** | **Dividing a Polynomial by a Monomial**

Divide the polynomial by the monomial.

1. $15x^3 + 3x^2$ by x
2. $42m^6 - 2m^3$ by $2m$
3. $36d^5 - 6d^2$ by $3d$
4. $22x^2 + 11y^5$ by 11
5. $48c^4 + 36c^5$ by $12c^2$
6. $27n^3 - 9n^2$ by $-3n$
7. $39p^2 + 52p^3$ by $-13p^2$
8. $40a^5 - 20a^2$ by $10a$
9. $-25x^3 - 15x^2$ by $5x$
10. $-55a^3d + 22a^2$ by $-a^2$
11. $-bm^3n^2 - 3bm^2n$ by $-m^2n$
12. $-5a^5b + 5ab^2$ by $-ab$
13. $8a^3b^2 + 4a^2b^3$ by $4ab^2$
14. $10x^3y - 5xy^4$ by $-5xy$
15. $x^2y^3z - xy^4z^2$ by $-xy^2z$
16. $16a^3bc^2 + 12a^2b^5c$ by $-4abc$
17. $m^2n^2 + m^3n^2 - m^2n^3$ by $-mn$
18. $p^5q^2 - p^2q^5 - p^3q^3$ by p^2q^2
19. $x^3y^3 - x^4y + xy^4$ by $-xy$
20. $-a^3b^3 - a^2b^2 - ab$ by $-ab$
21. $c^3 - 4c^2d^2 + d^3$ by cd^2
22. $r^4s^3 - r^2s^2 + r^4s^2$ by $-r^2s$
23. $a^4 + 2a^2b^2 - b^4$ by a^2b^2
24. $m^5n^2 + m^2n^2 - m^2n^4$ by $-m^2n$
25. $4x^3z + 2xz^2 - 3z^4$ by $-xz$
26. $ab^3 + a^3c^2 - b^2c^4$ by $-abc$
27. $p^3q^3 + pq^2r^3 - p^2r^4$ by $-p^2r$
28. $7m^4n^2 + 7m^3n^3 - 7m^2n^4$ by $-7m^2n^3$
29. $4c^4d + 3c^2d^3 - cd^5$ by $-cd^2$
30. $3a^2x + 5a^3x^3 - 2ax^2$ by $-a^2x^2$
31. $8b^2c^4 + 4b^2c - 12b^3c^3$ by $-b^3c^2$

9-3   QUOTIENT OF TWO POLYNOMIALS

To divide one polynomial by another polynomial, follow these steps.

1. Write the divisor and the dividend in the order of descending powers of the variable.

2. Supply any missing terms, using coefficients of zero.
3. Set up the division in long-division form, as in the following example.

Note that this method is used only for polynomials, expressions in which the exponents are all positive integers.

Example 14

Divide $(4x + x^2 + 3)$ by $(x + 1)$.

Solution:

1. Write the dividend in descending order of the powers.

$$x^2 + 4x + 3$$

2. There are no missing terms, so we go on to the next step.
3. Set up in long-division format.

$$(x + 1)\overline{\smash{)}x^2 + 4x + 3}$$

4. Divide the first term in the dividend (x^2) by the first term in the divisor (x). The result (x) is written above the dividend, in line with the term having the same power. It is the first term of the quotient.

$$
\begin{array}{r}
x \\
(x+1)\overline{\smash{)}x^2 + 4x + 3}
\end{array}
$$

5. Multiplying the divisor by the first term of the quotient. Write the result below the dividend. Subtract it from the dividend.

$$
\begin{array}{r}
x \\
(x+1)\overline{\smash{)}x^2 + 4x + 3} \\
\underline{-\ (x^2 +\ x)} \\
3x + 3
\end{array}
$$

6. Repeat Steps 4 and 5 until the degree of the remainder is *less* than the degree of the divisor.

$$
\begin{array}{r}
x + 3 \\
(x+1)\overline{\smash{)}x^2 + 4x + 3} \\
\underline{-\ (x^2 +\ x)} \\
3x + 3 \\
\underline{-\ (3x + 3)} \\
0
\end{array}
$$

The result is written

$$\frac{x^2 + 4x + 3}{x + 1} = x + 3$$

We now try a harder example.

Example 15

Divide $(2w^2 + 6w^4 - 2)$ by $(w + 1)$.

Solution:

1. Write the dividend in descending order of the powers.

$$6w^4 + 2w^2 - 2$$

2. Supply the missing terms with coefficients of zero.

$$6w^4 + 0w^3 + 2w^2 + 0w - 2$$

3. Set up in long-division format.

$$(w + 1)\overline{)6w^4 + 0w^3 + 2w^2 + 0w - 2}$$

4. Divide the first term in the dividend $(6w^4)$ by the first term in the divisor (w). The result $(6w^3)$ is written above the dividend, in line with the term having the same power. It is the first term of the quotient.

$$\begin{array}{r} 6w^3 \\ (w + 1)\overline{)6w^4 + 0w^3 + 2w^2 + 0w - 2} \end{array}$$

5. Multiply the divisor by the first term of the quotient. Write the result below the dividend. Subtract it from the dividend.

$$\begin{array}{r} 6w^3 \\ (w + 1)\overline{)6w^4 + 0w^3 + 2w^2 + 0w - 2} \\ -\,(6w^4 + 6w^3) \\ \hline -\,6w^3 + 2w^2 + 0w - 2 \end{array}$$

6. Repeat Steps 4 and 5, each time using the new dividend obtained, until the degree of the remainder is less than the degree of the divisor.

$$\begin{array}{r} 6w^3 - 6w^2 + 8w - 8 \\ (w + 1)\overline{)6w^4 + 0w^3 + 2w^2 + 0w - 2} \\ -\,(6w^4 + 6w^3) \\ \hline -\,6w^3 + 2w^2 + 0w - 2 \\ -\,(-6w^3 - 6w^2) \\ \hline 8w^2 + 0w - 2 \\ -\,(8w^2 + 8w) \\ \hline -\,8w - 2 \\ -\,(-8w - 8) \\ \hline 6 \end{array}$$

The result is written

$$\frac{6w^4 + 2w^2 - 2}{w + 1} = 6w^3 - 6w^2 + 8w - 8 + \frac{6}{w + 1}$$

COMMON ERROR Errors are often made during the subtraction step (Step 5 in the preceding example).

$$\begin{array}{r} 6w^3 \\ (w + 1)\overline{)6w^4 + 0w^3 + 2w^2 + 0w - 2} \\ -\,(6w^4 + 6w^3) \\ \hline 6w^3 + 2w^2 + 0w - 2 \end{array}$$

No.   Should be $-6w^3$

Example 16

Divide $(2x^2y - y^3 + x^3 - 2xy^2)$ by $(x - y)$.

Solution: Let us write each expression in the order of descending powers of x, that is, $x^3, x^2, \ldots$, and proceed as before.

$$
\begin{array}{r}
x^2 + 3xy + y^2 \\
x - y \overline{\smash{\big)}\ x^3 + 2x^2y - 2xy^2 - y^3} \\
\underline{-\ (x^3 -\ x^2y)} \\
3x^2y - 2xy^2 \\
\underline{-\ (3x^2y - 3xy^2)} \\
xy^2 - y^3 \\
\underline{-\ (xy^2 - y^3)} \\
0
\end{array}
$$

SECTION 9-3 Exercises Quotient of Two Polynomials

Divide each polynomial by the given binomial.

1. $a^2 + 15a + 56$ by $a + 7$
2. $27x^3 - 8y^3$ by $3x - 2y$
3. $a^2 + a - 56$ by $a - 7$
4. $4x^2 + 23x + 15$ by $4x + 3$
5. $2x^2 + 11x + 5$ by $2x + 1$
6. $6x^2 - 7x - 3$ by $2x - 3$
7. $a^2 - 15a + 56$ by $a - 7$
8. $a^2 - a - 56$ by $a + 7$
9. $3x^2 - 4x - 4$ by $2 - x$
10. $a^4 + 3a^2 + 2$ by $a^2 + 1$
11. $a^8 - 3a^4 + 2$ by $a^4 - 1$
12. $1 - x^3y^3$ by $1 - xy$
13. $a^3 - 8a - 3$ by $a - 3$

Chapter 9 REVIEW EXERCISES

Divide.

1. $p^2q^3r - pq^4r^2$ by $-pq^2r$
2. $16x^3yz^2 + 12x^2y^5z$ by $-4xyz$
3. $m^2 - m - 56$ by $m + 7$
4. $3n^2 - 4n - 4$ by $2 - n$
5. $p^2 - 2pq + q^2$ by $p - q$
6. $a^2b^6c^2$ by a^2c^3
7. $(u - v)^6$ by $(u - v)^3$
8. $a^2b^2 + a^3b^2 - a^2b^3$ by $-ab$
9. $x^5y^2 - x^2y^5 - x^3y^3$ by x^2y^2
10. $75abc \div 5a^3b^4c$
11. $-44ab^2 \div 4a^2b$
12. $1 - m^3n^3$ by $1 - mn$
13. $p^2 + 3pq + 2q^2$ by $p + q$
14. $m^2 - 2m - 3$ by $m - 3$
15. $x^2 - xy - 2y^2$ by $x + 2y$
16. $p^2 - 3q^2 + 2pq$ by $p - q$
17. $a^4 + a^3 - 3a^2 + a$ by $a^2 - a$
18. $x^4 + 3x^2 + 2$ by $x^2 + 1$
19. $z^8 + z^4 + 1$ by z^4
20. $p^3 - 4p^2q^2 + p$ by pq^2
21. $x^4y^3 - x^2y^2 + x^4y^2$ by $-x^2y$

22. In simplifying a fraction we are careful not to change its value. If that's the case, and the new fraction is "the same" as the old, then why bother? Write a few paragraphs explaining why you think it's valuable to simplify fractions (and other expressions as well) or if there is some advantage to just leaving them alone.

23. In the following chapter we will solve simple equations, something you must have done before. Without peeking ahead, write down in your own words whatever you remember about solving equations. You may give a list of steps, or give a description in paragraph form.

TEAM PROJECT

24. Write an hour exam that will test for the most important skills covered in this chapter. Be sure that it is not too long and hard, and not too short and easy. Then work the test and produce an answer key, to be handed in with the exam.

Simple Equations 10

An *equation* is obtained when two mathematical expressions are set equal to each other. For example, when the mathematical expression $2(x - 3)$ is set equal to the expression $5x + 3$, we get the equation

$$2(x - 3) = 5x + 3$$

In this chapter we will first learn how to substitute numerical values into an equation, and then go on to solve simple equations.

It is important to be able to solve equations because they are used to describe so many things in the physical world. In the following chapter we'll learn how to write an equation to describe a simple situation, but for now we'll just learn how to solve a given equation. In later chapters we will solve more complex equations.

10-1   SUBSTITUTING INTO EQUATIONS AND FORMULAS

Before learning how to substitute into and to solve equations, we will first learn some terms that we will use in talking about equations.

Equations

An equation has two sides and an equal sign.

$$2x^2 - 5 = 3x + 2$$

left side $\quad \perp \quad$ right side

equal sign

Identities

An equation that is true for *any value* of the variable is called an *identity*.

Example 1

$$x(x - 3) = x^2 - 3x$$

is an identity, because it is true for any value of *x*.

Conditional Equations

A *conditional equation* is one whose sides are equal only for certain values of the variable.

Example 2

The equation

$$x + 1 = 4$$

is a conditional equation because the sides are equal only when $x = 3$.

When we say *equation* in this textbook, we will mean a *conditional equation*.

Substituting into Equations

To *substitute into an equation* means to replace the letter quantities in an equation with their given numerical values, and to perform the indicated computation. This is different from, and usually simpler than, solving an equation.

Example 3

Substitute the values

$$a = 4, \quad b = 2, \quad \text{and} \quad c = 6$$

into the equation

$$x = \frac{2a + b}{c}$$

Solution: Substituting, we obtain

$$x = \frac{2(4) + 2}{6} = \frac{10}{6} = 1\frac{2}{3}$$

When substituting integers, as in the preceding example, treat them as exact numbers and give your answer in exact form. When substituting approximate numbers, be sure to round your answer to the proper number of digits. Recall the rules for rounding from Chapter 3:

> When we *add* or *subtract* approximate numbers, we round our answer to the same number of *decimal places* as in the original number having the fewest decimal places.
>
> When we *multiply*, *divide*, *raise to powers*, or *take roots*, we round our answer to the same number of *significant digits* as in the original number having the fewest significant digits.

Example 4

Substitute the values

$$r = 12.8, \quad m = 1.04, \quad \text{and} \quad p = 33.7$$

into the equation

$$x = \frac{p + r}{m} + \frac{p - r}{7}$$

Solution: We substitute, treating the 7 as an exact number,

$$x = \frac{33.7 + 12.8}{1.04} + \frac{33.7 - 12.8}{7}$$

$$= \frac{46.5}{1.04} + \frac{20.9}{7}$$

$$= 44.7 + 2.99 = 47.7$$

Note that in the last step of the addition we were justified in keeping only one decimal place.

Substituting into Formulas

The equation for the area of a circle of radius r:

$$A = \pi r^2$$

is also called a *formula* for the area of a circle. A formula is an equation which expresses some general mathematical or physical fact. To substitute into a formula we use the same procedure as for substituting into equations. But now we carry units along with the numerical values. Sometimes we will need a conversion factor to make the units cancel properly, so that the answer will be in the desired units.

Example 5

Find the area of a circle of radius 3.75 cm.

Solution: We substitute 3.75 cm into the formula for the area of a circle given above, and get

$$A = \pi r^2 = \pi (3.75 \text{ cm})^2$$

$$= 44.2 \text{ cm}^2$$

Here we have used an accurate value of π from our calculator, so we round our answer to the number of digits in the given radius.

Often when substituting into a formula, some conversion of units must be done to make those units cancel properly. Such conversions may be done after substituting into the formula or before, as shown in the next example.

Example 6

The velocity v of a moving object starting from rest is equal to the acceleration a times the elapsed time t.

$$v = at$$

Find the velocity, in feet per second, for an object that starts from rest and accelerates at 5.74 m/s² for 0.138 minutes.

Solution: In this example we will convert units *before* substituting. Since 1 m = 3.281 ft we get,

$$\frac{5.74\,\text{m}}{\text{s}^2} \cdot \frac{3.281\,\text{ft}}{\text{m}} = 18.8\,\text{ft/s}^2$$

and since 1 minute = 60 seconds, we get,

$$0.138\,\text{minute} \cdot \frac{60\,\text{seconds}}{\text{minute}} = 8.28\,\text{seconds}$$

Substituting into the given formula,

$$v = at = \frac{18.8\,\text{ft}}{\text{s}^2} \cdot 8.28\,\text{s} = 156\,\text{ft/s}$$

rounded to three significant digits.

COMMON ERROR Students often neglect to include units when substituting into a formula, with the result that the units often do not cancel properly.

In our next example, the units to be used in a certain formula are specified. Here it is necessary to convert all quantities to those specified units before substituting into the formula.

Example 7

An engineering handbook gives the formula for the flow over a dam (Fig. 10-1) as

$$Q = 0.727\,Lh\,\sqrt{2gh} \quad \text{ft}^3/\text{s}$$

where

Q = volumetric flow rate, ft^3/s
L = width of dam in ft
h = height of water over dam, in ft
g = 32.2 ft/s^2

Find the flow rate over a 4.72-m-wide dam, if the height of water is 8.75 in.

■ **FIGURE 10-1**
Water over a dam.

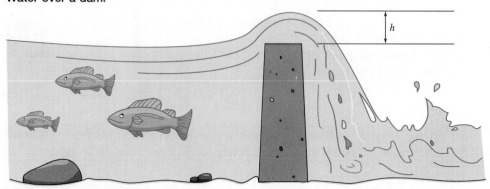

Solution: To use the given formula we must keep the units specified. Thus

$$L = 4.72 \text{ m} \frac{3.281 \text{ ft}}{\text{m}} = 15.5 \text{ ft}$$

$$h = 8.75 \text{ in.} \frac{\text{ft}}{12 \text{ in.}} = 0.729 \text{ ft}$$

Substituting,

$$Q = 0.727 \, Lh \sqrt{2gh}$$
$$= 0.727(15.5 \text{ ft})(0.729 \text{ ft}) \sqrt{2(32.2 \text{ ft/s}^2)(0.729 \text{ ft})}$$
$$= 56.3 \text{ ft}^3/\text{s}$$

SECTION 10-1 Exercises Substituting into Equations and Formulas

Substituting

Substitute the given integers into each equation and evaluate y. Do not round your answer.

1. $y = 3x + 3$ $(x = 4)$
2. $y = 3m^2 + 2m - 6$ $(m = -3)$
3. $y = 4a^3 - 3x$ $(a = -2, x = -3)$
4. $y = 4x^3 + 5x^2 - x - 12$ $(x = -2)$
5. $y = 4c + 3d^2 - 5e^3$ $(c = 5, d = -4, e = 2)$
6. $y = \dfrac{x^2}{z} - \dfrac{w^2}{x} + \dfrac{x}{z^2}$ $(x = -3, w = 2, z = 5)$

Substitute the given approximate numbers into each equation and evaluate y. Assume that all the given integers are exact numbers. Round your answer to the proper number of digits.

7. $y = 6 - 4x$ $(x = -3.445)$
8. $y = -4d^3 - 2e^2$ $(d = -6.375, e = 7.995)$
9. $y = a^3 - a + 9$ $(a = 8.01)$
10. $y = \sqrt[3]{7x - 5z}$ $(x = 5.39, z = -3.27)$
11. $y = \sqrt[3]{x^3 + 5x}$ $(x = 4.52)$
12. $y = (2z + 3x)^{2.3}$ $(z = 3.7, x = 4.8)$

Applications

13. An amount of money P deposited for t years at a simple interest rate r (expressed as a decimal) will grow to a new amount A where, $A = P(1 + rt)$. Find the amount to which \$2500 will accumulate in 4 years at a simple interest rate of 5.5%.

14. In time t, an object having initial velocity v_0 and acceleration a will travel a distance s, where $s = v_0 t + \dfrac{1}{2} at^2$. Find the distance traveled after 2.2 seconds of a body thrown downward with a velocity of 10.0 ft/second. Use $a = 32.2$ ft/s^2.

15. To convert a temperature C given in degrees Celsius to a temperature F given in degrees Fahrenheit, we use the formula $F = (9/5)C + 32$. Convert 48.0°C to degrees Fahrenheit.

16. The torque T delivered by a motor of horsepower P rotating at N rev/min is

$$T = \frac{33{,}000P}{2\pi N} \quad \text{ft-lb}$$

Find the torque delivered by a $\dfrac{2}{3}$-hp motor rotating at 1500 rev/min.

17. Find the torque delivered by a 2.75-hp motor rotating at 120 rad/s.

18. A bar with length L equal to 30.4 m and a cross-sectional area a of 32.8 cm^2 is subject to a tensile load P of 19,000 N. The elongation e is 2.45 mm. Use the equation

$$E = \frac{PL}{ae}$$

    to find the modulus of elasticity E in newtons per square centimeter.

19. Use the equation, $A = P(1 + r)^t$, to find the amount A obtained when an amount $P = \$7950$ is allowed to accumulate for $t = 4$ years at a compound interest rate of $r = 5\frac{1}{4}\%$.

20. The formula for the pressure loss h in a pipe is

$$h = \frac{6270\,fLQ^2}{D^5}\,\text{ft}$$

    Where f is the friction factor, L the length of the pipe in feet, Q the flow rate in cubic feet per second, and D the pipe diameter in inches. Compute the pressure drop in a 3.25-in.-diameter, 250-ft-long pipe, where $f = 0.019$ and the flow rate is 122 gal/min.

21. A metal having a resistance R_1 at temperature t_1 will have a new resistance R at temperature t, where $R = R_1[1 + \alpha(t - t_1)]$, and α is the temperature coefficient of resistance. If the resistance of a copper coil is 775 Ω at 20°C, and the temperature coefficient of resistance is 0.00393 at 20°C, find the resistance at 80°C.

10-2   SOLVING SIMPLE EQUATIONS

The Solution of an Equation

A value of the variable that makes the sides of an equation equal is called a *solution* of that equation.

┌─── **Example 8** ───────
The solution of the equation

$$x + 1 = 4$$

is $x = 3$. The value 3 is also called the *root* of the equation. We also say that it *satisfies* the equation, since it makes the equation true.

Checking

We can check a solution by substituting it back into the original equation.

┌─── **Example 9** ───────
Is 12 a solution of the equation $3(x - 4) = 2x$?

Solution: Substituting 12 for x in the equation, we get

$$3(12 - 4) \stackrel{?}{=} 2(12)$$

$$3(8) \stackrel{?}{=} 24$$

$$24 = 24 \text{ checks}$$

So the value $x = 12$ checks, and 12 is a solution of the given equation.

First-Degree Equations

Recall that a first-degree term is one in which the variable is raised to the power 1. An equation in which all terms containing the variable are of first degree is called a *first-degree equation*. A first-degree equation is also called a *linear* equation.

Example 10

The equations

$$5 + 9x = 4 - 2x, \quad 5x - 5z = 7z, \quad \text{and} \quad \frac{x}{3} - 8 = 12 - 3x$$

are all of the first degree.

The equation $x^2 - 3x + 2 = 0$ is not of the first degree since it contains a second-degree term (x^2).

Solving an Equation

When we *solve* an equation, our object is to get the variable standing alone on one side of the equation. To accomplish this, *we perform the same mathematical operation to both sides of the equation.* That is, we may add the same quantity to both sides, subtract the same quantity from both sides, multiply both sides by the same quantity, and so forth. This will be made clear by examples.

Example 11

Solve the equation $5x = 4x + 7$.

Solution: Subtracting $4x$ from both sides, we obtain

$$5x - 4x = 7 + 4x - 4x$$

Combining like terms yields

$$x = 7$$

Check:
Substituting 7 for x in the original equation gives

$$5(7) \overset{?}{=} 4(7) + 7$$
$$35 \overset{?}{=} 28 + 7$$
$$35 = 35 \quad \text{checks}$$

Example 12

Solve the equation $2x - 3 = 7 - 3x$.

Solution: Adding $3x + 3$ to both sides,

$$2x - 3 = 7 - 3x$$
$$\underline{3x + 3 \quad\quad 3 + 3x}$$
$$5x \quad = \quad 10$$

Dividing both sides by 5, we obtain

$$x = 2$$

Check:

Substituting 2 for x in the original equation yields

$$2(2) - 3 \overset{?}{=} 7 - 3(2)$$
$$4 - 3 \overset{?}{=} 7 - 6$$
$$1 = 1 \quad \text{checks}$$

Symbols of Grouping

We showed earlier that parentheses, brackets, and braces are often used to group quantities within an expression. When the equation contains such symbols of grouping, remove them early in the solution.

Example 13

Solve the equation $2 - 4(x + 2) = 5 - 3(2x + 1)$.

Solution: Removing the parentheses, we obtain

$$2 - 4x - 8 = 5 - 6x - 3$$

Combining like terms,

$$-4x - 6 = -6x + 2$$

Adding $6x + 6$ to both sides,

$$\frac{6x + 6 = \quad 6x + 6}{2x \quad = \quad\quad 8}$$

Dividing by 2,

$$x = 4$$

Check:

$$2 - 4(4 + 2) \overset{?}{=} 5 - 3(8 + 1)$$
$$2 - 24 \overset{?}{=} 5 - 27$$
$$-22 = -22 \quad \text{checks}$$

Simple Fractional Equations

An equation that contains one or more fractions is called a *fractional equation*.

Example 14

Some fractional equations are

$$\frac{x}{5} = 3x \quad \text{and} \quad \frac{2x}{5} = \frac{2}{7}$$

Here we will solve some very simple fractional equations, and in Chapter 18 we will learn how to solve more complex types.

If an equation contains a single fraction, that fraction can be eliminated by multiplying both sides by the denominator of the fraction.

---- **Example 15** ----

Solve

$$\frac{x}{2} - 4 = 7$$

Solution: Multiplying both sides by 2,

$$2\left(\frac{x}{2} - 4\right) = 2(7)$$

$$x - 8 = 14$$

Adding 8 to both sides,

$$x = 8 + 14 = 22$$

Check:

$$\frac{22}{2} - 4 \stackrel{?}{=} 7$$

$$11 - 4 = 7 \quad \text{checks}$$

When an equation has two or more fractions, we can clear the fractions by multiplying both sides by the least common denominator.

---- **Example 16** ----

Solve

$$\frac{x}{2} - 3 = \frac{x}{5}$$

Solution: Multiplying by 10, the product of the denominators 2 and 5,

$$10\left(\frac{x}{2} - 3\right) = 10\left(\frac{x}{5}\right)$$

$$10\left(\frac{x}{2}\right) - 10(3) = 10\left(\frac{x}{5}\right)$$

$$5x - 30 = 2x$$

Rearranging,

$$5x - 2x = 30$$

$$3x = 30$$

Dividing both sides by 3

$$x = 10$$

Check:

$$\frac{10}{2} - 3 \stackrel{?}{=} \frac{10}{5}$$

$$5 - 3 \stackrel{?}{=} 2$$

$$2 = 2 \quad \text{checks}$$

Strategy

Remember that our object when solving an equation is to get the variable by itself on one side of the equation. To do this we use any valid mathematical operation, applied to both sides of the equation. It is not possible to give a procedure that will work for every equation, but the following tips should help. You may have to do these operations in a different order than that shown.

TIPS
1. Eliminate any fractions by multiplying both sides by the least common denominator.
2. Remove any parentheses by performing the indicated multiplication.
3. Move all terms containing x to one side of the equation, and move all other terms to the other side.
4. Combine like terms on the same side of the equation at any stage of the solution.
5. Remove any coefficient of x by dividing both sides by that coefficient.
6. Check the answer by substituting it back into the original equation.

COMMON ERROR The mathematical operations you perform must be done to both sides of the equation in order to preserve the equality.

Example 17

Solve

$$\frac{x-2}{3} = \frac{2x-4}{2}$$

Solution: We eliminate the fractions by multiplying by 6,

$$6\left(\frac{x-2}{3}\right) = 6\left(\frac{2x-4}{2}\right)$$

$$2(x-2) = 3(2x-4)$$

Removing parentheses,

$$2x - 4 = 6x - 12$$

We get all x terms on one side by adding 12 and subtracting $2x$ from both sides,

$$-4 + 12 = 6x - 2x$$

Combining like terms and switching sides,

$$4x = 8$$

Dividing by the coefficient of x,

$$x = 2$$

Check:

$$\frac{2-2}{3} \stackrel{?}{=} \frac{2(2)-4}{2}$$

$$0 = 0 \quad \text{checks}$$

Solve and check each equation. Treat the constants in these equations as exact numbers. Leave your answers in fractional, rather than decimal, form.

1. $x + 9 = 16$ **2.** $3x - 2 = 10$

3. $30 + 5x = 20x$ **4.** $7x - 29 = 6$

5. $4x + 9 = 11x - 3x$ **6.** $20 - y = 13$

7. $x + 9 = 5$ **8.** $-6y - 4 = 2y$

9. $5x + 8 = 9x$ **10.** $4x = 3x - 6$

11. $10 - 5y = 1 - y$ **12.** $x + 3 = 10$

13. $x - 5 = 6$ **14.** $8y - 6 = 5y$

15. $7x - 3 = 5x + 1$ **16.** $25x - 5 = 3x + 6$

17. $y - 4 = 0$ **18.** $21x = -8 + 5x$

19. $4x - 5 = 10x - 2$ **20.** $7x + 15 = 8$

21. $6x + 4 = 3x + 19$ **22.** $16y - 10 = 10y + 14$

23. $\dfrac{1}{3a} = 7$ **24.** $\dfrac{y}{5} = 4$

25. $17 - 14x = 8 - 11x$ **26.** $3y - 15 + 4y = 6$

27. $5x - 10 + 13 = 18$ **28.** $44 - 11x + 4x = -5$

29. $49 - 5y = 3y - 7$ **30.** $5 + y = 1 + 15y$

31. $4x - 111 = 21 + 14x$ **32.** $47x - 84 = 2x + 6$

33. $16 = x + 3$ **34.** $15x - 9 = 7x - 5$

35. $3 - (2b + 4) = 5b - 2$ **36.** $x + 16 = 1 - 4(3x + 1)$

37. $6x - 25 = 4x + 7$ **38.** $80 - 7x = 5x - 16$

39. $3(y - 5) + 2(3 + y) = 21$ **40.** $5(3x - 13) = 10$

41. $4(y - 5) = 2(y - 10)$ **42.** $2x + 1 = -2(2x + 5)$

43. $3(15 - x) = 3(5 + x)$ **44.** $4(3x + 20) = 2(x - 5)$

45. $3(2x + 1) = 7$ **46.** $3(x - 5) = 4(x - 1)$

47. $5x + 6 = 6(x - 3) + 2$ **48.** $6(2x - 3) + 2(6x + 2) = 5(x + 1)$

49. $19y - 60 = 4y + 15$ **50.** $5(2x - 8) = 2(10 - 5x)$

51. $5 + 3(x - 2) = 58$ **52.** $3(3x + 13) = 5(x - 1)$

53. $4(x - 5) - 2(6x + 3) = 22$ **54.** $5(y - 1) + 4(3y + 3) = 3(4y - 6)$

55. $2(4x - 3) = 3(5x + 2)$ **56.** $3(4x - 2) = 9x$

57. $5a = -2(13 - 9a)$ **58.** $5 - 3(x - 2) = 4(3x - 1)$

59. $6x + 5 = 7(x - 3)$ **60.** $3(7 - 2x) + 2 = 3(4x - 1)$

61. $\dfrac{x}{4} - 3 = 11$ **62.** $5 + \dfrac{7x}{3} = \dfrac{x}{2}$

63. $\dfrac{5x}{4} = 2x - 3$ **64.** $\dfrac{y}{5} - \dfrac{y}{2} + \dfrac{y}{4} = 5$

65. $\dfrac{2x - 4}{7} = \dfrac{2 - 2x}{4}$

SIMPLE EQUATIONS WITH APPROXIMATE COEFFICIENTS AND LITERALS

Equations with Approximate Numbers

In technical work we usually must solve equations that have approximate numbers, rather than coefficients and constants that are integers. The method is no different than before. Just remember to round your answer to the proper number of significant digits.

─── **Example 18** ───

Solve for x: $4.23x + 1.33 = 2.16x + 5.82$.

Solution: Collecting terms,

$$4.23x - 2.16x = 5.82 - 1.33$$
$$2.07x = 4.49$$

Dividing,

$$x = \frac{4.49}{2.07} = 2.17$$

rounded to three significant digits.

─── **Example 19** ───

Solve for x: $2.93(x + 4.28) = 24.2 - 5.82x$.

Solution: Removing parentheses gives

$$2.93x + 12.5 = 24.2 - 5.82x$$

Collecting terms,

$$2.93x + 5.82x = 24.2 - 12.5$$
$$8.75x = 11.7$$

Dividing,

$$x = \frac{11.7}{8.75} = 1.34$$

rounded to three significant digits.

Simple Literal Equations

An equation in which some or all of the coefficients are represented by letters is called a *literal* equation.

─── **Example 20** ───

The equation

$$5x + 3c = d$$

is a literal equation.

We often want to isolate a particular literal quantity on one side of the equal sign. When we do this we say we are *solving* for that quantity. To solve for a literal quantity, we use the same methods as before.

--- **Example 21** ---

Solve the equation of Example 20 for x.

Solution: Subtracting $3c$ from both sides,

$$5x = d - 3c$$

Dividing by 5 gives

$$x = \frac{d - 3c}{5}$$

We will solve more complex literal equations in Chapter 18.

SECTION 10-3 Exercises Simple Equations with Approximate Coefficients and Literals

Approximate Coefficients

Solve for x. Round your answer to the proper number of significant digits.

1. $8.27x = 4.82$
2. $24.8x - 28.4 = 0$
3. $2.84x + 2.83 = 83.7$
4. $3.82 = 29.3 + 3.28x$
5. $3.82x - 3.28 = 5.29x + 5.82$
6. $382x + 827 = 625 - 846x$
7. $2.94(x + 8.27) = 3.27$
8. $9.38(5.82 + x) = 23.8$
9. $2.34(x - 4.27) = 5.27(x + 3.82)$
10. $92.1(x - 2.34) = 82.7(x - 2.83)$
11. $2.83x + 12.1(x - 6.54) = 2.47(x - 12.43)$
12. $1.58(8.22 + x) + 42.8 = 6.44(x - 6.17) - 1.87(x + 9.22)$

Literal Equations

Solve for x.

13. $dx + 5 = 8$
14. $8 + bx = d$
15. $ax + b = c + d$
16. $dx - 5 = cx - 2$
17. $a(x + 1) = b(x - 2)$
18. $3(bx + c) = 5$

■ Chapter 10 REVIEW EXERCISES

Solve and check each equation. Treat the constants in these equations as exact numbers. Leave your answers in fractional, rather than decimal, form.

1. $x + 3 = 15$
2. $2x - 6 = 12$
3. $10 + 2x = 30x$
4. $5x - 19 = 7$
5. $3x + 9 = 12x$
6. $25 - y = 17$
7. $x + 2 = 51$
8. $-8y - 2 = 2y$
9. $3x + 8 = 5x$
10. $2x = 3x - 7$
11. $12 - 5y = 6$
12. $x + 8 = 16$
13. $x - 2 = 61$
14. $3y - 5 = 5y$
15. $2x - 6 = 3x + 2$
16. $15x - 3 = 5x$
17. $y - 3 = 8$
18. $11x = -8 + 2x$

19. $5x - 5 = 16x$ **20.** $4x + 18 = 83$ **21.** $3x + 8 = 4x + 12$

22. $2x + 2 = 8(x - 4)$ **23.** $6(2 - 2x) + 6 = 7(3x - 1)$

24. $\dfrac{x}{2} - 6 = 21$ **25.** $2 + \dfrac{3x}{4} = \dfrac{8x}{3}$

26. $\dfrac{2x}{3} = 5x - 3$ **27.** $\dfrac{y}{3} - \dfrac{y}{2} + \dfrac{y}{4} = 7$

Substitute the given integers into each equation. Do not round your answer.

28. $y = 4x + 7\ (x = 8)$ **29.** $y = 2m^2 + 6m - 2\ (m = -7)$

30. $y = 2a^3 - 7x\ (a = -3, x = -7)$ **31.** $y = 6x^3 + 2x^2 - x - 1\ (x = -3)$

32. $y = 7c + 2d^2 - 8e\ (c = 2, d = -1, e = 6)$

Substitute the given approximate numbers into each equation. Round your answer to the proper number of digits.

33. $y = 4 - 8x\ (x = -1.34)$

34. $y = -2d^3 - 5e^2\ (d = -2.335, e = 3.195)$

35. $y = a^3 + a + 2\ (a = 4.61)$

Use the formulas given in Section 10-1 Exercises for problems 36–44.

36. Find the amount to which \$7420 will accumulate in 6 years at a simple interest rate of 3.8%.

37. Find the displacement after 5.6 seconds of a body thrown downward with a speed of 22 ft/s.

38. Convert 26°C to degrees Fahrenheit.

39. Find the torque delivered by a 1.50-hp motor rotating at 2740 rev/min.

40. Find the torque delivered by a 1.55-hp motor rotating at 225 rad/s.

41. A bar 51.4-m long having a cross-sectional area of 24.5 cm² is subject to a tensile load of 21,500 N. The elongation is 1.55 mm. Find the modulus of elasticity in newtons per square centimeter. (See Section 10-1 Exercises, Problem 18.)

42. Find the amount S obtained when an amount $P = \$3354$ is allowed to accumulate for $t = 7$ years at a compound interest rate of $n = 6.75\%$.

43. Compute the pressure drop in a 2.84-in.-diameter, 124-ft-long pipe, where $f = 0.022$ and the flow rate is 184 gal/min.

44. The resistance of a copper coil is 827 Ω at 18°C and the temperature coefficient of resistance is 0.00411 at 18°C. find the resistance at 75°C.

Solve for x.

45. $8.24(x + 1.17) = 6.67$ **46.** $2.48(3.12 + x) = 53.2$

47. $4.24(x - 6.77) = 2.67(x + 7.72)$ **48.** $12.1(x - 5.34) = 22.7(x - 5.83)$

49. $5.23x + 62.1(x - 2.14) = 7.37(x - 22.13)$

50. $4.18(3.22 + x) + 21.8 = 3.84(x - 2.47) - 4.17(x + 3.62)$

51. $cx + 5 = d$ **52.** $d + ax = c$

53. $mx + b = a + b$ **54.** $cx - d = 2x - a$

55. You probably have some idea what field you later plan to enter: business, engineering, and so forth. Look at a book from that field and find at least one formula commonly used. Write it out, describe what it is used for, and explain all the quantities it contains, and their units.

Solving Word Problems

In this chapter we are asked to read a short problem statement (the "word problem") and to decide what is being asked for. We then write an equation which enables us, by solving that equation, to find the quantity we seek.

Keep in mind that we are not solving these problems for their own sake. We don't really care how long it takes runner A to reach city Q. What we are trying to do is to help you to read a technical statement, to extract the important information, and to use simple mathematics to find a missing quantity. These skills will also help you to deal with the mountains of technical material in written form with which we are all faced: instruction manuals, textbooks, specifications, contracts, insurance policies, building codes, handbooks, and so forth.

11-1   SOLVING WORD PROBLEMS

It is not possible to give a step-by-step procedure that will enable someone to solve *any* word problem, but we can give some good tips that are almost always helpful. We first give some general tips on how to approach any word problem and later show how to set up specific kinds of problems.

Study the Problem

Read the problem carefully, more than once if necessary. Be sure you understand the words used. A technical problem will often have unfamiliar terms. Look up their meanings in a dictionary, handbook, or textbook.

Example 1

Word problems may contain words or phrases like

simply supported beam	*angular velocity*	*groundspeed*
concentrated load	*center of gravity*	*kinetic energy*

It goes without saying that you must know the meanings of these words before you can solve a problem in which they appear.

Picture the Problem

Try to visualize the situation described in the problem. Form a picture in your mind. Draw a diagram showing as much of the given information as possible.

Example 2

Suppose that part of a word problem states that "a car goes from city P to city Q, which is 236 miles away, at an average rate of 55 miles per hour."

At this point you should see not just dead words on paper, but a car moving along a road (perhaps a red sports car cruising down the interstate, stereo playing, you at the wheel). Make a sketch something like Fig. 11-1, showing the road as a line with cities P and Q at either end, 236 miles apart. Show the car with an arrow giving the direction of travel and the speed.

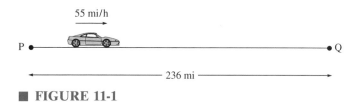

■ **FIGURE 11-1**

Of course, we have not solved the problem yet (it isn't even fully stated), but at least we have a good grip on the information given so far.

Identify the Unknown(s)

Look for a sentence that asks a question (it may end in a question mark). Find a sentence that starts with *Find . . .* or *Calculate . . .* or *What is . . .* or *How much . . .* or similar phrases. Such a sentence will usually contain the unknown quantity, the thing we want to solve for. Then label this unknown with a statement such as:

Let x = cost of each part, dollars

Be sure to include units of measure when defining the unknown.

Example 3

Suppose one sentence in a word problem says, "Find the speed of train *B*." You may be tempted to label the unknown in one of the following ways:

Let x = train B	This is too vague. Are we talking about the speed of train B or the time it takes for train B or the distance traveled by train B or something else?
Let x = speed of the train	Which train?
Let x = speed of train B	Units are missing.
Let x = speed of train B, mi/h	Good

Estimate the Answer

A valuable skill to develop is the ability to *estimate* the answer before starting the problem. Then you will be able to see if the answer you get is reasonable or not.

One way to estimate an answer is to *make simplifying assumptions.*

Example 4

Suppose we want to find the volume of the spindle in Fig. 11-2. To get an estimate, we can find the volume of a cylinder of the same length as the spindle, but has a diameter somewhere between the largest and smallest diameters of the spindle, say, 17 mm. This will not give us the exact answer, of course, but will put us in the ballpark. If our final answer is very different than our estimate, we would check our work.

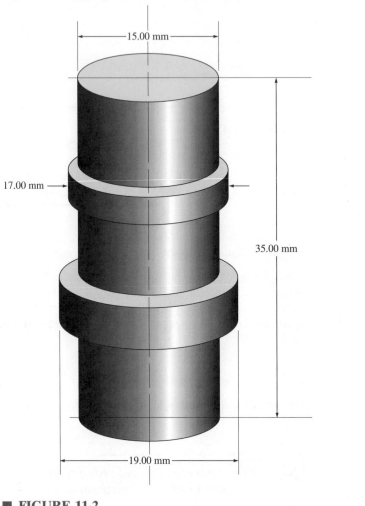

■ FIGURE 11-2
A spindle.

Another technique is to find two numbers that *bracket* the answer.

Example 5

We can *bracket* the volume of the spindle by noting that it must be less than that of a cylinder 19.00 mm in diameter and 35.00 mm in length, but greater than that of a cylinder 15.00 mm in diameter and 35.00 mm in length.

Estimation is not always easy, and it takes some time to develop this skill, but it is well worth it. We will show estimation in most of the word problems in this chapter.

Write and Solve an Equation

We must now write an equation to relate the given quantities to the unknown quantity. Sometimes the equation will be a formula from mathematics, such as the relationship between the volume of a sphere and its radius. At other times you will need a formula from technology, such as the one relating the strength of a steel rod to its diameter. The relationships you will need for the problems in this book can be found in Appendix A, "Summary of Facts and Formulas." Problems that use some of these formulas are treated later in this chapter. For now, we will solve simple number problems, in which the formula is given verbally right in the problem statement.

When you have an answer, label it with the same words you used in defining the unknown.

Don't Give Up

Be persistent. If you don't "get it" at first, keep trying. Take a break and return to the problem later. Read the problem again. Perhaps try a different approach. You know that these "school problems" you encounter in a mathematics class all have solutions, so keep at it.

Number Problems

These problems may not have much practical value, but they have the great value in giving us practice in setting up and solving word problems. In such problems, mathematical operations are often indicated with words such as

> . . . *the difference of* . . .
> . . . *is equal to* . . .
> . . . *the quotient of* . . .

Locate any such words and replace them with mathematical symbols.

Example 6

If two times a number is increased by seven, the result is equal to five less than four times the number. Find the number.

Solution: Let x = the number. Then,

$$2x = \text{two times the number}$$

From the problem statement,

$$2x + 7 = 4x - 5$$

Solving for x,

$$7 + 5 = 4x - 2x$$
$$2x = 12$$
$$x = 6$$

Check Your Answer

First see if your answer is reasonable. Did you come up with a bridge cable 1 mm in diameter or a person walking at a rate of 54 mi/h? Does your answer agree, within reason, with your estimate? Then you should check your answer in the original problem *statement*. Do *not* check the answer in your equation, which may already contain an error.

Let us check the answer from Example 6. Using the words of the problem statement, we get,

Twice six increased by seven is: $2(6) + 7 = 12 + 7 = 19$

Five less than four times six is: $4(6) - 5 = 24 - 5 = 19$ checks

> **COMMON ERROR** Checking an answer by substituting into the equation is *not good enough*. The equation may already contain an error. Check your answer in the problem statement.

Define Other Unknowns

If there is a second unknown, you will often be able to define it in terms of the original unknown, rather than to introduce a new symbol.

────── **Example 8** ──────

Find two numbers whose sum is 50 and whose difference is 30.

Solution: This problem has two unknowns. We could choose to label each with a separate symbol,

$$\text{Let } x = \text{larger number}$$

and

$$y = \text{smaller number}$$

but a better way would be to label the second unknown in terms of the first:

$$\text{Let } x = \text{larger number}$$

then,

$$50 - x = \text{smaller number}$$

Solving the problem, we get the difference by subtracting the smaller number from the larger.

$$x - (50 - x)$$

We are told that this equals 30, so our equation is,

$$x - (50 - x) = 30$$

Removing parentheses gives

$$x - 50 + x = 30$$

Collecting like terms we get,

$$2x - 50 = 30$$

Adding 50 to both sides gives

$$2x = 80$$

$$x = 40 = \text{the larger number}$$

$$50 - x = 10 = \text{the smaller number}$$

The usual steps followed in solving a word problem are summarized here.

> 1. **Study the Problem.** Look up unfamiliar words. Make a sketch. Try to visualize the situation in your mind.
> 2. **Identify the Unknown(s).** Give it a symbol, such as *let x* = ... If there is more than one unknown, try to label the others in terms of the first. Include units.
> 3. **Estimate the Answer.** Make simplifying assumptions or try to bracket the answer.
> 4. **Write and Solve an Equation.** Look for a relationship between the unknown and the known quantities that will lead to an equation. Write and solve that equation for the unknown. Include units in your answer. If there is a second unknown be sure to find it also.
> 5. **Check Your Answer.** See if the answer looks reasonable. See if it agrees with your estimate. Be sure to do a numerical check in the problem statement itself.

SECTION 11-1   Exercises   Solving Word Problems

Identify an unknown and rewrite each expression as an algebraic expression.

1. Ten more than three times a number.
2. Two numbers whose sum is 17.
3. Two numbers whose difference is 42.
4. The amounts of antifreeze and of water in 4 gallons of an antifreeze-water solution.
5. A fraction whose denominator is 4 more than 6 times its numerator.
6. The angles in a triangle, if one angle is three times the other.
7. The number of gallons of antifreeze in a radiator containing x gallons of a mixture that is 11% antifreeze.
8. The distance traveled in x hours by a car going 78 km/h.
9. If x equals the length of a rectangle, write an expression for the width of the rectangle if (a) the perimeter is 64; (b) the area is 320.

Solve each problem for the required quantity.

10. Four less than 6 times a number is 32. Find the number.
11. Find a number such that the sum of 16 and that number is 3 times that number.
12. Six less than 5 times a number is 19. Find the number.
13. The sum of 45 and some number equals 6 times that number. Find it.
14. Ten less than three times a certain number is 29. Find the number.
15. When three is added to seven times a number, the result is equal to the same as when we subtract seven from nine times that number. Find that number.
16. The denominator of a fraction is 2 less than 4 times its numerator, and the reduced value of the fraction is $\frac{4}{15}$. Find the numerator.

11-2   UNIFORM MOTION PROBLEMS

Number problems were good for a warm-up, but now we will do some types of problems that have more application to technology. We will start with uniform motion problems. The ideas here should be familiar; you already know that if you walk at a rate of 3 miles per hour for 2 hours you will travel 6 miles. The problems in this section are based on that one simple idea.

Uniform Motion

Motion is called *uniform* when the speed does not change. The distance traveled at constant speed is related to the speed (the *rate* of travel) and the elapsed time by

$$\text{rate} \times \text{time} = \text{Distance}$$

or

$$RT = D$$

Be careful not to use this formula for anything but uniform motion.

— **Example 9** —

A plane flies for 2.45 h at 325 km/h. How far does it travel?

Solution: From the equation above,

$$D = 325 \,(2.45) = 796 \text{ km}$$

We now do a typical motion problem, in which we will organize our information in table form, and use the equation (rate × time = distance) to help in the solution.

— **Example 10** —

A truck leaves a city traveling at a speed of 72.5 km/h, and a car leaves the same city 1.25 hour later to overtake the truck. If the car's speed is 108.0 km/h, how long will it take for the car to overtake the truck?

Estimate

The truck has a head start of 1.25 hour and goes at 72.5 km/h, so it is about 91 km ahead, when the car starts. But the car goes about 36 km/h faster than the truck, so each hour the truck's lead is cut by 36 km. Thus it would take between 2 and 3 hours to reduce the 91-km lead to zero.

Solution: Let us follow the list of steps suggested for solving any word problem.

Make a Sketch

A simple sketch, Fig. 11-3, shows the city, the road, the meeting point, and the rates of the truck and the car.

Identify the Given Information

Let us write down the formula for uniform motion,

$$\text{rate} \times \text{time} = \text{Distance}$$

■ **FIGURE 11-3**

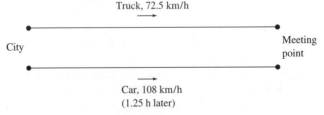

Truck, 72.5 km/h

City   Meeting point

Car, 108 km/h
(1.25 h later)

Under each of the three quantities we write the given value of that quantity, both for the truck and for the car. Reading the problem statement, we see that the speed of the truck and of the car are given.

	Rate (km/h)	× Time (h)	= Distance (km)
Truck	**72.5**		
Car	**108.0**		

Define the Unknown

What are we looking for? We seek the time for the car to overtake the truck, so we write

Let t = time traveled by car (hours)

and enter this information into our table

	Rate (km/h)	× Time (h)	= Distance (km)
Truck	72.5		
Car	108.0	t	

Define the Other Unknowns

We have three empty boxes in our table. First, the time for the truck can be written in terms of t by noting that the truck travels for 1.25 hour longer than the car.

	Rate (km/h)	× Time (h)	= Distance (km)
Truck	72.5	$t + \mathbf{1.25}$	
Car	108.0	t	

We complete the table by noting that distance = rate × time.

	Rate (km/h)	× Time (h)	= Distance (km)
Truck	72.5	$t + 1.25$	$\mathbf{72.5(t + 1.25)}$
Car	108.0	t	$\mathbf{108.0}t$

Write and Solve an Equation

An equation says that something is equal to something else. What quantities are equal in this problem? Since, at the instant that the car overtakes the truck, they have both gone the same distance, we set the distance gone by the truck equal to the distance gone by the car.

$$108.0t = 72.5(t + 1.25)$$

Solve the Equation

Removing the parentheses,

$$108.0t = 72.5t + 72.5(1.25)$$

$$108.0t - 72.5t = 90.63$$

$$35.5t = 90.63$$

$$t = 2.55 \text{ hours} = \text{time for car}$$

Check Your Answer

We first note that the answer (2.55 h for the car) agrees with our estimate, which was between 2 and 3 hours. Next, for a more accurate check, let's find the distance traveled by the truck and by the car.

$$\text{Time for truck} = t + 1.25 = 3.80 \text{ h}$$

$$\text{Truck distance} = (72.5 \text{ km/h})(3.80\text{h}) = 275.5 \text{ km}$$

$$\text{Car distance} = (108.0 \text{ km/h})(2.55 \text{ h}) = 275.4 \text{ km}$$

These distances agree to three significant digits, the precision to which we are working in this problem. Since most of our original numbers are known only to three significant digits, we cannot expect our answers to check to more than three significant digits.

In the preceding example we got our equation from the fact that both vehicles traveled the same distance. This, of course, will not be true for *every* motion problem, and you may have to look for other relationships in order to write an equation.

SECTION 11-2 Exercises Uniform Motion Problems

1. Two planes start from the same city at the same time. One travels at 252 mi/h and the other at 266 mi/h, in the opposite direction. How long will it take for them to be 1750 miles apart?

2. The pointer of a certain meter can travel to the right at the rate of 10.0 cm/s. What must be the return rate if the total time for the pointer to traverse the full 12.0-cm scale and return to zero must not exceed 2.00 seconds?

3. A train travels from P to Q at a rate of 22.5 km/h. After it has been gone 2.75 hours, an express train leaves P for Q traveling at 85.5 km/h, and reaches Q 1.50 hours ahead of the first train. Find the distance from P to Q, and the time taken by the express train.

4. A freight train leaves A for B, 175 miles distance and travels at the rate of 31.5 mi/h. After 1.50 hours, a train leaves B for A, traveling at 21.5 mi/h. How many miles from B will they meet?

5. Two submarines start from the same spot and travel in opposite directions, one at 115 km/day and the other at 182 km/day. How long will it take for the submarines to be 1470 km apart?

6. A bus travels 87.5 km to another town at a speed of 72.0 km/h. What must be its return rate if the total time for the round trip is to be 2.50 hours?

7. An oil slick from a runaway offshore oil well is advancing toward a beach 354 miles away at the rate of 10.5 mi/day. Two days after the spill, cleanup ships leave the beach and steam toward the slick at a rate of 44.6 mi/day. At what distance from the beach will they reach the slick?

8. A certain shaper has a forward cutting speed of 115 ft/min and a stroke of 10.5 in. It is observed to make 429 cuts (and returns) in 4.0 minutes. What is the return speed?

9. Spacecraft A is over Houston at noon on a certain day and traveling at a rate of 275 km/h. Spacecraft B, attempting to overtake and dock with A, is over Houston at 1:15 P.M. and is traveling in the same direction as A, at 444 km/h. At what time will B overtake A? At what distance from Houston?

FINANCIAL PROBLEMS

We all deal with money, so word problems involving money are probably the easiest for us to understand. We will need to use percentage often here, so you may want to review Chapter 5 quickly before starting these problems.

Example 11

A roofer needs $1700 worth of materials to roof a certain house. He estimates that 8.0% of the materials he buys will be waste. Find the cost of materials, to the nearest ten dollars, if he buys enough to allow for waste.

Estimate

We know that he must buy more than $1700 worth of materials, but not too much more. Let's guess at $2000 worth. The waste would then be 8% of $2000 or $160 worth, leaving $1840 actually used. This is *more* than is needed. So our answer must be between $1700 and $2000.

Solution: We will let x stand for the quantity we seek, so

$$\text{Let } x = \text{cost of roofing material, dollars}$$

From that amount we subtract the 8.0% of that amount for waste, and the result is the $1700 needed. That gives us our equation

$$x - 0.080x = 1700$$

Solving for x,

$$0.920x = 1700$$

$$x = \frac{1700}{0.920} = \$1850$$

This number lies within the values we bracketed in our estimate.

Example 12

A federal income tax table states that if your taxable income is over $55,100 but not over $115,000, your tax is $12,470 plus 31% of the amount over $55,100. What was your taxable income if your tax bill was $26,834? Work to the nearest dollar.

Estimate

For an income of $55,100, the tax is $12,470. For an income of $115,000, the tax is $12,470 + 0.31(115,000 − 55,100) = $31,039. Our tax ($26,834) lies between these values, but closer to the higher value than the lower. To get closer, let's guess at an income of $100,000, with a tax of $12,470 + 0.31(100,000 − 55,100), or $26,389, a bit lower than our actual tax. Thus, our income must be a little over $100,000 but not more than $115,000.

Solution:

$$\text{Let } x = \text{taxable income}$$

Your income then exceeds $55,100 by $(x - 55{,}100)$, so your tax is

$$\text{Tax} = 12{,}470 + 31\% \text{ of } (x - 55{,}100)$$

But your tax bill was $26,834 so our equation is

$$12{,}470 + 0.31(x - 55{,}100) = 26{,}834$$

Solving for x,

$$0.31(x - 55{,}100) = 26{,}834 - 12{,}470$$
$$0.31x - 0.31(55{,}100) = 14{,}364$$
$$x - 55{,}100 = 46{,}335$$
$$x = 46{,}335 + 55{,}100 = \$101{,}435$$

This agrees well with our estimate.

Example 13

A person had $24,924.85 in a bank, part in a savings account that earned an interest of 3.55% per year, and the remainder in a money market account earning 5.75% per year. (Assume that both are *simple interest;* that is, the interest is computed only once per year.) The interest earned from both accounts combined was $1125.84. How much was in each account?

Estimate

Let's assume that *half the money* (about $12,500) was put into each account. The interest at 3.55% would be $444, and at 5.75% would be $719, for a total of $1163. This is more than was actually earned. We conclude that *more than half* was invested at the lower rate.

Solution: Here we are looking for *two* things, the amount in each of two accounts. Let us set x equal to one of them. So,

Let x = amount deposited in the savings account

Then

$24{,}924.85 - x$ = amount deposited in the money market account

The interest earned on each account is then

3.55% of x = interest earned by savings account

and

5.75% of $(24{,}924.85 - x)$ = interest earned by money market account

Setting the sum of the two earnings equal to $1125.84 gives us our equation.

$$0.0355x + 0.0575(24{,}924.85 - x) = 1125.84$$

Solving for x,

$$0.0355x + 1433.18 - 0.0575x = 1125.84$$
$$1433.18 - 1125.84 = 0.0575x - 0.0355x$$
$$0.0220x = 307.34$$
$$x = \$13{,}970.00$$

That is the amount in the savings account. In the money market account we have

$$\$24{,}924.85 - x = \$10{,}954.85$$

We see that more than half the money was put into the lower-interest savings account, as we predicted in our estimate.

1. How much must a person earn to have $45,824 after paying 28% in taxes?

2. A carpenter estimates that a certain deck needs $4285 worth of lumber, if there were no waste. How much should she buy, if she estimates the waste at 7%?

3. The labor costs for a certain brick wall were $1118 per day for 10 masons and helpers. If a mason earns $125/day and a helper $92/day, how many masons were on the job?

4. According to a tax table, for your filing status, if your taxable income is over $38,100 but not over $91,850, your tax is $5700 plus 28% of the amount over $38,100. If your tax bill was $8126, what was your taxable income? Work to the nearest dollar.

5. A company has $173,924 in bonds and from them earns $13,824 in simple interest annually. Part of the money is invested at 6.75% and the remainder at 8.24%. How much is invested at each rate?

6. A company had $528,374 invested, part in stocks that earned 9.45% per year, and the remainder in bonds that earned 6.12% per year, both simple interest. The amount earned from both investments combined was $42,852. How much was in each investment?

7. A student sold a computer and printer for a total of $995, getting $1\frac{1}{2}$ times as much for the printer as for the computer. What was the price of each?

11-4   MIXTURE PROBLEMS

Before starting a typical mixture problem, let's look at some basic ideas about mixtures.

Basic Relationships

If a mixture contains two ingredients, A and B, the total amount of the mixture is equal to the sum of the amounts of the ingredients:

Total amount of mixture = amount of A + amount of B

If there are more than two ingredients, the total amount, of course, equals the sum of the individual amounts. If more of a particular ingredient, say A, is added to the mixture and/or some is removed, then,

Final amount of A = initial amount of A + amount of A added − amount of A removed

---
Example 14

A soldering machine for circuit boards contains 124 lb of solder, half lead and half zinc. From this, 22 lb of solder is removed, then 35 lb of lead is added. How much lead is contained in the final mixture?

Solution:

Initial weight of lead = 0.5(124) = 62 lb

Amount of lead removed = 0.5(22) = 11 lb

Amount of lead added = 35 lb

By the equation above,

Final amount of lead = 62 − 11 + 35 = 86 lb

Percent Concentration

Recall from Sec. 5-4 that the percent concentration of each ingredient is

$$\text{Percent concentration of ingredient } A = \frac{\text{amount of } A \text{ in mixture}}{\text{total amount of mixture}} \times 100 \qquad \boxed{16}$$

--- **Example 15** ---

The final percent concentration of lead in the preceding example is, by Eq. 16

$$\text{Percent lead} = \frac{86}{137} \times 100 = 63\%$$

--- **Example 16** ---

A tank contains 88.6 liters of gasohol that is 12.2% alcohol. How many liters of alcohol are in the tank?

Solution: Again using Eq. 16, but in the form,

Amount of A in mixture = percent concentration of A × total amount of mixture

$$\text{Liters of alcohol in tank} = 0.122 \text{ of } 88.6 \text{ liters}$$
$$= 10.8 \text{ liters}$$

--- **Example 17** ---

Find the percent concentration of alcohol in the tank of Example 16 if 2.6 liters of pure alcohol are added.

Solution: The amount of alcohol in the tank increases by 2.6 liters, so

$$\text{Total amount of alcohol} = 10.8 + 2.6 = 13.4 \text{ liters}$$

The total amount of mixture also increases by 2.6 liters, so

$$\text{Total amount of mixture} = 88.6 + 2.6 = 91.2 \text{ liters}$$

Then by Eq. 16,

$$\text{Percent concentration of alcohol} = \frac{13.4}{91.2} \times 100 = 14.7\%$$

Two Mixtures

When two mixtures are combined into a third mixture, the amount of any ingredient A in the final mixture is

Final amount of A = amount of A in first mixture + amount of A in second mixture

--- **Example 18** ---

A vat of 125 liters of gasohol containing 11.2% alcohol is mixed with 225 liters of gasohol containing 7.75% alcohol. How many liters of alcohol are in the final mixture?

Solution: By the equation above,

$$\text{Final amount of alcohol} = 11.2\% \text{ of } 125 + 7.75\% \text{ of } 225$$
$$= 0.112(125) + 0.0775(225)$$
$$= 31.4 \text{ liters}$$

─── **Example 19** ───

How much steel containing 5.25% nickel must be combined with another steel containing 2.84% nickel to make 3.25 tons of steel containing 4.15% nickel?

Estimate

The final steel needs 4.15% of 3.25 tons of nickel, about 0.135 tons. If we assume that *equal amounts* of each steel were used, the amount of nickel would be 5.25% of 1.625 tons, or 0.085 tons, from the first alloy, and 2.84% of 1.625 tons, or 0.046 tons, from the second alloy. This gives a total of 0.131 tons of nickel in the final steel. This is not enough (we need 0.135 tons). Thus, *more than half* of the final alloy must come from the higher-nickel alloy, or between 1.625 tons and 3.25 tons.

Solution: Let x = tons of 5.25% steel needed. Fig. 11-4 shows the three alloys, and the amount of nickel in each, with the nickel drawn as if it were separated from the rest of the steel. By the equation above, the weight of the 2.84% steel is $3.25 - x$. The weight of nickel it contains is, by Eq. 16,

$$0.0284(3.25 - x)$$

The weight of nickel in x tons of 5.25% steel is

$$0.0525x$$

■ **FIGURE 11-4**

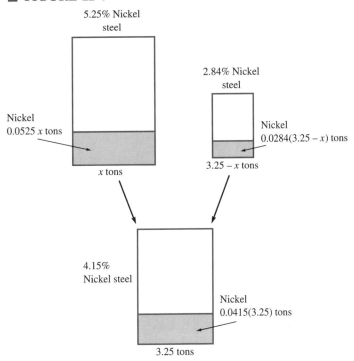

The sum of these must give the weight of nickel in the final mixture:

Final amount of A =

amount of A in first mixture and amount of A in second mixture

$$0.0525x + 0.0284(3.25 - x) = 0.0415(3.25)$$

Clearing parentheses,

$$0.0525x + 0.0923 - 0.0284x = 0.1349$$
$$0.0241x = 0.0426$$
$$x = 1.77 \text{ tons of } 5.25\% \text{ steel}$$
$$3.25 - x = 1.48 \text{ tons of } 2.84\% \text{ steel}$$

Check

First we see that more than half the final alloy comes from the higher-nickel steel, as predicted in our estimate. Now let's see if the final mixture has the proper percentage of nickel.

$$\text{Final tons of nickel} = 0.0525(1.77) + 0.0284(1.48)$$
$$= 0.135 \text{ tons}$$
$$\text{Percent nickel} = \frac{0.135}{3.25} \times 100 = 0.0415$$

or 4.15%, as required.

Alternate Solution: This problem can also be set up in table form, as follows.

	Percent Nickel	×	Amount (tons)	=	Amount of Nickel (tons)
Steel with 5.25% Ni	5.25%		x		0.0525x
Steel with 2.84% Ni	2.84%		3.25 − x		0.0284(3.25 − x)
Final Steel	4.15%		3.25		0.0415(3.25)

We then equate the sum of the amounts of nickel in the original steels with the amount in the final steel, getting

$$0.0525x + 0.0284(3.25 - x) = 0.0415(3.25)$$

as before.

COMMON ERROR If you wind up with an equation that looks like this,

( )lb nickel + ( )lb iron = ( )lb iron

you know something must be wrong. When using the equation

Final amount of A = amount of A in first mixture + amount of A in second mixture

all the terms must be for the same ingredient.

1. From 55.5 gallons of wine containing 12.0% alcohol, 10.0 gallons is removed. Then 5.50 gallons of alcohol is added. How many gallons of alcohol are contained in the final mixture?

2. A casting weighs 875 kg and is made of a brass that contains 87.5% copper. How many kg of copper are in the casting?

3. Find the percent concentration of copper in the casting of the preceding problem, if it is melted with 125 kg of pure copper.

4. There are 235 gallons of fuel for a two-cycle engine that contains 3.15% oil. To this is added 186 gallons of fuel containing 5.05% oil. How many gallons of oil are in the final mixture?

5. How much steel containing 1.15% chromium must be combined with another steel containing 1.50% chromium to make 8.00 tons of steel containing 1.25% chromium?

6. Two gasohol mixtures are available: one with 6.35% alcohol and the other with 11.28% alcohol. How many gallons of the 6.35% mixture must be added to 25.0 gallons of the other mixture to make a final mixture containing 8.00% alcohol?

7. How many tons of tin must be added to 2.75 tons of bronze to raise the percentage of tin from 10.5% to 16.0%?

8. How many pounds of nickel silver containing 12.5% zinc must be melted with 248 lb of nickel silver containing 16.4% zinc to make a new alloy containing 15.0% zinc?

9. How many pounds of candy costing $3.55 per pound must be mixed with 10 pounds of candy costing $5.25 per pound to make a mixture that costs $4.00 per pound?

10. How many liters of olive oil costing $4.86 per liter must be mixed with 136 liters of corn oil costing $2.75 per liter to make a blend costing $3.00 per liter?

■ Chapter 11 REVIEW EXERCISES

1. A company has $223,812 invested in two separate accounts and earns $14,824 in simple interest annually from these investments. Part of the money is invested at 5.94% and the remainder at 8.56%. How much is invested at each rate?

2. An airplane left an airport at 1:00 P.M. traveling at a rate of 242 mi/h. After traveling for some distance, one engine failed, and the airplane returned to the field at a reduced rate of 115 mi/h, arriving at 3:15 P.M. At what distance from the field did it turn around?

3. From 12.5 liters of coolant containing 13.4% antifreeze, 5.50 liters is removed. If 5.50 liters of antifreeze is then added, how much antifreeze will be contained in the final mixture?

4. A person had $125,815 invested, part in a mutual fund that earned 10.25% per year and the remainder in a bank account that earned 4.25% per year, both simple interest. The amount earned from both investments combined was $11,782. How much was in each investment?

5. How much must a person earn to have $50,000 after paying 27% in taxes?

6. In a group of 105 persons, composed of men, women, and children, there are four times as many children as men, and twice as many women as men. How many are there of each?

7. The labor costs for a certain project were $3971 per day for 15 technicians and helpers. If a technician earns $325/day and a helper $212/day, how many technicians were on the job?

8. A person walks a certain distance at a rate of 3.8 mi/h. Another person takes 2.1 hours longer to walk the same distance, at a rate of 3.2 mi/h. What distance did they walk?

9. If $5x + 3$ stands for 38, for what number does $2x - 4$ stand?

10. A certain submarine can travel at a rate of 31.5 km/h submerged and 42.3 km/h on the surface. How far can it travel submerged, and return to the starting point on the surface, if the total round trip is to take 8.25 hours?

11. A mixture is made where 15.6 gallons of cleaner containing 17.8% acid is added to 25.5 gallons of cleaner containing 10.3% acid. How many gallons of acid are in the final mixture?

12. A certain space probe traveled at a speed of 1280 km/h, and then, by firing retro-rockets, slowed to 950 km/h. The total distance traveled was 75,300 km in a time of 63.5 hours. Find the distance from the starting point at which the retro-rockets were fired.

13. How many pounds of copper are in an ingot that weighs 1550 lb and is made of a brass that contains 73.2% copper?

14. A ship leaves port at a rate of 24.5 km/h. After 8.25 hours a launch leaves the same port to overtake the ship and travels at a rate of 35.2 km/h. Find the time it will take for the launch to overtake the ship and the distance from port at which they will meet.

15. How much brass containing 75.5% copper must be combined with another brass containing 86.3% copper to make 5250 kg of brass containing 80.0% copper?

WRITING

16. Make up a word problem. It can be similar to one given in this chapter or, better, very different. Try to solve it yourself. Then swap with a classmate and solve each other's problem. Note down where the problem may be unclear, unrealistic, or ambiguous. Finally, each of you should rewrite your problem if needed.

Functions and Graphs

In this chapter, we introduce functions and functional notation. We use functions to solve applications including the volume of a cylinder and the velocity of a uniformly accelerating object. Then we introduce the rectangular coordinate system and study the graph of a function. We pay particular attention to linear and quadratic functions. We explore the use of a graphing calculator, with its ability to trace and zoom in on the graph of a function.

12-1 FUNCTIONS

The equations we solved in the last two chapters were equations in one variable, such as

$$3(x - 2) = 5(x + 1) - 4$$

But many equations and formulas involve two or more variables. For example, the formula for the area of a circle, $A = \pi r^2$, involves two variables, area A and radius r. As the radius r of the circle changes, the area A changes also (Fig. 12-1). The variables r and A are related. In this section we study equations that express relationships between variables.

The idea of a *function* provides us with a different way of speaking about mathematical relationships. It may become apparent as you study this chapter that these same problems could be solved without ever introducing the idea of a function. Does the function concept, then, merely give us new jargon for the same old ideas? Not really. A new way of *speaking* about something can lead to a new way of *thinking* about it, and so it is with functions. It also will lead to powerful and convenient functional notation, which will be especially useful as you continue your study of mathematics.

In this section, we define a function and study the different forms of a function. Many formulas are functions and express functional relationships between variables.

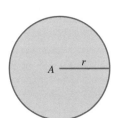

■ **FIGURE 12-1**
Area and radius of a circle.

■ **FIGURE 12-2**
Area and side of a square.

Example 1

The equation

$$A = s^2$$

is the formula for the area A of a square with side s (Fig. 12-2). This equation expresses the relationship between the two variables, s and A. It expresses one quantity, A, in terms of another quantity, s. In fact, if $s = 3$ in.,

then $A = 9$ square in.

We say that *the area A is a function of the side s*. Given any side s of a square, we can determine the area A of that square, using the equation above.

Definition of a Function

If we have an equation that enables us to find exactly one y for any given x, then we say y *is a function of* x. In Example 1, A is a function of s, since for each value of s, there is exactly one value of A.

Example 2

In the equation $y = 2x - 3$,

$$\text{if } x = 4$$

then, by substitution,

$$y = 2(4) - 3$$
$$= 8 - 3$$
$$= 5$$

Similarly,

$$\text{if } x = 5$$

then

$$y = 2(5) - 3 = 7$$

In fact, for each value of x, there is exactly one value of y, and y is a function of x.

Example 3

In the equation $x^2 + y^2 = 25$,

$$\text{if } x = 3$$

then

$$3^2 + y^2 = 25$$
$$y^2 = 25 - 9$$
$$y^2 = 16$$

or

$$y = \pm 4$$

Since y has *two* values for $x = 3$, the equation $x^2 + y^2 = 25$ is *not* a function.

Explicit Form

An equation is said to be in *explicit form* when one variable is isolated on one side of the equal sign. In the equation $y = 2x - 3$, the variable y is expressed *explicitly* in terms of x. That is, y has been isolated on one side of the equal sign.

Example 4

The equations

$$y = x^2 - 1, \quad A = s^2, \quad \text{and} \quad x = 3y + z$$

are all functions in explicit form.

Implicit Form

When a variable is *not* isolated on one side of the equal sign, the equation is said to be in *implicit form*.

Example 5

The equations

$$x + y = 1, \quad x = y^2 + 3x, \quad \text{and} \quad x^2 + y^2 = 25$$

are all in implicit form.

Dependent and Independent Variables

In the explicit equation $y = 2x - 3$, the variable y has been isolated on the left. If we know any value of x, then we can easily determine the value of y by substitution. In this equation, the variable y is called the *dependent* variable, because its value *depends* on the value of the other variable x. The variable x is called the *independent* variable. Only equations in explicit form have dependent and independent variables.

Example 6

In the equation

$$y = x - 3$$

y is the dependent variable and x is the independent variable. If we solve that same equation for x, we get

$$x = y + 3$$

Now x is the dependent variable and y is the independent variable. We could also write that same equation as

$$x - y = 3$$

This is an implicit form of the equation and neither x nor y is dependent or independent.

Types of Functions

There are many different types of functions and many have special names. For example,

$y = 2x - 3$	is called a *linear* function.
$y = 4x^2 - 3x + 5$	is called a *quadratic* function.
$y = x^3$	is called a *power* function.
$y = \cos 3x$	is called a *trigonometric* function.
$y = 2^x$	is called an *exponential* function.
$y = \log x$	is called a *logarithmic* function.

We will study all these types of functions in this book.

Functions in Verbal Form

A function may be expressed in verbal form by a sentence. We may then express that sentence in equation form.

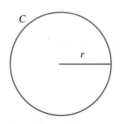

■ FIGURE 12-3
Circumference and
radius of a circle.

Example 7 ──────────

The circumference of a circle is twice the product of π times the radius. Express the circumference as a function in equation form.

Solution: Letting C represent the circumference of the circle and r represent the radius (Fig. 12-3), the equation form is

$$C = 2\pi r$$

A Function in the Form of a Formula

Many formulas are functions. The most common formulas are listed in Appendix A, "Summary of Facts and Formulas," beginning on page 677.

Example 8 ──────────

Express the volume V of a sphere as a function of its radius r.

Solution: This is another way of saying: "Write an equation for the volume of a sphere in terms of its radius." The formula for the volume of a sphere is found in Appendix A, "Summary of Facts and Formulas," Eq. 95, on page 681.

$$V = \frac{4}{3}\pi r^3$$

A Function as a Set of Ordered Pairs

We must often deal with pairs of numbers rather than with single values. An *ordered pair* of numbers is a pair of numbers in which the order of the numbers is important. An ordered pair of numbers is written with the numbers separated by a comma and enclosed within parentheses. Since order counts, the ordered pair (2, 3) is different from the ordered pair (3, 2). A set of ordered pairs is a function if for each first number there is only *one* second number.

Example 9 ──────────

The set of ordered pairs

$$(-2, 4), \quad (-1, 1), \quad (0, -2), \quad (1, -5), \quad (2, -8)$$

is a function, because each first number has exactly one second number associated with it.

This function could also be described by a table of values:

First	−2	−1	0	1	2
Second	4	1	−2	−5	−8

Example 10 ──────────

The set of ordered pairs

$$(2, -4), \quad (0, 0), \quad (1, 1), \quad (2, 4)$$

is not a function, because there are two second numbers, −4 and 4, for the first number 2.

Example 11

In this table, the quantity r represents the radius (in cm) of a circle and C represents the circumference (in cm) of the circle.

r	2.3	3.5	4.0	5.6	8.1	9.9
C	14.5	22.0	25.1	35.2	50.9	62.2

This table of values defines a function, since for each value of the radius, there is only one value of the circumference. We say that the circumference C is a function of the radius r.

This function may also be expressed as the set of ordered pairs (r, C): $(2.3, 14.5)$, $(3.5, 22.0)$, $(4.0, 25.1)$, $(5.6, 35.2)$, $(8.1, 50.9)$, and $(9.9, 62.2)$.

SECTION 12-1 Exercises Functions

Definition of a Function

Find the value y of the given function for the specified value of x.

1. $y = 3x - 2$, $x = 1$ **2.** $y = 2x^2 + x - 1$, $x = 3$

3. $y = x^3$, $x = -4$ **4.** $y = 2^x$, $x = 3$

State whether the equation is in implicit or explicit form. If explicit, name the dependent and the independent variables.

5. $y = 4x - 1$ **6.** $x = xy + 2$

7. $x^2 + y^2 = 25$ **8.** $s = 16.1t^2$

Functions in Verbal Form

Express in equation form.

9. y is equal to 5 times the square of x.

10. y is the amount by which 3 exceeds x.

11. y is the sum of x and its reciprocal.

12. y is twice the square root of x.

13. The surface area A of a sphere is 4π times the square of the radius r.

14. The area A of a circle is π times the square of the radius r.

15. The velocity v is the product of the acceleration a and time t.

16. The power P dissipated in a resistor is the quotient of the square of the voltage V divided by the resistance R.

A Function in the Form of a Formula

Refer to Appendix A to find the formula, if necessary.

17. A car is traveling on the highway at 55 mi/h. Express the distance d traveled as a function of the time t traveled.

18. If work is the product of force times distance, express the amount of work W done in lifting a 50-kg block as a function of the distance d lifted. A force of 50 kg is necessary to lift a 50-kg block.

19. The first minute of a telephone call costs $.75 and each additional minute is $.53. Express the cost c as a function of time t.

20. An empty spool of cable weighs 146 lb. Each yard of cable weighs 28 lb. Find the weight w of the filled spool as a function of the length x of cable on it.

21. Express the area A of a triangle with a base of 10 cm as a function of its height h.

22. Express the base b of a rectangle as a function of the height h, if the area A is 4.7 cm².

23. Express the area A of a square as a function of its perimeter p.

24. Express the circumference C of a circle as a function of its diameter D.

25. Express the area A of a circle as a function of its circumference C.

26. Express the volume V of a cube as a function of its side s.

A Function as a Set of Ordered Pairs

27. Express this function as a set of ordered pairs:

x	-2	-1	0	1	2
y	-5	-2	1	4	7

28. Make a table of values for this function: $(-1, 1)$, $(0, 0)$, $(1, 1)$, $(2, 4)$, $(3, 9)$.

12-2   FUNCTIONAL NOTATION

In this section we learn to use functional notation, determine the domain and range of a function, and find the value of a function by substitution.

Functional Notation

Sometimes we use a special notation with functions, a notation which specifies the rule, or operations on a variable. If y is a function of x, we write

$$y = f(x)$$

and we read this as *y equals f of x*.

Just as we use the symbol x to represent a number without saying which number we are specifying, we use the notation $y = f(x)$ to represent a function f without having to specify which particular function we are referring to. The letter f is used as the *name* of the function, to identify the rule or operations to be performed on the independent variable x.

COMMON ERROR The equation $y = f(x)$ does *not* mean "y equals f times x."

$$y \neq f \times x$$

Example 12

If $y = 2x - 3$, then y is a function of x. We write $y = f(x)$, where $f(x) = 2x - 3$. In other words,

$$y = 2x - 3 \quad \text{and} \quad f(x) = 2x - 3$$

are two *different* ways of describing the *same* function.

■ **FIGURE 12-4**
Volume and side of a cube.

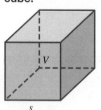

Example 13

The volume V of a cube is a function of the length s of a side (Fig. 12-4), such that

$$V = s^3$$

In functional notation, we write

$$V = f(s) \quad \text{or} \quad f(s) = s^3$$

Note that the independent variable is s and the dependent variable is V.

Example 14

The distance traveled by a freely falling body is a function of the elapsed time. We can express this relationship in functional notation by the equation

$$s = f(t)$$

where s represents the distance fallen, and t represents the elapsed time. From physics we learn that

$$s = v_0 t + \frac{1}{2} at^2$$

where v_0 is the initial velocity and a is the acceleration due to gravity. Note that the independent variable is t, v_0 and a are constants, and s is the dependent variable.

The letter f is usually used to represent a function, but other letters such as g and h may be used. Subscripts, as in f_1 and f_2, are also used to distinguish one function from another. For example, we could say $y = g(x)$, $y = h(x)$, $y = f_1(x)$, or $y = f_2(x)$ to describe different functions.

Domain and Range

The *domain* of the function $y = f(x)$ is the set of all x values, and the *range* of the function is the set of all y values of the function. If the domain of a function is not given, then it is assumed to be the set of *all possible real values* of x that yield *real values* of y.

Example 15

Find the domain and range of the function

$$y = x^2$$

Solution: We can square any real number so x can be any real number. Thus the domain is all real numbers. Any real number squared will be nonnegative, so the range is all nonnegative real numbers.

> Domain: all real numbers x
> Range: $y \geq 0$

Example 16

Find the domain and range of the function

$$y = \sqrt{x - 2}$$

Solution: Since we cannot take the square root of a negative number, then x must be greater than or equal to 2:

$$x \geq 2$$

Since the square root of any number is nonnegative, then

$$y \geq 0$$

So, we get

> Domain: $x \geq 2$
> Range: $y \geq 0$

In terms of dependent and independent variables, the domain is the set of all possible values of the independent variable, and the range is the set of all possible values of the dependent variable.

Example 17

A body, falling freely from rest, falls a distance s feet in t seconds according to the formula

$$s = 16t^2$$

Find the domain and range of this function.

Solution: Since $s = 16t^2$, t is the independent variable and s is the dependent variable. Since t represents the elapsed time, t cannot be negative, so

$$t \geq 0$$

The distance s is the product of 16 times a nonnegative number, t^2, so s is also nonnegative,

$$s \geq 0$$

So, we have

Domain: $t \geq 0$
Range: $s \geq 0$

Substituting into Functions

Functional notation is often used to designate the value of the function for a specific value of the independent variable. The value of the function $f(x)$ when $x = a$ is denoted as $f(a)$. Thus, $f(3)$ is the value of the function $f(x)$ when $x = 3$.

Example 18

Given $f(x) = x^2 - 5$, find $f(3)$ and $f(-2)$.

Solution: The notation $f(3)$ represents the value of the function $f(x)$ when $x = 3$. Therefore, we substitute 3 for x in the equation of the function.

$$f(x) = x^2 - 5$$
$$\updownarrow \quad \updownarrow$$
$$f(3) = 3^2 - 5$$
$$= 9 - 5$$
$$= 4$$

Similarly,

$$f(-2) = (-2)^2 - 5$$
$$= 4 - 5$$
$$= -1$$

We may substitute a literal value, such as a, into the equation of the function in the same way.

Example 19

Given $g(x) = 2x - 3$, find $g(a)$ and $g(-5b)$.

Solution: Substituting a for x in the equation of the function, we get

$$g(a) = 2(a) - 3$$
$$= 2a - 3$$

Substituting $-5b$ for x, we get

$$g(-5b) = 2(-5b) - 3$$
$$= -10b - 3$$

Example 20

If $g(x) = 2x - 3$, find $3g(-4)$.

Solution: The notation $3g(-4)$ means 3 times $g(-4)$. First we find $g(-4)$.

$$g(-4) = 2(-4) - 3$$
$$= -8 - 3$$
$$= -11$$

Now we multiply by 3:

$$3g(-4) = 3(-11)$$
$$= -33$$

SECTION 12-2   Exercises   Functional Notation

1. Rewrite "y is a function of x such that $y = 2x - 1$" using functional notation.
2. Rewrite "v is a function of t, such that $v = 24 + 32t$" using functional notation.
3. The Fahrenheit temperature F is 32 more than $\frac{9}{5}$ of the Celsius temperature C. Express this in functional notation.
4. Work W is defined as force F times distance d. Write work as a function of distance.

Domain and Range

Find the domain and range of each function.

5. $y = 3x - 2$ 6. $f(x) = x^2 + 1$ 7. $g(x) = \dfrac{1}{x}$ 8. $y = \sqrt{x - 1}$

9. The work W done by a constant force F acting on an object and moving it through a distance d is given by

$$W = Fd$$

10. If a body has a constant acceleration of 9.8 m/s² and initial velocity of 1.4 m/s then its velocity v after time t is given by

$$v = 1.4 + 9.8t$$

Substituting into Functions

11. Given $f(x) = 3x - 2$, find $f(1)$, $f(-2)$, $f(a)$, and $2f(3)$.
12. Given $g(x) = 5x + 2$, find $g(2)$, $g(-3)$, $g(b)$, and $3g(-1)$.

13. Given $h(x) = 3x^2 + 1$, find $h\left(\dfrac{1}{2}\right)$, $h(0)$, $h(c)$, and $h(1) - h(2)$.

14. Given $f_1(x) = 4 - x^2$, find $f_1(0.2)$, $f_1(0)$, $f_1(p)$, and $\dfrac{f_1(1)}{f_1(3)}$.

Round to three significant digits.

15. The velocity of a uniformly accelerating body at time t is given by

$$f(t) = v_0 + at,$$

where a is the acceleration and v_0 is the initial velocity. If v_0 is 8.00 ft/s and $a = 32.2$ ft/s^2, find $f(5.00)$, $f(10.0)$, and $f(20.0)$.

16. The power (in watts) dissipated in a resistor is given by

$$g(R) = \frac{V^2}{R}$$

where V is the voltage in volts and R is the resistance in ohms. If the voltage is 115 volts, find $g(7.00)$, $g(72.0)$, and $g(4750)$.

12-3   RECTANGULAR COORDINATES

In this section we define the rectangular coordinate system, and learn to graph points in rectangular coordinates. Graphing is a skill that is indispensable in mathematics. We often collect data, and it is essential for us to be able to graph that data and interpret the graph. Most physical phenomena can be represented algebraically as functions (equations), and the graph is a "picture" of that function. Graphs can be used to show the relationship between two quantities, such as time and velocity, diameter and area, and voltage and current. Lastly, we must be able to read information from a graph.

The Rectangular Coordinate System

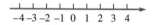

■ **FIGURE 12-5**
Real number line.

In Chapter 1 we saw that we can graphically represent every real number as a point on a line. We called that line the *real number line* (Fig. 12-5).

But how can we graph an *ordered pair* of numbers? To graph an ordered pair, we need two number lines, placed at right angles to each other, intersecting each other at the zero mark. We call this a *rectangular coordinate system.*

The horizontal number line is called the *x axis* and the vertical number line is called the *y axis.* They intersect at a point called the *origin.* These two axes divide the plane into four regions called *quadrants.* The quadrants are numbered counterclockwise, using Roman numerals, as shown in Fig. 12-6.

■ **FIGURE 12-6**
Rectangular coordinate system.

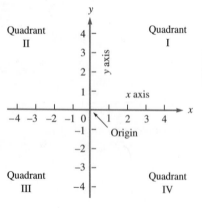

— Example 21 —

In Fig. 12-7, name the quadrant in which each point lies.

■ **FIGURE 12-7**

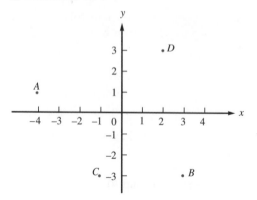

Solution: A: quadrant II; B: quadrant IV; C: quadrant III; D: quadrant I.

Graph of a Point

Each ordered pair of numbers can be graphed as a point. Fig. 12-8 shows a point *P* in quadrant I. Its horizontal distance from the origin, called the *x coordinate* of the point, is 2 units; thus, $x = 2$. Its vertical distance from the origin, called the *y coordinate* of the point, is 3 units; thus, $y = 3$. The numbers in the ordered pair (2, 3) are called the *rectangular coordinates,* or simply *coordinates,* of the point *P*. They are always written in the same order, with the *x* coordinate first, and the *y* coordinate second. A letter identifying the point is sometimes written before the coordinates, as in *P*(2, 3).

To plot any ordered pair (*h*, *k*), simply place a point at a distance *h* units from the *y* axis and *k* units from the *x* axis. Points with positive *x* coordinates are located to the *right* of the *y* axis and points with negative *x* coordinates are located to the *left* of the *y* axis. Points with positive *y* coordinates are located *above* the *x* axis and points with negative *y* coordinates are located *below* the *x* axis.

Example 22

Graph the points

  (a) *P*(3, 4)
  (b) *Q*(−3, 2)
  (c) *R*(4, −3)
  (d) *S*(−2, −4)

Solution: We draw the coordinate axes, label each axis, and label the scale on each axis.

  (a) The point *P*(3, 4) is located 3 units to the right of the *y* axis and 4 units above the *x* axis. So, we start at the origin, move 3 units to the right (on the *x* axis), and then move 4 units straight up. We label the point *P*(3, 4), as in Fig. 12-9.

  (b) The point *Q*(−3, 2) is located 3 units to the left of the *y* axis and 2 units above the *x* axis.

  (c) The point *R*(4, −3) is located 4 units to the right of the *y* axis and 3 units below the *x* axis.

  (d) The point *S*(−2, −4) is located 2 units to the left of the *y* axis and 4 units below the *x* axis.

■ **FIGURE 12-8**
Graph of a point.

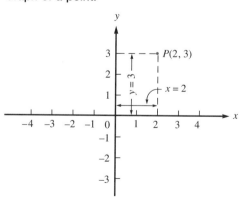

■ **FIGURE 12-9**

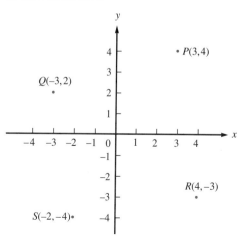

Graphing a Set of Ordered Pairs

To graph a *set* of ordered pairs, we simply plot each ordered pair. If the ordered pairs represent a function, we usually connect the points with a smooth curve, unless we have reason to believe that there are sharp corners, breaks, or gaps in the graph.

--- **Example 23** ---

Graph this set of ordered pairs and connect them in order with a smooth curve that seems to fit the curve best: $P_1(-4, 4)$, $P_2(-3, 0)$, $P_3(-2, -3)$, $P_4(-1, -4)$, $P_5(0, -3)$, $P_6(1, 0)$, $P_7(2, 4)$.

Solution: The points are graphed in Fig. 12-10 and then connected with a smooth curve.

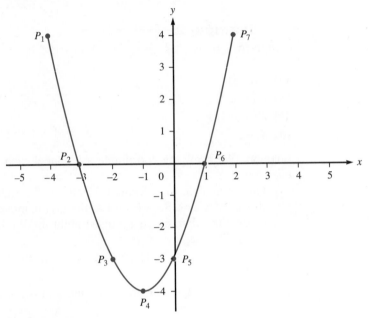

■ **FIGURE 12-10**
Graph of a set of ordered pairs.

Graphing Data

Ordered pairs of numbers may be given in a chart or a table. This information, called *data,* may be the result of an experiment or calculations. The ordered pairs are graphed as points, and the points connected by a smooth curve.

--- **Example 24** ---

Graph the following data, and connect the points with a smooth curve. The data shows the relationship between the diameter D and the circumference C of several circles.

D (in cm)	4.6	7.0	8.0	11.2	16.2	19.8
C (in cm)	14.5	22.0	25.1	35.2	50.9	62.2

Solution: We choose to graph the diameter D on the horizontal axis and the circumference C on the vertical axis. Thus, C is the dependent variable and D

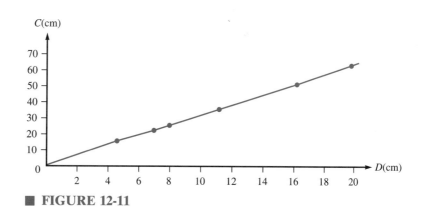

■ FIGURE 12-11

is the independent variable; C is being graphed as a function of D. We draw and label the axes as the D (horizontal) and C (vertical) axes. Since D ranges from 0 to 20, a convenient scale on the D axis is 2. Since C ranges from 0 to 70, a convenient scale on the C axis is 10. Each pair of numbers represents a point (D, C). The points (4.6, 14.5), (7.0, 22.0), (8.0, 25.1), (11.2, 35.2), (16.2, 50.9), and (19.8, 62.2) are plotted and connected, as shown in Fig. 12-11. In this case the graph is a straight line.

SECTION 12-3 Exercises Rectangular Coordinates

Graph of a Point

Write the coordinates of each point in Fig. 12-12.

1. A **2.** B **3.** C **4.** D

5. E **6.** F **7.** G **8.** H

Graph the following points.

9. $(3, 4)$ **10.** $(5, 3)$ **11.** $(-1, 2)$ **12.** $(-2, 4)$

13. $(4, -3)$ **14.** $(5, -2)$ **15.** $(-2, -3)$ **16.** $(-3, -4)$

17. $(-3, 0.5)$ **18.** $(1.6, -3)$ **19.** $(0, -2)$ **20.** $(-3, 0)$

21. $(-4, 0)$ **22.** $(0, -4)$ **23.** $\left(-\frac{1}{2}, 3\frac{3}{4}\right)$ **24.** $\left(-2\frac{1}{2}, -\frac{3}{4}\right)$

■ FIGURE 12-12

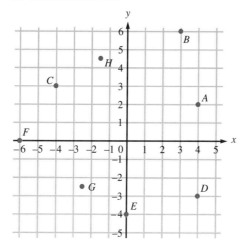

25. $(-3.7, -2.3)$ **26.** $(1.4, -3.7)$

27. Which quadrant contains points having a positive x coordinate and a negative y coordinate?

28. In which quadrant does a point (h, k) lie if $h < 0$ and $k < 0$?

29. In which quadrants is the x coordinate negative?

30. In which quadrants is the y coordinate positive?

31. What is the x coordinate of any point on the y axis?

32. What is the y coordinate of any point on the x axis?

33. In which quadrants is the y coordinate divided by the x coordinate negative?

34. In which quadrants is the x coordinate divided by the y coordinate positive?

Graphing a Set of Ordered Pairs

Graph each set of points. Connect them in order with line segments and identify the geometric figure formed.

35. $P_1(-3, 1)$, $P_2(2, 2)$, $P_3(5, -2)$, $P_4(0, -3)$

36. $P_1(-3.6, 4.4)$, $P_2(0, 6.2)$, $P_3(3.1, 0)$, $P_4(-0.5, -1.8)$

Graph each set of points. Connect them in order with a smooth curve that seems to fit the data best.

37. $P_1(-4, -9)$, $P_2(-3, -2)$, $P_3(-2, -1)$, $P_4(-1, 0)$, $P_5(0, 7)$

38. $P_1(-1, -7)$, $P_2(0, -2)$, $P_3(1, 1)$, $P_4(2, 2)$, $P_5(3, 1)$, $P_6(4, -2)$, $P_7(5, -7)$

Graphing Data

Graph the following data. Label the graph and connect the points with a smooth curve.

39. C represents the temperature in degrees Celsius, and F represents the temperature in degrees Fahrenheit.

C	-20.5	0	20.5	49.6	82.7	100
F	-4.9	32	68.9	121.3	180.9	212

40. Let t represent the time (in years) a sum of money is left in a savings account, and let I represent the interest (in dollars) earned during that time.

t	0.5	1	2.5	3.5	5	7.5	10
I	27.50	55.00	137.50	192.50	275.00	412.50	550.00

12-4   LINEAR FUNCTIONS

In this section we study linear functions. We learn to graph a linear function, find its intercepts and slope, and solve some applications involving linear functions.

Linear Function

A *linear function* is a function defined by an equation having the form

$$y = mx + b$$

where m and b are constants.

┌─── **Example 25** ───────
The equation $y = 3x - 2$ defines a linear function in which $m = 3$ and $b = -2$.

Graph of a Function

The graph of a function defined by an equation consists of all the points whose coordinates *satisfy* the equation—that is, the equation is true when the coordinate values of the point are substituted in.

Example 26

Show that the point (2, 4) satisfies the equation

$$y = 3x - 2$$

Solution: If $x = 2$ and $y = 4$, the equation becomes

$$4 = 3(2) - 2$$

or

$$4 = 4$$

Since this is a true equation, the coordinates of the point (2, 4) satisfy the equation. Therefore the point (2, 4) is on the graph of the equation $y = 3x - 2$, as we will see in Example 27.

Table of Values

The graph of the function

$$y = f(x)$$

consists of *all* points whose coordinates (x, y) satisfy the equation. Since the value of y depends on the value of x, then y is the dependent variable and x is the independent variable. Some points can be determined by choosing specific values for x and calculating y for each value of x.

To organize our work we may make a table of values that consists of the Oordinates of several points on the graph of the function. To graph the function, we plot those points and connect them with a curve.

Example 27

Make a table of values for the linear function

$$y = 3x - 2$$

for x from -2 to 2 and graph the function.

Solution: We choose integral values for x and calculate y by substituting these x values into the equation. The table of values is shown on the right.

$$y = 3x - 2$$

		x	y
if $x = -2$	then $y = 3(-2) - 2 = -8$	-2	-8
if $x = -1$	then $y = 3(-1) - 2 = -5$	-1	-5
if $x = 0$	then $y = 3(0) - 2 = -2$	0	-2
if $x = 1$	then $y = 3(1) - 2 = 1$	1	1
if $x = 2$	then $y = 3(2) - 2 = 4$	2	4

Note: A horizontal arrangement of this table looks like this:

x	-2	-1	0	1	2
y	-8	-5	-2	1	4

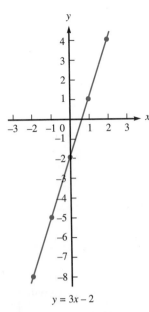

$y = 3x - 2$

■ FIGURE 12-13

Each ordered pair represents one point of the graph. The horizontal axis is the axis of the independent variable x, and the vertical axis is the axis of the dependent variable y. We plot these points and connect them with a smooth curve (Fig. 12-13). As we see, the graph of the equation $y = 3x - 2$ is a straight line.

In fact, *the graph of every linear function is a straight line.* If we know any *two* points on a line, then we can draw that line. So our table of values can be *very small* for a linear function. We can "get by" with plotting only two points, but it's wise to plot an extra point as a check.

Intercepts

The point where the graph crosses the x axis is called the *x intercept.* To find the x intercept, we note that *the y coordinate of any point on the x axis is 0.* So, to find the x intercept, we substitute 0 for y in the equation and solve for x.

The point where the graph crosses the y axis is called the *y intercept.* To find the y intercept, we note that *the x coordinate of any point on the y axis is 0.* So, to find the y intercept, we substitute 0 for x in the equation and solve for y.

■ FIGURE 12-14
Intercepts.

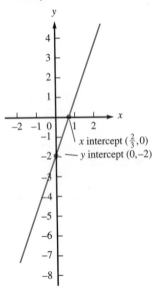

$y = 3x - 2$

Example 28

Find the intercepts of the function

$$y = 3x - 2$$

Solution: To find the x intercept, we substitute 0 for y in the equation to get

$$0 = 3x - 2$$

and solve for x:

$$x = \frac{2}{3}$$

So the x intercept is $\left(\frac{2}{3}, 0\right)$, as shown in Fig. 12-14. Sometimes we refer to the x coordinate as the x intercept; so we may say *the x intercept is $\frac{2}{3}$,* instead of saying *the x intercept is $\left(\frac{2}{3}, 0\right)$.*

To find the y intercept, we substitute 0 for x in the equation to get

$$y = 3(0) - 2$$

and solve for y:

$$y = -2$$

So the y intercept is $(0, -2)$, as shown in Fig. 12-14. Sometimes we refer to the y coordinate as the y intercept; so we may say *the y intercept is -2,* instead of saying *the y intercept is $(0, -2)$.*

Example 29

Find the y intercept of the general linear function

$$y = mx + b$$

Solution: To find the y intercept, we substitute 0 for x in the equation:

$$y = m(0) + b$$

or

$$y = b$$

So, *the y intercept of the linear function $y = mx + b$ is the constant b.*

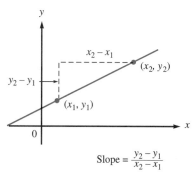

Slope = $\dfrac{y_2 - y_1}{x_2 - x_1}$

■ **FIGURE 12-15**
Slope of a line.

Slope

A very important concept in the study of linear functions is *slope*. Slope is a measure of the direction and steepness of a line. We can calculate the slope of a line if we know the coordinates of two points. If (x_1, y_1) and (x_2, y_2) are two points on a line (Fig. 12-15), the slope is defined as

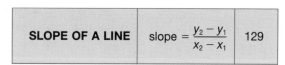

SLOPE OF A LINE	slope = $\dfrac{y_2 - y_1}{x_2 - x_1}$	129

Example 30

In Example 27, we found that the points $(1, 1)$ and $(2, 4)$ are on the graph of

$$y = 3x - 2$$

Use these points to calculate the slope of the line.

Solution: We use Eq. 129 to find the slope of the line, with $(x_1, y_1) = (1, 1)$ and $(x_2, y_2) = (2, 4)$ (Fig. 12-16). So, we let $x_1 = 1$, $y_1 = 1$, $x_2 = 2$, and $y_2 = 4$,

$$\text{Slope} = \frac{4 - 1}{2 - 1}$$

$$= \frac{3}{1}$$

$$= 3$$

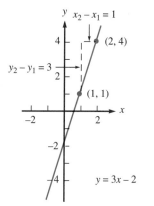

■ **FIGURE 12-16**

So the slope of the line $y = 3x - 2$ is 3, which is also the coefficient of x.

In fact, *the slope of the line $y = mx + b$ is always the coefficient m.*

Slope-Intercept Form

The form of a linear function

$$y = mx + b$$

$$\qquad \updownarrow \qquad \updownarrow$$

$$\qquad \text{slope} \quad y \text{ intercept}$$

is called the *slope-intercept form,* because m is the slope of the line and b is the y intercept.

SLOPE-INTERCEPT FORM	$y = mx + b$	134

12-4 Linear Functions   **211**

Example 31

Find the slope and the y intercept of the linear function

$$y = 2x + 3$$

Solution: The equation $y = 2x + 3$ is in slope-intercept form.

Since the slope m is the coefficient of the x term, then $m = 2$ and the slope is 2. The y intercept b is the constant term 3, so $b = 3$ and the y intercept is 3.

Example 32

Graph and find the equation of the line passing through the points $(-1, 3)$ and $(2, -3)$.

Solution: To graph the line, we plot the two points and draw the line through them (Fig. 12-17). To find the equation of the line, we need to know the slope and the y intercept of the line. We use Eq. 129 to find the slope of the line, with $(x_1, y_1) = (-1, 3)$ and $(x_2, y_2) = (2, -3)$,

$$m = \text{slope} = \frac{-3 - 3}{2 - (-1)}$$
$$= \frac{-6}{3}$$
$$= -2$$

So, substituting $m = -2$ into Eq. 134,

$$y = -2x + b$$

Since the point $(-1, 3)$ is on the line, its coordinates satisfy the equation. We let $x = -1$ and $y = 3$,

$$3 = -2(-1) + b$$

Solving for b,

$$b = 1$$

So

$$y = -2x + 1$$

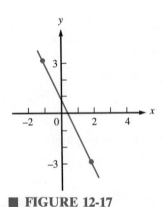

■ **FIGURE 12-17**

Horizontal and Vertical Lines

Now we consider some properties of horizontal and vertical lines.

Example 33

Find the slope and the y intercept of the linear function

$$y = 2$$

Graph the function.

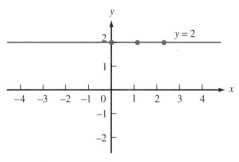

■ **FIGURE 12-18**
Graph of a horizontal line.

Solution: We write the equation in slope-intercept form:

$$y = 0x + 2$$

So the slope m is 0 and the y intercept is 2. We make a table of values to include at least three points, in which y is always 2:

x	0	1	2
y	2	2	2

We graph these points in Fig. 12-18 and get a horizontal line.

The slope of every horizontal line is 0. The equation of the horizontal line with y intercept b is

HORIZONTAL LINE	$y = b$	132

── **Example 34** ──
Find the equation of the vertical line that passes through (4, 3) and (4, −1).

Solution: We plot these two points and draw the line through them (Fig. 12-19).

■ **FIGURE 12-19**
Graph of a vertical line.

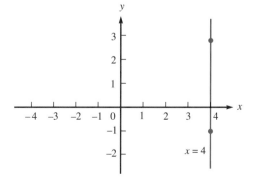

Using Eq. 129 to find the slope,

$$m = \frac{-1 - 3}{4 - 4}$$

$$= \frac{-4}{0}$$

Now $\frac{-4}{0}$ is undefined, so *the slope is undefined.* From the graph, we observe that there is no y intercept. This equation cannot be expressed in slope-intercept form. We note that the x coordinate of every point on the graph is 4, so the equation of this line is

$$x = 4$$

This equation, however, does *not* describe a *function*, since all points have the same x coordinate.

The slope of a vertical line is undefined. The equation of the vertical line with x intercept a is

VERTICAL LINE	$x = a$	133

Applications

Many applications involve formulas that are linear functions. A *formula* is an equation in which letters represent physical or geometric quantities. These letters may be either variables or constants. We graph a formula the same way we graph other functions.

Example 35

The velocity v of an object t seconds after being thrown down from the top of a building with an initial velocity v_0, is given by the formula

$$v = v_0 + gt$$

where g is the acceleration due to gravity, 9.8 m/s^2. If a ball is thrown with an initial downward velocity of 5.0 m/s, the equation becomes

$$v = 5.0 + 9.8t$$

Graph this equation for t from 0.0 to 10.0 seconds.

Solution: In this equation, t is the independent variable and v is the dependent variable. To make the table of values, we calculate the values for v by substituting the t values in the given equation and make a table of values

t	0.0	2.0	4.0	6.0	8.0	10.0
v	5.0	24.6	44.2	63.8	83.4	103.0

Since this is a linear function, we know the graph is a line. Since t goes from 0.0 to 10.0, we choose a scale of 2 for the t axis. Since v goes from 5.0 to 103.0, we choose a scale of 25 for the v axis. We use different scales for the two axes, since the ranges are so different. We plot the points and connect them with a straight line (Fig. 12-20).

■ **FIGURE 12-20**

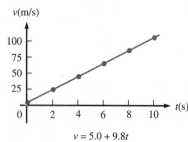

$v = 5.0 + 9.8t$

214   CHAPTER 12 / Functions and Graphs

For each equation in exercises 1-10,

(a) complete this table of values;

$$\begin{array}{c|ccccc} x & -2 & -1 & 0 & 1 & 2 \\ \hline y & & & & & \end{array}$$

(b) graph the equation;

(c) find the x and y intercepts; and

(d) find the slope.

1. $y = x$ **2.** $y = x + 2$ **3.** $y = 2x + 3$ **4.** $y = 3x + 1$

5. $y = 1 - 3x$ **6.** $y = 3 - 2x$ **7.** $y = \dfrac{x}{2} - 1$ **8.** $y = 2 - \dfrac{x}{3}$

9. $y = 3$ **10.** $y = -1$

For each equation in exercises 11 and 12,

(a) graph the equation;

(b) find the x and y intercepts; and

(c) find the slope.

11. $x = -1$ **12.** $x = 2$

Graph and find the equation of the linear function passing through the points.

13. (1, 2) and (3, 6)   **14.** (−1, 4) and (3, −4)

15. (−1, −1) and (4, −2)   **16.** (−2, −3) and (2, −1)

17. The circumference C of a circle is given by

$$C = 2\pi r$$

Applications

where r is the radius of the circle. Graph this function for values of r from 0 to 10 cm.

18. If an amount of money p (the principal) is invested at an annual interest rate r for t years, the simple interest earned is given by the formula

$$I = prt$$

Write the equation for the interest if $5000 is invested at the rate of 6.5%. Graph the function for t from 0 to 5 years.

19. The voltage V in a circuit with constant resistance R is a linear function of the current I, given by

$$V = IR$$

In a certain circuit the resistance is 4.5 ohms. Write the equation for the voltage as a function of the current. Graph the function for I from −2.5 to 2.5 amps.

20. The work W done by a constant force F acting on an object is a linear function of the distance d moved by the object, given by

$$W = Fd$$

A force of 100 lb is applied to move a block. Write the equation for the work W done on the block as a function of the distance d. Graph the function for d from 0 to 12 ft.

21. The weight W of an object of density D is a linear function of its volume V, given by

$$W = DV$$

The density of iron is 7.87 g/cm³. Write the equation for the weight W of an iron object as a function of its volume V. Graph the function for V from 0 to 400 cm³.

22. The velocity v of sound (in m/s) is a linear function of the temperature T (in °C), given by

$$v = 0.607T + 332$$

Graph the velocity v of sound as a function of the temperature T for T from −30°C to 30°C.

23. The total cost C of producing x units is a linear function:

$$C = mx + C_0$$

where m is the cost per unit and C_0 is the fixed (or initial) cost. A computer company determines that it costs $100 to produce one keyboard and that the fixed costs were $250,000. Express the cost C as a function of x and graph the function for x from 0 to 1000 keyboards.

24. The total cost C of producing x units is a linear function:

$$C = mx + C_0$$

where m is the cost per unit and C_0 is the fixed (or initial) cost. A new company invests $750,000 for equipment to produce laser disks. The unit cost per laser disk is $35. Express the cost C as a function of x and graph the function for x from 0 to 5000 disks.

25. A moving company buys a 35-ft van, which has a useful lifetime of 6 years. After x months of use, the value y of the van is estimated by the equation

$$y = 37000 - 500x$$

Graph y as a function of x for x from 0 to 6 years.

12-5   QUADRATIC AND OTHER FUNCTIONS

In this section, we study quadratic functions. We learn to graph a quadratic function, find its vertex, and solve some applications of quadratic functions. We then graph some other functions that are neither linear nor quadratic.

Quadratic Function

A quadratic function has the form

$$y = ax^2 + bx + c$$

where a, b, and c are constants, and $a \neq 0$.

―――― **Example 36** ――――
The equation

$$y = x^2 - 2x - 3$$

defines a quadratic function in which $a = 1$, $b = -2$, and $c = -3$.

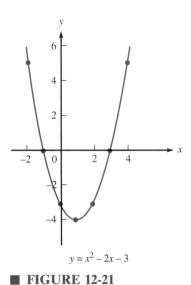

$$y = x^2 - 2x - 3$$

■ FIGURE 12-21

Graph of a Quadratic Function

The graph of a quadratic function has some very special properties which we now consider.

Example 37

Graph the quadratic function

$$y = x^2 - 2x - 3$$

for integral values of x from -2 to 4.

Solution: By substitution, we obtain the table of values,

x	-2	-1	0	1	2	3	4
y	5	0	-3	-4	-3	0	5

These ordered pairs are plotted and connected with a smooth curve in Fig. 12-21. Note that there are *two* x intercepts, $(-1, 0)$ and $(3, 0)$. The y intercept is $(0, -3)$.

> **COMMON ERROR** Be especially careful when substituting negative values into an equation. It is easy to make an error.

Parabola

The graph of a quadratic function is a curve called a *parabola*. Every parabola has a point that is either a high point on the curve or a low point. This point is called the *vertex*. Low points on a curve are called *minimum points*. High points on a curve are called *maximum points*. The vertex is either a minimum or a maximum point, depending on whether the curve opens up or down.

The graph of a quadratic function generally includes the vertex. Without proof we give this equation for the x coordinate of the vertex:

x COORDINATE OF THE VERTEX	$x = \dfrac{-b}{2a}$	136

Once we know the x coordinate of the vertex, we can calculate the y coordinate from the equation.

Example 38

Find the coordinates of the vertex of

$$y = x^2 - 2x - 3$$

Solution: Using Eq. 136 for the x coordinate, with $a = 1$ and $b = -2$,

$$x = \frac{-(-2)}{2(1)}$$
$$= 1$$

Substituting 1 for x in the equation,

$$y = 1^2 - 2(1) - 3 = -4$$

So the vertex of the parabola is $(1, -4)$. This is verified in Fig. 12-21, in which $(1, -4)$ is the lowest point of the graph.

── **Example 39** ──

Find the vertex and graph the function

$$y = -x^2 - 4x - 2$$

Solution: Using Eq. 136 for the x coordinate of the vertex, with $a = -1$ and $b = -4$,

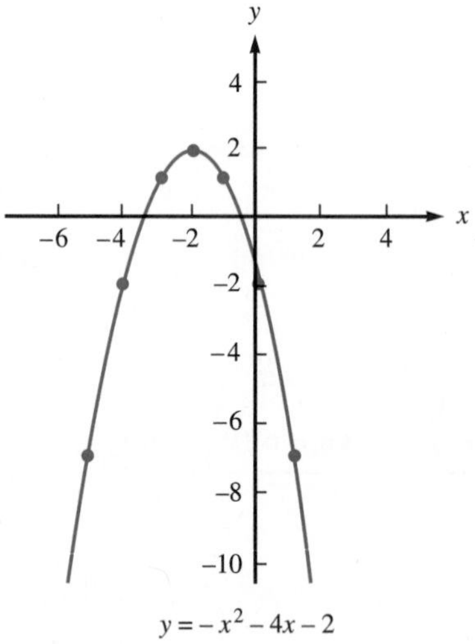

$$x = \frac{-(-4)}{2(-1)} = -2$$

We make a table of values, choosing several x values to the left of -2 and several to right of -2.

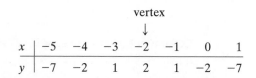

			vertex ↓			
x	-5	-4	-3	-2	-1	0	1
y	-7	-2	1	2	1	-2	-7

These points are plotted and connected with a smooth curve in Fig. 12-22.

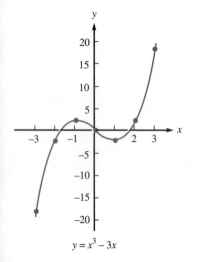

$$y = -x^2 - 4x - 2$$

■ **FIGURE 12-22**

Graphs of Other Functions

So far, we have graphed only functions that are linear or quadratic. There are also many other important types of functions. Some of these, such as trigonometric and exponential functions, will be covered in later sections. Here we consider a few other examples of functions.

■ **FIGURE 12-23**

── **Example 40** ──

Graph the function

$$y = x^3 - 3x$$

for x from -3 to 3.

Solution: By substitution, we obtain the table.

x	-3	-2	-1	0	1	2	3
y	-18	-2	2	0	-2	2	18

We plot these points, and connect them with a smooth curve to get the graph (Fig. 12-23).

$$y = x^3 - 3x$$

── **Example 41** ──

Graph the function

$$y = \frac{1}{x}$$

for x from -3 to 3.

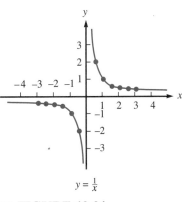

$y = \frac{1}{x}$

FIGURE 12-24

Solution: By substitution, we obtain the table, taking x increments of $\frac{1}{2}$.

x	-3	$\frac{-5}{2}$	-2	$\frac{-3}{2}$	-1	$\frac{-1}{2}$	0	$\frac{1}{2}$	1	$\frac{3}{2}$	2	$\frac{5}{2}$	3
y	$\frac{-1}{3}$	$\frac{-2}{5}$	$\frac{-1}{2}$	$\frac{-2}{3}$	-1	-2	undef.	2	1	$\frac{2}{3}$	$\frac{1}{2}$	$\frac{2}{5}$	$\frac{1}{3}$

We graph this function in Fig. 12-24. Note that something very special is happening near $x = 0$. First, x cannot equal zero, since $\frac{1}{0}$ is not defined; thus 0 is not in the domain of the function. But, for positive values of x, as x gets smaller (closer to 0), y gets larger. The graph approaches the y axis but never touches it! And, as negative values of x approach 0, y gets smaller (more negative). Again, the graph approaches the y axis but never touches it! A line which the graph approaches but never touches is called an *asymptote* of the graph. Here, the y axis is an asymptote. Similarly, we observe that the x axis is another asymptote of this graph—on the left and the right, the graph approaches the x axis, but never touches it.

Example 42

Graph the function

$$y = \sqrt{x}$$

for x from 0 to 9.

Solution: By substitution, we make the table of values, choosing perfect square values of x.

x	0	1	4	9
y	0	1	2	3

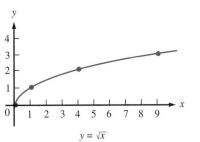

$y = \sqrt{x}$

FIGURE 12-25

We plot these points and connect them in Fig. 12-25. For the function $y = \sqrt{x}$, the domain and range are

Domain: $x \geq 0$
Range: $y \geq 0$

So the graph of the function lies entirely in the first quadrant.

SECTION 12-5 Exercises Quadratic and Other Functions

The Graph of a Quadratic Function

Fill in the missing values and graph the function.

1. $y = x^2 - 3$

x	-3	-2	-1	0	1	2	3
y							

2. $y = 6x - x^2 - 5$

x	0	1	2	3	4	5	6
y							

Find the vertex and graph the function.

3. $y = -x^2$
4. $y = 1 - x^2$
5. $y = x^2 + 2x + 3$
6. $y = -x^2 + 4x - 1$
7. $y = \frac{x^2}{4}$
8. $y = 5x - x^2$

Fill in the missing values and graph the function.

9. $y = x^3$

x	-2	-1	$-\frac{1}{2}$	0	$\frac{1}{2}$	1	2
y							

10. $y = \dfrac{x^3}{3}$

x	-2	-1	$-\frac{1}{2}$	0	$\frac{1}{2}$	1	2
y							

For each function, make a table of ordered pairs, using integral values of x from -3 to 3, rounding y values to tenths. Graph the function.

11. $y = \dfrac{-x^3}{3}$ **12.** $y = 2x^3$

13. $y = x^3 + 3x^2$ **14.** $y = 3x - x^3$

15. $y = -\dfrac{1}{x}$ **16.** $y = \dfrac{1}{x-1}$

17. $y = \dfrac{1}{x+1}$ **18.** $y = 1 + \dfrac{1}{x}$

For each function, make a table of ordered pairs, and graph the function.

19. $y = -\sqrt{x}$ **20.** $y = \sqrt{x+1}$

21. $y = \sqrt{1-x}$ **22.** $y = 1 + \sqrt{x}$

23. It has been shown that blood near the edge of an artery flows more slowly than blood near the center of the artery. The velocity v (in cm/s) of the blood x cm from the center of an artery is given by

$$v = 1000(0.04 - x^2)$$

Graph v as a function of x, for x from 0.0 to 0.2 cm.

24. The revenue R received by a company for selling a CD player at a price of p dollars is given by the revenue equation:

$$R = 8000p - 40p^2$$

Graph the revenue R as a function of the price p, for p from 0 to 200 dollars.

25. The farm population in the United States as a percentage of the total population can be expressed as a function:

$$y = \frac{200}{2t - 3895}$$

where y represents the percentage and t represents the year. Graph y as a function of t from the year 1950 to 2000.

12-6   READING GRAPHS

In this section we learn how to interpret a graph so that we can read information from it.

Example 43

An object is thrown down from the top of a cliff with an initial velocity of 6.5 m/s. Its velocity v, after t seconds, is given by the equation

$$v = 6.5 + 9.8t$$

(a) Graph the function for t from 0.0 to 10.0 seconds.

(b) From the graph, estimate the velocity of the object after $t = 7.0$ seconds.

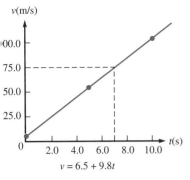

$$v = 6.5 + 9.8t$$

FIGURE 12-26

Solution: First we make a table of values for the function

$$v = 6.5 + 9.8t$$

Using t values from 0.0 to 10.0, we calculate the v values from the equation. Since this is a linear function, we need to plot only three points.

t (seconds)	0.0	5.0	10.0
v (m/s)	6.5	55.5	104.5

We plot the three points and draw the line (Fig. 12-26). To estimate the velocity from the graph we need to draw some extra lines. First, locate the point on the horizontal (time) axis where $t = 7.0$. Draw a vertical line up to where it meets the graph. Next, from the point where the vertical line intersects the graph, draw a horizontal line to the vertical v axis and read the value of the velocity. We estimate the velocity is about 75 m/s.

─── **Example 44** ───

If an amount of money a is invested at an annual interest rate r for t years, and the interest is compounded annually, the accumulated amount y is given by the formula

$$y = a(1 + r)^t$$

Suppose you have $100 which you invest at an annual interest rate of 5.5%, compounded annually.

(a) Write the formula for the accumulated amount.

(b) Graph the formula for $t = 0$ to 10 years.

(c) Estimate how many years it would take to accumulate to $150.

Solution:

(a) The interest rate r, expressed as a decimal, is

$$r = 5.5\% = 0.055$$

Substituting $a = 100$ and $r = 0.055$ in the given equation,

$$y = 100(1 + .055)^t$$

or

$$y = 100(1.055)^t$$

(b) We make the table of values, and substitute into the equation to find y, rounding to the nearest dollar.

FIGURE 12-27

$y(\$)$

$y = 100(1.055)^t$

t (years)	0	2	4	6	8	10
y (dollars)	100	111	124	138	153	171

We plot these points and connect them with a smooth curve (Fig. 12-27).

(c) To estimate how many years it would take to accumulate to $150, we draw a horizontal line at $y = 150$. Then we draw the vertical line from the point of intersection to the t axis and estimate that the time when the accumulated amount will be $150 is about 7.6 years.

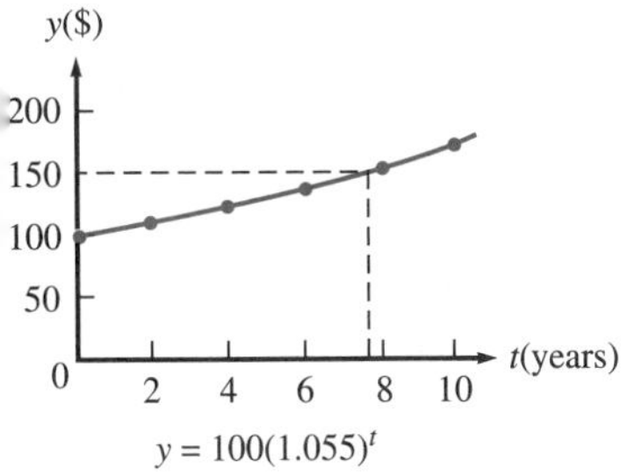

Example 45

The height s (in ft) of an object t seconds after it is thrown from the top of a building is given by

$$s = -16t^2 + 32t + 128$$

(a) Graph s as a function of t, for t from 0.0 to 4.0 seconds.

(b) Find the time at which the object hits the ground.

(c) From the graph, estimate the time at which it reaches its greatest height.

(d) Estimate the greatest height the object reaches.

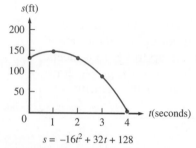

$s = -16t^2 + 32t + 128$

■ **FIGURE 12-28**

Solution:

(a) Since s is a quadratic function of t, the graph is a parabola. We make the table of values, for t from 0.0 to 4.0 seconds.

t (sec)	0.0	1.0	2.0	3.0	4.0
s (ft)	128	144	128	80	0

We plot these points (Fig. 12-28) and draw the parabola.

(b) The object hits the ground when $s = 0$. From the table, we see that $s = 0$ when $t = 4.0$. So, the object hits the ground after 4.0 seconds.

(c) The object reaches its greatest height at the vertex. From the graph, we can estimate that the vertex is (1, 144). We check it analytically, using Eq. 136 for the x coordinate of the vertex, with $a = -16$ and $b = 32$,

$$x = \frac{-32}{2(-16)} = 1$$

The object reaches its greatest height at $t = 1.0$ seconds.

(d) Its greatest height is the s coordinate of the vertex, 144 ft.

Example 46

The force F acting on a spring is a function of the distance x the spring is stretched. This data was taken in an experiment.

x (in.)	0.5	1.0	1.5	2.0	2.5	3.0
F (lb)	6.8	13.6	20.4	27.2	34.0	40.8

(a) Graph the force as a function of the displacement.

(b) From the graph, estimate the force required to stretch the spring 4.0 in.

(c) Estimate the amount of stretch when the spring is acted upon by a 25-lb force.

■ **FIGURE 12-29**

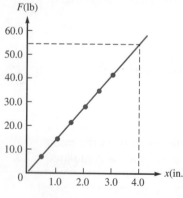

Solution:

(a) We plot the points in the data table. They are linear, so we draw the line through the points (Fig. 12-29).

(b) Although we don't have an equation for this function, we can estimate values from the graph. To estimate the force required to stretch the spring 4.0 in., we must extend the graph to the right to $x = 4.0$. We draw the vertical line at $x = 4.0$ up to the function line, draw the horizontal line over to the F axis and read the F value. We estimate it's about at $F = 54$. So the necessary force is about 54 lb.

(c) Similarly, to estimate the distance the spring is stretched when a 25-lb force is applied, we find $F = 25$ on the vertical F axis, draw the horizontal line across to the function line, draw the vertical line down to the horizontal x axis, and read the x value. We estimate it's about $x = 1.8$. So a 25-lb force would stretch this spring about 1.8 in.

Graphical Solution of Equations

We can use our knowledge of graphing functions to *solve equations* of the form $f(x) = 0$. The graph of the function $y = f(x)$ touches the x axis at an x intercept (Fig. 12-30), where the y coordinate is zero. Substituting $y = 0$ into the equation $y = f(x)$, we get $f(x) = 0$. So the x value of an x intercept is also a *solution* (or *root*) of the equation $f(x) = 0$.

Thus, to solve an equation graphically, we put the equation into the form

$$f(x) = 0$$

and graph the function

$$y = f(x)$$

The x intercepts are the solutions of the equation $f(x) = 0$. These solutions can be *approximated* from the graph.

The graphical solution of equations is a very powerful method of solving equations. It can be used to solve *any* equation, as long as the associated function can be graphed.

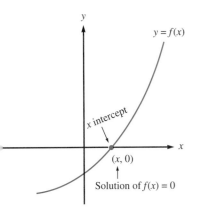

■ FIGURE 12-30
Graphical solution of the equation $f(x) = 0$.

Example 47

Solve the equation graphically. Approximate the solution(s) to the nearest tenth.

$$6x^2 - 4 = 5x$$

Solution: We put the equation into the form

$$f(x) = 0$$

by bringing all terms to one side of the equal sign:

$$6x^2 - 5x - 4 = 0$$

So,

$$f(x) = 6x^2 - 5x - 4$$

We make a table of values,

x	-2	-1	0	1	2
y	30	7	-4	-3	10

graph the function (Fig. 12-31), and read the x intercepts from the graph. We estimate (to the nearest tenth) that the intercepts are at $x = -0.5$ and $x = 1.3$. So the approximate solutions of the equation

$$6x^2 - 4 = 5x$$

are -0.5 and 1.3.

Check: We substitute into the original equation,

$$6(-0.5)^2 - 4 \stackrel{?}{=} 5(-0.5) \qquad 6(1.3)^2 - 4 \stackrel{?}{=} 5(1.3)$$

$$-2.5 = -2.5 \quad \text{check} \qquad 6.14 = 6.5 \quad \text{check}$$

We see from the check that the solution -0.5 is exact, and that the solution 1.3 is approximately correct (the check is accurate to 1 significant digit, if we round 6.5 back to 6).

■ FIGURE 12-31

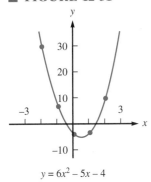

$y = 6x^2 - 5x - 4$

Estimate to two significant digits.

1. The displacement s for an object thrown upward with an initial velocity v_0 after t seconds is given by

$$s = v_0 t - \frac{1}{2} g t^2$$

where g is 32 ft/s². An object is thrown upward with an initial velocity of 160 ft/s.

   (a) Write the equation for displacement as a function of time.
   (b) Graph the function for t from 0 to 10 seconds.
   (c) From the graph, estimate the displacement after 7.5 seconds.

2. According to Ohm's law, the current I (in amps) is equal to the voltage V (in volts) divided by the resistance R (in ohms),

$$I = \frac{V}{R}$$

The voltage in a certain circuit is 100 volts.

   (a) Write the equation for the current in this circuit as a function of resistance.
   (b) Graph the function for R from 10 to 300 ohms.
   (c) From the graph, estimate the resistance when the current is 5 amps.

3. The surface area A of a cylinder with radius r and height h is given by

$$A = 2\pi r h + 2\pi r^2$$

A cylinder has a radius of 6.0 cm.

   (a) Write an equation for the surface area as a function of the height.
   (b) Graph the function for h from 0.0 to 8.0 cm.
   (c) From the graph, estimate the height if the surface area is 500 cm².
   (d) Estimate the surface area if the height is 4.0 cm.

4. The surface area A of a cylinder with radius r and height h is given by

$$A = 2\pi r h + 2\pi r^2$$

The height of the cylinder is 3.0 m.

   (a) Write the equation for the surface area as a function of the radius.
   (b) Graph the function for r from 0.0 to 5.0 m.
   (c) From the graph, estimate the radius when the surface area is 75 m².

5. The velocity v of a rocket after t seconds is given by

$$v = v_0 + at$$

where v_0 is the initial velocity and a is the acceleration. A rocket has an initial velocity of 5.0 m/s and an acceleration of 2.0 m/s².

   (a) Write an equation for its velocity v as a function of time t.
   (b) Graph the function for t from 0.0 to 10.0 seconds.
   (c) From the graph, estimate the velocity after 5.5 seconds.
   (d) Estimate the time at which the velocity is 20.0 m/s.

6. The volume of a cone with radius r and height h is given by

$$V = \frac{1}{3}\pi r^2 h$$

A certain conical sand pile is 2.7 m high.

(a) Write the equation for the volume of this sand pile as a function of the radius.

(b) Graph the function for r from 0.0 to 3.0 m.

(c) From the graph, estimate the radius of the sand pile if the volume is 8.0 m³.

7. The height s (in ft) of an object t seconds after it is thrown from the top of a building is given by

$$s = -16t^2 - 40t + 200$$

(a) Graph s as a function of t, for t from 0.0 to 3.0 seconds.

(b) Estimate the time at which the object hits the ground.

8. The height s (in feet) of an object t seconds after it is thrown upward from the ground is given by

$$s = 112t - 16t^2$$

(a) Graph the function for t from 0.0 to 8.0 seconds.

From the graph, estimate

(b) the time when the object hits the ground;

(c) the time when the object reaches its maximum height;

(d) the maximum height of the object; and

(e) the time(s) when the object is 100 feet above the ground.

9. The length l of a spring is a linear function of the force F applied to it. This data was taken in an experiment.

Force (lb)	0	2	4	6	8	10
Length (in.)	5.00	5.14	5.28	5.42	5.56	5.70

(a) Graph the length l as a function of the force F.

(b) Estimate the force necessary to stretch the spring to 6 in.

10. The following is a table of data which relates the force F required to pull a block along a rough surface and the normal force N.

N (lb)	500	600	700	800	900
F (lb)	225	270	315	360	405

(a) Graph the force F as a function of the normal force N.

(b) Estimate the force required to pull a block with normal force of 500 lb.

11. The velocity v of a projectile fired upward is a linear function of time t. This data was taken in an experiment.

t (seconds)	2	4	6
v (ft/s)	32	−32	−96

(a) Graph v as a function of t.

(b) The velocity of a projectile when $t = 0$ is called the initial velocity. Find the initial velocity.

(c) Estimate the time at which the velocity is 0.

12. The linear velocity v of a point on a rotating circle of radius r is a function of its angular velocity ω. The following data were taken in an experiment.

ω (rev/s)	1600	1800	2000	2200	2400
v (ft/s)	64000	72000	80000	88000	96000

   (a) Graph the linear velocity v as a function of the angular velocity ω.

   (b) Estimate the angular velocity of a point moving with a linear velocity of 70000 ft/s.

   (c) Estimate the linear velocity of a point moving with an angular velocity of 1000 rev/s.

13. The simple interest I earned on a savings account is a linear function of time t. After 1 year, the account earned \$35 in interest, and after 5 years, it earned \$175 in interest.

   (a) Graph I as a function of t, for t from 0 to 5 years.

   (b) From the graph estimate the interest after 4 years.

14. The volume V of a cone is a linear function of the height h if the radius is constant. For this function, $V = 1.57$ cm^3 when $h = 1.50$ cm and $V = 2.36$ cm^3 when $h = 2.25$ cm.

   (a) Graph V as a function of h, for h from 0.00 to 3.00 cm.

   (b) From the graph estimate the volume when $h = 2.00$ cm.

Graphical Solution of Equations

Solve the equation graphically. Approximate the solution(s) to the nearest tenth.

15. $x^2 - 3 = 4x$ **16.** $x^2 + x = 3$ **17.** $-x^2 = 4 + 6x$

18. $1 - x^2 = 3x$ **19.** $x^3 - 4x = 4$ **20.** $x^3 + x = 1$

12-7   GRAPHING CALCULATORS

A graphing calculator helps us to graph functions quickly. Since there are many graphing calculators on the market, our discussion is general, and not calculator-specific. In this section we graph functions using a graphing calculator. We use the TRACE and ZOOM features of the graphing calculator to determine the x and y intercepts and the maximum and minimum points of a graph.

Graph of a Function

We set the calculator *graph mode* for graphing functions in the rectangular coordinate system. We set the *window* (the displayed portion of the coordinate plane) by defining the minimum and maximum values of x and y that will appear in the window. We enter the function and graph. The steps are simple!

Graphing a Function on a Graphing Calculator

1. Set the window.
2. Enter the equation.
3. Graph.

Example 48

Use a graphing calculator to graph the function

$$y = x^2$$

for $-10 \le x \le 10$ and $0 \le y \le 100$.

Solution: If we look at the screen on the calculator where the window variables are defined, we see a screen similar to this:

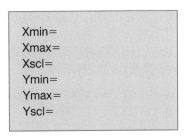

These six quantities are called the *window variables*. In this case, we're given that $-10 \le x \le 10$ and $0 \le y \le 100$, so we let Xmin = -10, Xmax = 10, Ymin = 0, and Ymax = 100. The values of Xscl and Yscl determine the spacing between the tick marks on the axes. We choose the scales so that there are neither too many nor too few tick marks on an axis. A scale of 5 on the x axis and a scale of 10 on the y axis are appropriate choices for this window. So we set Xscl = 5 and Yscl = 10. Our window variables look like this:

In the screen where the function is defined, we enter the function

$$y = x^2$$

Now we can display the graph (Fig. 12-32). We see that the graph is a parabola with the vertex at $(0, 0)$.

FIGURE 12-32
Graphing calculator screen.

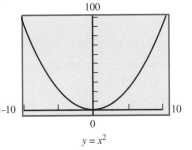

$y = x^2$

Note: When a figure includes a graphing calculator screen, *we will show the values of Xmin, Xmax, Ymin, and Ymax, although they do not appear on the screen.*

Complete Graph

We are often *given* the domain of a function we wish to graph. But if we're not told which x values to use, we must determine an appropriate domain. We want to include those x values that show the important features of the graph, such as asymptotes, intercepts, and maximum and minimum points. A graph that shows all the important features is called a *complete* graph.

If we don't know an appropriate domain and range which will display a complete graph of the function, we may need to *explore* with different windows.

On the graphing calculator, a good starting point may be a window in which $-5 \leq x \leq 5$ and $-3 \leq y \leq 3$. In this window, the origin is in the center of the screen and a scale of 1 is appropriate on both the x and y axes. Many graphing calculators have built-in window settings (standard, initial, decimal, trig, etc.), which use predefined domains and ranges. These settings are very handy and convenient to use. Check your calculator manual for specifics.

Example 49

Graph the function and estimate the important features of the graph:

$$y = x^3 - 3x^2 + 2$$

Solution: We set the window so that

$$-5 \leq x \leq 5 \quad \text{and} \quad -3 \leq y \leq 3$$

with x and y scales of 1. We enter and graph the function (Fig. 12-33). The complete graph of this function is displayed in the window. From the graph we estimate that the y intercept is 2, the x intercepts are about -0.7, 1, and 2.7, and maximum and minimum points are at about $(0, 2)$ and $(2, -2)$.

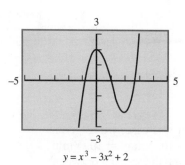

$$y = x^3 - 3x^2 + 2$$

■ **FIGURE 12-33**

Changing the Window on the Graphing Calculator

Sometimes the screen does not show a complete graph of the function, or it may not show the portion of the graph that we're interested in. After graphing with one window setting, if the important features of the graph are not shown on the screen, the window may be changed to a more appropriate one.

Example 50

Make a complete graph of the function

$$y = 2x^2 + 3x - 5$$

Solution: We really don't know what an appropriate window for this function is, so we try

$$-5 \leq x \leq 5 \quad \text{and} \quad -3 \leq y \leq 3$$

We enter and graph the function (Fig. 12-34). Since this is a quadratic equation, the graph should be a parabola. But the vertex is *not* on the screen. From the portion shown, we estimate that the vertex is at about $(-1, -6)$. We change the window so that

$$-4 \leq x \leq 2 \quad \text{and} \quad -8 \leq y \leq 2$$

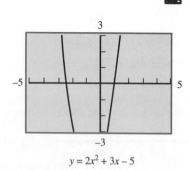

$$y = 2x^2 + 3x - 5$$

■ **FIGURE 12-34**

■ **FIGURE 12-35**

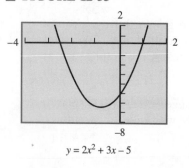

$$y = 2x^2 + 3x - 5$$

and graph the function again (Fig. 12-35). Since this graph shows the vertex of the parabola, and the x and y intercepts, it is now a complete graph. From the graph, we estimate that the x intercepts are -2.5 and 1, the y intercept is -5, and the vertex is at *about* $(-1, -6)$.

Using the TRACE Feature of the Graphing Calculator

The TRACE feature on a graphing calculator is used to move the cursor from one point to the next, while displaying the coordinates of the point on the screen.

Example 51

Find the approximate coordinates of the vertex of the parabola

$$y = 3x^2 - 7x + 5$$

Solution: Using a window so that $-5 \le x \le 5$ and $-3 \le y \le 3$, we graph the function

$$y = 3x^2 - 7x + 5$$

(Fig. 12-36). With the graph on the screen, we turn on the TRACE feature. A blinking point, called the *cursor,* appears on the graph. The coordinates of that point are shown at the bottom of the screen. We use the arrow keys to move the cursor along the curve until it is as close to the vertex as we can get. Since this parabola opens up, the vertex is the point which has the smallest y value. We read the coordinates of the point (Fig. 12-37). The coordinates of the vertex are about $x = 1.1702128$ and $y = 0.91670439$. *Not all of these digits are significant.* In fact, we can conclude only that the vertex is *approximately* at $(1.2, 0.9)$.

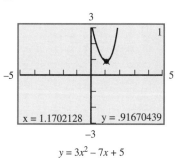

$$y = 3x^2 - 7x + 5$$

■ **FIGURE 12-36**

$x = 1.1702128$ $y = .91670439$

$$y = 3x^2 - 7x + 5$$

■ **FIGURE 12-37**
Tracing on a graph.

> **COMMON ERROR** When using the TRACE feature on a graphing calculator, the coordinates of the points are *approximations,* and many digits may *not* be significant.

Note: Many calculators have a feature to find the minimum value of a function. Using this feature, we find that the vertex (a minimum point) is at $(1.167, 0.917)$ rounded to thousandths place.

Using the ZOOM Feature of the Graphing Calculator

The ZOOM feature on a graphing calculator allows us to enlarge or reduce a portion of a graph. To enlarge a small portion of a graph, we can "zoom in" on a region around a point or we can "box" in a portion of the window. To display a larger portion of the graph on the screen, we can "zoom out"—this is useful if only part of the complete graph is shown. The extent of the enlargement or reduction depends on the "zoom factors."

■ **FIGURE 12-38**

Example 52

Graph the function

$$y = 5x - x^3$$

Solution: We clear the graph screen and set the window so that $-5 \le x \le 5$ and $-3 \le y \le 3$. We graph the function (Fig. 12-38). The maximum and minimum points are not shown on this graph, so we would like to see a larger portion of the graph. We set the zoom factors to 2 for both the x and the y axes—this *doubles* the length of both the x and the y axes when we zoom out. We "zoom out" to see a larger portion of the graph (Fig. 12-39). We use the TRACE feature to find that the x intercepts are at about $(-2.2, 0)$ and $(2.2, 0)$, and the minimum and maximum points are at about $(-1.3, -4.3)$ and $(1.3, 4.3)$.

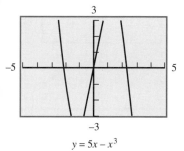

$$y = 5x - x^3$$

■ **FIGURE 12-39**
Zooming out.

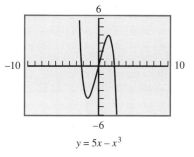

$$y = 5x - x^3$$

Sketching a Graphing Calculator Graph on Graph Paper

Often we need to sketch a graph on graph paper after we've graphed it on the calculator. To sketch the graph, *we must still plot points and connect them.* We use the graph on the calculator to determine the coordinates of important points, such as the x and y intercepts and maximum or minimum points. We plot these points and connect them with a smooth curve. If asymptotes exist, we show them. Finally we label the x and y axes and label the scales on each axis.

Graphical Solution of Equations by Graphing Calculator

In Sec. 12-6 we showed how to solve an equation graphically. Any equation can be solved graphically if we can graph the associated function, and the graphing calculator certainly simplifies this process. After we graph the function, we find the x intercepts using the root feature of the graphing calculator. The "roots" are the solutions of the original equation.

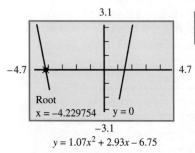

■ FIGURE 12-40
Finding a root.

— **Example 53** —

Use a graphing calculator to solve the equation graphically. Approximate the solutions to three significant digits.

$$1.07x^2 + 2.93x = 6.75$$

Solution: We bring all terms to one side of the equal sign,

$$1.07x^2 + 2.93x - 6.75 = 0$$

set the left side equal to y,

$$y = 1.07x^2 + 2.93x - 6.75$$

and graph the function. Using the root feature (Fig. 12-40), we find that one solution is

$$x = -4.23$$

Again, using the root feature (Fig. 12-41), we find that the other solution is

$$x = 1.49$$

Note that the y value at this x intercept is not *exactly* 0,

$$y = -2 \cdot 10^{-13}$$

but it is very close to 0! Again, we're only finding *approximations*, and this is close enough.

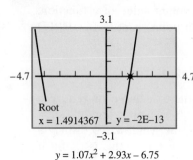

■ FIGURE 12-41
Another root.

SECTION 12-7 Exercises Graphing Calculators

Graph of a Function

Set the window so that $-5 \leq x \leq 5$ and $-3 \leq y \leq 3$ with a scale of 1 for both x and y. Clear the graph screen and graph each function on the calculator. Sketch the graph on graph paper.

1. $y = 3x - 2$ **2.** $y = 1 - 2x$

3. $y = x^2 - 2x - 1$ **4.** $y = x^2 + 3x + 1$

5. $y = x^3 - x$ **6.** $y = 2x - x^3$

Changing the Window on the Graphing Calculator

Graph these functions for $-5 \leq x \leq 5$ and $-3 \leq y \leq 3$, or use a similar window on your calculator. The complete graphs of these functions may not appear on the screen. Change the window as needed, so that you display the complete graph. Sketch on graph paper, using the same x and y scales.

7. $y = 2x^2 + 3x - 6$ **8.** $y = 1 - 5x - 2x^2$

9. $y = 2 + 3x - x^3$ **10.** $y = x^3 - 3x^2 - 2$

Graph each function and use the TRACE feature to approximate important features of the graph, such as the x and y intercepts and maximum and minimum points. Round to the nearest tenth. Sketch.

11. $y = 1 - 2x - x^2$ **12.** $y = x^2 - 7x + 10$

13. $y = 3x^2 - 4 - x^3$ **14.** $y = 4 - 3x^2 - x^3$

**Using the ZOOM
Feature**

Graph each function and use the TRACE, ZOOM, and other calculator features to approximate important features of the graph. Round to the nearest hundredth. Sketch.

15. $y = -x^2 + 2x + 5$ **16.** $y = 5 - 4x - 3x^2$

17. $y = -x^3 + 2x^2 - 1$ **18.** $y = x^3 - 4x^2 - 2$

**Graphical Solution
of Equations**

Using a graphing calculator to solve the equation graphically. Approximate the solutions to three significant digits.

19. $1.32x^2 + 1.95x = 4.02$ **20.** $2.56x^2 - 6.75 = 1.17x$

21. $2.56x^3 - 6.75 = 7.25x^2$ **22.** $1.36x + 3.75x^3 = 2.89$

■ Chapter 12   REVIEW EXERCISES

Express in equation form:

1. y is 3 times the square of x.

2. y is the amount by which 4 exceeds x.

3. Find the domain and range of the function

$$y = \sqrt{x - 3}$$

4. Given $f(x) = 2x^2 - 1$, find $f(-1)$, $f(2)$, $f(a)$, and $3f(0)$.

5. Given $g(x) = x - x^3$, find $g(1)$, $g(-2)$, $g(b)$, and $2g\left(\frac{1}{2}\right)$.

6. In which quadrant does a point (h, k) lie if $h > 0$ and $k < 0$?

7. Fill in the missing values and graph the equation.

$$y = 3x - 2$$

x	−2	−1	0	1	2
y					

State whether the equation is in implicit form or explicit form. If explicit, name the dependent and the independent variables.

8. $y = x^2 - 1$ **9.** $2x + 3y = 5$

10. Graphically solve the equation

$$x^3 - x^2 = 2x - 4$$

on a graphing calculator, rounding to the nearest tenth. Sketch on graph paper.

11. Make a table of ordered pairs, taking values of x from −3 to 3, and graph the function

$$y = x^2 + x - 6$$

12. Graph the linear function

$$y = 4 - 2x$$

Find the x intercept, the y intercept, and the slope of the line.

13. Graph the function

$$y = 3 - 5x - 2x^2$$

on a graphing calculator. Use the TRACE and ZOOM features to approximate the x and y intercepts and maximum and minimum points of the graph, rounding to the nearest hundredth. Sketch on graph paper.

14. The force F required to pull a block along a rough surface with a coefficient of friction μ is a linear function of the normal force N,

$$F = \mu N$$

A block is being pulled along a rough surface with a coefficient of friction of 0.20. Graph the force F as a function of the normal force N, for N from 0 to 35 lb.

15. The power company determines your monthly charge as follows: The customer is charged \$6.65 for the first 200 kWh of electricity used and an additional \$0.16 per kWh for electricity used over 200 kWh. Express the monthly charge C as a function of the electricity used x.

16. The centripetal force F of a body of mass m revolving in a circle of radius r with velocity v is given by

$$F = \frac{mv^2}{r}$$

A 300-g mass is revolving around a circle of radius 20 cm. Graph F as a function of v, for v from 0 to 300 cm/s. From the graph, estimate the velocity of the mass when the centripetal force is 600000 g-cm/s^2.

17. When an amount of money a is invested at an annual interest rate r for t years, the accumulated amount is given by

$$y = a(1 + r)^t$$

if the interest is compounded annually. Write a formula for the amount accumulated when \$300 is invested, if the interest is compounded annually for two years. Find the accumulated amounts for interest rates of 6.5%, 7.5%, and 10.2%, rounding to the nearest dollar.

18. If an object is dropped from a height h (in m), its velocity v (in m/s) when it hits the ground is given by

$$v = \sqrt{2ah}$$

where a is 9.81m/s^2. Write the equation for velocity and graph v as a function of h, for h from 0.0 to 20.0 m.

19. The quantity r represents the radius of the base of a cone and V represents the volume of the cone with radius r. Data were collected in an experiment:

r (cm)	1.5	2.6	3.5	4.4
V (cm^3)	10.5	31.4	57.0	90.0

Graph the volume V as a function of the radius r.

20. The amount of work required to lift a weight is defined as the product of force (or weight) times the distance lifted. Graph the work required to lift a 75-lb weight from 0.0 to 4.00 ft in increments of 0.5 feet. From the graph, estimate the work required to lift the weight 3.8 feet.

21. The resistance R of a wire having resistivity ρ and cross-sectional area A is a linear function of the length L of the wire, given by

$$R = \rho \frac{L}{A}$$

An aluminum wire has a diameter of 10.0 mm. The resistivity of aluminum is 2.63×10^{-8} ohm-meters. Graph the resistance R in the wire as a function of its length L, for L from 0.0 to 8.0 m.

WRITING

22. State, in your own words, what is meant by a "function." What are the domain and range of a function and how are they found? Describe the graph of a function, list the important features, and how you find them. You may use examples to help describe these terms.

TEAM PROJECT

23. Use a graphing calculator to graph the function

$$y = x^3 - 2x^2 + x - 30$$

Find, to the nearest hundredth, the x and y intercepts and maximum and minimum points.

13

Plane Geometry

In this chapter we present the basics of plane geometry, and applications of these concepts. They are presented informally, without proof, and the emphasis is on their use in applications. We examine lines, angles, polygons, triangles, quadrilaterals, and circles.

13-1 ——————— LINES AND ANGLES

In this section we study lines and angles. We look at two ways of defining an angle and three ways of measuring angles. We examine different types of angles and different relationships between angles.

■ **FIGURE 13-1**
Line *AB*.

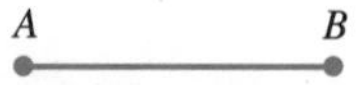

Lines, Segments, and Rays

A *line* (Fig. 13-1) is often denoted by a single letter, such as *L*. If *A* and *B* are two points on a line, then the line is denoted by the symbol $\overleftrightarrow{AB}$, or simply line *AB*.

A *line segment* (Fig. 13-2) is part of a line that lies between two points on the line. The two points are called the *endpoints* of the line segment. If *A* and *B* are the endpoints of the line segment, then the segment is sometimes denoted by the symbol $\overline{AB}$, or just segment *AB*. The *length* of the line segment is denoted by the symbol *AB*, where *AB* is a distance and represents a positive number.

■ **FIGURE 13-2**
Line segment *AB*.

■ **FIGURE 13-3**

┌─── **Example 1** ───
In Fig. 13-3, $\overline{PQ}$ is a line segment whose endpoints are *P* and *Q*. The length of the segment is *PQ*.

■ **FIGURE 13-4**
Ray *AB*.

A *ray* (Fig. 13-4) is part of a line that lies entirely on one side of a point. The point is called the *endpoint* of the ray. If *A* is the endpoint of the ray and *B* is any other point of the ray, then the ray is denoted by the symbol $\overrightarrow{AB}$, or simply ray *AB*, with the endpoint always given first.

■ **FIGURE 13-5**

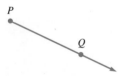

┌─── **Example 2** ───
In Fig. 13-5, $\overrightarrow{PQ}$ is a ray; its endpoint is *P*.

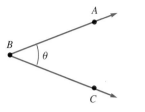

■ FIGURE 13-6
Angle.

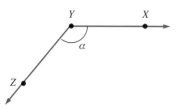

■ FIGURE 13-7

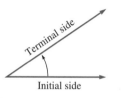

■ FIGURE 13-8
Angle formed by a rotation.

■ FIGURE 13-9
Positive and negative angles.

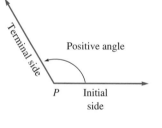

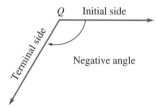

Angles

An *angle* (Fig. 13-6) is formed by two rays with a common endpoint. The common endpoint (the point of intersection of the two rays) is called the *vertex,* and the rays are called the *sides* of the angle. Often we give the angle a special name, such as θ or α (these are Greek letters). An angle can also be named using the angle symbol and the vertex, as in $\angle B$. If a point on each side of the angle is given, you may denote the angle by $\angle ABC$ or $\angle CBA$, taking special care to write the vertex as the middle point. So the angle in Fig. 13-6 can be written in any of the following ways:

$$\theta \quad \angle B \quad \angle ABC \quad \angle CBA$$

Example 3 ─────

Name the angle in Fig. 13-7 four different ways.

Solution: The angle can be written as α, $\angle Y$, $\angle XYZ$, or $\angle ZYX$.

An Angle Formed by a Rotation

Sometimes it is useful to think of an angle as being formed by the *rotation* of a ray. If we begin with a ray (Fig. 13-8) in some *initial position* (the *initial side*) and rotate the ray about its endpoint to a new position (the *terminal side*), then we have formed an angle. The angle is defined as the amount of rotation of the ray. We use a curved arrow to indicate the direction of the rotation. We usually consider angles formed by rotation in the *counterclockwise* direction as *positive,* and angles formed by *clockwise* rotation as *negative.*

Example 4 ─────

In Fig. 13-9, $\angle P$ is a positive angle, since it is formed by a counterclockwise rotation, and $\angle Q$ is a negative angle, since it is formed by a clockwise rotation.

Measure of an Angle

Angles are generally measured using one of three units: revolutions, degrees, or radians.

If we rotate a ray counterclockwise so that it returns to its initial position, we call the rotation one *revolution.*

Angles are also measured in degrees. We define one *degree* (1°) to be equal to $\frac{1}{360}$ of a revolution. Thus *there are 360 degrees in one revolution.*

We often use a *protractor* (Fig. 13-10) to measure angles in degrees. If the direction of rotation is not indicated, the angle is assumed to be positive.

■ FIGURE 13-10
Protractor.

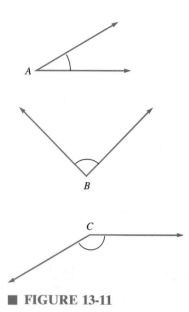

■ FIGURE 13-11

r r

1 rad

r

1 rad ≈ 57.3°

■ FIGURE 13-12
Angle of 1 radian.

┌─── **Example 5** ───────

Estimate the measure of the angles in Fig. 13-11 and then use a protractor to measure them to the nearest degree.

Solution: Since the direction of the rotation is not indicated, the angles are assumed to be positive. The measure of angle *A* is 30°. The measure of angle *B* is 90°. The measure of angle *C* is 150°.

Note: We will frequently omit the words "measure of angle" when discussing angle measurements. For example, we may say *A = 30°* instead of the more correct, but wordy, *the measure of angle A is 30°*.

Fractional parts of a degree may be expressed as decimal degrees (tenths, hundredths, thousandths, etc.) or as minutes and seconds. A *minute* (′) is equal to $\frac{1}{60}$ of a degree and a *second* (″) is equal to $\frac{1}{60}$ of a minute, or $\frac{1}{3600}$ of a degree. Surveyors must frequently be accurate to the nearest second.

In some fields, angles are measured in *radians.* The angle in Fig. 13-12 has a measure of 1 radian; its vertex is the center of a circle, and it "cuts off" a part of the circle equal in length to the radius. There are 2π radians in one revolution, where $\pi \approx 3.1416$. If the unit of an angle measurement is not given, the angle is assumed to be in radians. One radian is about 57.3°. (This topic will be covered in more detail in Sec. 24-1.)

You may find on your calculator a unit of angle measurement called a *grad.* There are 400 grads in one revolution. The grad is popular in Europe. We will not use grads in this book.

Conversions

We convert angles using the same method we used to convert other units in Chapter 4. When converting between units of measurement, *retain as many significant digits in the conversion as are given in the original measurement.* A table of conversions is given in Appendix B, starting on page 686. A summary of angle conversions follows:

ANGLE CONVERSIONS	1 rev = 360°	79
	1° = 60′	
	1′ = 60″	
	1 rev = 2π rad ≈ 6.28 rad	
	360° = 2π rad ≈ 6.28 rad	

┌─── **Example 6** ───────

Convert 34°23′56″ to decimal degrees.

Solution:

$$34°23'56'' = 34° + 23' + 56''$$

$$= 34° + 23'\left(\frac{1°}{60'}\right) + 56''\left(\frac{1'}{60''}\right)\left(\frac{1°}{60'}\right)$$

$$= \left(34 + \frac{23}{60} + \frac{56}{3600}\right)° = 34.3989°$$

We round to six significant digits, since there were six digits in the original measurement.

Many scientific calculators have special keys for converting degrees, minutes, and seconds into decimal degrees, and vice versa. Consult your calculator manual for specific instructions.

Example 7

Convert 67.2765° to degrees, minutes, and seconds.

Solution:

$$67.2765° = 67° + 0.2765°$$

Converting to degrees and minutes:

$$= 67° + 0.2765° \left(\frac{60'}{1°}\right)$$

$$= 67° + 16.59'$$

Converting to degrees, minutes, seconds:

$$= 67° + 16' + 0.59'$$

$$= 67° + 16' + 0.59' \left(\frac{60''}{1'}\right)$$

$$= 67° + 16' + 35'' \text{ (rounded)}$$

$$= 67°16'35''$$

Example 8

Convert to radians.
(a) 1.50 rev **(b)** 245°

Solution:

(a) $1.50 \text{ rev} = (1.50 \text{ rev}) \left(\frac{2\pi \text{ rad}}{1 \text{ rev}}\right)$

$$= 9.42 \text{ rad}$$

(b) $245° = 245° \left(\frac{2\pi \text{ rad}}{360°}\right) = 4.28 \text{ rad}$

Example 9

Convert to revolutions.
(a) 7.25 rad **(b)** 52°12'15''

Solution:

(a) $7.25 \text{ rad} = (7.25 \text{ rad}) \left(\frac{1 \text{ rev}}{2\pi \text{ rad}}\right) = 1.15 \text{ rev}$

(b) First we change to decimal degrees.

$$52°12'15'' = 52° + \frac{12°}{60} + \frac{15°}{3600} = 52.20417°$$

Converting to revolutions:

$$52°12'15'' = 52.20417° \left(\frac{1 \text{ rev}}{360°}\right)$$

$$= 0.145012 \text{ rev}$$

───── **Example 10** ─────

Convert to degrees.
(a) 1.72 rev   **(b)** 3.90 rad

Solution:

(a) $1.72 \text{ rev} = 1.72 \text{ rev} \left(\dfrac{360°}{1 \text{ rev}} \right) = 619°$

(b) $3.90 \text{ rad} = (3.90 \text{ rad}) \left(\dfrac{360°}{2\pi \text{ rad}} \right) = 223°$

Note that when an angle is considered a rotation, there is no limit to the size of an angle (Fig. 13-13). The measure of the angle is signed—it may be negative, zero, or positive, or even very large, like 900°.

Types of Angles

A *right* angle is defined as a 90° angle, or $\dfrac{1}{4}$ revolution. It is usually denoted by a small square at the vertex, as in Fig. 13-14. An *acute* angle is an angle between 0° and 90°. An *obtuse* angle is an angle between 90° and 180°.

───── **Example 11** ─────

In Fig. 13-11, angle A is acute, since $A = 30°$; angle B is a right angle, since $B = 90°$; angle C is an obtuse angle, since $C = 150°$.

■ **FIGURE 13-14**
Types of angles.

■ **FIGURE 13-13**

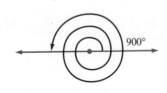

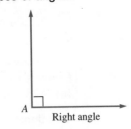
Right angle

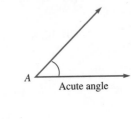
Acute angle

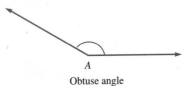
Obtuse angle

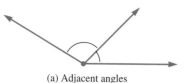

(a) Adjacent angles

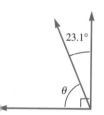

(b) Complementary angles

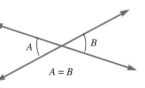

(c) Supplementary angles

FIGURE 13-15
Special relationships
between angles.

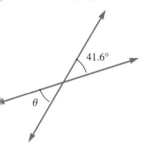

FIGURE 13-16

FIGURE 13-18
Vertical angles.

FIGURE 13-19

Special Relationships Between Angles

Two angles are called *adjacent* if they have a common side and a common vertex and they lie on opposite sides of their common side (Fig. 13-15). Two angles are called *complementary* if their sum is 90° (Fig. 13-15). Two angles are *supplementary* if their sum is 180° (Fig. 13-15). Note that complementary and supplementary angles are not always adjacent.

Example 12
In Fig. 13-16, find θ.

Solution: The two angles form a right angle and thus are complementary, so

$$\theta + 23.1° = 90°$$
$$\theta = 90° - 23.1° = 66.9°$$

Example 13
In Fig. 13-17, find θ.

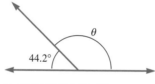

FIGURE 13-17

Solution: Since θ is supplementary to the 44.2° angle,

$$\theta = 180° - 44.2° = 135.8°$$

When two lines intersect they form two pairs of opposite angles. The angles within each pair are called *vertical* angles (Fig. 13-18). A very important theorem in geometry is

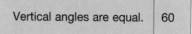

Vertical angles are equal.	60

Example 14
In Fig. 13-19, $\theta = 41.6°$, since vertical angles are equal.

Another very important theorem is

When two lines intersect, the adjacent angles are supplementary.	61

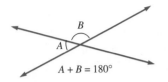

■ FIGURE 13-20
Supplementary adjacent angles.

Thus, when two lines intersect, the sum of adjacent angles is 180° (Fig. 13-20). So, $A + B = 180°$.

—— **Example 15** ——

In Fig. 13-21, find θ.

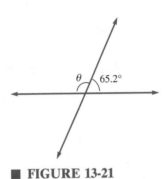

■ FIGURE 13-21

Solution: When two lines intersect, the adjacent angles are supplementary and their sum is 180°, so

$$\theta + 65.2° = 180°$$
$$\theta = 180° - 65.2° = 114.8°$$

Two lines that intersect at a right angle are called *perpendicular* (Fig. 13-22). Two lines in a plane that never intersect are called *parallel* (Fig. 13-23).

A *transversal* is a line that intersects two or more lines. In Fig. 13-24, two parallel lines, L_1 and L_2 are intersected by a transversal, T. Angles A, B, G, and H are called *exterior* angles, and angles C, D, E, and F are called *interior* angles. Angles A and E are called *corresponding* angles. Other corresponding pairs of angles are C and G, B and F, and D and H. Angles C and F are called *alternate*

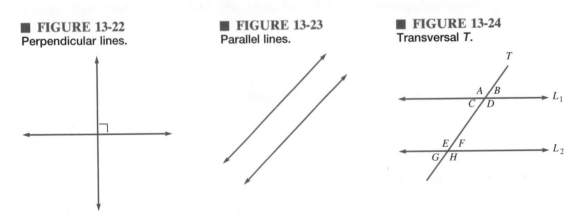

■ FIGURE 13-22
Perpendicular lines.

■ FIGURE 13-23
Parallel lines.

■ FIGURE 13-24
Transversal T.

interior angles, as are angles *D* and *E*. Refer to Fig. 13-24 to verify another important theorem in geometry:

> If two parallel lines are intersected by a transversal, corresponding angles are equal, and alternate interior angles are equal.
>
> **62**

Example 16

Find θ in Fig. 13-25.

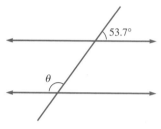

■ FIGURE 13-25   ■ FIGURE 13-26

Solution: We need to introduce another angle, α (Fig. 13-26), which is a corresponding angle to 53.7°. Thus, $\alpha = 53.7°$. Now, θ is supplementary to α, so

$$\theta = 180° - \alpha = 180° - 53.7° = 126.3°$$

SECTION 13-1   Exercises   Lines and Angles

Lines, Segments, and Rays

1. Draw a segment *RS*, so that *RS* = 3 cm.
2. Draw a ray *QP*. What is the endpoint of ray *QP*?

Angles

3. Name the angle in Fig. 13-27 four different ways.
4. Name the angle in Fig. 13-28 four different ways.

■ FIGURE 13-27   ■ FIGURE 13-28

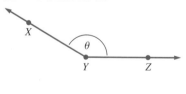

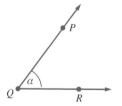

Measure of an Angle

5. Estimate the measure of each angle in Fig. 13-29 and then measure with a protractor to the nearest degree.

■ FIGURE 13-29

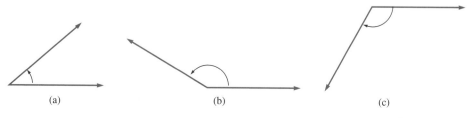

(a)   (b)   (c)

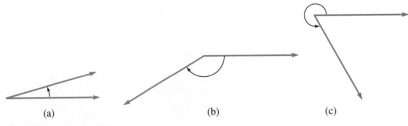

(a)   (b)   (c)

■ **FIGURE 13-30**

6. Estimate the measure of each angle in Fig. 13-30 and then measure with a protractor to the nearest degree.

7. What is a radian?

8. Draw an angle whose measure is 2 rads.

Conversions

Round all conversions appropriately.

Convert to decimal degrees.

| 9. 32°18′ | 10. 140°37′30″ | 11. 93°47′29″ |
| 12. 67°42′ | 13. 235°22′30″ | 14. 69°36′27″ |

Convert to degrees, minutes, and seconds.

| 15. 56.25° | 16. 172.125° | 17. 12.345° |
| 18. 74.7° | 19. 263.875° | 20. 164.985° |

Convert to radians.

| 21. 2.50 rev | 22. 15.0 rev | 23. 60.0° |
| 24. 75.0° | 25. 125°50′ | 26. 215°32′15″ |

Convert to revolutions.

| 27. 3.50 rad | 28. 1.25 rad | 29. 301° |
| 30. 335° | 31. 75°24′55″ | 32. 146°35′51″ |

Convert to decimal degrees.

33. 4.50 rev   34. 2.1345 rad

Convert to degrees, minutes, and seconds.

35. 5.123556 rad   36. 3.61000 rev

Sketch the following angles and show the rotation:

| 37. 135° | 38. −60° | 39. 300° | 40. −540° |
| 41. 225° | 42. −45° | 43. 150° | 44. −840° |

Special Relationships Between Angles

45. Draw two complementary angles that are not adjacent.

46. Draw two supplementary angles that are adjacent.

47. Find θ in each part of Fig. 13-31.

48. Find θ in each part of Fig. 13-32.

49. Find angles A, B, C, D, E, F, and G in Fig. 13-33, given that L_1 and L_2 are parallel.

50. Find angles A, B, C, D, E, F, and G in Fig. 13-34, given that L_1 and L_2 are parallel.

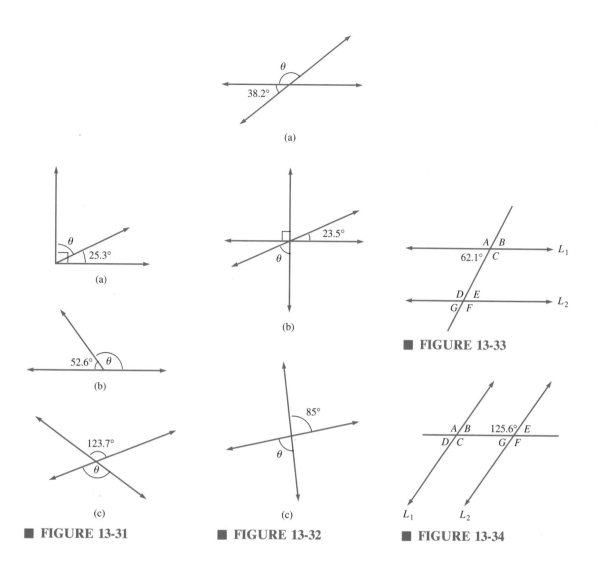

FIGURE 13-31

FIGURE 13-32

FIGURE 13-33

FIGURE 13-34

Applications

51. An escalator in a large store (Fig. 13-35) makes an angle of 34.8° with the first floor. Find the other angles A, B, and C.

52. A ray of light makes a 35.3° angle with the plane of a mirror. What angle does the ray of light make with a line perpendicular to the mirror?

53. A ship is traveling due west. If the captain turns 90° to the right, what is her new direction?

54. A support for a wall (Fig. 13-36) makes an angle of 32.5° with the floor at a corner. What angle does it make with the wall?

FIGURE 13-35

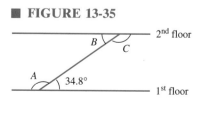

FIGURE 13-36

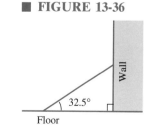

13-1 Lines and Angles   **243**

TRIANGLES

In this section we define the polygon and the triangle and look at different types of triangles. We discuss the concepts of perimeter and area and relate those concepts to the triangle. Lastly, we look at some properties of special triangles.

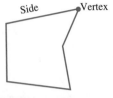

Polygons

A *polygon* (Fig. 13-37) is a plane figure formed by three or more line segments, called the *sides* of the polygon. The sides are joined at their endpoints to form a closed figure. A point where the two sides meet is called a *vertex* (plural, *vertices*). If all the sides and angles of a polygon are equal, the polygon is called a *regular* polygon.

■ **FIGURE 13-37**
Polygon.

Example 17

The polygons in Fig. 13-38 are all regular. The square and the equilateral triangle are the most common regular polygons. The polygons in Fig. 13-39 are *not* regular polygons, since their sides are not all equal.

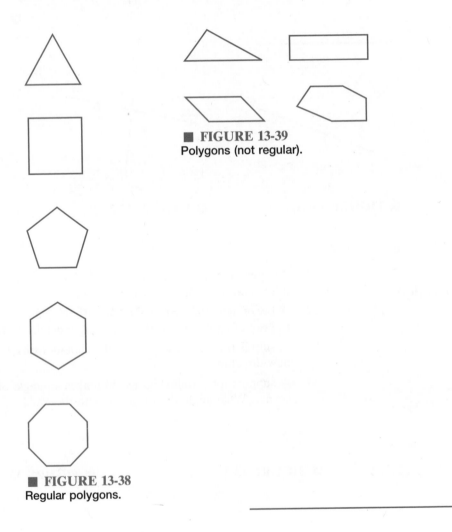

■ **FIGURE 13-39**
Polygons (not regular).

■ **FIGURE 13-38**
Regular polygons.

Perimeter of a Polygon

The *perimeter* of a polygon is simply the sum of its sides. It is the distance around the polygon.

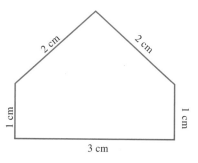

FIGURE 13-40

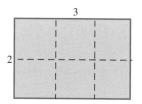

FIGURE 13-41

FIGURE 13-42
Types of triangles.

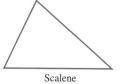

Scalene

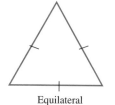

Equilateral

Isosceles

Example 18

The perimeter p of the polygon in Fig. 13-40 is the sum of its sides. We start at the lower left corner.

$$p = 1 + 2 + 2 + 1 + 3 = 9 \text{ cm}$$

Area of a Polygon

The *area* of a polygon is a measure of the interior of the polygon. It is usually expressed as a number of square units.

Example 19

The area A of the polygon in Fig. 13-41 can be found by counting the square units. Thus

$$A = 6 \text{ square units}$$

Types of Triangles

A *triangle* is a polygon with three sides. There are many different types of triangles (Fig. 13-42). A *scalene* triangle has no equal sides and an *equilateral* triangle has three equal sides. An *isosceles* triangle has two equal sides. In an isosceles triangle, the angle between the two equal sides is called *the vertex angle* and the side opposite the vertex angle is called the *base*.

   An *acute* triangle (Fig. 13-43) has three acute angles, and an *obtuse* triangle has one obtuse angle. A *right* triangle (Fig. 13-43) has one right angle, and an *oblique* triangle (Fig. 13-43) has no right angle.

Sum of the Angles

A very important theorem in geometry concerns the angles of a triangle: *The sum of the angles of any triangle is* 180° (Fig. 13-44).

SUM OF THE ANGLES OF A TRIANGLE	$A + B + C = 180°$	100

This enables us to find the third angle of a triangle if the other two angles are known.

FIGURE 13-43
Types of triangles.

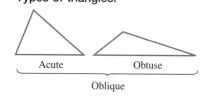

Acute   Obtuse

Oblique

Right

FIGURE 13-44
Sum of the angles of a triangle.

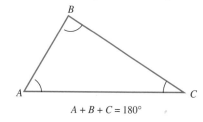

$A + B + C = 180°$

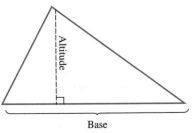

Base

Acute triangle

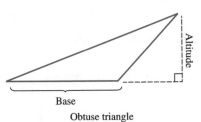

Base

Obtuse triangle

■ FIGURE 13-45
Altitude and base of a triangle.

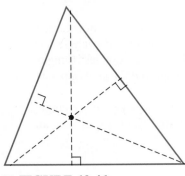

■ FIGURE 13-46
Altitudes.

■ FIGURE 13-48
Medians.

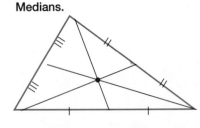

Example 20

Find angle A in a triangle if the other two angles are 53.2° and 46.5°.

Solution: By Eq. 100, the sum of the angles must be 180°. Therefore,

$$A = 180° - (53.2° + 46.5°)$$
$$A = 80.3°$$

Altitude and Base of a Triangle

An *altitude* of a triangle (Fig. 13-45) is a line segment which extends from a vertex perpendicular to the opposite side (called the *base*), or to an extension of that side. Every triangle has three altitudes, one from each vertex. *The three altitudes of a triangle intersect in a single point.*

Example 21

Draw the three altitudes in the given triangle and show that they intersect in a single point.

Solution: An altitude is drawn from each vertex (Fig. 13-46) so that it is perpendicular to the opposite side. We see that all three altitudes intersect in a single point.

Median

A *median* of a triangle (Fig. 13-47) is a line segment joining a vertex to the midpoint of the opposite side. In Fig. 13-47, the markings on the base indicate equal line segments.

Every triangle has *three* medians, one from each vertex. *The three medians intersect in a single point* (but not usually the same point at which the altitudes intersect).

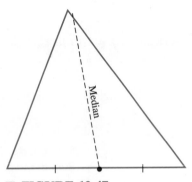

■ FIGURE 13-47
Median of a triangle.

Example 22

Draw the three medians in an acute triangle and show that they intersect in a single point.

Solution: A median is drawn from each vertex to the midpoint of the opposite side (Fig. 13-48). We see that the three medians intersect in a single point.

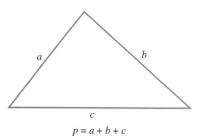

FIGURE 13-49
Perimeter of a triangle.

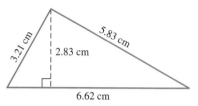

FIGURE 13-50

$A = \frac{1}{2}bh$

FIGURE 13-51
Area of a triangle.

Perimeter of a Triangle

The perimeter of a triangle having sides a, b, and c is the sum of a, b, and c (Fig. 13-49).

PERIMETER OF A TRIANGLE	$p = a + b + c$	97

Example 23

Find the perimeter of the triangle in Fig. 13-50.

Solution: To find the perimeter, use Eq. 97.

$$p = a + b + c$$
$$= 3.21 + 6.62 + 5.83$$
$$= 15.66 \text{ cm}$$

Area of a Triangle

One way to find the area of a triangle is to use the familiar formula: *The area of a triangle equals one-half the product of the base b and the altitude h to that base* (Fig. 13-51).

AREA OF A TRIANGLE WITH BASE b AND HEIGHT h	$A = \frac{1}{2}bh$	98

Example 24

Use Eq. 98 to find the area of the triangle in Fig. 13-50.

Solution: To find the area, use Eq. 98

$$A = \frac{1}{2}bh$$

$$= \frac{1}{2}(6.62)(2.83)$$

$$= 9.37 \text{ cm}^2$$

Hero's Formula

There is another method of finding the area of a triangle. If a, b, and c are the lengths of the sides of a triangle, then the area is given by *Hero's* (or *Heron's*) formula.

HERO'S FORMULA FOR THE AREA OF A TRIANGLE	$A = \sqrt{s(s-a)(s-b)(s-c)}$ where $s = \dfrac{a+b+c}{2}$	99

Example 25

Use Hero's formula to find the area of the triangle in Fig. 13-50.

Solution: Using Eq. 99,

$$s = \frac{3.21 + 6.62 + 5.83}{2} = 7.83$$

$$A = \sqrt{s(s-a)(s-b)(s-c)}$$
$$= \sqrt{7.83(7.83 - 3.21)(7.83 - 6.62)(7.83 - 5.83)} = 9.36 \text{ cm}^2$$

Note that this result agrees with the answer in Example 24, since there is always some uncertainty in the last significant digit.

We use Hero's formula to find the area of a triangle when we know all three sides of the triangle. If we know a base and a height, then we would use the familiar formula, $A = \frac{1}{2}bh$.

Example 26

A triangular parking lot is bounded by three streets (Fig. 13-52). Find the cost of paving the area, at \$9.50 per yd^2, if the triangle is 45.0 ft × 35.0 ft × 25.0 ft.

■ **FIGURE 13-52**

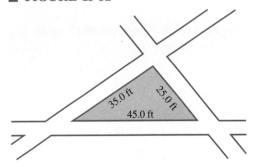

Solution: We calculate the area, using Hero's formula.

$$s = \frac{45.0 + 35.0 + 25.0}{2} = 52.5$$

$$A = \sqrt{52.5(7.5)(17.5)(27.5)} = 435.3 \text{ ft}^2$$

Converting ft^2 to yd^2,

$$A = 435.3 \text{ ft}^2 \left(\frac{1 \text{ yd}^2}{9 \text{ ft}^2}\right) = 48.37 \text{ yd}^2$$

Calculating the cost,

$$\text{Cost} = \left(\frac{\$9.50}{\text{yd}^2}\right)(48.37 \text{ yd}^2) = \$459$$

Types of Triangles

In Exercises 1 through 6 (Fig. 13-53), (a) find the measure of angle *A*, and (b) name all the terms from this list that apply to the triangle: scalene, isosceles, equilateral, right, acute, obtuse, oblique.

■ **FIGURE 13-53**

1.

2.

3.

4.

5.

6.

7. Sketch a right triangle, an equilateral triangle, and a scalene triangle. Measure the three angles of each triangle with a protractor. What is the sum of the angles of each triangle?

8. Sketch an isosceles triangle, an acute triangle, and an obtuse triangle. Measure the three angles of each triangle with a protractor. What is the sum of the angles of each triangle?

Sum of the Angles

Find the third angle in the triangle when the other two angles are given.

9. 62.7°, 75.9° 10. 32.6°, 87.4°

11. 47°20′, 67°35′ 12. 95°45′, 38°30′

Altitude

13. Draw an obtuse triangle and its three altitudes. Do they intersect in a single point?

14. Draw an acute triangle and its three altitudes. Do they intersect in a single point?

Median

15. Draw an acute triangle and its three medians. Do they intersect in a single point?

16. Draw an obtuse triangle and its three medians. Do they intersect in a single point?

17. Draw an isosceles triangle and the median from the vertex angle. Is the vertex angle bisected? Is the base bisected? Is the median perpendicular to the base?

18. Draw an equilateral triangle and the medians. What can you say about the angles? What can you say about the medians and sides?

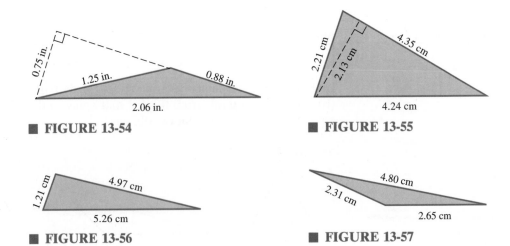

FIGURE 13-54

FIGURE 13-55

FIGURE 13-56

FIGURE 13-57

19. Find the perimeter and area of the triangle in Fig. 13-54.

20. Find the perimeter and area of the triangle in Fig. 13-55.

21. Find the perimeter and area of the triangle in Fig. 13-56.

22. Find the perimeter and area of the triangle in Fig. 13-57.

Applications

23. A wedge is in the shape of an isosceles triangle. If the vertex angle is 22.8°, what are the two base angles?

24. A vertical flagpole is supported by a wire that forms an angle of 42.5° with the pole. Find the angle that the wire makes with the ground.

25. Three stakes, A, B, and C, are placed in the ground so that $\angle ABC$ is 43.8°, and $\angle BCA$ is 62.3°. Make a sketch and find $\angle CAB$.

26. Two angles of a triangular lot measure 48.5° and 74.6°. Find the third angle.

Round to the nearest dollar.

27. Find the cost of fencing in a triangular pasture, whose sides are 625 ft, 455 ft, and 376 ft, if fence costs $1.50 per yard.

28. A garden is made in the shape of a regular pentagon, 12.0 m on a side. Find the cost of putting a brick border around the garden, if the brickwork costs $18.75 per meter.

29. What is the cost of a triangular piece of land whose base is 455 ft and altitude is 225 ft, at $975 an acre? (An acre is 43,560 ft².)

30. A certain field needs to be fertilized at the rate of 1000 lb of fertilizer per acre. A 50.0-lb bag of fertilizer costs $7.60. Find the cost of fertilizing a triangular field whose base is 575 feet and altitude is 489 feet. (An acre is 43,560 ft².)

31. At $4.50 a square yard, find the cost of paving a triangular court whose sides are 35.0 ft, 30.0 ft, and 25.0 ft.

32. If indoor-outdoor carpeting costs $24.00 a square yard, what is the cost of carpeting a triangular deck whose sides are 24.0 ft, 15.0 ft, and 15.0 ft?

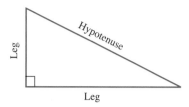

FIGURE 13-58
Right triangle.

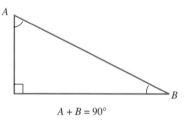

$A + B = 90°$

FIGURE 13-59
Acute angles of a right triangle.

13-3 RIGHT TRIANGLES

In this section, we study the right triangle, which you recall is a triangle with one right angle. We define the hypotenuse and legs and learn how to find them by using the Pythagorean Theorem. Finally, we consider properties of two special triangles: the 30-60-90 right triangle and the 45-45-90 right triangle.

Definitions

In a right triangle (Fig. 13-58), the side that is opposite the right angle is called the *hypotenuse*. The two other sides are called *legs*. Note that the hypotenuse is always the longest side of a right triangle.

Sum of the Angles

The right angle of a right triangle is, of course, 90°. Since the sum of the angles of any triangle is 180°, then *the sum of the two nonright angles in a right triangle is 90°* (Fig. 13-59) and the acute angles are complementary. Each one must be less than 90°, and therefore both are acute.

The acute angles of a right triangle are complementary.	$A + B = 90°$	103

Example 27

Find angle A in Fig. 13-60.

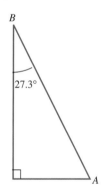

27.3°

FIGURE 13-60

Solution: Since angle A and angle B are the acute angles of a right triangle, they are complementary.

$$A + B = 90°$$

So,

$$A = 90° - B = 90° - 27.3° = 62.7°$$

Pythagorean Theorem

The sides of a right triangle are related by the Pythagorean Theorem. This theorem was named for the Greek mathematician and philosopher, Pythagoras, who lived

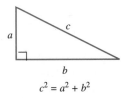

FIGURE 13-61
Pythagorean Theorem.

about 2500 years ago. Pythagoras proved that *the square of the hypotenuse of a right triangle is equal to the sum of the squares of the two legs* (Fig. 13-61).

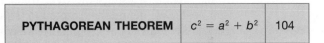

PYTHAGOREAN THEOREM	$c^2 = a^2 + b^2$	104

We use the Pythagorean Theorem to find a missing side of a right triangle if we know the other two sides. If the missing side is the hypotenuse c, then we take the square root of both sides of Eq. 104 to get

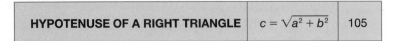

HYPOTENUSE OF A RIGHT TRIANGLE	$c = \sqrt{a^2 + b^2}$	105

Example 28

A right triangle has legs of length 4.37 cm and 6.59 cm. Find the length of the hypotenuse.

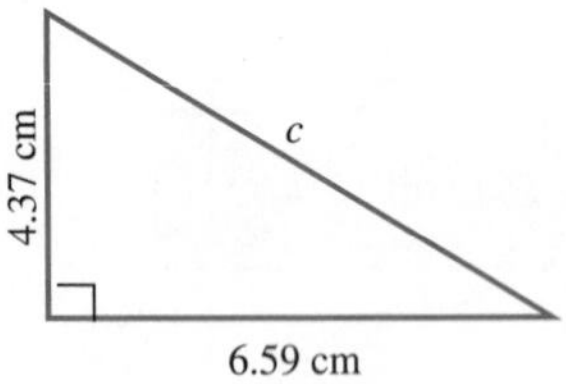

Solution: We sketch the triangle in Fig. 13-62 and let $a = 4.37$ and $b = 6.59$ in Eq. 105. Then c is the length of the hypotenuse.

$$c = \sqrt{4.37^2 + 6.59^2}$$
$$c = \sqrt{62.525} = 7.91 \text{ cm}$$

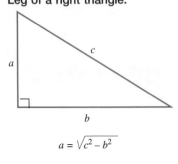

FIGURE 13-62

COMMON ERROR Remember that the Pythagorean Theorem applies only to *right triangles*. We need trigonometry (Chapter 22) to find the sides and angles of other (oblique) triangles.

Finding a Leg of a Right Triangle

FIGURE 13-63
Leg of a right triangle.

Suppose you need to find a leg, say a, of a right triangle (Fig. 13-63). We will solve Eq. 104 for the leg a.

$$c^2 = a^2 + b^2$$

Interchanging sides,

$$a^2 + b^2 = c^2$$

Subtracting b^2 from both sides,

$$a^2 = c^2 - b^2$$

Taking the square root of both sides,

$$a = \sqrt{c^2 - b^2}$$

LEG OF A RIGHT TRIANGLE	$a = \sqrt{c^2 - b^2}$	106

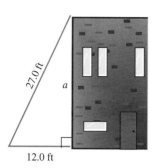

FIGURE 13-64

27.0 ft

a

12.0 ft

┌─── **Example 29** ───────────

A ladder 27.0 ft long reaches to the top of a building when its foot stands 12.0 feet from the building (Fig. 13-64). How high is the building?

Solution: We see that the unknown side of the right triangle is a leg. We use Eq. 106, letting a = the length of the missing leg.

$$a = \sqrt{27.0^2 - 12.0^2} = 24.2 \text{ ft}$$

───────────────────────────┘

┌───┐
│ **COMMON ERROR** You *add* the squares of the two legs to get the square of the hypotenuse. If the missing side is a leg, then you must *subtract* the square of the given leg from the square of the hypotenuse. │
└───┘

The 30-60-90 Right Triangle

When we draw a median in an equilateral triangle, the median bisects the angle and is perpendicular to the opposite side (Fig. 13-65). We consider just one of these

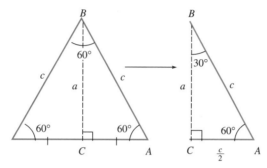

FIGURE 13-65

right triangles. Since $\angle B$ was 60° and this angle was bisected, then $\angle ABC$ in the right triangle is 30°. Angle C is a right angle, and $\angle A$ is still 60°. The three angles of the triangle are 30°, 60°, and 90°. Therefore, we call this triangle a *30-60-90 right triangle*. Since the base c of the equilateral triangle is bisected, the base of the right triangle is $\frac{c}{2}$. We use Eq. 106 to find the missing leg a.

$$a = \sqrt{c^2 - \left(\frac{c}{2}\right)^2} = \sqrt{\frac{4c^2}{4} - \frac{c^2}{4}} = \sqrt{\frac{3c^2}{4}} = \frac{c}{2}\sqrt{3}$$

So, in a 30-60-90 right triangle (Fig. 13-66), *the shorter leg is half the length of the hypotenuse and the longer leg is $\sqrt{3}$ times the shorter leg.*

┌───┐
│ **30-60-90** | |
│ **RIGHT TRIANGLE RELATIONSHIPS** | 107 |
│ 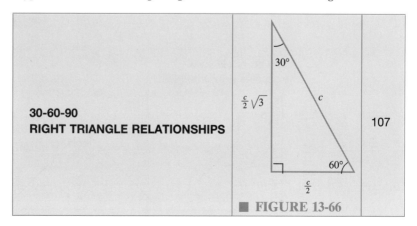 │
│ **FIGURE 13-66** │
└───┘

When we solve problems involving 30-60-90 right triangles, we assume the angle measurements are exact, and so we round the results to the number of significant digits in the given side.

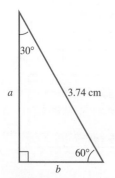

■ FIGURE 13-67

Example 30

The hypotenuse of a 30-60-90 right triangle is 3.74 cm long. Find the two legs.

Solution: We sketch the triangle in Fig. 13-67.

$$b = \frac{3.74}{2} = 1.87 \text{ cm}$$

$$a = (1.87)(\sqrt{3}) = 3.24 \text{ cm}$$

Example 31

The altitude of an equilateral triangle is 7.46 cm. Find the side c.

Solution: We sketch the equilateral triangle in Fig. 13-68, and draw an altitude to the base. Angle B is 60°, since all angles of an equilateral triangle are 60°. Thus, the altitude is the longer leg of a 30-60-90 right triangle. The side c of the equilateral triangle is the hypotenuse of the 30-60-90 right triangle. We are given the longer leg and need to find the hypotenuse c.

$$\frac{c}{2}\sqrt{3} = 7.46$$

Solving for c,

$$c = \frac{(2)(7.46)}{\sqrt{3}} = 8.61 \text{ cm}$$

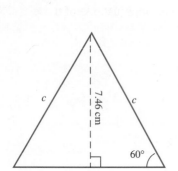

■ FIGURE 13-68

The 45-45-90 Right Triangle

A 45-45-90 right triangle (Fig. 13-69) has two 45° angles and one right angle. It is also called an *isosceles right triangle*. In an isosceles right triangle, the two legs are equal, and the third side is the hypotenuse. One angle is a right angle and the other two angles are equal, since the base angles of an isosceles triangle are equal. Therefore each base angle must be 45°.

If we are given the leg a of an isosceles right triangle (Fig. 13-70), we can find the hypotenuse c using Eq. 105

$$c = \sqrt{a^2 + a^2} = \sqrt{2a^2} = a\sqrt{2}$$

■ FIGURE 13-69
Isosceles right triangles.

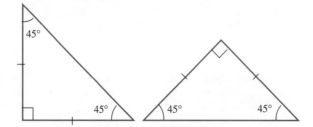

Thus, in an isosceles right triangle, the hypotenuse c is the length of the leg a times $\sqrt{2}$.

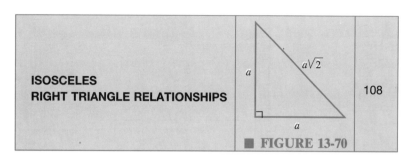

**ISOSCELES
RIGHT TRIANGLE RELATIONSHIPS**

108

■ **FIGURE 13-70**

┌─── **Example 32** ───

Find the hypotenuse of an isosceles right triangle whose legs are 5.00 cm each.

Solution:

$$c = a\sqrt{2} = (5.00)(\sqrt{2}) = 7.07 \text{ cm}$$

┌─── **Example 33** ───

A baseball diamond is a square. If it is 127 ft from home plate to second base (Fig. 13-71), how far is it between the bases? (Round to the nearest foot.)

Solution: The diagonal of a square separates the square into two isosceles right triangles. The diagonal is the hypotenuse of each triangle. So,

$$a\sqrt{2} = 127$$

$$a = \frac{127}{\sqrt{2}} = 90 \text{ ft}$$

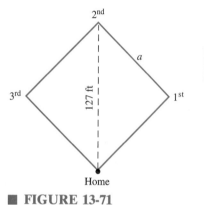

■ **FIGURE 13-71**

SECTION 13-3 **Exercises** **Right Triangles**

**Finding the Sides
of a Right Triangle**

1. What is wrong with this definition: "The hypotenuse is the longest side of a triangle"?
2. Name the hypotenuse and the legs of the triangles in Fig. 13-72.

■ **FIGURE 13-72**

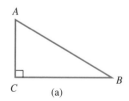

(a)

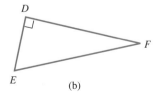

(b)

3. Find the missing sides in Fig. 13-73.

4. Find the missing sides in Fig. 13-74.

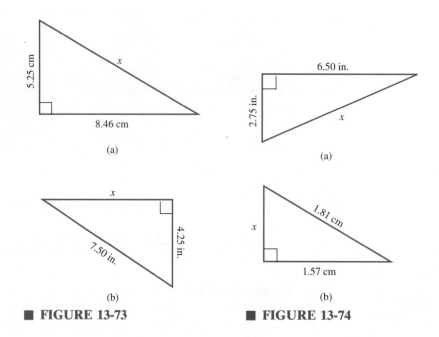

(a)

(b)

■ **FIGURE 13-73**

(a)

(b)

■ **FIGURE 13-74**

30-60-90 and 45-45-90 Right Triangles

5. Find the missing sides in Fig. 13-75.

6. Find the missing sides in Fig. 13-76.

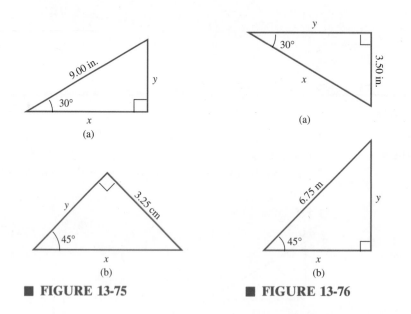

(a)

(b)

■ **FIGURE 13-75**

(a)

(b)

■ **FIGURE 13-76**

7. Find the altitude and the area of an equilateral triangle 4.75 inches on a side.

8. Find the side of an equilateral triangle whose altitude is 8.50 in.

9. Find the diagonal of a square whose side is 56.0 cm.

10. Find the diagonal of a square whose perimeter is 13.0 in.

11. Find the perimeter of a square whose diagonal is 18.25 in.

12. Find the side of a square whose diagonal is 3.56 cm.

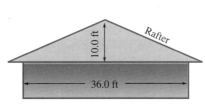

FIGURE 13-77

0.750 in.

FIGURE 13-78

13. The length of a rectangle is 54.2 mm. The diagonal is 67.3 mm long. Find the perimeter and area of the rectangle.

14. Find the perimeter and area of a 30-60-90 right triangle whose longer leg is 11.0 in.

15. A vertical pole 35.0 ft high is supported by a guy wire attached to the top. The wire reaches the ground 45.0 ft from base of the pole. How long is the wire?

16. How high is a loading platform if the bottom of a 15.0-ft ramp is 13.0 ft from the base of the platform?

17. A ladder 14.0 m long reaches to the top of a building when its foot stands 5.00 m from the building. How high is the building?

18. Two streets, one 15.6 m wide and the other 29.8 m wide, cross at right angles. What is the distance between opposite corners?

19. A rectangular room is 24.0 ft by 17.0 ft. What is the distance from one corner to its opposite corner?

20. A ladder 28.0 ft long stands flat against the side of a building. How many feet must it be pulled out at the bottom so that the top may be lowered 5.0 ft?

21. A house is 36.0 ft wide, the ridge is 10.0 ft higher than the side walls, and the rafters project 2.00 ft beyond the sides of the house (Fig. 13-77). How long are the rafters?

22. A carpenter must build a ramp for disabled persons to enter a door 3.50 ft above the ground level. How long will the ramp be if it must begin to rise 25.0 ft from the base of the building?

23. A hex head bolt measures 0.750 in. across the flats (Fig. 13-78). Find the distance between opposite corners. (Hint: Draw 6 equilateral triangles using diagonals, and then separate those into 12 equal right triangles.)

24. A room is 18.0 ft long, 15.0 ft wide, and 8.00 ft high. What is the distance from one of the lower corners to the opposite upper corner?

13-4

QUADRILATERALS

FIGURE 13-79
Quadrilateral.

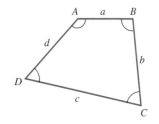

A *quadrilateral* (Fig. 13-79) is a polygon that has 4 sides. The trapezoid, parallelogram, rectangle, rhombus, and square are quadrilaterals.

Sum of the Angles

The sum of the angles of a quadrilateral is 360°.

SUM OF THE ANGLES OF A QUADRILATERAL	$A + B + C + D = 360°$	63

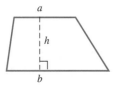

■ FIGURE 13-80
Trapezoid.

Trapezoid

A *trapezoid* (Fig. 13-80) is a quadrilateral in which only one pair of opposite sides are parallel. The two parallel sides are called the *bases, a* and *b*. The perpendicular distance between the bases is called the *altitude, h*. (The altitude is sometimes called the *height.*)

The perimeter *p* of a trapezoid is the sum of its sides, just as it is for any polygon.

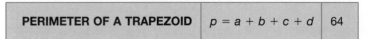

| PERIMETER OF A TRAPEZOID | $p = a + b + c + d$ | 64 |

The area *A* of a trapezoid is the product of the average of the bases, $\frac{a+b}{2}$, times the altitude *h*.

$$A = \left(\frac{a+b}{2}\right)h$$

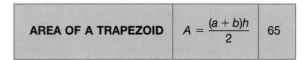

| AREA OF A TRAPEZOID | $A = \frac{(a+b)h}{2}$ | 65 |

If the two nonparallel sides are equal, the trapezoid is called an *isosceles* trapezoid (Fig. 13-81). In the figure, the marks on the two nonparallel sides indicate that the sides are equal.

■ FIGURE 13-81
Isosceles trapezoid.

■ FIGURE 13-82

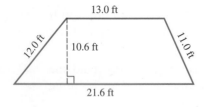

── **Example 34** ──

Find the cost of flooring the trapezoidal deck in Fig. 13-82, if the flooring costs $19.99 per square yard. Round to the nearest dollar.

Solution: Using Eq. 65 to find the area of the deck,

$$A = \frac{(a+b)h}{2} = \frac{(21.6 + 13.0)10.6}{2}$$

$$A = 183.4 \text{ ft}^2$$

Converting to yd²,

$$A = 183.4 \text{ ft}^2 \left(\frac{1 \text{ yd}^2}{9 \text{ ft}^2}\right) = 20.38 \text{ yd}^2$$

So,

$$\text{Cost} = (20.38 \text{ yd}^2)(\$19.99/\text{yd}^2) = \$407$$

■ FIGURE 13-83
Parallelogram.

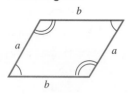

Parallelogram

A *parallelogram* (Fig. 13-83) is a quadrilateral in which both pairs of opposite sides are parallel. In a parallelogram, both pairs of opposite sides are equal, and opposite angles are equal. In the figure, equal angles are marked with the same number of arcs.

FIGURE 13-84
Diagonals of a parallelogram.

A *diagonal* of a parallelogram is a line segment joining two opposite vertices of the parallelogram. Every parallelogram has two diagonals and the diagonals bisect each other (Fig. 13-84).

The perimeter of a parallelogram is the sum of its four sides (Fig. 13-83).

$$p = a + b + a + b$$

Combining terms,

$$p = 2a + 2b$$

Factoring out a 2 on the right, we get the formula,

PERIMETER OF A PARALLELOGRAM	$p = 2(a + b)$	66

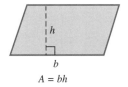

$A = bh$

FIGURE 13-85
Area of a parallelogram.

The area A of a parallelogram is equal to the product of its base b and its altitude h (Fig. 13-85).

AREA OF A PARALLELOGRAM	$A = bh$	67

Example 35

Find (a) the perimeter and (b) the area of the parallelogram in Fig. 13-86.

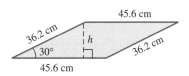

FIGURE 13-86

Solution:

(a) Using Eq. 66 for the perimeter of a parallelogram,

$$p = 2(a + b) = 2(36.2 + 45.6)$$
$$p = 163.6 \text{ cm}$$

(b) To find the area, first we draw an altitude to complete a 30-60-90 triangle. We see that the altitude is the short leg of the 30-60-90 right triangle and that the 36.2 cm side is the hypotenuse. The short leg is half the hypotenuse, so the altitude is 18.1 cm. Using Eq. 67 to find the area,

$$A = bh = (45.6)(18.1)$$
$$A = 825 \text{ cm}^2$$

FIGURE 13-87
Rectangle.

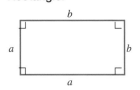

Rectangle

A *rectangle* (Fig. 13-87) is a parallelogram with four right angles. The diagonals of a rectangle bisect each other and are equal (Fig. 13-88).

In a rectangle, we call the base b and the height a. To find the perimeter, we use Eq. 66 to get the familiar formula,

PERIMETER OF A RECTANGLE	$p = 2(a + b)$	68

FIGURE 13-88
Diagonals of a rectangle.

To find the area of a rectangle, we use Eq. 67 to get the familiar formula,

AREA OF A RECTANGLE	$A = ab$	69

———— **Example 36** ————

How much can a farmer expect to earn from a rectangular corn field, 625 ft by 875 ft, if an acre yields corn worth \$225 on the average?

Solution: First we find the area of the field. Using Eq. 69,

$$A = ab = (625)(875)$$
$$A = 546900 \text{ ft}^2$$

Now we convert ft² to acres, using the conversion, 1 acre = 43560 ft².

$$A = 546900 \text{ ft}^2 = (546900 \text{ ft}^2) \left(\frac{1 \text{ acre}}{43560 \text{ ft}^2} \right)$$

$$A = 12.56 \text{ acres}$$

Multiplying the number of acres by \$225/acre, we have

$$\text{Earnings} = (12.56 \text{ acres})(\$225/\text{acre}) = \$2830$$

So the farmer could expect to earn about \$2830 from the corn in the field.

Rhombus

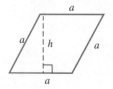

■ FIGURE 13-89
Rhombus.

A *rhombus* (Fig. 13-89) is a parallelogram in which all four sides are equal. The diagonals of a rhombus bisect each other, are perpendicular to each other, and bisect the angles of the rhombus (Fig. 13-90).

Since all four sides in a rhombus are equal, the perimeter is just four times the length of a side a.

PERIMETER OF A RHOMBUS	$p = 4a$	70

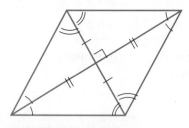

■ FIGURE 13-90
Diagonals of a rhombus.

The area of a rhombus is the product of the base a and the altitude h (Fig. 13-89).

AREA OF A RHOMBUS	$A = ah$	71

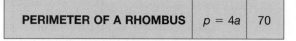

■ FIGURE 13-91

———— **Example 37** ————

Find (a) the perimeter and (b) the area of the rhombus in Fig. 13-91.

Solution:

(a) To find the perimeter of the rhombus, we use Eq. 70 with $a = 4.37$,

$$p = 4a = 4(4.37)$$
$$p = 17.48 \text{ cm}$$

(b) To find the area, we must first find the altitude h. Using Eq. 107 for the long leg of a 30-60-90 triangle, we have

$$h = \left(\frac{4.37}{2} \right) \sqrt{3} = 3.785 \text{ cm}$$

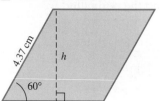

Thus, the area of the rhombus (Eq. 71) is

$$A = ah = (4.37)(3.785)$$
$$A = 16.5 \text{ cm}^2$$

Square

A *square* (Fig. 13-92) is a rectangle with four equal sides. The diagonals of a square bisect each other, are equal, are perpendicular to each other, and bisect the angles of the square (Fig. 13-93).

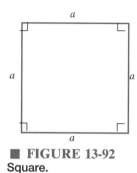

■ **FIGURE 13-92**
Square.

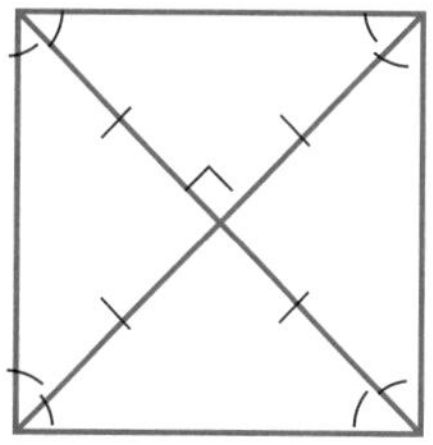

■ **FIGURE 13-93**
Diagonals of a square.

Since all four sides in a square are equal, the perimeter is just four times the length of a side a.

PERIMETER OF A SQUARE	$p = 4a$	72

The area A of a square is equal to the side a squared (Fig. 13-92).

AREA OF A SQUARE	$A = a^2$	73

┌─── **Example 38** ────────

The hectare is a metric unit for land area. It is used in Europe today. An area of one hectare equals the area of a square 100 m on each side. Find the area of a square plot of land 352 m on a side, and express it in hectares (ha).

Solution: Using Eq. 73 for the area of a square,

$$A = a^2 = (352)^2$$
$$A = 123900 \text{ m}^2$$

Converting to hectares,

$$1 \text{ hectare} = 1 \text{ ha} = (100 \text{ m})(100 \text{ m}) = 10000 \text{ m}^2$$

So,

$$A = 123900 \text{ m}^2 = 123900 \text{ m}^2 \left(\frac{1 \text{ ha}}{10000 \text{ m}^2} \right)$$

$$A = 12.4 \text{ ha}$$

13-4 Quadrilaterals   **261**

Name	Figure	Perimeter	Area
Trapezoid	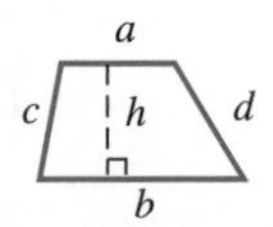	$p = a + b + c + d$	$A = \dfrac{(a+b)h}{2}$
Parallelogram	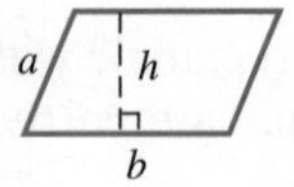	$p = 2(a + b)$	$A = bh$
Rectangle	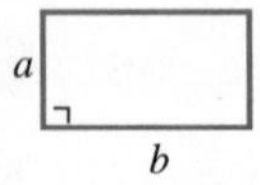	$p = 2(a + b)$	$A = ab$
Rhombus	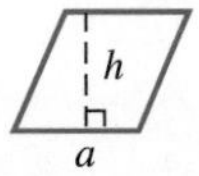	$p = 4a$	$A = ah$
Square	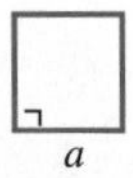	$p = 4a$	$A = a^2$

■ **FIGURE 13-94**
Quadrilaterals.

Since rectangles, rhombi, and squares are all parallelograms, the perimeter and area formulas for the parallelogram may be used for the rectangle, rhombus, and square.

Information on the quadrilaterals is summarized in Fig. 13-94.

SECTION 13-4   Exercises   Quadrilaterals

Quadrilaterals

1. Draw a parallelogram. Measure the angles with a protractor and add them. Is the sum close to 360°?

2. Draw a trapezoid. Measure the angles with a protractor and add them. Is the sum close to 360°? Do the same for a rhombus.

3. Draw a parallelogram and its two diagonals. State four properties concerning parallelograms and/or their diagonals.

4. Draw a rectangle and its two diagonals. State six properties concerning rectangles and/or their diagonals.

5. Draw a rhombus and its two diagonals. State six properties concerning rhombi and/or their diagonals.

6. Draw a square and its two diagonals. State seven properties concerning squares and/or their diagonals.

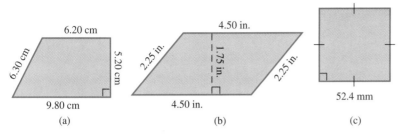

FIGURE 13-95

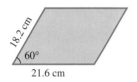

FIGURE 13-96

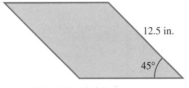

FIGURE 13-97

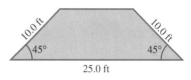

FIGURE 13-98

7. Find the perimeter and area of the quadrilaterals in Fig. 13-95.

8. Find the perimeter and area of the parallelogram in Fig. 13-96.

9. Find the perimeter and area of the rhombus in Fig. 13-97.

10. Find the perimeter and area of the trapezoid in Fig. 13-98.

Round to the nearest dollar.

Applications

11. Find the cost of shingling a two-sided roof, 39 ft 6 in. by 19 ft 6 in. on each side, at $1.25 per square foot.

12. What will be the cost of paving a sidewalk 285 ft long and 5.50 ft wide, at $14.50 per yd^2?

13. How many tiles 8.00 in. square will cover a floor 24.0 ft by 36.0 ft?

14. What will it cost to cement a cellar floor 35.0 ft wide by 31 ft 6 in. long, at 45¢ per ft^2?

15. Find the cost of paving 12.5 miles of road 37.4 ft wide at a cost of $1.15 per square yd.

16. What part of an acre is a baseball diamond, which is a square, 90.0 ft on a side? (1 acre = 43,560 ft^2)

17. Find the area of a garden, in acres, that is 150 ft by 75.0 ft.

18. Find the cost of painting a vertical wall of a house 32.0 ft by 36.0 ft, if a gallon of paint covers 375 ft^2 and costs $12.95.

19. What is the cost of plastering the walls and ceiling of a room 15.2 ft long, 11.8 ft wide, and 8.75 ft high, at $9.50/yd^2, deducting 124 ft^2 for doors and windows?

20. Find the cost of lining a topless rectangular tank 54.3 in. long, 48.9 in. wide, and 36.5 in. deep, with zinc, weighing 5.21 lb per sq. ft, at $1.65 per lb installed.

13-5

CIRCLES

In this section we discuss the circle and the parts of a circle. We show how to calculate the diameter, circumference, and area of a circle.

Definitions

A *circle* (Fig. 13-99) is the set of all points in a plane that are equidistant from a given point C, called the *center* of the circle. The distance from the center to any

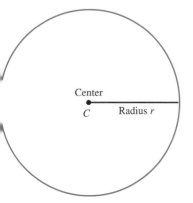

FIGURE 13-99
Circle.

■ FIGURE 13-100
Compass.

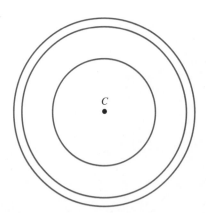

■ FIGURE 13-101
Concentric circles.

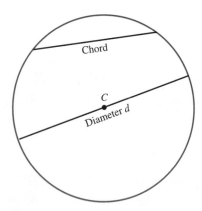

■ FIGURE 13-102
Chord and diameter of a circle.

point on the circle is called the *radius*. The term *radius* may also refer to a line segment whose endpoints are the center and a point on the circle. All the *radii* (plural of *radius*) of a circle, of course, are equal.

A circle may be drawn using an instrument called a *compass* (Fig. 13.-100). Circles that have the same center (Fig. 13-101) are called *concentric*.

A *chord* (Fig. 13-102) is a line segment whose endpoints are on the circle. If the chord passes through the center of the circle, then the chord is called a *diameter*. The diameter *d* of a circle is equal to twice its radius *r*.

DIAMETER AND RADIUS OF A CIRCLE	$d = 2r$ or $r = \dfrac{d}{2}$	74

■ FIGURE 13-103

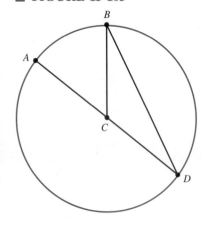

Example 39

If *C* is the center of the circle in Fig. 13-103, name the diameter, three radii, and two chords. If the diameter is 3.72 cm, find the radius.

Solution: The diameter is $\overline{AD}$; three radii are $\overline{AC}, \overline{CD}$, and $\overline{BC}$; two chords are $\overline{AD}$ and $\overline{BD}$. Using Eq. 74, with $d = 3.72$,

$$r = \frac{d}{2} = \frac{3.72}{2} = 1.86 \text{ cm}$$

Circumference

The perimeter of a circle is the distance around a circle and is called its *circumference*. The ratio (quotient) of the circumference *C* of a circle to (divided by) its diameter *d* is always a constant. This constant is called *pi* (a Greek letter pronounced *pie*), and is written π.

$$\frac{\text{Circumference}}{\text{Diameter}} = \frac{C}{d} = \pi$$

The constant π is an irrational number and has no exact fractional or decimal representation. It is approximately equal to 3.14, rounded to 2 decimal places. The value of π, rounded to 25 decimal places, is

$$\pi \approx 3.1415926535897932384626434 \text{ (rounded)}$$

All scientific calculators have a special key for π. You use the π key on your calculator when π is in a calculation, because the value the calculator gives for π has *many* significant digits and you won't be introducing rounding error into the calculations.

We see that the circumference C of a circle is equal to π times the diameter d. We can also express the circumference in terms of the radius:

$$C = \pi d$$

Since $d = 2r$

$$C = \pi(2r) = 2\pi r$$

CIRCUMFERENCE OF A CIRCLE	$C = \pi d$ or $C = 2\pi r$	75

─── **Example 40** ───

Find the circumference of a circular water main 94.6 cm in diameter.

Solution: By Eq. 75,

$$C = \pi d = \pi(94.6 \text{ cm}) = 297 \text{ cm}$$

Area of a Circle

The area A of a circle is equal to π times the square of the radius r.

AREA OF A CIRCLE	$A = \pi r^2$	76

COMMON ERROR Some students have trouble distinguishing the formulas for the circumference and area of a circle. Remember that units of area are square units and the formula for the area of a circle has a square (πr^2) in it.

─── **Example 41** ───

Find the cross-sectional area of a pipe 3.26 cm in diameter.

Solution: By Eq. 74,

$$r = \frac{d}{2} = \frac{3.26 \text{ cm}}{2} = 1.63 \text{ cm}$$

By Eq. 76,

$$A = \pi r^2 = \pi(1.63 \text{ cm})^2 = 8.35 \text{ cm}^2$$

Arcs, Sectors, Segments, and Tangents

Any two points on a circle divide the circle into two parts, called *arcs* (Fig. 13-104). The shorter of the two arcs is named by its two endpoints A and B, and is denoted by the symbol $\overset{\frown}{AB}$, or simply arc AB. A *semicircle* is an arc which is "cut off" by a diameter of the circle. The endpoints of the semicircle are the endpoints of the diameter. In Fig. 13-104, $\overset{\frown}{DE}$ is a semicircle. A semicircle is half a circle.

■ **FIGURE 13-104**
Arcs of a circle.

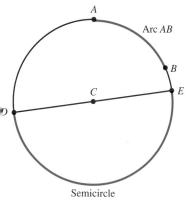

Arc *AB*

Semicircle

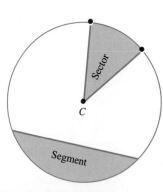

■ FIGURE 13-105
Sector and segment of a circle.

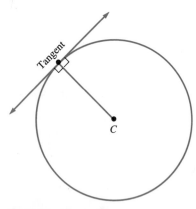

■ FIGURE 13-106
Tangent of a circle.

A *sector* of a circle (Fig. 13-105) is a region bounded by two radii and an arc. It is a pie-shaped region.

A *segment* of a circle (Fig. 13-105) is a region bounded by a chord and an arc.

A *tangent* (Fig. 13-106) is a line that intersects a circle in exactly one point. A tangent is always perpendicular to the radius drawn to the point of contact. We can often use that fact to solve problems involving tangents of circles.

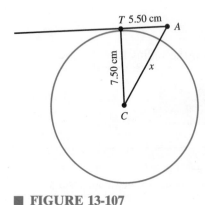

■ FIGURE 13-107

--- **Example 42** ---

From a point A outside a circle of radius 7.50 cm, a tangent is drawn to the circle (Fig. 13-107). The line segment $\overline{TA}$ is 5.50 cm long. Find the distance from A to the center of the circle.

Solution: Since $\overline{AT}$ is a tangent, it is perpendicular to the chord $\overline{CT}$. Since triangle ATC is a right triangle, we can use the Pythagorean Theorem to find the hypotenuse.

$$x = \sqrt{7.50^2 + 5.50^2} = 9.30 \text{ cm}$$

So, the point A is 9.30 cm from the center of the circle.

SECTION 13-5 Exercises Circles

Definitions

1. Identify each of the following in Fig. 13-108: center, radius, diameter, chord, arc, and semicircle.

2. Use a compass to draw a circle with all of the following and label each item: center, radius, diameter, chord, arc, semicircle, and sector.

Circumference and Area

3. Find the circumference and area of a circle whose radius is 3.60 cm.

4. Find the circumference and area of a circle whose diameter is 7.56 feet.

5. Find the area of a circle whose circumference is 12.75 in.

6. Find the circumference of a circle whose area is 15.4 m².

7. Draw a circle with a tangent. Draw the radius to the point of contact. What can you conclude?

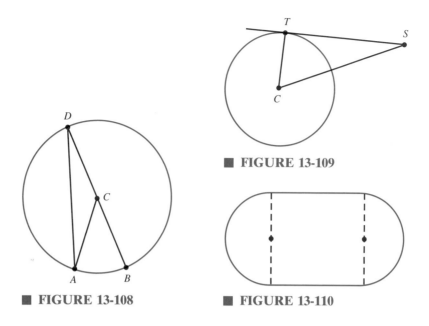

■ FIGURE 13-109

■ FIGURE 13-108

■ FIGURE 13-110

8. Find the distance TS in Fig. 13-109, if the diameter is 12.2 inches, SC is 18.6 inches, $\overline{CT}$ is a radius, and $\overline{ST}$ is a tangent.

9. What is the circumference of a circular pond 25.0 m in diameter?

10. What length of fence is needed to enclose a circular garden 15.8 ft in radius?

11. The orbit of the earth around the sun is nearly circular. Find the length of that orbit, assuming the earth is 92,900,000 miles away from the sun.

12. A certain car tire is 30.0 inches in diameter. How far will the car move forward with one revolution of the wheel?

13. Find the area of the face of a cylindrical piston 3.75 inches in diameter.

14. A circular fountain in a park has a basin that is 17.5 m in circumference. What is the diameter of the basin? What is the area of the basin?

15. The area of the bottom of a circular plate is 325 cm². Find its diameter.

16. Find the cost of cleaning up a circular oil spill 1.75 km in diameter, if it costs $6050/km² for cleanup.

17. A running track (Fig. 13-110) consists of a square with a semicircle at each end. Find the length of the track if the radius of the semicircles is 42.8 yards.

18. A bicycle tire has a radius of 12.0 in. How many revolutions will it make in going a mile?

19. An automobile tire is 26.4 inches in diameter and the wheel makes 8.00 revolutions per second. How far (in miles) will the car move in 1.00 hour?

20. Two ventilating pipes, each 20.0 cm in diameter, are joined to form one single large pipe having the same capacity (cross-sectional area) as the combined capacity of the two single pipes. What is the diameter of the single large pipe?

21. A circle 16.0 inches in diameter has a chord 9.00 inches long. Find the perpendicular distance from the chord to the center of the circle.

22. A circular path is 1.20 m wide and surrounds a circular garden 214.0 m in diameter. Find the area of the path.

23. A round washer has an outside diameter of 31.4 mm. It has a round hole in the center 12.0 mm in diameter. Find the area of the face of the washer.

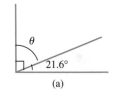

(a)

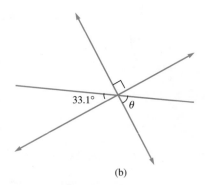

(b)

■ **FIGURE 13-111**

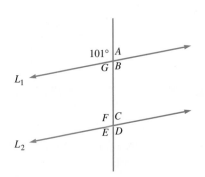

■ **FIGURE 13-112**

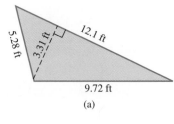

(a)

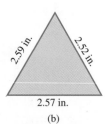

(b)

■ **FIGURE 13-113**

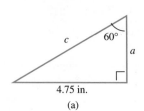

(a)

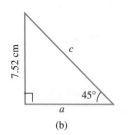

(b)

■ **FIGURE 13-114**

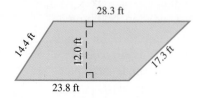

■ **FIGURE 13-115**

■ **FIGURE 13-116**

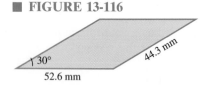

1. Find θ in each part of Fig. 13-111.

2. Find angles A, B, C, D, E, F, and G in Fig. 13-112, given that L_1 and L_2 are parallel.

3. Three stakes, A, B, and C, are placed in the ground so that $\angle ABC$ is 55.3° and $\angle ACB$ is 43.9°. Make a sketch and find $\angle BAC$.

4. Find the perimeter and area of the triangles in Fig. 13-113.

5. A vertical pole is supported by a cable that forms an angle of 38.7° with the pole. Find the angle that the cable makes with the ground.

6. Find the cost of a triangular plot of land 525 ft by 356 ft by 325 ft, at $1050 an acre. (An acre is 43,560 ft².)

7. The rafters of a certain roof form an isosceles triangle with the horizontal beam. What are the two base angles, if the vertex angle is 124°?

8. Find the missing sides of the triangles in Fig. 13-114.

9. Find the altitude and area of an equilateral triangle 6.54 cm on a side.

10. The perimeter of a square floor is 53.5 feet. Find the distance from one corner to the opposite corner.

11. Find the perimeter and area of a 30-60-90 right triangle whose shortest leg is 7.67 cm.

12. A ladder 30.0 feet long stands flat against the side of a building. How many feet must it be pulled out at the bottom so that the top may be lowered 6.00 feet?

13. Draw a parallelogram, a rectangle, a rhombus, and a square. Draw their diagonals. For each figure, state several properties that are true.

14. Find the perimeter and area of the shape in Fig. 13-115.

15. Find the perimeter and area of the parallelogram in Fig. 13-116.

16. What will it cost to carpet a rectangular floor, 6.50 m by 5.40 m, at $12.50 per square meter?

17. A rectangular field is 65.8 rods long and 52.9 rods wide. How many acres does it contain? (1 acre = 160 rod^2)

18. Find the circumference and area of a circle whose diameter is 12.4 cm.

19. A ranger measured the distance around a circular lake to be 2.15 miles. Find its diameter in ft and its area in acres. (1 acre = 43,560 ft^2)

20. Find the diameter of a circular garden whose area is 1.00 acre. (1 acre = 43,560 ft^2)

21. A circular water main 28.0 inches in diameter branches off into four smaller equal circular mains with the same total combined capacity as the larger main. Find the diameter of the smaller mains.

WRITING

22. Plane geometry is everywhere in the real world! Write an essay on plane geometry in either art, nature, or architecture.

TEAM PROJECT

23. In "The Musgrave Ritual" Sherlock Holmes calculates the length of the shadow of an elm tree that is no longer standing. He does know that the elm was 64 ft high and that the shadow was cast at the instant that the sun was grazing the top of a certain oak tree. Holmes held a 6.0-ft fishing rod vertical and measured the length of its shadow at the proper instant. It was 9.0 feet long. He then said, "Of course the calculation now is a simple one. If a rod of six feet threw a shadow of nine, a tree of sixty-four feet would throw one of _____." How long was the shadow of the tree?

Now use Sherlock Holmes' method to find the height of a flagpole, building, or tree on your campus. Find a way to check your result.

14 Solid Geometry

In this chapter we introduce you to the study of solid geometric figures. We define the prism, cylinder, pyramid, cone, and sphere and find the volumes and surface areas of these solids.

Students need to have a working knowledge of solid geometry. Auto manufacturers and consumers use the volume of cylinders to indicate the power of an engine; technicians calculate volumes of tanks and pipes, weights of steel balls, surface areas of domes, and so forth. All these involve concepts of solid geometry.

14-1 PRISMS

In this section we consider the prism. We define various terms, find the volume and surface area of prisms, and solve applications.

Definitions

■ **FIGURE 14-1**
Right prism.

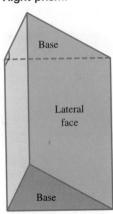

A *right prism* (Fig. 14-1) is a solid that is bounded by two equal parallel polygons (called the *bases*) and a number of rectangles (called the *lateral faces*). The lateral faces are perpendicular to the bases. In Fig. 14-1, the two triangles are the bases; they are equal and parallel. The three rectangles are the lateral faces. In this book, the term *prism* refers to a right prism.

Prisms occur in many forms, depending on the shape of the base. Figure 14-2 shows several examples of prisms: a brick with a rectangular base, an optical prism with a triangular base, and an I beam with an I-shaped base.

Volume of a Prism

The volume of a solid is a measure of the space it occupies or encloses. It is measured in cubic units (cm³, ft³, m³, etc.). The volume of a prism is equal to the product of the area B of the base and the altitude h (Fig. 14-3).

VOLUME OF A PRISM	$V = Bh$	80

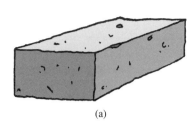
(a)

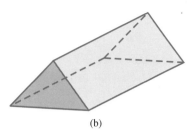
(b)

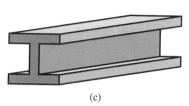
(c)

■ **FIGURE 14-2**
Examples of right prisms.

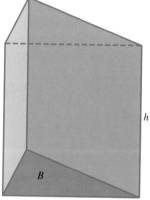

$V = Bh$

■ **FIGURE 14-3**
Volume of a prism.

Example 1

Find the volume of the triangular prism in Fig. 14-4.

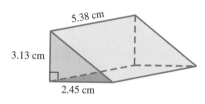

5.38 cm

3.13 cm

2.45 cm

■ **FIGURE 14-4**

Solution: In Fig. 14-4, the bases of the prism are the two triangles on the left and right ends of the prism. We find the area of the triangular base B,

$$B = \frac{1}{2}(2.45)(3.13)$$

$$B = 3.834 \text{ cm}^2$$

Using Eq. 80 to find the volume of the prism, with $h = 5.38$,

$$V = Bh = (3.834)(5.38)$$

Rounding to three significant digits,

$$V = 20.6 \text{ cm}^3$$

Lateral Area of a Prism

■ **FIGURE 14-5**
Lateral area of a prism.

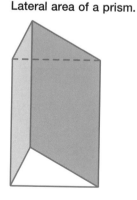

The area of all the lateral faces is called the *lateral area* (Fig. 14-5). To find the lateral area, we just add the areas of the rectangular lateral faces.

Example 2

Find the lateral area of the prism in Fig. 14-4.

Solution: The prism in Fig. 14-4 has three rectangular lateral faces. To calculate these areas, we first need to find the hypotenuse c of the right triangle:

$$c = \sqrt{3.13^2 + 2.45^2} = 3.975$$

Now we add the areas of the three rectangular faces:

$$L = (3.13)(5.38) + (2.45)(5.38) + (3.975)(5.38)$$

Rounding to three significant digits,

$$L = 51.4 \text{ cm}^2$$

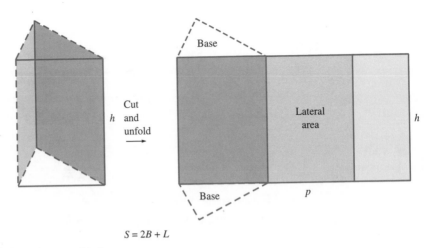

$$S = 2B + L$$

■ **FIGURE 14-6**
Surface area of a prism.

Now let's take a look at an alternative way to calculate the lateral area L. To visualize this, we cut along the dotted edges of the prism (Fig. 14-6) and "unfold" the prism. The "unfolded" lateral area is a rectangle whose length is the perimeter p of the base and whose width is the height h of the rectangular solid. Thus, the lateral area L of a prism is the product of the perimeter p of the base and the height h.

LATERAL AREA OF A PRISM	$L = ph$	81

Example 3

Use Eq. 81 to find the lateral area of the prism in Fig. 14-4.

Solution: In Example 2, we found that the hypotenuse of the triangular base was 3.975 cm. Now we can find the perimeter p of the triangular base:

$$p = 3.13 + 2.45 + 3.975$$
$$p = 9.555 \text{ cm}$$

Substituting into Eq. 81 with $h = 5.38$,

$$L = ph = (9.555)(5.38)$$
$$L = 51.4 \text{ cm}^2$$

This agrees with our answer in Example 2—we just used a different method to find it.

Surface Area of a Prism

The surface area of a solid is the total area of the entire surface of the solid. The surface area S of a prism (Fig. 14-6) is the sum of the area of the two bases, $2B$, and the lateral area, L.

SURFACE AREA OF A PRISM	$S = 2B + L$	82

Example 4

Find the surface area of the prism in Fig. 14-4.

Solution: In Example 1, we found the area B of the base,

$$B = 3.834 \text{ cm}^2$$

In Example 3, we found the lateral area L,

$$L = 51.4 \text{ cm}^2$$

Substituting into Eq. 82,

$$S = 2B + L = 2(3.834) + 51.4$$
$$S = 59.1 \text{ cm}^2$$

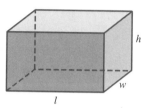

■ FIGURE 14-7
Rectangular solid.

Rectangular Solid

A *rectangular solid,* or box (Fig. 14-7), is a prism with a rectangular base. There are simple formulas for the volume and surface area of a rectangular solid.

To find the volume, we use the area B of the rectangular base:

$$B = lw$$

Substituting into Eq. 80,

$$V = Bh = (lw)h = lwh$$

VOLUME OF A RECTANGULAR SOLID	$V = lwh$	83

The surface of a rectangular solid consists of six rectangles, two of each size: l by w, h by w, and l by h. Thus,

SURFACE AREA OF A RECTANGULAR SOLID	$S = 2(lw + hw + lh)$	84

■ FIGURE 14-8

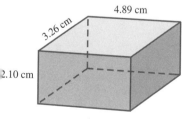

Example 5

Find the volume and surface area of the rectangular solid in Fig. 14-8.

Solution: We use Eq. 83 for the volume of a rectangular solid, letting $l = 4.89$, $w = 3.26$, and $h = 2.10$,

$$V = lwh = (4.89)(3.26)(2.10)$$
$$V = 33.5 \text{ cm}^3$$

We use Eq. 84 for the surface area of a rectangular solid,

$$S = 2(lw + hw + lh) = 2[(4.89)(3.26) + (2.10)(3.26) + (4.89)(2.10)]$$

Rounding to three significant digits,

$$S = 66.1 \text{ cm}^2$$

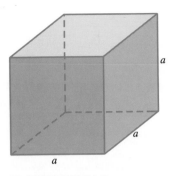

FIGURE 14-9
Cube.

Cube

A *cube* (Fig. 14-9) is a prism, in which all six faces are equal squares. There are simple formulas for its volume and surface area.

To find the volume of a cube with side a, we use the area B of the square base

$$B = a^2$$

Substituting into Eq. 80 with $h = a$,

$$V = Bh = (a^2)a$$

VOLUME OF A CUBE	$V = a^3$	85

The surface of a cube with side a consists of six squares, each having area a^2. Thus

SURFACE AREA OF A CUBE	$S = 6a^2$	86

--- **Example 6** ---

Find the volume and surface area of a cube with side 1.75 cm.

Solution: To find the volume, we use Eq. 85 with $a = 1.75$,

$$V = a^3 = (1.75)^3$$

$$V = 5.36 \text{ cm}^3$$

To find the surface area, we use Eq. 86 with $a = 1.75$,

$$S = 6a^2 = 6(1.75)^2$$

$$S = 18.4 \text{ cm}^2$$

FIGURE 14-10

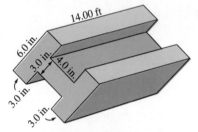

Applications

Applications involve finding areas and volumes of tanks, beams, boxes, rooms, wedges, and so forth.

FIGURE 14-11

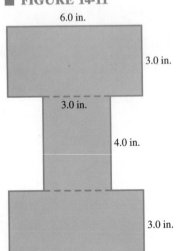

--- **Example 7** ---

Find the number of ft³ of steel necessary to make the I beam in Fig. 14-10.

Solution: The base of the I beam is shaped like an I. We sketch the I, and separate the region into rectangles so we can calculate its area (Fig. 14-11). The area B is

$$B = 6.00(3.00) + 3.00(4.00) + 6.00(3.00) = 48 \text{ in.}^2$$

The beam is 14 ft long, so $h = 14$ ft which we convert to in.:

$$h = 14 \text{ ft} = (14 \text{ ft}) \left(\frac{12 \text{ in.}}{1 \text{ ft}} \right) = 168 \text{ in.}$$

Using Eq. 80 for the volume,

$$V = Bh = (48)(168)$$

$$V = 8064 \text{ in.}^3$$

Converting to ft³,

$$V = 8064 \text{ in.}^3 \left(\frac{1 \text{ ft}^3}{12^3 \text{ in.}^3} \right) = 4.7 \text{ ft}^3 \quad \text{(rounding to two significant digits)}$$

This table summarizes the formulas found in this section.

	Lateral Area	Surface Area	Volume
Prism	$L = ph$	$S = 2B + L$	$V = Bh$
Rectangular solid		$S = 2(lw + hw + lh)$	$V = lwh$
Cube		$S = 6a^2$	$V = a^3$

SECTION 14-1 Exercises Prisms

Prisms

Find the volume and surface area of the following prisms.

1.

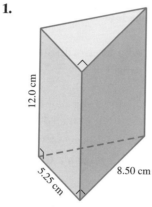

12.0 cm
5.25 cm
8.50 cm

■ FIGURE 14-12

2.

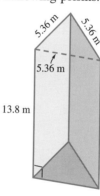

5.36 m
5.36 m
5.36 m
13.8 m

■ FIGURE 14-13

3.

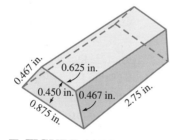

0.467 in.
0.625 in.
0.450 in. 0.467 in.
0.875 in. 2.75 in.

■ FIGURE 14-14

4.

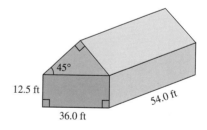

45°
12.5 ft
54.0 ft
36.0 ft

■ FIGURE 14-15

Rectangular Solids

Find the volume and surface area of the rectangular solids with these dimensions.

5. $l = 4.60$ cm, $w = 5.80$ cm, $h = 2.50$ cm
6. $l = 1.55$ ft, $w = 2.75$ ft, $h = 4.25$ ft
7. $l = 5.25$ in., $w = 2.56$ in., $h = 4.58$ in.
8. $l = 3.24$ m, $w = 5.38$ m, $h = 1.47$ m

Cubes

Find the volume and surface area of the cubes with the given side.

9. $a = 3.52$ ft   **10.** $a = 4.58$ cm
11. $a = 0.549$ m   **12.** $a = 0.375$ in.

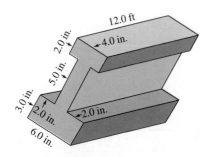

■ FIGURE 14-16

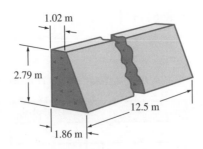

■ FIGURE 14-17

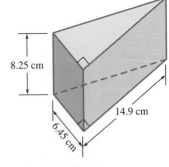

■ FIGURE 14-18

13. A brick measures 8.85 cm by 5.75 cm by 21.1 cm. Find the volume of the brick.

14. A classroom is 45.0 ft long, 28.0 ft wide, and 12.0 feet high. If there are 36 people in the room, find the number of cubic feet of air for each person in the room.

15. White light is separated into a spectrum by passing through a prism which is 4.25 cm in height and whose base is a right isosceles triangle whose equal sides are 2.68 cm long. Find its volume.

16. Find the volume of a gas tank that is a rectangular prism 13.0 in. by 15.5 in. by 36.0 in.

17. A rectangular tank is to be lined with stainless steel sheets. Find the area to be lined (bottom and sides) if the tank is 12.5 ft long, 8.75 ft wide, and 4.25 ft high.

18. A rectangular storage shed is to be painted. Find the area to be painted (top and sides) if the shed is 5.20 m long, 4.50 m wide, and 6.50 m high.

19. Find the number of ft³ of steel necessary to make the I beam in Fig. 14-16.

20. How many cubic yards of gravel can a truck carry if the trailer is 6.35 yd long, 2.75 yd wide, and 1.45 yd high?

21. How many loads of gravel will be needed to cover a driveway 75.2 yd long and 12.5 ft wide, to a depth of 15.0 in., if one truckload contains 8.00 yd³ of gravel?

22. Find the number of cubic meters of concrete required to make the concrete retaining wall in Fig. 14-17. Find the total cost of the concrete at $87.50 per cubic meter.

23. Find the volume of the solid steel wedge in Fig. 14-18. Find its weight (in kg) if the steel weighs 7.82 grams per cubic centimeter.

24. A cube has a surface area of 57.8 cm². Find the length of its side and its volume.

25. A rectangular prism with a square base has a volume of 892 cubic inches. The height is 8.00 inches. Find the total surface area.

14-2   CYLINDERS

In this section we define the cylinder, and find its volume and surface area. We study liquid capacity and conversion of units, and solve applications involving automobile cylinders, pipes, and cylindrical tanks.

A *circular cylinder* (Fig. 14-19) is a geometric solid which has two equal parallel *circular* bases. It is similar to a prism and can be thought of as a "circular prism." Although there are non-circular cylinders, all the cylinders that we deal with in this book are circular cylinders. Whenever we say "cylinder" we are referring to a circular cylinder.

FIGURE 14-19
Cylinder.

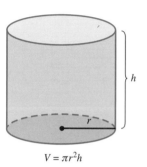

$V = \pi r^2 h$

FIGURE 14-20
Volume of a cylinder.

FIGURE 14-21

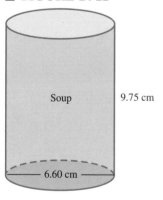

Soup   9.75 cm

6.60 cm

FIGURE 14-22
Surface area of a cylinder.

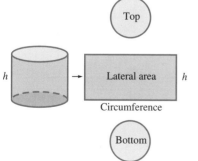

Volume of a Cylinder

Since a cylinder can be thought of as a circular prism, we use Eq. 80 for the volume of a prism to determine the equation of the volume of a cylinder having height h and base radius r (Fig. 14-20):

$$V = Bh$$

Since the base is a circle, its area B is

$$B = \pi r^2$$

Substituting, we get

$$V = (\pi r^2)h$$

VOLUME OF A CYLINDER	$V = \pi r^2 h$	87

Example 8

Find the volume of a cylindrical soup can (Fig. 14-21) whose diameter is 6.60 cm and height is 9.75 cm.

Solution: We use Eq. 87 for the volume,

$$V = \pi r^2 h$$

First we need the radius,

$$r = \frac{d}{2} = \frac{6.60}{2} = 3.30 \text{ cm}$$

So,

$$V = \pi(3.30)^2(9.75) = 334 \text{ cm}^3$$

Surface Area of a Cylinder

The surface area of a cylinder can be visualized by "unfolding" the surface of the cylinder. Think of a soup can; if we remove the top and bottom "lids" and then slit the can up one side and spread out the lateral area (Fig. 14-22), we get two circles and a rectangle. By Eq. 76, the area B of each circle is

$$B = \pi r^2$$

The lateral area is a rectangle, whose length is the circumference C of the base and whose width is the height h of the can. (Note that this also follows from Eq. 81 that the lateral area of a prism is the perimeter of the base times the height.)

So,

$$L = Ch$$

Since

$$C = 2\pi r$$

We substitute and get

$$L = (2\pi r)h = 2\pi rh$$

Substituting in Eq. 82, we get

$$S = 2B + L = 2(\pi r^2) + 2\pi rh$$

SURFACE AREA OF A CYLINDER	$S = 2\pi r^2 + 2\pi rh$	88

Example 9

Find the surface area of the can in Example 8 (Fig. 14-21).

Solution: We use Eq. 88

$$S = 2\pi r^2 + 2\pi rh$$

Since $r = 3.30$ cm and $h = 9.75$ cm, we substitute and get

$$S = 2\pi(3.30)^2 + 2\pi(3.30)(9.75)$$
$$= 271 \text{ cm}^2$$

Liquid Capacity

In solving volume problems, we are often asked to find the volume of liquids in gallons, liters, quarts, and so forth. This type of volume is called *liquid capacity*. These problems are solved by finding the volume in cubic units and then converting to liquid capacity units.

Example 10

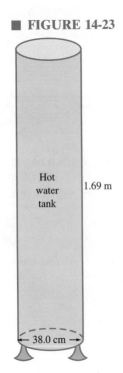

■ FIGURE 14-23

A hot water tank (Fig. 14-23) has a diameter of 38.0 cm and a height of 1.69 m. How many liters does it hold?

Solution: We find the radius from the diameter

$$r = \frac{d}{2} = \frac{38.0}{2} = 19.0 \text{ cm}$$

The units of r and h must agree, and so we convert $h = 1.69$ m to centimeters.

$$1.69 \text{ m} = (1.69 \text{ m})\left(\frac{100 \text{ cm}}{1 \text{ m}}\right) = 169 \text{ cm}$$

Using Eq. 87 to calculate V

$$V = \pi r^2 h = \pi(19.0)^2(169)$$
$$V = 191700 \text{ cm}^3$$

We must now convert cm³ to liters. One liter (ℓ) is 1000 cm³.

So,

$$V = (191700 \text{ cm}^3)\left(\frac{1 \ell}{1000 \text{ cm}^3}\right)$$
$$V = 192 \ell$$

Hollow Cylinders

A *hollow cylinder* is the shell between two concentric cylinders, like a pipe or tube. To find the volume of a hollow cylinder, we subtract the volume of the small cylinder from the volume of the large cylinder. If the radius of the small cylinder is r and the radius of the large cylinder is R, then

$$V = \pi R^2 h - \pi r^2 h$$

Example 11

Find the volume of metal in a 4.75-m length of the hollow iron pipe shown in Fig. 14-24.

Solution: The large radius R is

$$R = \frac{4.86}{2} = 2.43 \text{ cm}$$

The small radius r is

$$r = \frac{3.64}{2} = 1.82 \text{ cm}$$

The length of the cylinder h, in cm, is

$$h = 4.75 \text{ m} = 475 \text{ cm}$$

So,

$$
\begin{aligned}
V &= \pi R^2 h - \pi r^2 h \\
&= \pi(2.43)^2(475) - \pi(1.82)^2(475) \\
&= 3870 \text{ cm}^3
\end{aligned}
$$

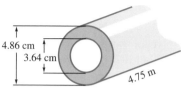

4.86 cm
3.64 cm
4.75 m

FIGURE 14-24

SECTION 14-2 Exercises Cylinders

Volume and Surface Area

1. Find the volume of a cylinder whose radius is 3.46 ft and whose height is 9.52 ft.
2. Find the surface area of the cylinder in Exercise 1.
3. Find the surface area of a cylinder whose diameter is 5.65 cm and whose height is 8.75 cm.
4. Find the volume of the cylinder in Exercise 3.

Liquid Capacity

5. Find the liquid capacity, in gallons, of the tank in Fig. 14-25. (1 gal $\approx$ 231 in.^{3})
6. Find the liquid capacity, in mℓ, of a can 6.75 cm in diameter and 10.3 cm high. (1 mℓ = 1 cm^3)

Hollow Cylinders

7. Find the volume of the hollow cylinder in Fig. 14-26.
8. Find the volume of a hollow cylinder with an inside radius of 5.25 in., an outside radius of 7.75 in., and a height of 6.50 in.

FIGURE 14-25

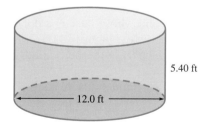

5.40 ft

12.0 ft

FIGURE 14-26

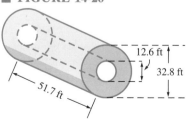

12.6 ft
32.8 ft
51.7 ft

9. Find the amount of steel in a steel shaft 35.0 cm long with a diameter of 2.35 cm.

10. Find the volume of silage that can be stored in a cylindrical silo 15.0 ft in diameter and 315.0 ft high.

11. Find the volume of copper necessary to make a mile of copper wire 0.3125 inch in diameter.

12. Find the amount of wood (in m³) in a cylindrical cedar post 12.5 cm in diameter and 2.50 m long.

13. A stainless steel milk tank 5.25 ft in diameter and 315 ft long is being constructed. Find the amount, in ft², of stainless steel needed, allowing 10.0% for seams and waste.

14. The cylindrical tank of an electric hot water heater is 21.5 inches in diameter and 58.5 inches in height. How many square feet of insulation are needed to wrap the lateral area and top?

15. In automotive engineering, the volume (or displacement) of a cylinder in a piston engine is measured in cubic centimeters (cc). The diameter of the cylinder is called the bore and the height is called the stroke. What is the displacement of a cylinder with a bore of 72.0 mm and a stroke of 85.0 mm?

16. What is the engine displacement (in cc) of an 8-cylinder engine if each cylinder has a bore of 65.0 mm and a stroke of 88.0 mm? (Refer to Exercise 15.)

17. An oil tank is 24.0 ft long and has a diameter of 10.5 ft. How many gallons will it hold? (1 ft³ ≈ 7.48 gal)

18. An artesian well driller had to drill down 287 ft before striking water. If the pipe is 3.00 inches in diameter, how much (in gallons) is the "head" (water stored in the pipe)?

19. A certain bushing is a hollow cylinder with an outer diameter of 16.0 mm, an inner diameter of 10.0 mm, and a length of 24.0 mm. Compute the volume of the bushing in cm³.

■ FIGURE 14-27

20. Find the amount of iron in a 10.0-ft length of the hollow pipe in Fig. 14-27.

1.625 in.
1.250 in.

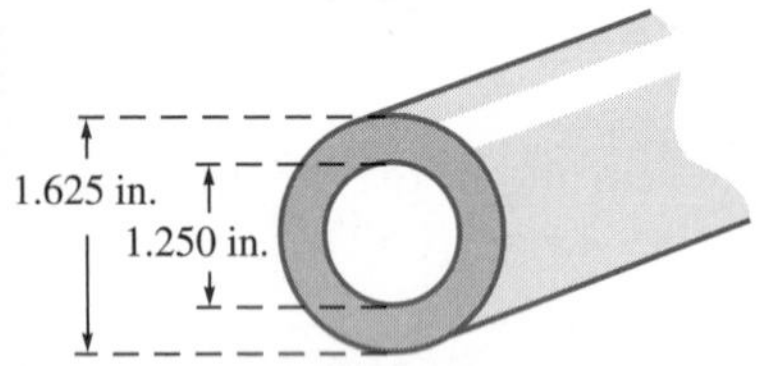

21. A gasoline storage tank has a diameter of 20.0 m and a height of 8.58 m. When the tank is half full, how many liters of gasoline remain? (1 ℓ = 1000 cm³)

22. A water main is 16.0 inches in diameter. If the water flows at 1.56 ft per second, how much water (in gallons) flows in one hour?

23. An engine piston has a height of 9.80 cm and a volume of 352 cubic centimeters. Find the diameter of the piston.

24. A cylindrical quart can is 4.50 in. high. What is its diameter? (1 gal ≈ 231 in.³)

14-3 PYRAMIDS AND CONES

In this section we define pyramids and cones and find their volumes and surface areas. We also discuss weight, density, and other applications.

Pyramid

A *pyramid* (Fig. 14-28) is a solid whose base is a polygon and whose lateral faces are triangles with a common vertex. Figure 14-28 shows a pyramid whose base is a square. The *altitude h* (or *height*) is the perpendicular distance from the vertex to the base.

A *regular* pyramid is a pyramid whose base is a regular polygon and whose vertex is centered over the base. All pyramids we consider in this chapter will be regular pyramids. The faces of a regular pyramid are equal isosceles triangles.

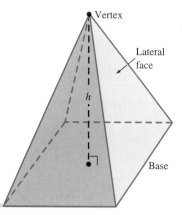

FIGURE 14-28
Pyramid.

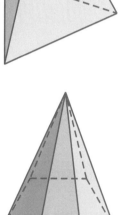

FIGURE 14-29
Pyramids.

Pyramids are named by the shape of their base. Figure 14-28 shows a square pyramid, and Fig. 14-29 shows a triangular pyramid and a hexagonal pyramid.

Volume of a Pyramid

The volume of a pyramid is *one-third* the volume of a prism of the same base and height. Recall from Section 14-1 that the volume of a prism is the product of the area of the base B and the height h. So, the volume V of a pyramid (Fig. 14-30) is one-third the area of the base B times the height h.

VOLUME OF A PYRAMID	$V = \dfrac{1}{3}Bh$	89

Example 12

Find the volume of a square pyramid (Fig. 14-30) if the base is 23.8 cm on a side and the height is 32.5 cm.

Solution: For this pyramid,

$$h = 32.5 \text{ cm}$$

Since the base is a square, its area B is

$$B = (23.8)^2 = 566.4 \text{ cm}^2$$

Substituting into Eq. 89, we get

$$V = \frac{1}{3}Bh = \frac{1}{3}(566.4)(32.5)$$

$$V = 6140 \text{ cm}^3$$

Lateral and Surface Area of a Regular Pyramid

In a regular pyramid (Fig. 14-31), each lateral face is an isosceles triangle; the altitude s of the isosceles triangle is called the *slant height* of the pyramid. The area A of one lateral face (one of the isosceles triangles) is one-half the product of the base b of the triangle times the slant height s.

$$A = \frac{1}{2}bs$$

FIGURE 14-30
Volume of a pyramid.

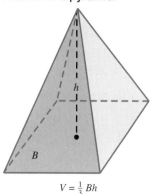

$V = \frac{1}{3}Bh$

FIGURE 14-31
Regular pyramid.

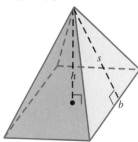

If we multiply this area A by the number n of lateral faces, then we have the lateral area L of the pyramid.

$$L = nA$$

Substituting,

$$L = n\left(\frac{1}{2}bs\right)$$

Rearranging the factors,

$$L = \frac{1}{2}(nb)s$$

We note that nb is the perimeter p of the base of the pyramid, so,

$$nb = p$$

Substituting,

$$L = \frac{1}{2}ps$$

So, the lateral area of a pyramid is one-half the perimeter p of the base times the slant height s of the pyramid.

LATERAL AREA OF A PYRAMID	$L = \frac{1}{2}ps$	90

The surface area of a pyramid includes both the area B of the base and the lateral area L.

SURFACE AREA OF A PYRAMID	$S = B + L$	91

Example 13

Find the lateral area and the surface area of a square pyramid with a base 4.76 cm on a side and a slant height of 6.23 cm.

Solution: We sketch the pyramid in Fig. 14-32. For this pyramid, the base is a square of side 4.76, so

$$p = 4(4.76) = 19.04 \text{ cm}$$

and

$$s = 6.23 \text{ cm}$$

Substituting into Eq. 90 to find the lateral area of a pyramid,

$$L = \frac{1}{2}ps = \frac{1}{2}(19.04)(6.23)$$

$$L = 59.31 \text{ cm}^2$$

To find the surface area, we need to calculate the area B of the square base,

$$B = s^2 = (4.76)^2$$

$$B = 22.66 \text{ cm}^2$$

Substituting into Eq. 91 for the surface area of the pyramid,

$$S = B + L = 22.66 + 59.31$$

$$S = 82.0 \text{ cm}^2$$

■ **FIGURE 14-32**

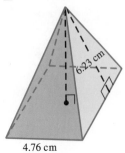

6.23 cm

4.76 cm

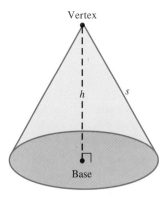

FIGURE 14-33
Right circular cone.

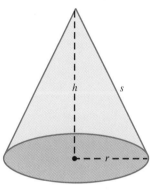

$V = \frac{1}{3}\pi r^2 h$

FIGURE 14-34
Volume of a cone.

FIGURE 14-35

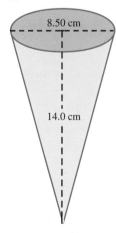

8.50 cm

14.0 cm

Cone

A *circular cone* (Fig. 14-33) is a geometric solid that has a circular base and a lateral surface which tapers from the base to a point called the *vertex*. The *altitude h* (or *height*) of a cone is the perpendicular distance from the vertex to the base. The *slant height* of a cone is the distance from the vertex to a point on the circular base. In a *right* circular cone, the line from the vertex to the center of the base is perpendicular to the base.

Volume of a Cone

We can think of a cone as a "circular pyramid." Since the volume of a pyramid is one-third the volume of the prism with the same dimensions, then the volume of a cone is one-third the volume of the cylinder which has the same radius and height (Fig. 14-34), or

$$V = \frac{1}{3}(\pi r^2 h)$$

where r is the radius of the base and h is the height of the cone.

VOLUME OF A CONE	$V = \frac{1}{3}\pi r^2 h$	92

Example 14

Find the volume of a cone whose diameter is 8.50 cm and height is 14.0 cm.

Solution: We sketch the cone in Fig. 14-35. For this cone, $d = 8.50$ cm and $h = 14.0$ cm. We find the radius,

$$r = \frac{d}{2} = \frac{8.50}{2} = 4.25 \text{ cm}$$

Substituting into Eq. 92 for the volume of the cone,

$$V = \frac{1}{3}\pi r^2 h = \frac{1}{3}\pi(4.25)^2(14.0)$$

$$V = 265 \text{ cm}^3$$

Lateral and Surface Area of a Cone

The lateral area of a cone is determined using Eq. 90

$$L = \frac{1}{2}ps$$

where s is the slant height of the cone and the perimeter p is the circumference of the circular base. Thus,

$$p = 2\pi r$$

Substituting into Eq. 90, we have

$$L = \frac{1}{2}ps = \frac{1}{2}(2\pi r)s$$

or

$$L = \pi rs$$

LATERAL AREA OF A CONE	$L = \pi rs$	93

The total surface area is

$$S = B + L$$

Substituting $B = \pi r^2$ and $L = \pi rs$, we have

$$S = \pi r^2 + \pi rs$$

SURFACE AREA OF A CONE	$S = \pi r^2 + \pi rs$	94

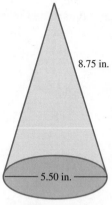

■ **FIGURE 14-36**

─── **Example 15** ───

Find the lateral and surface areas of a cone (Fig. 14-36) with diameter 5.50 inches and slant height 8.75 inches.

Solution: For this cone, $d = 5.50$ in. and $s = 8.75$ in. First we find r,

$$r = \frac{5.50}{2} = 2.75 \text{ in.}$$

Substituting into Eq. 93 to find the lateral area L,

$$L = \pi rs = \pi(2.75)(8.75)$$
$$L = 75.6 \text{ in.}^2$$

Substituting into Eq. 94 to find the surface area S,

$$S = \pi r^2 + \pi rs = \pi(2.75)^2 + 75.6$$
$$S = 99.4 \text{ in.}^2$$

Weight

Often we need to find the weight of objects, and weight depends on volume. To calculate the weight of an object, we must know the weight per unit volume of the material. The weight per unit volume of a material is called its *density*. The weight W of an object is the density d of the material times the volume V of the object.

$$\text{Weight} = (\text{density})(\text{Volume}) \qquad \text{or} \qquad W = dV$$

The densities of common materials can be found in a density table found in most engineering handbooks. In this book, we provide the densities as needed.

─── **Example 16** ───

Find the weight of the cone in Example 14 (Fig. 14-35), if it is made of bronze weighing 9.20 grams per cm³.

Solution: The density is the weight per unit volume, so it's given that the density of bronze is

$$d = 9.20 \text{ g/cm}^3$$

Using the volume from Example 14,

$$V = 265 \text{ cm}^3$$

Substituting into the weight equation,

$$W = dV = \left(9.20 \frac{\text{g}}{\text{cm}^3}\right)(265 \text{ cm}^3)$$

$$W = 2440 \text{ g}$$

Volume

1. Find the volume of a square pyramid 6.21 cm on a side and 7.36 cm in height.

2. Find the volume of a regular triangular pyramid 3.52 inches on a side with an altitude of 6.05 inches.

3. Find the volume of a cone whose radius is 35.2 mm and height is 87.6 mm.

4. Find the volume of a cone whose diameter is 0.0781 m and height is 0.0459 m.

Surface Area

5. Find the surface area of a regular triangular pyramid whose base is 6.56 in. on a side and slant height is 9.25 in.

6. Find the surface area of a square pyramid whose base is 14.25 in. on a side and slant height is 8.75 in.

7. Find the surface area of a cone whose diameter is 15.0 in. and slant height is 8.50 in.

8. Find the surface area of a cone whose radius is 2.50 ft and slant height is 4.75 ft.

9. A cone has a diameter of 12.0 in., a height of 18.0 in., and a slant height of 19.0 in. Find the surface area of the cone.

10. A square pyramid has a base 24.0 cm on a side, a height of 35.0 cm, and a slant height of 37.0 cm. Find the surface area of the pyramid.

Weight

11. Find the weight of the pyramid in Exercise 1 if it is made of gold weighing 19.3 grams per cm^3.

12. Find the weight of the pyramid in Exercise 2 if it is made of bronze, which weighs 548 pounds per cubic foot.

13. Find the weight of the cone in Exercise 3 if it is made of silver. The density of silver is 10.5 g/cm^3.

14. Find the weight of the cone in Exercise 4 if it is made of steel whose density is 7.84 g/cm^3.

Applications

15. A sand pile is in the shape of a cone, 8.54 ft in diameter and 5.56 ft in height. Find the weight of the sand pile if the sand weighs 102 lb per cubic foot.

16. A granite monument is in the shape of a tall square pyramid. It is 27.0 m tall and the base is 2.25 m on a side. Find the weight (in metric tons) of the monument if the granite weighs 5.95 g/cm^3.

17. Find the weight (in kg) of a conical brass casting which is 46.7 cm high and 12.9 cm in diameter. The density of brass is 8.65 g/cm^3.

18. A drinking cup is in the shape of a cone, 3.00 inches in diameter and 3.50 inches high. Find its volume. Will the drinking cup hold *more* than or *less* than $\frac{1}{2}$ cup of water? (1 pint $\approx$ 28.9 in.^{3})

19. A canvas tent is in the shape of a square pyramid. It is 4.50 ft tall with a slant height of 5.55 ft, and the base is 6.50 ft on a side. How much canvas was needed to make the tent (floor and sides)?

20. A steeple, in the shape of a hexagonal pyramid, is to be painted. It is 30.0 m high with a slant height of 30.8 m, and its base is 4.00 m on a side. A liter of paint will cover 11.8 m^2. How much will the paint cost to paint the sides of the steeple with two coats of paint, if a 5.00-liter can of paint costs $18.95?

21. The tin cover of a maple syrup sap bucket is conical. It is 15.0 inches in diameter and 2.50 inches in height with a slant height of 7.91 inches. Find the lateral area of the cover. Then find the weight of the cover, allowing 5% for overlap, if tin sheeting weighs 2.55 pounds per square foot.

22. A grain bin on a combine is conical. It is 58.0 inches high and 54.0 inches in diameter. Find the amount of metal (in ft²) necessary to make the open bin and the volume (in ft³) of grain it will hold. (Hint: Use the Pythagorean Theorem to find the slant height.)

23. A concrete monument, in the shape of a square pyramid, weighs 7250 kg. The concrete weighs 2.21 grams per cm³. If the base of the monument is 1.00 m on a side, how tall is the monument?

24. A cone, 12.0 cm in slant height, has a lateral surface area of 226 cm². Find its volume.

25. The volume V of a *frustum* (Fig. 14-37) of a cone is

$$V = \frac{1}{3}\pi h(R^2 + r^2 + Rr)$$

where h is the height and R and r are the radii of the two bases. Find the capacity in gallons of a pail that is 1.25 feet high, 9.00 inches wide at the bottom, and 12.0 inches wide at the top. How much would a full pail of water weigh, if the density of water is 62.4 lb/ft³? (1 ft³ ≈ 7.48 gal)

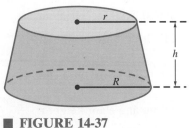

■ **FIGURE 14-37**
Frustum of a cone.

14-4   SPHERES

In this section we define a sphere and find its volume and surface area. These are essential skills in dealing with applications such as ball bearings, geodesic domes, planets, and satellites.

Definitions

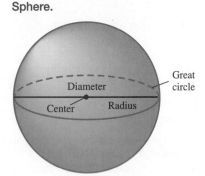

■ **FIGURE 14-38**
Sphere.

A *sphere* (Fig. 14-38) is a geometric solid in which every point of the sphere is equidistant from a given point, called the *center* of the sphere. The *radius* of a sphere is the distance from the center to a point of the sphere. The *diameter* of a sphere is the length of a line segment through the center with its endpoints on the sphere. Thus, the diameter is twice its radius, or the radius is half the diameter,

$$r = \frac{d}{2}$$

A *great circle* of a sphere (Fig. 14-38) is a circle cut by a plane which passes through the center of the sphere. The radius of a great circle is the radius r of the sphere. A great circle separates the sphere into two equal half-spheres, called *hemispheres.*

Volume of a Sphere

The volume V of a sphere is the product of $\frac{4}{3}\pi$ times the cube of the radius r.

VOLUME OF A SPHERE	$V = \frac{4}{3}\pi r^3$	95

Surface Area of a Sphere

The surface area S of a sphere is 4 times the area of the great circle of the sphere. Since the radius of a great circle is the radius r of the sphere, then the area A of a great circle is

$$A = \pi r^2$$

and the surface area S of a sphere is

$$S = 4A$$

Substituting,

$$S = 4(\pi r^2)$$

SURFACE AREA OF A SPHERE	$S = 4\pi r^2$	96

Example 17

A baseball has a diameter of 2.96 in. Find its volume and surface area.

Solution: To find the volume, we first find the radius, using $d = 2.96$ in.,

$$r = \frac{d}{2} = \frac{2.96}{2} = 1.48 \text{ in.}$$

Substituting into Eq. 95 for the volume,

$$V = \frac{4}{3} \pi r^3 = \frac{4}{3} \pi (1.48)^3$$

$$V = 13.6 \text{ in.}^3$$

To find the surface area, we use Eq. 96,

$$S = 4\pi r^2 = 4\pi (1.48)^2$$

$$S = 27.5 \text{ in.}^2$$

SECTION 14-4　Exercises　Spheres

Volume and Surface Area of a Sphere

Find the volume and surface area of each of the following spheres.

1. Tennis ball, radius = 3.32 cm
2. Beach ball, radius = 11.2 in.
3. Earth, diameter = 7920 miles
4. Ball bearing, diameter = 0.925 mm

Applications

5. A weather balloon is filled with helium, to a diameter of 2.50 m. Find its volume.
6. Oil comes out of a sprayer in spherical droplets 1.005 mm in diameter. Find the volume (in cm³) of 1 million droplets.
7. A glass globe is filled with water. Find its capacity in gallons if the diameter of the globe is 7.61 inches.
8. A hemispherical vat has a diameter of 21.3 inches. Find its capacity in gallons.
9. A spherical gas storage tank, 35.4 m in diameter, is to be painted. How much paint is needed for two coats, if a 4.00-liter can of paint covers 50.0 m²?
10. A company manufactures plastic hemispherical covers for plants. The covers are 42.5 cm in diameter. Find the cost of making 125,000 covers, if the plastic costs $4.75 per square meter.
11. A hemispherical dome on a state house has a radius of 72.5 ft. Find the cost of applying a thin layer of gold to the dome at $1.93 per square ft.

12. Two fuel storage tanks each have the same capacity. One is spherical with a diameter of 3.00 ft. The other is cylindrical with the same diameter and a height of 2.00 ft. Which of the two tanks required more surface area material? How much more?

13. Find the weight of a golden sphere 7.00 inches in diameter. (The density of gold is 1204 lb per cubic foot.)

14. Find the weight of a brass sphere 3.25 ft in diameter. Brass weighs 511 lb per cubic foot.

15. For the shot put, a contestant throws (puts) an iron sphere (a shot). Find the weight of a shot 5.00 inches in diameter, if it is made of iron whose density is 491 lb/ft^3.

16. Find the weight of 1000 steel ball bearings 2.50 mm in diameter, if the density of the steel is 7.79 g/cm^3.

17. A certain sphere holds 1.00 cubic foot of water. What is its diameter? What is its surface area? Hint: $r = \sqrt[3]{\dfrac{3V}{4\pi}}$

18. A certain sphere holds 1.00 liter of water. What is its diameter? What is the surface area?

19. Find the volume of a sphere whose surface area is 1.00 m^2.

■ Chapter 14 REVIEW EXERCISES

1. Find the volume and surface area of the rectangular prism whose length is 6.75 cm, width is 4.52 cm, and height is 5.88 cm.

2. Find the volume and surface area of the cube whose side is 3.75 inches.

3. Find the volume and surface area of the prism in Fig. 14-39.

■ FIGURE 14-39

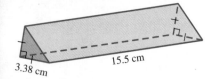

15.5 cm

3.38 cm

4. An excavation for a building foundation is 50.0 ft long, 32.5 ft wide, and 12.0 ft deep. How many cubic yards of soil are removed? How many truckloads is that if the truck can haul 6.50 yd^3 per load?

5. A certain silo is in the form of a prism with octagonal bases. Each side of the octagonal base is 8.55 ft and the silo is 32.0 feet high. How much paint will be needed to paint the lateral sides of the silo, if one gallon of paint covers 257 ft^2?

■ FIGURE 14-40

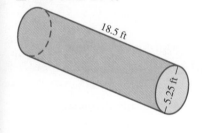

18.5 ft

5.25 ft

6. Find the volume and surface area of the cylinder in Fig. 14-40.

7. A steam boiler has 18 cylindrical flues, each 3.00 inches in diameter and 8.00 feet long. Find the total area of heating surface (the lateral area) of the 18 flues.

8. A brass rod is 0.360 inch in diameter and 6.50 feet long. Find its volume in cubic inches.

9. The Alaskan pipeline is 122 cm in diameter. How many kiloliters of crude oil will a mile section hold?

10. A hollow concrete tube is 24.5 ft long. The inside diameter of the tube is 28.6 in. and the outside diameter is 36.7 inches. Find the number of cubic feet of concrete in the tube.

11. Find the volume of a triangular pyramid 34.5 mm high and whose base is 27.5 mm on each side.

12. Find the total surface area of a cone whose height is 3.75 ft, slant height is 4.37 ft, and radius is 2.25 ft.

13. One of the colossal Egyptian pyramids, the great pyramid of Cheops in Gizeh, originally rose to a height of 481 ft and had a slant height of 611 ft. Its square

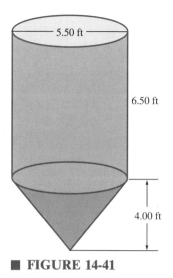

■ FIGURE 14-41

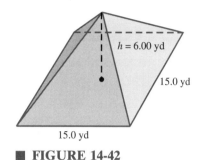

■ FIGURE 14-42

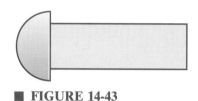

■ FIGURE 14-43

base was 750.0 ft on each side. Find the volume and the total surface area of the great pyramid of Cheops.

14. Find the volume of the sawdust collector in Fig. 14-41.

15. The roof of a building (Fig. 14-42) is in the shape of a square pyramid and must be shingled. Find the lateral area of the roof. How many boxes of shingles are needed, if each box covers 50.0 ft²?

16. A pendulum weight is a solid gold cone, 3.00 inches in diameter and 5.00 inches in height. Find its value if gold sells for $390.00 an ounce. (The density of gold is 1204 lb/ft³.)

17. Find the surface area, volume, and weight of an iron ball, 3.50 inches in diameter. (The density of iron is 491 pounds/ft³.)

18. A hemispherical radome protects a radar antenna on an aircraft. If the radome is 5.85 m in diameter, what volume is enclosed? If the radome is constructed of sheet material weighing 1.81 kg/m², find the weight of the radome.

19. Find the weight of the steel round-head rivet in Fig. 14-43. The shaft is 1.32 cm in diameter and 3.87 cm in length. The head is hemispherical and is 2.35 cm in diameter. The steel weighs 7.79 g/cm³.

WRITING

20. In this chapter, we studied geometric solids, including the prism, rectangular solid, cube, cylinder, pyramid, cone, and sphere. In your own words, describe each of these geometric solids. Which solid is your favorite? Give an example of your favorite solid from real life.

TEAM PROJECT

21. Which weighs more, a million dollars in $1 bills or a million dollars in gold?

Right Triangle Trigonometry

One of the most important topics in mathematics is trigonometry. In this chapter, we lay the foundation for solving applications involving right triangles using trigonometry.

Trigonometry is the branch of mathematics that deals mostly with the solution of triangles and the functions derived from right triangles. The six trigonometric functions are introduced here and are used to solve right triangles. We consider the inverse trigonometric functions and some of the many applications involving right triangle trigonometry, such as problems in surveying, navigation, angles of depression and elevation, and indirect measurement.

15-1

SINE, COSINE, AND TANGENT FUNCTIONS

In this section we define an angle in standard position and define the three most useful trigonometric functions. We calculate the trigonometric functions, for a given angle or for a given point on the terminal side of the angle. We will use this in Sec. 15-4 to solve applications involving right triangles.

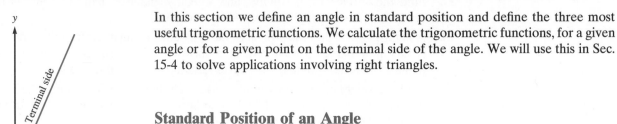

Standard Position of an Angle

In Sec. 13-1 we discussed the definition of an angle formed by a rotating ray. In this section, we extend that concept with angles drawn on the coordinate axes. If we draw an angle on the coordinate axes, with its vertex at the origin and one side along the positive x axis (initial side), then the angle is said to be in *standard position* (Fig. 15-1). All the angles referred to in this chapter will be positive acute angles and their terminal sides will be in quadrant I.

■ **FIGURE 15-1**
Standard position of an angle.

■ **FIGURE 15-2**

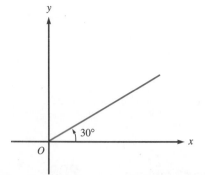

┌─ **Example 1** ─────
Sketch a 30° angle in standard position.

Solution: The vertex of the angle is at the origin and the one side is along the positive x axis (Fig. 15-2). The terminal side is drawn in quadrant I, so that a 30° angle is formed. The position of the terminal side may be found using a protractor if accuracy is important.

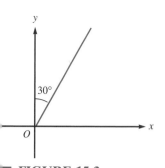

FIGURE 15-3

Angle *not* in standard position.

Example 2

Sketch the positive angle θ in standard position for which the point (3, 4) is on the terminal side. Measure the angle.

Solution: We draw the vertex of the angle at the origin and draw one side along the positive *x* axis. We plot the point (3, 4) and draw a line from the vertex through the point (Fig. 15-4). Using a protractor, we find that the angle θ is about 53°.

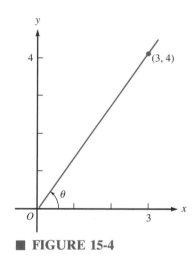

FIGURE 15-4

Trigonometric Functions

FIGURE 15-5

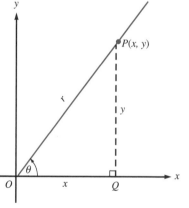

Now let's consider an angle θ in standard position (Fig. 15-5). We select a point *P* on the terminal side of θ, and let the coordinates of *P* be (*x*, *y*). If we drop a perpendicular from *P* to a point *Q* on the *x* axis, we have a right triangle. The hypotenuse of the right triangle is $\overline{OP}$, and we call its length *r* (for *radius*). The leg opposite θ is $\overline{PQ}$ and has length *y*. The leg adjacent to θ is $\overline{OQ}$ and has length *x*.

The hypotenuse *r* is found by the Pythagorean Theorem.

HYPOTENUSE	$r = \sqrt{x^2 + y^2}$	109

The right triangle has three sides (*x*, *y*, and *r*) which are related to each other by Eq. 109 and are functions of the angle θ. The quotients (ratios) of any two sides $\left(\dfrac{y}{r}, \dfrac{x}{r}, \text{and} \dfrac{y}{x}\right)$ are very important numbers. If we had chosen to place point *P* farther out on the terminal side of angle θ, then the lengths *x*, *y*, and *r* would all be greater than they are now. But the quotient of any two sides would be the same, regardless of where we placed *P*. The quotients change only if we change the angle θ. That is, the quotients of two sides are functions of the angle θ. They are called the *trigonometric functions*.

The most useful trigonometric functions are the sine, cosine, and tangent functions. The symbols for these functions are given below with their definitions.

TRIGONOMETRIC FUNCTIONS	sine $\theta = \sin \theta = \dfrac{y}{r}$	110
	cosine $\theta = \cos \theta = \dfrac{x}{r}$	111
	tangent $\theta = \tan \theta = \dfrac{y}{x}$	112

── Example 3 ──

Find the sine, cosine, and tangent of the angle in Fig. 15-4, rounded to three significant digits.

Solution: For this angle θ, $x = 3$ and $y = 4$
Using Eq. 109 to calculate r,

$$r = \sqrt{3^2 + 4^2} = \sqrt{25} = 5$$

Using Eqs. 110–112,

$$\sin \theta = \frac{y}{r} = \frac{4}{5} = 0.800$$

$$\cos \theta = \frac{x}{r} = \frac{3}{5} = 0.600$$

$$\tan \theta = \frac{y}{x} = \frac{4}{3} = 1.33$$

COMMON ERROR When writing the trigonometric functions, *do not omit the angle.*
If sin θ = 0.800,   NEVER write   sin = 0.800

■ **FIGURE 15-6**

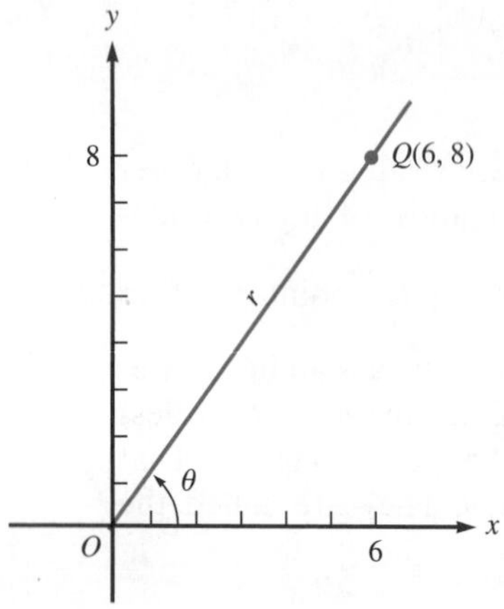

You will notice that the trigonometric functions are *pure* numbers; that is, they have no units. If the sides of a triangle have units, the units cancel out in the division.

── Example 4 ──

Sketch the angle in standard position for which the point $Q(6, 8)$ is on the terminal side. Find the sine, cosine, and tangent of the angle, rounded to three significant digits.

Solution: We sketch the angle (Fig. 15-6) so that the point $Q(6, 8)$ is on the terminal side. For point Q, $x = 6$ and $y = 8$.
Using Eq. 109 to calculate r,

$$r = \sqrt{6^2 + 8^2} = \sqrt{100} = 10$$

By Eqs. 110–112,

$$\sin \theta = \frac{y}{r} = \frac{8}{10} = 0.800$$

$$\cos \theta = \frac{x}{r} = \frac{6}{10} = 0.600$$

$$\tan \theta = \frac{y}{x} = \frac{8}{6} = 1.33$$

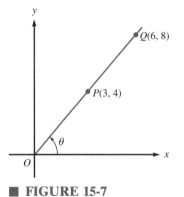

■ FIGURE 15-7

We see that the values of the trigonometric functions in Example 4 are the same as in Example 3. That is because the points (3, 4) and (6, 8) are both on the same terminal side (Fig. 15-7). The angle is the same for both examples.

So the trigonometric functions are not functions of the chosen point on the terminal ray. The specific point doesn't matter. Any point on the terminal side of the angle will do. The trigonometric functions are functions of the angle, and not of a particular point on the terminal side.

Trigonometric Functions by Calculator

We generally use a scientific calculator to find the trigonometric functions of a given angle. First, the calculator must be in the proper angle unit mode, either "degree" or "radian," depending on the units of the given angle. Refer to your calculator manual to determine how to change modes on your calculator and how to enter the trigonometric functions.

Example 5

Use your calculator to evaluate the cosine of a 30° angle, to three significant digits.

Solution: With our calculator in degree mode, we find that

$$\cos 30° = 0.8660254038$$

Rounding to three significant digits,

$$\cos 30° = 0.866$$

Example 6

Evaluate tan 1.25 to three significant digits.

Solution: We're to find the tangent of an angle whose measure is 1.25. Recall that an angle with no units is assumed to be in radians. So, with our calculator in radian mode, we find that

$$\tan 1.25 = 3.009569674$$

Rounding to three significant digits,

$$\tan 1.25 = 3.01$$

COMMON ERROR It is easy to forget to change to degree or radian mode as necessary. Be sure to check it each time.

**Standard Position
of an Angle**

Sketch each angle in standard position, showing the rotation. Then check your accuracy with a protractor.

1. 45°	**2.** 60°	**3.** 80°	**4.** 20°
5. 30°	**6.** 75°	**7.** 10°	**8.** 40°

Trigonometric Functions

The given point is a point on the terminal side of a positive angle θ in standard position. Sketch each angle and give the sine, cosine, and tangent for each angle rounded to three significant digits.

9. (5, 12)	**10.** (6, 8)	**11.** (15, 8)	**12.** (7, 24)
13. (2, 3)	**14.** (1, 5)	**15.** $(\sqrt{3}, 1)$	**16.** $(5, \sqrt{16})$
17. (34.6, 79.1)	**18.** (502, 289)		

Using a protractor, sketch the given angle in standard position on coordinate axes. Drop a perpendicular from any point on the terminal side and measure the distances, x, y, and r in cm. Use these measured distances to determine the sine, cosine, and tangent of the angle. Round to three significant digits. Check your results with the calculator. They should match approximately.

19. 60°	**20.** 30°	**21.** 45°	**22.** 75°

Plot the given point on coordinate axes and draw a line segment to the origin. With a protractor, measure the angle formed with the positive x axis to the nearest degree. Find the tangent of the angle two ways: (a) find the tangent of the measured angle by calculator, and (b) calculate the ratio $\frac{y}{x}$. They should be approximately equal.

23. (4, 3)	**24.** (2, 5)	**25.** (1, 6)	**26.** (4, 1)

27. Draw a 45° angle in standard position. Drop a perpendicular from a point on the terminal side, to form a 45-45-90° right triangle. Label the lengths of the legs 1 and 1, and the hypotenuse $\sqrt{2}$. Use these lengths to determine the sine, cosine, and tangent of 45°. Verify these functions using your calculator.

28. Draw a 60° angle in standard position. Drop a perpendicular from a point on the terminal side, to form a 30-60-90° right triangle. Label the lengths of the two legs $\frac{\sqrt{3}}{2}$ and $\frac{1}{2}$, and the hypotenuse 1. Use these lengths to determine the functions of sine, cosine, and tangent of 60°. Verify these functions using your calculator.

Evaluate, rounding to three significant digits.

29. sin 52.1°	**30.** cos 84.3°
31. tan 12.5°	**32.** sin 46.9°
33. cos 36.8°	**34.** tan 18.3°
35. sin 78.6°	**36.** cos 54.3°
37. tan 67°12′	**38.** sin 29°35′
39. cos 43°40′18″	**40.** sin 6°52′42″
41. tan (1.54 rad)	**42.** cos (0.367 rad)
43. cos (0.0256 rad)	**44.** sin (1.27 rad)

15-2   INVERSE TRIGONOMETRIC FUNCTIONS

In this section we study the inverse trigonometric functions. This involves finding the angle when a trigonometric function (sine, cosine, or tangent) is given. We also learn to use the special inverse function notation. We need this for applications involving right triangles.

Finding the Angle When the Trigonometric Function Is Given

Suppose we were given:

$$\sin \theta = 0.256$$

What is θ? To begin, θ is an angle; specifically, it is an angle whose sine is 0.256. We call the angle θ the *inverse sine* of 0.256, and we write

$$\theta = \sin^{-1} 0.256$$

The angle θ is also called the *arcsine* of 0.256, and may also be written

$$\theta = \arcsin 0.256$$

Inverse Functions by Calculator

You can use your calculator to find an angle, given one of its trigonometric functions. Consult your calculator manual for specifics on the inverse trigonometric functions.

Example 7

Use the calculator to find the angle θ such that

$$\sin \theta = 0.256$$

Solution: First we rewrite the equation

$$\sin \theta = 0.256$$

as

$$\theta = \sin^{-1} 0.256$$

With our calculator in degree mode, we find that

$$\sin^{-1} 0.256 = 14.83284792$$

Rounding to the nearest tenth, we get

$$\theta = 14.8°$$

Example 8

Show the meaning of the equation $\cos \theta = 0.345$ by completing this sentence:
_____ is the _____ whose _____ is _____.

Solution: $\underline{\theta}$ is the angle whose cosine is 0.345.

Example 9

Find θ to the nearest tenth of a degree if

$$\tan \theta = 3.75$$

Solution: We rewrite the equation as

$$\theta = \tan^{-1} 3.75$$

With the calculator in degree mode, we find that

$$\tan^{-1} 3.75 = 75.06858282$$

Rounding,

$$\theta = 75.1°$$

Example 10

Evaluate $\cos^{-1} 0.345$, in radians, rounded to the nearest hundredth.

Solution: With the calculator in radian mode, we find that

$$\cos^{-1} 0.345 = 1.218557542$$

Rounding,

$$\cos^{-1} 0.345 = 1.22$$

TIP If θ is an angle and a is the sine of that angle, then these equations all mean the same thing:

$$\sin \theta = a$$
$$\sin^{-1} a = \theta$$
$$\arcsin a = \theta$$

They all mean "θ is the angle whose sine is a."

There may be some confusion with the special notation used to indicate the inverse trigonometric function. In the expression

$$\sin^{-1} a$$

the raised -1 is *not* an exponent, but indicates the inverse sine (arcsin) of the number a.

$$\sin^{-1} a = \arcsin a$$

However, in the expression

$$(\sin a)^{-1}$$

the -1 *is* an exponent, and we saw in Chapter 7 that

$$(\sin a)^{-1} = \frac{1}{\sin a}$$

COMMON ERROR $\qquad\qquad \sin^{-1} a \neq (\sin a)^{-1}$

The $\sin^{-1} a$ is *not* the reciprocal of a.

The same is true for the other trigonometric functions.

Example 11

In radian mode, show that

$$\cos^{-1} 0.123 \neq (\cos 0.123)^{-1}$$

Solution: We see that $\cos^{-1} 0.123$ is the *angle* whose cosine is 0.123. With the calculator in radian mode, we find that

$$\cos^{-1} 0.123 = 1.45$$

Since $(\cos 0.123)^{-1}$ is the *reciprocal* of cos 0.123, we find that

$$(\cos 0.123)^{-1} = \frac{1}{\cos 0.123} = 1.01$$

Since $1.45 \neq 1.01$,

$$\cos^{-1} 0.123 \neq (\cos 0.123)^{-1}$$

Finding the Angle When the Trigonometric Function Is Given

Show the meaning of these equations by completing the sentence: _____ is the _____ whose _____ is _____.

1. $\sin \theta = 0.636$
2. $\cos \alpha = 0.398$
3. $\arctan 3.45 = \alpha$
4. $\arcsin 0.035 = \theta$
5. $\cos^{-1} a = 30°$
6. $\tan^{-1} b = \dfrac{\pi}{3}$
7. $\tan \dfrac{\pi}{2} = x$
8. $\sin 3 = y$
9. $\sin^{-1} a = B$
10. $\cos^{-1} x = Z$
11. $\sin 1 = 0.8415$
12. $\tan 0.5 = 0.5463$
13. $\arccos 0.399 = 66.5°$
14. $\arctan 0.222 = 12.5°$

Write the equation two other ways.

15. $\sin \alpha = 0.634$
16. $\cos \theta = 0.294$
17. $\tan \theta = 4.67$
18. $\sin^{-1} a = 45°$
19. $\cos^{-1} b = 60°$
20. $\tan^{-1} b = 30°$
21. $\arcsin 0.268 = \alpha$
22. $\arccos 0.839 = \theta$
23. $\arccos x = 0.785$
24. $\tan 27° = y$
25. $\tan 89.0° = 57.3$
26. $\arcsin 0.738 = 0.830$

Solve to the nearest tenth of a degree.

27. $\cos A = 0.707$
28. $\sin B = 0.866$
29. $\tan \theta = 4.87$
30. $\cos A = 0.197$
31. $\sin \alpha = 0.5$
32. $\tan \theta = 2$
33. $\tan \alpha = 1$
34. $\cos \alpha = 0.5$
35. $\cos B = 0.866$
36. $\sin C = 0.707$

Solve to the nearest hundredth of a radian.

37. $\sin \theta = 0.382$
38. $\tan A = 25.6$
39. $\tan B = 50.1$
40. $\cos \alpha = 0.999$

Evaluate to the nearest tenth of a degree.

41. $\arctan 6.20$
42. $\arccos 0.707$
43. $\sin^{-1} 0.707$
44. $\tan^{-1} 1$
45. $\cos^{-1} 0.5$
46. $\arcsin 0.866$

Evaluate to the nearest hundredth of a radian.

47. $\tan^{-1} 10$
48. $\arctan 35$
49. $\arccos 0.866$
50. $\sin^{-1} 0.5$

Inverse Functions by Calculator

51. Show that $\sin^{-1} 0.738 \neq (\sin 0.738)^{-1}$, in radian mode.
52. Show that $\tan^{-1} 3.00 \neq (\tan 3.00)^{-1}$, in radian mode.

15-3

SOLUTION OF RIGHT TRIANGLES

In Sec. 15-1, we defined the sine, cosine, and tangent of an angle, and saw that these functions could give us the quotient of two sides of a right triangle. But we considered only angles that were in standard position on the coordinate axes. In this section, we will see how the sine, cosine, and tangent can be used with a right triangle that is not on coordinate axes.

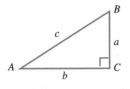

■ FIGURE 15-8
Labeling a right triangle.

Right Triangles

Figure 15-8 shows a right triangle that has been labeled in the conventional way. We will usually label a right triangle as shown in Fig. 15-8. We label the angles (vertices) with the capital letters A, B, and C, with C being the right angle. We label the sides with the lower case letters a, b, and c. Side a is opposite angle A; side b is opposite angle B; and side c (the hypotenuse) is opposite angle C (the right angle). The formulas for right triangles that we have already learned are repeated here.

	$c^2 = a^2 + b^2$	104
PYTHAGOREAN THEOREM	$c = \sqrt{a^2 + b^2}$	105
	$a = \sqrt{c^2 - b^2}$	106
SUM OF THE ACUTE ANGLES	$A + B = 90°$	103

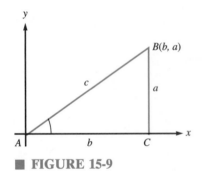

■ FIGURE 15-9

When we solve a right triangle, it is convenient to express the trigonometric functions of the acute angles in terms of the sides. If we place triangle ABC (Fig. 15-8) on the coordinate axes, so that A is at the origin, B is in Quadrant I, and C is on the positive x axis (Fig. 15-9), then angle A is in standard position, and we can find the trigonometric functions of angle A. Using Eqs. 110–112 with $\theta = A$, $x = b$, $y = a$, and $r = c$,

$$\sin A = \frac{y}{r} = \frac{a}{c} = \frac{\text{leg opposite } A}{\text{hypotenuse}}$$

$$\cos A = \frac{x}{r} = \frac{b}{c} = \frac{\text{leg adjacent } A}{\text{hypotenuse}}$$

$$\tan A = \frac{y}{x} = \frac{a}{b} = \frac{\text{leg opposite } A}{\text{leg adjacent } A}$$

We can generalize for any acute angle θ in any right triangle:

	$\sin \theta = \dfrac{\text{opposite leg}}{\text{hypotenuse}}$	110
RIGHT TRIANGLE TRIGONOMETRY	$\cos \theta = \dfrac{\text{adjacent leg}}{\text{hypotenuse}}$	111
	$\tan \theta = \dfrac{\text{opposite leg}}{\text{adjacent leg}}$	112

COMMON ERROR Eqs. 110–112 above are true only in a *right* triangle.

TIP Equations 110–112 are very important since they are used so often. They should be memorized. One way to memorize these is to remember

"SOH-CAH-TOA"

which comes from

$$\text{Sin } \theta = \frac{\text{Opposite leg}}{\text{Hypotenuse}}$$

$$\text{Cos } \theta = \frac{\text{Adjacent leg}}{\text{Hypotenuse}}$$

$$\text{Tan } \theta = \frac{\text{Opposite leg}}{\text{Adjacent leg}}$$

Note: These formulas are true for a right triangle in *any* position. The angle θ need not be in standard position on the coordinate axes.

─── **Example 12** ───

Find the hypotenuse, the tangent of angle A, and angle A in Fig. 15-10.

Solution: We use the formula derived from the Pythagorean Theorem (Eq. 105) to find the hypotenuse

$$c = \sqrt{a^2 + b^2} = \sqrt{12.0^2 + 5.00^2}$$
$$c = 13.0 \text{ in.}$$

(Note that we round to as many significant digits as are in the given numbers.)

Hint: Always make sure your answers are reasonable as you go. Does the hypotenuse look like 13 in. (to scale)? Yes, it does.

We use Eq. 112 to find the tangent

$$\tan A = \frac{\text{opposite leg}}{\text{adjacent leg}} = \frac{12.0}{5.00}$$
$$\tan A = 2.40$$

To find angle A, we solve the equation above for A, using the inverse tangent function.

$$A = \tan^{-1} 2.40$$

With the calculator in degree mode, we find that

$$A = 67.4°$$

Once again, make sure your answers are reasonable as you go. Does angle A look like it is about 67°? Yes, it does.

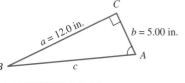

FIGURE 15-10

Solving Right Triangles

To *solve* a triangle means to find all the missing sides and angles. In most applications we will need to find only one angle or one side. But first we will practice by finding all the missing sides and angles. We can solve any right triangle if we know at least one side and either another side or an angle. So we will know three parts (including the right angle), and we will need to find the remaining three parts.

Steps to Solve a Right Triangle

1. Make an accurate sketch (to scale) and label the known quantities.
2. Solve for the missing angles (by either subtracting a known angle from 90°, or by using an inverse trigonometric function).
3. Solve for the missing sides (by using the trigonometric ratios or the Pythagorean Theorem).
4. Check.

Solving a Right Triangle When One Side and One Angle Are Known

If one side and one angle (and the right angle) are known, then we must find the other two sides and one angle.

Example 13

Solve right triangle ABC if $B = 32.5°$ and $a = 12.1$ cm.

Solution: First we make a sketch in Fig. 15-11. We make our sketch as accurate as possible (to scale), so that we can verify our results as we go. We label the known quantities, and the unknown sides and angles by letters. Next, we find the missing angle, A.

$$A = 90° - 32.5°$$
$$A = 57.5°$$

Does it look like A is about 57°? Of course.

Next, we find the missing sides, using the trigonometric ratios. We need to find b and c, and we want to use only angle B, not angle A. After all, we may have made an error in finding A, and if we used an incorrect A to find the sides, then they would be wrong also. So we want to use only the *given* quantities, B and a.

We must determine which trigonometric function to use to find b. The unknown side b is *opposite* angle B. The given side a is the side *adjacent* to angle B. The trigonometric function which relates the opposite side to the adjacent side is the tangent.

$$\tan B = \frac{\text{opposite leg}}{\text{adjacent leg}}$$

Substituting,

$$\tan 32.5° = \frac{b}{12.1}$$

Solving for b,

$$b = 12.1(\tan 32.5°)$$

With the calculator in degree mode, we find that

$$12.1(\tan 32.5°) = 7.708550156$$

So, rounding to three significant digits,

$$b = 7.71 \text{ cm}$$

(Does it look right?)

We now find side c. One way of finding c is to use the Pythagorean Theorem. The problem with that is that we would need to use b, which was not given, but which we calculated. Thus, a possible error in b would be carried to c. So

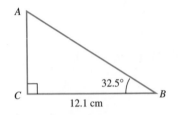

■ **FIGURE 15-11**

A

32.5°

C

12.1 cm

B

we choose not to use the Pythagorean Theorem now. (We will use it later in the check.)

We will solve for c using a trigonometric function. We note that side c is the hypotenuse. And we will once again use the given adjacent side, a. The obvious choice for the trigonometric function is the cosine.

$$\cos B = \frac{\text{adjacent leg}}{\text{hypotenuse}}$$

Substituting,

$$\cos 32.5° = \frac{12.1}{c}$$

Solving for c, and remembering to *divide* this time,

$$c = \frac{12.1}{\cos 32.5°}$$

$$c = 14.3 \text{ cm}$$

Check:

Does it look like the hypotenuse is about 14 cm, to scale? Yes it does. We can check numerically, using the formula derived from the Pythagorean Theorem:

$$\sqrt{12.1^2 + 7.71^2} \overset{?}{=} 14.3$$

$$14.3 = 14.3 \quad \text{checks}$$

COMMON ERROR Remember to use only the *given* information for each computation, rather than some quantity previously calculated. This way, any errors in the early computations will not be carried along.

Solving a Right Triangle When Two Sides Are Known

If two sides are known, then we must find one side and two angles.

┌─── **Example 14** ───
Solve right triangle ABC if $a = 42.0$ cm and $b = 48.0$ cm.

Solution: We sketch the triangle in Fig. 15-12, labeling all parts. We find the angles, A and B. To find angle A, we note that side a is opposite angle A, and side b is adjacent to angle A. We must use the tangent function.

$$\tan A = \frac{\text{opposite leg}}{\text{adjacent leg}}$$

Substituting,

$$\tan A = \frac{42.0}{48.0}$$

Solving for A,

$$A = \tan^{-1}\frac{42.0}{48.0} = 41.2°$$

Similarly,

$$\tan B = \frac{48.0}{42.0}$$

$$B = \tan^{-1}\frac{48.0}{42.0} = 48.8°$$

FIGURE 15-12

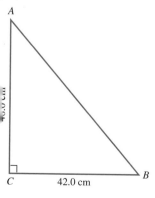

Do the angles look like they are about 41° and 49°? An intermediate check is that

$$41.2° + 48.8° = 90.0°$$

as it should.

Next, we find side c. We use Eq. 105:

$$c = \sqrt{42.0^2 + 48.0^2} = 63.8 \text{ cm}$$

We round to three significant digits, since the given numbers had three significant digits.

Check:

As a quick check, does it look like the hypotenuse is about 64 cm long, to scale? Yes, it does. For a numerical check, we can use one of the other trigonometric functions, say the sine, of one of the angles.

$$\sin 41.2° \overset{?}{=} \frac{42.0}{63.8}$$

$$0.659 = 0.658 \quad \text{checks}$$

Important: This *apparent* discrepancy in the check is due to the rounding, where there is always a little uncertainty in the last digit.

SECTION 15-3 **Exercises** **Solution of Right Triangles**

Right Triangles

Find the third side, the tangent of angle A, and angle A.

1.

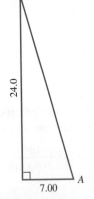

■ **FIGURE 15-13**

2.

■ **FIGURE 15-14**

3. ■ **FIGURE 15-15**

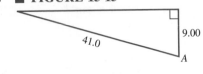

4. ■ **FIGURE 15-16**

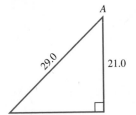

5.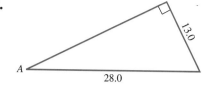

■ **FIGURE 15-17**

6.

■ **FIGURE 15-18**

7.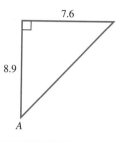

■ **FIGURE 15-19**

8.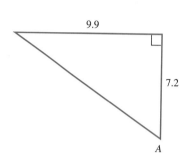

■ **FIGURE 15-20**

9.

■ **FIGURE 15-21**

10.

■ **FIGURE 15-22**

Solving a Right Triangle When One Side and One Angle Are Known

Sketch each right triangle using the conventional labeling, and find all the missing parts.

11. $B = 48.0°$, $c = 35.0$ in.

12. $A = 35.0°$, $c = 42.6$ cm

13. $A = 69.0°$, $a = 4.32$ m

14. $B = 54.0°$, $b = 8.37$ cm

15. $B = 25.2°$, $a = 56.8$ cm

16. $A = 75.4°$, $b = 28.5$ in.

17. $A = 42°15'$, $a = 1.65$ m

18. $B = 64°24'$, $b = 3.75$ m

Solving a Right Triangle When Two Sides Are Known

Sketch each right triangle using the conventional labeling, and find all missing parts. Express the angles in decimal degrees.

19. $a = 4.00$ in., $b = 7.00$ in.

20. $b = 5.00$ cm, $c = 8.00$ cm

21. $a = 4.60$ cm, $c = 6.90$ cm

22. $a = 1.50$ in., $b = 7.50$ in.
23. $b = 35.70$ m, $c = 41.30$ m
24. $a = 85.6$ m, $c = 101$ m

15-4   APPLICATIONS

There are many applications involving right triangle trigonometry. In this section, we solve problems in geometry, angles of elevation and angles of depression, indirect measurements, navigation, and shop applications. In all of these applications, the key to the solution is in determining the appropriate trigonometric function to use. Each trigonometric function involves three quantities: one angle and two sides (opposite, adjacent, or hypotenuse). First we make a sketch, and determine which of the sides is given and which is needed. If they are opposite and adjacent, we use the tangent; if they are opposite and hypotenuse, we use the sine; if they are adjacent and hypotenuse, we use the cosine. Remember that each result is rounded according to the number of significant digits in the given quantities.

Geometry

Geometric problems involving polygons other than right triangles are often solved using trigonometry. In the following examples we use right triangle trigonometry to solve geometric problems.

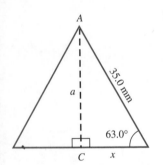

■ **FIGURE 15-23**

■ **FIGURE 15-24**

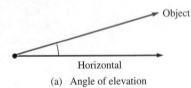

(a) Angle of elevation

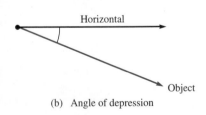

(b) Angle of depression

┌─ **Example 15** ────────
Two of the sides of an isosceles triangle have a length of 35.0 mm and the base angles are each 63.0°. Find the altitude and the base of the triangle.

Solution: We sketch the triangle in Fig. 15-23, and note that this is not a right triangle, so we cannot use the trigonometric functions directly. We draw the altitude a to the base and create two right triangles. In right triangle ABC, a is opposite the 63.0° angle and 35.0 mm is the hypotenuse, so we use the sine.

$$\sin 63.0° = \frac{a}{35.0}$$

Solving for a,

$$a = 35.0(\sin 63.0°)$$
$$a = 31.2 \text{ mm}$$

We let x be the side adjacent to the 63.0° angle. The hypotenuse is 35.0, so we use the cosine.

$$\cos 63.0° = \frac{x}{35.0}$$
$$x = 35.0(\cos 63.0°) = 15.9$$

So, the base is

$$\text{base} = 2x = 31.8 \text{ mm}$$

So, the altitude is 31.2 mm and the base is 31.8 mm.

Angles of Elevation and Depression

Many right triangle trigonometry applications involve an angle made with a horizontal line (Fig. 15-24). For an object *above* the horizontal, the angle formed between the horizontal line and the line of sight from an observer to the object is called the *angle of elevation*. If the object is *below* the horizontal line, then the angle formed between the horizontal line and the line of sight from an observer to the object is called the *angle of depression*.

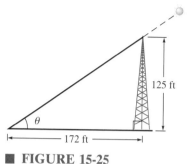

■ FIGURE 15-25

┌─── **Example 16** ───

A 125-ft tower casts a shadow 172 ft long. Find the angle of elevation of the sun.

Solution: In Fig. 15-25, 125 is opposite θ and 172 is adjacent to θ. Therefore we use the tangent function:

$$\tan \theta = \frac{125}{172}$$

$$\theta = \tan^{-1}\frac{125}{172} = 36.0°$$

Indirect Measurements

Distances that are difficult or impossible to measure directly are often found indirectly using trigonometry. A surveyor, for example, uses a transit to measure angles between locations, and then applies trigonometry to calculate the distances.

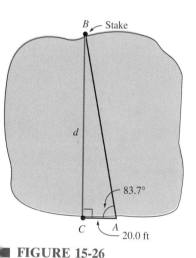

■ FIGURE 15-26

┌─── **Example 17** ───

A surveyor needs to measure the distance across a pond (Fig. 15-26). First she places a stake C on one shore. With a transit directly above C, she sights a stake B on the opposite shore. She then turns the transit 90° and places a third stake along that line of sight at A, 20.0 ft from C. At A she measures the angle between the two stakes B and C to be 83.7°. What is the distance d across the pond?

Solution: Since d is the side opposite the 83.7° angle and 20.0 ft is the adjacent side, we use the tangent.

$$\tan 83.7° = \frac{d}{20.0}$$
$$d = 20.0(\tan 83.7°)$$
$$d = 181 \text{ ft}$$

So, the distance across the pond is 181 ft.

■ FIGURE 15-27

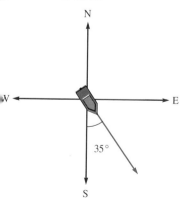

Navigation

Problems in navigation often require the use of trigonometry. Before doing a navigation problem we show how compass directions are commonly designated. For example, a ship that is traveling in a compass direction of *35° east of south* is illustrated in Fig. 15-27. The direction 35° east of south means that the angle of travel is 35° in an easterly direction from due south.

┌─── **Example 18** ───

An observer at a point P (Fig. 15-28) on a coast sights a ship S in a direction of 42.0° west of north. At the same time, the ship is directly west of a point Q, which is 12.0 km due north of P. Find the distance x of the ship from point P.

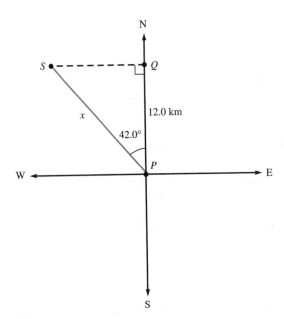

■ FIGURE 15-28

Solution: First we note how the angle is drawn. From point P, a line is drawn due north. We draw the 42.0° angle on the west side of the north line. We note that x is the hypotenuse of the right triangle SPQ and 12.0 is the side adjacent to the 42.0° angle. We use the cosine:

$$\cos 42.0° = \frac{12.0}{x}$$

$$x = \frac{12.0}{\cos 42.0°}$$

$$x = 16.1 \text{ km}$$

So the distance from the ship to P is 16.1 km.

■ FIGURE 15-29

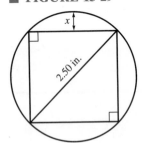

Shop Trigonometry

There are many uses of trigonometry in machine and tool shops.

Example 19

Determine the depth of cut x (to the nearest hundredth of an inch) required to mill a square on the end of a 2.50-in. circular shaft (Fig. 15-29).

Solution: The radius of the circle is 1.25 in. We draw another radius, which is the perpendicular bisector of the top of the square. Recall from Sec. 13-4 that the diagonals of a square bisect the 90° angles of the square. So the angle θ is 45°. (We treat 45° as an exact number.) We first solve for y, and then use y to find x.

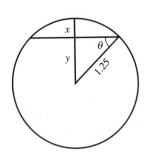

$$\sin 45° = \frac{y}{1.25}$$

$$y = 1.25(\sin 45°) = 0.884$$

$$x = 1.25 - 0.884 = 0.366$$

$$x = 0.37 \text{ in.}$$

We round this result to the hundredths place, since 1.25 is accurate only to the hundredths place. So the cut must be 0.37 in. deep on each side of the shaft.

Geometry

Round answers to three significant digits.

1. Find the length of the altitude of an isosceles triangle with a 65° base angle and a base 145 in. long (Fig. 15-30).

2. Find the base of an isosceles triangle with a 54° vertex angle and equal sides of 43.8 mm each.

3. Find the measure of the base angles of an isosceles triangle with equal legs of 7.00 in. each and a base of 5.00 in.

4. Find the base of an isosceles triangle with 72° base angles and an altitude 58.4 in. long.

5. The sides of a rectangle are 467 mm and 638 mm. Find the angle θ between a diagonal and the shorter side (Fig. 15-31).

6. The short side of a certain rectangle is 32.0 cm and the diagonal is 54.0 cm long. Find the length of the long side and the angle between the diagonal and the short side.

Angles of Elevation and Depression

7. The shadow of a tree is 45.0 ft long when the angle of elevation of the sun is 68.7° (Fig. 15-32). Find the height of the tree.

8. How long a shadow will be cast by a monument that is 429 ft tall, when the angle of elevation of the sun is 42.6°?

9. An airplane, at an altitude of 12,500 ft, is approaching an airport. If the angle of depression from the airplane to the airport is 20.3°, find the horizontal distance from the airplane to the airport.

10. The Washington Monument is 555 ft high. If a person stood one quarter of a mile away and looked at the top of the monument, what would the angle of elevation be? (Fig. 15-33)

11. A woman at a window 18.0 ft above a street is looking at a street light 13.0 ft above the street. The angle of depression is 15.8°. How far is she from the street light?

■ **FIGURE 15-30**

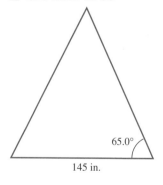

145 in.

65.0°

■ **FIGURE 15-31**

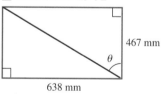

467 mm

638 mm

θ

■ **FIGURE 15-32**

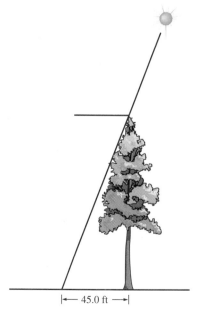

|← 45.0 ft →|

■ **FIGURE 15-33**

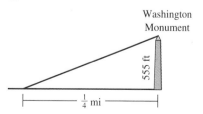

Washington Monument

555 ft

$\frac{1}{4}$ mi

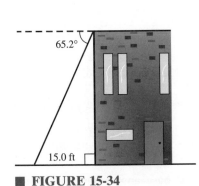

■ **FIGURE 15-34**

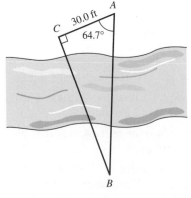

■ **FIGURE 15-35**

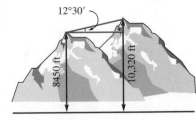

■ **FIGURE 15-36**

12. From a boat at sea, the angle of elevation to the top of a 375 ft cliff is 28.3°. How far is the boat from the base of the cliff?

13. From the top of a building, the angle of depression to a person on the street, 15.0 ft in front of the building, is 65.2° (Fig. 15-34). How tall is the building?

14. From the top of a 125-ft sea cliff, the angle of depression to a small boat sailing toward the cliff was 26.3°. Two hours later the angle of depression to the boat was 35.8°. How far had the boat traveled?

Indirect Measurements **15.** A surveyor needs to measure the distance between two points *B* and *C* on opposite sides of a river (Fig. 15-35). She places a transit at a point *C* and sights point *B* across the river. She then turns the transit 90° and places a stake along that line of sight at point *A* 30.0 ft from C. Angle *CAB* is measured with the transit to be 64.7°. What is the distance between points *B* and *C*?

16. To measure the height of a cliff, a surveyor measures a point 125 m from the base of the cliff and measures the angle of elevation to the top of the cliff to be 28.5°. How high is the cliff?

17. The angle of depression of a ship from the bridge of an aircraft carrier 237 ft above water level is 10.5°. How far is the ship from the waterline of the carrier?

18. The angle of elevation from one mountain peak to the adjacent higher peak is 12°30′ (Fig. 15-36). What is the air distance between the peaks if their heights are 8450 ft and 10,320 ft?

19. From a point 125 ft from the base of a building, a surveyor uses a transit to sight an angle of 36.2° to the top of a building. What is the height of the building if the sight of the transit is 5 ft from the ground?

20. A woman wants to measure the height of an antenna on the top of a 125-ft building. From a point in front of the building, she measures the angle of elevation to the top of the building to be 68.2° and the angle of elevation to the top of the antenna to be 71.0°. How tall is the antenna?

Navigation **21.** An observer at a point *A* on a coast sights a ship in a direction of 37.0° east of south. At the same time, the ship is directly east of a point *B*, 19.0 km due south of *A* (Fig. 15-37). Find the distance of the ship from point *A*.

22. A ship at sea was 3.50 km due west of a lighthouse at noon one day. The ship was traveling due north, and later that day the captain sighted the lighthouse in a direction of 27° east of south. How far had the ship traveled since noon?

23. A ship traveled from a port in the direction of 57° west of north for 348 miles. How far north and how far west of the port is the ship located?

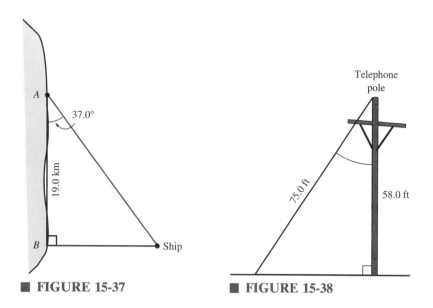

■ FIGURE 15-37

■ FIGURE 15-38

Structures

24. A ship traveled from a port in the direction of 43° west of south at 26.0 knots for 3.00 hours. How far south and how far west of the port was the ship located? (A knot is about 1.151 mi/h.)

25. A guy wire 75.0 ft long is stretched from the ground to the top of a telephone pole 58.0 ft high (Fig. 15-38). Find the angle between the wire and the pole.

26. The roof of a building (Fig. 15-39) slopes up at an angle of 29.0° with the horizontal. The span (horizontal distance) of the roof is 32.0 ft. Find the rise and the length of a roof rafter.

27. Find the angle a rafter makes with the horizontal if the rise is 10.0 ft and the span is 50.0 ft (Fig. 15-39). Then find the length of the rafter.

28. A stair has a rise of 5.00 in. for every 6.00 in. of run (Fig. 15-40). What is the angle of inclination θ of the stair?

29. A shelf (Fig. 15-41) is supported by a 27.0-cm brace attached to the wall 14.0 cm below the shelf. Find the angle between the brace and the wall.

30. An entrance ramp for the handicapped must be built to a door 1.45 m above ground level. If the maximum angle the ramp can make with the ground is 8°, how far from the base of the building must they begin the ramp?

■ FIGURE 15-39   **■ FIGURE 15-40**   **■ FIGURE 15-41**

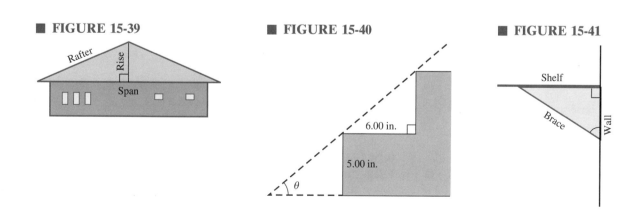

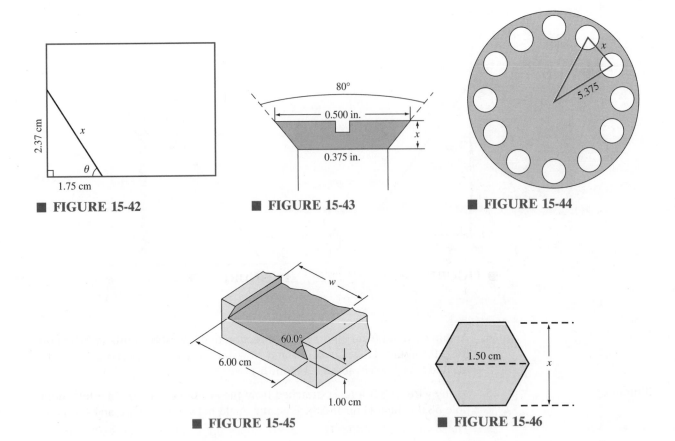

■ FIGURE 15-42

■ FIGURE 15-43

■ FIGURE 15-44

■ FIGURE 15-45

■ FIGURE 15-46

Shop Trigonometry

31. The corner of the rectangular metal plate in Fig. 15-42 is to be cut off. Find x and θ.

32. Determine the missing dimension in the head of the screw in Fig. 15-43.

33. A machinist must make a bolt circle of radius 5.375 in. with 12 equally spaced bolt holes (Fig. 15-44). What is the straight line distance between the holes?

34. Sixteen rivets are equally spaced on the circumference of a circular plate. If the center-to-center distance between two adjacent rivets is 5.75 cm, what is the radius of the circle?

35. Figure 15-45 is a sketch of a dovetail wedge. Find the upper width, w, of the wedge.

36. The distance across the corners of a regular hexagonal bolt is 1.50 cm (Fig. 15-46). Find the distance across the flats.

15-5   COSECANT, SECANT, AND COTANGENT FUNCTIONS

In this section we define the cosecant, secant, and cotangent of an angle and then use the calculator to calculate them. We use these functions to express some right triangle relationships in a simpler form and also to simplify some trigonometric formulas. We didn't cover these functions earlier because we didn't need them for solving right triangles; all the calculations could be done using only the sine, cosine, and tangent functions.

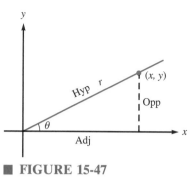

FIGURE 15-47

Cosecants, Secants, and Cotangents as Functions

In Sec. 15-1 we defined the sine, cosine, and tangent of an angle as a ratio of coordinates or as a quotient of sides of a right triangle (for acute angles). There are three additional trigonometric functions that are similarly defined, using the triangle in Fig. 15-47.

THREE MORE TRIGONOMETRIC FUNCTIONS	$\text{cosecant } \theta = \csc \theta = \dfrac{r}{y} = \dfrac{\text{hypotenuse}}{\text{opposite leg}}$	113
	$\text{secant } \theta = \sec \theta = \dfrac{r}{x} = \dfrac{\text{hypotenuse}}{\text{adjacent leg}}$	114
	$\text{cotangent } \theta = \cot \theta = \dfrac{x}{y} = \dfrac{\text{adjacent leg}}{\text{opposite leg}}$	115

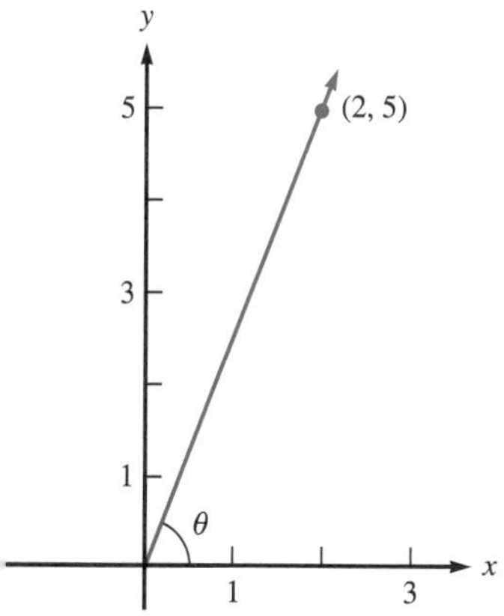

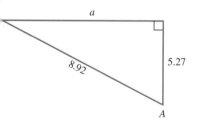

FIGURE 15-48

Example 20

The point $(2, 5)$ is on the terminal side of a positive angle θ in standard position (Fig. 15-48). Find the cosecant, secant, and cotangent of θ, rounded to three significant digits.

Solution: We calculate r, using Eq. 109, with $x = 2$ and $y = 5$,

$$r = \sqrt{x^2 + y^2} = \sqrt{2^2 + 5^2}$$
$$r = \sqrt{29}$$

We leave r as $\sqrt{29}$, without approximating it as a decimal, so that we don't introduce error at this point! Now we use Eqs. 113–115 to calculate the trigonometric functions.

$$\csc \theta = \frac{r}{y} = \frac{\sqrt{29}}{5} = 1.08$$

$$\sec \theta = \frac{r}{x} = \frac{\sqrt{29}}{2} = 2.69$$

$$\cot \theta = \frac{x}{y} = \frac{2}{5} = 0.400$$

FIGURE 15-49

Example 21

Find the cosecant, secant, and tangent of angle A in Fig. 15-49. Round to three significant digits.

Solution: Using the formula derived from the Pythagorean Theorem (Eq. 106),

$$a = \sqrt{8.92^2 - 5.27^2} = \sqrt{51.79} = 7.197$$

Here we approximate $\sqrt{51.79}$ but carry one extra digit for substituting into Eqs. 113–115,

$$\csc A = \frac{\text{hypotenuse}}{\text{opposite leg}} = \frac{8.92}{7.197} = 1.24$$

$$\sec A = \frac{\text{hypotenuse}}{\text{adjacent leg}} = \frac{8.92}{5.27} = 1.69$$

$$\cot A = \frac{\text{adjacent leg}}{\text{opposite leg}} = \frac{5.27}{7.197} = 0.732$$

Reciprocal Relationships

How do you determine the cosecant of a specific angle, say csc 52.6°, by calculator? You may have noticed that there is no "csc" key on your calculator.

Since

$$\csc \theta = \frac{\text{hypotenuse}}{\text{opposite leg}} \quad \text{and} \quad \sin \theta = \frac{\text{opposite leg}}{\text{hypotenuse}}$$

we can see that the cosecant and the sine are reciprocals of each other.

Therefore,

$$\csc \theta = \frac{1}{\sin \theta}$$

Similarly, the secant is the reciprocal of the cosine, and the cotangent is the reciprocal of the tangent.

RECIPROCAL RELATIONSHIPS	$\csc \theta = \dfrac{1}{\sin \theta}$	116(a)
	$\sec \theta = \dfrac{1}{\cos \theta}$	116(b)
	$\cot \theta = \dfrac{1}{\tan \theta}$	116(c)

So, to find the cosecant, secant, or cotangent of an angle, first find the sine, cosine, or tangent of the angle and then take the reciprocal.

Example 22

Find csc 52.6°, rounded to three significant digits.

Solution: Since $\csc 52.6° = \dfrac{1}{\sin 52.6°}$, we evaluate with the calculator. With the calculator in degree mode, we find that

$$\frac{1}{\sin 52.6°} = 1.258788514$$

So csc 52.6° = 1.26.

Finding the Angle When the Trigonometric Function Is Given

Here we reverse the previous process. We must now use both the reciprocal and the inverse keys on the calculator.

Example 23

Given sec θ = 2.25, find θ, and round to the nearest tenth of a degree.

Solution: The secant is the reciprocal of the cosine, so

$$\sec \theta = \frac{1}{\cos \theta} \quad \text{so} \quad \frac{1}{\cos \theta} = 2.25$$

Taking reciprocals,

$$\cos \theta = \frac{1}{2.25}$$

Taking the inverse,

$$\theta = \cos^{-1}\left(\frac{1}{2.25}\right)$$

The final equation shows θ expressed in terms of an inverse trigonometric function $(\cos^{-1})$ and a reciprocal $\left(\frac{1}{2.25}\right)$. With the calculator in degree mode, we find that

$$\cos^{-1}\left(\frac{1}{2.25}\right) = 63.61220004$$

So, $\theta = 63.6°$.

Example 24

Evaluate $\cot^{-1} 2.75$ in radians, and round to three significant digits.

Solution: Recall that $\cot^{-1} 2.75$ is the angle whose cotangent is 2.75. That is,

$$\cot^{-1} 2.75 = \theta$$
$$\cot \theta = 2.75$$
$$\frac{1}{\tan \theta} = 2.75$$
$$\tan \theta = \frac{1}{2.75}$$
$$\theta = \tan^{-1}\left(\frac{1}{2.75}\right)$$

With the calculator in radian mode, we find that

$$\tan^{-1}\left(\frac{1}{2.75}\right) = 0.3487710036$$

So, $\cot^{-1} 2.75 = 0.349$ rad.

SECTION 15-5 Exercises Cosecant, Secant, and Cotangent Functions

Cosecants, Secants, and Cotangents as Functions

The given point is a point on the terminal side of a positive angle θ in standard position. Sketch each angle and give the cosecant, secant, and cotangent for each angle, rounded to three significant digits.

1. $(3, 4)$ 2. $(12, 5)$ 3. $(24, 7)$ 4. $(6, 8)$
5. $(1, 3)$ 6. $(3, 6)$ 7. $(3, \sqrt{2})$ 8. $(\sqrt{6}, 4)$
9. $(27.3, 83.7)$ 10. $(12.5, 93.3)$

Find the cosecant, secant, and cotangent of angle A (rounded to three significant digits), and the degree measure of A (rounded to the nearest tenth).

11.

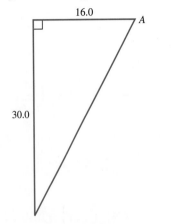

16.0

30.0

■ **FIGURE 15-50**

12.

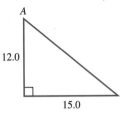

12.0

15.0

■ **FIGURE 15-51**

13.

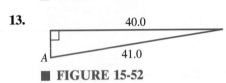

40.0

41.0

■ **FIGURE 15-52**

14.

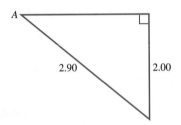

2.90   2.00

■ **FIGURE 15-53**

15.

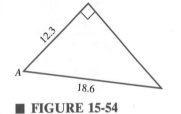

12.3

18.6

■ **FIGURE 15-54**

16.

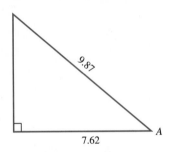

9.87

7.62

■ **FIGURE 15-55**

Reciprocal Relationships

Evaluate, and round to three significant digits.

17. cot 63.3°	**18.** csc 74.5°	**19.** sec 28.1°
20. cot 89.6°	**21.** csc 2.39°	**22.** sec 3.58°
23. cot 52°45′	**24.** csc 73°54′	**25.** sec (1.25 rad)
26. csc (0.75 rad)		

Finding the Angle When the Trigonometric Function Is Given

Find θ, and round to the nearest tenth of a degree.

27. sec $\theta = 1.15$	**28.** csc $\theta = 2.25$
29. cot $\theta = 12.6$	**30.** sec $\theta = 1.05$

Solve to the nearest thousandth of a radian.

31. csc $\theta = 4.50$	**32.** cot $\theta = 8.60$

Evaluate to the nearest tenth of a degree.

33. sec^{-1} 1.75	**34.** arccsc 3.50

Evaluate to the nearest thousandth of a radian.

35. arccsc 5.30	**36.** cot^{-1} 4.25

Given a point on the terminal side of an angle in standard position,

    (a) Sketch the angle;

    (b) Give the six trigonometric functions for the angle, to three significant digits; and

    (c) Find the angle, to the nearest tenth of a degree.

1. (0.950, 4.70)   **2.** (32.5, 74.2)

Evaluate to three significant digits.

3. $\sin 76.8°$ **4.** $\cos 5°18'$

5. $\cot (1.25 \text{ rad})$ **6.** $\csc (2.37 \text{ rad})$

7. $\tan (12.6°)$ **8.** $\sec 35°42'$

Find the angle to the nearest tenth of a degree.

9. $\cos A = 0.866$ **10.** $\cot \alpha = 1.50$

11. $\tan \theta = 4.75$ **12.** $\csc \theta = 5.71$

Find the angle to the nearest thousandth of a radian.

13. $\sin \alpha = 0.567$ **14.** $\sec \theta = 1.15$

15. Evaluate $\sin^{-1} 0.56$ to the nearest tenth of a degree.

16. Evaluate arcsec 1.41 to the nearest thousandth of a radian.

Find the six trigonometric functions of angle A, and the degree measure of A.

17.

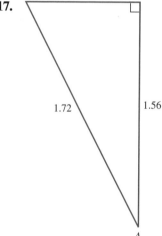

■ **FIGURE 15-56**

18.

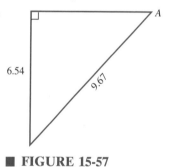

■ **FIGURE 15-57**

■ **FIGURE 15-58**

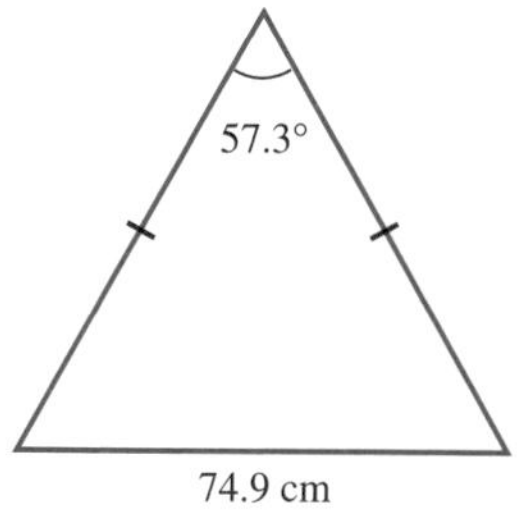

Sketch each right triangle and find all the missing parts.

19. $A = 37.6°$, $c = 65.7$ cm

20. $c = 4.625$ in., $b = 3.875$ in.

21. $B = 0.825$ rad, $a = 3.25$ cm

22. $c = 63.3$ in., $B = 67.8°$

23. Find the altitude of an isosceles triangle with a 57.3° vertex angle and a base 74.9 cm long (Fig. 15-58).

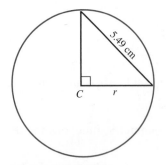

■ **FIGURE 15-59**

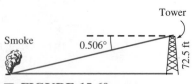

■ **FIGURE 15-60**

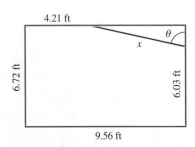

■ **FIGURE 15-61**

24. The sides of a rectangle are 15.0 and 25.0 inches. Find the angle between a diagonal and the longer side.

25. Find the area of a parallelogram whose sides are 48.0 cm and 36.0 cm, if one angle is 73.5°.

26. Find the diameter of a circle inscribed in a hexagon whose sides are 15.0 in. each.

27. Find the radius of the circle in Fig. 15-59.

28. How long is the shadow of a 56.0-ft tree, when the angle of elevation of the sun is 32.8°?

29. A ranger spots a fire from the top of a 72.5-ft fire tower. The angle of depression to the fire is 0.506° (Fig. 15-60). How far away from the base of the tower is the fire?

30. To measure the height of a cliff, a surveyor measures a point 275 m from the base of the cliff and measures the angle of elevation to the top of the cliff to be 16.4°. How high is the cliff?

31. A 20.0-ft ladder leans against a vertical wall so that the base of the ladder is 4.56 ft away from the base of the wall. How high up the wall does the ladder go, and what angle does the ladder make with the ground?

32. The corner of the rectangular metal plate in Fig. 15-61 is to be cut off. Find x and θ.

WRITING

33. Write an essay on how you might use right triangle trigonometry in real-life situations on the job, in a hobby, or around the house.

34. Make a device for measuring angles in the horizontal plane (a "transit"), using a sighting tube of some sort and a protractor. Use your transit and a rope of known length to find some distance on your campus, such as the distance across a ball field, using the method of Example 17 in Sec. 15-4. Compare your result with the actual taped distance.

16 Factoring

In Sec. 8-3 we learned how to multiply factors to get a product. In this chapter, we do the reverse: We begin with the product and find those quantities (factors) which, when multiplied together, give the original product. To *factor* expressions easily, we must be very familiar with the *special products* of algebraic multiplication. We will look at these special products first in each section.

But why must we learn to factor? We learn to factor because factoring is an essential step in simplifying expressions and solving equations, especially quadratic equations, literal equations, and formulas. Factoring is also used extensively in simplifying fractions. We will consider many factorable expressions which appear in equations that arise from applications.

16-1   COMMON FACTORS

In this section we learn to *factor* certain polynomials by factoring out a monomial which is a factor of each term of the polynomial. Recall that a monomial has only one term and a polynomial has more than one term.

Factoring an Expression

The *factors* of an expression are those quantities (numbers and variables) whose product is the original expression. In this chapter, we consider only factors with integral coefficients.

> ### Example 1
> Since
>
> $$3a - 6b = 3(a - 2b)$$
>
> then   3 and $(a - 2b)$ are factors of $3a - 6b$.

Prime Factors

If an expression has no factors, other than 1 and itself, then it is called *prime*.

> ### Example 2
> Two factors of 70 are 7 and 10. The 7 is prime because it has no factors other than 1 and 7. However, the 10 is not prime, because it can be factored into 2 times 5 (both 2 and 5 are prime). Thus the prime factors of 70 are 2, 5, and 7.

Example 3

The expressions x, $x - 2$, and $x^2 - x - 1$ are all prime, because they have no factors other than 1 and themselves.

Example 4

The expressions $x^2 - 4$ and $x^2 - 5x + 6$ are not prime, because they can be factored, as we shall see later in this chapter. We say they are *factorable*.

It's important to note that *not all* expressions are factorable. Many expressions, as in Example 3, are prime.

Common Factors

In Sec. 1-3, we discussed the distributive law

$$a(b + c) = ab + ac$$

This gives us the product of a monomial times a polynomial. If we reverse the sides of this equation, then we are *factoring* the left side.

$$ab + ac = a(b + c)$$
$$\text{common factor}$$

In this equation, the quantity a is called a *common factor,* since it is a factor *common* to both terms. It may be factored out, using the reverse of the distributive law (Eq. 5).

FACTORING OUT A COMMON FACTOR	$ab + ac = a(b + c)$	5

Example 5

Factor.

(a) $5p + 5q = 5(p + q)$
(b) $6a - 12b =$

$$6a - 6(2b) = 6(a - 2b)$$
$$\text{common factor}$$

The common factor 6 is *not* prime. That is, it can be factored

$$6 = 2(3)$$

However, in general, *we do not factor the numerical coefficient into primes.* So we are finished factoring the expression.

Example 6

$$6x - 2xy = 2x(3) - 2x(y) = 2x(3 - y)$$
$$\text{common factor}$$

Sometimes the common factor is not as obvious as it is in the examples above.

Example 7

Factor $12x^2 + 8x$.

Solution: Since 4 is a factor of both 12 and 8 and there is a common x in both terms, then $4x$ is a common factor of $12x^2 + 8x$.

$$12x^2 + 8x = \underbrace{4x}(3x) + \underbrace{4x}(2) = \underbrace{4x}(3x + 2)$$

common factor

Note that $4x$ is only *one* common factor of $12x^2 + 8x$. Other common factors are 2, 4, x, and $2x$. So the expression $12x^2 + 8x$ may be factored several different ways:

$$12x^2 + 8x = 2(6x^2 + 4x)$$

or

$$= 4(3x^2 + 2x)$$

or

$$= x(12x + 8)$$

or

$$= 2x(6x + 4)$$

But the preferred factoring is

$$12x^2 + 8x = 4x(3x + 2)$$

because $4x$ is the *greatest* common factor, as described below.

Factoring Completely

We say that an expression is factored *completely* if it is expressed as a product of its prime factors. One way to factor an expression completely is to factor out its "greatest common factor." The *greatest common factor* (also called the *gcf*) is the common factor that has the greatest numerical coefficient and the greatest degree. (Recall that the degree of a term is the power to which the variable is raised.)

But how do we find the gcf? Generally, the gcf is found by *inspection*. That is, we look at (inspect) the terms and say, "Ah ha, the gcf is $2x^2$." Specifically, the numerical coefficient of the gcf is the gcf of the coefficients of the terms, and the variable part is the highest power of each variable that is a factor of *each term*.

Example 8

Factor $10x^3 - 6x^2y$ completely.

Solution: First the gcf of the expression must be determined. The gcf of 10 and 6 is 2, the highest power of x contained in each term is x^2, and no power of y is contained in the first term. Thus, the gcf of the expression is $2x^2$.

$$10x^3 - 6x^2y = 2x^2(5x) - 2x^2(3y)$$
$$= 2x^2(5x - 3y)$$

Another method of factoring completely is to factor the expression, and keep factoring, until all the factors are prime.

Example 9

$$10x^3 - 6x^2y = 2(5x^3 - 3x^2y)$$
$$= 2x^2(5x - 3y)$$

Factors of 1

In factoring expressions, you must be especially careful with factors of 1.

Example 10 ───────────

Factor $12x^2 + 6x$ completely.

Solution: By inspection we find the gcf is $6x$. Factoring $6x$ out of each term:

$$12x^2 - 6x = 6x(2x) - 6x(1)$$
$$= 6x(2x - 1)$$

Students are sometimes careless about the 1 in the factorization above. It's a common error to forget the 1, but it must be included!

COMMON ERROR If the gcf of an expression is the same as one of its terms, we must include the 1 in the factored form.

Common Expressions

The common factors we have dealt with thus far have been monomials. But sometimes the common factor may be an expression, such as a binomial or a polynomial. Again, use the reverse of the distributive property:

$$ab + ac = a(b + c)$$

except that the common factor a will be an expression. Suppose the common factor was the expression $x + y$. Then the equation above would be:

$$\underbrace{(x + y)}b + \underbrace{(x + y)}c = \underbrace{(x + y)}(b + c)$$
common factor

Note that this is just a special case of the distributive property.

Example 11 ───────────

Factor completely:

$$(x - 2)x + (x - 2)3$$

Solution: The common factor is the expression $(x - 2)$. Factoring out the expression $(x - 2)$, we get

$$(x - 2)x + (x - 2)3 = (x - 2)(x + 3)$$

Example 12 ───────────

Factor completely:

$$2(p - 3) - x(p - 3)$$

Solution: The common factor is the expression $(p - 3)$. Factoring out $(p - 3)$, we get

$$2(p - 3) - x(p - 3) = (2 - x)(p - 3)$$

Factoring by Grouping

When the expression to be factored has four or more terms, these terms can sometimes be arranged in smaller groups, which can be factored separately.

─── **Example 13** ───

Factor $ab + ac + 2b + 2c$.

Solution: First we notice that there is no factor common to all four terms. However, the first two terms have a common factor, a, and the second two terms have a common factor, 2. We will try grouping the first two terms together and the last two terms together.

$$ab + ac + 2b + 2c = (ab + ac) + (2b + 2c)$$

Now factor out the common factor in each group:

$$= a(b + c) + 2(b + c)$$

common factor

Both terms now have the common factor $(b + c)$. We now treat $(b + c)$ as a single expression and factor it out using the reverse of the distributive property. So,

$$ab + ac + 2b + 2c = (a + 2)(b + c)$$

Factoring by grouping can be quite tricky. Sometimes you must rearrange terms, factor out negative factors, and in general, use *trial and error* to see if any arrangement gives you another common factor. Sometimes no arrangement works, and the expression is not factorable.

Checking

To check if factoring has been done correctly, simply multiply the factors together. If the product is the original expression, then you were correct. This check does not, however, tell you if the expression is factored *completely*. You check that by inspecting the factors to make sure they cannot be factored further.

─── **Example 14** ───

Factor and check:

$$6x^2 - 3x$$

Solution: Factoring,

$$6x^2 - 3x = 3x(2x - 1)$$

To check we multiply the factors,

$$3x(2x - 1) = 3x(2x) - 3x(1)$$
$$= 6x^2 - 3x$$

So, it checks, and the factors are correct.

Factoring Completely

Factor each expression completely and check your result.

1. $xy + xz$
2. $pr - qr$
3. $3a - 3b$
4. $7x - 7y^2$
5. $x^2 - xy$
6. $ab + b^2$
7. $6x + 18$
8. $16x - 4y$
9. $24x^2 - 3$
10. $36ax + 3a$
11. $ac + a$
12. $4x - 2$
13. $4x^3 - 3x^2$
14. $5b^2 - 6b^3$
15. $24x^3y + 15x^2y^2 - 40xy$
16. $15p^3 - 10p^2 + 20p^5$
17. $4x^2y + 6x^3 - 2x$
18. $6y^3 - 12y^2 + 3y$
19. $6a^2b^3 + 4a^3b^3 - 10a^2b^4$
20. $6a^2b^3c - 9a^2bc^3 + 12a^3b^3c^2$

Common Expressions

21. $2(x + y) + x(x + y)$
22. $4(a - b) - b(a - b)$
23. $(p - 1)q - (p - 1)p$
24. $(m + n)m + (m + n)n$
25. $(c - d)d + (c - d)$
26. $(x - 3)x - (x - 3)$
27. $(x - y)^2 + (x - y)$
28. $4(a - b) - 2(a - b)^2$
29. $(3 + a)^3 - (3 + a)^2$
30. $(p - q)^3 - (p - q)^2 + 5(p - q)$

Factoring by Grouping

Group the terms in pairs and factor completely.

31. $xy + xz + 3y + 3z$
32. $pq - pr + 5q - 5r$
33. $xy - 2x + 3y - 6$
34. $ab - 3a + 4b - 12$
35. $x^2 + xy - 2x - 2y$
36. $2a - 2b - a^2 + ab$
37. $m^3 - 5m^2 + 3m - 15$
38. $1 - x + x^2 - x^3$

39. The total surface area of a cylinder is given by the formula

$$A = 2\pi r^2 + 2\pi rh$$

where r is the radius of the cylinder and h is the height (Fig. 16-1). Factor the right side of this equation.

Applications

40. If p dollars is invested for t years at an annual rate of simple interest r, the principal p would accumulate to y dollars, given by the formula:

$$y = p + prt$$

Factor the right side of this equation.

■ **FIGURE 16-1**

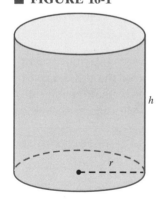

41. The total surface area of a cone is given by the formula

$$A = \pi r^2 + \pi rs$$

where r is the radius of the base and s is the slant height (Fig. 16-2). Factor the right side of the equation.

42. The total surface area A of a rectangular solid is given by the formula

$$A = 2lw + 2lh + 2wh$$

where l is the length, w is the width, and h is the height of the solid. Factor the right side of this equation.

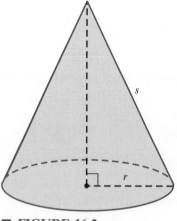

■ FIGURE 16-2
Cone.

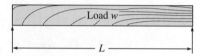

■ FIGURE 16-3
Simply supported beam.

43. An object is thrown upward with an initial velocity of 32 ft/s from a building 64 ft above the ground. The height S of the object above the ground at any time t is given by

$$S = 64 + 32t - 16t^2$$

Factor the right side of this equation.

44. In a battery, the sum of the voltage-drops across the resistors R_1, R_2, and R_3 with current i must equal the battery voltage E:

$$E = iR_1 + iR_2 + iR_3$$

Factor the right side of this equation.

45. For a simply supported beam of length L having a distributed load of w (Fig. 16-3), the bending moment M at any distance x from one end is given by

$$M = \frac{1}{2}wLx - \frac{1}{2}wx^2$$

Factor the right side of this equation.

16-2   DIFFERENCE OF TWO SQUARES

In this section we consider products that give the difference of two squares, such as

$$x^2 - y^2$$

and then we factor the difference of two squares. We learn to recognize the *pattern* for the difference of two squares, and use the pattern to factor the difference of any two even powers of expressions. We learn to group terms so that they fit into the pattern. Finally we consider applications of this type of factoring.

The Sum Times the Difference of Two Numbers

In Sec. 8-3 we saw in Eq. 31 that the sum of two numbers times the difference of the same numbers was equal to the difference of their squares.

$$\overset{\text{sum}}{\underset{\downarrow}{}} \quad \overset{\text{difference}}{\underset{\downarrow}{}}$$
$$(x + y)(x - y) = x^2 - y^2$$
$$\underset{\underset{\text{difference of two squares}}{\uparrow}}{}$$

Example 15

$$(x + 3)(x - 3) = x^2 - 9$$

Example 16

$$(2x - 5)(2x + 5) = (2x)^2 - 5^2$$
$$= 4x^2 - 25$$

Factoring the Difference of Two Squares

If we reverse the sides of Eq. 31, we have

FACTORING THE DIFFERENCE OF TWO SQUARES	$x^2 - y^2 = (x + y)(x - y)$	31

Once we recognize that an expression is the difference of two squares

$$x^2 - y^2$$

we can fit it into the pattern of Eq. 31.

Example 17

Factor $y^2 - 16$.

Solution: First we recognize that $y^2 - 16$ is a difference of two squares, since both y^2 and 16 are perfect squares. The square root of 16 is 4, so we write

$$y^2 - 16 = y^2 - 4^2$$

Thus, using Eq. 31,

$$y^2 - 16 = (y + 4)(y - 4)$$

Example 18

Factor $9x^2 - 4$.

Solution: Is $9x^2 - 4$ a difference of two squares? Does it fit into the pattern of $(x^2 - y^2)$? Yes, both $9x^2$ and 4 are perfect squares.

$$9x^2 - 4 = (3x)^2 - (2)^2$$

We fit these quantities into the pattern of the right side of Eq. 31:

$$9x^2 - 4 = (3x)^2 - (2)^2 = (3x \quad)(3x \quad)$$

Same!

$$9x^2 - 4 = (3x)^2 - (2)^2 = (3x \quad 2)(3x \quad 2)$$

Same!

$$9x^2 - 4 = (3x + 2)(3x - 2)$$

sum   difference

━━━━━ **Example 19** ━━━━━

Factor completely:

$$2x^2 - 18$$

Solution: First we recognize that this is a difference but not of two squares. Neither $2x^2$ nor 18 are perfect squares. But there *is* a common factor, 2. We factor it out first:

$$2x^2 - 18 = 2(x^2 - 9)$$

Now we see that $(x^2 - 9)$ is the difference of two squares and can be factored further.

$$2x^2 - 18 = 2(x^2 - 9) = 2(x + 3)(x - 3)$$

TIP Always factor out the gcf *first*, before doing other types of factoring.

Difference of Two Even Powers

The difference of any two even powers can be factored as the difference of two squares, since every even power is a perfect square.

━━━━━ **Example 20** ━━━━━

Factor completely:

$$x^4 - y^4.$$

Solution: The expression $x^4 - y^4$ is the difference of two even powers, and therefore the difference of two squares:

$$x^4 - y^4 = (x^2)^2 - (y^2)^2$$
$$= (x^2 + y^2)(x^2 - y^2)$$

Can either of these factors be factored further? The first factor $(x^2 + y^2)$ is prime, and cannot be factored. But the second factor $(x^2 - y^2)$ is the difference of two squares, and can be factored further.

$$x^4 - y^4 = (x^2 + y^2)(x^2 - y^2)$$
$$= (x^2 + y^2)(x + y)(x - y)$$

None of these factors can be factored again, so the expression is factored completely.

In Example 20, we noted that the sum

$$x^2 + y^2$$

was prime. It *cannot* be factored, since the *sum* of two squares is prime.

Squared Expressions

Sometimes one (or both) of the perfect squares will be squares of expressions.

Example 21

Factor completely:

$$(a + 1)^2 - b^2$$

Solution: In this example, the expression $(a + 1)$ corresponds to the x in Eq. 31, and b corresponds to the y.

$$x^2 \quad - y^2 = \quad [x \quad + y] \quad [x \quad - y]$$

Replacing x by $(a + 1)$, $(a + 1)^2 - b^2 = [(a + 1) + b][(a + 1) - b]$

Removing the inner parentheses on the right:

$$(a + 1)^2 - b^2 = (a + 1 + b)(a + 1 - b)$$

SECTION 16-2 Exercises Difference of Two Squares

Factor completely.

1. $m^2 - n^2$
2. $c^2 - d^2$
3. $x^2 - 16$
4. $9 - x^2$
5. $25 - p^2$
6. $y^2 - 4$
7. $9y^2 - 1$
8. $36x^2 - y^2$
9. $4a^2 - 9b^2$
10. $9b^2 - 25a^2$
11. $p^4 - 4$
12. $4m^2 - n^6$
13. $3x^2 - 12y^2$
14. $18x^2 - 8y^2$
15. $4a^2b^2 - 4c^2$
16. $20x^2y^4 - 5$

Difference of Two Even Powers

17. $a^4 - b^4$
18. $x^6 - y^4$
19. $p^6 - q^8$
20. $49m^2 - 9n^4$
21. $100x^4 - 81$
22. $a^2b^4 - 4c^6$
23. $x^8 - 1$
24. $x^4 - 1$

Squared Expressions

25. $(a + b)^2 - c^2$
26. $(x - y)^2 - z^2$
27. $p^2 - (q + r)^2$
28. $m^2 - (n - p)^2$
29. $1 - (a - b)^2$
30. $(x - y)^2 - (a + b)^2$

Applications

31. To find the difference in elevation, h, of a slope (Fig. 16-4), a surveyor uses the equation

$$h^2 = s^2 - d^2$$

Factor the right side of this equation.

■ **FIGURE 16-4**

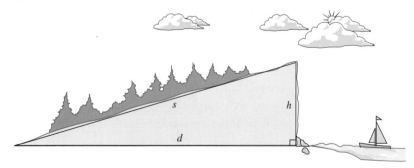

■ FIGURE 16-5

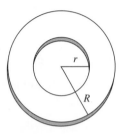

■ FIGURE 16-6

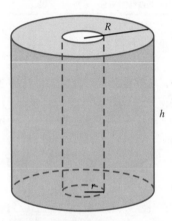

■ FIGURE 16-7

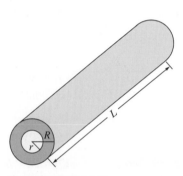

■ FIGURE 16-8

32. The area of a square picture frame (Fig. 16-5) is given by the formula

$$A = s_2^2 - s_1^2$$

Factor the right side of this equation.

33. The surface area of a washer (Fig. 16-6) is given by the formula

$$A = \pi R^2 - \pi r^2$$

Factor the right side of this equation.

34. The volume of a hollow cylinder (Fig. 16-7) is given by the formula

$$V = \pi R^2 h - \pi r^2 h$$

Factor the right side of this equation.

35. The weight of an iron pipe (Fig. 16-8) is given by the formula

$$W = \pi R^2 LD - \pi r^2 LD$$

where D is the density of iron. Factor the right side of this equation.

36. The "work-energy" principle states that in any finite motion, the work of the resultant force on a body equals the change in its kinetic energy:

$$W = \frac{1}{2} mV^2 - \frac{1}{2} mv^2$$

Factor the right side of this equation.

SIMPLE TRINOMIALS

In this section we factor quadratic trinomials of the form $ax^2 + bx + c$, in which $a = 1$. In the next section, we factor the general quadratic trinomial which has a leading coefficient different from 1. In all cases the coefficients in this chapter will be integers. We also solve applications that require us to factor trinomials.

Recall from Sec. 8-2 that a *trinomial* is a polynomial that has three terms. A *quadratic trinomial* in x has an x^2 term, an x term, and a constant term. The coefficient of the x^2 term in a quadratic trinomial is called the *leading coefficient* and the x term is sometimes called the *middle term*.

Example 22

The expression $3x^2 - 2x + 5$ is a quadratic trinomial. The leading coefficient is 3, the coefficient of the x term is -2, and the constant term is 5.

Trinomials with a Leading Coefficient of 1

We get a quadratic trinomial with a leading coefficient of 1 when we multiply two binomials, if the coefficients of the x terms in the binomials are also 1. In Sec. 8-3, we saw that we can multiply two binomials together using the FOIL method.

Example 23

Recall how we multiply $(x + 2)(x + 3)$:

$$F + O + I + L$$
$$(x + 2)(x + 3) = x^2 + \underbrace{3x + 2x} + 6$$

So,
$$(x + 2)(x + 3) = x^2 + 5x + 6$$

On the right, the coefficient of the x term is 5 (the sum of 2 and 3), and the constant term is 6 (the product of 2 and 3):

$$(x + 2)(x + 3) = x^2 + 5x + 6$$

In general, by Eq. 35

$$(x + a)(x + b) = x^2 + (a + b)x + ab$$

If we reverse Eq. 35, then it shows us a method to factor quadratic trinomials.

FACTORING A TRINOMIAL WITH LEADING COEFFICIENT 1	$x^2 + (a + b)x + ab = (x + a)(x + b)$	35

Note that in the given trinomial, the leading coefficient is 1, the coefficient of the x term is a sum, and the constant term is a product. The key to factoring is to *find the two numbers, a and b, which have that sum and product.*

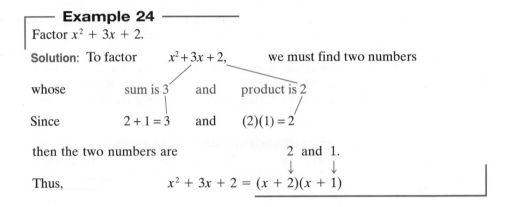

Example 24

Factor $x^2 + 3x + 2$.

Solution: To factor $\quad x^2 + 3x + 2,\quad$ we must find two numbers

whose $\quad\quad$ sum is 3 $\quad$ and $\quad$ product is 2

Since $\quad\quad\quad 2 + 1 = 3 \quad$ and $\quad (2)(1) = 2$

then the two numbers are $\quad\quad\quad\quad\quad$ 2 and 1.

Thus, $\quad\quad\quad\quad\quad x^2 + 3x + 2 = (x + 2)(x + 1)$

It's important to note that *not all quadratic trinomials can be factored,* using only factors with integral coefficients.

Example 25

The quadratic trinomial $x^2 + x + 1$ is not factorable, since

$$x^2 + x + 1 = x^2 + 1x + 1$$
$$\text{sum} \quad \text{product}$$

and there are no numbers, a and b, whose sum is 1:

$$a + b = 1$$

and whose product is 1:

$$ab = 1$$

So, the quadratic trinomial $x^2 + x + 1$ is prime.

Using the Signs to Aid Factoring

The signs of the terms of the trinomial indicate the signs of the factors.

Example 26

Factor $x^2 + 11x + 18$.

Solution: If we let $\quad x^2 + 11x + 18 = (x + a)(x + b)$
$$\text{sum} \quad \text{product}$$

then we must find two numbers, a and b, whose sum is +11 and whose product is +18:

$$a + b = +11 \quad \text{and} \quad ab = +18$$

Since both the sum +11 and the product +18 are positive, then both a and b must be positive. We consider products first because there are fewer pairs of numbers with a given product than a given sum. So, we begin with positive pairs of numbers (factors) whose products are $+18$. We list these pairs of factors and calculate their sums. A table is useful to organize our work:

Positive Factors of 18		Product	Sum
a	b	ab	$a + b$
$+1$	$+18$	$+18$	$+19$
$+2$	$+9$	$+18$	$+11$
$+3$	$+6$	$+18$	$+9$

The correct pair is $+2$ and $+9$, since the sum

$$+2 + 9 = +11$$

gives us the coefficient of the x term, $+11$.

Thus,
$$x^2 + 11x + 18 = (x + 2)(x + 9)$$

Check:

Just multiply $(x + 2)(x + 9)$ to check:

$$(x + 2)(x + 9) = x^2 + 11x + 18 \quad \text{checks}$$

Example 27

Factor $x^2 - 11x + 18$.

Solution: This trinomial is similar to the one in Example 26, except that the x term is negative. In this case we need to find a pair of numbers, a and b, whose product is $+18$ and whose sum is -11:

$$ab = +18 \quad \text{and} \quad a + b = -11$$

Since the product is positive, then either both numbers are positive, or both are negative. Since the sum is negative, then both must be *negative*. We make a table:

Negative Factors of $+18$		Product	Sum
a	b	ab	$a + b$
-1	-18	$+18$	-19
-2	-9	$+18$	-11
-3	-6	$+18$	-9

Since the sum
$$-2 + -9 = -11$$

and -11 is the coefficient of the x term, the result is

$$x^2 - 11x + 18 = (x - 2)(x - 9)$$

You can check by multiplying the factors, to get the original trinomial.

Example 28

Factor $t^2 - 4t - 12$.

Solution: We are looking for two numbers, a and b, whose product is -12 and whose sum is -4:

So,
$$t^2 - 4t - 12 = (t + a)(t + b)$$

where
$$ab = -12 \quad \text{and} \quad a + b = -4$$

The product is negative, so the numbers have opposite signs, one positive and one negative. There are many ways to factor -12, so we make a table to organize our work:

Factors of -12		Product	Sum
a	b	ab	$a + b$
$+1$	-12	-12	-11
$+2$	-6	-12	-4
$+3$	-4	-12	-1
$+4$	-3	-12	$+1$
$+6$	-2	-12	$+4$
$+12$	-1	-12	$+11$

All the pairs of factors give a product of -12, but only one pair gives the required sum of -4:

$$+2 + -6 = -4$$

Thus the numbers are $+2$ and -6.

So,

$$t^2 - 4t - 12 = (t + 2)(t - 6)$$

SECTION 16-3 Exercises Simple Trinomials

Trinomials with a Leading Coefficient of 1

Factor completely and check by multiplying.

1. $x^2 + 8x + 7$
2. $x^2 + 7x + 10$
3. $a^2 - a - 6$
4. $c^2 - 3c - 10$
5. $u^2 + 7u - 18$
6. $k^2 + k - 12$
7. $y^2 - 7y + 6$
8. $a^2 - 6a + 8$
9. $14 + 9k + k^2$
10. $21 - 10y + y^2$
11. $x^2 - 8xy + 15y^2$
12. $m^2 + 14mn + 24n^2$
13. $18 - 3x - x^2$
14. $20 - y - y^2$
15. $a^2 - 2ab - 8b^2$
16. $s^2 - 5st - 24t^2$

Applications

17. To find the width of a sheet of metal needed to make a box with a specific volume, one must solve the following equation:

$$w^2 - 11w - 60 = 0$$

Factor the left side of this equation.

18. To find the width of a brick border on a certain patio one must solve the equation

$$w^2 - 26w + 69 = 0$$

Factor the left side of this equation.

19. To find the dimensions of a rectangular field that has a perimeter of 724 m and an area of 32,400 m², one must solve the equation

$$x^2 - 362x + 32400 = 0$$

Factor the left side of this equation.

20. To find the dimensions of a rectangular field which is 12 m longer than it is wide and contains 448 m², one must solve the equation

$$w^2 + 12w - 448 = 0$$

Factor the left side of this equation.

21. To find the width of a certain picture frame, we must solve the equation

$$x^2 - 11x + 10 = 0$$

Factor the left side of this equation.

22. An object is thrown upward with an initial velocity of 30 ft/s from a building 64 ft above the ground. The height s of the object above the ground at any time t is given by

$$s = 64 + 30t - t^2$$

Factor the right side of this equation.

Factor completely and check.

23. $(x + y)^2 - 5(x + y) + 4$ **24.** $(y - 3)^2 - 3(y - 3) + 2$

25. $t^4 - 2t^2 + 1$ **26.** $x^4 - 10x^2 + 9$

16-4   GENERAL TRINOMIALS

In this section we learn how to factor the general quadratic trinomial where the leading coefficient is not 1. In Sec. 8-3, we saw that we get such a trinomial when we multiply two binomials, Eq. 36:

$$(ax + b)(cx + d) = acx^2 + (ad + bc)x + bd$$

In the trinomial on the right,

ac = the leading coefficient,
$ad + bc$ = the x term coefficient, and
bd = the constant term

┌─── **Example 29** ───────

If we multiply the binomials below, we can see the relationship among the coefficients. We know that

$$(2x - 1)(3x + 4) = \ 6x^2 + \quad 5x \quad - \ 4$$

We compare this to

$$(ax + b)(cx + d) = acx^2 + (ad + bc)x + bd$$

In the trinomial,

the leading coefficient 6 is the product ac, $2(3) = 6$

the x coefficient 5 is the sum, $ad + bc$, $(2)(4) + (-1)(3) = 5$

the constant -4 is the product bd $(-1)(4) = -4$

When we factor, we reverse the process of multiplication. Reversing Eq. 36, we get

FACTORING A GENERAL TRINOMIAL	$acx^2 + (ad + bc)x + bd = (ax + b)(cx + d)$	36

To factor a trinomial, we look for the four numbers a, b, c, and d, which satisfy the equation above.

───── **Example 30** ─────

Factor $2x^2 + 7x + 3$.

Solution: The leading coefficient is 2 and the constant is 3. We can "set up" the problem like this:

$$2x^2 + 7x + 3 = \quad (\_x + \_)(\_x + \_)$$

Product of coefficients must be 2.

↓

The only factors of 2 are 2 and 1.

Product of constants must be 3.

↓

The only factors of 3 are 3 and 1.

Since all the coefficients are positive, we need only consider positive factors. We "fill in the blanks," listing all the possibilities.

$$2x^2 + 7x + 3 \stackrel{?}{=} (2x + 3)(x + 1)$$

Reversing the constants, $\qquad\qquad \stackrel{?}{=} (2x + 1)(x + 3)$

Reversing the coefficients, $\qquad\quad \stackrel{?}{=} (x + 3)(2x + 1)$

Reversing the constants, $\qquad\qquad \stackrel{?}{=} (x + 1)(2x + 3)$

Note that the third possibility is identical to the second and that the fourth is identical to the first. So we can eliminate the third and fourth possibilities. In fact, when you consider possible factorizations, *you need not reverse the factors of the leading coefficient as long as you do reverse the factors of the constant.*

Now we check the middle terms to see which, if either, give us the required $7x$.

Factors	Middle term	
$(2x + 3)(x + 1)$	$2x + 3x$	$= 5x$
$(2x + 1)(x + 3)$	$6x + x$	$= 7x$

The second possibility gives the correct middle term, so

$$2x^2 + 7x + 3 = \underline{(2x + 1)(x + 3)}$$

Example 30 was simple because all terms were positive and the numbers had few factors. We will now factor some more "interesting" trinomials.

───── **Example 31** ─────

Factor $2x^2 + 5x - 12$.

Solution: We write down the setup:

$$2x^2 + 5x - 12 = \quad (\_x \quad \_)(\_x \quad \_)$$

Product must be 2 $\qquad\qquad$ Product must be -12

In general, the leading coefficient is usually positive, and we need consider only its positive factors. In this case, the leading coefficient is 2, and the only positive factors of 2 are 2 and 1. So, we can fill in the coefficients of x:

$$2x^2 + 5x - 12 = (\underline{2}x \quad \_) \; (\underline{1}x \quad \_)$$

The constant term is -12. Since it is negative, the constant factors have opposite signs, one positive and one negative. The positive factors of 12 are

$$1 \text{ and } 12 \qquad 2 \text{ and } 6 \qquad 3 \text{ and } 4$$

For each pair of factors, we use opposite signs, reverse the signs, reverse the position of the factors, and reverse those signs! For the factors 1 and 12, we have 4 possible factorizations:

$$(2x + 1)(x - 12)$$

Reversing the signs:

$$(2x - 1)(x + 12)$$

Reversing the factors:

$$(2x + 12)(x - 1)$$

Reversing the signs again:

$$(2x - 12)(x + 1)$$

A similar set of possibilities is found for the factors 2 and 6, and another set of 4 possibilities is found for the factors 3 and 4. This list shows all possibilities:

$$2x^2 + 5x - 12 \overset{?}{=} (2x + 1)(x - 12)$$
$$(2x - 1)(x + 12)$$
$$(2x + 12)(x - 1)$$
$$(2x - 12)(x + 1)$$
$$(2x + 2)(x - 6)$$
$$(2x - 2)(x + 6)$$
$$(2x + 6)(x - 2)$$
$$(2x - 6)(x + 2)$$
$$(2x + 3)(x - 4)$$
$$(2x - 3)(x + 4)$$
$$(2x + 4)(x - 3)$$
$$(2x - 4)(x + 3)$$

By trial and error we determine which combination (if any) will produce a middle term of $+5x$. *We calculate the middle term for each pair of factors* until we find that $(2x - 3)(x + 4)$ has the required middle term of $+5x$. So, the solution is

$$2x^2 + 5x - 12 = \underline{(2x - 3)(x + 4)}$$

Note: You will rarely have to test *all* the possibilities. In fact, with a little practice, you will, in most cases, quickly find the right combination by trial and error.

Test for Factorability

As we noted before, not all quadratic trinomials can be factored. We test for factorability as follows.

TEST FOR FACTORABILITY	The trinomial $ax^2 + bx + c$ (where a, b, and c are constants) is factorable if $b^2 - 4ac$ is a perfect square.	34

Example 32

Test these trinomials for factorability:

(a) $2x^2 + 5x - 12$ **(b)** $3x^2 - 2x + 1$

Solution:

(a) This quadratic is from Example 31, so we already know it is factorable. We verify this by calculating $b^2 - 4ac$, where $a = 2$, $b = 5$, and $c = -12$:

$$b^2 - 4ac = 5^2 - 4(2)(-12) = 121$$

The number 121 is a perfect square, so the quadratic $2x^2 + 5x - 12$ is factorable.

(b) For $3x^2 - 2x + 1$, we calculate $b^2 - 4ac$, where $a = 3$, $b = -2$, and $c = 1$:

$$b^2 - 4ac = (-2)^2 - 4(3)(1) = -8$$

Since -8 is not a perfect square, the quadratic $3x^2 - 2x + 1$ is *not* factorable.

Factoring Trinomials by Grouping

Factoring becomes more complicated when the leading coefficient can also be factored several ways. But there is an alternative method of factoring trinomials that reduces the amount of trial and error and uses the grouping method from Sec. 16-1. We will illustrate this method in the next example.

Example 33

Factor, using the grouping method:

$$6x^2 - 5x - 4$$

Solution: Letting

$$6x^2 - 5x - 4 = ax^2 + bx + c$$

then

$$a = 6, \quad b = -5, \quad \text{and} \quad c = -4$$

To use the grouping method, we must find two numbers whose product is ac and whose sum is b.

Since

$$ac = 6(-4) = -24$$

and

$$b = -5$$

then we need to find two numbers whose product is -24 and whose sum is -5. (Sound familiar?) The positive factors of 24 are

$$1 \text{ and } 24 \quad 2 \text{ and } 12 \quad 3 \text{ and } 8 \quad 4 \text{ and } 6$$

Since the product is negative, we use opposite signs. Calculating sums to find a sum of -5, we find

$$+3 + -8 = -5$$

So, using these factors as coefficients of x:

$$+3x - 8x = -5x$$

In the original trinomial,

$$6x^2 \qquad - 5x \quad - 4$$

We substitute $3x - 8x$ for $-5x$,

$$= 6x^2 + 3x - 8x - 4$$

Grouping terms,

$$= (6x^2 + 3x) + (-8x - 4)$$

Factoring,

$$= 3x(2x + 1) - 4(2x + 1)$$

Factoring out the $(2x + 1)$,

$$= (3x - 4)(2x + 1)$$

So,

$$6x^2 - 5x - 4 = (3x - 4)(2x + 1)$$

This grouping method is useful because it eliminates most of the trial and error aspect of factoring trinomials.

Common Factors

If the trinomial has any common factors, it is wise to factor these common factors out first, because the coefficients will then be smaller and more manageable.

┌─── **Example 34** ───
Factor $3x^2 - 21x + 30$.

Solution: The trinomial $3x^2 - 21x + 30$ has a common factor 3. Factor out the 3 first, and then factor the new trinomial.

$$3x^2 - 21x + 30 = 3(x^2 - 7x + 10)$$
$$= 3(x - 5)(x - 2)$$

┌─── **Example 35** ───
Factor $12x^2 + 14x - 6$.

Solution: The trinomial $12x^2 + 14x - 6$ has a common factor 2. Factor out the 2 and then factor the new trinomial.

$$12x^2 + 14x - 6 = 2(6x^2 + 7x - 3)$$
$$= 2(3x - 1)(2x + 3)$$

Trinomials with More than One Variable

Two (or more) variables may appear in some factoring problems. The process of factoring is the same, but remember to include all the variables in the factors.

┌─── **Example 36** ───
Factor $9x^2 - 15xy - 14y^2$.

Solution:

$$9x^2 - 15xy - 14y^2 = (3x - 7y)(3x + 2y)$$

Remember the y's!

— **Example 37** —

Factor $4a^3 - 10a^2b + 6ab^2$.

Solution: We see that the trinomial has a common factor, $2a$, and we factor that out first:

$$4a^3 - 10a^2b + 6ab^2 = 2a(2a^2 - 5ab + 3b^2)$$

Now we factor the new trinomial by trial and error:

$$= 2a(2a - 3b)(a - b)$$
$$\downarrow \qquad \downarrow$$
Remember the b's!

Thus,

$$4a^2 - 10a^2b + 6ab^2 = 2a(2a - 3b)(a - b)$$

SECTION 16-4   Exercises   General Trinomials

Factor completely.

1. $2x^2 + 3x + 1$
2. $2y^2 + 5y + 3$
3. $5y^2 - 7y + 2$
4. $6n^2 - 11n + 3$
5. $6x^2 - 13x - 5$
6. $4y^2 - 4y - 15$
7. $3a^2 + 7a - 6$
8. $8r^2 + 2r - 3$
9. $15 + 11x - 12x^2$
10. $12 - y - 6y^2$
11. $6t^2 + 25st + 14s^2$
12. $36u^2 - 5uv - 24v^2$
13. $3b^2 - 15b - 18$
14. $7x^2 - 14x - 21$
15. $2ab^3 - 14ab^2 + 24ab$
16. $3x^3y - 21x^2y + 30xy$
17. $15k^2 - 6k - 21$
18. $10mn^2 - 6mn - 4m$

Applications

19. An object is thrown upward with an initial velocity of 16 ft/s from a building 96 ft above the ground. The height s of the object above the ground at any time t is given by

$$s = 96 + 16t - 16t^2$$

Factor the right side of this equation.

20. An object is thrown into the air with an initial velocity of 48 ft/s. To find the time it takes for the object to reach a height of 35 ft, we must solve the quadratic equation

$$16t^2 - 48t + 35 = 0$$

Factor the left side of this equation.

■ **FIGURE 16-9**

21. An open box is made from a rectangular sheet of cardboard by cutting out square corners and folding up its sides (Fig. 16-9). To find the dimensions of the box, we must solve the equation

$$6x^2 - 54x - 540 = 0$$

Factor the left side of this equation.

■ **FIGURE 16-10**

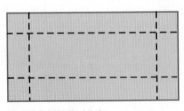

22. A rectangular patio (Fig. 16-10) is surrounded by a brick border of width w. To find the width of the border, we must solve the equation

$$4w^2 - 104w + 276 = 0$$

Factor the left side of this equation.

23. $6(y - 2)^2 + (y - 2) - 2$ **24.** $8(x - y)^2 - 14(x - y) - 15$

25. $15 - 8(x - y) + (x - y)^2$ **26.** $6a^2 + a(b + c) - (b + c)^2$

27. $6(m + n)^2 + 7(m + n)(m - n) + (m - n)^2$

16-5   PERFECT SQUARE TRINOMIALS

In this section we define the perfect square trinomial, show which products give a perfect square trinomial, and learn how to factor a perfect square trinomial.

Perfect Square Trinomials

When a binomial is squared, we get a *perfect square trinomial.*

Example 38

$$(3x - 4)^2 = (3x - 4)(3x - 4)$$
$$= 9x^2 - 12x - 12x + 16$$
$$= 9x^2 - 24x + 16 \leftarrow \text{perfect square trinomial}$$

In a perfect square trinomial, the first and last terms of the trinomial are the squares of the first and last terms of the binomial,

$$(3x - 4)^2$$

$$\text{square} \quad \text{square}$$

$$9x^2 \quad\quad 16$$

and the middle term is twice the product of the terms of the binomial.

$$(3x - 4)^2$$

twice the product $2(3x)(-4) = -24x$

Thus, $(3x - 4)^2 = 9x^2 - 24x + 16$

Recall the formulas, Eqs. 37 and 38, for the perfect square trinomials from Sec. 8-5,

$$(x + y)^2 = x^2 + 2xy + y^2$$
$$(x - y)^2 = x^2 - 2xy + y^2$$

Factoring a Perfect Square Trinomial

To factor a perfect square trinomial you must first *recognize the form of the trinomial as a perfect square:* The first and last terms must both be positive perfect squares and the middle term must be twice the product of the square roots of the first and last terms. Then to factor, simply reverse the process described above. Reversing Eqs. 37 and 38, we get

FACTORING PERFECT SQUARE TRINOMIALS	$x^2 + 2xy + y^2 = (x + y)^2$	37
	$x^2 - 2xy + y^2 = (x - y)^2$	38

Example 39

Factor $x^2 + 10x + 25$.

Solution: First recognize the form.
First and last terms are positive perfect squares:

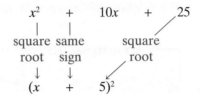

The middle term is twice the product of x and 5:

So, factoring

$$x^2 \quad + \quad 10x \quad + \quad 25$$

square same square
root sign root

We get,

$$(x \quad + \quad 5)^2$$

Example 40

Factor $9x^2 - 24x + 16$.

Solution: Is this a perfect square trinomial? The first and last terms are positive perfect squares and the middle term is twice the product of the square roots of the first and last terms:

$$9x^2 - 24x + 16$$

$$(3x)^2 \quad (4)^2$$

$$2(3x)(4) = 24x$$

So, yes, this is a perfect square trinomial. Since the middle term is negative, we refer to Eq. 38 and fit the trinomial into the pattern in Eq. 38:

$$9x^2 - 24x + 16 = (3x)^2 - 2(3x)(4) + (4)^2$$
$$= (3x - 4)^2$$

Of course, you would get the same result using the methods in Sec. 16-4 to factor a general trinomial. But, if you learn to recognize a perfect square trinomial, the factoring will be easier for you.

Note that *a perfect square trinomial must have a middle term.* The sum of two squares *cannot* be factored as a perfect square.

COMMON ERROR $x^2 + y^2 \neq (x + y)^2$

SECTION 16-5 Exercises Perfect Square Trinomials

Factoring a Perfect Square Trinomial

Factor completely and check.

1. $x^2 - 4x + 4$

2. $25 + 10x + x^2$

3. $y^2 + 6y + 9$

4. $b^2 - 8b + 16$

5. $36 - 12s + s^2$

6. $1 - 4x + 4x^2$

7. $9a^2 + 12ab + 4b^2$

8. $25x^2 - 60xy + 36y^2$

9. $1 - 22y + 121y^2$

10. $9a^2 - 42a + 49$

11. $100 + 60x + 9x^2$

12. $4x^2 + 20x + 25$

13. $81y^2 + 72y + 16$

14. $9x^2 - 24x + 16$

15. $16t^2 - 24tuv + 9u^2v^2$

16. $4x^2y^2 - 12xyz + 9z^2$

17. $7y^3 + 14y^2 + 7y$

18. $24x + 24x^2 + 6x^3$

19. $20xy^2 - 60xy + 45x$

20. $4x^4 - 8x^2y^2 + 4y^4$

21. $(x + y)^2 + 2(x + y) + 1$

22. $9 - 6(a - b) + (a - b)^2$

23. $4 - 4(a - b) + (a - b)^2$

24. $4(x - 2)^2 - 12(x - 2)(y - 1) + 9(y - 1)^2$

25. $16(m + n)^2 + 40(m + n)(m - n) + 25(m - n)^2$

16-6　SUM OR DIFFERENCE OF TWO CUBES

In this section we define the sum of two cubes, $x^3 + y^3$, and the difference of two cubes, $x^3 - y^3$. We find out what factors will give a product that is the sum or the difference of two cubes, and then reverse the process to find the factors of the sum or the difference of two cubes.

If we divide the sum of two cubes, $x^3 + y^3$, by $x + y$, we find that there is no remainder and therefore $x + y$ is a factor of $x^3 + y^3$:

$$
\begin{array}{r}
x^2 - xy + y^2 \\
x + y \overline{) x^3 \qquad\qquad + y^3} \\
\underline{x^3 + x^2y} \\
-x^2y \\
\underline{-x^2y - xy^2} \\
+xy^2 + y^3 \\
\underline{+xy^2 + y^3}
\end{array}
$$

Thus, the other factor of $x^3 + y^3$ is $x^2 - xy + y^2$.

FACTORING THE SUM OF TWO CUBES	$x^3 + y^3 = (x + y)(x^2 - xy + y^2)$	32

So the sum of two cubes, $x^3 + y^3$, is always divisible by the sum, $x + y$. Similarly, we can show that the *difference* of two cubes, $x^3 - y^3$, is divisible by the *difference*, $x - y$. By division, we find

FACTORING THE DIFFERENCE OF TWO CUBES	$x^3 - y^3 = (x - y)(x^2 + xy + y^2)$	33

The two equations above should be learned, if you don't want to have to divide it out each time!

Factoring the Sum or Difference of Two Cubes

Once we recognize an expression as the sum or the difference of two cubes, we apply Eq. 32 or Eq. 33.

Example 41

Factor $p^3 + 8$.

Solution: First we recognize this as the sum of two cubes:

$$p^3 + 8 = p^3 + 2^3$$

Now we apply Eq. 32:

$$p^3 + 2^3 = (p + 2)(p^2 - 2p + 2^2)$$

same sign   opposite signs   always positive

So,

$$p^3 + 8 = (p + 2)(p^2 - 2p + 4)$$

Example 42

Factor $8x^3 - 27y^3$.

Solution: First we recognize this as the difference of two cubes:

$$8x^3 - 27y^3 = (2x)^3 - (3y)^3$$

Applying Eq. 33:

$$(2x)^3 - (3y)^3 = [2x - 3y][(2x)^2 + (2x)(3y) + (3y)^2]$$
$$= [2x - 3y][4x^2 + 6xy + 9y^2]$$

So,

$$8x^3 - 27y^3 = (2x - 3y)(4x^2 + 6xy + 9y^2)$$

same sign   opposite signs   always positive

COMMON ERROR The middle term of the trinomials in Eq. 32 and Eq. 33 is often *mistaken* as 2xy.

$$x^3 + y^3 \neq (x + y)(x^2 - 2xy + y^2)$$
$$\downarrow$$
$$\text{no!}$$

SECTION 16-6   Exercises   Sum or Difference of Two Cubes

Factor completely.

1. $a^3 - 1$	**2.** $x^3 + 1$	**3.** $x^3 + 27$
4. $a^3 - 8$	**5.** $b^3 - 64$	**6.** $y^3 - 27$
7. $1 - 8a^3$	**8.** $27p^3 + 125$	**9.** $1000 + 27x^3$
10. $343 - 8c^3$	**11.** $64a^3 - 125b^3$	**12.** $729c^3 + 64d^3$
13. $3y^3 - 24$	**14.** $4x^3 + 32$	**15.** $54p^4 + 2p$
16. $250q - 2q^4$	**17.** $(a + b)^3 - x^3$	**18.** $(x - y)^3 + 27z^3$

19. $8x^3 - (a - b)^3$ **20.** $64a^3 + (x + y)^3$ **21.** $(a - b)^3 + (x + y)^3$

22. $(a - 1)^3 - (b - 2)^3$ **23.** $x^6 + 1$ **24.** $1 - y^6$

25. $a^9 - 1$ **26.** $27x^9 + 512$

27. The volume of a hollow spherical shell having an inside radius of r and an outside radius of R is

$$V = \frac{4}{3}\pi R^3 - \frac{4}{3}\pi r^3$$

Factor the right side of this equation.

28. A concrete container for holding radioactive material is in the shape of a hollow cube whose inside dimension is s and whose outside dimension is S. If it is made of concrete of density d, its weight is

$$W = dS^3 - ds^3$$

Factor the right side of this equation.

29. Factor $a^6 - b^6$ in two different ways:

     (a) As the difference of two cubes, and

     (b) As the difference of two squares.

Which way gives you the complete factorization more easily?

■ Chapter 16   REVIEW EXERCISES

Factor completely.

1. $15y^2 - 27y$ **2.** $25xy + 5y$

3. $15a^2 - 21ab + 3ac$ **4.** $(y - 2)y - (y - 2)$

5. $3x - 3y + xy - y^2$

6. An item costing P dollars is reduced in price by 12%. The resulting price C is

$$C = P - 0.12P$$

Factor the right side of this equation.

7. $4x^2 - 25$ **8.** $9x^2 - 16a^2b^2$

9. $27 - 12y^2$ **10.** $(c - d)^2 - e^2$

11. $m^2 - (n - 1)^2$ **12.** $x^4 - 1$

13. Due to a drop in temperature, a spherical weather balloon shrinks from radius R to radius r. The change in surface area is given by the formula

$$S = 4\pi R^2 - 4\pi r^2$$

Factor the right side of this equation.

14. $x^2 + 8x - 9$ **15.** $m^2 + m - 90$

16. $y^2 - 4y - 12$ **17.** $b^2 - 11b + 28$

18. $x^2 + x - 30$ **19.** $b^2 - 20b + 36$

20. $6a^2 + 19a + 15$ **21.** $24y^2 + 17y - 20$

22. $4y^2 - 25y + 6$ **23.** $15x^2 + x - 40$

24. $4a^2 - 12a + 9$ **25.** $y^2 + 16y + 64$

26. $49 - 14p + p^2$ **27.** $16x^2 + 72x + 81$

28. An object is thrown into the air with an initial velocity of 82 ft/s. To find the time it takes for the object to reach a height of 45 ft, we must solve the quadratic equation

$$16t^2 - 82t + 45 = 0$$

Factor the left side of this equation. (Extra: Show where the equation comes from.)

Factor completely.

29. $1 - y^3$

30. $a^3 - 125$

31. $27x^3 + 8y^3$

32. $2a^4 - 16ab^3$

33. $a^3 - (b - 2)^3$

34. $8 + a^6$

35. We have studied the factoring of several different types of expressions in this chapter. List at least six types and give an example of each. State in words how to recognize each type and how to tell one from the other. List two other expressions that are not examples of the given types.

36. The area of a rectangle is the product of its sides, so we can think of the product

$$(x + 1)(x + 2) = x^2 + 3x + 2$$

as the area of a rectangle of sides $(x + 1)$ and $(x + 2)$ (Fig. 16-11).

Similarly, to factor an expression using areas of rectangles, we can cut rectangles of areas equal to the terms in the trinomial. You may use graph paper and choose any length for x, but avoid making it an integer value. The expression $x^2 + 3x + 2$ can be represented by rectangles of area x^2, $3x$, and 2 (Fig. 16-12). Now arrange the individual rectangles to form one large rectangle, as in Fig. 16-11. There will be only one way to do this. The sides $(x + 1)$ and $(x + 2)$ of this rectangle will be the factors of the trinomial. Use this method to factor these trinomials:

(a) $x^2 + 5x + 6$

(b) $2x^2 + 5x + 2$

(c) $x^2 + 2x - 3$

■ **FIGURE 16-11**   ■ **FIGURE 16-12**

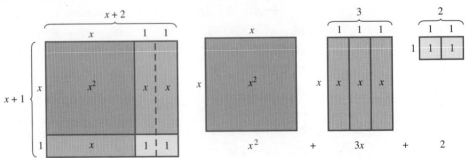

344   CHAPTER 16 / Factoring

Algebraic Fractions

Because the calculator and the metric system both use decimal notation, the importance of common fractions has been somewhat reduced. However, common fractions are still widely used in everyday situations and algebraic fractions are just as important as ever. We must be able to handle algebraic fractions in order to solve fractional equations and the applications from which they arise.

In this chapter we study equivalent fractions. We learn to add, subtract, multiply, and divide fractions. We simplify complex fractions and solve fractional equations, formulas with fractions, and applications involving fractions.

17-1   EQUIVALENT FRACTIONS

In this section we define algebraic fractions, manipulate fractions with signs, and express fractions in equivalent forms by using the fundamental principle of fractions.

Algebraic Fractions

An *algebraic fraction* is a fraction whose numerator and/or denominator contain literal quantities (variables).

Example 1

Some examples of algebraic fractions are

$$\frac{1}{x}, \quad \frac{x+y}{2}, \quad \frac{y}{x-1}, \quad \frac{5x^2b}{7y^5}, \quad \text{and} \quad \frac{x^2-3x+4}{x^2-9}$$

Manipulation with Signs

A fraction may have as many as three signs associated with it: the sign of the numerator, the sign of the denominator, and the sign of the fraction itself. Any two of the three signs of a fraction may be changed without changing the value of the fraction.

Example 2

The fraction $\frac{a}{b}$ may be written as $+\frac{+a}{+b}$ with three positive signs. It can also be expressed as $+\frac{-a}{-b}$ or $-\frac{+a}{-b}$ or $-\frac{-a}{+b}$ by changing two of the three positive signs to negative signs.

Fractions preceded by a negative sign are treated similarly.

Example 3

The fraction $-\dfrac{a}{b}$ may be written as $-\dfrac{+a}{+b}$ with one negative sign. It can also be expressed as

$$-\frac{-a}{-b} \text{ or } +\frac{+a}{-b} \text{ or } +\frac{-a}{+b}$$

by changing two of the three signs.

Generally, if a number is preceded by a + sign, then the + sign is omitted.

Example 4

The fraction $+\dfrac{-a}{+b}$ is generally written as $\dfrac{-a}{b}$.

Examples 2, 3, and 4 can be summarized as the rules of signs.

RULES OF SIGNS	$\dfrac{a}{b} = \dfrac{-a}{-b} = -\dfrac{a}{-b} = -\dfrac{-a}{b}$	11
	$-\dfrac{a}{b} = -\dfrac{-a}{-b} = \dfrac{a}{-b} = \dfrac{-a}{b}$	12

It is useful to eliminate as many negative signs as possible when you simplify fractions.

Example 5

Use the rules of signs to simplify each of the following fractions:

(a) $\quad -\dfrac{-x}{y}$ (b) $\quad -\dfrac{-p}{-q}$

Solution:

(a) $\quad -\dfrac{-x}{y} = \dfrac{x}{y}$

(b) $\quad -\dfrac{-p}{-q} = -\dfrac{p}{q}$

A negative sign in front of the quantity can be interpreted as a factor of -1. So,

$$-(a - b) = -1(a - b) = -a + b = b - a$$

$-(a - b) = b - a$	8

This is often used to simplify fractions or to express a fraction differently.

Example 6

Show that $\dfrac{x-3}{y-x} = \dfrac{3-x}{x-y}$.

Solution: Using the rules of signs,

$$\frac{x-3}{y-x} = \frac{-(x-3)}{-(y-x)}$$

Using Eq. 8,

$$= \frac{3-x}{x-y}$$

Example 7

Simplify the fraction $-\dfrac{p-1}{q-p}$.

Solution: Using the rules of signs, Eq. 12,

$$-\frac{p-1}{q-p} = \frac{-(p-1)}{q-p}$$

Using Eq. 8,

$$= \frac{1-p}{q-p}$$

Alternate Solution:

$$-\frac{p-1}{q-p} = \frac{p-1}{-(q-p)} = \frac{p-1}{p-q}$$

So, we see that the given fraction can be simplified by associating the negative sign with either the numerator or the denominator. The resulting simplified fractions look different, but both are equal to the original equation.

Equivalent Fractions

Recall that two fractions are *equivalent* if they have the same value. When we calculate with a fraction, we frequently have to use a different yet equivalent form of that fraction. We use the *fundamental principle of fractions* to find equivalent fractions.

FUNDAMENTAL PRINCIPLE OF FRACTIONS	If the numerator and the denominator of a fraction are multiplied or divided by the same (nonzero) number, then the value of the fraction remains unchanged.

In symbols, if $\dfrac{a}{b}$ is a fraction, and c is any nonzero number, then the fundamental principle of fractions gives us:

EQUIVALENT FRACTIONS	$\dfrac{a}{b} = \dfrac{ac}{bc}$	39
	$\dfrac{a}{b} = \dfrac{a \div c}{b \div c}$	40

Example 8

Multiply the numerator and denominator of the fraction $\dfrac{3a}{5x}$ by $2x$ to obtain an equivalent fraction.

Solution:
$$\frac{3a}{5x} = \frac{(3a)(2x)}{(5x)(2x)} = \frac{6ax}{10x^2}$$

Fractions Involving Polynomials

The fundamental principle of fractions also applies to fractions involving polynomials.

Example 9

Multiply the numerator and denominator of the fraction $\dfrac{1}{x-2}$ by $(x + 2)$ to obtain an equivalent fraction.

Solution:
$$\frac{1}{x-2} = \frac{1(x+2)}{(x-2)(x+2)} = \frac{x+2}{x^2-4}$$

COMMON ERROR We may *not add or subtract* the same quantity in the numerator and denominator of a fraction to obtain an equivalent fraction, because this will change the value of the fraction. For example,

$$\frac{x}{y} \neq \frac{x+1}{y+1}$$

Similarly,

COMMON ERROR We may *not take the square root* of both the numerator and denominator of a fraction to obtain an equivalent fraction, because this will change the value of the fraction. For example,

$$\frac{9a^2}{4b^2} \neq \frac{3a}{2b}$$

Reducing Fractions to Lowest Terms

You will recall from Sec. 2-2 that a common fraction is reduced to lowest terms, or simplified, when it has the smallest possible denominator. An algebraic fraction is simplified when the numerator and the denominator have no common factor other than 1. We can cancel common factors in algebraic fractions, just like we did with common fractions.

REDUCING FRACTIONS	$\dfrac{ac}{bc} = \dfrac{a\cancel{c}^{\,1}}{b\cancel{c}_{\,1}} = \dfrac{a \cdot 1}{b \cdot 1} = \dfrac{a}{b}$	39

Example 10

Reduce $\dfrac{9x^2y}{6x^3y}$ to lowest terms.

Solution: We cancel common factors, one by one, until there are no common factors left.

$$\frac{9x^2y}{6x^3y} = \frac{\overset{3}{\cancel{9}} \cdot \overset{1}{\cancel{x^2}} \cdot \overset{1}{\cancel{y}}}{\underset{2}{\cancel{6}} \cdot \underset{x}{\cancel{x^3}} \cdot \underset{1}{\cancel{y}}} = \frac{3}{2x}$$

Alternate Solution: Sometimes it is easier to factor the greatest common factor out of the numerator and the denominator and cancel. The gcf is $3x^2y$. We factor and cancel.

$$\frac{9x^2y}{6x^3y} = \frac{\overset{1}{\cancel{(3x^2y)}}(3)}{\underset{1}{\cancel{(3x^2y)}}(2x)} = \frac{3}{2x}$$

Example 11

Reduce the fraction $\dfrac{2(x-3)}{(x-3)(x+3)}$ to lowest terms.

Solution: A common factor is $(x - 3)$. We cancel $(x - 3)$ in the numerator and in the denominator:

$$\frac{2(x-3)}{(x-3)(x+3)} = \frac{2}{x+3} \cdot \frac{\overset{1}{\cancel{x-3}}}{\underset{1}{\cancel{x-3}}} = \frac{2}{x+3}$$

To reduce a fraction in which the numerator and/or denominator are not factored, we must first factor them and then cancel any factors common to both.

Example 12

Reduce to lowest terms:

$$\frac{3x-6}{4x-8}$$

Solution:

$$\frac{3x-6}{4x-8} = \frac{3\overset{1}{\cancel{(x-2)}}}{4\underset{1}{\cancel{(x-2)}}} = \frac{3}{4}$$

Reducing more complicated fractions involves more factoring.

Example 13

Reduce $\dfrac{6a^2-a-2}{3+5a-2a^2}$ to lowest terms.

Solution:
$$\frac{6a^2-a-2}{3+5a-2a^2} = \frac{(3a-2)\overset{1}{\cancel{(2a+1)}}}{\underset{1}{\cancel{(1+2a)}}(3-a)} = \frac{3a-2}{3-a}$$

Note that

$$2a + 1 = 1 + 2a$$

and that the terms of identical factors need not be written in the same order. But take care to cancel only common *factors*.

> **COMMON ERROR** Remember to cancel only common *factors*. You cannot cancel common *terms*. For example,
>
> $$\frac{\overset{1 \text{ no!}}{\cancel{3}x} - 1}{\underset{2}{\cancel{6}x} - 5} \neq \frac{x-1}{2x-5} \quad \text{and} \quad \frac{\overset{1 \text{ no!}}{\cancel{x^2}} - 4}{\underset{1}{\cancel{x^2}} - 9} \neq \frac{4}{9}$$

Simplifying Fractions by Changing Signs

Sometimes fractions can be reduced by manipulating the algebraic signs.

Example 14

Reduce to lowest terms: $\dfrac{y-x}{x-y}$

Solution: Since, by Eq. 8,

$$y - x = -(x - y)$$

we have

$$\frac{y-x}{x-y} = \frac{-\overset{1}{\cancel{(x-y)}}}{\underset{1}{\cancel{(x-y)}}} = \frac{-1}{1} = -1$$

Example 15

Reduce $\dfrac{9-x^2}{x^2-5x+6}$ to lowest terms.

Solution: First we factor,

$$\frac{9-x^2}{x^2-5x+6} = \frac{(3+x)(3-x)}{(x-3)(x-2)}$$

Replacing $(x-3)$ with $-(3-x)$,

$$= \frac{(3+x)\overset{1}{\cancel{(3-x)}}}{-\underset{1}{\cancel{(3-x)}}(x-2)}$$

Canceling the common factors,

$$= \frac{3+x}{-(x-2)}$$

Simplifying,

$$\frac{9-x^2}{x^2-5x+6} = \frac{3+x}{2-x}$$

SECTION 17-1 Exercises Equivalent Fractions

Algebraic Fractions and Manipulation with Signs

Use the rules of signs to simplify the fractions.

1. $\dfrac{-r}{-s}$ **2.** $-\dfrac{-m}{n}$ **3.** $-\dfrac{x}{-2}$ **4.** $-\dfrac{-3}{-y}$

Write an equivalent fraction.

5. $\dfrac{a-b}{b-1}$ **6.** $\dfrac{p-q}{p-3}$ **7.** $-\dfrac{r-s}{r-1}$ **8.** $-\dfrac{x-2}{x-y}$

Equivalent Fractions

Multiply the numerator and denominator of the given fractions by the given factor to obtain an equivalent fraction.

9. $\dfrac{3x}{5y}$, by $4x$

10. $\dfrac{2}{p-3}$, by $(p+3)$

11. $\dfrac{a+b}{a-b}$, by $(a-b)$

12. $\dfrac{2x-1}{3x-2}$, by $-3x$

Reducing Fractions to Lowest Terms

Reduce to lowest terms, without using negative exponents.

13. $\dfrac{12x}{8y}$

14. $\dfrac{6a^2}{4a^6}$

15. $\dfrac{18x^4y}{15x^2y^3}$

16. $\dfrac{15ab^4}{21a^5b}$

17. $\dfrac{81p^6qr^4}{45pq^5r^4}$

18. $\dfrac{49xy^3z^5}{56x^6y^3z}$

19. $\dfrac{4(x-2)}{(x-2)(x+2)}$

20. $\dfrac{(a+1)(a-3)}{(a+4)(a-3)}$

21. $\dfrac{8(x-1)(x-2)(x-3)}{6(x-1)(x+2)(x-3)}$

22. $\dfrac{12(p+2q)(3p+q)(2p-q)}{15(2p-q)(p+2q)(3p-q)}$

23. $\dfrac{5x-15}{3x-9}$

24. $\dfrac{4+2y}{5y+10}$

25. $\dfrac{x^2-16}{x^2-7x+12}$

26. $\dfrac{x^2-25}{x^2-10x+25}$

27. $\dfrac{x^2+14x+49}{x^2-49}$

28. $\dfrac{4x^2-9}{2x^2-9x+9}$

29. $\dfrac{x^2-5x+6}{x^2-x-6}$

30. $\dfrac{6x^2-11x+4}{9x^2-9x-4}$

31. $\dfrac{1+4x+4x^2}{6x^2-x-2}$

32. $\dfrac{9x^2-4}{9x^2-12x+4}$

33. $\dfrac{x^3+y^3}{x^2-y^2}$

34. $\dfrac{a^3-1}{a^2-1}$

35. $\dfrac{(x-y)^3}{x^3-y^3}$

36. $\dfrac{(a+b)^3}{a^3+b^3}$

Simplifying Fractions by Changing Signs

Reduce to lowest terms.

37. $\dfrac{a-b}{b-a}$

38. $\dfrac{2m-2n}{3n-3m}$

39. $-\dfrac{q-3p}{3p-q}$

40. $-\dfrac{2x-3y}{3y-2x}$

41. $\dfrac{(x-y)(a-b)}{y-x}$

42. $\dfrac{a-b}{(b-a)(b-2)}$

43. $\dfrac{(2x-1)(x-3)}{(3-x)(2+x)}$

44. $\dfrac{(1-x)(2-x)(3-x)}{(x-3)(x+2)(x-1)}$

45. $\dfrac{6-5x+x^2}{x^2-x-6}$

46. $\dfrac{4a-8a^2}{2a-1}$

47. $\dfrac{3x^2-9x}{9-x^2}$

48. $\dfrac{y^2-4y+3}{1-y^2}$

49. $\dfrac{2xy-y^2}{y^2-4x^2}$

50. $\dfrac{a^2-6a+8}{12-3a}$

51. $\dfrac{x^3-8}{4-x^2}$

52. $\dfrac{p^3-q^3}{(q-p)^3}$

53. $\dfrac{6x^2y-3x^3}{x^3-8y^3}$

54. $\dfrac{6y^2-2y}{1-27y^3}$

The following expressions are used in the given application. Simplify each expression.

55. Velocity of a bullet:

$$\frac{mv^2 + Mv^2}{mv^2}$$

56. Electricity:

$$\frac{E^2R^2 - E^2r^2}{(R^2 + 2Rr + r^2)^2}$$

57. Machine design:

$$\frac{R^3 - r^3}{R^2 - r^2}$$

58. Mechanical advantage of a machine:

$$\frac{4F^2 - 4f^2}{2F - 2f}$$

17-2 MULTIPLYING AND DIVIDING ALGEBRAIC FRACTIONS

In this section, we multiply and divide with algebraic fractions.

Multiplying Algebraic Fractions

Recall from Sec. 2-3 how we multiply two common fractions:

Example 16

$$\left(\frac{2}{3}\right)\left(\frac{4}{5}\right) = \frac{2 \cdot 4}{3 \cdot 5} = \frac{8}{15}$$

So, *the product of two or more fractions is a fraction whose numerator is the product of the original numerators and whose denominator is the product of the original denominators.* This rule applies to algebraic fractions as well. In symbols,

MULTIPLICATION OF FRACTIONS	$\dfrac{a}{b} \cdot \dfrac{c}{d} = \dfrac{ac}{bd}$	41

Example 17

$$\left(\frac{2a}{b}\right)\left(\frac{x}{3y}\right) = \frac{(2a)(x)}{(b)(3y)} = \frac{2ax}{3by}$$

If there are common factors in the numerator and denominator, it is recommended that *we cancel the common factors before we multiply.* If this is done first, the resulting fraction will be in lowest terms.

$$\frac{3x}{4y} \cdot \frac{6y}{7z} = \frac{3x}{\overset{}{\underset{2}{4y}}} \cdot \frac{\overset{3}{6y}}{7z} = \frac{9x}{14z}$$

We use the same process when multiplying fractions in which the numerator or denominator is a polynomial.

┌─── **Example 19** ───────

$$\frac{x+y}{x-y} \cdot \frac{x^2-y^2}{(x+y)^2} = \frac{(x+y)(x^2-y^2)}{(x-y)(x+y)^2}$$

Factoring and canceling,

$$= \frac{\overset{1}{\cancel{(x+y)}}\,\overset{1}{\cancel{(x+y)}}\,\overset{1}{\cancel{(x-y)}}}{\underset{1}{\cancel{(x-y)}}\,\underset{1}{\cancel{(x+y)}}\,\underset{1}{\cancel{(x+y)}}} = 1$$

Dividing Algebraic Fractions

We recall from Sec. 2-4 how we divided fractions:

┌─── **Example 20** ───────

$$\left(\frac{2}{5}\right) \div \left(\frac{3}{4}\right) = \frac{2}{5} \cdot \frac{4}{3} = \frac{8}{15}$$

So, to divide by a fraction, we multiply by its reciprocal. The reciprocal of a fraction is found by inverting the fraction. Thus, the reciprocal of $\frac{c}{d}$ is $\frac{d}{c}$.

DIVISION OF FRACTIONS	$\dfrac{a}{b} \div \dfrac{c}{d} = \dfrac{a}{b} \cdot \dfrac{d}{c} = \dfrac{ad}{bc}$	42

Thus, *to divide by a fraction (the divisor), we invert the divisor and multiply!*

┌─── **Example 21** ───────
Divide

$$\frac{3x}{4y} \div \frac{2a}{3} = \frac{3x}{4y} \cdot \frac{3}{2a} = \frac{9x}{8ay}$$

After we invert the divisor, if there are common factors, we cancel these common factors before we multiply.

Example 22

$$\frac{2a^2}{3b} \div \frac{5x}{6y^2} = \frac{2a^2}{3b} \cdot \frac{\overset{2}{\cancel{6}}y^2}{5x} = \frac{4a^2y^2}{5bx}$$

Example 23

$$\frac{x^2 - x - 6}{3x - 3} \div \frac{x^2 - 4}{x - 1} = \frac{x^2 - x - 6}{3x - 3} \cdot \frac{x - 1}{x^2 - 4}$$

Factoring and canceling,

$$= \frac{\overset{1}{\cancel{(x+2)}}(x-3)}{3\cancel{(x-1)}} \cdot \frac{\overset{1}{\cancel{(x-1)}}}{\cancel{(x+2)}(x-2)}$$

$$= \frac{x-3}{3(x-2)}$$

Generally, we leave the result in its simplified factored form.

When dividing a fraction by a polynomial, just multiply by the reciprocal of the polynomial.

Example 24

$$\frac{a^2 + 6a + 8}{a^2} \div (3a^2 + 6a) = \frac{a^2 + 6a + 8}{a^2} \cdot \frac{1}{3a^2 + 6a}$$

Factoring and canceling,

$$= \frac{\overset{1}{\cancel{(a+2)}}(a+4)}{a^2} \cdot \frac{1}{3a\cancel{(a+2)}}$$

Simplifying,

$$= \frac{a+4}{3a^3}$$

Combined Multiplication and Division

In problems involving three or more multiplications and divisions, invert only the fraction *immediately following* the division sign.

Example 25

$$\frac{6a}{7b} \div \frac{12a^2}{35c^2} \cdot \frac{5b^2}{c^4} = \frac{\overset{1 \cdot 1}{\cancel{6}\,\cancel{a}}}{\underset{1 \cdot 1}{\cancel{7}\,\cancel{b}}} \cdot \frac{\overset{5 \cdot 1}{\cancel{35}\,\cancel{c^2}}}{\underset{2 \cdot a}{\cancel{12}\,\cancel{a^2}}} \cdot \frac{\overset{b}{5\cancel{b^2}}}{\underset{c^2}{\cancel{c^4}}} = \frac{25b}{2ac^2}$$

Multiplying Algebraic Fractions

Multiply and simplify.

1. $\dfrac{x}{y} \cdot \dfrac{2a}{3b}$

2. $\dfrac{a}{5} \cdot \dfrac{2b}{3c}$

3. $\dfrac{5x}{2y} \cdot \dfrac{8y}{x^2}$

4. $\dfrac{3p^2}{2q} \cdot \dfrac{q^5}{12p^4}$

5. $\dfrac{24a^2b^3}{5c^2} \cdot \dfrac{15c}{36ab^3}$

6. $\dfrac{5x^2}{6a^2y^3} \cdot \dfrac{27a^3y}{35x}$

7. $\dfrac{a}{a+b} \cdot \dfrac{b}{a-b}$

8. $\dfrac{p-q}{p} \cdot \dfrac{p+q}{q}$

9. $\dfrac{x-y}{a^2} \cdot \dfrac{2a}{x-y}$

10. $\dfrac{a+b}{6} \cdot \dfrac{2a}{3(a+b)}$

11. $\dfrac{3a+2b}{6a} \cdot \dfrac{3a}{2a+3b}$

12. $\dfrac{2x}{x-2y} \cdot \dfrac{2x-y}{4x}$

13. $\dfrac{x^2+x-2}{x+1} \cdot \dfrac{x+1}{x^2-3x+2}$

14. $\dfrac{x-2}{x^2+2x-3} \cdot \dfrac{x^2+4x+3}{x-2}$

Dividing Algebraic Fractions

Divide and simplify.

15. $\dfrac{5a}{6b} \div \dfrac{3x}{7y}$

16. $\dfrac{x^2}{y^3} \div \dfrac{2a}{b^2}$

17. $\dfrac{3x^3}{2y^2} \div \dfrac{x}{y}$

18. $\dfrac{a^2}{3b} \div \dfrac{a}{6b^2}$

19. $6 \div \dfrac{3}{x^2}$

20. $9a^2 \div \dfrac{a}{b^2}$

21. $\dfrac{x+2}{3} \div \dfrac{2}{x-1}$

22. $\dfrac{3}{x-3} \div \dfrac{2}{x+2}$

23. $\dfrac{a+5}{3} \div (6-a)$

24. $\dfrac{7-x}{y} \div (x+y)$

25. $\dfrac{2x-6}{x+1} \div \dfrac{3x-9}{2x+2}$

26. $\dfrac{6x-4}{3} \div \dfrac{3x-2}{2}$

27. $\dfrac{6(x+y)^2}{x-y} \div 2(x+y)$

28. $\dfrac{8(a-b)(a+b)}{a^2} \div 4(a-b)$

29. $2(a+c) \div \dfrac{ab+bc}{c}$

30. $(x-y) \div \dfrac{xz-yz}{x+y}$

31. $\dfrac{3a^2-6ab}{2ab+4b^2} \div \dfrac{9a^2-18ab}{4ab+8b^2}$

32. $\dfrac{4x^2+2xy}{6xy-3y^2} \div \dfrac{10x^2+5xy}{12xy-6y^2}$

33. $\dfrac{2x^2+3x-2}{3xy} \div \dfrac{x^3+8}{6xy}$

34. $\dfrac{3x^2-x-2}{6x^3} \div \dfrac{2x^2-5x+3}{4x^2}$

35. $\dfrac{3x^2+x-2}{2x^2-7x+6} \div \dfrac{3x^2-8x+4}{2x^2-x-3}$

36. $\dfrac{12x^2+5x-3}{6x^2+7x+2} \div \dfrac{4x^2-x-3}{2x^2-x-1}$

37. $\dfrac{15x^2-x-2}{2x-6} \div (10x-4)$

38. $\dfrac{8x^2+10x-3}{3x+3} \div (8x+12)$

Perform the indicated operations and simplify.

39. $\dfrac{5x}{6y} \div \dfrac{2a^2}{3x} \cdot \dfrac{4y^3}{a^2}$

40. $\dfrac{2a}{3b^3} \div \dfrac{5x^2}{7b} \cdot \dfrac{5a}{6x}$

41. $\dfrac{25a^2b}{8xy} \div \dfrac{10ab^2}{3cx^2y} \cdot \dfrac{4by}{9ac}$

42. $\dfrac{14x^2y}{15ab^2} \div \dfrac{9cx}{10ab} \cdot \dfrac{21ac}{8y^2}$

43. $\dfrac{2x^2 + x - 1}{2x^2 - x - 1} \cdot \dfrac{4x^2 - 1}{3x^2 + 2x - 1} \div \dfrac{4x^2 - 4x + 1}{9x^2 - 6x + 1}$

44. $\dfrac{2a^2 - 2ab}{a^2 + 2ab + b^2} \cdot \dfrac{2a^2 + 2ab}{3ab - 3b^2} \div \dfrac{4ab - 4b^2}{3a + 3b}$

45. $\dfrac{(x - 1)^3}{3x + 3} \div \dfrac{x^2 - 1}{6x - 12} \cdot \dfrac{x^2 + 2x + 1}{x^3 - 1}$

46. $\dfrac{x^2 - 4x - 5}{3x^4 - 27x^2} \div \dfrac{3x^2 + 4x + 1}{x^3 + x^2 - 6x} \cdot \dfrac{9x^3 - 18x^2 - 27x}{x^2 - 7x + 10}$

47. $\dfrac{x^2 - 2x - 3}{2x^4 - 72x^2} \div \dfrac{x^2 - 4x + 3}{x^3 - 7x^2 + 6x} \div \dfrac{2x^2 + 5x + 2}{4x^3 + 32x^2 + 48x}$

48. $\dfrac{x^3 - 5x^2 + 4x}{6x^2 + 30x + 36} \div \left(\dfrac{x^4 - 16x^2}{3x^2 - 6x - 45} \cdot \dfrac{x^2 - 6x + 5}{x^3 + 8x^2 + 16x} \right)$

Applications

49. The power P dissipated in a resistor is given by

$$P = i^2 R$$

where i is the current and R is resistance. Find the formula for P in terms of the time t, if

$$i = \frac{3t + 1}{2t^2} \quad \text{and} \quad R = \frac{2t}{3t + 1}$$

50. The resistance in a simple electric circuit is the voltage divided by the current. Find the formula for the resistance R of a circuit in which the voltage V and current i are functions of time, as follows:

$$V = \frac{4t + 12}{2t + 3} \quad \text{and} \quad i = \frac{6t}{2t^2 + 9t + 9}$$

51. The density of an object is its mass divided by its volume. Write and simplify an expression for the density of a cone (Fig. 17-1) having a mass m and a volume equal to $\dfrac{\pi r^2 h}{3}$.

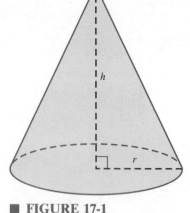

■ **FIGURE 17-1**

17-3   ADDING AND SUBTRACTING ALGEBRAIC FRACTIONS

In Sec. 2-5, we learned how to add and subtract common fractions. In this section we learn how to add and subtract algebraic fractions. We find the least common denominator, add and subtract mixed expressions, and solve applications.

Adding and Subtracting Like Fractions

Recall that like, or similar, fractions are fractions that have the same (common) denominator. *The sum of like fractions is the sum of the numerators divided by the denominator.* When we add or subtract fractions (or other quantities), we may say that we *combine* them. As usual, we simplify the result if possible.

ADDITION AND SUBTRACTION OF FRACTIONS WITH THE SAME DENOMINATOR	$\dfrac{a}{c} + \dfrac{b}{c} = \dfrac{a+b}{c}$ $\dfrac{a}{c} - \dfrac{b}{c} = \dfrac{a-b}{c}$	43

Example 26

$$\frac{2}{x} + \frac{3}{x} = \frac{5}{x}$$

Sometimes the numerators can be combined into a single sum or difference, as above. If the numerators cannot be combined, then the sum or difference must be indicated using the + and − symbols.

Example 27

$$\frac{2x}{a^2b} - \frac{3}{a^2b} = \frac{2x-3}{a^2b}$$

Special care must be taken when subtracting fractions involving polynomials.

Example 28

Combine and simplify:

$$\frac{2y}{y-1} - \frac{2+y}{y-1}$$

Solution: We use parentheses to show that we are subtracting the entire second numerator, $2 + y$:

$$\frac{2y}{y-1} - \frac{2+y}{y-1} = \frac{2y - (2+y)}{y-1}$$

Subtracting,

$$= \frac{2y - 2 - y}{y-1}$$

Combining,

$$= \frac{y-2}{y-1}$$

COMMON ERROR When a fraction is subtracted, be sure to subtract the entire numerator.

We use Eq. 8

$$-(a - b) = b - a$$

to combine certain fractions, as shown in the following example.

Example 29

Combine and simplify:

$$\frac{1}{x-2} + \frac{x}{2-x}$$

Solution: First we rewrite the fractions with a common denominator: Using Eq. 8,

$$\frac{1}{x-2} + \frac{x}{2-x} = \frac{1}{x-2} + \frac{x}{-(x-2)}$$

Using the rules of signs,

$$= \frac{1}{x-2} - \frac{x}{x-2}$$

Combining,

$$= \frac{1-x}{x-2}$$

COMMON ERROR You may combine numerators only when adding or subtracting fractions with like denominators. NEVER add or subtract denominators:

$$\frac{a}{b} + \frac{a}{c} \neq \frac{a}{b+c} \quad \text{and} \quad \frac{a}{b} + \frac{c}{d} \neq \frac{a+c}{b+d}$$

Example 30

$$\frac{1}{x} + \frac{1}{y} \neq \frac{1}{x+y} \quad \text{and} \quad \frac{1}{x} + \frac{y}{2} \neq \frac{1+y}{x+2}$$

Least Common Denominator

To add or subtract fractions with different denominators, we must rewrite the fractions as equivalent fractions that do have the same common denominator. Any common denominator will work. However, the *least common denominator (LCD)* is the most convenient and useful. The LCD is the simplest expression that is exactly divisible by each of the denominators. The LCD is also the least common multiple of the denominators and must contain all the prime factors of each of the denominators.

Steps to Find the LCD

1. Factor each denominator into its prime factors.
2. Take the highest power of each prime factor occurring in any denominator.
3. The LCD is the product of these highest powers.

Example 31

Find the LCD of $\dfrac{5}{6x^2y}$ and $\dfrac{4}{9xy^3}$.

Solution: Factoring the denominators into primes, we have

$$6x^2y = 2 \cdot 3x^2y$$

---

and

$$9xy^3 = 3^2xy^3$$

Note that with monomials, only the numerical coefficients must be factored, since the variable part is already factored. The prime factors of these two denominators are 2, 3, x, and y.
The highest power of 2 that occurs is 2^1 (or 2).
The highest power of 3 that occurs is 3^2.
The highest power of x is x^2, and the highest power of y is y^3.
The LCD is the product of these powers:

$$LCD = 2 \cdot 3^2 \cdot x^2y^3 = 18x^2y^3$$

In many problems, you may be able to find the LCD by inspection. The process given above is one that we can use when the LCD is not obvious.

Example 32 ─────────

Find the LCD of the fractions

$$\frac{2}{4x^2}, \quad \frac{3x}{2x^2 - 8}, \quad \text{and} \quad \frac{5}{x^2 - 2x}$$

Solution: Factoring the denominators into primes, we have

$$4x^2 = 2^2x^2$$

$$2x^2 - 8 = 2(x^2 - 4) = 2(x + 2)(x - 2)$$

and

$$x^2 - 2x = x(x - 2)$$

The LCD is the product of the highest powers of the factors 2, x, $x + 2$, and $x - 2$, so

$$LCD = 2^2x^2(x + 2)(x - 2)$$
$$= 4x^2(x + 2)(x - 2)$$

We could multiply this to find the product, but the LCD is more useful for addition and subtraction in its factored form.

TIP It is preferable to leave the least common denominator in its factored form for ease in addition and subtraction.

Adding and Subtracting Algebraic Fractions with Different Denominators

To add or subtract algebraic fractions with different denominators, we must rewrite the fractions as equivalent fractions that do have the same common denominator. Any common denominator will work. If the denominators have no factors in common, then the common denominator is the product of the denominators.

─── **Example 33** ───

Combine $\dfrac{a}{b} + \dfrac{c}{d}$.

Solution: We multiply the two denominators to get the common denominator, bd. Expressing each fraction over bd:

$$\frac{a}{b} + \frac{c}{d} = \frac{a(d)}{b(d)} + \frac{c(b)}{d(b)}$$

$$= \frac{ad}{bd} + \frac{bc}{bd}$$

$$= \frac{ad + bc}{bd}$$

This brings us to the rules for adding and subtracting fractions with different denominators.

ADDITION AND SUBTRACTION OF FRACTIONS WITH DIFFERENT DENOMINATORS	$\dfrac{a}{b} + \dfrac{c}{d} = \dfrac{ad + bc}{bd}$ $\dfrac{a}{b} - \dfrac{c}{d} = \dfrac{ad - bc}{bd}$	44

─── **Example 34** ───

Combine and simplify:

$$\frac{x}{2y} - \frac{3}{x}$$

Solution: By inspection, the LCD is $2xy$. We use the fundamental principal of fractions to find equivalent fractions.

Expressing each fraction over $2xy$:

$$\frac{x}{2y} - \frac{3}{x} = \frac{x(x)}{2y(x)} - \frac{3(2y)}{x(2y)}$$

$$= \frac{x^2}{2xy} - \frac{6y}{2xy}$$

Combining,

$$= \frac{x^2 - 6y}{2xy}$$

Since the numerator is not factorable, this is already simplified.

Steps to Add or Subtract Fractions with Different Denominators

1. Find the LCD.
2. Change each fraction to an equivalent one whose denominator is the LCD by multiplying the numerator and denominator of the fraction by that expression which will make its denominator equal to the LCD.

3. Add the equivalent fractions by adding the numerators and writing that sum over the LCD.

4. Simplify the result if possible.

─── **Example 35** ───

Combine and simplify:

$$\frac{5}{6x^2y} - \frac{4}{9xy^3}$$

Solution: We found the LCD of these fractions in Example 31. It is $18x^2y^3$. We find equivalent fractions:

$$\frac{5}{6x^2y} - \frac{4}{9xy^3} = \frac{5}{6x^2y} \cdot \frac{3y^2}{3y^2} - \frac{4}{9xy^3} \cdot \frac{2x}{2x}$$

$$= \frac{15y^2 - 8x}{18x^2y^3}$$

─── **Example 36** ───

Combine and simplify:

$$\frac{a}{a-b} + \frac{b}{a^2 - b^2}$$

Solution: We factor the second denominator:

$$a^2 - b^2 = (a+b)(a-b)$$

Thus the LCD is $(a+b)(a-b)$.

$$\frac{a}{a-b} + \frac{b}{a^2 - b^2} = \frac{a}{(a-b)} \cdot \frac{(a+b)}{(a+b)} + \frac{b}{(a+b)(a-b)}$$

$$= \frac{a(a+b) + b}{(a+b)(a-b)}$$

$$= \frac{a^2 + ab + b}{(a+b)(a-b)}$$

Since the numerator is not factorable, this result is in simplest form.

Mixed Expressions

A mixed expression is an expression in which some, but not all, terms are fractions. To add or subtract mixed expressions, we must still find the LCD and express each term with that denominator.

─── **Example 37** ───

Combine and simplify:

$$x - \frac{2}{x-1}$$

Solution: First we change the non-fraction part to fractional form:

$$x = \frac{x}{1}$$

and proceed as above. The LCD is $x - 1$. (The factor of 1 need not be written.)

$$x - \frac{2}{x-1} = \frac{x}{1} - \frac{2}{x-1}$$

$$= \frac{x}{1} \cdot \frac{(x-1)}{(x-1)} - \frac{2}{x-1}$$

$$= \frac{x(x-1) - 2}{x-1}$$

$$= \frac{x^2 - x - 2}{x-1}$$

$$= \frac{(x+1)(x-2)}{x-1}$$

Since no factors cancel, the expression is in simplest form.

SECTION 17-3 **Exercises** **Adding and Subtracting Algebraic Fractions**

Adding and Subtracting Like Fractions

Combine and simplify.

1. $\dfrac{5}{y} + \dfrac{2}{y}$

2. $\dfrac{7}{x} - \dfrac{3}{x}$

3. $\dfrac{6}{ab} - \dfrac{2}{ab} + \dfrac{3}{ab}$

4. $\dfrac{5}{x^2} + \dfrac{2}{x^2} - \dfrac{3}{x^2}$

5. $\dfrac{2x}{a^2b^3} - \dfrac{5y}{a^2b^3} + \dfrac{6x}{a^2b^3}$

6. $\dfrac{2a}{x^2y} + \dfrac{3}{x^2y} - \dfrac{4a}{x^2y}$

7. $\dfrac{5a}{x-y} - \dfrac{a}{x-y}$

8. $\dfrac{2x}{a+b} + \dfrac{3x}{a+b}$

9. $\dfrac{a}{a-2} - \dfrac{a-3}{a-2}$

10. $\dfrac{q}{p+1} - \dfrac{q+1}{p+1}$

11. $\dfrac{c^2}{c-1} - \dfrac{1}{c-1}$

12. $\dfrac{x^2}{x+3} + \dfrac{3x}{3+x}$

13. $\dfrac{x^2}{x+1} - \dfrac{3+2x}{1+x}$

14. $\dfrac{6}{y-2} - \dfrac{5y-y^2}{y-2}$

15. $\dfrac{1}{m-n} + \dfrac{2}{n-m}$

16. $\dfrac{3}{x-y} - \dfrac{5}{y-x}$

17. $\dfrac{a}{a-b} - \dfrac{b}{b-a} + \dfrac{a+b}{a-b}$

18. $\dfrac{3}{1-p} + \dfrac{p}{p-1} - \dfrac{1+p}{1-p}$

Adding and Subtracting Algebraic Fractions with Different Denominators

Combine and simplify.

19. $\dfrac{1}{x} + \dfrac{1}{y}$

20. $\dfrac{p}{10} - \dfrac{2p}{15}$

21. $\dfrac{7}{a} - \dfrac{5}{a^2}$

22. $\dfrac{x}{y} + \dfrac{3}{xy}$

23. $\dfrac{2a}{3x} + \dfrac{a}{2x}$

24. $\dfrac{y}{4x} - \dfrac{5y}{6x^2}$

25. $\dfrac{x}{4} - \dfrac{3y}{2} + \dfrac{2z}{3}$

26. $\dfrac{1}{r_1} + \dfrac{1}{r_2} + \dfrac{1}{r_3}$

27. $\dfrac{1}{x} - \dfrac{1}{y} + \dfrac{1}{z}$

28. $\dfrac{a-b}{4} - \dfrac{a+b}{6}$

29. $\dfrac{1}{2} - \dfrac{5}{6ab^2} + \dfrac{3}{8a^2b}$

30. $\dfrac{2}{3x^2} + \dfrac{1}{6y} - \dfrac{3}{xy^3}$

31. $\dfrac{2}{x+y} - \dfrac{3}{x-y}$

32. $\dfrac{a+1}{a-1} - \dfrac{a-1}{a+1}$

33. $\dfrac{p-1}{p+2} + \dfrac{p+2}{p-3}$

34. $\dfrac{1}{x-1} + \dfrac{2}{x-2} - \dfrac{3}{x-3}$ **35.** $\dfrac{2}{b+3} - \dfrac{3}{b-2} + \dfrac{1}{b+1}$ **36.** $\dfrac{a}{a+b} - \dfrac{2b^2}{a^2-b^2}$

37. $\dfrac{2}{x+1} - \dfrac{x-3}{x^2-1}$ **38.** $\dfrac{1}{2x} - \dfrac{x}{3x-6}$ **39.** $\dfrac{3}{x^2-x} + \dfrac{5}{2x-2}$

40. $\dfrac{4y}{4y^2-1} - \dfrac{3y+1}{1-2y} + \dfrac{2y}{1+2y}$

Mixed Expressions

Combine and simplify.

41. $a + \dfrac{1}{a}$ **42.** $\dfrac{3}{y} - y$ **43.** $c + \dfrac{3}{c-2}$

44. $\dfrac{d}{d+1} - d$ **45.** $\dfrac{2}{x} - 3 + \dfrac{1}{x^3}$ **46.** $\dfrac{1}{a+2} - a + 3$

47. $\dfrac{2x-3}{3x-2} - \dfrac{2}{x} - 1$ **48.** $\dfrac{2a-3}{a-2} - \dfrac{4}{a} - 3a - 4$

Applications

The following expressions are used with the given application. Combine and simplify.

49. Forces acting on a concrete slab:

$$1 - \frac{4c}{\pi l} + \frac{c^3}{3l^3}$$

50. The time t required to make a round trip by boat:

$$\frac{d}{V-v} - \frac{d}{V+v}$$

51. Resistance of a circuit (Fig. 17-2):

$$R_1 + \frac{R_2 R_3}{R_2 + R_3}$$

52. Focal length of a thin lens (Fig. 17-3):

$$\frac{1}{f} = \frac{1}{p} + \frac{1}{q} \qquad \text{Combine the right side.}$$

■ **FIGURE 17-2**

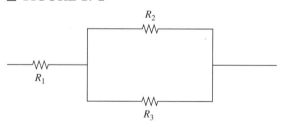

■ **FIGURE 17-3**

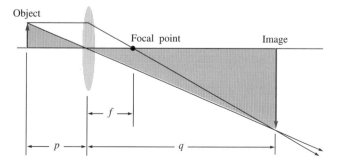

COMPLEX FRACTIONS

In this section we learn to simplify a complex fraction. Recall from Sec. 2-6 that a *complex fraction* is a fraction in which the numerator or denominator, or both, contains a fraction.

Example 38

Examples of complex fractions are

$$\frac{\dfrac{x}{y}}{z}, \quad \frac{\dfrac{p}{q}}{r}, \quad \frac{\dfrac{a}{b}}{\dfrac{c}{d}}, \quad \frac{1-\dfrac{1}{x}}{2}, \quad \frac{1-\dfrac{1}{x}}{1+\dfrac{1}{x}}, \quad \text{and} \quad \frac{x+\dfrac{1}{x-y}}{x-\dfrac{1}{x+y}}$$

The main fraction line is longer and in boldface type.

To simplify a complex fraction, we perform the indicated operations to obtain a simple fraction in which there is only one fraction line. In a fraction, the fraction line indicates division, so we express the fraction as a division problem, divide, and simplify the result, if possible.

Example 39

Simplify:

$$\frac{\dfrac{x}{3}}{y}$$

Solution: First we express the fraction as a division problem:

$$\frac{\dfrac{x}{3}}{y} = x \div \frac{3}{y}$$

Inverting and multiplying,

$$= \frac{x}{1} \cdot \frac{y}{3}$$

$$= \frac{xy}{3}$$

If the numerator and/or denominator are more complicated, just express them as simple fractions. Then divide the fraction in the numerator by the fraction in the denominator. Recall that a fraction line is also a symbol of grouping, so all quantities grouped by a fraction line must be combined before dividing.

Example 40

Simplify:

$$\frac{y-\dfrac{1}{5}}{5-\dfrac{1}{y}}$$

Solution: Express the numerator and denominator as simple fractions:

$$\frac{y-\dfrac{1}{5}}{5-\dfrac{1}{y}} = \frac{\dfrac{5y-1}{5}}{\dfrac{5y-1}{y}}$$

Divide:

$$= \frac{5y-1}{5} \div \frac{5y-1}{y}$$

Invert and multiply:

$$= \frac{\overset{1}{\cancel{5y-1}}}{5} \cdot \frac{y}{\cancel{5y-1}}_{1}$$

$$= \frac{y}{5}$$

Alternate Solution: Multiply the numerator and denominator of the given fraction by the LCD of the fractions that appear in the numerator or denominator, and simplify. The LCD of the denominators is $5y$. We multiply the numerator and denominator by $5y$:

$$\frac{y - \frac{1}{5}}{5 - \frac{1}{y}} = \frac{\left(y - \frac{1}{5}\right)5y}{\left(5 - \frac{1}{y}\right)5y}$$

Distributing the $5y$ and multiplying,

$$= \frac{y(5y) - \frac{1}{5}(5y)}{5(5y) - \frac{1}{y}(5y)} = \frac{5y^2 - y}{25y - 5}$$

Factoring and canceling,

$$= \frac{y\overset{1}{\cancel{(5y-1)}}}{5\cancel{(5y-1)}_{1}} = \frac{y}{5}$$

We see that the two methods yield the same result, $\frac{y}{5}$.

Example 41

Simplify:

$$\frac{\frac{1}{9} - \frac{a}{3}}{\frac{a}{6} - \frac{1}{18}}$$

Solution: The LCD is 18. We multiply the numerator and denominator by 18:

$$\frac{\frac{1}{9} - \frac{a}{3}}{\frac{a}{6} - \frac{1}{18}} = \frac{\left(\frac{1}{9} - \frac{a}{3}\right)18}{\left(\frac{a}{6} - \frac{1}{18}\right)18}$$

Distributing,

$$= \frac{\frac{1}{9}(18) - \frac{a}{3}(18)}{\frac{a}{6}(18) - \frac{1}{18}(18)}$$

$$= \frac{2 - 6a}{3a - 1}$$

$$= \frac{2(1 - 3a)}{3a - 1}$$

$$= \frac{-2\overset{1}{\cancel{(3a-1)}}}{\cancel{(3a-1)}_{1}}$$

$$= -2$$

Simplify.

1. $\dfrac{\dfrac{a}{b}}{2}$ **2.** $\dfrac{\dfrac{a}{b}}{2}$

3. Show that $\dfrac{x}{\dfrac{y}{z}} \neq \dfrac{\dfrac{x}{y}}{z}$

4. Show that $\dfrac{2}{\dfrac{x}{3}} \neq \dfrac{\dfrac{2}{x}}{3}$

Simplify.

5. $\dfrac{\dfrac{a}{b}}{\dfrac{c}{d}}$ **6.** $\dfrac{\dfrac{3}{x}}{\dfrac{x}{4}}$ **7.** $\dfrac{x-4}{2-\dfrac{8}{x}}$ **8.** $\dfrac{x-12}{1-\dfrac{12}{x}}$

9. $\dfrac{b-\dfrac{1}{4}}{4-\dfrac{1}{b}}$ **10.** $\dfrac{\dfrac{1}{2}-x}{2-\dfrac{1}{x}}$ **11.** $\dfrac{3+\dfrac{2}{y}}{9y-\dfrac{4}{y}}$ **12.** $\dfrac{\dfrac{9}{y}-16y}{1+\dfrac{3}{4y}}$

13. $\dfrac{\dfrac{3a}{2}-1}{6a-\dfrac{8}{3a}}$ **14.** $\dfrac{\dfrac{y}{3}+\dfrac{y}{4}}{\dfrac{5y}{12}+1}$ **15.** $\dfrac{1+\dfrac{a}{b}}{1-\dfrac{a^2}{b^2}}$ **16.** $\dfrac{\dfrac{1}{a}+\dfrac{1}{b}}{\dfrac{1}{a}-\dfrac{1}{b}}$

17. $\dfrac{1-\dfrac{7a+4}{2a^2}}{a-\dfrac{16}{a}}$ **18.** $\dfrac{1-\dfrac{7b+6}{3b^2}}{b-\dfrac{9}{b}}$ **19.** $\dfrac{\dfrac{x}{x-2y}-1}{1-\dfrac{x}{x+2y}}$ **20.** $\dfrac{2-\dfrac{b}{2a-b}}{\dfrac{b}{2a+b}-1}$

These complex fractions are used with the given application. Simplify.

Applications

21. Resistance of two resistors in parallel (Fig. 17-4):

$$\dfrac{1}{\dfrac{1}{R_1}+\dfrac{1}{R_2}}$$

■ FIGURE 17-4

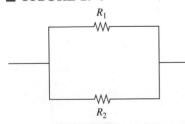

R_1

R_2

22. Focal length of a thin lens:

$$\dfrac{1}{\dfrac{p+q}{pq}}$$

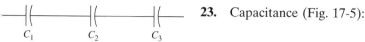

FIGURE 17-5

23. Capacitance (Fig. 17-5):

$$\frac{1}{\dfrac{1}{C_1} + \dfrac{1}{C_2} + \dfrac{1}{C_3}}$$

24. Present value of a stock:

$$\frac{\dfrac{1}{1+i}}{1 - \dfrac{1}{1+i}}$$

■ Chapter 17 REVIEW EXERCISES

Reduce to lowest terms.

1. $\dfrac{x^2 - 36}{x^2 - 12x + 36}$

2. $\dfrac{x^3 + 1}{(x+1)^3}$

3. $\dfrac{a^2 - 4b^2}{2b - a}$

4. $\dfrac{9a^2 - 24a + 16}{9a^2 - 16}$

5. $\dfrac{y^3 - x^3}{(x - y)^2}$

6. $\dfrac{a^3 + b^3}{a^2 - b^2}$

Perform the indicated operations and simplify.

7. $\dfrac{3x^3 - 75x}{2x^2 - 2x - 24} \cdot \dfrac{4x^2 - 28x + 48}{x^4 - 8x^3 + 15x^2}$

8. $\dfrac{a^3 - 2a^2b + ab^2}{4a + 4b} \div \dfrac{ab - a^2}{2a + 2b}$

9. $\dfrac{4 - n^2}{n^3 - 8} \div \dfrac{n^3 - 2n^2 - 8n}{4n^2 - 16n}$

10. $\dfrac{x^2 - 3x - 4}{3x^4 - 75x^2} \div \dfrac{2x^2 + 3x + 1}{x^3 + 3x^2 - 10x} \cdot \dfrac{6x^3 - 24x^2 - 30x}{x^2 - 6x + 8}$

Combine and simplify.

11. $\dfrac{1}{1 - a} - \dfrac{a^2}{1 - a}$

12. $\dfrac{x^2}{x + 2} - \dfrac{2 - x}{2 + x}$

13. $\dfrac{x}{x - y} + \dfrac{y}{y - x} - \dfrac{x + y}{x - y}$

14. $\dfrac{x}{12} - \dfrac{7x}{8}$

15. $\dfrac{x}{4} + \dfrac{5x}{6} - \dfrac{3x}{8}$

16. $\dfrac{x + y}{2} - \dfrac{x - y}{3}$

17. $\dfrac{x + 5}{x^2 - 16} + \dfrac{x - 3}{4 - x}$

18. $\dfrac{a - 4}{2a} + \dfrac{a + 2}{a + 1} - \dfrac{3a + 2}{a^2 + a}$

19. $\dfrac{a}{5} - 3$

20. $x - \dfrac{2}{x - 1} + 1$

21. $\dfrac{3p - 5}{p - 3} - \dfrac{5}{p} - 9$

22. $\dfrac{2x - 3}{x - 2} - \dfrac{5}{x} - 2$

Simplify.

23. $\dfrac{\dfrac{1}{3} - x}{3 - \dfrac{1}{x}}$ **24.** $\dfrac{\dfrac{1}{4} - \dfrac{x}{2}}{\dfrac{x}{6} - \dfrac{1}{12}}$ **25.** $\dfrac{1 - \dfrac{1}{x}}{x - \dfrac{1}{x^2}}$ **26.** $\dfrac{1 - \dfrac{3x + 2}{2x^2}}{x - \dfrac{4}{x}}$

The following expressions are used with the given application. Simplify.

27. Displacement of a vibrating string:

$$\frac{t}{T} - \frac{x}{\lambda}$$

28. Focal length of a lens:

$$\frac{1}{\dfrac{1}{p} + \dfrac{1}{q}}$$

WRITING

29. We have seen many applications of algebraic fractions in this chapter. How are algebraic fractions used in science or in business? Give three examples, explaining what each letter represents and the units associated with each letter.

TEAM PROJECT

30. A small computer company has five employees: Ms. Adams, Ms. Black, Mr. Clark, Mr. Downing, and Mr. Ewing. They hold the positions of technician, analyst, mathematician, programmer, and manager, although not necessarily in that order. Here is some additional information about the employees:

The analyst and the manager were same-sex roommates in college.
The technician is a bachelor.
Mr. Ewing and Ms. Adams have had only business contacts with each other.
Mrs. Clark was greatly disappointed when her husband told her that the manager had refused to give him a raise.
Mr. Downing is going to be best man when the mathematician and the analyst are married.

What position does each person hold in the company? Explain how you came to your conclusions.

Fractional Equations

Many equations in applied mathematics involve fractions. In this chapter we solve fractional equations and formulas involving fractions. We then solve some applications with fractional equations.

18-1   SOLUTION OF FRACTIONAL EQUATIONS

In this section we solve fractional equations in which the denominators are either constants or variables.

Solving Fractional Equations

A *fractional equation* is an equation in which one or more terms is a fraction.

Example 1

Some examples of fractional equations are

$$\frac{2x}{3} = 13 - \frac{x}{5} \quad \text{and} \quad \frac{3x}{x+2} - 2 = \frac{2x-3}{2x-1}$$

A good way to begin to solve a fractional equation is to *first multiply each side of the equation by the LCD of all the fractions.* The resulting equation will contain no fractions, and can be solved using the methods described in Chapter 10. As always, it is wise to *check* solutions by substituting them into the original equation.

Example 2

Solve and check:

$$\frac{2x}{3} = 13 - \frac{x}{5}$$

Solution: The LCD of these fractions is 15. Multiplying each side of the equation by the LCD, we have

$$\left(\frac{2x}{3}\right)15 = \left(13 - \frac{x}{5}\right)15$$

$$\left(\frac{2x}{3}\right)15 = (13)15 - \left(\frac{x}{5}\right)15$$

Simplifying,

$$10x = 195 - 3x$$
$$13x = 195$$
$$x = 15$$

Check:

$$\frac{2(15)}{3} \stackrel{?}{=} 13 - \left(\frac{15}{5}\right)$$
$$10 = 10 \quad \text{checks}$$

Note: In Example 2, we could have eliminated one step by just multiplying each *term* of the equation by the LCD.

If a fraction with a polynomial in it is preceded by a minus sign, then the entire fraction must be subtracted. The wise use of parentheses is very important.

Example 3

Solve and check:

$$\frac{7y - 5}{10} - \frac{y + 1}{3} = 1$$

Solution: We multiply, term by term, by the LCD, 30,

$$\left(\frac{7y - 5}{10}\right)30 - \left(\frac{y + 1}{3}\right)30 = (1)30$$
$$(7y - 5)3 - (y + 1)10 = 30$$
$$21y - 15 - 10y - 10 = 30$$
$$11y = 55$$
$$y = 5$$

Check:

$$\frac{7(5) - 5}{10} - \frac{5 + 1}{3} \stackrel{?}{=} 1$$
$$1 = 1 \quad \text{checks}$$

Equations with the Variable in the Denominator

The procedure is the same for solving an equation in which the variable is in one or more denominators. Multiply both sides of the equation by the LCD, which in this case contains the variable. Solve the resulting equation and check.

Example 4

Solve and check:

$$\frac{1}{4x} - \frac{3}{4} = \frac{7}{x}$$

Solution: The LCD is $4x$. Multiplying term by term by $4x$, we get

$$\frac{1}{4x}(4x) - \frac{3}{4}(4x) = \frac{7}{x}(4x)$$

Simplifying

$$1 - 3x = 28$$
$$x = -9$$

Check: Substituting $x = -9$ in the original equation:

$$\frac{1}{4(-9)} - \frac{3}{4} \overset{?}{=} \frac{7}{-9}$$

$$-\frac{7}{9} = -\frac{7}{9} \quad \text{checks}$$

Example 5

Solve and check:

$$\frac{3x}{x+2} - 2 = \frac{2x-3}{2x-1}$$

Solution: The LCD is $(x + 2)(2x - 1)$. Multiplying term by term by the LCD,

$$\left(\frac{3x}{x+2}\right)(x+2)(2x-1) - 2(x+2)(2x-1) = \left(\frac{2x-3}{2x-1}\right)(x+2)(2x-1)$$

Simplifying,

$$3x(2x-1) - 2(2x^2 + 3x - 2) = (2x-3)(x+2)$$
$$6x^2 - 3x - 4x^2 - 6x + 4 = 2x^2 + x - 6$$
$$-10x = -10$$
$$x = 1$$

Check:

$$\frac{3}{1+2} - 2 \overset{?}{=} \frac{2-3}{2-1}$$

$$-1 = -1 \quad \text{checks}$$

Alternate Solution: Recall that all equations can be solved graphically. To solve this equation, we rewrite it as

$$\frac{3x}{x+2} - 2 - \frac{2x-3}{2x-1} = 0$$

and graph the function

$$y = \frac{3x}{x+2} - 2 - \frac{2x-3}{2x-1}$$

on a graphing calculator. Using the "root" or "zero" feature of the calculator (Fig. 18-1), we find that there is an x intercept at $x = 1$, so that the solution of the given equation is

$$x = 1$$

■ FIGURE 18-1

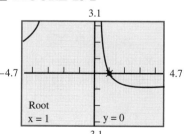

When solving equations in which the variable is in the denominator, *the check is essential.* In the process of multiplying both sides by a variable expression (the

LCD), we may introduce a false "solution," a value that does not make the original equation true. We call such a value *extraneous*. It becomes clear that a value is extraneous only in the check. If the check produces a denominator of zero in the given equation, then that value is extraneous and must be discarded. There may be *no solution*. Since we cannot predict which equations produce extraneous values, we must check all solutions.

COMMON ERROR It is *essential* to check the solution of an equation in which the variable appears in a denominator, for the value may be extraneous.

Example 6

Solve and check:

$$\frac{2}{a-2} = \frac{a}{a-2} - 2$$

Solution: Multiplying by the LCD, $a - 2$,

$$\left(\frac{2}{a-2}\right)(a-2) = \left(\frac{a}{a-2}\right)(a-2) - 2(a-2)$$

$$2 = a - 2(a-2)$$

$$2 = a - 2a + 4$$

$$a = 2$$

Check:

$$\frac{2}{0} \stackrel{?}{=} \frac{2}{0} - 2$$

Since we have division by zero, the value 2 is not a solution; it is extraneous. Thus, there is no solution.

Fractional Solutions

If a solution is a fraction, checking by substitution may not be simple. A quick check to see if it is extraneous can be made by inspecting the denominators. A check by calculator may be easier, or you may choose to rework the problem, rather than doing a regular check.

Example 7

Check that $x = \dfrac{12}{5}$ is a solution to the equation

$$\frac{6}{x^2 - 4} - 1 = \frac{4}{x+2} - \frac{x-3}{x-2}$$

Solution: A check by hand would be laborious. A quick check of the denominator does not indicate division by zero, so the value is not extraneous. We check by evaluating each side of the equation *by calculator:*

$$\frac{6}{\left(\frac{12}{5}\right)^2 - 4} - 1 \stackrel{?}{=} \frac{4}{\frac{12}{5} + 2} - \frac{\frac{12}{5} - 3}{\frac{12}{5} - 2}$$

$$2.4090909 = 2.4090909 \quad \text{checks by calculator}$$

Solving Fractional Equations

Solve and check.

1. $\dfrac{2x}{3} = 12$

2. $\dfrac{11}{5} = \dfrac{b}{15}$

3. $a - 3 = \dfrac{2a}{5}$

4. $n + 6 = \dfrac{3n}{5}$

5. $4 - x = \dfrac{5 + x}{8}$

6. $x = \dfrac{3x}{8} - 5$

7. $\dfrac{x}{3} - \dfrac{x}{4} = 5$

8. $\dfrac{7x}{2} - \dfrac{9x}{4} = x + 6$

9. $\dfrac{x}{2} - x = \dfrac{x}{3} - 15$

10. $\dfrac{5c}{6} - \dfrac{3c}{4} = 3 - \dfrac{2c}{3}$

11. $\dfrac{4 - 5a}{3} - 2 = \dfrac{5 - 9a}{7}$

12. $\dfrac{2y - 1}{4} = \dfrac{2y + 3}{5} - 3$

13. $\dfrac{5 - a}{7} - \dfrac{a - 1}{14} = 1$

14. $\dfrac{3x - 1}{4} - \dfrac{2x + 1}{7} = 1$

15. $2 - \dfrac{x - 7}{12} = \dfrac{x + 2}{6}$

16. $\dfrac{3b - 1}{4} - \dfrac{b + 5}{3} = 1$

17. $\dfrac{5 - y}{9} - \dfrac{y + 2}{16} = \dfrac{1}{12}$

18. $\dfrac{3a - 2}{5} - \dfrac{2a - 7}{4} = 2$

19. $\dfrac{2y - 4}{3} - \dfrac{y + 1}{5} = 5$

20. $\dfrac{a - 5}{12} - \dfrac{1 - 3a}{2} = \dfrac{2}{3}$

21. $\dfrac{4y + 3}{5} - \dfrac{y + 4}{3} - \dfrac{2y - 1}{15} = 2$

22. $\dfrac{5 - 2a}{2} - \dfrac{1 - 3a}{3} = \dfrac{2a - 3}{6} + 4$

Equations with the Variable in the Denominator

Solve and check.

23. $\dfrac{1}{2x} = \dfrac{1}{6}$

24. $\dfrac{2}{3x} = \dfrac{2}{9}$

25. $\dfrac{1}{x} - \dfrac{1}{3x} = \dfrac{1}{3}$

26. $\dfrac{3x - 6}{x} + 1 = \dfrac{14}{x}$

27. $\dfrac{y - 3}{y + 3} = 7$

28. $\dfrac{x - 2}{x + 8} = 6$

29. $\dfrac{8}{x - 2} = \dfrac{4x}{x - 2}$

30. $\dfrac{2}{x - 1} = 0$

31. $\dfrac{y}{y - 5} - 2 = \dfrac{5}{y - 5}$

32. $\dfrac{x + 1}{2x - 6} + \dfrac{3}{3 - x} = \dfrac{3}{2}$

33. $\dfrac{2}{x - 2} - \dfrac{9}{x} = \dfrac{1}{x}$

34. $\dfrac{4}{a - 1} = \dfrac{7}{a - 1} - 2$

35. $\dfrac{2}{x + 1} + \dfrac{1}{x} = \dfrac{1}{x^2 + x}$

36. $\dfrac{3}{a + 2} = \dfrac{1}{a - 2} - \dfrac{4}{a^2 - 4}$

37. $\dfrac{3y + 4}{y + 2} - \dfrac{3y - 5}{y - 4} = \dfrac{12}{y^2 - 2y - 8}$

38. $\dfrac{5y}{y + 3} - \dfrac{3y + 2}{y - 4} = \dfrac{y - 3}{y^2 - y - 12} + 2$

39. $1.83x = \dfrac{x}{2.75} + 8.14$

40. $87.4 = \dfrac{35.6}{x} + 45.7$

41. $\dfrac{0.429}{x - 0.326} - 0.567 = \dfrac{0.395}{x - 0.326}$

42. $\dfrac{2.34}{x - 3.89} + 4.56 = 8.46 - \dfrac{6.39}{x - 3.89}$

LITERAL EQUATIONS AND FORMULAS

Recall that a literal equation is an equation in which some or all of the constants are represented by letters (literal) and that a formula is a literal equation that relates two or more mathematical or physical quantities.

Many important formulas in applications involve fractions. We look at a few of them here, and solve for different variables. The process of solving is the same as in the last section, except that now the solution will be a literal expression rather than a number.

Example 8

Solve for x:

$$\frac{x}{a} - b = \frac{x}{c}$$

Solution: We combine the two x terms on the left and bring the constant b to the right:

$$\frac{x}{a} - \frac{x}{c} = b$$

We multiply both sides of the equation by the LCD:

$$ac\left(\frac{x}{a} - \frac{x}{c}\right) = ac(b)$$

Distributing, we get

$$ac\left(\frac{x}{a}\right) - ac\left(\frac{x}{c}\right) = acb$$

Simplifying,

$$cx - ax = abc$$

Factoring,

$$(c - a)x = abc$$

Dividing,

$$x = \frac{abc}{c - a}$$

Example 9

The kinetic energy E of a body with mass m and velocity v is

$$E = \frac{1}{2}mv^2$$

Solve this equation for m.

Solution: Multiplying both sides by the LCD, 2, we have:

$$2E = mv^2$$

Dividing both sides by v^2, and changing sides:

$$m = \frac{2E}{v^2}$$

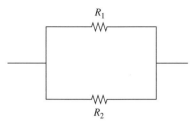

R_1

R_2

■ **FIGURE 18-2**

Example 10 ────────────

If two resistors R_1 and R_2 are connected in parallel (Fig. 18-2), the total resistance R is found using the formula:

$$\frac{1}{R} = \frac{1}{R_1} + \frac{1}{R_2}$$

Solve for R_1.

Solution: The LCD is RR_1R_2. Multiplying term by term:

$$\frac{1}{R} \cdot RR_1R_2 = \frac{1}{R_1} \cdot RR_1R_2 + \frac{1}{R_2} \cdot RR_1R_2$$

Simplifying,

$$R_1R_2 = RR_2 + RR_1$$

Combining all terms involving R_1,

$$R_1R_2 - RR_1 = RR_2$$

Factoring out R_1,

$$R_1(R_2 - R) = RR_2$$

Solving for R_1,

$$R_1 = \frac{RR_2}{R_2 - R}$$

Be sure that the variable you have solved for does not also appear on the other side of the equation.

> **COMMON ERROR** When solving for a variable, be sure to combine all terms involving that variable before solving for the variable. It is *not useful to solve for a variable in terms of itself.*

Substituting into Formulas

In many problems we are given a formula and values for all but one variable in the formula. We substitute all known values in the formula and then solve for the unknown variable.

Example 11 ────────────

The formula for the displacement s of a freely falling body having an initial velocity v and acceleration g is

$$s = vt + \frac{1}{2}gt^2$$

Find the initial velocity of an object thrust downward off the top of a 195-ft building, if it hits the ground after 3.00 seconds. (Use $g = 32.2$ ft/s².)

Solution: Substituting into the formula, with $s = 195$, $g = 32.2$, and $t = 3.00$, we have:

$$195 = v(3) + \frac{1}{2}(32.2)(3^2)$$

$$195 = 3v + 144.9$$

Solving,

$$v = 16.7 \text{ ft/s}$$

So, the initial velocity is 16.7 ft/s.

Solve each literal equation for x.

1. $\dfrac{x}{a} + b = c$

2. $a - \dfrac{b}{x} = c$

3. $\dfrac{ax}{b} = c$

4. $a = \dfrac{bc}{x}$

5. $\dfrac{x}{a} + \dfrac{x}{b} = \dfrac{1}{c}$

6. $\dfrac{a}{x} = \dfrac{b}{x} - c$

7. $a = \dfrac{b + x}{c}$

8. $a = \dfrac{b}{x + c}$

9. $\dfrac{a}{b} = \dfrac{c}{x} + 1$

10. $a = \dfrac{bx}{b + x}$

Solve each formula for the letter in parentheses on the right.

11. $A = \dfrac{ab}{2}$ (a)

12. $V = \dfrac{1}{3}\pi r^2 h$ (h)

13. $T^2 = 4\pi^2 \dfrac{L}{g}$ (g)

14. $v^2 = \dfrac{2GM}{r}$ (r)

15. $I = \dfrac{E + e}{R}$ (e)

16. $I = \dfrac{E}{R + r}$ (r)

17. $\dfrac{1}{R} = \dfrac{1}{R_1} + \dfrac{1}{R_2} + \dfrac{1}{R_3}$ (R_1)

18. $\dfrac{1}{f} = \dfrac{1}{p} + \dfrac{1}{q}$ (p)

19. $\dfrac{E}{e} = \dfrac{R + r}{r}$ (r)

20. $\dfrac{W_1}{W_2} = \dfrac{L_2}{L_1}$ (L_1)

21. $\dfrac{P_1 V_1}{T_1} = \dfrac{P_2 V_2}{T_2}$ (T_1)

22. $C = \dfrac{C_1 C_2}{C_1 + C_2}$ (C_1)

23. $s = vt - \dfrac{1}{2}gt^2$ (g)

24. $C = \dfrac{5}{9}(F - 32)$ (F)

25. $A = \dfrac{h}{2}(a + b)$ (a)

26. $I = \dfrac{E}{r + \dfrac{R}{n}}$ (R)

Applications

27. A formula for calculating a child's dosage for medication is

$$\text{Child's dose} = \frac{\text{age of child}}{\text{age of child} + 12} \times \text{adult dose}$$

If a 10-year-old child's dose of a particular medication is 15 mg, what is the adult's dose?

28. A formula for calculating an infant's dosage for medication is

$$\text{Infant's dose} = \frac{\text{age of infant in months}}{150} \times \text{adult dose}$$

If a 12-month-old infant is to receive 8.5 mg of medication, what is the equivalent adult dose (to the nearest mg)?

29. The formula in Exercise 16 is used to find the current I (in amperes) in an electric circuit when voltage E (in volts), external resistance R (in ohms), and internal resistance r (in ohms) are known. Find R when $E = 30.5$ volts, $r = 0.756$ ohms, and $I = 1.67$ amperes.

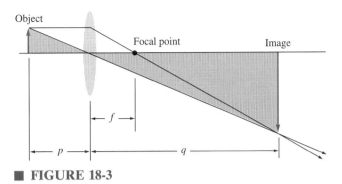

Object Focal point Image

f

p q

■ **FIGURE 18-3**

30. The formula in Exercise 21 shows the relation among the pressure P, volume V, and temperature T (in the Kelvin temperature scale) of a gas. If the volume of oxygen is 2.50 ft^3 when the pressure is 2150 lb/in.^{2} and the temperature is 27.5°C, find the volume of oxygen when the pressure is 1150 lb/in.^{2} and the temperature is 34.5°C. (Hint: To convert °C to Kelvin units, add 273 to the degrees Celsius.)

31. The formula in Exercise 18 expresses the relation among the focal length f of a lens, the distance p to an object, and the distance q to the image (Fig. 18-3). Find the image distance for a lens with a focal length of 50.6 cm, if the object is 325 cm from the lens.

32. What resistor must be wired in parallel with a 125-ohm resistor if the total resistance is 20.5 ohms? (Refer to Example 10).

C_1 C_2

■ **FIGURE 18-4**

33. The formula in Exercise 22 is used to find the capacitance C of a circuit containing two capacitances, C_1 and C_2, in series. A circuit contains two capacitances in series, one of which is 6.50 μF (Fig. 18-4). Find the second capacitance if the capacitance of the circuit is 3.50 μF.

34. The formula in Exercise 25 is used to find the area A of a trapezoid when the altitude h and the bases a and b are known. If one base is 14.4 cm, the altitude is 9.50 cm, and the area is 108 cm^2, find the second base.

18-3

APPLICATIONS

Many types of applications involve equations with fractions. These include work and uniform motion problems. The method for solving these problems is the same as that given in Chapter 11.

Work

To solve work problems, we use an equation that we know from everyday experience:

$$(\text{rate of work}) \cdot (\text{time worked}) = \text{Amount of work done}$$

or,

$$\text{rate} \cdot \text{time} = \text{Work}$$

Example 12

Suppose a person can mow a lawn in 3 hours. To find his rate of work, we solve for the rate:

$$\text{rate} \cdot \text{time} = \text{Work}$$

$$\text{rate} = \frac{\text{Work}}{\text{time}}$$

In this case, the "Work" is mowing "1 lawn" and the time worked is 3 hours. So the rate is

$$\text{rate} = \frac{1 \text{ lawn}}{3 \text{ hours}} \text{ or } \frac{1}{3} \text{ lawn/h}$$

Now we look at a problem in which two people are working together to complete a job (the work).

Example 13

One person can mow a lawn in 2.75 hours, and a second person takes 3.50 hours to mow the same lawn. How long would it take them to mow the lawn, working together?

Solution: We make a table. But first we need to find the individual rates of work for each person:

$$\text{For the 1st person: } r_1 = \frac{1 \text{ lawn}}{2.75 \text{ hours}} = \frac{1}{2.75} \text{ lawn/h}$$

$$\text{For the 2nd person: } r_2 = \frac{1 \text{ lawn}}{3.50 \text{ hours}} = \frac{1}{3.50} \text{ lawn/h}$$

We make a table:

	Rate (lawn/h)	× Time (h)	= Work (lawn)
1st person	$\frac{1}{2.75}$	t	$\frac{1}{2.75}t$
2nd person	$\frac{1}{3.50}$	t	$\frac{1}{3.50}t$
Together			1

The equation for the problem comes from the last column of the table. Their work is 1, meaning 1 lawn, because they mowed the whole (1) lawn.

$$\text{1st person's work} + \text{2nd person's work} = \text{their work together}$$

$$\frac{1}{2.75}t + \frac{1}{3.50}t = 1$$

Multiplying through by the LCD 2.75(3.50), we get

$$2.75(3.50)\left(\frac{1}{2.75}t\right) + 2.75(3.50)\left(\frac{1}{3.50}t\right) = 2.75(3.50)(1)$$

$$3.50t + 2.75t = 9.625$$

$$6.25t = 9.625$$

$$t = 1.54 \text{ h}$$

So, it would take them 1.54 hours to mow the lawn together.

─── **Example 14** ───

One technician can assemble 9 instruments in 4.25 hours. A second technician can assemble 7 instruments in 3.75 hours. If they worked together, how long would it take them to assemble 25 instruments?

Solution: First we find the technicians' individual rates. Working alone, the first technician works at the rate

$$r_1 = \frac{9 \text{ instruments}}{4.25 \text{ hours}} = \frac{9}{4.25} \text{ instruments/h}$$

Working alone, the second technician works at the rate

$$r_2 = \frac{7 \text{ instruments}}{3.75 \text{ hours}} = \frac{7}{3.75} \text{ instruments/h}$$

We make a table:

	Rate (ins/h) ×	Time (h) =	Work (ins)
1st technician	$\frac{9}{4.25}$	t	$\frac{9}{4.25}t$
2nd technician	$\frac{7}{3.75}$	t	$\frac{7}{3.75}t$
Together			25

Here, the technicians are assembling 25 instruments, so the total work to be completed is 25 (instruments). As in Example 13, the sum of their individual works is the total work:

$$\frac{9}{4.25}t + \frac{7}{3.75}t = 25$$

The LCD is 4.25(3.75), so multiplying by the LCD,

$$4.25(3.75)\frac{9}{4.25}t + 4.25(3.75)\frac{7}{3.75}t = 4.25(3.75)25$$

Simplifying,

$$33.75t + 29.75t = 398.4$$
$$63.5t = 398.4$$
$$t = 6.27 \text{ h}$$

So it will take them about 6.27 hours to assemble 25 instruments, working together.

Uniform Motion

Certain problems of uniform motion involve equations with fractions. In the following navigational example, we need to know that the airspeed of a plane is its speed relative to the air and the groundspeed is its speed relative to the ground. The groundspeed is equal to the sum of the airspeed and the wind velocity.

$$\text{Groundspeed} = \text{airspeed} + \text{wind velocity}$$

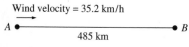

Wind velocity = 35.2 km/h

$A \bullet$ ————————— $\bullet B$
485 km

■ **FIGURE 18-5**

───── **Example 15** ─────

A pilot flies 485 km from airport A to airport B with a tailwind of 35.2 km/h. On her return flight, she flies directly into the same wind and flies only 415 km in the same amount of time. What is the plane's airspeed?

Solution: First we make a diagram (Fig. 18-5). Let r = the plane's airspeed (in km/h). On the first flight (with the wind),

$$\text{The plane's groundspeed} = \text{airspeed} + \text{wind velocity}$$
$$= r + 35.2$$

On the return flight (against the wind),

$$\text{The plane's groundspeed} = \text{airspeed} - \text{wind velocity}$$
$$= r - 35.2$$

We make a table, filling in the time column for each flight using the rate and distance columns, and

$$\text{Time} = \frac{\text{distance}}{\text{rate}}$$

	Rate (km/h)	×	Time (h)	=	Distance (km)
1st flight	$r + 35.2$		$\dfrac{485}{r + 35.2}$		485
Return flight	$r - 35.2$		$\dfrac{415}{r - 35.2}$		415

Since the times are equal, we write the equation from the time column:

$$\frac{485}{r + 35.2} = \frac{415}{r - 35.2}$$

$$485(r - 35.2) = 415(r + 35.2)$$

$$485r - 17072 = 415r + 14608$$

$$70r = 31680$$

$$r = 453 \text{ km/h}$$

So the plane is flying with an airspeed of 453 km/h.

───────────────────────────────

SECTION 18-3 **Exercises** **Applications**

───────────────────────────────

Solve and check. Round to three significant digits.

1. A rectangle is $\dfrac{3}{4}$ as wide as it is long. If the length is increased by 3.00 cm and the width decreased by 2.00 cm, then the area is 25.0 cm² more than the original area. Find the dimensions of the original rectangle.

2. The width of a rectangle is $\dfrac{3}{5}$ of its length and the perimeter is 48.0 cm. Find the dimensions of the rectangle.

3. One side of a triangular lot is $\frac{1}{2}$ the length of the second side. The third side is $\frac{5}{6}$ the length of the first side. Find the sides if the perimeter of the lot is 184 m.

4. After spending $\frac{3}{5}$ of her money, a woman had $24.00 left. How much did she have before?

5. A man spent $\frac{3}{4}$ of his paycheck, and then $\frac{1}{3}$ of the remainder. If he had $54.00 left, how much was his paycheck?

6. Three children were left an inheritance of which the eldest received $\frac{1}{2}$ and the second received $\frac{1}{3}$. The third child received the remainder, which was $2500. How much was the total inheritance?

Work

7. A woman can mow the lawn in 5.25 hours and her husband can mow the lawn in 4.75 hours. If each had their own mowers and mowed the lawn together, how long would it take them?

8. A holding tank can be drained through one pipe in 8.50 hours, or through another pipe in 6.75 hours. If both pipes were used, how long would it take to drain the tank?

9. One road crew can construct 525 ft of road in 3.50 days. A second crew can construct 1050 ft in 7.25 days. If the crews worked together, how long would it take them to construct a road one mile long?

10. One machine can make 25 parts in 6.75 hours. Another machine can make 18 parts in 5.25 hours. If both machines are working, how long would it take to make 100 parts?

11. A swimming pool can be filled in 15.5 hours through one opening at the top. It can be filled in 12.6 hours through a second opening. How long will it take to fill the pool if both openings are used?

12. A 50.0-gallon tank can be filled in 8.25 minutes. It can be drained in 12.5 minutes. Five (5.00) minutes after they began filling the tank, a technician accidentally opened the drain. How much longer does it take for the tank to fill?

Uniform Motion

13. A biker rides 15.0 miles down a long straight road with a wind of 8.25 mi/h at his back. He turns around and finds that he can go only 7.25 miles in the same time, with the wind at his front. What is his bike velocity, without wind?

14. The current in a certain stream is 4.47 km/h. A canoeist can go 7.28 km upstream in the same time it takes her to go 15.8 km downstream. What is her rate of canoeing in still water?

FIGURE 18-6

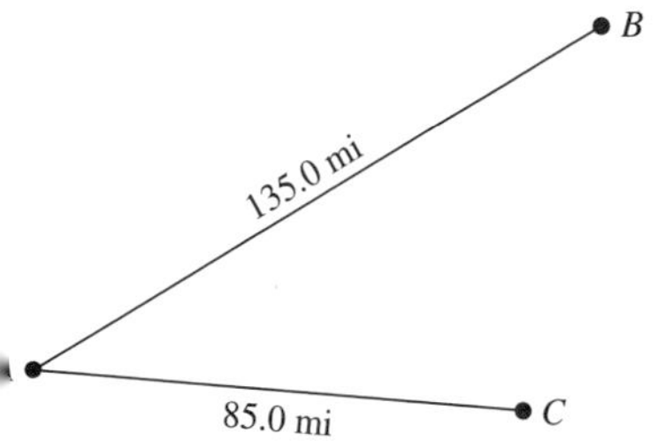

15. A salesman drove 185 miles to a convention. Then he drove 132 miles to a city at the same rate in 1.17 hours less than his first trip. How fast did the salesman drive? How long did it take him to drive to the convention?

16. Two ships leave port A at the same time and travel at the same rate. One ship sailed to port B 135.0 miles away and the other to port C 85.0 miles away (Fig. 18-6). The second ship arrived at her port 2.25 hours before the first ship arrived at her port. How fast were the ships traveling?

17. A baseball player's batting average is found by dividing h, the number of hits, by b, the number of times at bat:

$$\frac{h}{b}$$

If the player gets a hit, then the new batting average is

$$\frac{h+1}{b+1}$$

If the player does not get a hit, then the new batting average is

$$\frac{h}{b+1}$$

Will a player's batting average always change after the player goes to bat? Under what conditions will the batting average remain the same?

■ Chapter 18 REVIEW EXERCISES

Solve and check.

1. $y - 3 = \dfrac{y+6}{7}$

2. $\dfrac{2a}{5} - \dfrac{a}{2} = a$

3. $\dfrac{2x}{5} - \dfrac{3x}{10} = 2 + \dfrac{4x}{15}$

4. $\dfrac{2b}{3} - \dfrac{3b-1}{8} = 1$

5. $\dfrac{7-y}{24} + \dfrac{3y+4}{16} = 2$

6. $\dfrac{n-5}{n-4} = \dfrac{3n}{3n+1}$

7. $2 + \dfrac{1}{b-1} = \dfrac{b}{b-1}$

8. $\dfrac{5x}{x-2} - 3 = \dfrac{4x}{2x-7}$

Solve for the variable in parentheses on the right.

9. $F = k\dfrac{M_1 M_2}{d^2}$ (M_1)

10. $\dfrac{V_1}{T_1} = \dfrac{V_2}{T_2}$ (T_1)

11. $F = \dfrac{9}{5}C + 32$ (C)

12. The formula,

$$I = \frac{E}{R+r}$$

is used to find the current I in an electric circuit when voltage E, external resistance R, and internal resistance r are known. Find R when $E = 33.5$ volts, $r = 0.525$ ohms, and $I = 1.25$ amperes.

13. The width of a certain rectangle is $\dfrac{2}{3}$ of the length. If the width is decreased by 3.00 in. and the length increased by 3.00 in., the area is 36.0 in.² less than the original rectangle. Find the dimensions of the original rectangle.

14. One person can build a shed in 5.25 days and another person can build an identical shed in 4.50 days. How long should it take them to build the shed working together?

15. One mason can build 5.00 ft of wall in a day, and another mason can build 6.00 ft of wall in a day. How many days will it take them to build 120.0 ft of wall together?

16. Two planes, flying at the same rate, are scheduled to arrive at Chicago's O'Hare Airport at the same time. One plane left its origin, 3125 miles from O'Hare, 3.00 hours before the other plane left its origin, 1250 miles from O'Hare. What was their groundspeed and how many hours was the longer flight?

17. Sid can type 7 pages in 2.00 hours. Shirley can type 13 pages in 3.00 hours. After Shirley has been typing a 35-page document for 2.00 hours, Sid joins her. How much longer must they work together to complete the job?

WRITING

18. When we solve a fractional equation algebraically, we sometimes get a false "solution," which is called *extraneous*. Make up a fractional equation that has an extraneous solution, solve the equation, and show that it has an extraneous solution. Where does this extraneous solution come from? How did you discover that it is extraneous?

TEAM PROJECT

19. An alphametic is an algebra puzzle in which digits are replaced by letters of the alphabet. In each puzzle each letter stands for a particular, but different, digit. For example, a familiar alphametic is

$$\begin{array}{r} \text{SEND} \\ \text{MORE} \\ \hline \text{MONEY} \end{array}$$

The solution is:

S = 9	M = 1	Check:	9567
E = 5	O = 0		1085
N = 6	R = 8		10652
D = 7	Y = 2		

Solve the alphabetic

$$\frac{\text{SEVEN}}{\text{TWO}} = \text{TWO}$$

in which

$$\begin{array}{r} \text{TWO} \\ \text{TWO)}\overline{\text{SEVEN}} \\ \text{BOB} \\ \hline \text{JOE} \\ \text{OVV} \\ \hline \text{VESN} \\ \text{VESN} \\ \hline \end{array}$$

Systems of Linear Equations

Sometimes a physical situation can be described by a single equation. For example, the motion of a car being driven down a highway at 55 mi/h can be described by the equation

$$d = 55t$$

where d is the distance (in miles) driven during time t (in hours). However, a more complicated situation may require two or more equations for a complete description. For example, if there is a *second* car, we need *two* equations to completely describe the motion of the two cars. If the second car travels the same distance as the first car, but at 60 mi/h and in two hours less time, then

$$d = 55t \qquad \text{for the first car}$$

and also

$$d = 60(t - 2) \quad \text{for the second car}$$

In this chapter we learn how to solve systems of two or more linear equations using several different methods. We show graphical and algebraic solutions. We also include applications for systems of linear equations, and solve systems involving fractional and literal equations.

19-1   GRAPHICAL METHOD

In this section we learn how to solve a system of two linear equations graphically.

Linear Equations

A *linear equation* is an equation in which the terms containing the variable(s) are all of first degree. Recall from Sec. 7-1 that the degree of a term is the power to which the variable in the term is raised. A linear equation in two variables can be written as

LINEAR EQUATION IN TWO VARIABLES	$Ax + By = C$	131

where A, B, and C are constants, and A and B cannot both be zero. In a linear equation, no variable has an exponent other than 1.

Example 1

The equation

$$x + y = 8$$

is a linear equation in two variables, in which $A = 1$, $B = 1$, and $C = 8$ in Eq. 131.

Suppose we had to *solve* the linear equation in Example 1. We need to find all values of the variables that make the equation true. If $x = 1$, for example, then $y = 7$ would make the equation true. Denoting this as an ordered pair, we say that $(1, 7)$ is *a solution* of the equation. Similarly, $x = 2$ and $y = 6$ would also make the equation true, so $(2, 6)$ is a solution, and so forth. In fact, there are an *infinite* number of solutions.

Graph of a Linear Equation

We saw in Chapter 12 that these solutions can be displayed graphically, by plotting the ordered pairs as points on the coordinate plane. As you might expect, the graph of a linear equation in two variables is a *line*.

Example 2

Graph the equation:

$$x + y = 8$$

Solution: To graph the equation, we'll plot some points and connect them. To make a table of values, we solve the equation for y, so that we can easily calculate the y values from the x values:

$$y = -x + 8$$

Note that this puts the equation into the slope-intercept form of a linear function. Since the graph is a line, we will plot three points. We already have two points from Example 1, so we make a table and calculate a third point.

x	y
1	7
2	6
3	5

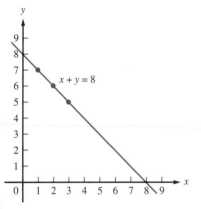

FIGURE 19-1
Graph of a linear equation.

We plot these points and see that the graph is a line (Fig. 19-1).

System of Equations

If two or more equations that contain the same variables are considered together, they are called a *system of equations*. (Sometimes they are referred to as *simultaneous* equations.)

Example 3

The equations

$$x + y = 8$$
$$x - y = 4$$

are a system of two linear equations in two variables.

Solution of a System of Equations

The *solution* of a system of equations is the set of *all* values of the variables that make every equation in the system true.

Example 4

The solution of the system in Example 3 is the set of all ordered pairs (x, y) which make *both* equations true. The ordered pair $(6, 2)$ is a solution, since the sum, $x + y$, of the numbers is 8 and the difference, $x - y$, is 4:

$$6 + 2 = 8 \quad \text{and} \quad 6 - 2 = 4$$

In fact, the ordered pair $(6, 2)$ is the only solution, since they are the only numbers whose sum is 8 and difference is 4.

Example 5

Determine if the ordered pair $(5, -4)$ is a solution to this system:

$$2x + y = 6$$
$$3x + 4y = 4$$

Solution: We must substitute the ordered pair $(5, -4)$ into both equations. Letting $x = 5$ and $y = -4$ in the first equation, we have

$$2(5) + (-4) \stackrel{?}{=} 6$$
$$6 = 6$$

Thus the pair $(5, -4)$ satisfies the first equation. Now we substitute the values in the second equation:

$$3(5) + 4(-4) \stackrel{?}{=} 4$$
$$-1 \neq 4$$

The substitution fails in the second equation, so $(5, -4)$ is not a solution to the system.

COMMON ERROR When checking, be sure to substitute the apparent solution in *both* equations.

Graphical Solution of a System

To solve a system of two linear equations graphically, graph both equations on the same coordinate axes and find the point of intersection. The solution is given by the coordinates of the point of intersection. Any graphical solution is approximate and *must be checked in both equations.*

Example 6

Solve this system graphically and check.

$$2x - 3y = 12$$
$$x + y = 1$$

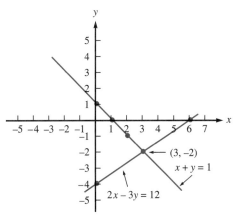

■ FIGURE 19-2

Solution: To graph the equations, we make a table of values for each equation. We solve each equation for y, and calculate at least three ordered pairs.

$$2x - 3y = 12 \qquad\qquad x + y = 1$$

$$y = \frac{2x - 12}{3} \qquad\qquad y = 1 - x$$

x	y
0	-4
3	-2
6	0

x	y
0	1
1	0
2	-1

We plot the points and draw the lines (Fig. 19-2). It looks like the lines intersect at about the point $(3, -2)$. The apparent solution is $(3, -2)$, but these values must be checked.

Check:
We substitute $x = 3$ and $y = -2$ into *both* of the original equations:

$$2(3) - 3(-2) \stackrel{?}{=} 12 \qquad\qquad 3 + -2 \stackrel{?}{=} 1$$

$$6 + 6 = 12 \qquad\qquad 1 = 1 \quad \text{checks}$$

$$12 = 12 \quad \text{checks}$$

The values check in both equations and therefore the ordered pair $(3, -2)$ is the solution.

Again, a graphical solution to any problem is *approximate,* and must always be checked. If the graphical approximation happens to be exact, as in Example 6 above, the check will also be exact. Sometimes, however, the coordinates of the point of intersection are not integers, and must be approximated, say, to the nearest tenth. If the values we check are not exact, then they will not satisfy the equations exactly. The approximated values will be satisfactory, if, in the check, both sides of the equations are approximately equal.

┌─── **Example 7** ───────

Solve this system graphically. Approximate the solution to the nearest tenth and check:

$$3x - y = 8$$
$$2x - 5y = 15$$

Solution: In this example, we graph the lines from the slope-intercept form of each equation. We solve each equation for y:

$$3x - y = 8 \qquad 2x - 5y = 15$$
$$y = 3x - 8 \qquad y = \frac{2}{5}x - 3$$

For the first equation, we graph a line whose slope is 3 and whose y intercept is -8. For the second, we graph a line whose slope is $\frac{2}{5}$ and whose y intercept is -3. From the graph (Fig. 19-3), we approximate the solution to be $(1.9, -2.2)$.

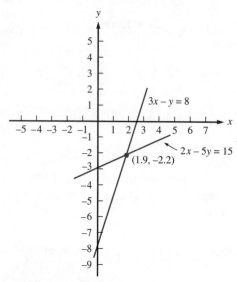

■ FIGURE 19-3

Check:

We check in the original equations:

$$3(1.9) - (-2.2) \overset{?}{=} 8 \qquad 2(1.9) - 5(-2.2) \overset{?}{=} 15$$
$$7.9 \approx 8 \quad \text{checks} \qquad 14.8 \approx 15 \quad \text{checks}$$

The sides of each equation are approximately equal, so the solution $(1.9, -2.2)$ is a good approximation.

┌─── **Example 8** ───────

Solve this system graphically, and check:

$$x - y = 2$$
$$-3x + 3y = 9$$

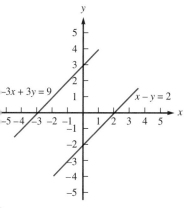

◀ FIGURE 19-4

Solution: We rewrite the equations in slope-intercept form:

$$y = x - 2$$
$$y = x + 3$$

and graph the equations in Fig. 19-4. We see that the lines are parallel, and there is *no solution.*

Example 9

Solve this system graphically, and check:

$$-2x + y = 5$$
$$4x - 2y = -10$$

Solution: We rewrite the equations in slope-intercept form:

$$y = 2x + 5$$
$$y = 2x + 5$$

and note that the two equations are the same. So the graphs of the equations (Fig. 19-5) are really the same line. Since all the points on the line are solutions of the system, the *solution is infinite,* and consists of

all ordered pairs (x, y) such that $-2x + y = 5$

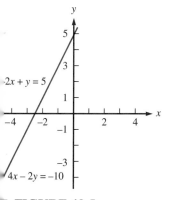

◀ FIGURE 19-5

Graphing Calculator Solution

It's very simple to solve a system of equations using a graphing calculator. Just graph the two equations on the same screen. If the point of intersection is not on the screen, zoom out or change the window until the point of intersection is on the screen. Now use the "intersect" feature to find the coordinates of the point of intersection.

Example 10

Use a graphing calculator to solve the system:

$$1.24x - 1.65y = -2.39$$
$$3.06x + 2.47y = -2.59$$

accurate to the nearest hundredth.

Solution: We solve each equation for y, so that we can enter the function in the calculator.

$$y = \frac{1.24x + 2.39}{1.65}$$

$$y = \frac{-3.06x - 2.59}{2.47}$$

◀ FIGURE 19-6

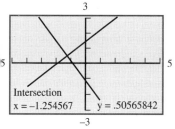

We use the range $-5 \leq x \leq 5$ and $-3 \leq y \leq 3$. We graph both equations on the same screen and use the "intersect" feature to find the point of intersection (Fig. 19-6). We see that the coordinates are $(-1.25, 0.51)$, rounded to hundredths. Thus the solution is $x = -1.25$ and $y = 0.51$, rounded to hundredths.

**Solution of a
System of Equations**

Determine if the given ordered pair is a solution to the system of equations.

1. $x + 2y = -7$
$5x - y = 9$ $(1, -4)$

2. $2x - y = 5$
$x - 3y = 5$ $(2, -1)$

3. $x - 2y = -3$
$3x + y = 5$ $(-1, 1)$

4. $x + y = 3$
$2x - 3y = 11$ $(2, 1)$

5. $2x - 3y = -5$
$x + 6y = 2$ $(-1.6, 0.6)$

6. $2x - 5y = 8$
$6x + y = 0$ $(0.25, -1.5)$

For each pair of lines in Fig. 19-7, name the coordinates of the point of intersection. If the coordinates are not integers, round to the nearest tenth.

7. a and b **8.** a and c **9.** a and d

10. b and c **11.** b and d **12.** c and d

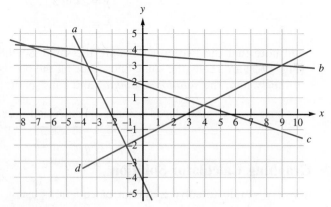

■ **FIGURE 19-7**

**Graphical Solution
of a System**

Solve these systems graphically. If the solution is an ordered pair, approximate the solution to the nearest tenth and check it.

13. $x = -2$
$y = 1$

14. $x = 3$
$y = -2$

15. $2x - y = -5$
$x + y = 2$

16. $3x - 2y = -4$
$x + 2y = -4$

17. $2x + 5y = 18$
$3x + 4y = 7$

18. $x + y = 3$
$4x - 6y = 7$

19. $x - 2y = 6$
$2x - 4y = 12$

20. $2x + y = -3$
$-4x - 2y = 6$

21. $2x - 3y = 3$
$x - 2y = -2$

22. $x - 2y = 4$
$3x + 2y = 4$

23. $9x - 8y = 1$
$6x + 12y = 5$

24. $4x + y = 11$
$2x + 3y = 4$

25. $2x - 5y = 10$
$-4x + 10y = 30$

26. $x - y = 6$
$2x - 2y = 4$

27. $5x + 2y = 3$
$3x - 3y = 4$

28. $5x - y = 7$
$2x + y = 4$

29. $x + y = 2$
$3x - y = 10$

30. $2x + y = 6$
$3x + 4y = 4$

31. $2x - 5y = -2$
$-4x + 10y = 4$

32. $4x - 3y = -2$
$12x - 9y = 1$

33. $6.25x - 5.56y = 14.0$
$4.18x + 1.04y = 8.56$

34. $5.35x + 1.46y = 10.5$
$5.48x + 3.27y = 15.7$

35. $2x + y = -3$
$6x + 3y = 2$

36. $-2x + 3y = 4$
$4x - 6y = -8$

37. $1.67x - 0.76y - 3.18 = 0$
$2.56x + 3.82y - 11.5 = 0$

38. $2.45x + 1.73y + 1.95 = 0$
$2.38x - 1.25y - 3.45 = 0$

39. $0.2x + 0.1y - 0.1 = 0$
$-0.3x + 0.1y - 0.1 = 0$

40. $3.1x + 1.2y + 3.9 = 0$
$1.8x - 3.0y + 1.1 = 0$

Graphing Calculator Solution

Use a graphing calculator to solve the system, rounding to hundredths.

41. $x - y = 1$
$x + y = 5$

42. $x - y = 3$
$2x + 3y = 11$

43. $-2x + y = -3$
$4x - 2y = 2$

44. $-2x + 3y = 4$
$6x - 5y = 1$

45. $5.24x - 3.45y = 7.82$
$2.14x + 1.58y = 4.36$

46. $2.74x + 1.69y = -2.45$
$-6.55x - 2.33y = 4.50$

47. $2.36x + 3.47y = 4.27$
$8.46x - 12.0y = -16.7$

48. $5.36x + 1.49y = 10.4$
$9.63x + 1.03y = -20.4$

19-2   ADDITION METHOD

In the last section we learned how to approximate the solution to a system of two linear equations using a graphical method. But when the problem requires an *exact* solution, we may need to use an algebraic method to solve the system.

Algebraic Solution by the Addition Method

With any algebraic method of solving a system of equations in two variables, we try to *eliminate* one of the variables, so that we have a single equation in one variable. In the *addition method,* we eliminate one of the variables by adding the two equations together; that is, we add the left sides together, add the right sides together, and get a new equation. The new equation is in one variable and we can solve for that variable. We find the other variable by substituting back into one of the original equations.

Example 11

Solve by the addition method, and check:

$$2x - y = 3$$

$$3x + y = 2$$

Solution: If we add the two equations together, we eliminate the y terms:

$$2x - y = 3$$
$$\underline{3x + y = 2}$$

Adding the two equations, we get $5x + 0 = 5$

Solving for x,

$$x = 1$$

Substituting $x = 1$ into the first of the original equations,

$$2(1) - y = 3$$

Solving for y,

$$y = -1$$

So the solution is $x = 1$ and $y = -1$. We write the solution as the ordered pair $(1, -1)$.

Check:
We substitute the values into *both* original equations.

$$2(1) - (-1) \stackrel{?}{=} 3 \quad \text{and} \quad 3(1) + (-1) \stackrel{?}{=} 2$$
$$3 = 3 \quad \text{checks} \qquad\qquad 2 = 2 \quad \text{checks}$$

In Example 11, y was eliminated when we added the original equations, because the coefficients of y in the two equations (-1 and 1) were equal in absolute value and opposite in sign. Often this is not the case. If one coefficient of a variable is a *multiple* of the other coefficient of that variable, then we may multiply each side of one equation by a factor that will make the coefficients in both equations equal in absolute value and opposite in sign.

Example 12

Solve by the addition method, and check:

$$2x + 3y = 6$$
$$x + 2y = 5$$

Solution: No pair of coefficients are equal in absolute value and opposite in sign, but one coefficient of x is a multiple of the other coefficient of x. If we multiply each side of the second equation by -2, then the x coefficients will be equal in absolute value and opposite in sign:

$$-2(x + 2y) = -2(5)$$
$$-2x - 4y = -10$$

So the new system becomes

$$2x + 3y = 6$$
$$\underline{-2x - 4y = -10}$$

Adding, we get $\qquad\qquad\qquad -y = -4$

Solving for y,

$$y = 4$$

Substituting $y = 4$ into the original second equation,

$$x + 2(4) = 5$$

Solving for x,

$$x = -3$$

So the solution is $(-3, 4)$.

Check:
As in Example 11, we check by substituting $x = -3$ and $y = 4$ into both equations.

$$2(-3) + 3(4) \stackrel{?}{=} 6 \qquad -3 + 2(4) \stackrel{?}{=} 5$$
$$6 = 6 \qquad\qquad\qquad 5 = 5 \quad \text{checks}$$

If one coefficient of a variable is not a multiple of the other coefficient of that variable, then we may have to multiply each side of *both* equations by factors that will make the coefficients of one variable opposites.

Example 13

Solve by the addition method, and check:

$$3x + 2y = 6$$
$$8x - 3y = 1$$

Solution: We choose to eliminate y, because the y coefficients are smaller than the x coefficients and they are already opposite in sign. To eliminate y, we multiply the first equation by 3 and the second by 2:

$$9x + 6y = 18$$
$$16x - 6y = 2$$

Adding, we get

$$25x = 20$$

Solving for x,

$$x = \frac{20}{25} = \frac{4}{5}$$

Substituting into the first equation,

$$3\left(\frac{4}{5}\right) + 2y = 6$$

$$2y = 6 - \frac{12}{5} = \frac{18}{5}$$

$$y = \frac{9}{5}$$

So the solution is $\left(\frac{4}{5}, \frac{9}{5}\right)$.

Check:

We substitute the solution into both equations:

$$3\left(\frac{4}{5}\right) + 2\left(\frac{9}{5}\right) \stackrel{?}{=} 6 \qquad 8\left(\frac{4}{5}\right) - 3\left(\frac{9}{5}\right) \stackrel{?}{=} 1$$

$$6 = 6 \qquad\qquad\qquad 1 = 1 \quad \text{checks}$$

Example 14

Solve, round to three significant digits, and check:

$$-3.26x + 2.04y = 15.2$$
$$5.75x + 6.42y = 3.97$$

Solution: Since it doesn't matter which variable we choose to eliminate, we choose x because its coefficients have opposite signs. We eliminate x by multiplying the first equation by 5.75 and the second equation by 3.26. We carry four (or more) significant digits and round the final answer to three significant digits at the end.

$$-18.75x + 11.73y = 87.40$$
$$\underline{18.75x + 20.93y = 12.94}$$

Adding, $\qquad\qquad\qquad 32.66y = 100.34$

$$y = 3.072$$

Substituting $y = 3.072$ into the first equation,

$$-3.26x + 2.04(3.072) = 15.2$$

Solving for x,

$$x = -2.740$$

We round to get the solution $x = -2.74$ and $y = 3.07$, and express the solution as the ordered pair $(-2.74, 3.07)$.

Check:

$$-3.26(-2.74) + 2.04(3.07) \stackrel{?}{=} 15.2 \qquad 5.75(-2.74) + 6.42(3.07) \stackrel{?}{=} 3.97$$
$$15.2 = 15.2 \qquad\qquad\qquad\qquad\quad 3.95 \approx 3.97 \quad \text{checks}$$

Note: Since we're dealing with rounded numbers, there's some uncertainty in the last digit.

Systems with No Solution

When solving certain systems algebraically, both variables may be eliminated. If we get an equation that is *never* true, like

$$0 = 2$$

then there is no solution. In the graph of such systems, the lines are parallel and there is no point of intersection.

Example 15

Solve by the addition method:

$$2x + y = 2$$
$$-6x - 3y = 4$$

Solution: To eliminate x, we multiply each side of the first equation by 3:

$$6x + 3y = 6$$

Adding the second equation, $\qquad \underline{-6x - 3y = 4}$

we get $\qquad\qquad\qquad\qquad\qquad 0 = 10$

This equation is never true, and there is no solution.

Systems with Infinite Solutions

When solving certain systems algebraically, we may get an equation that is *always true*, like

$$0 = 0 \quad \text{or} \quad 2 = 2$$

In this case, there are an infinite number of solutions. In the graph of such systems, the two lines are the same, and there are an infinite number of points on the line.

Example 16

Solve by the addition method:

$$2x + 3y = 4$$
$$-6x - 9y = -12$$

Solution: To eliminate x, we multiply the first equation by 3:

$$6x + 9y = 12$$

Adding the second equation, $\quad \dfrac{-6x - 9y = -12}{0 = 0}$

This equation is always true, so the system has an infinite number of solutions. We express the solution as:

all ordered pairs (x, y) such that $\quad 2x + 3y = 4$

Solution of a System of Linear Equations by the Addition Method

This procedure may be followed to solve a system of two linear equations in two variables using the addition method:

1. If necessary, multiply each side of one or both equations by a factor, or factors, so that the coefficients of one variable are equal in absolute value and opposite in sign.
2. Add the equations, to eliminate that variable.
3. Solve the resulting equation for the remaining variable.
4. Substitute the solution for the variable found in Step 3 into an equation and solve for the remaining variable.
5. Check by substituting the solution into both original equations.

SECTION 19-2 Exercises Addition Method

Algebraic Solution by the Addition Method

Solve by the addition method, and check.

1. $x + y = 7$
   $x - y = 3$

2. $x + y = 9$
   $4x - y = 6$

3. $2x - 3y = 5$
   $4x + 3y = 7$

4. $4x + 3y = -9$
   $-4x + y = 3$

5. $-3x + 2y = 9$
   $2x + y = 1$

6. $3x - 2y = -8$
   $-x + 4y = 6$

7. $3x + y = 8$
   $x - 5y = 0$

8. $3x + 5y = 4$
   $x - y = -6$

9. $2x + 3y = 5$
   $-3x + 2y = 12$

10. $8x + 3y = 1$
    $6x + 2y = 0$

11. $5.2x + 2.3y = 3.8$
    $3.8x - 3.4y = 4.7$

12. $-5.6x + 2.8y = -4.9$
    $3.7x + 3.5y = 5.8$

13. $12x - 3y = 8$
$4x = y + 1$

14. $2x = 3y + 6$
$14x - 21y = -3$

15. $8x = 4y + 12$
$2x - y = 3$

16. $x - 3y = 5$
$2x = 6y + 10$

17. $2x + 3y = 13$
$-3x + 2y = 0$

18. $3x - 2y = 19$
$5x + 3y = 19$

19. $x + 5 = 3x - 2y$
$4 - x = 2x + 4y$

20. $9 + (x - y) = 3x$
$2(x + y) = 6x - 18 + 5y$

21. $y - 10 = -3x$
$6x - 5 = -2x$

22. $3y = 9 + 6x$
$2x = y - 3$

23. $1.25x - 2.24y = 1.43$
$7.54x + 5.72y = 50.4$

24. $1.53x - 0.82y = 2.75$
$0.97x - 1.19y = -1.42$

25. $23.6x - 37.1y = -43.7$
$12.7x + 18.2y = 31.8$

26. $54.1x - 36.7y = 190$
$75.8x + 41.7y = 24.1$

27. $\dfrac{1}{3}x - \dfrac{1}{2}y = -\dfrac{5}{6}$
$\dfrac{1}{2}x + \dfrac{1}{3}y = 2$

28. $\dfrac{1}{2}x - \dfrac{3}{4}y = 6$
$-\dfrac{2}{3}x + \dfrac{1}{4}y = 2$

29. $2ax + by = 3$
$ax + 3by = 2$

30. $2x - 3y = a$
$3x + y = b$

31. $ax + by = c$
$dx + ey = f$

19-3

SUBSTITUTION METHOD

A second method of algebraically solving two linear equations in two variables is the *substitution method*.

Algebraic Solution by the Substitution Method

First solve either equation for one variable in terms of the other variable. Substitute this expression into the other equation to eliminate the first variable. Finish as with the addition method by solving the resulting equation for the remaining variable. Substitute that value into one of the equations and solve for the remaining variable.

> **Example 17**
>
> Solve by the substitution method, and check:
>
> $$3x - 2y = 1$$
> $$2x + y = 10$$

Solution: We may choose to solve either equation for either variable. Since the coefficient of y in the second equation is 1, this is the best choice:

$$y = 10 - 2x$$

We substitute the expression for y in the other (first) equation:

$$3x - 2(10 - 2x) = 1$$

Solving this equation for x:

$$3x - 20 + 4x = 1$$
$$7x = 21$$
$$x = 3$$

Now we solve for y by substituting $x = 3$ into either equation. Since y has already been solved for in the second equation, we use the equation:

$$y = 10 - 2x = 10 - 2(3)$$
$$y = 4$$

So the solution is $(3, 4)$.

Check:

$$3(3) - 2(4) \stackrel{?}{=} 1 \qquad 2(3) + 4 \stackrel{?}{=} 10$$
$$1 = 1 \qquad\qquad 10 = 10 \quad \text{checks}$$

In general, we wish to avoid substituting a fractional expression. However, it is sometimes impossible to do so.

―― **Example 18** ――――

Solve by substitution, and check:

$$5x + 3y = 1$$
$$3x + 4y = -6$$

Solution: In this system, no variable can be solved for without introducing a fractional expression. We choose to solve for x in the first equation:

$$x = \frac{1 - 3y}{5}$$

We substitute the right side of this equation for x into the second equation:

$$3\left(\frac{1 - 3y}{5}\right) + 4y = -6$$

To eliminate the fraction, we multiply both sides of the equation by the LCD, 5:

$$3(1 - 3y) + 20y = -30$$

Solving for y:

$$3 - 9y + 20y = -30$$
$$11y = -33$$
$$y = -3$$

Substituting $y = -3$ into the equation for x:

$$x = \frac{1 - 3y}{5} = \frac{1 - 3(-3)}{5} = \frac{10}{5}$$
$$x = 2$$

So the solution is $(2, -3)$.

Check:

$$5(2) + 3(-3) \stackrel{?}{=} 1 \qquad 3(2) + 4(-3) \stackrel{?}{=} -6$$
$$1 = 1 \qquad\qquad -6 = -6 \quad \text{checks}$$

Solution of a System of Linear Equations by the Substitution Method

This procedure may be followed to solve a system of two linear equations in two variables using the substitution method:

1. Solve either equation for one variable in terms of the other variable.
2. Substitute this expression into the other equation.
3. Solve the resulting equation for the remaining variable.
4. Substitute the solution for the variable found in Step 3 into an equation, and solve for the remaining variable.
5. Check by substituting the solution into both original equations.

A Choice of Methods

When the method of solution for a system of linear equations is not specified, you may choose among the three methods: graphical, addition, or substitution. The addition method may be used easily with any system. The substitution method is easiest when the coefficient of one of the variables is 1 or -1. Graphical solutions are approximate and can be used when exact answers are not needed or to verify algebraic results.

SECTION 19-3 | **Exercises** | **Substitution Method**

Algebraic Solution
by the Substitution
Method

Solve by the substitution method, and check.

1. $4x - 3y = 8$
   $y = 4$

2. $x - 5 = 0$
   $3x - 2y = 1$

3. $3x - y = 10$
   $x + y = 2$

4. $-2x + 5y = 22$
   $2x + y = -10$

5. $x - y = 3$
   $2x + 3y = 11$

6. $2x + y = 2$
   $6x - y = 22$

7. $5x - 8y = 6$
   $x + 5y = -3$

8. $2x - y = 3$
   $2x + y = -1$

9. $-3x + 4y = 12$
   $3x + 5y = 15$

10. $5x - 3y = 21$
    $6x + 5y = 8$

11. $2x - 3y = 4$
    $3x - 2y = -2$

12. $2x - 3y = 3$
    $3x - 2y = -3$

13. $4x - 3y = 11$
    $5x + 4y = 6$

14. $5x - 9y = 12$
    $7x - 5y = 13$

15. $3x + 2y = -1$
    $-6x - 4y = 2$

16. $5y - 2 = 2x$
    $4x = 10y - 4$

17. $3x - 6y = 1$
    $5x - 10y = 2$

18. $3x - y = -4$
    $-3x + y = 5$

19. $y = 2x + 1$
    $y = -x + 3$

20. $y = -3x + 2$
    $y = 2x - 1$

21. $2x - y = 3a$
$\quad x - 2y = -2b$

22. $rx + sy = t$
$\quad 2rx - sy = 2t$

23. $0.9x - 0.8y = 0.1$
$\quad 0.6x + 1.2y = 0.5$

24. $1.5x - 2y = -1$
$\quad 3x + 2.5y = 3.5$

25. $\dfrac{x}{2} + 2y = 22$
$\quad x - \dfrac{y}{2} = 8$

26. $\dfrac{2x}{3} + y = 18$
$\quad x - \dfrac{y}{4} = 6$

A Choice of Methods

Solve by any convenient method and check.

27. $y = 3x - 2$
$\quad y = -2x + 3$

28. $3x + 4y = 1$
$\quad x + 4y = -5$

29. $5x - 3y = -13$
$\quad x + 4y = 2$

30. $y = \dfrac{1}{3}x - 1$
$\quad y = -\dfrac{1}{3}x - 3$

31. $5x + 7y = 1$
$\quad 4x - 2y = 16$

32. $3x - 2y = 7$
$\quad 3x + 5y = 14$

Solve for x and y, by making the substitutions:

$$m = \frac{1}{x} \quad \text{and} \quad n = \frac{1}{y}$$

33. $\dfrac{12}{x} + \dfrac{6}{y} = 5$
$\quad \dfrac{4}{x} + \dfrac{3}{y} = 2$

34. $\dfrac{1}{x} + \dfrac{1}{y} = \dfrac{5}{6}$
$\quad \dfrac{1}{x} - \dfrac{1}{y} = \dfrac{13}{6}$

19-4

APPLICATIONS OF SYSTEMS OF TWO EQUATIONS

In many applications, there are two or more unknown values. To solve such problems, we must write *as many equations as there are variables*. Otherwise, it is not possible to obtain a unique numerical solution.

In general, we use the same process we used to solve applications involving one equation, except that we need two equations, since we have two variables.

Solution of Applications Involving Systems of Two Equations in Two Variables

1. Understand the written problem: Read it carefully once for a general overview and again for the details. Draw a sketch if possible.
2. Identify the unknown quantities with variables.
3. Determine any formulas you need to solve the problem. Make a table if possible, headed by the formula.
4. Write two equations relating the two variables.
5. Solve the system.
6. Answer the question(s) asked, and label your answers with the proper units.
7. Check your answer(s) in the words of the problem.

─── **Example 19** ───

A technician ordered 7 transistors and 4 resistors, for a total price of $3.15. The next week, she ordered 5 transistors and 6 resistors, for a total price of $3.35. Find the price of each.

Solution: We first identify and label the unknown quantities:

Let t = the price of a transistor, in cents
and r = the price of a resistor, in cents

We need to use the relationship, which we know from everyday experience, that

$$\text{(No. of items)} \times \text{(price/item)} = \text{cost}$$

We use the first sentence of the problem statement to write the first equation:

$$7t + 4r = 315$$

We use the second sentence in the problem to write the second equation:

$$5t + 6r = 335$$

We now have a system of two equations in two variables. We choose to solve this system using the addition method. To eliminate r, we multiply both sides of the first equation by 3, and we multiply both sides of the second equation by -2:

$$
\begin{array}{rr}
21t + 12r = & 945 \\
-10t - 12r = & -670 \\
\hline
\end{array}
$$

Adding,
$$11t \qquad = \quad 275$$

Solving for t,

$$t = 25$$

Substituting $t = 25$ into the first equation,

$$7(25) + 4r = 315$$

Solving for r,

$$r = 35$$

So each transistor was 25¢ and each resistor was 35¢.

Check:
We check back in the words of the problem.

Seven transistors at 25¢ each comes to	$7(\$.25) = \$1.75$
and four resistors at 35¢ each comes to	$4(\$.35) = \$1.40$
for a total of	$\$3.15$ checks

Five transistors at 25¢ each comes to	$5(\$.25) = \$1.25$
and six resistors at 35¢ each comes to	$6(\$.35) = \$2.10$
for a total of	$\$3.35$ checks

You will notice that some problems in this section are similar to problems presented in Chapters 11 and 18. In these problems there are two unknown quantities, but earlier we were able to use one equation in one variable by expressing the second variable in terms of the first. It may be more natural for you to solve these problems using two equations in two variables.

Example 20

A solution that is 34% silver nitrate is to be mixed with a solution that is 4% silver nitrate to form 100 cc of solution that is 7% silver nitrate. How much of each is needed?

Solution:

Let x = amount of the 34% solution (in cc)

and y = amount of the 4% solution (in cc)

We use a form of Eq. 16:

% concentration of A × amount of solution = amount of A in the solution

A table is very helpful:

	% of silver nitrate (as a decimal) ×	Amount of solution (cc) =	Amount of silver nitrate (cc)
	0.34	x	$0.34x$
	0.04	y	$0.04y$
New solution	0.07	100	0.07(100)

We calculate the third column by multiplying the first and second columns. Then we use columns two and three to write a system of two equations:

$$x + y = 100$$
$$0.34x + 0.04y = 0.07(100)$$

We choose to solve this system using the substitution method.

Solving for y,

$$y = 100 - x$$

Substituting into the second,

$$0.34x + 0.04(100 - x) = 0.07(100)$$

To clear the equation of decimals, we multiply both sides by 100,

$$34x + 4(100 - x) = 7(100)$$

Solving for x,

$$34x + 400 - 4x = 700$$
$$30x = 300$$
$$x = 10$$

Substituting $x = 10$ into the first equation,

$$10 + y = 100$$
$$y = 90$$

So we need to mix 10 cc of the 34% solution with 90 cc of the 4% solution to make 100 cc of a 7% solution.

Check:

We check back in the words of the problem:

We have 100 cc since

$$10 + 90 = 100 \quad \text{checks}$$

We have a 7% solution since

$$(34\% \text{ of } 10) + (4\% \text{ of } 90) \overset{?}{=} 7\% \text{ of } 100$$

$$3.4 + 3.6 = 7 \quad \text{checks}$$

Example 21

A motorboat travels downstream for a distance of 42.6 miles in 2.75 hours and makes the return trip in 3.25 hours. Find the rate of the motorboat in still water and the rate of the current in the river. Round to the nearest tenth of a mi/h.

Solution:

$$\text{Let} \quad m = \text{the rate of the motorboat in still water}$$
$$\text{and} \quad c = \text{the rate of the current in the river}$$

Then,

$$\text{Downstream rate} = m + c$$

$$\text{Upstream rate} = m - c$$

Using the equation

$$\text{rate} \times \text{time} = \text{Distance}$$

Once again we make a table:

	Rate (mi/h)	×	Time (h)	=	Distance (mi)
Downstream	$m + c$		2.75		$(m + c)\,2.75$
Upstream	$m - c$		3.25		$(m - c)\,3.25$

We multiply the first and second columns to find the third column. Now, using the given fact that the distance downstream is 42.6 mi and the distance upstream is 42.6 mi, we get a system of two equations:

$$(m + c)\,2.75 = 42.6$$

$$(m - c)\,3.25 = 42.6$$

We simplify the equations by dividing both sides of the first equation by 2.75 and both sides of the second by 3.25:

$$m + c = 15.49$$

$$m - c = 13.11$$

We can eliminate c by adding the equations,

$$2m = 28.60$$

Solving for m,

$$m = 14.30$$

Substituting to find c,

$$14.30 + c = 15.49$$

$$c = 15.49 - 14.30 = 1.19$$

Rounding to tenths, $m = 14.3$ and $c = 1.2$. So the rate of the motorboat is 14.3 mi/h in still water and the rate of the current is 1.2 mi/h.

Check:
Checking in the words of the problem:

Going downstream, the rate is

$$14.3 + 1.2 = 15.5 \text{ mi/h}$$

After 2.75 hours, the distance traveled is

$$d = rt = (15.5)(2.75) = 42.6 \text{ mi} \quad \text{checks}$$

Going upstream, the rate is

$$14.3 - 1.2 = 13.1 \text{ mi/h}$$

After 3.25 hours, the distance traveled is

$$d = 13.1(3.25) = 42.6 \text{ mi} \quad \text{checks}$$

SECTION 19-4 Exercises Applications of Systems of Two Equations

Write two equations in two variables, solve, and check. Round approximate answers to three significant digits.

1. A piece of pipe is 21.0 ft long. It must be cut so that one piece is 3.00 ft longer than the other. What are the lengths of the two pieces of pipe?

2. The sum of the number of teeth on two gears is 73, and their difference is 17. How many teeth are on each gear?

3. A salesman sells TVs and stereos. One day he sold 3 TVs and 4 stereos for $2530. The next day he sold 4 TVs and 3 stereos for $2510. What are the prices of a TV and a stereo?

4. While wiring a new house, an electrician and her assistant worked 25.0 and 17.0 hours, respectively, and earned $671. In wiring a second house, the electrician and her assistant worked 21.0 and 13.0 hours, respectively, and earned $547. What were the hourly wages of the electrician and the assistant?

5. A man takes a trip of 360 miles, part of the way by car at an average of 48.0 mi/h, and the rest of the way by train at an average of 64.0 mi/h. If the total travel time is 6.00 hours, how far does he travel by each mode of travel?

6. A student, traveling 655 miles on his way back to school, takes a train at an average of 70.0 mi/h part of the way, and a bus, at an average of 55.0 mi/h for the rest of the way. If the total trip was 10.0 hours, how many hours and how many miles did he travel by each method of transportation?

7. A ski lift carries a skier up a slope at the rate of 120 ft/min and she skis down from the top to the bottom on a path parallel to the lift at an average rate of 1080 ft/min. If the round trip traveling time is 20.0 minutes, how many minutes does it take her to ski down? How long is the lift?

8. An airplane on a search mission flies due east from an airport, turns and flies due west back to the airport. The plane cruises at 200 mi/h when flying east and 250 mi/h when flying west. What is the farthest point from the airport the plane can reach if it can remain in the air for 9.00 hours?

9. A canoeist paddles downstream for a distance of 12.5 miles in 4.25 hours and returns in 6.25 hours. Find the rate of the river current.

10. A fish swims 12.0 miles downstream in 2.45 hours and returns in 6.75 hours. How fast does it swim in still water?

11. An airplane travels 1860 miles in 2.00 hours with the wind and then returns in 2.50 hours traveling against the wind. Find the speed of the airplane with respect to the air and the velocity of the wind.

12. A cyclist rode 5.75 miles in 15.0 minutes with the wind, and returned in 20.0 minutes against the same wind. Find her speed (in mi/h) without a wind.

13. The velocity v of a uniformly accelerated particle is linearly related to the elapsed time t by the equation

$$v = v_0 + at$$

where v_0 is the initial velocity and a is the acceleration. In one trial, the velocity of a particle was 23.0 ft/s after 5.00 seconds, and in another trial, the velocity was 40.5 ft/s after 10.0 seconds. Find the initial velocity and the acceleration.

14. An automobile company is conducting tests on a new model of car. Starting from rest, the car accelerates at a constant rate for 11.0 s to a certain velocity. In a second test, starting at 18.0 km/h, the car accelerates at the same constant rate and reaches that certain velocity in only 8.00 s. Find that velocity and the constant acceleration (in km/h/s). (Refer to Exercise 13 if necessary.)

15. A metallurgist wishes to make 1000 kg of an alloy that is 75% copper, using a 78% copper alloy and a 63% copper alloy. How much of each does he need?

16. A creamery has a vat of milk with 2.5% butterfat, and a vat of milk with 30% butterfat. How much of each is needed to make 100 gallons of milk with 3.6% butterfat?

17. Water must be added to a solution of 75% sulfuric acid in water to make 150 cc of a 20% solution of sulfuric acid in water. How much of the 75% solution should be used and how much water should be added?

18. Water must be added to a solution of 80% nitric acid to make 840 cc of a 2.0% solution. How much of the 80% solution should be used and how much water should be added?

19. A vat contains 440 liters of gasohol (gasoline and alcohol) which is 11% alcohol. How much gasohol must be drained from the vat and replaced with gasoline to produce a gasohol mixture which is 6.5% alcohol?

20. A certain brass alloy contains 60% copper, 34% zinc, and 6% lead. How many kg of this brass alloy and of zinc must be added to 20 kg of lead to produce a new brass alloy that is 40% zinc and 8% lead?

21. A man has $8000 invested, part at 6.0% and the remainder at 10.5%. If his total income from these two investments is $678 per year, how much has he invested at each rate?

22. A sum of $20,000 is invested, part at 8.5% and part at 7.5%. If the total interest from these two investments is $1620 per year, how much is invested at each rate?

23. A woman invested $4200, part of it in bonds bearing 9.0% interest and the remainder in stocks bearing 12.0% in dividends. If she received the same income from each investment, how much was invested in each?

24. A sum of $10,000 is invested, part at 7.0% and part at 10%. The annual income from the 7.0% investment is $65 less than from the 10% investment. How much was invested in each?

25. A carpenter and a helper can do a certain job in 12 days. If the carpenter works twice as fast as the helper, how long would it take each to do the job, working alone?

26. One person can mow a lawn twice as fast as another. Together they can mow the lawn in 6.0 hours. How long would it take the slower person to mow the lawn alone?

27. During one week two machines produce a total of 180 parts, with the faster machine working for 30 hours and the slower for 20 hours. During another week, they produce 495 parts, with the faster machine working 60 hours and the slower working for 85 hours. How many parts can each produce in an hour working alone?

28. A wind generator operates for 18 hours one day and a hydro unit operates for 20 hours; together they produce 860 kWh. Another day the wind generator operated for 12 hours and the hydro unit operated for 15 hours; together they produced 615 kWh that day. Find the number of kW produced by each in one hour.

19-5   SYSTEMS OF THREE EQUATIONS

Some applications need a system of three equations in three variables to describe them. A linear equation in three variables is an equation in three variables in which no variable is of a degree greater than 1.

Example 22

The equation

$$x + 2y + 3z = 6$$

is a linear equation in three variables, since no variable has an exponent other than 1.

Example 23

The equations

$$x - y + z = 6$$
$$x + y + 2z = 5$$
$$2x + y - z = -3$$

are a *system* of three linear equations in three variables.

Algebraic Solution

To solve a system of three equations in three variables, we first eliminate one variable and get a system of two equations in two variables. Then we eliminate a variable from *that* system to get one equation in one variable. We solve that equation for the variable, and substitute back to find the second variable. Finally we substitute both these values into one of the preceding equations and solve for the last variable. You should always check your solution in all three equations.

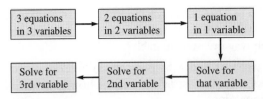

■ **FIGURE 19-8**

Solution of a System of Three Linear Equations in Three Variables

1. Choose a variable to be eliminated, and eliminate it from any pair of equations.
2. Eliminate the same variable from another pair of equations.
3. Solve this system of two equations in two variables.
4. Solve for the third variable by substituting these values into one of the preceding equations.
5. Check by substituting the solution into all three original equations.

A flowchart for the process is given in Fig. 19-8.

─── **Example 24** ───

Solve and check:

$$x - y + z = 6 \tag{1}$$
$$x + y + 2z = 5 \tag{2}$$
$$2x + y - z = -3 \tag{3}$$

Solution: We choose to eliminate y, since the coefficients of y in all three equations are 1 and -1. We eliminate y from equations (1) and (2) by adding them together:

$$
\begin{array}{r}
x - y + z = 6 \\
x + y + 2z = 5 \\
\hline
2x \qquad + 3z = 11
\end{array} \tag{4}
$$

We eliminate y from equations (2) and (3) by multiplying both sides of the second equation by -1, and adding the equations:

$$
\begin{array}{r}
-x - y - 2z = -5 \\
2x + y - z = -3 \\
\hline
x \qquad - 3z = -8
\end{array} \tag{5}
$$

Now we solve the system consisting of two new equations (4) and (5) in x and z:

$$2x + 3z = 11$$
$$x - 3z = -8$$

We solve this system by adding the equations to get:

$$3x = 3$$

Solving for x:

$$x = 1$$

We substitute $x = 1$ into equation (4):

$$2(1) + 3z = 11$$

Solving for z:

$$z = 3$$

We substitute $x = 1$ and $z = 3$ into equation (1):

$$1 - y + 3 = 6$$

Solving for y:

$$y = -2$$

So the solution is $(1, -2, 3)$.

Check:

We substitute back in all three original equations:

$$1 - (-2) + 3 \stackrel{?}{=} 6 \qquad 1 + (-2) + 2(3) \stackrel{?}{=} 5 \qquad 2(1) + (-2) - 3 \stackrel{?}{=} -3$$

$$6 = 6 \quad \text{checks} \qquad 5 = 5 \quad \text{checks} \qquad -3 = -3 \quad \text{checks}$$

Applications

Some applications require three equations in three variables.

■ FIGURE 19-9

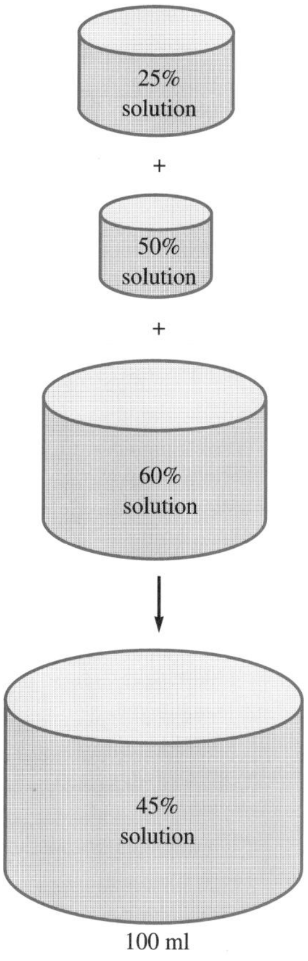

25% solution

+

50% solution

+

60% solution

↓

45% solution

100 ml

Example 25

Solutions containing 25%, 50%, and 60% sulfuric acid are mixed to produce 100 mℓ of a 45% solution of sulfuric acid (Fig. 19-9). If four times as much of the 25% solution is used as the 50% solution, how much of each is used?

Solution:

$$\text{Let} \quad x = \text{number of m}\ell \text{ of the 25\% solution}$$
$$y = \text{number of m}\ell \text{ of the 50\% solution}$$
$$z = \text{number of m}\ell \text{ of the 60\% solution}$$

The equations come from three parts of the problem statement.

1. There is a total of 100 mℓ:

$$x + y + z = 100$$

2. The sum of the acid in the parts is the same as the acid in the final solution:

$$.25x + .50y + .60z = .45(100)$$

or simplifying,

$$25x + 50y + 60z = 4500$$

3. There is four times as much 25% solution as 50% solution:

$$x = 4y$$

So the system of three equations in three unknowns is:

$$x + y + z = 100 \tag{1}$$
$$25x + 50y + 60z = 4500 \tag{2}$$
$$x = 4y \tag{3}$$

To solve, first we use (3) to substitute $4y$ for x in (1):

$$4y + y + z = 100$$

Simplifying:

$$5y + z = 100 \tag{4}$$

Now substitute $4y$ for x in (2):

$$25(4y) + 50y + 60z = 4500$$

Simplifying:

$$150y + 60z = 4500 \tag{5}$$

We now have a system of two equations in two unknowns, (4) and (5).

Solving for z in (4), we get

$$z = 100 - 5y \tag{6}$$

Substituting in (5), we get

$$150y + 60(100 - 5y) = 4500$$

Solving,

$$y = 10$$

Substituting in (3),

$$x = 4(10) = 40$$

Substituting in (6),

$$z = 100 - 5(10) = 50$$

So we need to use 40 mℓ of the 25% solution, 10 mℓ of the 50% solution, and 50 mℓ of the 60% solution to make 100 mℓ of the 45% solution.

Check:
Checking back in the words of the problem:

We "produce 100 mℓ":

$$40 + 10 + 50 = 100$$

There is "four times as much of the 25% solution" as the 50%:

$$4(10) = 40$$

It makes a "45% solution of sulfuric acid":

$$25\%(40) + 50\%(10) + 60\%(50) \overset{?}{=} 45$$

$$= 45\%(100) \quad \text{checks}$$

SECTION 19-5 Exercises Systems of Three Equations

Solve and check.

1. $x + y + z = 6$
$x - y + z = 2$
$2x - y + 3z = 6$

2. $2x + y + z = 13$
$x + 2y + z = 11$
$x + 3y + 3z = 19$

3. $x + y - z = 4$
$2x - 3y + z = 1$
$x - 4y - 2z = -7$

4. $2x - 3y + 3z = 7$
$3x + y - 2z = -11$
$5x - 2y + 4z = 11$

5. $x + 2y - z = 2$
$3x - y + 3z = -2$
$-2x - 4y + 2z = -1$

6. $3x + y - z = -1$
$2x - 4y + 2z = 1$
$-x + 2y - z = -3$

7. $x + y - 3z = 7$
$x + y + 4z = 0$
$5x + 2z = 8$

8. $x - y = -6$
$x + z = -9$
$3y + z = 1$

9. $x + 3z = -2$
$2x + y = 2$
$7y + 4z = -4$

10. $x - y = 1$
$y + z = 3$
$x - z = 2$

11. $y + z = 2$
$x - z = -1$
$x - y = 3$

12. $x - z = 1$
$x + y + z = 2$
$x + y = 1$

13. $x - y + 4z = 1$
$2x + y + 6z = 3$
$3x + 2y - 2z = 2$

14. $x - 2y - 3z = -4$
$2x + 3y + 4z = 8$
$3x + 4y - 5z = 16$

Write three equations in three variables, solve, and check.

Applications

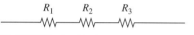

$R_1 \quad R_2 \quad R_3$

■ **FIGURE 19-10**

15. Three solutions contains 20%, 30%, and 40% nitric acid are mixed together to produce 100 mℓ of 25% solution. Twice as much of the 40% solution is used as the 30% solution. How many mℓ of each are used?

16. The three resistors in Fig. 19-10 have a total resistance of 2740 ohms. The second resistance is 200 more than the first resistance, and the third resistance is 100 less than twice the second resistance. Find the resistances.

17. In a certain electric circuit, the total of three voltages in series is 340 volts. The first voltage is 25 volts less than twice the second. The third is equal to the sum of the other two. Find each voltage.

18. A woman wants to invest $16,000, part in a checking account, part in a money market account, and the rest in a certificate of deposit. She wants twice as much in the certificate of deposit as in her checking account. She also wants $1000 more in the money market account than in the certificate of deposit. How much should be placed in each account?

19. A man has $8000 invested in three investments. On the first, he receives 8% interest, on the second, 9%, and the third, 12%. He has twice as much invested at 9% than at 8%, and receives $860 annually in interest. How much does he have in each of the three investments?

20. An assembly of a nut, a bolt, and a washer weighs 80 g. Two nuts and one bolt weigh 90 g. Three nuts and bolts with double washers weigh 270 g. Find the weight of a nut, a bolt, and a washer.

21. A dietician is planning a meal using three foods. Each ounce of food A contains 10% of the daily requirements for carbohydrates, 5% for protein, and 0% for sodium. Each ounce of food B contains 10% of the daily requirements for carbohydrates, 10% for protein, and 15% for sodium. Each ounce of food C contains 10% of the daily requirement for carbohydrates, 25% for protein, and 10% for sodium. How many ounces of each food should be used to supply 100% of the daily requirements for each nutrient?

22. An athlete enters a triathlon and runs at the rate of 8 mi/h, canoes at 2 mi/h, and bikes at 15 mi/h. She finishes the race in 3 hours 35 minutes. Her friend, who is doing the triathlon just for fun, runs at 6 mi/h, canoes at 1 mi/h, bikes at 10 mi/h, and finishes the race in 5 hours and 40 minutes. How many miles was each part (running, canoeing, biking), if the total course was 32 miles long?

19-6

SOLUTION OF SYSTEMS OF TWO LINEAR EQUATIONS BY DETERMINANTS

We have seen that systems of linear equations may be solved graphically or algebraically, using the addition method or the substitution method. But systems may also be solved using *determinants*. In this section we introduce determinants and learn how to evaluate them. Then we use determinants to solve a system of two linear equations.

To begin, our goal is to solve a general system of two linear equations:

$$a_1x + b_1y = c_1$$
$$a_2x + b_2y = c_2$$

Using the addition method, we multiply the first equation by b_2 and the second by $-b_1$:

$$a_1b_2x + b_1b_2y = b_2c_1$$
$$\underline{-a_2b_1x - b_1b_2y = -b_1c_2}$$

Adding: $\qquad (a_1b_2 - a_2b_1)x = b_2c_1 - b_1c_2$

Dividing by $a_1b_2 - a_2b_1$ gives

$$x = \frac{b_2c_1 - b_1c_2}{a_1b_2 - a_2b_1}$$

Similarly, we can solve for y although we don't show the work for this. These results can be summarized as follows:

SOLUTION OF TWO LINEAR EQUATIONS $\quad a_1x + b_1y = c_1 \quad a_2x + b_2y = c_2$	$x = \dfrac{b_2c_1 - b_1c_2}{a_1b_2 - a_2b_1}$ $y = \dfrac{a_1c_2 - a_2c_1}{a_1b_2 - a_2b_1}$ where $a_1b_2 - a_2b_1 \neq 0$	52

These equations can be used to solve any system of two linear equations.

Definitions

The denominator, $a_1b_2 - a_2b_1$, of the fractions above can be expressed using a special arrangement of the coefficients a_1, b_1, a_2, and b_2 called a *determinant*. It is denoted by the symbol

$$\begin{vmatrix} a_1 & b_1 \\ a_2 & b_2 \end{vmatrix}$$

Each number in the determinant is called an *element*. This determinant has two *rows* and two *columns* and is called a *second-order determinant*. Every second-order determinant has two diagonals. The diagonal from the upper left to lower right is called the *principal diagonal*. The diagonal from the lower left to the upper right is called the *secondary diagonal*.

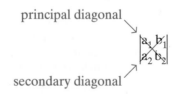
principal diagonal

secondary diagonal

Value of a Second-Order Determinant

The *value* of a second-order determinant is equal to the product of the elements on the principal diagonal minus the product of the elements on the secondary diagonal.

SECOND-ORDER DETERMINANT	$\begin{vmatrix} a_1 & b_1 \\ a_2 & b_2 \end{vmatrix} = a_1b_2 - a_2b_1$	54

Example 26

In the determinant $\begin{vmatrix} 2 & -3 \\ 1 & 5 \end{vmatrix}$ each of the numbers 2, 1, -3, and 5 are the elements.

The first row contains the elements 2 and -3; the second row contains the elements 1 and 5. The first column contains the elements 2 and 1; the second column contains the elements -3 and 5. The value of this determinant is equal to the product of the elements on the principal diagonal, 2(5), minus the product of the elements on the secondary diagonal, 1(-3).

$$\begin{vmatrix} 2 & -3 \\ 1 & 5 \end{vmatrix} = 2(5) - 1(-3) = 10 + 3 = 13$$

COMMON ERRORS Students sometimes *add* the numbers on a diagonal instead of *multiplying*. Don't do that.

Don't just ignore a zero element. It causes the product along its diagonal to be zero.

Example 27

$$\begin{vmatrix} 0 & 4 \\ 2 & -3 \end{vmatrix} = 0(-3) - 2(4) = -8$$

Solving a System of Two Equations by Determinants

We now return to the solution of the general system of two linear equations:

$$a_1x + b_1y = c_1$$
$$a_2x + b_2y = c_2$$

Recall that the solution for x is

$$x = \frac{b_2c_1 - b_1c_2}{a_1b_2 - a_2b_1}$$

We can replace the denominator in this expression by the determinant of the coefficients:

$$\begin{vmatrix} a_1 & b_1 \\ a_2 & b_2 \end{vmatrix} = a_1b_2 - a_2b_1$$

Now look at the numerator $b_2c_1 - b_1c_2$. This can be expressed as a determinant also:

$$\begin{vmatrix} c_1 & b_1 \\ c_2 & b_2 \end{vmatrix} = c_1b_2 - c_2b_1$$

$$= b_2c_1 - b_1c_2$$

So the numerator can also be expressed as a determinant. Note that the determinant for the numerator is the same as the determinant of the coefficients (the denominator) except that the column of x coefficients, a_1 and a_2, have been replaced by the column of constants, c_1 and c_2. So our solution for x can be expressed as

$$x = \frac{\begin{vmatrix} c_1 & b_1 \\ c_2 & b_2 \end{vmatrix}}{\begin{vmatrix} a_1 & b_1 \\ a_2 & b_2 \end{vmatrix}}$$

Similarly, we can develop an expression for y. We don't show the work for this, but we find that

$$y = \frac{\begin{vmatrix} a_1 & c_1 \\ a_2 & c_2 \end{vmatrix}}{\begin{vmatrix} a_1 & b_1 \\ a_2 & b_2 \end{vmatrix}}$$

We can now express the solution of a system of two linear equations in terms of determinants. This is called Cramer's Rule, after the Swiss mathematician Gabriel Cramer (1704–1752).

CRAMER'S RULE	$x = \dfrac{\begin{vmatrix} c_1 & b_1 \\ c_2 & b_2 \end{vmatrix}}{\begin{vmatrix} a_1 & b_1 \\ a_2 & b_2 \end{vmatrix}}$ and $y = \dfrac{\begin{vmatrix} a_1 & c_1 \\ a_2 & c_2 \end{vmatrix}}{\begin{vmatrix} a_1 & b_1 \\ a_2 & b_2 \end{vmatrix}}$ where $\begin{vmatrix} a_1 & b_1 \\ a_2 & b_2 \end{vmatrix} \neq 0$	56

Thus, the solution for any variable is a fraction whose denominator is the determinant of the coefficients, and whose numerator is that same determinant with the column of coefficients for the variable replaced by the column of constants.

Of course the determinant of the coefficients cannot be zero or we would get division by zero. This would indicate that the system has no unique solution.

─── **Example 28** ───

Solve by determinants:

$$3x - 2y = 4$$
$$x + 3y = 2$$

Solution: Using Eq. 56 for x and y, we get

$$x = \frac{\begin{vmatrix} 4 & -2 \\ 2 & 3 \end{vmatrix}}{\begin{vmatrix} 3 & -2 \\ 1 & 3 \end{vmatrix}} = \frac{4(3) - 2(-2)}{3(3) - 1(-2)} = \frac{16}{11}$$

and

$$y = \frac{\begin{vmatrix} 3 & 4 \\ 1 & 2 \end{vmatrix}}{\begin{vmatrix} 3 & -2 \\ 1 & 3 \end{vmatrix}} = \frac{3(2) - 1(4)}{11} = \frac{2}{11}$$

Note that we don't need to calculate the denominator a second time—we already know it's 11.

Check:

$$3\left(\frac{16}{11}\right) - 2\left(\frac{2}{11}\right) \stackrel{?}{=} 4 \qquad\qquad \frac{16}{11} + 3\left(\frac{2}{11}\right) \stackrel{?}{=} 2$$

$$\frac{48}{11} - \frac{4}{11} \stackrel{?}{=} 4 \qquad\qquad \frac{16}{11} + \frac{6}{11} \stackrel{?}{=} 2$$

$$\frac{44}{11} \stackrel{?}{=} 4 \qquad\qquad \frac{22}{11} \stackrel{?}{=} 2$$

$$4 = 4 \quad \text{checks} \qquad\qquad 2 = 2 \quad \text{checks}$$

COMMON ERROR Be sure to arrange the equations into the form

$$a_1x + b_1y = c_1$$
$$a_2x + b_2y = c_2$$

before writing the determinants.

Calculator Solution

Many calculators can also evaluate determinants. Check your calculator manual for specific instructions. On many calculators, you first enter the *dimensions* of the determinant (the dimensions are the number of rows and columns). Then enter the elements of the determinant, and press the appropriate key to evaluate the determinant.

Example 29

Use a calculator to solve this system to three significant digits.

$$2.37x - 3.54y - 4.49 = 0$$
$$3.16x + 2.34 = 2.75y$$

Solution: First we rewrite the equations so they are in the correct form, with the *x* and *y* terms on the left of the equal sign and the constants on the right:

$$2.37x - 3.54y = 4.49$$
$$3.16x - 2.75y = -2.34$$

Then we enter the determinant of the coefficients in the calculator (Fig. 19-11), and evaluate the determinant (Fig. 19-12). Now we find the numerator for *x* by entering the constants in the first column, and evaluating that determinant by calculator.

$$\begin{vmatrix} 4.49 & -3.54 \\ -2.34 & -2.75 \end{vmatrix} = -20.6311$$

By Cramer's Rule,

$$x = \frac{-20.6311}{4.6689} = -4.42$$

Now we find the numerator for *y* by entering the determinant,

$$\begin{vmatrix} 2.37 & 4.49 \\ 3.16 & -2.34 \end{vmatrix} = -19.7342$$

■ **FIGURE 19-11**

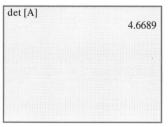

```
MATRIX[A] 2 × 2
[ 2.37    -3.54          ]
[ 3.16    -2.75          ]
```

■ **FIGURE 19-12**

```
det [A]
              4.6689
```

Once again, by Cramer's Rule

$$y = \frac{-19.7342}{4.6689} = -4.23$$

rounded to three significant digits. So the solution is $(-4.42, -4.23)$.

SECTION 19-6 Exercises Solution of Systems of Two Linear Equations by Determinants

Value of a Second-Order Determinant

Evaluate each determinant.

1. $\begin{vmatrix} 2 & 1 \\ 3 & 4 \end{vmatrix}$
2. $\begin{vmatrix} 3 & -4 \\ -1 & 7 \end{vmatrix}$
3. $\begin{vmatrix} -3 & -2 \\ -5 & -3 \end{vmatrix}$
4. $\begin{vmatrix} -2 & 1 \\ -3 & 2 \end{vmatrix}$

5. $\begin{vmatrix} 2.4 & 3.7 \\ -5.3 & 1.5 \end{vmatrix}$
6. $\begin{vmatrix} 5.67 & -3.21 \\ -3.49 & -1.38 \end{vmatrix}$
7. $\begin{vmatrix} 2.43 & 0 \\ -1.25 & -3.34 \end{vmatrix}$
8. $\begin{vmatrix} 0 & 1.67 \\ 0 & 3.93 \end{vmatrix}$

Solving a System of Two Equations by Determinants

Solve each system using determinants. Check.

9. $2x - y = 9$
   $3x + y = 11$
10. $4x - 3y = 6$
    $2x - 3y = 12$
11. $5x - 3y = 7$
    $2x - 4y = 9$

12. $x - 6y = -20$
    $2x - 7y = -17$
13. $9x - y = 7$
    $5x - 3y = -11$
14. $x - 4y = -17$
    $2x - 3y = -9$

15. $5x + 2y - 20 = 0$
    $3x - 2y - 12 = 0$
16. $4 - x + 3y = 0$
    $1 + 2x + 3y = 0$

17. $y = 3 - x$
    $y = 11 - 3x$
18. $x = 2y + 3$
    $2x = 9 + 3y$

Calculator Solution

Solve each system using determinants with a calculator. Round to three significant digits.

19. $2.45x - 5.62y = 5.71$
    $3.62x - 4.49y = 12.6$
20. $5.34x = 20.5 + 4.67y$
    $2.87y = 6.37 - 3.21x$

21. $1.25x + 3.58y = 5.41$
    $6.32x - 2.90y = 1.66$
22. $4.65x - 7.75y = 9.65$
    $6.37x - 8.61y = 1.58$

Applications

Solve by determinants. Check.

23. In a certain electric circuit, two resistors, R_1 and R_2, have a total resistance of 87 ohms when connected in series. Find the resistance of the two resistors if R_1 has a resistance of 19 ohms less than R_2.

24. The sum of two capacitors is 65 microfarads and the difference between them is 29 microfarads. What is the size of each capacitor?

25. How many pounds each of 95¢-a-pound candy and 75¢-a-pound candy should be mixed to yield 32 pounds of 90¢-a-pound candy?

26. A tea merchant prepares a blend of 24 lb of tea to sell for $2.15 per lb. For the blend, he uses two types of tea, one selling at $1.95 per lb, the other at $2.25 per lb. How much of each type should he use for the blend?

27. A dietician in a nursing home is to plan a special diet composed of two foods, A and B. Each ounce of food A contains 6 units of calcium and 2 units of iron. Each ounce of food B contains 5 units of calcium and 3 units of iron. How many ounces of foods A and B should be used to obtain a food mix that has 52 units of calcium and 24 units of iron?

28. A certain zoo animal is kept under a strict diet. He is to receive, among other things, 40 grams of protein and 12 grams of fat. The zookeeper is able to purchase two food mixes of the following compositions: Mix M has 10% protein and 6% fat; mix N has 20% protein and 2% fat. How many grams of each mix should be used to obtain the right diet for this animal?

19-7

SOLUTION OF SYSTEMS OF THREE LINEAR EQUATIONS BY DETERMINANTS

In the last section we solved a system of two linear equations using determinants. In this section we solve a system of three linear equations using determinants.

Definitions

If we algebraically solved the system of equations:

$$a_1x + b_1y + c_1z = k_1$$

$$a_2x + b_2y + c_2z = k_2$$

$$a_3x + b_3y + c_3z = k_3$$

we would get the solution:

SOLUTION OF A SYSTEM OF THREE EQUATIONS	$x = \dfrac{b_2c_3k_1 + b_1c_2k_3 + b_3c_1k_2 - b_2c_1k_3 - b_3c_2k_1 - b_1c_3k_2}{a_1b_2c_3 + a_3b_1c_2 + a_2b_3c_1 - a_3b_2c_1 - a_1b_3c_2 - a_2b_1c_3}$ $y = \dfrac{a_1c_3k_2 + a_3c_2k_1 + a_2c_1k_3 - a_3c_1k_2 - a_1c_2k_3 - a_2c_3k_1}{a_1b_2c_3 + a_3b_1c_2 + a_2b_3c_1 - a_3b_2c_1 - a_1b_3c_2 - a_2b_1c_3}$ $z = \dfrac{a_1b_2k_3 + a_3b_1k_2 + a_2b_3k_1 - a_3b_2k_1 - a_1b_3k_2 - a_2b_1k_3}{a_1b_2c_3 + a_3b_1c_2 + a_2b_3c_1 - a_3b_2c_1 - a_1b_3c_2 - a_2b_1c_3}$ where $$a_1b_2c_3 + a_3b_1c_2 + a_2b_3c_1 - a_3b_2c_1 - a_1b_3c_2 - a_2b_1c_3 \neq 0$$	53

Notice that the three denominators are identical.

The denominator

$$a_1b_2c_3 + a_3b_1c_2 + a_2b_3c_1 - a_3b_2c_1 - a_1b_3c_2 - a_2b_1c_3$$

can be expressed as the *determinant of the coefficients* of the three given equations, just as in Sec. 19-6. It is commonly expressed as

$$\begin{vmatrix} a_1 & b_1 & c_1 \\ a_2 & b_2 & c_2 \\ a_3 & b_3 & c_3 \end{vmatrix}$$

This determinant has three rows and three columns and is called a *third-order determinant*. The principal diagonal consists of the elements a_1, b_2, and c_3. The secondary diagonal consists of the elements a_3, b_2 and c_1.

Value of a Third-Order Determinant

An easy way to evaluate a third-order diagonal is to use its diagonals. Write the determinant and copy the first and second columns to the right of the determinant. Now there are two more diagonals parallel to the principal diagonal to its right. There are also two more diagonals parallel to the secondary diagonal to its right.

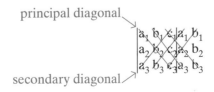

principal diagonal

secondary diagonal

To find the value of the determinant, first add the products of the elements of the principal diagonal and the products of the two parallel diagonals. Then from the sum, subtract the products of the elements of the secondary diagonal and the products of the two parallel diagonals.

THIRD-ORDER DETERMINANT	$\begin{vmatrix} a_1 & b_1 & c_1 \\ a_2 & b_2 & c_2 \\ a_3 & b_3 & c_3 \end{vmatrix} = \begin{aligned} & a_1b_2c_3 + a_3b_1c_2 + a_2b_3c_1 \\ & - a_3b_2c_1 - a_1b_3c_2 - a_2b_1c_3 \end{aligned}$	55

Example 30

Evaluate the determinant

$$\begin{vmatrix} 1 & -2 & 3 \\ -4 & 5 & 6 \\ 7 & 8 & -9 \end{vmatrix}$$

Solution: We rewrite the determinant with the first two columns written to the right, and draw all six diagonals.

$$\begin{vmatrix} 1 & -2 & 3 & 1 & -2 \\ -4 & 5 & 6 & -4 & 5 \\ 7 & 8 & -9 & 7 & 8 \end{vmatrix}$$

Adding the subtacting the products of the diagonals, we get a value of

$$(1)(5)(-9) + (-2)(6)(7) + (3)(-4)(8) - (7)(5)(3) - (8)(6)(1) - (-9)(-4)(-2)$$
$$= -45 - 84 - 96 - 105 - 48 + 72$$
$$= -306$$

Determinants by Calculator Solution

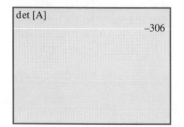

This determinant can also be evaluated by using a calculator. Check your calculator manual for specific instructions.

Example 31

Evaluate the determinant in Example 30 using a calculator.

Solution: As before, we enter the *dimensions* of the determinant (this is a 3-by-3 determinant). Then enter the elements of the determinant (Fig. 19-13), and evaluate the determinant (Fig. 19-14).

```
det [A]
                    -306
```

Solving a System of Three Equations by Determinants

We can solve a system of three linear equations by determinants just as we solved a system of two linear equations by determinants. We return to the equations for the solutions of the general system in Eq. 53. The numerator of each solution can be obtained from the determinant of the coefficients by replacing the coefficients of the variable in question with the constants k_1, k_2, k_3. The solution by determinants now becomes Cramer's Rule.

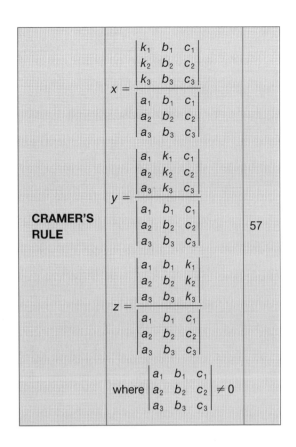

CRAMER'S RULE

$$x = \frac{\begin{vmatrix} k_1 & b_1 & c_1 \\ k_2 & b_2 & c_2 \\ k_3 & b_3 & c_3 \end{vmatrix}}{\begin{vmatrix} a_1 & b_1 & c_1 \\ a_2 & b_2 & c_2 \\ a_3 & b_3 & c_3 \end{vmatrix}}$$

$$y = \frac{\begin{vmatrix} a_1 & k_1 & c_1 \\ a_2 & k_2 & c_2 \\ a_3 & k_3 & c_3 \end{vmatrix}}{\begin{vmatrix} a_1 & b_1 & c_1 \\ a_2 & b_2 & c_2 \\ a_3 & b_3 & c_3 \end{vmatrix}}$$

$$z = \frac{\begin{vmatrix} a_1 & b_1 & k_1 \\ a_2 & b_2 & k_2 \\ a_3 & b_3 & k_3 \end{vmatrix}}{\begin{vmatrix} a_1 & b_1 & c_1 \\ a_2 & b_2 & c_2 \\ a_3 & b_3 & c_3 \end{vmatrix}}$$

where $\begin{vmatrix} a_1 & b_1 & c_1 \\ a_2 & b_2 & c_2 \\ a_3 & b_3 & c_3 \end{vmatrix} \neq 0$

57

Example 32

Solve by determinants, and use a calculator to evaluate the determinants.

$$3x - 7y + 3z = 6$$
$$3x + 3y + 6z = 1$$
$$5x - 5y + 2z = 5$$

Solution: In the calculator, we enter the determinant of the coefficients (Fig. 19-13) and evaluate it (Fig. 19-14).

By Cramer's Rule, the solution is

$$x = \frac{\begin{vmatrix} 6 & -7 & 3 \\ 1 & 3 & 6 \\ 5 & -5 & 2 \end{vmatrix}}{\begin{vmatrix} 3 & -7 & 3 \\ 3 & 3 & 6 \\ 5 & -5 & 2 \end{vmatrix}} = \frac{-40}{-150} = \frac{4}{15}$$

$$y = \frac{\begin{vmatrix} 3 & 6 & 3 \\ 3 & 1 & 6 \\ 5 & 5 & 2 \end{vmatrix}}{-150} = \frac{90}{-150} = -\frac{3}{5}$$

$$z = \frac{\begin{vmatrix} 3 & -7 & 6 \\ 3 & 3 & 1 \\ 5 & -5 & 5 \end{vmatrix}}{-150} = \frac{-50}{-150} = \frac{1}{3}$$

So the solution is $x = \dfrac{4}{15}$, $y = -\dfrac{3}{5}$, and $z = \dfrac{1}{3}$.

Check:
We use a calculator to check in all three equations.

Value of a Third-Order Determinant

Evaluate each determinant.

1.
$$\begin{vmatrix} 1 & 2 & 2 \\ 3 & -1 & 1 \\ -2 & 4 & -1 \end{vmatrix}$$

2.
$$\begin{vmatrix} -2 & 1 & 3 \\ 5 & -1 & 2 \\ -3 & 2 & 4 \end{vmatrix}$$

3.
$$\begin{vmatrix} 0 & 1 & -2 \\ 1 & 0 & -1 \\ -1 & 2 & 0 \end{vmatrix}$$

4.
$$\begin{vmatrix} 2 & 0 & -1 \\ 0 & 1 & 3 \\ -1 & -3 & -2 \end{vmatrix}$$

Evaluate by calculator. Round to tenths.

5.
$$\begin{vmatrix} 1.3 & 2.3 & -3.2 \\ -3.4 & 1.7 & -2.5 \\ 5.4 & -2.4 & 0.6 \end{vmatrix}$$

6.
$$\begin{vmatrix} 4.5 & -3.1 & 2.7 \\ 1.9 & -2.7 & -8.5 \\ -3.2 & 0.7 & 1.5 \end{vmatrix}$$

Determinants by Calculator Solution

Solve each system using determinants with a calculator. Check.

7.
$$6x - 4y - 7z = 17$$
$$9x - 7y - 16z = 29$$
$$10x - 5y - 3z = 23$$

8.
$$2x + 4y - 3z = 22$$
$$4x - 2y + 5z = 18$$
$$6x + 7y - z = 63$$

Solve by determinants. Round to tenths.

9.
$$2.4x - 1.7y - 3.2z = 4.5$$
$$1.3x - 5.4y + 3.8z = 6.4$$
$$-3.4x + 4.9y - 2.6z = 3.7$$

10.
$$-9.8x + 7.8y + 5.7z = 9.4$$
$$8.4x - 4.9y - 6.5z = 0.7$$
$$5.8x - 5.8y + 3.7z = 8.5$$

Solving a System of Three Equations by Determinants

Solve each system using determinants.

11.
$$x + y + z = 18$$
$$x - y + z = 6$$
$$x + y - z = 4$$

12.
$$x + y + z = 35$$
$$x - 2y + 3z = 15$$
$$x - y - z = 5$$

13.
$$x + y = 25$$
$$x + z = 30$$
$$y + z = 35$$

14.
$$x - y = 6$$
$$x - z = 10$$
$$y + z = 14$$

15.
$$2x + 3y + z = 2$$
$$-x + 2y + 3z = -1$$
$$-3x - 3y + z = 0$$

16.
$$2x - 2y + 3z = 5$$
$$2x + y - 2z = -1$$
$$4x - y - 3z = 0$$

Applications

Solve by determinants.

17. The three resistors in Fig. 19-15 have a total resistance of 2600 ohms when connected in series. If the resistance of R_1 is 420 ohms more than 3 times R_2, and the resistance of R_3 is 480 ohms more than R_2, find each resistance.

18. Three solutions containing 4% acid, 28% acid, and 60% acid, are to be mixed to get 1000 cc of a new solution, which is 30% acid. If the new solution is to contain twice as much of the 4% solution as of the 28% solution, how much of each should be mixed together?

■ **FIGURE 19-15**

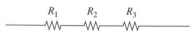
R_1 R_2 R_3

1. Determine if $(2, -1)$ is a solution of the system

$$3x - y = 7$$
$$2x + 3y = -1$$

2. Determine if $\left(\dfrac{2}{13}, -\dfrac{14}{13}\right)$ is a solution of this system:

$$5x - 3y = 4$$
$$x + 2y = -2$$

Solve these systems graphically. If the solution is an ordered pair, estimate the solution to the nearest tenth and check it.

3. $3x + y = -3$
   $x - 3y = -1$

4. $2x + 5y = 18$
   $3x + 4y = 7$

5. $2x + 3y = 5$
   $4x + 6y = 7$

6. $2x - y = 3$
   $4x - 2y = 6$

Solve by the addition method, and check.

7. $x + y = 2$
   $3x - y = 10$

8. $3x + y = -3$
   $x - 3y = -1$

9. $3x - 4y = 2$
   $5x - 7y = -5$

10. $2x + 3y = 3$
    $5x - 4y = 1$

11. $y = 5 - 2x$
    $4x = 10 - 2y$

12. $6x = 3y - 9$
    $y = 2x + 4$

Solve by the substitution method, and check.

13. $2x + y = 13$
    $3x + y = 17$

14. $2x - y = 4$
    $x + y = 5$

15. $y + 2 = 4x$
    $8x - 2y = -1$

16. $3y - 2 = 4x$
    $12x - 9y = -6$

17. $2x + 3y = 6$
    $3x + 4y = 8$

18. $3x - 5y = 7$
    $5x - 7y = 8$

Evaluate the determinant.

19. $\begin{vmatrix} -4 & 2 \\ -3 & -2 \end{vmatrix}$

20. $\begin{vmatrix} -2.6 & -3.4 \\ 5.3 & 4.6 \end{vmatrix}$

Solve by determinants, and check.

21. $x + 2y = 2$
    $3x - 2y = -10$

22. $y = x + 1$
    $y = -2x + 4$

23. $5x - 2y = 4$
    $10x - 4y = 8$

24. $4.3x - 3.2y = -6.0$
    $1.5x + 3.8y = -9.7$

Solve by any convenient method, and check:

25. $3x - 2y = 4$
    $2y = 3x - 6$

26. $2x + 7y = 1$
    $2x + 5y = 3$

27. $2x - 5y = 12$
    $2x - 3y = 12$

28. $3x + 4y = 4$
    $x - 2y = 0$

Write two equations in two variables, solve, and check.

29. The sum of two voltages is 86 volts and the difference is 14 volts. Find the two voltages.

30. A dietician in a hospital is to plan a special diet composed of the following: Each ounce of Food A contains 9 units of calcium and 4 units of iron, each ounce of food B contains 5 units of calcium and 4 units of iron, and the mix of A and B must have exactly 81 units of calcium and 48 units of iron.

31. An investor bought 80 acres of land, part at $550 per acre and part at $650 per acre. If he paid a total of $49,000, how much land was there in each part?

32. A cyclist rode 60 miles in 3 hours with the wind, and returned in 4 hours against the same wind. Find her speed without the wind and the speed of the wind.

33. How much solder containing 30% tin and 70% lead must be combined with another solder containing 60% tin and 40% lead to make 120 pounds of solder containing 50% tin and 50% lead?

34. How many cc of water must be added to a solution of 85% carbolic acid to make 850 cc of a 1.5% solution?

35. How many pounds of tea costing $0.70 per pound should be mixed with tea costing $1.50 per pound to make 32 lb of tea that sells for $1.30 a pound?

Solve and check.

36. $2x + y - 3z = -2$
$x + 3y - 2z = -5$
$3x + 2y - z = 7$

37. $x + y = -1$
$y + z = -2$
$2x - z = 4$

Evaluate the determinant.

38. $\begin{vmatrix} 3 & -2 & 4 \\ 0 & 5 & -1 \\ 2 & 0 & -3 \end{vmatrix}$

39. $\begin{vmatrix} 3.5 & -6.5 & -3.8 \\ -1.6 & 2.9 & 3.7 \\ 0 & -4.8 & -3.9 \end{vmatrix}$

Solve by determinants, and round to tenths.

40. $1.2x + 2.3y + 3.4z = 9.9$
$7.8x - 2.3y - 3.5z = 1.6$
$4.7x + 3.8y - 7.2z = 0.4$

Write three equations in three variables, solve, and check.

41. A lab technician must mix three solutions of 10%, 30%, and 60% alcohol, to make 1000 cc of a 39% solution. If the new solution must contain three times as much of the 30% solution as of the 10% solution, how much of each must she mix?

WRITING

42. We have learned four methods of solving a system of linear equations. Describe each method, including its advantages and/or disadvantages. Make up a system of two linear equations and use it to illustrate all four methods of solution. Which do you prefer, and why?

43. A certain nickel silver alloy contains:

55.90% copper	0.10% lead
31.25% zinc	12.00% nickel
0.50% tin	0.25% manganese

How many pounds of zinc, tin, lead, nickel, and manganese must be added to 400.0 kg of this alloy to make a new "leaded nickel silver" with the following composition:

44.50% copper	1.00% lead
42.00% zinc	10.00% nickel
0.50% tin	2.00% manganese

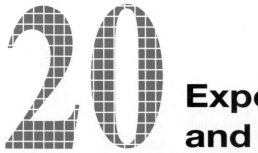

Exponents and Radicals

In this chapter we study fractional exponents, rules of radicals, simplification of radical expressions, addition, subtraction, multiplication, and division of radical expressions, and radical equations. Radical expressions need to be studied because they are solutions to many equations derived from applications.

20-1   FRACTIONAL EXPONENTS

In this section we define a quantity raised to a fractional exponent and express such a quantity in both exponential and radical form. We evaluate numerical expressions with fractional exponents. Finally, we extend the laws of integral exponents from Sec. 7-3 to fractional exponents so that we can simplify algebraic expressions with fractional exponents.

To refresh your memory, we restate the laws of integral exponents:

PRODUCT	$x^m x^n = x^{m+n}$	19
QUOTIENT	$\dfrac{x^m}{x^n} = x^{m-n}$	20
POWER	$(x^m)^n = x^{mn}$	21
PRODUCT RAISED TO A POWER	$(xy)^n = x^n y^n$	22
QUOTIENT RAISED TO A POWER	$\left(\dfrac{x}{y}\right)^n = \dfrac{x^n}{y^n}$	23
ZERO EXPONENT	$x^0 = 1$	24
NEGATIVE EXPONENT	$x^{-n} = \dfrac{1}{x^n}$	25

These laws, which we developed for integral exponents, also apply to fractional exponents, providing the bases, x and y, are positive real numbers. Therefore, in this chapter, *we will assume that variable bases represent positive real numbers.*

Radical and Exponential Forms

A radical expression can be written as a quantity raised to a fractional exponent. We make the following definition:

RADICAL AND EXPONENTIAL FORMS	$\sqrt[r]{x} = x^{1/r}$	26

The expression on the left is in *radical* form. The expression on the right is in *exponential* form.

Example 1

By Eq. 26,

$$\sqrt[3]{x} = x^{1/3}$$

$$\sqrt[4]{x} = x^{1/4}, \text{ and so forth}$$

Example 2

Express $\sqrt{x}$ in exponential form.

Solution: Recall that

$$\sqrt{x} = \sqrt[2]{x}$$

Then by Eq. 26,

$$\sqrt{x} = x^{1/2}$$

Example 3

Change $y^{-1/2}$ to radical form.

Solution: By Eqs. 25 and 26,

$$y^{-1/2} = \frac{1}{y^{1/2}} = \frac{1}{\sqrt{y}}$$

The negative sign in the exponent indicates a reciprocal. Even though the exponent is negative, the expression is not negative.

COMMON ERROR If a fractional exponent is negative, it does *not* mean that the expression is negative.

Example 4

Write $x^{2/3}$ in radical form.

Solution: By Eq. 21,

$$x^{2/3} = (x^{1/3})^2 = (\sqrt[3]{x})^2$$

Or looking at it another way:

$$x^{2/3} = (x^2)^{1/3} = \sqrt[3]{x^2}$$

We see that there are two equivalent ways of writing $x^{2/3}$:

$$x^{2/3} = (\sqrt[3]{x})^2 = \sqrt[3]{x^2}$$

In general,

FRACTIONAL EXPONENTS	$x^{p/r} = (\sqrt[r]{x})^p = \sqrt[r]{x^p}$	27

---- **Example 5** ----

Express $x^{3/4}$ in two equivalent radical forms.

Solution:

$$x^{3/4} = (\sqrt[4]{x})^3 = \sqrt[4]{x^3}$$

It follows from Eq. 27 that

RADICAL FORMS OF x	$\sqrt[r]{x^r} = x$ and $(\sqrt[r]{x})^r = x$	28

---- **Example 6** ----

$$\sqrt[3]{x^3} = x \quad \text{and} \quad (\sqrt[3]{x})^3 = x$$

Evaluation of Expressions

Numerical expressions with fractional exponents can be evaluated algebraically using the laws of exponents or numerically using a calculator. We show examples of both methods of evaluation. First we show how to evaluate an expression algebraically using the laws of exponents. This yields an exact answer, with irrational answers expressed in radical form.

---- **Example 7** ----

Evaluate $64^{1/2}$ algebraically.

Solution:

$$64^{1/2} = \sqrt{64} = \sqrt{8^2} = 8$$

---- **Example 8** ----

Evaluate $(-8)^{1/3}$ algebraically.

Solution:

$$(-8)^{1/3} = \sqrt[3]{-8} = \sqrt[3]{(-2)^3} = -2$$

This illustrates that an odd root of a negative number is always negative.

---- **Example 9** ----

$$100000^{1/5} = \sqrt[5]{100000} = \sqrt[5]{10^5} = 10$$

Example 10

$$8^{2/3} = (\sqrt[3]{8})^2 = 2^2 = 4$$

Although an expression like this, of the form $x^{p/r}$, can be evaluated using either $(\sqrt[r]{x})^p$ or $\sqrt[r]{x^p}$, it is generally easier to *take the root first* and then raise the result to the power, using $(\sqrt[r]{x})^p$.

Example 11

$$16^{-3/4} = \frac{1}{16^{3/4}} = \frac{1}{(\sqrt[4]{16})^3} = \frac{1}{2^3} = \frac{1}{8}$$

Evaluation by Calculator

Many numerical expressions with fractional exponents represent irrational numbers and these numbers can be approximated using a calculator. Powers are generally entered using the $\boxed{\wedge}$ key. Calculators give answers in decimal form, so an irrational answer can be expressed as a decimal approximation. (Some calculators also give answers in exact radical and fraction form.) When using a calculator, the use of parentheses around the fraction is very important.

Example 12

Express in radical form and approximate to three significant digits:

$$6^{2/3}$$

Solution: Changing to radical form,

$$6^{2/3} = \sqrt[3]{6^2} = \sqrt[3]{36}$$

We use a calculator to approximate the number,

$$6^{2/3} \approx 3.30$$

rounded to three significant digits.

COMMON ERROR Remember to use parentheses around a fractional exponent when entering it into the calculator.

Laws of Exponents

As stated earlier in this section, the laws of exponents also apply to fractional exponents, provided that the base is a positive real number. The following examples illustrate these laws.

Example 13

Simplify $x^{1/2}x^{1/3}$.

Solution: By Eq. 19, we add the exponents:

$$x^{1/2}x^{1/3} = x^{1/2+1/3} = x^{5/6}$$

20-1 Fractional Exponents　　**425**

Example 14

Simplify $\dfrac{y^{1/2}}{y^{1/3}}$.

Solution: By Eq. 20, we subtract the exponents:

$$\frac{y^{1/2}}{y^{1/3}} = y^{1/2-1/3} = y^{1/6}$$

Example 15

Simplify $(x^{1/2})^{4/5}$.

Solution: By Eq. 21, we multiply the exponents:

$$(x^{1/2})^{4/5} = x^{4/10} = x^{2/5}$$

Example 16

Expand $(3y)^{1/4}$.

Solution: By Eq. 22, each factor is raised to the $\dfrac{1}{4}$ power:

$$(3y)^{1/4} = 3^{1/4}y^{1/4}$$

Example 17

Expand $\left(\dfrac{x}{6}\right)^{1/3}$.

Solution: By Eq. 23, both the numerator and denominator are raised to the 1/3 power.

$$\left(\frac{x}{6}\right)^{1/3} = \frac{x^{1/3}}{6^{1/3}}$$

SECTION 20-1   Exercises   Fractional Exponents

Radical and Exponential Forms

Change to radical form. Express in two different ways if possible.

1. $x^{1/2}$	**2.** $y^{3/4}$	**3.** $y^{1/3}$	**4.** $x^{2/3}$
5. $a^{-1/2}$	**6.** $b^{-2/3}$	**7.** $(x-y)^{1/2}$	**8.** $(a+b)^{2/3}$
9. $(-c)^{2/3}$	**10.** $(-d)^{-3/5}$	**11.** $-c^{2/3}$	**12.** $-d^{-3/5}$

Change to exponential form.

13. $\sqrt{x}$	**14.** $\sqrt[3]{a}$	**15.** $\sqrt[4]{y}$	**16.** $\sqrt{z}$
17. $\sqrt[3]{b^2}$	**18.** $\sqrt[5]{x^3}$	**19.** $\dfrac{1}{\sqrt{y}}$	**20.** $\dfrac{1}{\sqrt[4]{x^3}}$
21. $\sqrt{x+y}$	**22.** $\sqrt[3]{a-b}$		

Evaluation of Expressions

Evaluate algebraically, without using a calculator.

23. $16^{1/2}$	**24.** $49^{1/2}$	**25.** $64^{1/3}$	**26.** $125^{1/3}$
27. $(-27)^{1/3}$	**28.** $(-1)^{1/3}$	**29.** $1^{1/5}$	**30.** $(10000)^{1/4}$
31. $(0.01)^{1/2}$	**32.** $(0.001)^{1/3}$	**33.** $25^{3/2}$	**34.** $27^{2/3}$
35. $81^{-1/2}$	**36.** $8^{-1/3}$	**37.** $64^{-2/3}$	**38.** $100^{-3/2}$
39. $(-32)^{4/5}$	**40.** $-32^{4/5}$	**41.** $-81^{-3/4}$	**42.** $-(-216)^{-2/3}$

Evaluation by Calculator

Evaluate using a calculator, and round to three significant digits.

43. $9^{3/2}$	**44.** $16^{3/4}$	**45.** $(-8)^{2/3}$	**46.** $(-64)^{2/3}$
47. $27^{-2/3}$	**48.** $125^{-2/3}$	**49.** $-16^{3/2}$	**50.** $-81^{-3/4}$
51. $5^{1/2}$	**52.** $(-7)^{1/3}$	**53.** $(-6)^{1/5}$	**54.** $8^{1/6}$
55. $9^{2/3}$	**56.** $12^{3/4}$	**57.** $4^{-2/3}$	**58.** $3^{-3/2}$
59. $(-2)^{-3/4}$	**60.** $-2^{-3/4}$		

Laws of Exponents

Simplify using the laws of exponents. Assume all variable bases are positive. Leave your answers in exponential form, with positive exponents.

61. $a^{1/3}a^{1/4}$	**62.** $y^{1/2}y^{2/3}$	**63.** $(b^{1/2})^{6/7}$	**64.** $(x^{2/3})^{1/2}$
65. $(4a^2)^{1/2}$	**66.** $(x^3y)^{2/3}$	**67.** $\dfrac{x^{2/3}}{x^{1/6}}$	**68.** $\dfrac{b^{1/2}}{b^{1/4}}$
69. $\left(\dfrac{x}{y^2}\right)^{3/2}$	**70.** $\left(\dfrac{a^2}{b}\right)^{1/2}$	**71.** $a^{-1/3}$	**72.** $c^{-2/3}$
73. $x \cdot x^{1/2}$	**74.** $a^{2/3} \cdot a$	**75.** $\dfrac{a}{a^{1/3}}$	**76.** $\dfrac{xy}{x^{2/3}}$
77. $\dfrac{a(a^{3/2})^2}{a^{2/3}}$	**78.** $\dfrac{(x^{-1/2})^3x}{x^{1/2}}$		

79. A gear in the shape of Fig. 20-1 can be described by the equation

$$x^{2/3} + y^{2/3} = a^{2/3}$$

Write this equation with radicals instead of fractional exponents.

80. The tensile strength S (in pounds) of a wire varies with temperature T (in °F) according to the equation

$$S = 784 - 0.0500T^{3/2}$$

Find the tensile strength when the temperature is 85.0° F.

81. If 500 bacteria double their number every hour, the equation for the number of bacteria n after t hours is

$$n = 500 \cdot 2^t$$

Evaluate the number of bacteria after $\dfrac{2}{5}$ of an hour.

82. A yeast manufacturer finds that the yeast will triple every hour. If the manufacturer starts with 100 pounds of yeast, the number of pounds of yeast n after t hours is

$$n = 100 \cdot 3^t$$

Evaluate the number of pounds of yeast after $\dfrac{3}{4}$ hour.

Applications

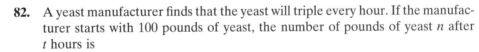

■ **FIGURE 20-1**

20-2   SIMPLIFICATION OF RADICALS

In this section, we use two rules of radicals, the root of a product and the root of a quotient, to simplify radical expressions.

Root of a Product

The root of a product is equal to the product of the roots:

ROOT OF A PRODUCT	$\sqrt[n]{xy} = \sqrt[n]{x}\sqrt[n]{y}$	29

Example 18

It follows from Eq. 29 that

$$\sqrt{xy} = \sqrt{x}\sqrt{y}$$

and similarly,

$$\sqrt[3]{xy} = \sqrt[3]{x}\sqrt[3]{y}, \text{ and so forth.}$$

Simplest Form of a Radical Expression

A radical expression is said to be *in simplest form* when

1. The radicand has been made as small as possible, and
2. There are no radicals in the denominator and no fractional radicands.

We now use the rule for the root of a product (Eq. 29) to simplify radical expressions. First we factor the radicand into factors that are rth powers, and then use Eq. 28 to eliminate these factors from the radicand. For example, to simplify a *square* root, we factor perfect *squares* from the radicand, apply the rule, and simplify.

Example 19

Write $\sqrt{18}$ in simplest form.

Solution: Since this is a *square* root, we factor a perfect *square* from the radicand, 18.

$$\sqrt{18} = \sqrt{9 \cdot 2}$$

By Eq. 29,

$$= \sqrt{9} \cdot \sqrt{2}$$

Simplifying,

$$= 3\sqrt{2}$$

Note that we could have factored 18 differently:

$$\sqrt{18} = \sqrt{6 \cdot 3}$$

Then, by Eq. 29,

$$\sqrt{18} = \sqrt{6} \cdot \sqrt{3}$$

But this cannot be simplified, since neither 6 nor 3 is a perfect square. In order to use Eq. 29 to simplify square roots, you must factor a *perfect square* from the radicand.

We also use the rule for the root of a product to simplify square roots of variable expressions, recalling from Eq. 28 that

$$\sqrt{x^2} = x$$

for all nonnegative x. We also use the fact that all even powers of x are perfect squares:

$$x^2 = (x)^2, \quad x^4 = (x^2)^2, \quad x^6 = (x^3)^2, \quad x^8 = (x^4)^2, \quad \text{and so forth}$$

Example 20

Simplify $\sqrt{x^8}$.

Solution:

$$\sqrt{x^8} = \sqrt{(x^4)^2} = x^4$$

―― **Example 21** ――

Simplify $\sqrt{8x^9}$.

Solution: Factoring out the perfect squares,

$$\sqrt{8x^9} = \sqrt{4 \cdot 2 \cdot x^8 \cdot x}$$

Rearranging, and using Eq. 29,

$$= \sqrt{4} \cdot \sqrt{x^8} \cdot \sqrt{2x}$$

Simplifying,

$$= 2 \cdot x^4 \cdot \sqrt{2x}$$
$$= 2x^4\sqrt{2x}$$

To simplify a cube root, we factor out the perfect cubes:

$$8, \quad 27, \quad 64, \quad 125, \quad 256, \quad \dots$$

The perfect cubes of x are

$$x^3, \quad x^6, \quad x^9, \quad x^{12}, \quad x^{15}, \quad \dots \quad \text{(exponents are all multiples of 3)}$$

Then we apply Eq. 28, using the fact that

$$\sqrt[3]{x^3} = x$$

for all x, and simplify.

―― **Example 22** ――

Simplify $\sqrt[3]{8x^6}$.

Solution:

$$\sqrt[3]{8x^6} = \sqrt[3]{8} \cdot \sqrt[3]{x^6}$$
$$= 2x^2$$

―― **Example 23** ――

Simplify $\sqrt[3]{16x^8}$.

Solution: Factoring out the perfect cubes, we have

$$\sqrt[3]{16x^8} = \sqrt[3]{8 \cdot 2 \cdot x^6 \cdot x^2}$$
$$= \sqrt[3]{8} \cdot \sqrt[3]{x^6} \cdot \sqrt[3]{2x^2}$$
$$= 2x^2\sqrt[3]{2x^2}$$

―― **Example 24** ――

Simplify $\sqrt[5]{32x^{11}}$.

Solution: Factoring out the fifth powers, we have

$$\sqrt[5]{32x^{11}} = \sqrt[5]{2^5 \cdot x^{10} \cdot x}$$
$$= \sqrt[5]{2^5} \cdot \sqrt[5]{x^{10}} \cdot \sqrt[5]{x}$$
$$= 2 \cdot \sqrt[5]{(x^2)^5} \cdot \sqrt[5]{x}$$
$$= 2 \cdot x^2 \cdot \sqrt[5]{x}$$
$$= 2x^2\sqrt[5]{x}$$

Rationalizing the Denominator

Simplifying radical expressions also requires that we write a given expression without any radicals in the denominator. This is called *rationalizing the denominator.* To do this, we use the rule for the root of a quotient.

Root of a Quotient

The root of a quotient is equal to the quotient of the roots:

ROOT OF A QUOTIENT	$\sqrt[r]{\dfrac{x}{y}} = \dfrac{\sqrt[r]{x}}{\sqrt[r]{y}}$	30

Note that if r is even, x must be nonnegative and y must be positive.

Example 25

$$\sqrt[3]{\frac{x}{y}} = \frac{\sqrt[3]{x}}{\sqrt[3]{y}}$$

Example 26

Simplify $\sqrt[3]{\dfrac{5}{8x^3}}$.

Solution: By Eq. 30,

$$\sqrt[3]{\frac{5}{8x^3}} = \frac{\sqrt[3]{5}}{\sqrt[3]{8x^3}} = \frac{\sqrt[3]{5}}{2x}$$

Since the denominator in the example above was a perfect cube, it was rationalized immediately after we applied Eq. 30. But frequently the radicand is *not* a perfect power, and we must make it a perfect power to rationalize it. If the denominator is a square root, multiply the numerator and denominator of the fraction by a quantity that will make the radicand in the denominator a perfect square.

Example 27

Simplify $\dfrac{3}{\sqrt{2}}$.

Solution:

$$\frac{3}{\sqrt{2}} = \frac{3}{\sqrt{2}} \cdot \frac{\sqrt{2}}{\sqrt{2}} = \frac{3\sqrt{2}}{2}$$

Here we multiply the numerator and denominator by $\sqrt{2}$ to make the denominator rational.

---

430 CHAPTER 20 / Exponents and Radicals

Example 28

Simplify $\sqrt{\dfrac{2}{3}}$.

Solution:

$$\sqrt{\frac{2}{3}} = \sqrt{\frac{2 \cdot 3}{3 \cdot 3}} = \frac{\sqrt{6}}{\sqrt{9}} = \frac{\sqrt{6}}{3}$$

Here we multiply both numerator and denominator *under* the radical sign by the smallest quantity that will make the denominator rational.

Example 29

Simplify:

$$\sqrt{\frac{1}{2x}} = \sqrt{\frac{1 \cdot 2x}{2x \cdot 2x}} = \frac{\sqrt{2x}}{2x}$$

To rationalize the denominator of a cube root, multiply by the *smallest* quantity that will make the denominator a perfect cube.

Example 30

$$\sqrt[3]{\frac{2a}{9b}} = \sqrt[3]{\frac{2a \cdot 3b^2}{9b \cdot 3b^2}} = \frac{\sqrt[3]{6ab^2}}{\sqrt[3]{27b^3}} = \frac{\sqrt[3]{6ab^2}}{3b}$$

In a more complicated problem we may be required to both remove square factors *and* rationalize the denominators. It is best to remove the square factors first, and then rationalize the denominator.

Example 31

$$\sqrt{\frac{y}{12x^3}} = \sqrt{\frac{y}{4 \cdot 3 \cdot x^2 \cdot x}} = \frac{1}{2x}\sqrt{\frac{y}{3x}}$$

$$= \frac{1}{2x}\sqrt{\frac{y \cdot 3x}{3x \cdot 3x}} = \frac{\sqrt{3xy}}{2x \cdot 3x} = \frac{\sqrt{3xy}}{6x^2}$$

It is important to note that while we have rules for the root of a product and of a quotient, there are *no rules* for the root of a *sum* or *difference*.

> **COMMON ERROR** The root of a sum is *not* equal to the sum of the roots.
>
> $$\sqrt[n]{x+y} \neq \sqrt[n]{x} + \sqrt[n]{y}$$

Example 32

(a) $\sqrt{x+y} \neq \sqrt{x} + \sqrt{y}$

(b) $\sqrt[3]{x-y} \neq \sqrt[3]{x} - \sqrt[3]{y}$

(c) $\sqrt{x^2 - 4} \neq x - 2$

---- **Example 33** ----

Show that

$$\sqrt{16 + 9} \neq \sqrt{16} + \sqrt{9}$$

Solution: The left side,

$$\sqrt{16 + 9} = \sqrt{25} = 5$$

but the right side

$$\sqrt{16} + \sqrt{9} = 4 + 3 = 7$$

Since $5 \neq 7$, then $\sqrt{16 + 9} \neq \sqrt{16} + \sqrt{9}$.

SECTION 20-2 Exercises Simplification of Radicals

Root of a Product

Write in simplest form.

1. $\sqrt{12}$	**2.** $\sqrt{75}$	**3.** $\sqrt{72}$	**4.** $\sqrt{128}$
5. $\sqrt[3]{24}$	**6.** $\sqrt[3]{32}$	**7.** $\sqrt[5]{64}$	**8.** $\sqrt[4]{162}$
9. $\sqrt{9x^4}$	**10.** $\sqrt{25a^6}$	**11.** $\sqrt{24b^5c^2}$	**12.** $\sqrt{40m^7n^4}$
13. $\sqrt[3]{27x^4}$	**14.** $\sqrt[3]{64x^8}$	**15.** $\sqrt[3]{16a^7}$	**16.** $\sqrt[3]{54b^5}$
17. $\sqrt[4]{81x^5}$	**18.** $\sqrt[5]{243x^7}$	**19.** $\sqrt[6]{64x^7}$	**20.** $\sqrt[4]{16x^9y}$

Root of a Quotient

Write in simplest form.

21. $\sqrt{\dfrac{5}{9}}$	**22.** $\sqrt[3]{\dfrac{3}{8}}$	**23.** $\dfrac{2}{\sqrt{3}}$	**24.** $\dfrac{2}{\sqrt[3]{9}}$
25. $\sqrt{\dfrac{4}{5}}$	**26.** $\sqrt{\dfrac{5}{6}}$	**27.** $\sqrt[3]{\dfrac{2}{3}}$	**28.** $\sqrt[3]{\dfrac{7}{9}}$
29. $\sqrt{\dfrac{7}{4x^2}}$	**30.** $\sqrt{\dfrac{5}{9y^4}}$	**31.** $\sqrt[3]{\dfrac{2y}{27x^3}}$	**32.** $\sqrt[3]{\dfrac{a^2}{64}}$
33. $\dfrac{-3}{\sqrt{5x}}$	**34.** $\dfrac{1}{\sqrt[3]{4x}}$	**35.** $\sqrt{\dfrac{1}{3x}}$	**36.** $\sqrt{\dfrac{2}{5a}}$
37. $\sqrt{\dfrac{5y}{6}}$	**38.** $\sqrt{\dfrac{3a}{2}}$	**39.** $\sqrt[3]{\dfrac{2}{25x}}$	**40.** $\sqrt[3]{\dfrac{3}{4a^2}}$
41. $\sqrt{\dfrac{a}{28b^3}}$	**42.** $\sqrt{\dfrac{7}{18x^5}}$	**43.** $\sqrt{\dfrac{x^3}{50y}}$	**44.** $\sqrt{\dfrac{a^6}{200}}$
45. $\sqrt{\dfrac{2y^4}{3x^3}}$	**46.** $\sqrt{\dfrac{9x^5}{5y^3}}$	**47.** $\sqrt{\dfrac{4a^2c^3}{3b}}$	**48.** $\sqrt{\dfrac{25x^6}{6y^5}}$

Applications

49. The radius of a circle having area A is

$$\sqrt{\dfrac{A}{\pi}}$$

Simplify this expression.

50. The radius of a sphere having volume V is

$$\sqrt[3]{\dfrac{3V}{4\pi}}$$

Simplify this expression.

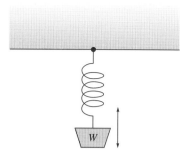

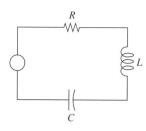

■ **FIGURE 20-2**　　　　　■ **FIGURE 20-3**

51. The natural frequency of a body under simple harmonic motion (Fig. 20-2) is

$$\frac{1}{2\pi}\sqrt{\frac{kg}{W}}$$

where W is the weight of the body and g is the acceleration due to gravity. Simplify this expression.

52. The resonant frequency for the circuit in Fig. 20-3 is

$$\sqrt{\frac{1}{LC}}$$

where L is the inductance and C is the capacitance of the circuit. Simplify this expression.

53. Factor the radicand and simplify.

　(a) $\sqrt{x^4 + x^2}$　　　**(b)** $\sqrt{a^2 + 2ab + b^2}$

54. Simplify by changing to exponential form and reducing the index. Express the result in radical form.

　(a) $\sqrt[4]{x^2}$　　　**(b)** $\sqrt[6]{4y^2}$

20-3　ADDING AND SUBTRACTING RADICALS

In this section we learn how to add and subtract similar radical expressions.

Similar Radical Expressions

Only *similar* (or *like*) radical expressions can be added or subtracted, just as with any algebraic expressions. Two radical expressions are *similar* if they have the same radicand and the same index.

Example 34 ───────────

(a) $\sqrt{3}$ and $\sqrt{2}$ are *not* similar expressions, because they have different radicands.

(b) $3\sqrt{5}$ and $-4\sqrt{5}$ are similar expressions, since they have the same radicand, 5, and the same index, 2.

(c) $\sqrt[3]{5}$ and $\sqrt{5}$ are not similar expressions, since they have different indices.

(d) $-4\sqrt[3]{a}$ and $\sqrt[3]{a}$ are similar expressions.

Combining Radical Expressions

As with algebraic expressions, we add or subtract similar radical expressions by *combining the coefficients* of the radicals.

Example 35

$$3\sqrt{2} + 4\sqrt{2} - 5\sqrt{2} = (3 + 4 - 5)\sqrt{2} = 2\sqrt{2}$$

Note that this is just an application of the distributive property; we factor out the similar radical parts. Since this is a numerical expression, we could approximate it using a calculator:

$$2\sqrt{2} \approx 2.83$$

but in general we leave the result in its radical form.

Example 36

$$6\sqrt[3]{x} - 2\sqrt[3]{x} + \sqrt[3]{x} = 6\sqrt[3]{x} - 2\sqrt[3]{x} + 1\sqrt[3]{x}$$
$$= (6 - 2 + 1)\sqrt[3]{x}$$
$$= 5\sqrt[3]{x}$$

It's very important to note that $\sqrt[3]{x} = 1\sqrt[3]{x}$.

Not all radical expressions can be combined algebraically.

Example 37

Combine, if possible, and approximate to three significant digits:

$$\sqrt{3} + \sqrt{2}$$

Solution: These radicals are not similar, and cannot be combined algebraically. Remember that

$$\sqrt{3} + \sqrt{2} \neq \sqrt{3 + 2}$$

The algebraic sum can only be *indicated,* as given. The terms cannot be combined into a single term. However, this numerical expression can be approximated using a calculator:

$$\sqrt{3} + \sqrt{2} \approx 3.15$$

rounded to three significant digits.

COMMON ERROR Remember that the sum of two roots is *not* equal to the root of the sum!

$$\sqrt[n]{x} + \sqrt[n]{y} \neq \sqrt[n]{x + y}$$

Example 38

Combine $2\sqrt{3x} - 5\sqrt{2x} - 7\sqrt{3x} + 8\sqrt{2x}$.

Solution: Rearranging the terms, we get

$$2\sqrt{3x} - 5\sqrt{2x} - 7\sqrt{3x} + 8\sqrt{2x} = (2\sqrt{3x} - 7\sqrt{3x}) + (-5\sqrt{2x} + 8\sqrt{2x})$$

Combining,
$$= -5\sqrt{3x} + 3\sqrt{2x}$$

Radicals that do not appear to be similar at first may turn out to be similar after simplification. So, we express each radical in simplest form, and then combine any similar terms.

Example 39

Combine algebraically.

$$2\sqrt{48} - \sqrt{80} + \sqrt{108}$$

Solution:

$$2\sqrt{48} - \sqrt{80} + \sqrt{108} = 2\sqrt{16 \cdot 3} - \sqrt{16 \cdot 5} + \sqrt{36 \cdot 3}$$
$$= 2(4)\sqrt{3} - 4\sqrt{5} + 6\sqrt{3}$$
$$= 8\sqrt{3} - 4\sqrt{5} + 6\sqrt{3}$$

Combining,
$$= 14\sqrt{3} - 4\sqrt{5}$$

Example 40

Simplify and combine:

$$4a\sqrt[3]{16a} - 2\sqrt[3]{54a^4}$$

Solution:

$$4a\sqrt[3]{16a} - 2\sqrt[3]{54a^4} = 4a\sqrt[3]{8 \cdot 2a} - 2\sqrt[3]{27 \cdot 2 \cdot a^3 \cdot a}$$
$$= 4a(2)\sqrt[3]{2a} - 2(3)(a)\sqrt[3]{2a}$$
$$= 8a\sqrt[3]{2a} - 6a\sqrt[3]{2a}$$
$$= 2a\sqrt[3]{2a}$$

To simplify radical expressions with fractions, first rationalize the denominators of each term, and then combine terms using the least common denominator.

Example 41

Combine algebraically.

$$4\sqrt{\frac{2}{3}} - 5\sqrt{\frac{1}{6}}$$

Solution: First we rationalize denominators:

$$4\sqrt{\frac{2}{3}} - 5\sqrt{\frac{1}{6}} = 4\sqrt{\frac{2}{3} \cdot \frac{3}{3}} - 5\sqrt{\frac{1}{6} \cdot \frac{6}{6}}$$
$$= \frac{4\sqrt{6}}{3} - \frac{5\sqrt{6}}{6}$$

Getting a common denominator of 6,

$$= \frac{4\sqrt{6}}{3} \cdot \frac{2}{2} - \frac{5\sqrt{6}}{6}$$

Simplifying,

$$= \frac{8\sqrt{6}}{6} - \frac{5\sqrt{6}}{6}$$
$$= \frac{3\sqrt{6}}{6}$$
$$= \frac{\sqrt{6}}{2}$$

┌── **Example 42** ──────────────┐

Combine:

$$\sqrt{\frac{5}{2x}} - \sqrt{\frac{2}{3x}} = \sqrt{\frac{5}{2x} \cdot \frac{2x}{2x}} - \sqrt{\frac{2}{3x} \cdot \frac{3x}{3x}}$$

Rationalizing each denominator,

$$= \frac{\sqrt{10x}}{2x} - \frac{\sqrt{6x}}{3x}$$

Getting the common denominator of $6x$,

$$= \frac{\sqrt{10x}}{2x} \cdot \frac{3}{3} - \frac{\sqrt{6x}}{3x} \cdot \frac{2}{2}$$

Combining,

$$= \frac{3\sqrt{10x}}{6x} - \frac{2\sqrt{6x}}{6x}$$

$$= \frac{3\sqrt{10x} - 2\sqrt{6x}}{6x}$$

In summary,

To Add or Subtract Radicals

1. Simplify each radical expression.
2. Add or subtract similar radicals by combining their coefficients.

SECTION 20-3 | **Exercises** | **Adding and Subtracting Radicals**

Combine algebraically. Express the answer in radical form.
1. $5\sqrt{3} - 2\sqrt{3} + \sqrt{3}$
2. $6\sqrt{5} + 3\sqrt{5} - \sqrt{5}$
3. $2\sqrt[3]{2} - \sqrt[3]{2} + 3\sqrt[3]{2}$
4. $\sqrt[4]{3} - 5\sqrt[4]{3} + 7\sqrt[4]{3}$
5. $\sqrt{8} - \sqrt{18} + \sqrt{32}$
6. $\sqrt{12} - \sqrt{48} - \sqrt{27}$
7. $\sqrt{75} + 3\sqrt{72} - 2\sqrt{27}$
8. $4\sqrt{80} - 2\sqrt{32} - 3\sqrt{125} + \sqrt{72}$
9. $5\sqrt{150} - 2\sqrt{500} - 3\sqrt{180} - \sqrt{600}$
10. $\sqrt[3]{108} - 2\sqrt[3]{32}$
11. $8 + \sqrt{8} + \sqrt[3]{8} + \sqrt[4]{8}$
12. $16 + \sqrt{16} + \sqrt[3]{16} + \sqrt[4]{16}$
13. $\sqrt{\frac{1}{2}} + \sqrt{\frac{1}{8}}$
14. $\sqrt{\frac{2}{3}} - \sqrt{\frac{2}{27}}$

Combine algebraically.
15. $5\sqrt{x} - 7\sqrt{x} + 3\sqrt{x}$
16. $-2\sqrt[3]{a} + 9\sqrt[3]{a} - 4\sqrt[3]{a}$
17. $2\sqrt{3x} - 4\sqrt{2x} - 5\sqrt{3x} + 7\sqrt{2x}$
18. $\sqrt{10a} + \sqrt{15a} - 2\sqrt{10a} + 3\sqrt{15a}$
19. $2\sqrt{50x} + 4\sqrt{45x} - \sqrt{98x}$
20. $3a\sqrt{98a} - 5\sqrt{24a^3} + 2a\sqrt{72a} - 4\sqrt{54a^3}$
21. $\sqrt{10x} + \sqrt{100x} + \sqrt{1000x} + \sqrt{10000x}$
22. $2\sqrt[3]{24x^4} - x\sqrt[3]{81x}$
23. $x\sqrt{3x^3} + x^2\sqrt{12x} - \sqrt{27x^5}$
24. $\sqrt{2y^2} - y\sqrt{18} + \sqrt{50y}$
25. $3\sqrt{\frac{2x}{5}} - \sqrt{\frac{x}{10}} + 2\sqrt{\frac{5x}{2}}$
26. $4\sqrt{\frac{y}{6}} - 3\sqrt{\frac{2y}{3}} - 2\sqrt{\frac{3y}{2}}$
27. $\sqrt[3]{\frac{1}{2y}} - \sqrt[3]{\frac{1}{16y}}$
28. $\sqrt[3]{\frac{3}{4x}} + \sqrt[3]{\frac{2}{9x}}$
29. $27^{1/3} - 16^{1/4} + 25^{-1/2} + 8^{2/3}$
30. $4^{-3/2} + 8^{1/3} - 81^{1/4} + 49^{1/2}$

MULTIPLYING RADICALS

In this section we learn how to multiply radicals with the same index. Then we multiply multinomials with radicals.

Multiplying Radicals with the Same Index

Radicals having the *same index r* can be multiplied using Eq. 29:

$$\sqrt[r]{x} \cdot \sqrt[r]{y} = \sqrt[r]{xy}$$

Example 43

(a) $\sqrt{2} \cdot \sqrt{3} = \sqrt{6}$

(b) $\sqrt[3]{5} \cdot \sqrt[3]{x} = \sqrt[3]{5x}$

(c) $\sqrt{x} \cdot \sqrt{x} = (\sqrt{x})^2 = x$

Radical expressions having the same index may be multiplied just like algebraic expressions. The coefficient of the product is the product of the coefficients. The radicands are multiplied and the resulting radical expression should be simplified if possible.

Example 44

(a) $2\sqrt{3} \cdot 3\sqrt{5} = 2(3)\sqrt{3 \cdot 5} = 6\sqrt{15}$

(b) $5\sqrt{3} \cdot 2\sqrt{3} = 5(2)(\sqrt{3})^2 = 5(2)3 = 30$

Example 45

(a) $3\sqrt{a} \cdot 5\sqrt{b} = 3(5)\sqrt{a \cdot b} = 15\sqrt{ab}$

(b) $2x\sqrt{5y} \cdot 3x\sqrt{2y} = 2x \cdot 3x \cdot \sqrt{5y \cdot 2y}$
$$= 6x^2\sqrt{10y^2}$$
$$= 6x^2\sqrt{10}\sqrt{y^2}$$
$$= 6x^2y\sqrt{10}$$

It is very important to note that in the examples above, we are multiplying two radicals with the *same index*. Eq. 29 holds only for radicals with the *same index*.

Multiplying Multinomials Containing Radicals

Recall from Sec. 8-2 that a *multinomial* is an algebraic expression having more than one term. Multinomials containing radicals are multiplied like other algebraic expressions, using the distributive property. So, we follow the same general procedure that we used in multiplying polynomials with variables.

Example 46

$$\sqrt{2}(4 - 5\sqrt{3}) = \sqrt{2}(4) - \sqrt{2}(5\sqrt{3})$$
$$= 4\sqrt{2} - 5\sqrt{6}$$

Example 47

$$2\sqrt{3x}(5\sqrt{6} - 4\sqrt{x}) = 2\sqrt{3x}(5\sqrt{6}) - 2\sqrt{3x}(4\sqrt{x})$$
$$= 10\sqrt{18x} - 8\sqrt{3x^2}$$
$$= 10\sqrt{9\cdot 2x} - 8\sqrt{x^2\cdot 3}$$
$$= 10\cdot 3\sqrt{2x} - 8x\sqrt{3}$$
$$= 30\sqrt{2x} - 8x\sqrt{3}$$

The product of two binomials may be found by multiplying vertically or horizontally, as illustrated below.

Example 48

Multiply $(\sqrt{x} + 5\sqrt{y})(4\sqrt{x} - \sqrt{y})$.

Solution: Multiplying vertically:

$$\begin{array}{r} \sqrt{x} + 5\sqrt{y} \\ 4\sqrt{x} - \sqrt{y} \\ \hline -\sqrt{xy} - 5\cdot y \\ 4\cdot x + 20\sqrt{xy} \\ \hline 4x + 19\sqrt{xy} - 5y \end{array}$$

Alternate Solution: Multiplying horizontally:

$$(\sqrt{x} + 5\sqrt{y})(4\sqrt{x} - \sqrt{y}) = \sqrt{x}\cdot 4\sqrt{x} - \sqrt{x}\cdot\sqrt{y} + 5\sqrt{y}\cdot 4\sqrt{x} - 5\sqrt{y}\cdot\sqrt{y}$$
$$= 4\cdot x - \sqrt{xy} + 20\sqrt{xy} - 5\cdot y$$
$$= 4x + 19\sqrt{xy} - 5y$$

Recall from Sec. 8-3 that this method is also called the FOIL method.

Example 49

Multiply $(3 + \sqrt{a})(3 - \sqrt{a})$.

Solution: Using FOIL,

$$(3 + \sqrt{a})(3 - \sqrt{a}) = 9 - 3\sqrt{a} + 3\sqrt{a} - a$$
$$= 9 - a$$

Conjugates

In Example 49, the product of two binomials with radicals was an expression without radicals. This happened because $(3 + \sqrt{a})$ and $(3 - \sqrt{a})$ are alike except for the sign of one term. Expressions such as these are called *conjugates* of each other. A conjugate of a binomial is a binomial that differs from the original binomial only in the sign of one term, usually the second term.

Example 50

The conjugate of $a + b$ is $a - b$,

The conjugate of $x - y$ is $x + y$, and

The conjugate of $\sqrt{a} + \sqrt{b}$ is $\sqrt{a} - \sqrt{b}$.

When we multiply two conjugates, the middle terms drop out, and the product has no radical in it!

Example 51
$$(\sqrt{a} + \sqrt{b})(\sqrt{a} - \sqrt{b}) = (\sqrt{a})^2 - \sqrt{a}\sqrt{b} + \sqrt{b}\sqrt{a} - (\sqrt{b})^2$$
$$= a - b$$

Example 52

Multiply $\sqrt{3} - \sqrt{x}$ by its conjugate.

Solution: The conjugate of $\sqrt{3} - \sqrt{x}$ is $\sqrt{3} + \sqrt{x}$. Multiplying, we have
$$(\sqrt{3} - \sqrt{x})(\sqrt{3} + \sqrt{x}) = \sqrt{3}^2 - \sqrt{x}^2$$
$$= 3 - x$$

SECTION 20-4 Exercises Multiplying Radicals

Multiplying Radicals with the Same Index

Multiply and simplify.

1. $\sqrt{3} \cdot \sqrt{5}$
2. $\sqrt{7} \cdot \sqrt{2}$
3. $(\sqrt{x})(-\sqrt{2})$
4. $\sqrt{5} \cdot \sqrt{2y}$
5. $\sqrt[3]{2} \cdot \sqrt[3]{3x^2}$
6. $(\sqrt[4]{3x})(\sqrt[4]{5})$
7. $\sqrt{3y} \cdot \sqrt{3y}$
8. $\sqrt{7x} \cdot \sqrt{7x}$
9. $3\sqrt{2} \cdot 4\sqrt{3}$
10. $5\sqrt{5x} \cdot 4\sqrt{2}$
11. $2\sqrt{3x} \cdot 5\sqrt{3x}$
12. $3\sqrt{2y} \cdot 4\sqrt{2y}$
13. $\sqrt[3]{6} \cdot \sqrt[3]{4}$
14. $2\sqrt[3]{18} \cdot \sqrt[3]{3}$
15. $(3\sqrt{5a})^2$
16. $(2\sqrt[3]{4y})^2$
17. $2\sqrt{10a} \cdot \sqrt{5a}$
18. $\sqrt{3x} \cdot 4\sqrt{6}$
19. $2\sqrt{15x} \cdot 3\sqrt{6}$
20. $5\sqrt{14x} \cdot 4\sqrt{10x}$
21. $3x\sqrt{2y} \cdot 2x\sqrt{5y}$
22. $4\sqrt{3x} \cdot 3y\sqrt{2x}$
23. $4\sqrt{6x} \cdot 3\sqrt{2x}$
24. $5x\sqrt{6x} \cdot 2\sqrt{15x}$
25. $(2\sqrt{5})(-4\sqrt{3})(3\sqrt{5})$
26. $(-3\sqrt{6})(2\sqrt{2})(5\sqrt{3})$

Multiplying Multinomials Containing Radicals

27. $\sqrt{3}(5 - 2\sqrt{2})$
28. $\sqrt{7}(\sqrt{2} + 3)$
29. $3\sqrt{2x}(4\sqrt{2} + 6\sqrt{x})$
30. $5\sqrt{3x}(2\sqrt{6} - 3\sqrt{2x})$
31. $2\sqrt{3}(4 - 2\sqrt{3} + \sqrt{6})$
32. $3\sqrt{5}(\sqrt{10} - 3 - 2\sqrt{5})$
33. $\sqrt{a}(\sqrt{2b} - \sqrt{ab})$
34. $\sqrt{3x}(\sqrt{6x} - 2\sqrt{xy})$
35. $(2x + \sqrt{3})(3x - \sqrt{3})$
36. $(\sqrt{2} - 5)(\sqrt{2} - 3)$
37. $(\sqrt{5} - \sqrt{2})(2\sqrt{5} + \sqrt{2})$
38. $(x\sqrt{3} + 2\sqrt{5})(4x\sqrt{3} + \sqrt{5})$
39. $(\sqrt{x} - \sqrt{3})(2\sqrt{x} - \sqrt{3})$
40. $(\sqrt{a} + \sqrt{b})(\sqrt{a} - 2\sqrt{b})$
41. $(\sqrt{5} + 2a)(\sqrt{5} - 2a)$
42. $(2\sqrt{3} - \sqrt{x})(2\sqrt{3} + \sqrt{x})$
43. $(\sqrt{x} - \sqrt{y})(\sqrt{x} + \sqrt{y})$
44. $(3\sqrt{2} + 4\sqrt{5})(3\sqrt{2} - 4\sqrt{5})$
45. $(\sqrt{2y} + \sqrt{3})^2$
46. $(\sqrt{5x} - \sqrt{7})^2$
47. $(\sqrt{3} - 2\sqrt{5})(\sqrt{2} + 3)$
48. $(3\sqrt{2} - 4\sqrt{3})(2\sqrt{5} - \sqrt{7})$

49. Multiply these monomials with different indices and different bases. (Hint: Change the following factors to exponential form, express the exponents with their LCD, use Eq. 21 to "factor" out the common denominator, and change back to radical form.)

   (a) $(\sqrt{6})(\sqrt[3]{5})$ **(b)** $(\sqrt[3]{5})(\sqrt[4]{2})$

DIVIDING RADICALS

In this section we learn how to divide radicals that have the same index. We also learn to divide multinomials with radicals by monomials and binomials with radicals.

Dividing Radicals with the Same Index

Radicals having the same index can be divided using Eq. 30.

$$\sqrt[r]{x} \div \sqrt[r]{y} = \frac{\sqrt[r]{x}}{\sqrt[r]{y}} = \sqrt[r]{\frac{x}{y}}$$

Just divide the radicands and keep the common index.

Example 53

Divide and simplify.

(a) $\sqrt{6} \div \sqrt{3} = \frac{\sqrt{6}}{\sqrt{3}} = \sqrt{\frac{6}{3}} = \sqrt{2}$

(b) $\frac{\sqrt{15xy}}{\sqrt{3x}} = \sqrt{\frac{15xy}{3x}} = \sqrt{5y}$

Note that division problems are often expressed as fractions, as in Example 53(b) above. It is common practice to rationalize the denominator in a division problem.

Example 54

$$\sqrt{14x} \div \sqrt{7y} = \frac{\sqrt{14x}}{\sqrt{7y}} = \sqrt{\frac{14x}{7y}}$$

$$= \sqrt{\frac{2x}{y}} = \sqrt{\frac{2x}{y} \cdot \frac{y}{y}}$$

$$= \frac{\sqrt{2xy}}{y}$$

Dividing a Multinomial by a Monomial Radical

If a multinomial is divided by a monomial, the division is done *term by term* (using the distributive property). Then each term is simplified.

Example 55

Divide and simplify:

$$\frac{\sqrt{2} + 4\sqrt{x}}{\sqrt{2}}$$

Solution:

$$\frac{\sqrt{2} + 4\sqrt{x}}{\sqrt{2}} = \frac{\sqrt{2}}{\sqrt{2}} + \frac{4\sqrt{x}}{\sqrt{2}}$$

$$= 1 + \frac{4}{1}\sqrt{\frac{x}{2}} = 1 + \frac{4}{1}\sqrt{\frac{x \cdot 2}{2 \cdot 2}}$$

$$= 1 + \frac{4}{2}\sqrt{2x} = 1 + 2\sqrt{2x}$$

Dividing by a Binomial Containing a Radical

An example of an expression divided by a binomial containing a radical is

$$4 \div (3 + \sqrt{2}) \quad \text{or} \quad \frac{4}{3 + \sqrt{2}}$$

We simplify the fraction by rationalizing the denominator. To rationalize a binomial denominator containing square roots, we multiply the numerator and denominator of the fraction by the *conjugate* of the denominator. This will remove all square roots from the denominator.

Example 56

Divide and simplify:

$$\frac{4}{3 + \sqrt{2}}$$

Solution: The conjugate of $3 + \sqrt{2}$ is $3 - \sqrt{2}$. So we multiply the numerator and denominator by $3 - \sqrt{2}$:

$$\frac{4}{3 + \sqrt{2}} = \frac{(4)(3 - \sqrt{2})}{(3 + \sqrt{2})(3 - \sqrt{2})} = \frac{12 - 4\sqrt{2}}{9 - 2} = \frac{12 - 4\sqrt{2}}{7}$$

Example 57

Divide and simplify:

$$\frac{3\sqrt{x} + \sqrt{2}}{\sqrt{x} - \sqrt{2}}$$

Solution: The conjugate of $\sqrt{x} - \sqrt{2}$ is $\sqrt{x} + \sqrt{2}$, so we multiply the numerator and denominator by $\sqrt{x} + \sqrt{2}$:

$$\frac{3\sqrt{x} + \sqrt{2}}{\sqrt{x} - \sqrt{2}} = \frac{(3\sqrt{x} + \sqrt{2})(\sqrt{x} + \sqrt{2})}{(\sqrt{x} - \sqrt{2})(\sqrt{x} + \sqrt{2})} = \frac{3x + 4\sqrt{2x} + 2}{x - 2}$$

Radical expressions should be reduced to lowest terms.

SECTION 20-5   Exercises   Dividing Radicals

Dividing Radicals with the Same Index

Divide and simplify. Express answers in radical form.

1. $\sqrt{6} \div \sqrt{2}$ **2.** $\sqrt{15} \div \sqrt{3}$ **3.** $\sqrt{10x} \div \sqrt{5}$

4. $\sqrt{21a} \div \sqrt{7}$ **5.** $\sqrt{21ab} \div \sqrt{3a}$ **6.** $\sqrt{55xy} \div \sqrt{5x}$

7. $\dfrac{5}{\sqrt{3}}$ **8.** $\dfrac{7}{\sqrt{5}}$ **9.** $\dfrac{x\sqrt{2}}{2\sqrt{7}}$

10. $\dfrac{y\sqrt{5}}{3\sqrt{3}}$ **11.** $\sqrt[3]{14xy} \div \sqrt[3]{2x}$ **12.** $\sqrt[4]{6ab} \div \sqrt[4]{3b}$

Dividing a Multinomial by a Monomial Radical

13. $(\sqrt{3} + 6\sqrt{2}) \div \sqrt{3}$ **14.** $(4\sqrt{5} - \sqrt{2}) \div \sqrt{2}$ **15.** $\dfrac{2 - \sqrt{x}}{\sqrt{x}}$

16. $\dfrac{2\sqrt{3} - 3}{\sqrt{3}}$ **17.** $\dfrac{2 + 4\sqrt{2}}{6\sqrt{2}}$ **18.** $\dfrac{1 - 6\sqrt{x}}{9\sqrt{x}}$

19. $3 \div (2 - \sqrt{3})$

20. $7 \div (\sqrt{2} + 1)$

21. $\dfrac{\sqrt{3}}{4\sqrt{2} + \sqrt{3}}$

22. $\dfrac{\sqrt{5}}{2\sqrt{3} - \sqrt{7}}$

23. $\dfrac{3 + \sqrt{x}}{1 - \sqrt{x}}$

24. $\dfrac{6 - \sqrt{y}}{3 + \sqrt{y}}$

25. $\dfrac{\sqrt{7} - \sqrt{5}}{\sqrt{7} + \sqrt{5}}$

26. $\dfrac{\sqrt{2} + \sqrt{3}}{\sqrt{2} - \sqrt{3}}$

27. $\dfrac{2\sqrt{a} - \sqrt{2}}{\sqrt{a} - 3\sqrt{2}}$

28. $\dfrac{\sqrt{m} + 2\sqrt{3}}{3\sqrt{m} + \sqrt{3}}$

29. $\dfrac{1 - \sqrt{2}}{\sqrt{3} - \sqrt{2}}$

30. $\dfrac{5 + \sqrt{3}}{\sqrt{2} - \sqrt{3}}$

31. $\dfrac{3 - \sqrt{x}}{2 + \sqrt{x}}$

32. $\dfrac{\sqrt{x} + 1}{\sqrt{x} - 1}$

33. $\dfrac{1 - \sqrt{5}}{3 + \sqrt{2} - \sqrt{5}}$

34. $\dfrac{3 - \sqrt{2}}{1 - \sqrt{2} - \sqrt{3}}$

35. A certain motion problem involves the expression

$$\frac{\sqrt{v}}{\sqrt{m} + \sqrt{v}}$$

Simplify this expression.

36. When studying the expansion of gases, we find the expression

$$\frac{\sqrt{V_2}}{\sqrt{V_2} - \sqrt{V_1}}$$

Simplify this expression.

37. When studying the theory of waves in wires, we find the expression

$$\frac{\sqrt{d_1} - \sqrt{d_2}}{\sqrt{d_1} + \sqrt{d_2}}$$

Simplify this expression.

38. Divide. (First rationalize the denominator, then change to exponential form, express the exponents with their LCD, use Eq. 21 to factor out the common denominator, and change back to radical form.)

(a) $\dfrac{\sqrt[3]{2}}{\sqrt{5}}$ (b) $\dfrac{\sqrt[4]{5}}{\sqrt[3]{4}}$

20-6   RADICAL EQUATIONS

In this section we define a radical equation, learn how to solve a radical equation, and learn how to recognize when a solution is extraneous.

Definition

A *radical equation* is an equation in which the variable is under a radical sign.

Example 58

$$\sqrt{x - 2} = 3$$

is a radical equation, but the equation

$$x - \sqrt{2} = 3$$

is not a radical equation, because the variable is not under a radical sign.

Solution of a Radical Equation

In Chapter 10, we learned to solve an equation by performing the same operation on both sides of the equation. That is, we could add, subtract, multiply, or divide both sides of an equation by the same (nonzero) number in order to isolate the variable on one side of the equation. But, if the equation has a radical in it, like

$$\sqrt{x} - 3 = 0$$

none of these operations eliminate the radical sign.

To solve a radical equation it is necessary to transpose the radical term so it is alone on one side of the equal sign and then to raise both sides to a power which will eliminate the radical. That power is indicated by the index of the radical. So, to solve the radical equation,

$$\sqrt[r]{x} = a$$

raise both sides of the equation to the rth power.

Example 59

Solve and check:

$$\sqrt{x} - 3 = 0$$

Solution: First we isolate the radical part by adding 3 to both sides:

$$\sqrt{x} = 3$$

This is a *square* root, so we *square* both sides of the equation.

$$(\sqrt{x})^2 = 3^2$$

What does the squaring do? In squaring both sides of this equation, we multiply the left side by $\sqrt{x}$ and the right side by 3. Since $\sqrt{x} = 3$, we are multiplying both sides by the same number (if there is a number x such that $\sqrt{x}$ is equal to 3). Squaring both sides, we get

$$x = 9$$

Check:
We check in the original equation:

$$\sqrt{9} - 3 \stackrel{?}{=} 0$$
$$3 - 3 \stackrel{?}{=} 0$$
$$0 = 0 \quad \text{checks}$$

So the solution of the equation $\sqrt{x} - 3 = 0$ is $x = 9$.

This method of solving an equation by squaring both sides produces a new equation. This new equation will have all of the roots of the original equation. However, in some cases, it will also have roots that are *not* roots of the original equation. Hence we must *always* check the roots obtained from the squaring process by substituting them into the *original* equation.

Example 60

Solve and check:

$$\sqrt{x} + 3 = 0$$

Solution: We isolate the radical part:

$$\sqrt{x} = -3$$

Squaring both sides,

$$(\sqrt{x})^2 = (-3)^2$$
$$x = 9$$

Check:
Substituting into the original equation,

$$\sqrt{9} + 3 \overset{?}{=} 0$$
$$3 + 3 \overset{?}{=} 0$$
$$6 \neq 0$$

The solution does *not* check. The solution $x = 9$ is an *extraneous* solution. Recall from Sec. 18-1 that an extraneous solution is a "false" solution—one that does *not* satisfy the original equation and therefore is not a solution. This extraneous solution was introduced at the squaring step, since squaring meant multiplying the left by $\sqrt{x}$ and the right by -3. We may do this only if $\sqrt{x} = -3$, and of course, a square root may never be negative. So there is *no real solution.*

COMMON ERROR The squaring process may introduce *extraneous* solutions. These are discovered only in the check. They must be discarded because they do not satisfy the original equation.

Example 61

Solve $\sqrt[3]{2x} = 4$ and check.

Solution: Cubing both sides:

$$(\sqrt[3]{2x})^3 = 4^3$$
$$2x = 64$$
$$x = 32$$

Check:
$$\sqrt[3]{2 \cdot 32} \overset{?}{=} 4$$
$$\sqrt[3]{64} \overset{?}{=} 4$$
$$4 = 4 \quad \text{checks}$$

Steps for Solving a Radical Equation with One Radical

1. Transpose the radical term until it is alone on one side of the equation.
2. Raise both sides of the resulting equation to the power indicated by the index of the radical.
3. Simplify the resulting equation and solve for the variable.
4. Check the solution in the original equation to see if it is extraneous.

Example 62 ─────────

Solve and check:

$$4 + \sqrt{2x - 5} = 7$$

Solution: Transposing, we have

$$\sqrt{2x - 5} = 3$$

Squaring both sides,

$$2x - 5 = 9$$

Solving for x,

$$2x = 14$$
$$x = 7$$

Check:

$$4 + \sqrt{2 \cdot 7 - 5} \stackrel{?}{=} 7$$
$$4 + \sqrt{9} \stackrel{?}{=} 7$$
$$7 = 7 \quad \text{checks}$$

Equations with Fractional Exponents

To solve an equation with fractional exponents, we change it to radical form and then follow the steps above.

Example 63 ─────────

Solve $1 + x^{1/3} = 5$.

Solution: Transposing, we have

$$x^{1/3} = 4$$

Changing to radical form,

$$\sqrt[3]{x} = 4$$

Cubing both sides,

$$(\sqrt[3]{x})^3 = 4^3$$
$$x = 64$$

Check:

Substituting into the original equation:

$$1 + 64^{1/3} \stackrel{?}{=} 5$$
$$1 + 4 \stackrel{?}{=} 5$$
$$5 = 5 \quad \text{checks}$$

Alternate Solution: We can also solve this equation in exponential form:

After transposing,

$$x^{1/3} = 4$$

we cube both sides,

$$(x^{1/3})^3 = (4)^3$$

so that we get x^1

$$x^1 = 64$$

or

$$x = 64$$

Equations with More Than One Square Root

If an equation has more than one square root radical, isolate one at a time and square both sides. Generally it's better to isolate the most complicated radical first.

Example 64

Solve $\sqrt{x} + \sqrt{x+2} = 2$.

Solution: We isolate the most complicated radical:

$$\sqrt{x+2} = 2 - \sqrt{x}$$

Squaring both sides,

$$(\sqrt{x+2})^2 = (2 - \sqrt{x})^2$$
$$x + 2 = 4 - 4\sqrt{x} + x$$

Combining terms,

$$-2 = -4\sqrt{x}$$
$$\sqrt{x} = \frac{1}{2}$$

Squaring both sides,

$$(\sqrt{x})^2 = \left(\frac{1}{2}\right)^2$$
$$x = \frac{1}{4}$$

Check:
Substituting into the original equation:

$$\sqrt{\frac{1}{4}} + \sqrt{\frac{1}{4} + 2} \stackrel{?}{=} 2$$

$$\frac{1}{2} + \frac{3}{2} \stackrel{?}{=} 2$$

$$2 = 2 \quad \text{checks}$$

Graphical Solution of Radical Equations

We learned how to solve equations graphically in Sec. 12-6. Recall that to solve an equation graphically, we bring all terms to one side of the equal sign, set those terms equal to y, and graph the function. The x intercepts are the solutions of the given equation. We can use a graphing calculator to graph the function, and use the "root" or "zero" feature to find the solutions of the equation.

Example 65

Solve the equation in Example 64 graphically:

$$\sqrt{x} + \sqrt{x+2} = 2$$

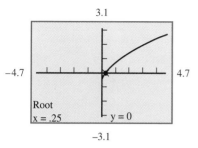

3.1

−4.7   4.7

Root
x = .25   y = 0

−3.1

■ **FIGURE 20-4**

Solution: We bring all terms to one side of the equal sign:

$$\sqrt{x} + \sqrt{x+2} - 2 = 0$$

set the left side equal to y and graph

$$y = \sqrt{x} + \sqrt{x+2} - 2$$

Using the "root" or "zero," we find that the solution is $x = 0.25$ (Fig. 20-4), which agrees with our answer of $\frac{1}{4}$ found in Example 64.

SECTION 20-6   Exercises   Radical Equations

Solve and check.

1. $\sqrt{x} = 2$ **2.** $\sqrt{x} = 5$ **3.** $\sqrt{x} - 4 = 0$

4. $\sqrt{x} + 1 = 0$ **5.** $\sqrt{x} + 6 = 0$ **6.** $\sqrt{x} - 7 = 0$

7. $\sqrt[3]{x} = 3$ **8.** $2 - \sqrt[3]{x} = 0$ **9.** $\sqrt[3]{4x} + 1 = -1$

10. $\sqrt[3]{2x} = 2$ **11.** $\sqrt{x} + 1 = 4$ **12.** $2 - \sqrt{x} = 1$

13. $3 - \sqrt{x} = 7$ **14.** $2 + \sqrt{x} = 1$ **15.** $3 + \sqrt{x} = 7$

16. $\sqrt{x} + 4 = 2$ **17.** $\sqrt{2x+3} = 5$ **18.** $4 = \sqrt{3x-2}$

19. $2 - \sqrt{1+5x} = 8$ **20.** $5 - \sqrt{4x-3} = 2$ **21.** $\sqrt{2x-1} = \sqrt{x+3}$

22. $\sqrt{3x-1} = \sqrt{2x+1}$ **23.** $\sqrt[3]{x-1} = 2$ **24.** $\sqrt[3]{x+3} = 1$

25. $\sqrt[3]{2x-2} = 4$ **26.** $\sqrt[3]{2x-5} = 5$ **27.** $\sqrt{x-2} = \dfrac{1}{\sqrt{x-2}}$

28. $\dfrac{2}{\sqrt{x-1}} = \sqrt{x-1}$ **29.** $\sqrt{x^2-3} = x + 3$ **30.** $x - 4 = \sqrt{x^2-8}$

31. $x^{1/3} = -2$ **32.** $x^{1/3} = 3$ **33.** $x^{3/2} - 8 = 0$

34. $2 - x^{3/2} = 1$ **35.** $\sqrt{12+x} = 2 + \sqrt{x}$ **36.** $\sqrt{x+7} = 7 - \sqrt{x}$

37. $\sqrt{3+2x} = \sqrt{2x} - 3$ **38.** $\sqrt{3x} - 1 = \sqrt{3x-5}$

39. $\sqrt{8+x} = 3 + \sqrt{x-1}$ **40.** $\sqrt{x-3} = 2 - \sqrt{x+1}$

Solve the equations graphically. Approximate the solution(s) to the nearest tenth.

41. $\sqrt{2x-1} - \sqrt{2x+6} = -1$ **42.** $\sqrt{3x+1} - \sqrt{3x-8} = 1$

43. $\sqrt{x^2-6.75} = 9.25 - x$ **44.** $\sqrt{x+2.25} = \sqrt{x+8.75}$

Applications

45. In simple harmonic motion, the undamped angular velocity (Fig. 20-5) is given by the equation

$$\omega = \sqrt{\dfrac{kg}{W}}$$

Solve this equation for k.

■ **FIGURE 20-5**

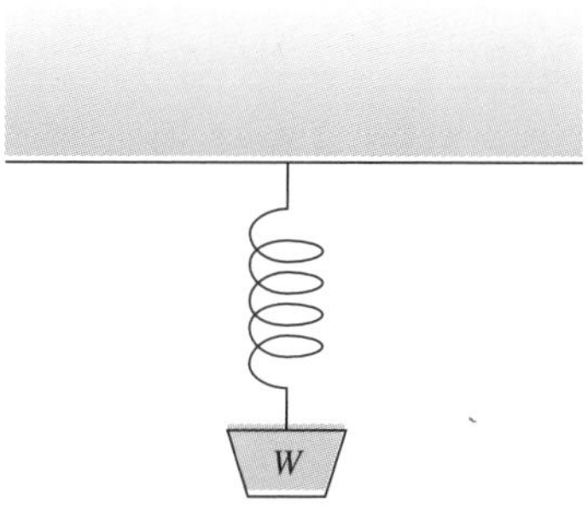

W

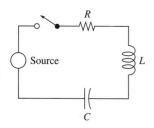

■ FIGURE 20-6

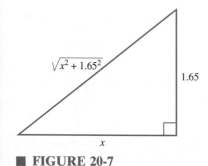

■ FIGURE 20-7

46. In a certain circuit (Fig. 20-6), the resonant frequency is given by the equation

$$\omega = \sqrt{\frac{1}{LC}}$$

Solve this equation for C.

47. The right triangle in Fig. 20-7 has one leg equal to 1.65 cm. The other leg is x and an expression for the hypotenuse is found using the Pythagorean Theorem. If the perimeter is 6.65 cm, then

$$x + 1.65 + \sqrt{x^2 + 1.65^2} = 6.65$$

Solve this equation for x.

■ Chapter 20 REVIEW EXERCISES

Change to radical form.

1. $x^{1/5}$ **2.** $5^{2/3}$ **3.** $b^{-1/3}$ **4.** $-7^{3/4}$

Change to exponential form.

5. $\sqrt{3y}$ **6.** $\sqrt[4]{x^3}$ **7.** $\dfrac{a}{\sqrt[3]{5}}$ **8.** $\sqrt{x-y}$

Evaluate without using a calculator.

9. $8^{1/3}$ **10.** $-9^{3/2}$ **11.** $49^{-1/2}$ **12.** $27^{-2/3}$

Evaluate using a calculator. Round to the nearest hundredth if necessary.

13. $64^{2/3}$ **14.** $-9^{-3/2}$ **15.** $6^{1/4}$ **16.** $5^{-2/3}$

Simplify using the laws of exponents. Assume all variable bases are positive, and leave your answers in exponential form, with positive exponents.

17. $(x^{1/2})^{1/4}$ **18.** $a^{-1/3}$ **19.** $(x^{2/3})^{9/10}$ **20.** $(8x^3)^{2/3}$

21. $\dfrac{a^{3/4}}{a^{1/3}}$ **22.** $\left(\dfrac{x^2}{y}\right)^{1/3}$

23. A pharmaceutical company, growing an organism to be used in a vaccine, finds that the organism grows at a rate so that it quadruples every day. If the company starts with 300 units one day, it will have n units in d days according to the equation

$$n = 300 \cdot 4^d$$

Evaluate the number of units after 7.00 hours $\left(\dfrac{7}{24}\text{ of a day}\right)$.

Write in simplest form.

24. $\sqrt{98x}$ **25.** $\sqrt{800y}$ **26.** $\sqrt[3]{54}$ **27.** $\sqrt[4]{32}$

28. $\sqrt{10000y^9}$ **29.** $\sqrt{18xy^5}$ **30.** $\sqrt[3]{48x^7}$ **31.** $\sqrt[4]{256x^9}$

32. $\sqrt{\dfrac{8}{9}}$ **33.** $\sqrt{\dfrac{11}{25a^6}}$ **34.** $\dfrac{5}{\sqrt{6x}}$ **35.** $\sqrt{\dfrac{3}{2}}$

36. $\sqrt[3]{\dfrac{3}{4}}$ **37.** $\sqrt{\dfrac{5}{6x}}$ **38.** $\sqrt[3]{\dfrac{5}{16x}}$ **39.** $\sqrt{\dfrac{y^3}{96x}}$

40. $\sqrt{\dfrac{15x^5y^8}{8z^3}}$

Combine algebraically and simplify. Leave the answer in radical form.

41. $5\sqrt{2} - 8\sqrt{2} - \sqrt{2}$ **42.** $\sqrt{45} - \sqrt{80} + \sqrt{20}$

43. $\sqrt[3]{250} - 2\sqrt[3]{54}$ **44.** $\sqrt[3]{\dfrac{5}{9}} - \sqrt[3]{\dfrac{3}{25}}$

45. $81 + \sqrt{81} + \sqrt[3]{81} + \sqrt[4]{81}$ **46.** $81^{1/3} - 27^{-1/2} - 12^{3/2}$

47. $3\sqrt{3}(2\sqrt{3})$ **48.** $3\sqrt{6}(2\sqrt{3})$

49. $\sqrt{2}(4\sqrt{3} - 5)$ **50.** $(\sqrt{7} - \sqrt{3})(2\sqrt{7} - \sqrt{3})$

51. $(3\sqrt{2} - 2\sqrt{3})^2$ **52.** $(\sqrt{5} - 4)(3\sqrt{2} + \sqrt{3})$

53. $\sqrt{15} \div \sqrt{5}$ **54.** $\dfrac{\sqrt{3}}{2\sqrt{5}}$

55. $\dfrac{6\sqrt{2} - \sqrt{5}}{\sqrt{5}}$ **56.** $\dfrac{\sqrt{3}}{\sqrt{2} - \sqrt{3}}$

57. $\dfrac{\sqrt{5} - 3\sqrt{2}}{2\sqrt{5} + \sqrt{2}}$ **58.** $\dfrac{2 - \sqrt{3}}{\sqrt{2} + \sqrt{5}}$

Combine algebraically and simplify.

59. $4x\sqrt[3]{2} - 3y\sqrt[3]{2}$ **60.** $-3\sqrt{y} + 7\sqrt{y} - 2\sqrt{y}$

61. $5\sqrt{7x} - 2\sqrt{6x} - 3\sqrt{7x} + \sqrt{6x}$ **62.** $5y^2\sqrt{28y} - 2y\sqrt{54y^3} + 3\sqrt{63y^5}$

63. $6\sqrt{28x} - 4\sqrt{12x} - 5\sqrt{63x} + 3\sqrt{75x}$ **64.** $\sqrt{5a^3} - a\sqrt{45a} + a\sqrt{20a^2}$

65. $\sqrt{\dfrac{3x}{5}} - \sqrt{\dfrac{3x}{20}}$ **66.** $3a\sqrt{\dfrac{3}{5}} - 2a\sqrt{\dfrac{1}{15}} + a\sqrt{\dfrac{5}{3}}$

67. $\sqrt[3]{3} \cdot \sqrt[3]{4x^2}$ **68.** $3\sqrt{2x} \cdot 6\sqrt{3}$

69. $(4x\sqrt{3y})(2x\sqrt{5y})$ **70.** $(2\sqrt[5]{a})(3\sqrt[3]{a})$

71. $4\sqrt{5x}(2\sqrt{x} - 3\sqrt{5})$ **72.** $(\sqrt{3} - 2x)(\sqrt{3} + x)$

73. $(\sqrt{x} + 2\sqrt{y})(3\sqrt{x} + \sqrt{y})$ **74.** $(4\sqrt{5a} - 2\sqrt{3})(4\sqrt{5a} + 2\sqrt{3})$

75. $\sqrt{6xy} \div \sqrt{3x}$ **76.** $3x \div \sqrt{7}$

77. $\dfrac{\sqrt[3]{22xy}}{\sqrt[3]{2x}}$ **78.** $\dfrac{2 - 6\sqrt{x}}{3\sqrt{x}}$

79. $\dfrac{3 + \sqrt{a}}{2 - \sqrt{a}}$ **80.** $\dfrac{3\sqrt{x} - 2\sqrt{y}}{\sqrt{x} - \sqrt{y}}$

Solve and check.

81. $\sqrt{x} - 2 = 0$ **82.** $\sqrt[3]{x} = -3$

83. $\sqrt{x + 5} + 4 = 0$ **84.** $\sqrt[3]{3 - x} = 5$

85. $\sqrt{3x - 2} = 7$ **86.** $2 - \sqrt{2x - 3} = 5$

87. $\sqrt{3x - 5} = \sqrt{x + 1}$ **88.** $x - 2 = \sqrt{x^2 - 4}$

89. $x^{1/3} - 1 = 3$ **90.** $\sqrt{x + 3} = 4 - \sqrt{x - 5}$

91. The diameter of a circle having area A is

$$\sqrt{\dfrac{4A}{\pi}}$$

Simplify this expression.

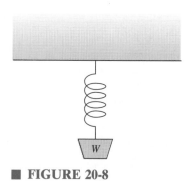

FIGURE 20-8

92. When studying the resistance in an electric circuit, we find the equation

$$\frac{\sqrt{R_1 R_2}}{\sqrt{R_1} + \sqrt{R_2}}$$

Simplify this expression.

93. The natural frequency f of a body under simple harmonic motion (Fig. 20-8) is given by the equation

$$f = \frac{1}{2\pi} \sqrt{\frac{kg}{W}}$$

Solve this equation for W.

WRITING

94. Explain how exponents and radicals are used to write two different forms of the same expression. Show examples of each form and explain why we need both forms.

TEAM PROJECT

95. If we know the perimeter and one leg of a right triangle, then we can find the other two sides. Suppose we know that the perimeter of a right triangle is 98.0 cm and that one leg is 35.0 cm. Cut a piece of string 98.0 cm long and mark a 35.0 cm length. Make the right triangle with these dimensions, being careful to create a precise right angle. Use this model to estimate the lengths of the other two sides, to three significant digits. Use a protractor to estimate the measures of the two acute angles to the nearest degree.

Verify your work algebraically. Use the formula for perimeter and the Pythagorean Theorem to write a radical equation that models this triangle. Solve the equation. Use right triangle trigonometry to find the acute angles. Calculate the percent error in your experimental results.

Quadratic Equations 21

In this chapter we solve quadratic equations by factoring and by using the quadratic formula. We then solve applications that involve quadratic equations. Finally, we graph the quadratic function.

21-1 INTRODUCTION TO QUADRATIC EQUATIONS

In this section we define a quadratic equation and its terminology. We learn about the different types of solutions and learn to solve a pure quadratic equation. Finally, we apply some formulas which are pure quadratic equations.

Quadratic Equation in One Variable

A *quadratic equation* (in x) is a polynomial equation of second degree. It must contain a second degree term (x^2), but no higher degree terms. It may (or may not) have a first-degree (x) term and a zero-degree term (constant). It cannot have a variable under a radical or in the denominator of a fraction.

Example 1

The equations
$$x^2 - 2x + 3 = 0, \quad 3x^2 - x = 5, \quad \text{and} \quad x^2 = 6$$
are all quadratic equations. The equations
$$x^3 - x = 2, \quad x - \sqrt{x} + 1 = 0, \quad \text{and} \quad \frac{1}{x^2} - x - 2 = 0$$
are *not* quadratic equations.

General Form of a Quadratic Equation

A quadratic equation is said to be in *general form* when it is written in the form
$$ax^2 + bx + c = 0$$
where a, b, and c are constants, with $a \neq 0$.

GENERAL FORM OF A QUADRATIC EQUATION	$ax^2 + bx + c = 0$	58

Quadratic equations in general form are usually written *without fractions* and with the first term *positive*. It is important to be able to identify a, b, and c for later use with the quadratic formula.

Example 2

The equation $2x^2 + 3x - 1 = 0$ is a quadratic equation in general form, with $a = 2$, $b = 3$, and $c = -1$.

Example 3

Write the equation $3x - x^2 = 5$ in general form and give the value of a, b, and c.

Solution: Transposing the 5, we have

$$-x^2 + 3x - 5 = 0$$

To make the coefficient of x^2 positive, we multiply both sides of the equation by -1:

$$x^2 - 3x + 5 = 0$$

This is the general form of the equation, with $a = 1$, $b = -3$, and $c = 5$.

Solution of a Quadratic Equation

Recall from Chapter 10 that a *solution* (or *root*) of an equation is a value of the variable that satisfies the equation; that is, it makes the equation true. A quadratic equation always has *two* solutions: They may be two *different real* solutions, or two *equal real* solutions, or two *non-real* solutions. We check the solutions of a quadratic equation by substituting them back into the original equation.

Example 4

Determine which, if any, of these values

$$x = 2, 1, \frac{1}{3}, \text{ or } -2$$

are solutions to the equation

$$3x^2 + 5x - 2 = 0$$

Solution: Substituting $x = 2$ into the equation:

$$3(2)^2 + 5(2) - 2 \overset{?}{=} 0$$
$$20 \neq 0$$

So, $x = 2$ is *not* a solution to the given equation.
Substituting $x = 1$,

$$3(1)^2 + 5(1) - 2 \overset{?}{=} 0$$
$$6 \neq 0$$

So $x = 1$ is *not* a solution to the given equation.
Substituting $x = \frac{1}{3}$:

$$3\left(\frac{1}{3}\right)^2 + 5\left(\frac{1}{3}\right) - 2 \overset{?}{=} 0$$
$$\frac{1}{3} + \frac{5}{3} - 2 \overset{?}{=} 0$$
$$0 = 0 \quad \text{checks}$$

So $x = \dfrac{1}{3}$ is a solution.

Substituting $x = -2$,

$$3(-2)^2 + 5(-2) - 2 \stackrel{?}{=} 0$$

$$12 - 10 - 2 \stackrel{?}{=} 0$$

$$0 = 0 \quad \text{checks}$$

So $x = -2$ is a solution.

We have now found the *two* solutions of the quadratic equation. Thus, the two solutions are

$$\dfrac{1}{3} \quad \text{and} \quad -2$$

Pure Quadratic Equation

A quadratic equation in which there is no first-degree (x) term is called a *pure quadratic* equation. In a pure quadratic equation, the coefficient of the x term, b, is always 0. Its general form is

$$ax^2 + c = 0$$

┌─── **Example 5** ───────

The equation

$$3x^2 - 12 = 0$$

is a pure quadratic equation, since there is no x term.

Solution of a Pure Quadratic Equation

One way to solve a pure quadratic equation is by first solving for x^2, and then taking the square root of both sides of the equation. We use this principle:

If

$$x^2 = a$$

then

$$x = \sqrt{a} \quad \text{or} \quad x = -\sqrt{a}$$

or

$$x = \pm\sqrt{a}$$

If x^2 is nonnegative, the solutions are real. If it is negative, the solutions are not real. The solutions should be simplified, if possible. As always, we check each solution by substituting back into the original equation.

┌─── **Example 6** ───────

Solve and check:

$$3x^2 - 12 = 0$$

Solution: Solving for x^2,

$$3x^2 = 12$$

$$x^2 = 4$$

Taking square roots,

$$x = 2 \qquad x = -2$$

(We may also write $x = \pm2$.)

So the solutions are 2 and -2.

Check: If $x = 2$,   If $x = -2$,

$$3(2)^2 - 12 \stackrel{?}{=} 0 \qquad\qquad 3(-2)^2 - 12 \stackrel{?}{=} 0$$

$$0 = 0 \quad \text{checks} \qquad\qquad\qquad 0 = 0 \quad \text{checks}$$

COMMON ERROR When taking the square root of both sides of an equation, be sure to write the positive and negative signs ($\pm$) on one side of the equation. There will always be *two* solutions to a quadratic equation.

In the example above, the solutions are two *different real (and rational)* numbers. But often the solutions to a quadratic equation are *irrational*.

Example 7

Solve $2x^2 - 3 = 0$.

Solution: Solving for x^2,

$$2x^2 = 3$$

$$x^2 = \frac{3}{2}$$

Taking square roots,

$$x = \pm\sqrt{\frac{3}{2}}$$

Rationalizing the denominator,

$$x = \pm\sqrt{\frac{3 \cdot 2}{2 \cdot 2}} = \pm\frac{\sqrt{6}}{2}$$

So the solutions are $\dfrac{\sqrt{6}}{2}$ and $-\dfrac{\sqrt{6}}{2}$.

For this equation, the solutions are two *different real (and irrational)* numbers.

Steps for the Solution of a Pure Quadratic Equation

1. Solve for x^2.
2. Take the square root of both sides of the equation.
3. Simplify the result, if necessary.
4. Check both solutions by substituting them into the original equation.

Quadratic Equations with No Real Solution

All quadratic equations have two solutions, but they may not be *real* numbers. The square root of a negative number is not a real number; we call it a *complex* number, and we will study complex numbers in Chapter 29. For now, it will suffice to know that the square root of a negative number is not a real number.

Example 8

Solve $2x^2 + 3 = 0$.

Solution: Solving for x^2,

$$x^2 = \frac{-3}{2} = -\frac{3}{2}$$

Taking the square root of both sides,

$$x = \pm\sqrt{-\frac{3}{2}}$$

We may stop at this point, since we must take the square root of a negative number, and simply state that there is *no real solution.*

Formulas

Many formulas in applications are quadratic equations. It will be helpful to be able to solve them.

Example 9

The formula for kinetic energy K of a moving body is given by

$$K = \frac{1}{2}mv^2$$

where m is the mass of a body and v is its velocity. Solve this formula for its velocity v.

Solution: Transposing, we have

$$\frac{1}{2}mv^2 = K$$

To solve for v^2, we multiply both sides by $\frac{2}{m}$,

$$\frac{2}{m}\left(\frac{1}{2}mv^2\right) = \frac{2}{m} \cdot K$$

$$v^2 = \frac{2K}{m}$$

Taking square roots,

$$v = \pm\sqrt{\frac{2K}{m}}$$

Rationalizing the denominator,

$$v = \pm\sqrt{\frac{2K \cdot m}{m \cdot m}} = \pm\frac{\sqrt{2Km}}{m}$$

Since v is a positive velocity, we choose only the *positive* root,

$$v = \frac{\sqrt{2Km}}{m}$$

Example 10

If current I is flowing through a resistance R, then the power dissipated is given by

$$P = I^2R$$

What current flows in a wire to a home electrical oven if the resistance is 10.0 ohms and the power is 5.30 kW?

Solution: Before substituting into the formula, we change 5.30 kW to watts:

$$5.30 \text{ kW} = 5.30 \text{ kW} \left(\frac{1000 \text{ W}}{1 \text{ kW}} \right) = 5300 \text{ W}$$

Substituting into the given equation,

$$5300 = I^2(10.0)$$
$$I^2 = 530$$
$$I = \pm\sqrt{530}$$

Taking only the positive current and approximating $\sqrt{530}$,

$$I = 23.0 \text{ amps}$$

SECTION 21-1 Exercises Introduction to Quadratic Equations

Quadratic Equation in One Variable

1. Which of the following are quadratic equations?

(a) $x^2 - x + 1 = 0$ (b) $3 - \dfrac{1}{x^2} = 2x$

(c) $4 = 3x^2$ (d) $5x^2 - \sqrt{x} = 3$

(e) $3x - 2 = 5$ (f) $\dfrac{1}{3}x^2 = 7x$

2. Which of the following are quadratic equations?

(a) $x^3 + x = 0$ (b) $3x^2 = 4x$

(c) $6x = 5$ (d) $x = \dfrac{3}{x^2} + 4$

(e) $\dfrac{3}{4}x^2 - 2x = \dfrac{5}{8}$ (f) $2\sqrt{x} - x^2 + 1 = 0$

General Form of a Quadratic Equation

Write in general form, with a positive leading coefficient and no fractions.

3. $2x^2 - 3x = 4$ **4.** $4x^2 - 5 = 2x$

5. $7x = 4x^2$ **6.** $5 = 2x^2$

7. $\dfrac{1}{3}x - 3 = \dfrac{5}{6}x^2$ **8.** $4 + \dfrac{1}{5}x^2 = \dfrac{3}{8}x$

9. $(3x + 1)(x - 2) = 1$ **10.** $(3x - 2)x = 4$

Solution of a Quadratic Equation

Determine which, if any, of the given values are solutions to the given equation.

11. $x^2 - 5x + 6 = 0; x = -1, 1, 2, 3$

12. $2x^2 - 5x = 3; x = -2, \dfrac{1}{2}, -3, 1$

13. $3x^2 - 2x - 1 = 0; x = -\dfrac{1}{3}, -1, 1, 2$

14. $(x - 2)(x - 1) = 6; x = 2, 4, 1, -1$

Pure Quadratic Equation

Solve and check. If a solution is irrational, write it in its simplified radical form. If there is no real number solution, write *no real solution*.

15. $x^2 - 25 = 0$　　　　16. $49 - x^2 = 0$

17. $12 - x^2 = 0$　　　　18. $x^2 - 20 = 0$

19. $5x^2 + 2 = 2$　　　　20. $4x^2 - 9 = 0$

21. $2x^2 - 5 = 0$　　　　22. $7 - 3x^2 = 7$

23. $x^2 + 3 = 0$　　　　24. $2x^2 = -1$

Formulas

Solve these formulas from science for the variable in parentheses. Write the solution in simplest radical form.

25. $e = mc^2$　　(c)　　　　26. $V = \pi r^2 h$　　(r)

27. $V = \dfrac{1}{3} \pi r^2 h$　　(r)　　　　28. $P = \dfrac{V^2}{R}$　　(V)

29. $I = \dfrac{k}{d^2}$　　(d)　　　　30. $W = \dfrac{1}{2} kx^2$　　(x)

31. $F = K \dfrac{M_1 M_2}{d^2}$　　(d)　　　　32. $F = \dfrac{Mv^2}{r}$　　(v)

Solve and round to three significant digits.

Applications

33. The volume V of a cylinder with radius r and height h is given by

$$V = \pi r^2 h$$

Find the radius of the cylinder in Fig. 21-1 if its volume is 7.95 cm³.

34. The volume of a cone with radius r and height h is given by

$$V = \dfrac{1}{3} \pi r^2 h$$

Find the diameter of a conical sand pile in Fig. 21-2 if it holds 8.50 yd³ of sand.

35. The power P (in watts) dissipated in an electrical load is given by

$$P = \dfrac{V^2}{R}$$

where V is the voltage (in volts) across the load and R is the resistance (in ohms) of the load. What is the voltage across a 100.0-watt light bulb whose resistance is 132 ohms?

■ **FIGURE 21-1**　　　　■ **FIGURE 21-2**

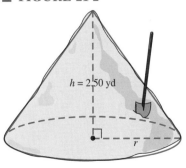

$h = 2.56$ cm

$h = 2.50$ yd

r

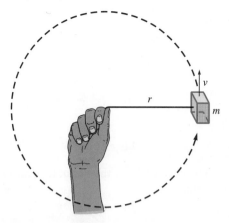

■ **FIGURE 21-3**

36. The power P (in watts) dissipated in a resistance R (in ohms) through which a current I (in amps) is flowing is given by

$$P = I^2R$$

What current flows in a wire of a home electrical circuit whose reistance is 150.0 milliohms and is dissipating 62.0 watts of power?

37. The centripetal force F of a body of mass m revolving in a circle of radius r with velocity v (Fig. 21-3) is

$$F = \frac{mv^2}{r}$$

Find the velocity of a 375-g object revolving around a circle of radius 21.5 cm with a centripetal force of 625,000 g-cm/s².

21-2 SOLUTION BY FACTORING

In this section we see how certain quadratic equations can be solved by factoring. In Chapter 16, we studied factoring expressions using integral coefficients. Here we also consider only integral coefficients. Not all quadratic equations can be solved by factoring using only integral coefficients, but many can.

Zero-Product Principle

When we can solve an equation by factoring, we use the *zero-product principle* to find the solutions:
If

$$ab = 0$$

then

$$\text{either} \quad a = 0 \quad \text{or} \quad b = 0 \quad \text{(or both)}$$

This principle says that if a product of two factors is equal to *zero,* then one (or both) of the factors must equal *zero.*

Solution of a Pure Quadratic Equation

Pure quadratic equations ($ax^2 + c = 0$) can often be solved by factoring the nonzero side of the equation.

Example 11

Solve and check:

$$9x^2 - 4 = 0$$

Solution: The right side of the equation is already zero, so we just factor the left:

$$(3x + 2)(3x - 2) = 0$$

Using the zero-product principle

$$3x + 2 = 0 \qquad\qquad 3x - 2 = 0$$

Solving for x.

$$x = -\frac{2}{3} \qquad\qquad x = \frac{2}{3}$$

So the solutions are $-\dfrac{2}{3}$ and $\dfrac{2}{3}$.

Check:

$$\text{If } x = -\frac{2}{3}, \qquad\qquad \text{If } x = \frac{3}{3},$$

$$9\left(-\frac{2}{3}\right)^2 - 4 \stackrel{?}{=} 0 \qquad\qquad 9\left(\frac{2}{3}\right)^2 - 4 \stackrel{?}{=} 0$$

$$0 = 0 \quad \text{checks} \qquad\qquad 0 = 0 \quad \text{checks}$$

Quadratic Equation with No Constant Term

A quadratic equation with no constant term has the form

$$ax^2 + bx = 0$$

To solve this equation, we write the equation in general form and factor the nonzero side of the equation. We factor out a common x and perhaps a coefficient, and then use the zero-product principle to set each factor equal to zero, and solve each resulting equation for x. Of course, it's wise to check each solution by substituting back into the original equation.

Example 12

Solve and check:

$$2x^2 = 8x$$

Solution: We write the equation in general form:

$$2x^2 - 8x = 0$$

Factoring, we get

$$2x(x - 4) = 0$$

Setting each factor equal to zero,

$$2x = 0 \qquad x - 4 = 0$$

Solving,

$$x = 0 \qquad\qquad x = 4$$

So the solutions are 0 and 4.

Check:

If $x = 0,$ If $x = 4,$

$2(0)^2 \overset{?}{=} 8(0)$ $2(4)^2 \overset{?}{=} 8(4)$

$0 = 0$ checks $32 = 32$ checks

In the last example, it would have been tempting to divide both sides of the given equation by $2x$. Let's look at what happens if we *do* divide by $2x$:

$$2x^2 = 8x$$

$$\frac{2x^2}{2x} = \frac{8x}{2x}$$

$$x = 4$$

We get *one* of the roots, $x = 4$, but we lose the other root, $x = 0$. (This error occurs because we divided both sides of the equation by zero, since $2x = 0$.)

COMMON ERROR In solving an equation, do *not* divide both sides by the variable or a variable expression. This improper step may cause a root to be lost.

Solution of the General Quadratic Equation

We can solve certain general quadratic equations by factoring. Just bring all terms to one side of the equal sign, and factor (if the expression is factorable using integral coefficients). Set each factor equal to zero, and solve.

Example 13

Solve $x^2 + x - 6 = 0$.

Solution: The right side is already zero, so we can begin with factoring the left side:

$$(x + 3)(x - 2) = 0$$

Using the zero-product principle,

$$x + 3 = 0 \qquad\qquad x - 2 = 0$$

Solving,

$$x = -3 \qquad\qquad x = 2$$

The solutions are -3 and 2.

Example 14

Solve $4x^2 + 6x - 4 = 0$.

Solution: If, when the equation is in general form, there is a greatest common *numerical* factor greater than 1, then we should first divide both sides of the equation by that factor so as to make the equation easier to factor. Dividing both sides of this equation by 2, we get:

$$2x^2 + 3x - 2 = 0$$

Factoring,

$$(2x - 1)(x + 2) = 0$$

Setting each factor equal to 0,

$$2x - 1 = 0 \qquad x + 2 = 0$$

$$x = \frac{1}{2} \qquad x = -2$$

The solutions are $\frac{1}{2}$ and -2.

If the given equation is not in general form, we must rewrite it in general form, with zero on one side, before we factor.

Example 15

Solve $12x = -32 + 9x^2$.

Solution: Rewriting in general form:

$$9x^2 - 12x - 32 = 0$$

Factoring,

$$(3x + 4)(3x - 8) = 0$$

$$3x + 4 = 0 \qquad 3x - 8 = 0$$

$$x = -\frac{4}{3} \qquad x = \frac{8}{3}$$

The solutions are $-\frac{4}{3}$ and $\frac{8}{3}$.

COMMON ERROR The zero-product principle can be applied only to an equation in which *one side is zero*. Always write the equation in general form before you factor and use the zero-product principle.

The procedure for solving quadratic equations by factoring is as follows:

Steps for the Solution of a Quadratic Equation by Factoring

1. Write the equation in general form.
2. Factor the nonzero side of the equation, if possible.
3. Set each factor equal to zero.
4. Solve for the variable in each new equation.
5. Check both solutions by substituting them into the original equation.

Example 16

Solve $2x^2 + 2x = 3$.

Solution: We write the equation in general form: $2x^2 + 2x - 3 = 0$.
We try to factor the left side:

$$(2x + 1)(x - 3)$$
$$(2x - 1)(x + 3)$$
$$(2x + 3)(x - 1)$$
$$(2x - 3)(x + 1)$$

but we never get a middle term of $2x$. None of these work! So, the equation cannot be solved by factoring. We'll see how we can solve it in the next section.

Formulas

A formula or equation for a certain application may often be solved by factoring. Since we frequently deal with physical problems, any root that does not make sense in the problem is rejected as unsuitable.

Example 17

The total surface area A of a cone with radius r and slant height s is given by Eq. 94,

$$A = \pi r^2 + \pi r s$$

Find the radius of the cone in Fig. 21-4 if its total surface area is 21π inches².

Solution: We substitute $s = 4$ and $A = 21\pi$ into the formula:

$$21\pi = \pi r^2 + \pi r(4)$$

Dividing both sides by π,

$$21 = r^2 + 4r$$

Putting in general form,

$$r^2 + 4r - 21 = 0$$

Factoring,

$$(r + 7)(r - 3) = 0$$
$$r + 7 = 0 \qquad\qquad r - 3 = 0$$
$$r = -7 \qquad\qquad\quad r = 3$$

We reject the solution $r = -7$, since the radius cannot be negative.

Check:
We check the solution $r = 3$:

$$21\pi \overset{?}{=} \pi(3)^2 + \pi(3)(4)$$
$$\overset{?}{=} 9\pi + 12\pi$$
$$= 21\pi \quad \text{checks}$$

So the radius is 3 inches.

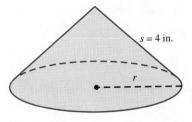

$s = 4$ in.

r

■ **FIGURE 21-4**

Solution of a Pure Quadratic Equation

Solve by factoring and check.

1. $x^2 - 9 = 0$
2. $4x^2 - 25 = 0$
3. $2x^2 - 8 = 0$
4. $3x^2 - 48 = 0$
5. $9x^2 = 64$
6. $25x^2 = 16$

Quadratic Equation with No Constant Term

7. $x(x - 2) = 0$
8. $3x(x + 1) = 0$
9. $x^2 + 2x = 0$
10. $3x^2 - 12x = 0$
11. $2x^2 - 6x = 0$
12. $5x^2 = 15x$

Solution of the General Quadratic Equation

13. $(x + 2)(x - 3) = 0$
14. $(2x - 1)(x + 3) = 0$
15. $x^2 + 2x - 3 = 0$
16. $x^2 - 6x = 7$
17. $x^2 + 2x + 1 = 0$
18. $x^2 + 25 = 10x$
19. $2x^2 + 9x + 10 = 0$
20. $10x^2 + 3 = 11x$
21. $2x^2 - 4x - 30 = 0$
22. $6x^2 - 3x = 18$
23. $x^2 + 9 = 6x$
24. $10x^2 = 18 - 3x$

Formulas

Solve by factoring.

25. The total surface area of a right circular cylinder is given by the formula

$$A = 2\pi r^2 + 2\pi rh$$

where r is the radius and h the height of the cylinder. Find the radius of the cylinder in Fig. 21-5 if the surface area is 8π m^2.

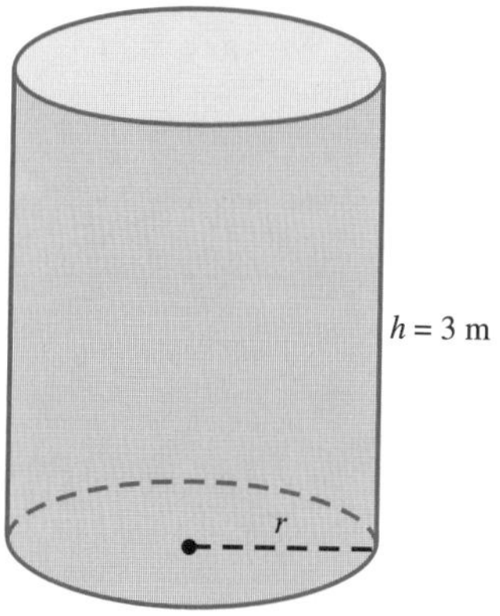

26. A variable electric current I (in amps) is given by the formula

$$I = t^2 - 9t + 23$$

where t is time in seconds. Find the time when the current is equal to

(a) 3 amps

(b) 5 amps

27. A variable voltage V (in volts) is given by the formula

$$V = t^2 - 12t + 35$$

where t is time in seconds. Find the time when the voltage is equal to

(a) 3 volts

(b) 24 volts

28. The cost c for manufacturing n items is approximated using the formula

$$c = -4n^2 + 5n + 625$$

Find the number of items made if the manufacturing cost is $550.

■ **FIGURE 21-5**

$h = 3$ m

r

21-3

QUADRATIC FORMULA

In the last section we saw that only certain quadratic equations could be solved by factoring. *All* quadratic equations can be solved using the quadratic formula. The quadratic formula is derived using a process called *completing the square*. We will demonstrate this process with an example, and then derive the quadratic formula. The quadratic formula is the most useful of all methods for solving quadratic equations.

Completing the Square

In the process of completing the square, we manipulate the given equation so that one side of the equation is a perfect square trinomial. Recall from Sec. 16-5 that in a perfect square trinomial, the first and last terms are perfect squares, and the middle term is twice the product of the square roots of the outer terms.

Example 18

Solve by completing the square:

$$2x^2 - 3x - 1 = 0$$

Solution: We add 1 to both sides,

$$2x^2 - 3x = 1$$

and divide both sides by the coefficient of x^2,

$$\frac{2x^2 - 3x}{2} = \frac{1}{2}$$

Simplifying,

$$x^2 - \frac{3}{2}x = \frac{1}{2}$$

We complete the square on the left by adding the square of half the coefficient of the x term to both sides. The coefficient of the x term is $\frac{-3}{2}$. We take half of $\frac{-3}{2}$,

$$\frac{1}{2} \cdot \frac{-3}{2} = \frac{-3}{4}$$

and square it, getting $\left(\frac{-3}{4}\right)^2$. Adding that to both sides, we have

$$x^2 - \frac{3}{2}x + \left(\frac{-3}{4}\right)^2 = \frac{1}{2} + \left(\frac{-3}{4}\right)^2$$

Factoring and simplifying,

$$\left(x - \frac{3}{4}\right)^2 = \frac{1}{2} + \frac{9}{16} = \frac{17}{16}$$

Taking the square root of both sides,

$$x - \frac{3}{4} = \pm\sqrt{\frac{17}{16}}$$

$$x - \frac{3}{4} = \pm\frac{\sqrt{17}}{4}$$

$$x = \frac{3}{4} \pm \frac{\sqrt{17}}{4} = \frac{3 \pm \sqrt{17}}{4}$$

So the solutions are $\dfrac{3 + \sqrt{17}}{4}$ and $\dfrac{3 - \sqrt{17}}{4}$.

Derivation of the Quadratic Formula

As you see from the example above, the method of completing the square is too cumbersome to be a practical tool for solving quadratic equations. We illustrated its use in Example 18 so we could use it to derive the quadratic formula.

We wish to find the roots of Eq. 58

$$ax^2 + bx + c = 0$$

by completing the square. We start by subtracting c from both sides,

$$ax^2 + bx = -c$$

We divide both sides by a,

$$x^2 + \frac{b}{a}x = -\frac{c}{a}$$

Completing the square, we add the square of half the coefficient of the x term, $\left(\frac{b}{2a}\right)^2$ to both sides,

$$x^2 + \frac{b}{a}x + \left(\frac{b}{2a}\right)^2 = -\frac{c}{a} + \left(\frac{b}{2a}\right)^2$$

Simplifying the right side,

$$x^2 + \frac{b}{a}x + \left(\frac{b}{2a}\right)^2 = \frac{b^2}{4a^2} - \frac{c}{a}$$

Factoring the left and simplifying the right,

$$\left(x + \frac{b}{2a}\right)^2 = \frac{b^2 - 4ac}{4a^2}$$

Taking the square root of both sides,

$$x + \frac{b}{2a} = \pm\sqrt{\frac{b^2 - 4ac}{4a^2}} = \pm\frac{\sqrt{b^2 - 4ac}}{2a}$$

Subtracting $\frac{b}{2a}$ from both sides,

$$x = -\frac{b}{2a} \pm \frac{\sqrt{b^2 - 4ac}}{2a}$$

Putting the right side over the common denominator, we have the quadratic formula:

QUADRATIC FORMULA	If $ax^2 + bx + c = 0,$ then $x = \dfrac{-b \pm \sqrt{b^2 - 4ac}}{2a}$	59

Solution of a Quadratic Equation by the Quadratic Formula

To use the quadratic formula to solve a quadratic equation, write the equation in general form, determine the coefficients a, b, and c and substitute them into the formula.

Example 19

Solve the equation of Example 18 by the quadratic formula:

$$2x^2 - 3x - 1 = 0$$

Solution: The equation is in general form, with $a = 2$, $b = -3$, and $c = -1$. Substituting these values in the quadratic formula, we have

$$x = \frac{-(-3) \pm \sqrt{(-3)^2 - 4(2)(-1)}}{2(2)}$$

$$x = \frac{3 \pm \sqrt{9 + 8}}{4} = \frac{3 \pm \sqrt{17}}{4}$$

So, we found the same solutions as before,

$$\frac{3 + \sqrt{17}}{4} \quad \text{and} \quad \frac{3 - \sqrt{17}}{4}$$

TIP

1. Always rewrite the quadratic equation in general form before trying to use the quadratic formula.

2. Use only the coefficients and not the variables to substitute into the quadratic formula.

3. Be careful with negative signs.

Example 20

Solve by the quadratic formula. Express the solutions in simplest radical form and also approximated to three significant digits.

$$x^2 - 8x - 4 = 0$$

Solution: The equation is in general form, with $a = 1$, $b = -8$, and $c = -4$. Substituting into the quadratic formula, we have

$$x = \frac{8 \pm \sqrt{64 - 4 \cdot 1(-4)}}{2 \cdot 1}$$

$$= \frac{8 \pm \sqrt{80}}{2}$$

Simplifying the radical,

$$= \frac{8 \pm 4\sqrt{5}}{2}$$

$$= 4 \pm 2\sqrt{5}$$

Approximating to three significant digits, the solutions are

$$-0.472 \quad \text{and} \quad 8.47$$

The quadratic formula can, of course, be used to solve quadratics with rational roots. If the roots are rational, then they can also be solved by factoring.

Example 21

Solve by the quadratic formula:

$$2x^2 - x = 6$$

Solution: First, we write the equation in general form:

$$2x^2 - x - 6 = 0$$

so that $a = 2$, $b = -1$, and $c = -6$.

Substituting into the quadratic formula,

$$x = \frac{-(-1) \pm \sqrt{(-1)^2 - 4(2)(-6)}}{2(2)}$$

$$x = \frac{1 \pm \sqrt{1 + 48}}{4}$$

$$x = \frac{1 \pm 7}{4}$$

$$x = \frac{1 + 7}{4} \quad \text{or} \quad x = \frac{1 - 7}{4}$$

$$x = \frac{8}{4} \qquad\qquad x = \frac{-6}{4}$$

$$x = 2 \qquad\qquad x = -\frac{3}{2}$$

So the solutions are 2 and $-\frac{3}{2}$.

Alternate Solution: We could have solved this equation by factoring,

$$2x^2 - x - 6 = 0$$
$$(2x + 3)(x - 2) = 0$$
$$2x + 3 = 0 \qquad\qquad x - 2 = 0$$
$$x = -\frac{3}{2} \qquad\qquad x = 2$$

These factors yield the identical solutions, using a much simpler method. So, if a quadratic is easily factorable, it is usually wise to use the method of factoring, rather than the quadratic formula. In the exercises, however, we offer practice using the formula, even when the solutions are rational and easy to find by factoring.

In applications, we often must solve quadratics with approximate decimal coefficients.

Example 22

Solve and round to three significant digits:

$$6.23x^2 - 5.82 = 4.93x$$

Solution: Writing the equation in general form:

$$6.23x^2 - 4.93x - 5.82 = 0$$

Substituting in the quadratic formula, we have

$$x = \frac{4.93 \pm \sqrt{4.93^2 - 4(6.23)(5.82)}}{2(6.23)}$$

We evaluate x by calculator, and round to three significant digits,

$$x \approx 1.44 \quad \text{and} \quad -0.649$$

Note: Some calculators have a built-in program for solving quadratic equations. Other calculators are programmable. Since quadratic equations are common, a program for solving them is very useful! Check your calculator manual for specific instructions.

Some quadratic equations do not have *real* solutions. When you use the quadratic formula, the radicand may be *negative*. A negative radicand indicates that the solutions will be complex numbers. Complex numbers will be studied in more detail in Chapter 29. For now, if the radicand is negative, then we say there is *no real solution*.

Example 23

Solve by the quadratic formula:

$$3x^2 - 3x + 5 = 0$$

Solution: The equation is in general form with $a = 3$, $b = -3$ and $c = 5$. Substituting into the quadratic formula,

$$x = \frac{-(-3) \pm \sqrt{(-3)^2 - 4(3)(5)}}{2(3)}$$

$$x = \frac{3 \pm \sqrt{9 - 60}}{6}$$

$$x = \frac{3 \pm \sqrt{-51}}{6}$$

So the radicand is negative and there is *no real solution*.

Example 24

Use the quadratic formula to solve for I in this formula for voltage in an electric circuit.

$$E = IR + \frac{P}{I}$$

Solution: To put the equation in general form, we first eliminate fractions by multiplying both sides by I,

$$EI = I^2R + P$$

Then,

$$RI^2 - EI + P = 0$$

Substituting into the quadratic formula,

$$I = \frac{E \pm \sqrt{E^2 - 4RP}}{2R}$$

As we have seen, the quadratic formula may be used to solve *any* quadratic equation. It is a very important and useful method and the formula should be *memorized*.

Solve these quadratic equations. Give irrational solutions in simplest radical form and approximate them to three significant digits. If there are no real number solutions, write *no real solution.*

Completing the Square

Solve by completing the square.

1. $x^2 + 6x - 2 = 0$ **2.** $x^2 - 8x - 3 = 0$

3. $3x^2 - 4x - 8 = 0$ **4.** $5x^2 - 2x - 1 = 0$

Solution of a Quadratic Equation by the Quadratic Formula

Solve by using the quadratic formula.

5. $x^2 + 5x - 1 = 0$ **6.** $x^2 - 4x - 2 = 0$

7. $x^2 - 3x + 4 = 0$ **8.** $x^2 + 2x + 6 = 0$

9. $2x^2 - 7x + 4 = 0$ **10.** $4x^2 - 2x - 1 = 0$

11. $2x^2 + 3x + 2 = 0$ **12.** $3x^2 - 5x + 2 = 0$

13. $x^2 + 2x = 1$ **14.** $x^2 - 7 = -2x$

15. $x^2 = 3$ **16.** $x^2 = 5x$

17. $x^2 + 5 = 2x$ **18.** $2x^2 = x + 6$

19. $3x^2 + 1 = 6x$ **20.** $5x^2 + 8x = -2$

Solve by factoring or by the quadratic formula.

21. $x^2 - 3x = 10$ **22.** $x^2 = 6x + 7$

23. $x^2 + 8 = -8x$ **24.** $x^2 + 2 = 6x$

25. $2x^2 + 6x = 3$ **26.** $5x^2 + 10x = -1$

27. $3x^2 + 5x = 2$ **28.** $2x^2 + x = 3$

29. $x^2 - \dfrac{4}{3}x = 1$ **30.** $\dfrac{1}{3}x^2 + \dfrac{3}{2}x = 3$

31. $\dfrac{2}{x} + 3x = 8$ **32.** $-\dfrac{1}{x+1} = \dfrac{3}{x} + 2$

Solve to three significant digits.

33. $5.21x^2 + 6.13x - 9.85 = 0$ **34.** $0.956x^2 + 0.852x = 12.1$

35. $3.67x^2 + 12.2x = 35.6$ **36.** $28.3x^2 + 29.0x - 7.52 = 0$

Solve these applications. Round to three significant digits.

Applications

37. The work w done in a circuit varies with time t (in milliseconds) according to the formula

$$w = 6t^2 - 8t + 14$$

Find the time(s) when the work is 12.5.

38. A circular highway curve of radius R and roadway width w has a visual obstacle beside the roadway. The sight distance C is given by the formula

$$C^2 = 4Rw - w^2$$

How wide would a roadway have to be to have a sight distance of 525 ft for a curve with a 2050-ft radius?

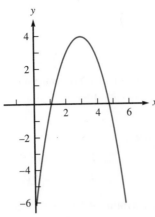

■ FIGURE 21-6

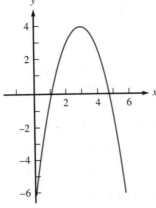

■ FIGURE 21-7

39. The path of an object following a parabolic curve (Fig. 21-6) is described by the following equation:

$$y = -1.12x^2 + 6.52x - 5.07$$

Find the x coordinate(s) when $y = 3.00$.

40. The total surface area A of a cone with radius r and slant height s is given by

$$A = \pi r^2 + \pi rs$$

Find the radius of the cone in Fig. 21-7 if the total surface area is 112 cm².

Write each equation in general form, and use the quadratic formula to solve for the variable in parentheses.

41. The surface area of a cylinder:

$$A = 2\pi r^2 + 2\pi rh \qquad (r)$$

42. Displacement:

$$s = vt + \frac{1}{2}gt^2 \qquad (t)$$

43. The sum of the first n positive integers:

$$S = \frac{1}{2}n(n + 1) \qquad (n)$$

44. The bending moment of a simply supported beam:

$$M = \frac{1}{2}wlx - \frac{1}{2}wx^2 \qquad (x)$$

21-4 APPLICATIONS OF QUADRATIC EQUATIONS

We now solve applications that require us to solve quadratic equations. You may wish to review the problem-solving techniques from Chapter 11. You should set these problems up as you did then. In this section, we solve geometric problems involving area and motion problems involving constant acceleration.

When we solve a quadratic equation, we have found that there are always two roots. We discard any roots that do not make sense in the physical problem (like a width of −3.62 meters). But do not be too hasty in rejecting solutions. Sometimes a second root will give an unexpected but equally good answer.

Area Problems

Quadratic equations are often derived from geometric problems involving areas.

Example 25

A rectangular plate of steel is needed in which the length is 3.06 m greater than twice its width and its area is 9.68 m². Find its dimensions.

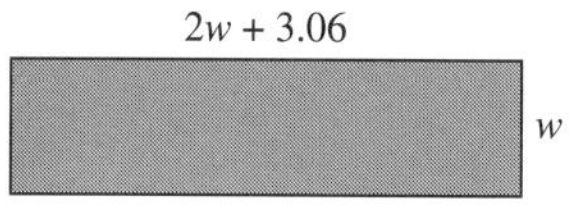

Solution: First we sketch and label the rectangle (Fig. 21-8). We let w = the width of the rectangle. The length l will be

$$l = 2w + 3.06$$

2w + 3.06

w

FIGURE 21-8

We use Eq. 69 for the area of a rectangle:

$$A = 1 \cdot w$$

Substituting,

$$9.68 = (2w + 3.06)(w)$$
$$9.68 = 2w^2 + 3.06w$$
$$2w^2 + 3.06w - 9.68 = 0$$

We solve the equation by the quadratic formula:

$$w = \frac{-3.06 \pm \sqrt{3.06^2 - 4(2)(-9.68)}}{2 \cdot 2}$$

By calculator, we get the two solutions,

$$w = 1.564 \quad \text{and} \quad w = -3.094$$

rounded to three significant digits. Since we cannot have a negative width, we reject the second solution, $w = -3.094$.
Substituting $w = 1.564$ into the equation

$$l = 2w + 3.06 = 2(1.564) + 3.06$$

we find that

$$l = 6.19$$

So the dimensions of the rectangle are

$$1.56 \text{ m by } 6.19 \text{ m}$$

Uniformly Accelerated Motion

Uniformly accelerated motion refers to the motion of an object which has a uniform (constant) acceleration. However, its velocity is constantly changing. If an object, with an initial velocity v_0, moves with uniform acceleration a, then the distance s after time t is given by

$$s = v_0 t + \frac{1}{2} a t^2$$

An object that is dropped or thrown is an example of an object with uniformly accelerated motion—the uniform (constant) acceleration is the acceleration due to gravity. The acceleration due to gravity is called g, where

$$g = \pm 32.2 \text{ ft/s}^2 \quad \text{or} \quad g = \pm 9.81 \text{ m/s}^2$$

depending on whether measurements are given in English or metric units.

Example 26

If a diver falls from a diving board 10.0 m above the water, how much later does he strike the water?

Solution: Since the diver falls, the initial velocity is 0 and the acceleration is due to gravity ($g = 9.81$ m/s^2).
Using the equation

$$s = v_0 t + \frac{1}{2} a t^2$$

with $v_0 = 0$ and $a = g$, we get

$$s = \frac{1}{2} g t^2$$

Substituting $s = 10.0$ and $g = 9.81$ into this equation, we have

$$10.0 = \frac{1}{2}(9.81)t^2$$

$$t = \pm \sqrt{\frac{20}{9.81}}$$

$$t = \pm 1.43 \text{ seconds}$$

Since a negative time makes no sense in this problem, we reject the negative root.

Check: We check the positive root, $t = 1.43$:

$$10.0 \stackrel{?}{=} \frac{1}{2}(9.81)(1.43)^2$$

$$10.0 = 10.0 \quad \text{checks}$$

So it takes about 1.43 seconds for the diver to fall 10.0 meters off the diving board.

Example 27

From the top of a 205-ft building, a ball is thrown toward the ground with an initial velocity of 42.5 feet per second. After how many seconds will the ball hit the ground?

Solution: Use the equation

$$s = v_0 t + \frac{1}{2} g t^2$$

Substituting $s = 205$, $v_0 = 42.5$, and $g = 32.2$, we get

$$205 = 42.5t + \frac{1}{2}(32.2)t^2$$

Simplifying,

$$16.1t^2 + 42.5t - 205 = 0$$

Using the quadratic formula to solve for t,

$$t = \frac{-42.5 \pm \sqrt{42.5^2 - 4(16.1)(-205)}}{2(16.1)}$$

Evaluating by calculator, we get the two approximate solutions,

$$t = 2.48 \qquad t = -5.12$$

Again, we reject the negative solution, $t = -5.12$ seconds, since it doesn't make sense in this problem. So the ball hits the ground after about 2.48 seconds.

With uniform motion problems, we must pay particular attention to the directions (up or down) in the problem. In Example 27, all directions were *down*—the ground was *below* the top of the building, the ball was thrown *down*, and gravity acts *down* (toward the earth). So we actually chose *down* as the positive direction. In the next example, some directions are up and some are down, so we must choose an appropriate positive direction.

Example 28 ──────────

A ball is thrown upward at 112 ft/s. In how many seconds will the ball be 125 ft above the ground?

Solution: The ball is thrown upward and the distance is 125 ft above the gound, so we choose up as the positive direction. But the force of gravity pulls the ball down (toward the ground), so that g must be negative ($g = -32.2$ ft/s^2). Using the equation

$$s = v_0 t + \frac{1}{2} g t^2$$

we substitute $s = 125$, $v_0 = 112$, and $g = -32.2$, and we get

$$125 = 112t + \frac{1}{2}(-32.2)t^2$$

Simplifying,

$$125 = 112t - 16.1t^2$$

$$16.1t^2 - 112t + 125 = 0$$

We use the quadratic formula:

$$t = \frac{112 \pm \sqrt{(-112)^2 - 4(16.1)(125)}}{2(16.1)}$$

Using a calculator, we find that the two approximate solutions are

$$t = 1.40 \quad \text{and} \quad t = 5.56$$

We find that both solutions check, so the ball passes 125 ft above ground at about 1.40 seconds (on the way up) and again at about 5.56 seconds (on the way down).

SECTION 21-4 Exercises Applications of Quadratic Equations

First write the equation or equations for each problem. Then solve each problem, using either factoring or the quadratic formula. Reject solutions that do not meet the conditions of the problem. Round answers appropriately.

Area Problems

1. A reception hall is three times as long as it is wide. Find its dimensions if the area is 1450 ft^2.

2. Find the length and width of a rectangular field containing 8.50 acres if the length is twice the width. (1 acre = 43,560 ft^2)

3. A small rectangular garden is to be 3.50 m longer than it is wide and to have an area of 28.0 m^2. Find its dimensions.

4. The area of a rectangular sheet of tin is 154 cm^2. If the sheet is 3.75 cm longer than it is wide, find the dimensions of the sheet.

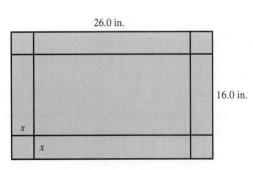

26.0 in.

16.0 in.

x

x

■ FIGURE 21-9

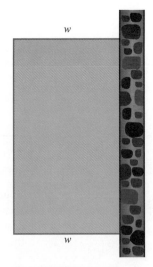

w

w

■ FIGURE 21-10

5. A technician must make an open pan by cutting squares from each corner of a 16.0-in. by 26.0-in. rectangular sheet of aluminum and folding up the sides (Fig. 21-9). If the base of the pan is to have an area of 200.0 in.^{2}, what should be the length of the side of each square?

6. A farmer makes a rectangular enclosure using a stone wall for one side and 200.0 yd of fence for the other three sides (Fig. 21-10). Find the dimensions of the enclosure if the area enclosed is 1.00 acre. (1 acre = 4840 yd^2)

7. A bricklayer has bricks to lay 100.0 m^2 of walkway around a square garden (Fig. 21-11). The area of the garden and the area of the walkway are to be the same. How wide should he make the walkway?

8. A portrait 45.0 cm by 60.0 cm is to be set in a frame of uniform width so that the area of the frame is equal to that of the portrait (Fig. 21-12). Find the uniform width of the frame.

Uniformly Accelerated Motion

9. An object is dropped off the top of a 104-ft building. Determine how long it takes for the object to hit the ground.

10. How long will it take a drop of water to reach the bottom of a 153-ft waterfall? (Assume $v_0 = 0$.)

■ FIGURE 21-11

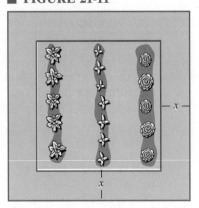

x

x

■ FIGURE 21-12

x

x

11. If supplies are dropped from an airplane 258 m above the ground, how much later do they hit the ground?

12. Gravity on the moon is about $\frac{1}{6}$ the gravity on earth. About how many seconds does it take an object to fall 4.00 m from the door of the lunar module to the surface of the moon?

13. An object is thrown downward with a velocity of 8.00 m/s from the top of a 185-m building. In how many seconds will it hit the ground?

14. From an airplane flying at an altitude of 2560 feet, an object is thrown downward at a rate of 96.0 feet per second. In how many seconds will it strike the ground?

15. A projectile is shot upward at 92.0 ft/s. After how many seconds will the projectile first be 125 feet above ground?

16. An object is projected upward at 98.6 m/s. In how many seconds will the object first be 275 m above the ground?

17. A missile is shot upward at 147 m/s. In how many seconds will the missile hit the ground?

18. A ball is thrown directly upward at 128 ft/s. How long will it take the ball to return to ground level?

19. An object is projected upward at 196.2 m/s.

    (a) In how many seconds will the object be 1962 m above the ground?

    (b) In how many seconds will the object be 1250 m above the ground? Why are there two answers?

    (c) In how many seconds will the object hit the ground?

20. A ball is thrown upward at 96.6 feet/s.

    (a) In how many seconds will the object be 144.9 feet above the ground?

    (b) In how many seconds will the ball be 65.5 feet above the ground? Why are there two answers?

    (c) In how many seconds will the ball hit the ground?

21-5   APPLICATIONS INVOLVING RATES

Many applications involving rates lead to quadratic equations. These include uniform motion problems, rates of pay, and rates of work.

Uniform Motion

Recall that if an object is traveling with uniform velocity, the distance D the object travels is given by the equation

$$D = r \cdot t$$

where r is the uniform velocity rate and t is the time.

Example 29

A woman makes a regular trip of 365 miles by car, always at the same average speed. One day, because of bad weather, her average speed was reduced by 9.50 mi/h. As a result, her trip took 2.00 hours longer. What was her regular average speed?

Solution: We make a table:

	r	$\times$	t	$=$	D
Regular run	r		t		365
Bad weather run	$r - 9.50$		$t + 2.00$		365

From this table we get the two equations:

$$r \cdot t = 365$$

and

$$(r - 9.50)(t + 2.00) = 365$$

We wish to find r, so we eliminate t. Solving for t in the first equation,

$$t = \frac{365}{r}$$

Substituting into the second equation,

$$(r - 9.50)\left(\frac{365}{r} + 2.00\right) = 365$$

Multiplying both sides by r,

$$(r - 9.50)(365 + 2.00r) = 365r$$

Simplifying,

$$365r + 2.00r^2 - 3467.5 - 19.0r = 365r$$

$$2.00r^2 - 19.0r - 3467.5 = 0$$

We solve this equation using the quadratic formula. At this point, a calculator program which solves quadratic equations in general form would be very useful. We find the solutions are

$$r = 46.7 \quad \text{and} \quad r = -37.2$$

We reject the negative answer, since it makes no sense in this problem. The only possible solution is $r = 46.7$ mi/h.

Check: We calculate how long it would take the woman to go 365 miles at 46.7 mi/h:

$$365 \div 46.7 = 7.82 \text{ hours}$$

In bad weather, she went 9.50 mi/h slower,

$$46.7 - 9.50 = 37.2 \text{ mi/h}$$

and it took her 2.00 hours longer,

$$7.82 + 2.00 = 9.82 \text{ hours}$$

The distance traveled then was

$$37.2 \cdot 9.82 = 365 \text{ checks}$$

So the average speed on her regular run is 46.7 mi/h.

Other Rate Problems

Other types of problems involving rates may also lead to quadratic equations. These problems may concern rates of pay, rates of cost, or rates of work.

Example 30

Last year a computer company divided $7200.00 in Christmas bonuses equally among its employees. This year the company employed 6 more people than last year, and therefore each employee received $60 less. How much did each employee receive last year?

Solution: First we note that an important equation in this problem is

$$\text{(bonus rate)} \cdot \text{(number of employees)} = \text{Total bonus}$$

Using this equation, we make a table,

	Bonus rate	× No. of employees	= Total bonus
Last year	r	n	7200
This year	$r - 60$	$n + 6$	7200

From the table we get the two equations,

$$r \cdot n = 7200$$

and

$$(r - 60)(n + 6) = 7200$$

We wish to find r, so we eliminate n. Solving for n in the first equation, we have

$$n = \frac{7200}{r}$$

Substituting into the second equation,

$$(r - 60)\left(\frac{7200}{r} + 6\right) = 7200$$

$$(r - 60)(7200 + 6r) = 7200r$$

$$(r - 60)(1200 + r) = 1200r$$

$$1200r + r^2 - 72000 - 60r = 1200r$$

$$r^2 - 60r - 72000 = 0$$

Using the quadratic formula, we find the solutions are

$$r = 300 \quad \text{and} \quad r = -240$$

(We could have factored this quadratic, but most quadratics from applications are not factorable.) Rejecting the negative root as inappropriate for this problem, the only possible solution is $r = \$300$.

Check: If the company paid $300 in bonuses last year, for a total of $7200, there must have been 24 employees ($7200 \div 300 = 24$). This year there are 6 more, or 30 employees, and the bonus rate went down $60 to $240 per employee. Does this amount to $7200?

$$30 \cdot \$240 = \$7200 \quad \text{checks}$$

So, each employee received $300 as a Christmas bonus last year.

First write the equation or equations for each problem. Then solve each problem. Reject solutions that do not meet the conditions of the problem. Round answers appropriately.

Uniform Motion

Solve by factoring or the quadratic formula.

1. A plane, making a 3000-mile flight, encountered a head wind of 100 mi/h for the entire trip, making it one hour late in arriving. What was its ground speed?

2. A man made a trip of 495 miles. On his return trip, the weather was bad, and his average speed was reduced by 10 mi/h. His total driving time was 20 hours. Find his average speed each way.

3. A motorboat takes 1 hour less to go 36 miles down a river than to return. If the river flows at 3 mi/h, find the rate at which the boat travels in still water.

4. A motorboat heads south from a dock at the same time that a sailboat heads west. The sailboat travels 7 mi/h slower than the motorboat. The boats are 13 miles apart at the end of one hour. Find the rate of each boat.

Solve by quadratic formula (or calculator program), and round to three significant digits.

5. A jet plane, making a routine 3250-mile cross-country flight, encountered a head wind of 112 mi/h for the entire trip, making it 1.00 hour late in arriving. What was its groundspeed?

6. A canoeist paddles 1670 yards across a lake. If he had paddled 15.0 yd/min faster, he would have arrived at the other side 10.5 minutes earlier. How fast was he paddling?

7. A woman made a trip of 575 miles. On her return trip, it was snowing, and her average speed was reduced by 10.4 mi/h. As a result, her return trip took 2.25 hours longer. Find her average speed each way.

8. A train goes 208 miles to city A. Then, with better track, it continues on to city B, 321 miles from city A, at a speed of 15.4 mi/h greater than on the first leg of the trip. The total travel time was 6.25 hours. Find the speed at which the train traveled to city A.

9. A man made a trip of 495 miles. On his return trip, it was icy, and his average speed was reduced by 9.75 mi/h. His total driving time was 19.5 hours. Find his average speed each way.

10. The average speed of a plane in still air is 425 mi/h. With a constant wind blowing, the plane flies 906 miles with a tail wind and returns with a head wind. If the round trip takes 4.25 hours of flying time, find the wind speed.

Other Rate Problems

Solve by factoring or quadratic formula.

11. The members of a technical organization decided to donate $2000 for a scholarship, and to share the cost of the contribution equally. If the organization had 20 more members, each would have had to contribute $5 less. How many members now belong to the organization?

12. A woman earned $1100 in a certain number of days. If her daily wage had been $10 more, she would have taken one less day to earn the same amount. How many days did she work at the lower rate?

13. A stonemason built 54 m of wall. If he had built 3 m less each day, it would have taken him 3 days longer. How many meters did he build each day, on the average?

14. A technician put together 48 computer boards. If she had put together 2 fewer boards each day, then it would have taken her 2 days longer to build those 48 boards. How many days did it take her to build the 48 boards?

15. Two resistors wired in series will give a total resistance R_T, where $R_T = R_1 + R_2$. Two resistors wired in parallel will give a total resistance R_T, where

$$\frac{1}{R_T} = \frac{1}{R_1} + \frac{1}{R_2}$$

What two resistors will give a total resistance of 500 ohms when wired in series and 100 ohms when wired in parallel? (Round to the nearest ohm.)

21-6

QUADRATIC FUNCTIONS

Recall from Sec. 12-5 that a quadratic function has the form

$$y = ax^2 + bx + c$$

However, quadratic functions from applications are not usually expressed with the variables x and y. Self-descriptive variables, such as t for time and r for radius, are generally used.

Example 31

These quadratic functions are examples from math, science, and technology.

(a) The area A of a circle is a quadratic function of the radius r:

$$A = \pi r^2$$

(b) The kinetic energy K of a moving body is a quadratic function of the velocity v:

$$K = \frac{1}{2} m v^2$$

(c) The displacement s of a freely falling body is a quadratic function of the time t:

$$s = vt + \frac{1}{2} g t^2$$

(d) The power P in a resistor is a quadratic function of the current I:

$$P = EI - RI^2$$

Graph of a Quadratic Function

Recall that the graph of a quadratic function is called a *parabola*. The parabola will open up or it will open down. A parabola is curved at all points (never straight). Every parabola has a maximum or a minimum point, called the *vertex*. The function may be graphed by making a table of values and plotting points, or by using a graphing calculator.

Applications

There are many examples of quadratic functions in math, science, and technology. We graph these functions, placing the independent variable along the horizontal axis and the dependent variable along the vertical axis.

Example 32 ────────

The kinetic energy K (in joules) of a moving object with mass m (in kg) and velocity v (in m/s) is given by

$$K = \frac{1}{2}mv^2$$

(a) On graph paper, graph the kinetic energy of a 2-g rifle bullet as a function of its velocity, up to $v = 500$ m/s.

(b) Graph again, using a graphing calculator.

(c) Use the calculator to find the kinetic energy of the bullet at 355 m/s.

Solution:

(a) The formula above works for mass in kilograms and velocity in m/s. The velocity is in m/s, and we must convert the mass in grams to kilograms,

$$2\,\text{g} = 0.002\,\text{kg}$$

Substituting $m = 0.002$ into the equation for K we have

$$K = \frac{1}{2}(0.002)v^2$$

$$K = (0.001)v^2$$

The graph of this equation will be a parabola. In this problem, only positive velocities are meaningful, so our graph will be a partial parabola. We make a table:

v	0	100	200	300	400	500
K	0	10	40	90	160	250

We choose appropriate scales for each axis, plot the points, and connect them with a smooth curve (Fig. 21-13).

(b) Using a graphing calculator, we let the independent variable v be x and the dependent variable K be y. We graph

$$y = 0.001x^2$$

with a window of $0 \le x \le 500$ and $0 \le y \le 250$. The graph is identical to Fig. 21-13.

(c) To find the kinetic energy when $v = 355$ m/s, we use the "value" feature of the calculator, letting $x = 355$, to find that

$$y = 126$$

So the kinetic energy of the bullet is about 126 joules at 355 m/s.

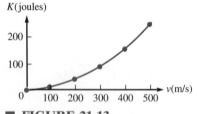

K(joules)

■ FIGURE 21-13

 Example 33 ────────

A ball is thrust upward at 112 ft/s. Express the distance s of the ball above the ground as a function of time t, and **(a)** graph the function by hand, **(b)** graph the function by graphing calculator, and **(c)** use the calculator to find how long it takes for the ball to reach its highest point, to three significant digits.

Solution: We use the equation

$$s = vt - \frac{1}{2}gt^2$$

with $v = 112$ ft/s and $g = 32.2$ ft/s². Making these substitutions,

$$s = 112t - 16.1t^2$$

or

$$s = -16.1t^2 + 112t$$

(a) We make a table, choosing integral values of t until we get a negative value for s. (A negative value of s indicates that the ball has hit the ground, and the equation no longer holds.)

t	0	1	2	3	4	5	6	7
s	0	95.9	159.6	191.1	190.4	157.5	92.4	−4.9

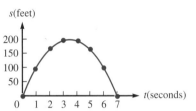

s(feet)

t(seconds)

■ **FIGURE 21-14**

We plot these points and connect them with a smooth curve, stopping between $t = 6$ and $t = 7$ where the graph goes below the t axis (Fig. 21-14).

(b) Use a graphing calculator to graph

$$y = -16.1x^2 + 112x$$

with $0 \le x \le 7$ and $0 \le y \le 200$. The graph is identical to Fig. 21-14.

(c) We need to find x (the time) at the vertex (the maximum point) of the parabola. On the graphing calculator, we use the "maximum" feature (Fig. 21-15) to find that:

$$x = 3.48 \qquad \text{at the highest point}$$

So it takes the ball about 3.48 seconds to reach its highest point.

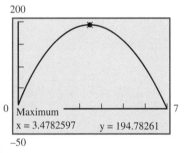

Maximum
x = 3.4782597   y = 194.78261

■ **FIGURE 21-15**

Graphical Solution of a Quadratic Equation

We can solve a quadratic equation graphically by graphing the quadratic function, and finding the x intercepts.

■ **FIGURE 21-16**

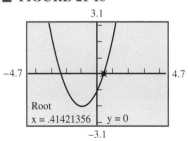

Root
x = .41421356   y = 0

─── **Example 34** ───

Solve graphically, rounding to three significant digits.

$$x^2 + 2x = 1$$

Solution: First we bring all terms to one side of the equals sign:

$$x^2 + 2x - 1 = 0$$

Now we graph the function

$$y = x^2 + 2x - 1$$

and use the "root" or "zero" feature to find that the x intercepts are -2.41 and 0.414 (Fig. 21-16).

Graph the function, placing the independent variable along the horizontal axis and the dependent variable along the vertical axis, using appropriate scales for both axes. If using the graphing calculator, determine the appropriate x and y ranges, graph the function, and sketch the graph on graph paper. Round answers to three significant digits.

1. The area of a circle with radius r is given by

$$A = \pi r^2$$

(a) Graph A as a function of r, for radii from 0 to 5 cm.

(b) Graphically find the area when the radius is 2.50 cm.

2. The power P (in watts) dissipated in an electrical load is given by

$$P = \frac{V^2}{R}$$

where V is the voltage (in volts) across the load and R is the resistance (in ohms) of the load.

(a) If the resistance is 100 ohms, graph P as a function of V for $0 \le V \le 30$ volts.

(b) Graphically find the power dissipated if the voltage across the load is 18.0 volts.

3. The equation for the height h (in m) of a certain parabolic road arch at a distance d (in m) from the center of the road (Fig. 21-17) is given by

$$h = 30 - \frac{1}{15}d^2$$

(a) Graph h as a function of d for $-21 \le d \le 21$.

(b) Find the height of the arch 6.70 m from the center of the road.

4. A projectile is shot upward at 96.6 ft/s.

(a) Graph its distance s above the ground as a function of time t, for t between 0 and 6 seconds.

(b) After how many seconds does it reach its highest point?

(c) What is its maximum height?

■ FIGURE 21-17

5. The cost c (in dollars) for manufacturing n items is approximated using the formula

$$c = -4n^2 + 5n + 625$$

   (a) Graph c as a function of n for up to 10 items.

   (b) Find the cost for manufacturing 6 items.

6. A variable electric current I (in amps) is given by the formula

$$I = t^2 - 9t + 23$$

   where t is the time in seconds.

   (a) Graph I as a function of t for $0 \le t \le 10$ s.

   (b) For what times is the current 12.5 amps?

7. The volume V of a cylinder with radius r and height h is given by

$$V = \pi r^2 h$$

   (a) Graph V as a function of r for cylinders with a constant height $h = 10$ in., for radii up to 8 in.

   (b) Find the radius when the volume is 1250 in³.

8. In a certain circuit, the power P (in watts) dissipated in the load resistor R (in ohms) with current I (in amps) is

$$P = EI - I^2 R$$

   (a) If the voltage E is 120 volts and $R = 4$ ohms, graph P as a function of I, for I between 0 and 30 amps.

   (b) Find the current when the power dissipated is 575 watts.

9. The stopping distance d (in feet) for a car traveling at velocity v (in mi/h) is given by

$$d = 0.044v^2 + 1.1v$$

   (a) Graph the stopping distance as a function of the velocity, for velocities from 0 to 100 mi/h.

   (b) If it took a car 165 ft to stop, find the car's speed when the brakes were first applied.

Solve each equation graphically. Round to three significant digits.

10. $2x^2 + 10 = 9x$ 11. $10x^2 + 3 = 11x$

12. $3x^2 + 2 = 8x$ 13. $3x^2 - 4x = 3$

■ **Chapter 21 REVIEW EXERCISES**

1. Which of the following are quadratic equations?

   (a) $x^3 - 3x^2 - x = 8$ **(b)** $5x - x^2 = 3$

   (c) $\dfrac{1}{2} + \dfrac{2}{x} = \dfrac{3}{x^2}$ **(d)** $\dfrac{1}{4}x^2 - 3x = 1$

   (e) $3x^2 - 2 = 0$ **(f)** $5 = 2x$

Write in general form, with no fractions and $a > 0$.

2. $3x - x^2 = 4$

3. $\frac{1}{2}x^2 = \frac{3}{4}$

4. $(2x - 3)(x + 2) = 1$

Determine which, if any, of the given values are solutions to the given equation.

5. $5x^2 + 9x - 2 = 0;\ x = -1, 1, -2, 2$

6. $(x - 1)(x + 2) = -4;\ x = 2, -2, 1, -3$

Solve and check.

7. $64 - x^2 = 0$

8. $3x^2 - 75 = 0$

9. $2x^2 + 3 = 3$

10. $49x^2 - 4 = 0$

11. $x^2 - 5 = 0$

12. $5.25x^2 - 3.75 = 0$

13. $3.46x^2 + 5.78 = 3.64$

Solve these formulas for the variable in parentheses. Write in simplest radical form.

14. $A = \pi r^2 \qquad (r)$

15. $s = \frac{1}{2}gt^2 \qquad (t)$

16. The power P (in watts) dissipated in an electrical load is given by

$$P = \frac{V^2}{R}$$

where V is the voltage (in volts) across the load and R is the resistance (in ohms) of the load. What is the voltage across a load whose power is 250 mW and resistance is 324 ohms?

Solve by factoring, and check.

17. $5x(x + 2) = 0$

18. $2x^2 = 10x$

19. $(2x + 1)(x - 1) = 0$

20. $x^2 + 15 = 8x$

21. $x^2 + 4x + 4 = 0$

22. $4x^2 + 18x + 20 = 0$

23. $6x^2 + x = 2$

24. $3x^2 + 3 = 6x$

25. $16x^2 - 9 = 0$

26. $8x^2 = 18$

27. $6x^2 - 6 = 5x$

28. $8x^2 + 3 = 14x$

29. $12 = x^2 - 11x$

30. A charge q (in coulombs) flows in a given circuit according to the formula

$$q = 2t^2 - 6t + 6$$

where t is time in microseconds. Find the time when the charge is equal to

(a) 2.0 coulombs

(b) 6.5 coulombs

Solve by completing the square. Give all irrational answers in simplest radical form and also approximate to three significant digits.

31. $x^2 - 7x + 2 = 0$

32. $2x^2 + 12x - 7 = 0$

Solve by using the quadratic formula. Give all irrational answers in simplest radical form and also approximate to three significant digits.

33. $x^2 - 4x + 1 = 0$

34. $2x^2 - 3x - 9 = 0$

35. $5x^2 + 6x + 4 = 0$

36. $2x^2 + 4x = 3$

37. $10 = 17x - 3x^2$

38. $2x^2 = 4x - 1$

39. $\frac{1}{6}x^2 + x = \frac{8}{3}$

40. $9x + \frac{25}{x} = 42$

41. $5.75x - 2.25x^2 = 1.25$

42. $3.01x^2 - 2.24 = 0.983x$

43. Write the equation in general form and use the quadratic formula to solve for r in this formula for the surface area A of a cone with slant height s:

$$A = \pi r^2 + \pi rs$$

44. In a certain circuit, the power P (in watts) dissipated in the load resistor R (in ohms) is

$$P = EI - I^2R$$

where E is the voltage (in volts) and I is the current (in amps). If the voltage E is 125 volts and $R = 103$ ohms, find the current I needed to produce a power of

(a) 20.0 watts

(b) 30.0 watts

45. The total surface area A of a cylinder with radius r and height h is given by

$$A = 2\pi r^2 + 2\pi rh$$

Find the radius of the cylindrical soup can in Fig. 21-18, if its surface area is 275 cm².

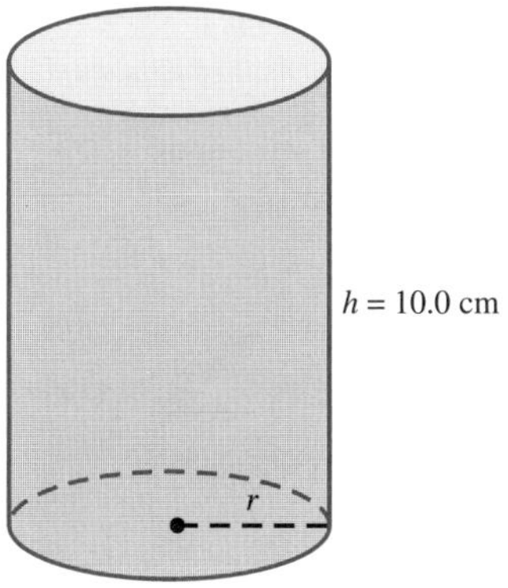

46. A rectangular field is fenced in with 725 m of fence. Find its dimensions if the area is 32,500 m².

47. If a flowerpot falls from a window and lands 78.4 m below, how many seconds does it fall?

48. An object is thrown downward from the top of a 285-ft building at a rate of 24.3 ft/s. In how many seconds does it hit the ground?

49. An object is projected upward at 39.24 m/s.

(a) In how many seconds will the object be 30.0 m above the ground?

(b) In how many seconds will the object hit the ground?

50. A jet flies a regular run of 7250 miles. With a tail wind of 103 mi/h, the trip takes 1.50 hours less than if there is no wind. How long does the regular (windless) run take?

51. A technician bought some transistors for $16.20. If she had received 3 more transistors for the same total price, then each transistor would have been 6 cents less. How much was each transistor?

Graph the function, placing the independent variable along the horizontal axis and the dependent variable along the vertical axis, using appropriate scales for both axes.

52. The volume V of a sphere with radius r is given by

$$V = \frac{4}{3}\pi r^3$$

(a) Graph V as a function of r for spheres for radii up to 8.00 cm.

(b) Find the volume for a sphere of radius 4.50 cm.

(c) Find the radius of a sphere of volume 125 cm³.

$h = 10.0$ cm

r

■ **FIGURE 21-18**

WRITING

53. Describe four different methods of solving a general quadratic equation. What are the advantages and/or disadvantages of each method? Make up a quadratic equation and use it to illustrate all four methods of solution. Which do you prefer, and why?

54. A baseball player hits a pop fly directly above home plate. The ball is 1.00 m above ground when the player hits the ball. The ball goes upward at an initial velocity of 40.0 m/s.

(a) What is the height of the ball after 3.00 seconds?

(b) When is the ball 30.0 m above the ground?

(c) When does the ball hit the ground, if the catcher misses it?

(d) How high is the ball at its highest point?

(e) On the moon, the force of gravity is about one-sixth that on earth. How high would the ball have gone if it had been hit on the moon?

Oblique Triangles

So far we have worked with only acute angles and right triangles. In this chapter we begin dealing with angles of any size and triangles that are not right triangles, *oblique triangles*. We extend the trigonometric functions of acute angles to angles of any size. We then derive the law of sines and the law of cosines, formulas that enable us to solve oblique triangles. Finally, we solve applications involving oblique triangles.

22-1   TRIGONOMETRIC FUNCTIONS OF ANY ANGLE

In this section, we extend our use of the six trigonometric functions to angles of any size. We examine the signs of the trigonometric functions in each quadrant and consider some special angles called *quadrantal* angles. Finally, we determine the trigonometric functions of any given angle, using the calculator.

The Trigonometric Functions of Any Angle

In Chapter 15, we defined the six trigonometric functions of an angle θ, in standard position (Fig. 22-1), with a point (x, y) on its terminal side, as follows:

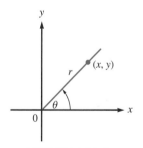

■ **FIGURE 22-1**
Standard position of an angle.

TRIGONOMETRIC FUNCTIONS		
	$\sin \theta = \dfrac{y}{r}$	110
	$\cos \theta = \dfrac{x}{r}$	111
	$\tan \theta = \dfrac{y}{x}$	112
	$\csc \theta = \dfrac{r}{y}$	113
	$\sec \theta = \dfrac{r}{x}$	114
	$\cot \theta = \dfrac{x}{y}$	115

where $r = \sqrt{x^2 + y^2}$. We used these functions for acute angles in quadrant I only but stated that the definitions were true for angles in any quadrant. Now we consider angles of any size and in any quadrant.

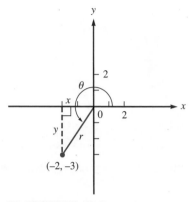

■ FIGURE 22-2

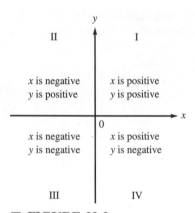

■ FIGURE 22-3
Signs of *x* and *y*.

■ FIGURE 22-4
Positive trigonometric
functions.

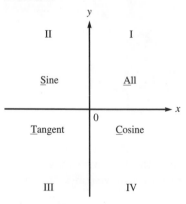

┌─── **Example 1** ───────────

The point $(-2, -3)$ is on the terminal side of a positive angle θ in standard position. Sketch the angle and write the six trigonometric functions of θ, rounded to four significant digits.

Solution: We plot the point $(-2, -3)$ and draw the terminal side of the angle from the origin through the point (Fig. 22-2), noting that this is a quadrant III angle. Since the angle is positive, we draw the *arc* showing a counterclockwise rotation from the positive *x* axis to the terminal side, and label the angle θ. Drawing the perpendicular to the *x* axis, we find the distance *r*

$$r = \sqrt{x^2 + y^2}$$
$$= \sqrt{(-2)^2 + (-3)^2}$$
$$= \sqrt{13}$$

We leave *r* in radical form, as $\sqrt{13}$, rather than using its decimal approximation, so that we don't introduce rounding error. So, by Eq. 110–115, with $x = -2$, $y = -3$, and $r = \sqrt{13}$,

$$\sin \theta = \frac{y}{r} = \frac{-3}{\sqrt{13}} = -0.8321$$

$$\cos \theta = \frac{x}{r} = \frac{-2}{\sqrt{13}} = -0.5547$$

$$\tan \theta = \frac{y}{x} = \frac{-2}{-3} = -0.6667$$

$$\csc \theta = \frac{r}{y} = \frac{\sqrt{13}}{-3} = -1.202$$

$$\sec \theta = \frac{r}{x} = \frac{\sqrt{13}}{-2} = -1.803$$

$$\cot \theta = \frac{x}{y} = \frac{-3}{-2} = 1.500$$

────────────────────────────────┘

Algebraic Signs of the Trigonometric Functions

In Chapter 15, we saw that the six trigonometric functions of acute (quadrant I) angles were always positive. From the preceding example, it is clear that some of the trigonometric functions of angles in quadrants II, III, and IV are negative, because *x* or *y* (or both) may be negative. (Remember that *r* is always positive.)

The trigonometric functions are defined in terms of *x*, *y*, and *r*. Since *r* is always positive, the signs of the functions depend on the signs of *x* and *y*. Figure 22-3 shows the signs of *x* and *y* in each of the four quadrants.

Now let's determine the quadrants in which the sine, cosine, and tangent are positive:

1. Since $\sin \theta = \frac{y}{r}$, the sine will be positive when *y* is positive: quadrants I and II.

2. Since $\cos \theta = \frac{x}{r}$, the cosine will be positive when *x* is positive: quadrants I and IV.

3. Since $\tan \theta = \frac{y}{x}$, the tangent will be positive when *x* and *y* have the same sign: quadrants I and III.

This is summarized in Fig. 22-4.

Some students learn this as the "ASTC" graph (using the first letter of the

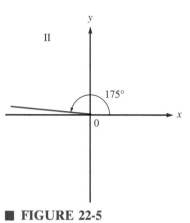

■ FIGURE 22-5

function in each quadrant). It may help to remember the words "All Students Take Calculus" for "ASTC." Or you may find it easier just to sketch the angle and note whether x or y is positive or negative. From this you can determine whether the function you want is positive or negative.

--- **Example 2** ---

Sketch the angle in standard position for each of the following and determine the quadrant of the terminal side. Then give the algebraic sign of each:

(a) sin 175°   **(b)** tan 310°   **(c)** cos 100°   **(d)** cot $(-120°)$

Solution:

(a) We sketch a 175° angle (Fig. 22-5) and see that it is a quadrant II angle. Since

$$\sin \theta = \frac{y}{r}$$

and y is positive in Quadrant II, then sin 175° is positive.

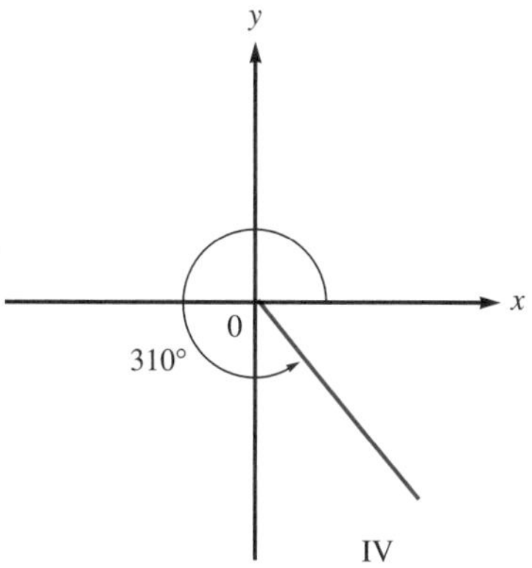

(b) We sketch a 310° angle (Fig. 22-6) and see that it is a quadrant IV angle. Since

$$\tan \theta = \frac{y}{x}$$

and x and y have opposite signs in quadrant IV, then tan 310° is negative.

(c) We sketch a 100° angle (Fig. 22-7) and see that it is a quadrant II angle. Since

$$\cos \theta = \frac{x}{r}$$

and x is negative in quadrant II, then cos 100° is negative.

(d) Since $-120°$ is negative, we sketch the angle with a clockwise rotation (Fig. 22-8), and see that it is a quadrant III angle. Since

$$\cot \theta = \frac{x}{y}$$

and x and y have the same sign in quadrant III, then cot $(-120°)$ is positive.

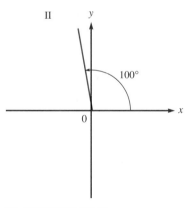

■ FIGURE 22-7

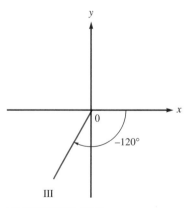

■ FIGURE 22-8

■ FIGURE 22-6

Quadrantal Angles

If the terminal side of an angle lies along one of the coordinate axes, the angle is called a *quadrantal angle.* The angles 0°, 90°, 180°, and 270° are the *primary* quadrantal angles. Examples of other quadrantal angles are 360°, $-180°$, and 540°.

On the x axis, $y = 0$

and
$$r = \sqrt{x^2 + y^2} = \sqrt{x^2 + 0^2} = \sqrt{x^2}$$
$$= |x| \qquad = x \text{ on the positive } x \text{ axis}$$
$$= -x \text{ on the negative } x \text{ axis}$$

On the y axis, $x = 0$

and
$$r = \sqrt{x^2 + y^2} = \sqrt{0^2 + y^2} = \sqrt{y^2}$$
$$= |y| \qquad = y \text{ on the positive } x \text{ axis}$$
$$= -y \text{ on the negative } x \text{ axis}$$

Since the trigonometric functions are all quotients of x, y, and r, then the sine, cosine, and tangent of quadrantal angles are 0, 1, -1, or undefined. We illustrate this in the next example.

┌─── **Example 3** ───────
│ Sketch each angle, and evaluate the trigonometric function without using a calculator:

(a) $\tan 0°$ **(b)** $\sin 90°$ **(c)** $\cos 180°$ **(d)** $\sec 270°$

Solution:

(a) We sketch the $0°$ angle (Fig. 22-9) and see that
$$y = 0$$

Using Eq. 112,
$$\tan 0° = \frac{y}{x}$$

Substituting $y = 0$,
$$\tan 0° = \frac{0}{x} = 0$$

(b) We sketch the $90°$ angle (Fig. 22-10) and see that
$$x = 0$$

and
$$r = |y| = y \qquad \text{(since } y \text{ is positive)}$$

Using Eq. 110,
$$\sin 90° = \frac{y}{r}$$

Substituting $r = y$,
$$\sin 90° = \frac{y}{y} = 1$$

■ **FIGURE 22-9**

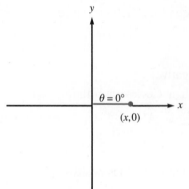

■ **FIGURE 22-10**

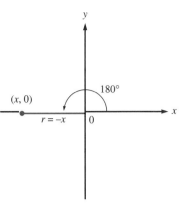

FIGURE 22-11

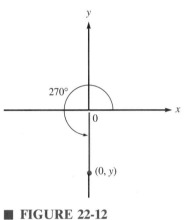

FIGURE 22-12

(c) We sketch the 180° angle (Fig. 22-11) and see that

$$y = 0$$

and

$$r = |x| = -x \qquad \text{(since } x \text{ is negative)}$$

Using Eq. 111,

$$\cos 180° = \frac{x}{r}$$

Substituting $r = -x$,

$$\cos 180° = \frac{x}{-x} = -1$$

(d) We sketch the 270° angle (Fig. 22-12) and see that

$$x = 0$$

Using Eq. 114,

$$\sec 270° = \frac{r}{x}$$

Substituting $x = 0$,

$$\sec 270° = \frac{r}{0} \text{ undefined!}$$

Trigonometric Functions of Any Angle by Calculator

Given an angle, any of the trigonometric functions can be found using a calculator. The calculator must be in the proper angle unit mode: "degree" if the angle is in degrees, or "radian" if the angle is in radians. Refer to your calculator manual to determine how to change modes.

Example 4

Find cos 205° to four significant digits.

Solution: First, we put the calculator in the degree mode.

We find that

$$\cos 205° = -0.9063$$

rounded to four significant digits.

Example 5

Find tan 5.62 to four significant digits.

Solution: Since the angle 5.62 has no written units, we assume it is in radians and put the calculator in the radian mode.

We find that

$$\tan 5.62 = -0.7812$$

rounded to four significant digits.

COMMON ERROR It is easy to forget to set the angle mode on your calculator. Be sure to set the appropriate mode (degrees or radians) before starting a problem in trigonometry!

Example 6

Find csc $(-121°)$ to four significant digits.

Solution: We put the calculator in the degree mode. As noted in Sec. 15-5, there is no cosecant key on the calculator, so we must evaluate the cosecant as the reciprocal of the sine.

Since

$$\csc(-121°) = \frac{1}{\sin(-121°)}$$

We find that

$$\frac{1}{\sin(-121°)} = -1.167$$

So,

$$\csc(-121°) = -1.167$$

rounded to four significant digits.

SECTION 22-1   Exercises   Trigonometric Functions of Any Angle

The Trigonometric Functions of Any Angle

The given point is on the terminal side of a positive angle θ. Sketch the angle θ in standard position, showing the rotation. Then write the six trigonometric functions of θ, rounded to four significant digits.

1. $(3, -4)$ **2.** $(-8, 15)$ **3.** $(-6, 9)$ **4.** $(-4, -8)$

5. $(0, -4)$ **6.** $(6, 0)$ **7.** $(7, 0)$ **8.** $(0, -7)$

The given point is on the terminal side of a positive angle θ in standard position. Determine the quadrant of the terminal side and write the six trigonometric functions of θ in decimal form rounded to four significant digits.

9. $(-2.43, -0.750)$ **10.** $(1.24, -0.575)$

11. $(7.17, -3.74)$ **12.** $(-9.59, 8.76)$

13. $(-2.50, -\sqrt{5.75})$ **14.** $(\sqrt{7.25}, -3.75)$

Algebraic Signs of the Trigonometric Functions

For each of the following, an angle θ is in standard position and has its terminal side in the given quadrant. Determine the algebraic sign of the sine, cosine, and tangent of θ.

15. quadrant II   **16.** quadrant III

17. quadrant IV   **18.** quadrant I

State in what quadrant or quadrants the terminal side of θ can lie if:

19. $\sin \theta$ is positive   **20.** $\cos \theta$ is positive

21. $\cos \theta$ is negative   **22.** $\sin \theta$ is negative

23. $\tan \theta$ is negative   **24.** $\tan \theta$ is positive

25. $\sin \theta$ is negative and $\cos \theta$ is positive

26. $\cos \theta$ is negative and $\tan \theta$ is negative

27. both $\sin \theta$ and $\cos \theta$ are negative

28. $\sin \theta$ is positive and $\cos \theta$ is negative

Sketch the given angle and determine the algebraic signs of the sine, cosine, and tangent of the following angles.

29. 130°	**30.** 225°	**31.** 290°	**32.** −350°
33. 255°	**34.** 310°	**35.** −150°	**36.** 175°

Quadrantal Angles

Without using a calculator, find the following:

37. cos 90°	**38.** tan 180°	**39.** tan 270°	**40.** sin 360°
41. sin 0°	**42.** cos 270°	**43.** cos (−180°)	**44.** sin (−90°)
45. sin 270°	**46.** tan 90°	**47.** cot 180°	**48.** sec 0°

Trigonometric Functions of Any Angle by Calculator

Using a calculator, find the trigonometric function to four significant digits.

49. sin 245°	**50.** cos 167°	**51.** tan 95.7°	**52.** sin 331°
53. cos (−85.3°)	**54.** tan (−198°)	**55.** sec 145°	**56.** csc 287°
57. cot 325°	**58.** sec 103°	**59.** sin (−175°)	**60.** cos (−305°)
61. cos 46°50′	**62.** sin 25°3′	**63.** cos 11.5	**64.** sin 8.75
65. sin 3.42	**66.** tan 6.17		

Write the sine, cosine, and tangent of θ in fraction form given the following.

67. $\sin \theta = \dfrac{-4}{5}$ and tan θ is negative

68. $\cos \theta = \dfrac{12}{13}$ and sin θ is negative

69. $\sec \theta = \dfrac{-17}{8}$ and cot θ is positive

70. $\csc \theta = \dfrac{41}{9}$ and tan θ is negative

22-2 FINDING ANGLES FROM TRIGONOMETRIC FUNCTIONS

In this section, we learn what a reference angle is. Then we use reference angles to determine the angle(s) when the value of a trigonometric function is given.

Angles with the Same Terminal Side

If we are given an angle θ in standard position, we can find another angle with the same terminal side by adding 360° (or a multiple of 360°) to θ (Fig. 22-13). We can find another angle which has the same terminal side by subtracting 360° (or a multiple of 360°) from θ (Fig. 22-14). Since these angles have the same terminal side, then the trigonometric functions of these angles have the same value.

■ **FIGURE 22-13**
Angles with the same terminal side.

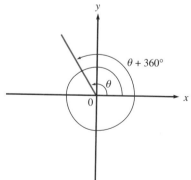

■ **FIGURE 22-14**
Angles with the same terminal side.

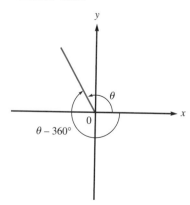

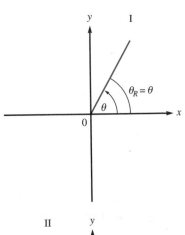

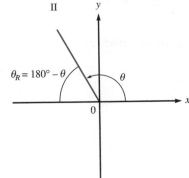

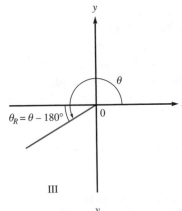

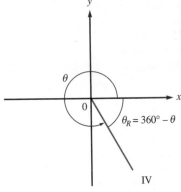

—— **Example 7** ——

Find a positive angle and a negative angle that have the same terminal side as the angle 145°, and show that the sine, cosine, and tangent of the three angles are the same.

Solution: To find another positive angle, just add 360° to the given angle: 145° + 360° = 505°.

To find a negative angle, just subtract 360° from the given angle:

$$145° - 360° = -215°$$

Now we'll compare the trigonometric functions. Using the calculator,

$$\sin 145° = 0.5736$$
$$\sin 505° = 0.5736$$
$$\sin(-215°) = 0.5736$$

Similarly,

$$\cos 145° = -0.8192 = \cos 505° = \cos(-215°)$$

and,

$$\tan 145° = -0.7002 = \tan 505° = \tan(-215°)$$

So the trigonometric functions of 145°, 505°, and −215° are the same.

Reference Angles

Finding the value of a trigonometric function of an angle, such as sin 35°, is no problem if we use a calculator—the calculator does all the work! This is not the case, however, when we are given the value of a trigonometric function, such as sin θ = 0.5, and asked to find the angle (or angles) that have that trigonometric function value. For this, we need to make use of the *reference angle*.

The *reference angle* of an angle θ in standard position is the acute angle that the terminal side of θ makes with the x axis. We call the reference angle θ_R. It is an unsigned angle (not a rotation). We sketch the reference angle by drawing the angle between the terminal side of the given angle and the x axis—either on the negative side or the positive side, whichever is closer (Fig. 22-15). This is summarized in the table:

θ	Quadrant	Reference Angle θ_R
$0° < \theta < 90°$	I	θ
$90° < \theta < 180°$	II	$180° - \theta$
$180° < \theta < 270°$	III	$\theta - 180°$
$270° < \theta < 360°$	IV	$360° - \theta$

—— **Example 8** ——

Find the reference angle for the following:

(a) 50°   **(b)** 135°

Solution: First we sketch the given angle, then sketch the reference angle.

(a) We sketch the angle θ = 50° in standard position (Fig. 22-16) and note that θ is a quadrant I angle. The reference angle, θ_R, for a quadrant I angle θ is θ itself. Thus,

$$\theta_R = \theta = 50°$$

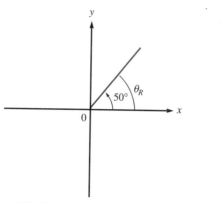

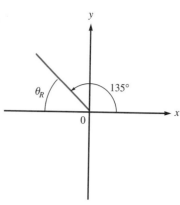

■ **FIGURE 22-16**   ■ **FIGURE 22-17**

(b) We sketch the angle $\theta = 135°$ and note that θ is a quadrant II angle (Fig. 22-17). The reference angle, θ_R, for a quadrant II angle θ is

$$\theta_R = 180° - \theta$$
$$\theta_R = 180° - 135° = 45°$$

Example 9

Find the reference angle for the following:

(a) 230°   **(b)** 345°   **(c)** −150°

Solution:

(a) $\theta = 230°$ is a quadrant III angle (Fig. 22-18). The reference angle is

$$\theta_R = \theta - 180°$$
$$\theta_R = 230° - 180° = 50°$$

(b) $\theta = 345°$ is a quadrant IV angle (Fig. 22-19). The reference angle is

$$\theta_R = 360° - \theta$$
$$\theta_R = 360° - 345° = 15°$$

(c) $\theta = -150°$ is a negative angle whose terminal side is in the third quadrant (Fig. 22-20). From the sketch we see that

$$\theta_R = 180° - 150° = 30°$$

■ **FIGURE 22-18**   ■ **FIGURE 22-19**   ■ **FIGURE 22-20**

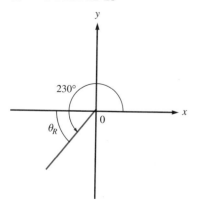

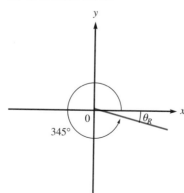

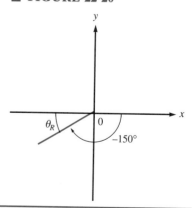

The reference angle is useful because the values of the trigonometric functions of an angle θ and of its reference angle θ_R *may* be different only in the algebraic signs, while the absolute values are always the same.

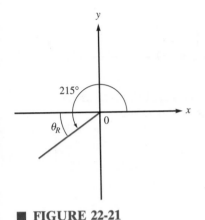

■ FIGURE 22-21

── **Example 10** ──

Find the reference angle θ_R for an angle of $\theta = 215°$, and compare the sines, cosines, and tangents of θ and θ_R.

Solution: First we sketch the angle θ and find its reference angle θ_R (Fig. 22-21). Since θ is a quadrant III angle,

$$\theta_R = 215° - 180° = 35°$$

So its reference angle is 35°. Comparing trigonometric functions, we find

$$\sin 215° = -0.5736 \quad \text{and} \quad \sin 35° = 0.5736$$
$$\text{so } \sin 215° = -\sin 35°$$
$$\cos 215° = -0.8192 \quad \text{and} \quad \cos 35° = 0.8192$$
$$\text{so } \cos 215° = -\cos 35°$$
$$\tan 215° = 0.7002 \quad \text{and} \quad \tan 35° = 0.7002$$
$$\text{so } \tan 215° = \tan 35°$$

We see that the sines and cosines of θ and of its reference angle θ_R differ only in algebraic sign, and that the tangents are the same.

Finding the Angle When the Value of a Trigonometric Function Is Given

In determining an angle when the value of a trigonometric function is given, we usually need only consider nonnegative angles less than 360°. These angles are called the *primary angles*. An angle θ is a primary angle if

$$0° \leq \theta < 360°$$

Recall from Sec. 15-2 that to find an angle that has a given trigonometric function value, we use an inverse trigonometric function key on the calculator. But the calculator gives us only *one* angle with the given trigonometric function value. In the next example, you will see that *there may be more than one angle having the same trigonometric function value,* and we show you how to find them. We limit our answers to *primary angles.*

── **Example 11** ──

Find all the primary angles θ for which $\sin \theta = 0.5$.

Solution: If $\sin \theta = 0.5$, then $\theta = \sin^{-1} 0.5$.

Using the calculator, we find one angle:

$$\theta = \sin^{-1} 0.5 = 30°$$

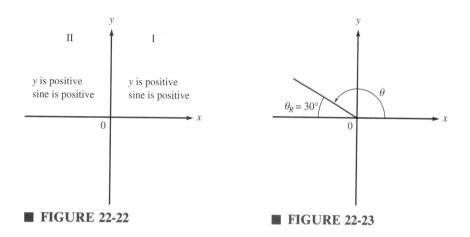

■ FIGURE 22-22

■ FIGURE 22-23

Are there any other primary angles for which the sine is 0.5? We must consider the algebraic sign of the sine function:

Since $\sin \theta = 0.5$, then the sine is positive.

Since $\sin \theta = \dfrac{y}{r}$, then $\dfrac{y}{r}$ must be positive.

Since r is always positive, then y must be positive.

And y is positive only in quadrants I and II (Fig. 22-22). Hence the sine is positive in only quadrants I and II.

We already found the quadrant I angle ($\theta = 30°$). The quadrant II angle is found using the reference angle, which is the angle itself; so $\theta_R = 30°$. We sketch the reference angle θ_R in quadrant II (Fig. 22-23) and sketch the positive angle θ, in standard position. We see

$$\theta = 180° - \theta_R$$
$$\theta = 180° - 30° = 150°$$

So, $\theta = 150°$ has a sine of 0.5. Checking on the calculator, we see that

$$\sin 150° = 0.5$$

The two primary angles which have a sine of 0.5 are 30° and 150°.

If the reference angle is known, we can find angle θ in a specified quadrant using this table:

Quadrant	θ
I	θ_R
II	$180° - \theta_R$
III	$180° + \theta_R$
IV	$360° - \theta_R$

In general, to find all the primary angles which have a given trigonometric function:

1. Use the sign of the function to determine which are the quadrants of the primary angles.
2. Find the reference angle by calculator.
3. Determine the primary angles from the quadrants and the reference angle.

─── **Example 12** ───

Find the primary angles, to the nearest tenth of a degree, that have a cosine of −0.3480.

Solution: We want to find θ, so that

$$\cos \theta = -0.3480$$

Since $\cos \theta = \dfrac{x}{r}$, and r is always positive, the cosine will be negative only if x is negative. And x is negative only in quadrants II and III (Fig. 22-24). Using a calculator, we find

$$\theta = \cos^{-1}(-0.3480) = 110.4°$$

This θ is a quadrant II angle. To find the quadrant III angle, we need the reference angle. We draw the 110.4° angle (Fig. 22-25) and see that the reference angle θ_R for a quadrant II angle is

$$\theta_R = 180° - \theta$$
$$\theta_R = 180° - 110.4° = 69.6°$$

Using a reference angle of 69.6° in quadrant III (Fig. 22-26) we calculate the other primary angle,

$$\theta = 180° + \theta_R$$
$$\theta = 180° + 69.6° = 249.6°$$

Checking on the calculator, we find that

$$\cos 249.6° = -0.3480$$

Thus, the primary angles that have a cosine of −0.3480 are 110.4° and 249.6°.

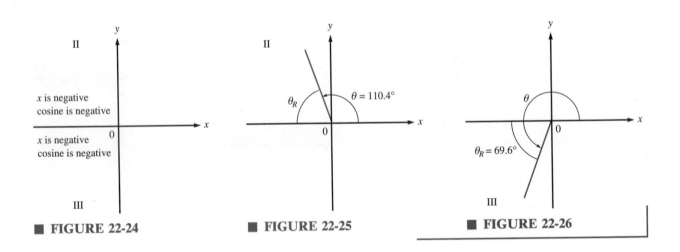

■ **FIGURE 22-24**   ■ **FIGURE 22-25**   ■ **FIGURE 22-26**

When the cotangent, secant, or cosecant is given, we make use of the reciprocal relationships, Eq. 116.

Example 13

Find all the primary angles, to the nearest tenth of a degree, that have a tangent of -2.5.

Solution: Since $\tan\theta = \dfrac{y}{x}$, then the tangent will be negative only if the signs of x and y are *different*. This occurs in quadrants II and IV (Fig. 22-27).

If

$$\tan\theta = -2.5$$

then

$$\theta = \tan^{-1}(-2.5)$$

By calculator, rounding to the nearest tenth of a degree,

$$\tan^{-1}(-2.5) = -68.2°$$

Sketching the angle to determine the reference angle (Fig. 22-28), we find the reference angle

$$\theta_R = 68.2°$$

So, in quadrant IV

$$\theta = 360° - 68.2° = 291.8°$$

and in quadrant II (Fig. 22-29)

$$\theta = 180° - 68.2° = 111.8°$$

So the primary angles that have a tangent of -2.5 are $111.8°$ and $291.8°$.

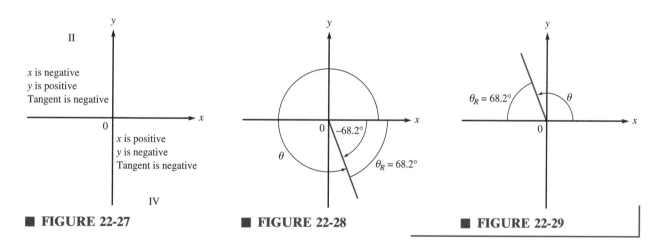

■ **FIGURE 22-27**   ■ **FIGURE 22-28**   ■ **FIGURE 22-29**

SECTION 22-2   Exercises   Finding Angles from Trigonometric Functions

Find a positive angle and a negative angle which have the same terminal side as the given primary angle, and show that the sine, cosine, and tangent of the three angles are the same.

1. $164°$ **2.** $261°$ **3.** $328°$ **4.** $82°$

Find the reference angle for each of the following angles.

5. 113°　　**6.** 317°　　**7.** −75°　　**8.** −128°

Find the reference angle for each of the following, and compare the sines, cosines, and tangents of both angles.

9. 219°　　**10.** 121°　　**11.** 307°　　**12.** 353°

**Finding the Angle
When the Value of a
Trigonometric Function
Is Given**

Find all the primary angles, to the nearest tenth of a degree, whose functions are given below.

13. $\cos \theta = 0.5$　　　　　　**14.** $\sin \theta = 0.7$

15. $\sin \theta = 0.1234$　　　　**16.** $\cos \theta = 0.5678$

17. $\tan \theta = 1$　　　　　　**18.** $\tan \theta = 1.5$

19. $\sin \theta = -0.5$　　　　　**20.** $\cos \theta = -0.5$

21. $\tan \theta = -2.348$　　　　**22.** $\tan \theta = -1$

23. $\cos \theta = 0.375$　　　　　**24.** $\sin \theta = 0.7525$

25. $\cot \theta = 1.75$　　　　　**26.** $\sec \theta = -2.75$

27. $\csc \theta = -3.25$　　　　**28.** $\sin \theta = -0.75$

29. $\cos \theta = 0$　　　　　　**30.** $\sin \theta = 0$

31. $\tan \theta = 0$　　　　　　**32.** $\cot \theta = 0$

22-3　　LAW OF SINES

In this section, we derive a new formula, the law of sines, to enable us to solve certain oblique triangles quickly. Then we use the law of sines to solve some oblique triangles and applications.

Oblique Triangles

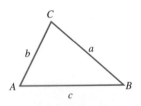

■ **FIGURE 22-30**
Oblique triangle.

An *oblique triangle* is a triangle that has no right angle. When we sketch an oblique triangle, if we label the three angles as A, B, and C, then we label the sides *opposite* these angles as a, b, and c, respectively (Fig. 22-30).

Recall from Sec. 15-3 that to *solve* a triangle means to find all the missing sides and angles. Any triangle can be solved if we know three parts of the triangle, at least one of which is a side. If the triangle is a right triangle, we can use the Pythagorean Theorem and the definitions of the six trigonometric functions. If the triangle is oblique (not a right triangle), you cannot use the Pythagorean Theorem or the definitions of the six trigonometric functions.

> **COMMON ERROR** The Pythagorean Theorem and the definitions of the six trigono-
> metric functions are used to solve only *right* triangles, and they *cannot* be used
> directly to solve *oblique* triangles.

Solving Oblique Triangles

Many oblique triangles can be solved using the *law of sines: In any triangle, the sides are proportional to the sines of the opposite angles.*

Mathematically, the law of sines can be stated:

LAW OF SINES	$\dfrac{a}{\sin A} = \dfrac{b}{\sin B} = \dfrac{c}{\sin C}$	101

Derivation of the Law of Sines

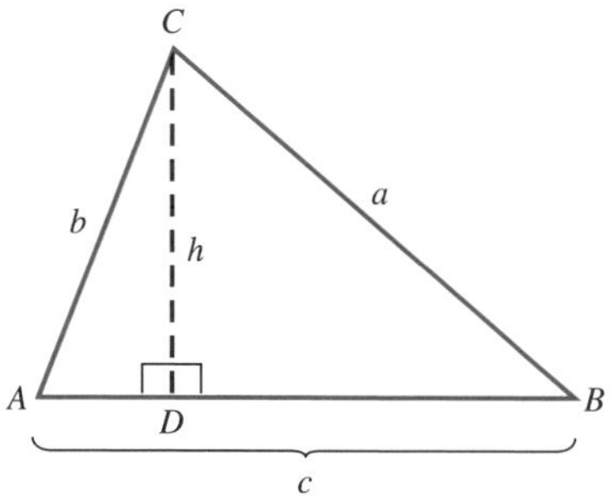

We will derive the law of sines for an oblique triangle in which all three angles are acute. (The derivation is similar if one of the angles is obtuse.)

We start by separating the given triangle into two right triangles by drawing altitude h to side AB (Fig. 22-31). In right triangle ACD,

$$\sin A = \frac{h}{b} \quad \text{or} \quad h = b \sin A$$

■ **FIGURE 22-31**

In right triangle BCD,

$$\sin B = \frac{h}{a} \quad \text{or} \quad h = a \sin B$$

So

$$b \sin A = a \sin B$$

Dividing by $(\sin A)(\sin B)$:

$$\frac{b \sin A}{(\sin A)(\sin B)} = \frac{a \sin B}{(\sin A)(\sin B)}$$

Or,

$$\frac{b}{\sin B} = \frac{a}{\sin A}$$

Similarly, by drawing the altitude to side AC, we find

$$\frac{a}{\sin A} = \frac{c}{\sin C}$$

Combining these two equations we obtain

$$\frac{a}{\sin A} = \frac{b}{\sin B} = \frac{c}{\sin C}$$

This is the law of sines.

Classification of Triangles

Triangles can be classified according to which three parts are known. In the following, A refers to an angle and S refers to a side.

> AAS: Two angles and a side *opposite* one of the angles
> ASA: Two angles and the *included* side
> SSA: Two sides and an angle *opposite* one of the sides
> SAS: Two sides and the *included* angle
> SSS: Three sides

The notations ASA, SSS, and the like remind us of the congruence theorems and postulates of plane geometry, which establish the *uniqueness* of the triangles in all but the SSA case. In learning to solve oblique triangles, we will consider each of these cases (AAS and ASA in this section, and SSA, SAS, and SSS in the next two sections).

Solving a Triangle with Two Angles and an *Opposite* Side Known (AAS)

To use the law of sines to solve an oblique triangle, we must have a known side *opposite* a known angle (AAS). The law of sines actually gives us three equations:

$$\frac{a}{\sin A} = \frac{b}{\sin B}$$

$$\frac{a}{\sin A} = \frac{c}{\sin C}$$

$$\frac{b}{\sin B} = \frac{c}{\sin C}$$

To use the law of sines, choose whichever equation has the three known parts and the one unknown part.

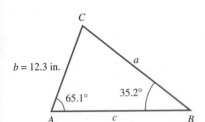

FIGURE 22-32

Example 14

Solve the triangle in which $A = 65.1°$, $B = 35.2°$, and $b = 12.3$ in. (Fig. 22-32).

Solution: We can use the law of sines, because we are given a side b and the opposite angle B. Referring to Fig. 22-32, and using the law of sines, Eq. 101:

$$\frac{a}{\sin 65.1°} = \frac{12.3}{\sin 35.2°}$$

Solving for a, we get

$$a = \frac{(\sin 65.1°)\,(12.3)}{\sin 35.2°} = 19.4$$

In this triangle, we are given two of the three angles, so we can determine C easily:

$$C = 180° - (65.1° + 35.2°) = 79.7°$$

Again using the law of sines, Eq. 101:

$$\frac{c}{\sin 79.7°} = \frac{12.3}{\sin 35.2°}$$

Solving for c, we get

$$c = \frac{(\sin 79.7°)\,(12.3)}{\sin 35.2°} = 21.0$$

So, the three angles of this triangle are

$$A = 65.1°, B = 35.2°, \text{ and } C = 79.7°$$

The three sides are

$$a = 19.4 \text{ in., } b = 12.3 \text{ in., and } c = 21.0 \text{ in.}$$

If you make a sketch more or less to scale, it can serve as an approximate check of your work. As another rough check, determine if the longest side is opposite the largest angle and if the shortest side is opposite the smallest angle.

Solving a Triangle with Two Angles and the *Included* Side Known (ASA)

In the preceding example, two angles and a side *opposite* one of the angles were given. We now consider an example in which two angles and the *included* side are given (ASA).

—— **Example 15** ——

Solve triangle ABC if $B = 123.6°$, $C = 38.3°$, and $a = 9.75$ cm.

Solution: First we sketch the triangle, more or less to scale (Fig. 22-33). In order to use the law of sines, we need a side and the angle opposite that side. The only side we have is a, and we don't have angle A. But we can find A easily:

$$A = 180° - (123.6° + 38.3°) = 18.1°$$

So now we can use the law of sines (Eq. 101):

$$\frac{9.75}{\sin 18.1°} = \frac{b}{\sin 123.6°}$$

Solving for b:

$$b = \frac{(\sin 123.6°)\,(9.75)}{\sin 18.1°} = 26.1$$

Again using the law of sines (Eq. 101):

$$\frac{9.75}{\sin 18.1°} = \frac{c}{\sin 38.3°}$$

Solving for c:

$$c = \frac{(\sin 38.3°)\,(9.75)}{\sin 18.1°} = 19.5$$

So, the missing angle and sides are

$$C = 18.1°, b = 26.1 \text{ cm, and } c = 19.5 \text{ cm}$$

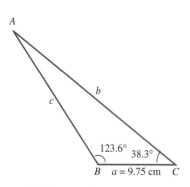

■ **FIGURE 22-33**

Applications

As with right triangles, oblique triangles have many applications. You should follow the same procedures for setting up these problems as for other word problems. If the resulting triangle has two angles and a side given, then you can solve the triangle using the law of sines.

—— **Example 16** ——

A vertical antenna 15.2 ft high stands on the top of a building on level ground. From a person on the street, whose eye level is 5.00 ft above ground level, the angle of elevation of the top of the antenna is 36.5° and the angle of elevation of the top of the building is 33.4°. How tall is the building?

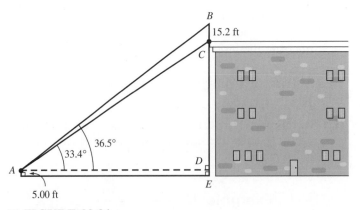

■ FIGURE 22-34

Solution: First we make a sketch, drawn more or less to scale (Fig. 22-34). We need to find the distance CE. If we knew AC, we could use right triangle trigonometry in triangle ACD. We use the law of sines in triangle ABC to solve for AC. We have a side ($BC = 15.2$ ft) and need two angles. We can find angle BAC by subtraction:

$$\angle BAC = 36.5° - 33.4° = 3.1°$$

We can also find angle B, since angle B and angle BAD are complementary:

$$B = 90° - 36.5° = 53.5°$$

So we now have two angles and an opposite side in oblique triangle ABC. Using the law of sines (Eq. 101), we find

$$\frac{AC}{\sin 53.5°} = \frac{15.2}{\sin 3.1°}$$

Solving for AC:

$$AC = \frac{(\sin 53.5°)\,(15.2)}{\sin 3.1°} = 225.9$$

Now we use AC in right triangle ACD:

$$\sin 33.4° = \frac{CD}{AC}$$

Solving for CD:

$$CD = (AC)\,(\sin 33.4°)$$
$$CD = (225.9)\,(\sin 33.4°) = 124.4$$

Adding 5.00 ft for the eye to ground distance,

$$124.4 + 5.00 = 129.4$$

Rounding to three significant digits, we find that the building is about 129 ft tall.

Round sides to three significant digits and angles to the nearest tenth of a degree.

Solving a Triangle with Two Angles and an Opposite Side Known (AAS)

Solve for the missing part of these triangles which is in parentheses.

1. $A = 37.0°$, $B = 87.0°$, $a = 54.0$ in.   (*b*)
2. $B = 68.0°$, $C = 47.0°$, $b = 72.5$ cm   (*c*)
3. $A = 61.0°$, $C = 104.0°$, $c = 13.5$ cm   (*a*)
4. $A = 114.0°$, $B = 51.0°$, $a = 26.7$ cm   (*b*)
5. $B = 38.9°$, $C = 62.7°$, $b = 75.3$ in.   (*c*)
6. $A = 78.5°$, $B = 54.1°$, $b = 6.75$ cm   (*a*)
7. $A = 100°30'$, $C = 43°54'$, $c = 33.2$ cm   (*a*)
8. $A = 0.900$ rad, $B = 1.50$ rad, $a = 56.5$ in.   (*b*)

Solving a Triangle with Two Angles and the Included Side Known (ASA)

Solve the triangles for all the missing parts.

9. $A = 73.0°$, $B = 83.0°$, $c = 4.50$ cm
10. $B = 48.0°$, $C = 105.0°$, $a = 58.6$ in.
11. $A = 35.7°$, $C = 115.4°$, $b = 78.4$ cm
12. $B = 59.5°$, $C = 68.5°$, $a = 529$ in.

Solve and round to three significant digits.

Applications

13. Two stakes, A and B, are 68.0 ft apart. A third stake C is placed so that angle ABC is 75.0° and angle BAC is 82.0°. Find the distance from C to each of the other stakes.

14. A surveyor needs to find the distance AB across a marsh (Fig. 22-35). A stake C is placed 155 m from A, so that angle CAB is 56.0° and angle ABC is 64.0°. Find the distance AB across the marsh.

15. Find the length of the shorter support of the shelf in Fig. 22-36, if the longer support is 37.5 cm long.

16. Two observers 1.45 km apart spot an airplane simultaneously. The angle of elevation of the airplane from observer A is 28.5°. The angle of elevation of the airplane from observer B is 47.9°. What is the height of the airplane?

■ **FIGURE 22-35**

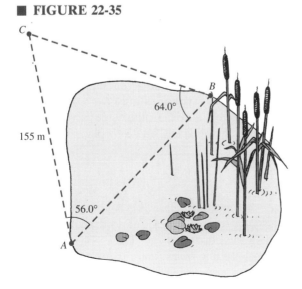

■ **FIGURE 22-36**

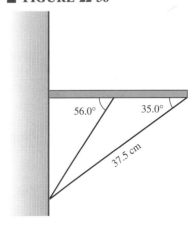

17. The pilot flying a plane can see the beginning and the end of a 1150-ft runway. The angle of depression of the beginning of the runway is 28.6°, and the angle of depression of the end of the runway is 22.5°. How far is the plane from the beginning of the runway?

18. A forest ranger in an observation tower at A spots a fire in the direction 37°15′ west of north. Another forest ranger in an observation tower 4.00 miles due west of A spots the fire in the direction 43°36′ east of north. How far is the fire from each tower?

19. A 125-ft lighthouse is on the edge of a sea cliff. From a ship, the angle of elevation of the bottom of the lighthouse is 36.5°, and the angle of elevation of the top of the lighthouse is 48.0°. How high is the cliff?

20. From a plane flying due north, the bearing of a radio station is 39.2° east of north (N 39.2° E). After flying for 4.50 km, the bearing of the radio station is 46.0° east of north. Find the distance of the plane from the station at the time of the second bearing.

21. Town A is 169 miles 18.5° west of south from town B. The bearing of town C from town B is 35.7° west of north. The bearing of town C from town A is 25.9° west of north. How far is town C from town A?

22-4   AMBIGUOUS CASE

In the last section, we learned to solve oblique triangles in which two angles and one side were known (AAS and ASA). We solved these triangles using the law of sines. We now consider triangles in which two sides and an angle *opposite* one of them are known (SSA). This case is called the *ambiguous* case, because sometimes there is one *triangle* and sometimes there are *two triangles*.

Solving a Triangle with Two Sides and an *Opposite* Angle Known (SSA)

The law of sines can be used to solve triangles in which two sides and an *opposite* angle are known.

■ FIGURE 22-37

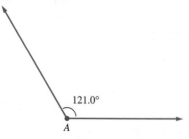

■ FIGURE 22-38

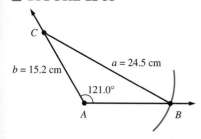

—— **Example 17** ——
Solve triangle ABC where $A = 121.0°$, $a = 24.5$ cm, and $b = 15.2$ cm.

Solution: First we will make an accurate sketch of the triangle; an accurate sketch (drawn to scale) will usually indicate whether we have one or two solutions. For consistency, we will always begin by drawing a horizontal ray with its endpoint on the left and construct the given angle ($A = 121.0°$) at the left end of the segment (Fig. 22-37). Two sides are given, one of which is opposite the given angle ($a = 24.5$ cm), and the other side is adjacent to the given angle ($b = 15.2$ cm). We use the adjacent side b to place C on the upper ray b units from A:

$$AC = b = 15.2$$

We construct side a (opposite angle A), beginning at point C. We open the compass to radius 24.5 (drawn to scale) and draw an arc and label point B at the intersection with the horizontal ray (Fig. 22-38). Since the arc intersects the horizontal ray in exactly one point, *there is exactly one triangle* with the given parts, triangle ABC.

We have an angle and the side opposite that angle, so we can use the law of sines:

$$\frac{24.5}{\sin 121.0°} = \frac{15.2}{\sin B}$$

Solving for $\sin B$:

$$\sin B = \frac{(\sin 121.0°)(15.2)}{24.5} = 0.5318$$

Taking the inverse sine,

$$B = \sin^{-1} 0.5318 = 32.13°$$

We find C by subtracting:

$$C = 180° - (121.0° + 32.13°) = 26.87°$$

To find side c, we again use the law of sines:

$$\frac{24.5}{\sin 121.0°} = \frac{c}{\sin 26.87°}$$

$$c = \frac{(\sin 26.87°)(24.5)}{\sin 121.0°} = 12.9$$

So, the missing angles and side are

$$B = 32.1°, C = 26.9°, \text{ and } c = 12.9 \text{ cm}$$

In the example above, the given angle was obtuse, and there was only one triangle that satisfied the given conditions. We can see from Fig. 22-38 that there can be *at most* one triangle if the given angle is obtuse. However, if the given angle is acute, then we must take special care when solving the triangle. We sometimes get a *second* solution (triangle) that may be just as important as the first.

■ **FIGURE 22-39**

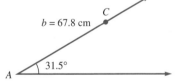

■ **FIGURE 22-40**

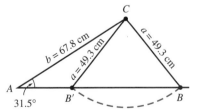

── **Example 18** ──
Solve triangle ABC in which $A = 31.5°$, $a = 49.3$ cm, and $b = 67.8$ cm.

Solution: As before, we sketch the given parts, starting with the given angle and the adjacent side (Fig. 22-39). With the compass point at C, we draw an arc with radius about 49.3 (to scale). We see that the arc intersects the horizontal ray in *two points*, B and B' (Fig. 22-40). So, there are *two different triangles* that have the same given parts! These triangles are overlapping in Fig. 22-40; we separate them in Fig. 22-41 so we can see them better. In the triangle on the left, angle B' (angle $AB'C$) is obtuse. In the triangle on the right, angle B (angle ABC) is acute. We can solve both triangles using the law of sines. We start with the acute triangle on the right.

■ **FIGURE 22-41**
Two triangles!

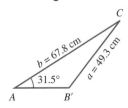
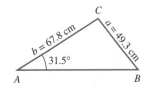

Triangle on the Right: Using the law of sines, we first find the acute angle B:

$$\frac{49.3}{\sin 31.5°} = \frac{67.8}{\sin B}$$

$$\sin B = \frac{(\sin 31.5°)(67.8)}{49.3} = 0.7186$$

By calculator,

$$B = \sin^{-1} 0.7186 = 45.94°$$

We find angle C by subtracting:

$$C = 180° - (31.5° + 45.94°) = 102.56°$$

We find side c using the law of sines again:

$$\frac{49.3}{\sin 31.5°} = \frac{c}{\sin 102.56°}$$

$$c = \frac{(49.3)(\sin 102.56°)}{\sin 31.5°} = 92.1$$

So, the missing angles and side of the triangle on the *right* are

$$B = 45.9°, C = 102.6°, \text{ and } c = 92.1 \text{ cm}$$

Triangle on the Left: To find the triangle on the left, we return to the equation above in which

$$\sin B = 0.7186$$

There are *two* angles which have a sine of 0.7186. One is in quadrant I and the other is in quadrant II. (Note that this is one place where we need that second angle!) The calculator gave us the first quadrant angle, $B = 45.94°$, and we used that angle to calculate the other parts of the first triangle. The quadrant II angle, B', is found using $45.94°$ as the *reference angle*.

In quadrant II,

$$B' = 180° - B$$

$$B' = 180° - 45.94° = 134.06°$$

When $B' = 134.1°$,

$$C = 180° - (31.5° + 134.06°) = 14.44°$$

Using the law of sines to find the new side c:

$$\frac{49.3}{\sin 31.5°} = \frac{c}{\sin 14.44°}$$

$$c = \frac{(49.3)(\sin 14.44°)}{\sin 31.5°} = 23.5$$

So, the missing angles and side of the triangle on the *left* are

$$B' = 134.1°, C = 14.4°, \text{ and } c = 23.5 \text{ cm}$$

COMMON ERROR In a problem like the preceding one, it is easy to forget the possibility of a second solution, especially since a calculator gives only one angle when taking the inverse sine.

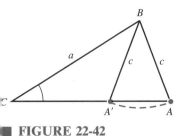

FIGURE 22-42

Whenever we are given two sides and an opposite angle, there are always three possibilities for the number of triangles that satisfy the given conditions: 0, 1, or 2. So how can we determine how many triangles to solve for? Well, there is a complex mathematical way of determining the number of solutions, but perhaps the *easiest* way is just to sketch the triangle fairly accurately, and see if it looks like there may be 2, 1, or 0 triangle solutions.

In all the examples above, the given parts were A, a, and b. Regardless of which parts are given, we suggest that you always start with sketching the angle on the left, labeling the adjacent side on the upper ray and then drawing the arc for the opposite side. For example, if C, a, and c were given, your sketch would be as in Fig. 22-42. You solve the triangle as shown above, using the new letters.

SECTION 22-4 Exercises Ambiguous Case

Sketch the triangle(s) with the given parts, and solve the triangle(s). Round sides to three significant digits and angles to the nearest tenth of a degree.

Solving a Triangle with Two Sides and an Opposite Angle Known (SSA)

1. $A = 126.0°$, $a = 25.2$ cm, $b = 20.8$ cm

2. $A = 135.0°$, $a = 30.5$ cm, $b = 25.9$ cm

3. $B = 36.0°$, $b = 127$ in., $c = 169$ in.

4. $C = 143.0°$, $b = 13.5$ in., $c = 18.3$ in.

5. $C = 152.0°$, $a = 74.8$ cm, $c = 89.2$ cm

6. $B = 95.0°$, $a = 53.8$ in., $b = 75.7$ in.

7. $B = 75.0°$, $a = 53.0$ cm, $b = 85.0$ cm

8. $C = 35.0°$, $b = 78.0$ in., $c = 49.0$ in.

9. $A = 16.7°$, $a = 10.3$ in., $c = 23.3$ in.

10. $B = 28.2°$, $b = 98.5$ cm, $c = 72.7$ cm

11. $C = 57.8°$, $b = 4.65$ m, $c = 4.25$ m

12. $A = 85.7°$, $a = 10.7$ m, $c = 9.25$ m

13. $C = 65.5°$, $b = 8.25$ in., $c = 9.75$ in.

14. $B = 30.7°$, $b = 44.7$ cm, $c = 87.4$ cm

Applications

15. A power pole on level ground is supported by two wires that run from the top of the pole to the ground. One wire is 12.5 m long and makes an angle of 48.6° with the ground, and the other wire is 11.0 m long. Find the angle the second wire makes with the ground.

16. A pole standing on level ground makes an angle of 84.3° with the horizontal. The pole is supported by a 25.5-ft prop whose base is 14.8 ft from the base of the pole. Find the angle made by the prop with the horizontal.

17. A ship leaves a port and travels due west. At a certain point the ship turns 62.5° toward the north and continues in that direction for 45.5 km to a point 84.0 km from the port. How far did the ship travel west before turning?

18. A man owns a triangular lot. Markers at corners A and B are found but the third marker at C cannot be located. The deed states that $AB = 152$ ft, BC 115 ft, and $A = 53.5°$. Is the deed correct?

LAW OF COSINES

Not all triangles can be solved using the law of sines. In this section, we derive another formula, the law of cosines, with which we can solve the remaining two types of triangles (SAS and SSS). We solve triangles of both types and do some applications.

Law of Cosines

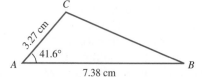

■ **FIGURE 22-43**

Can you use the law of sines to solve the triangle in Fig. 22-43? Since the side opposite angle A is not known, *you cannot use the law of sines*. We need to use a new law, the *law of cosines,* to solve this triangle. The law of cosines states that *the square of any side of a triangle is equal to the sum of the squares of the other two sides minus twice the product of these two sides and the cosine of their included angle.* This sounds a bit complicated, but it's really easy to use. Mathematically, the law of cosines tells us that in any triangle ABC (Fig. 22-44),

$$a^2 = b^2 + c^2 - 2bc \cos A$$

The same relationships hold for the other two sides.

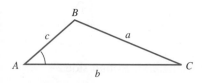

■ **FIGURE 22-44**

Derivation of the Law of Cosines

Consider the oblique triangle ABC in Fig. 22-45. We start by dividing the triangle into two right triangles by drawing an altitude h from vertex C to side AB. In right triangle BCD, we have

$$a^2 = h^2 + (c - m)^2$$

In right triangle ACD, we have

$$b^2 = h^2 + m^2$$

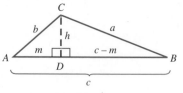

■ **FIGURE 22-45**

Expanding the first equation $\quad a^2 \qquad = h^2 + c^2 - 2cm + m^2$
Rewriting the second equation $\quad \dfrac{b^2 = h^2 \qquad\qquad\;\; + m^2}{}$
Subtracting the second equation $\quad a^2 - b^2 = \qquad c^2 - 2cm$

Solving for a^2:

$$a^2 = b^2 + c^2 - 2cm$$

But, in triangle ACD,

$$\cos A = \frac{m}{b}$$

Or,

$$m = b \cos A$$

Substituting for m:

$$a^2 = b^2 + c^2 - 2c(b \cos A)$$

or,

$$a^2 = b^2 + c^2 - 2bc \cos A$$

This is one form of the law of cosines. We could derive the other forms of the law of cosines by drawing altitudes to side AC and to side BC using the same process. So there are three forms of the law of cosines, one for each side of a triangle:

LAW OF COSINES	$a^2 = b^2 + c^2 - 2bc \cos A$ $b^2 = a^2 + c^2 - 2ac \cos B$ $c^2 = a^2 + b^2 - 2ab \cos C$	102

Solving a Triangle with Two Sides and the Included Angle Known (SAS)

We need the law of the cosines to solve this type of triangle. We will solve the triangle in Fig. 22-43 that we could not solve at the beginning of this section.

Example 19

Solve triangle ABC in which $A = 41.6°$, $b = 3.27$ cm, and $c = 7.38$ cm.

Solution: We refer back to Fig. 22-43 and use the law of cosines:

$$a^2 = b^2 + c^2 - 2bc \cos A$$
$$a^2 = 3.27^2 + 7.38^2 - 2(3.27)(7.38)(\cos 41.6°)$$

The right side can be evaluated by calculator:

$$a^2 = 29.06$$

Taking the square root of both sides:

$$a = \pm 5.391$$

We keep only the positive value, since a is a distance. Rounding to three significant digits,

$$a = 5.39 \text{ cm}$$

We could now use the law of cosines to find angles B and C, or we could use the law of sines (since we now have an angle and the side opposite that angle). If one of the unknown angles is acute and one obtuse, choose to *solve for the acute angle first,* since you don't need to use reference angles for acute angles. We will solve for the acute angle B, using the law of sines.

$$\frac{5.391}{\sin 41.6°} = \frac{3.27}{\sin B}$$

$$\sin B = \frac{(\sin 41.6°)(3.27)}{5.391} = 0.4027$$

Rounding to the nearest tenth of a degree:

$$B = \sin^{-1} 0.4027 = 23.7°$$

Finding C:

$$C = 180° - 41.6° - 23.7° = 114.7°$$

So, the missing side and angles of the triangle are

$$a = 5.39 \text{ cm}, B = 23.7°, \text{ and } C = 114.7°$$

Example 20

Solve triangle ABC in which $B = 135.0°$, $a = 25.0$ in., and $c = 37.0$ in.

Solution: We make a sketch (Fig. 22-46). By the law of cosines

$$b^2 = a^2 + c^2 - 2ac \cos B$$
$$b^2 = 25.0^2 + 37.0^2 - 2(25.0)(37.0)(\cos 135.0°)$$
$$b^2 = 3302.1$$
$$b = 57.46$$

Rounding to three significant digits, $b = 57.5$ in.

FIGURE 22-46

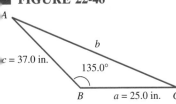

By the law of sines we find the angle C:

$$\frac{57.46}{\sin 135.0°} = \frac{37.0}{\sin C}$$

$$\sin C = \frac{(\sin 135.0°)(37.0)}{57.46} = 0.4553$$

$$C = 27.1°$$

We find angle A by subtraction:

$$A = 180° - 135.0° - 27.1° = 17.9°$$

So, the missing parts of the triangle are

$$b = 57.5 \text{ in., } C = 27.1°, \text{ and } A = 17.9°$$

COMMON ERROR The cosine of an obtuse angle is negative. Be sure to use the proper algebraic sign when applying the law of cosines to an obtuse angle.

Solving a Triangle with Three Sides Known (SSS)

When three sides of an oblique triangle are known, we can use the law of cosines to solve for one of the angles. A second angle is found using the sine law, and the third angle is found by subtracting the other two from 180°.

Example 21

Solve triangle ABC, in which $a = 195$ cm, $b = 243$ cm, and $c = 137$ cm.

Solution: We sketch the triangle in Figure 22-47. Using the law of cosines we find the acute angle A:

$$a^2 = b^2 + c^2 - 2bc \cos A$$

$$195^2 = 243^2 + 137^2 - 2(243)(137)(\cos A)$$

$$\cos A = \frac{195^2 - 243^2 - 137^2}{-2(243)(137)} = 0.5977$$

$$A = 53.30°$$

We find the acute angle C by the law of sines:

$$\frac{195}{\sin 53.30°} = \frac{137}{\sin C}$$

$$\sin C = \frac{(\sin 53.30°)(137)}{195} = 0.5633$$

$$C = 34.28°$$

Finally,

$$B = 180° - 53.30° - 34.28° = 92.42°$$

So, the angles of the triangle are $A = 53.3°$, $B = 92.4°$, and $C = 34.3°$.

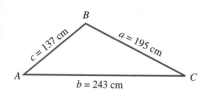

■ **FIGURE 22-47**

In the example above, we could have used the law of cosines to find the obtuse angle B, with no "ambiguity," since the cosine of an acute angle is positive, and the cosine of an obtuse angle is negative. Thus the law of cosines would tell us directly if the angle were acute or obtuse.

Sketch and solve each triangle ABC with the given parts. Round sides to three significant digits and angles to the nearest tenth of a degree.

Solving a Triangle with Two Sides and the Included Angle Known (SAS)

1. $A = 63.2°$, $b = 476$ cm, $c = 583$ cm

2. $B = 59.8°$, $a = 36.7$ in., $c = 45.5$ in.

3. $C = 35.8°$, $a = 175$ in., $b = 156$ in.

4. $A = 63.9°$, $b = 35.7$ cm, $c = 73.6$ cm

5. $B = 52.7°$, $a = 41.5$ cm, $c = 98.0$ cm

6. $C = 27.7°$, $a = 754$ ft, $b = 123$ ft

7. $C = 142°$, $a = 89.2$ mm, $b = 96.1$ mm

8. $B = 121°$, $a = 425$ cm, $c = 237$ cm

Solving a Triangle with Three Sides Known (SSS)

9. $a = 7.23$ m, $b = 7.64$ m, $c = 7.45$ m

10. $a = 546$ in., $b = 487$ in., $c = 528$ in.

11. $a = 329$ cm, $b = 651$ cm, $c = 442$ cm

12. $a = 98.3$ mm, $b = 158$ mm, $c = 67.4$ mm

13. $a = 309$ ft, $b = 128$ ft, $c = 182$ ft

14. $a = 28.6$ cm, $b = 20.5$ cm, $c = 10.4$ cm

Applications

15. To measure the width BC of a pond, a transit is placed at A, and angle CAB is measured to be $27.4°$ (Fig. 22-48). The distances AB and AC are measured and found to be 73.5 m and 84.2 m, respectively. Find the width of the pond.

16. A triangular lot measures 768 ft, 867 ft, and 935 ft along its sides. Find the angles between the sides.

17. Two boats are 68.5 km apart. Both are traveling toward the same port, which is 82.1 km from one of them and 96.4 km from the other. Find the angle at which their paths intersect.

18. Two straight roads intersect at an angle of $48.5°$. A car on one road is 145 ft from the crossing and a car on the other road is 179 ft from the crossing. Find the distance between the two cars.

19. A ship is 7.75 km directly west of a port. If the ship sails northwest ($45.0°$ west of north) for 2.14 km, how far will it be from the port?

20. From a point on level ground between two same-height telephone poles, cables are stretched to the top of each pole. One cable is 59.0 ft long and the other is 72.0 ft long. If the angle between the two cables is $135°$, find the distance between the poles.

21. A ship leaves a port traveling due west at 21.0 km/h. Three hours later the ship turns $18.0°$ north of west. How far is the ship from the port 5.00 hours after it left the port?

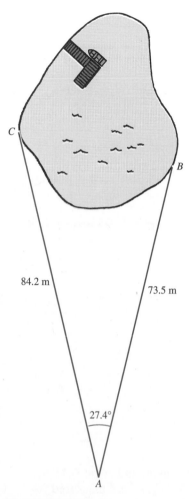

84.2 m

73.5 m

27.4°

A

C

B

■ **FIGURE 22-48**

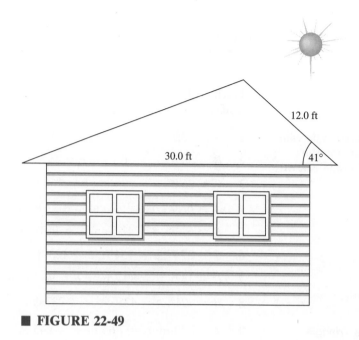

30.0 ft

12.0 ft

41°

■ **FIGURE 22-49**

θ

24.0 ft

24.0 ft

36.0 ft

■ **FIGURE 22-50**

■ **FIGURE 22-51**

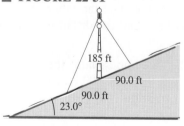

185 ft

90.0 ft

90.0 ft

23.0°

22. Solar panels are to be placed on the southern roof of a home. The panels are 12.0 ft long and make an angle of 41.0° with the horizontal (Fig. 22-49). How long must the rafter on the north be?

23. Find the angle *θ* between the rafters in Fig. 22-50.

24. A surveyor measures one angle of a triangular lot to be 68.0° and the two sides that meet at this corner to be 375 m and 256 m. Find the length of the third side.

25. A vertical radio tower is built on the side of a hill, which makes an angle of 23.0° with the horizontal. Two guy wires support the tower 185 ft up and are attached to the ground 90.0 ft downhill and 90.0 ft uphill from the tower (Fig. 22-51). What are the lengths of the guy wires?

1. Sketch the angle θ in standard position that has the point $(-4, -5)$ on the terminal side of the angle. Then write the six trigonometric functions of θ, rounded to four significant digits.

2. The point $(3.4, -2.6)$ is on the terminal side of an angle θ in standard position. Determine the quadrant of the terminal side, and write the six trigonometric functions of θ in decimal form with four significant digits.

3. An angle θ in standard position has its terminal ray in quadrant II. Determine the algebraic sign of the sine, cosine, and tangent of θ.

4. In what quadrant must the terminal side of θ lie if
   (a) the sine is positive and the cosine is negative?
   (b) both the cosine and the tangent are negative?

5. Without using a calculator, find the following functions of quadrantal angles:
   (a) $\tan 0°$ **(b)** $\cos 270°$ **(c)** $\sin 180°$ **(d)** $\sec 90°$

6. Find a positive angle and a negative angle which have the same terminal side as the primary angle $A = 154°$, and show that the sine, cosine, and tangent of the three angles are the same.

7. Find the reference angle for $A = 248°$, and compare the sines, cosines, and tangents of both angles.

8. Find the primary angles whose functions are given below. Round to the nearest degree.
   (a) $\cos \theta = -0.5878$ **(b)** $\sin \theta = -1$ **(c)** $\tan \theta = 0$ **(d)** $\csc \theta = -1.794$

Sketch the triangle and solve for all missing parts. Round all sides to three significant digits and angles to the nearest tenth of a degree.

9. $A = 68.3°$, $C = 75.2°$, $c = 5.37$ cm

10. $A = 21.0°$, $a = 7.20$ m, $c = 8.40$ m

11. $B = 82.0°$, $a = 47.0$ mm, $c = 75.0$ mm

12. $C = 31.0°$, $a = 14.5$ cm, $c = 43.7$ cm

13. $a = 6.30$ cm, $b = 7.50$ cm, $c = 5.90$ cm

14. $A = 72.3°$, $B = 31.7°$, $c = 56.8$ ft

15. Find the length of the rafters PR and QR for the roof in Fig. 22-52.

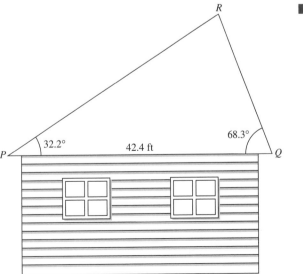

■ FIGURE 22-52

16. A woman owns a triangular lot. Markers at corners A and B are found, but the third marker at C cannot be located. The deed states that $AB = 187$ ft, $BC = 209$ ft, and $A = 38.7°$. How far is C from A?

17. A ranger leaves his station and walks due east 4.50 miles. He then turns and walks 3.50 miles in a direction 30.0° north of east. How far away from the station is he, and in what direction must he walk if he wants to go directly back to the station?

18. A drafting triangle has sides 10.2 cm, 24.5 cm, and 30.6 cm. Find the angles between the sides.

19. If $\tan \theta = -\dfrac{7}{24}$ and $\cos \theta$ is negative, write the sine and cosine of θ.

WRITING

20. Suppose you work for a company that builds bridges. You analyzed a proposed bridge truss made up of many triangles, sometimes using the law of sines and sometimes the more time-consuming law of cosines. Your client's accountant, angry over the size of your bill, has attacked your method of sometimes using the longer law of cosines when it is clear to him that the shorter law of sines is also "good for solving triangles." Write a letter to your client explaining why you sometimes had to use one law and sometimes the other.

TEAM PROJECT

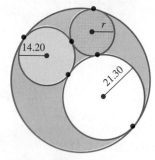

■ FIGURE 22-53

21. Four mutually tangent circles are shown in Fig. 22-53. Find the radius r of the green circle.

Vectors 23

In this chapter, we illustrate vectors graphically and use right triangle trigonometry to resolve vectors into components. We add vector quantities using two methods: addition by components and addition using the law of cosines. We look at many applications of vectors, including displacement, velocity, and force.

23-1   VECTORS AND COMPONENTS

In this section, we define a vector quantity and show the notation used for vectors. We show how to graphically represent a vector and the rectangular components of a vector.

Definitions

Many quantities can be described completely by giving their *magnitude* (a number indicating size). Such quantities are called *scalar* quantities.

> ### Example 1
> (a) Length is a scalar quantity, since it can be completely described by a number indicating its size: A certain board is 1.37 m long.
> (b) Temperature is a scalar quantity: The temperature of a steel forging is 1500° F.

Other quantities need to be described by giving not only the magnitude of the quantity but also its *direction*. Such quantities are called *vector quantities* (or vectors for short).

> ### Example 2
> (a) A plane takes off from Philadelphia and flies 79 miles. This is a distance, a scalar quantity. If we knew that the plane flew 79 miles from Philadelphia to New York City in the direction of 45° north of east, we would be specifying both magnitude and direction, or a vector quantity. This vector is called the *displacement* of an object. The displacement vector includes the direction of the displacement as well as the magnitude (the distance from the starting point).
> (b) A speed of 55 mi/h is a scalar quantity. If we say we are driving due south at 55 mi/h, we are specifying a vector, called *velocity*. The velocity vector has both a magnitude (speed) and a direction (of motion).
> (c) A *force* of 11 pounds is necessary to push a block *up* a 23° inclined plane. A force vector includes not only the magnitude of the force (11 pounds) but also the direction (23° from the horizontal).

Algebraic Representation of a Vector

Vectors are generally represented by boldface type, scalars by nonboldface type. Thus **V** (in boldface) represents a vector quantity, and *V* (in nonboldface) represents a scalar quantity. We use *V* to denote the *magnitude only* of the vector **V**. When writing by hand, it is customary to place an arrow over a vector quantity. So the vector **V** may be handwritten as $\vec{V}$.

■ **FIGURE 23-1**
Vector V.

Graphical Representation of a Vector

A vector is represented graphically by a line segment with an arrowhead at one end (a *directed line segment*). The length of the line segment represents the magnitude of the vector. The arrowhead is used to indicate the direction of the vector. The endpoint with the arrowhead is called the *tip* of the vector and the other endpoint is called the *tail* (Fig. 23-1).

When we draw a vector **V** on the coordinate axes, we generally start at the origin and draw the directed line segment at an angle *θ* specified by the direction of the vector (Fig. 23-2). The length of the directed line segment is drawn to scale, according to the magnitude of the vector. We can specify the vector **V** using the notation

$$V \angle \theta$$

which is read *V at an angle θ*. This is called the *polar form* of the vector.

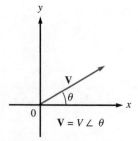

■ **FIGURE 23-2**

Example 3

The force vector in Example 2c is drawn on the coordinate plane as a directed line segment 11 units long (drawn to scale) at an angle of 23° from the positive *x* axis (Fig. 23-3). The vector **F** is written in polar form as

$$\mathbf{F} = 11 \angle 23°$$

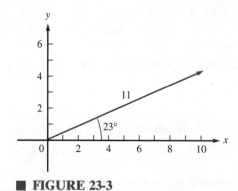

■ **FIGURE 23-3**

Rectangular Components of a Vector

Any vector can be replaced by two (or more) vectors which, acting together, exactly duplicate the effect of the original vector. These vectors are called the *components* of the original vector. We shall be concerned with components that are perpendicular to each other, called *rectangular components*. We sometimes use the word *component* to refer to only the magnitude of a component vector.

Resolving a Vector Graphically

To *resolve* a vector means to find its components. To resolve a vector graphically (to find its rectangular components), we sketch the vector and then draw a rectangle

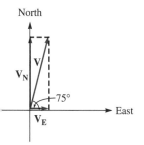

FIGURE 23-4
Velocity vectors.

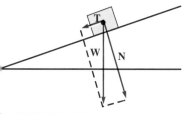

FIGURE 23-5
Force vector on an inclined plane.

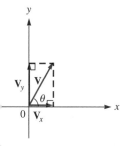

FIGURE 23-6
Components of a vector.

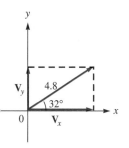

FIGURE 23-7

FIGURE 23-8
Resolving a vector.

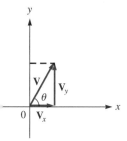

that has the given vector as one diagonal. From this rectangle, we can approximate the components of the vector.

Example 4

Figure 23-4 shows a velocity vector **V** of an airplane flying in a direction 75° north of east. From due east (the horizontal axis) we draw an angle of 75° toward the north. We resolve the vector **V** to find the east component **V**$_E$ and the north component **V**$_N$, by sketching a rectangle with a horizontal base that has **V** as its diagonal.

Example 5

The weight **W** (a force vector) of a block on an inclined plane (Fig. 23-5) is shown resolved into two perpendicular components, **N** and **T**. **N** is called the *normal* force, acting perpendicular to the plane (holding the block on the plane), and **T** is called the *tangential* force, acting parallel, or "tangent," to the plane (causing the block to slide down the plane).

Often we need to resolve a vector into its horizontal and vertical components. If **V** is vector, then the horizontal component is called the *x component* **V**$_x$ and the vertical component is called the *y component* **V**$_y$ (Fig. 23-6).

Example 6

Graphically resolve the vector

$$\mathbf{V} = 4.8 \angle 32°$$

into its vertical and horizontal components. Use the sketch to approximate V_x and V_y.

Solution: Using a protractor, we sketch a 32° angle in standard position. Next we draw the length of the vector to scale; since $V = 4.8$, we use 4.8 cm as the length of the directed line segment (Fig. 23-7). We draw a rectangle and use a cm-ruler to measure the length of **V**$_x$ and **V**$_y$ to be:

$$V_x \approx 4.1$$

and

$$V_y \approx 2.5$$

Resolving a Vector Algebraically

Any vector **V** can be resolved algebraically into a horizontal x component **V**$_x$ and a vertical y component **V**$_y$. In Fig. 23-6, the vector **V** is the diagonal of a rectangle. Since opposite sides of a rectangle are equal, the right side of this rectangle is equal to the vector **V**$_y$ (Fig. 23-8). The vector **V** is the hypotenuse of a right triangle in which its components **V**$_x$ and **V**$_y$ are the two legs. Using right triangle trigonometry, we have

$$\cos \theta = \frac{V_x}{V}$$

Solving for V_x:

$$V_x = V \cos \theta$$

And,

$$\sin \theta = \frac{V_y}{V}$$

Solving for V_y:

$$V_y = V \sin \theta$$

So, these are the algebraic equations needed to find the x and y components of a vector if the magnitude and direction of the vector are given:

	$V_x = V \cos \theta$	117
x AND y COMPONENTS OF A VECTOR V	$V_y = V \sin \theta$	118

Example 7

Algebraically find the components of the vector **V** in Example 6.

$$\mathbf{V} = 4.8 \angle 32°$$

Solution: Using Eqs. 117 and 118,

$$V_x = 4.8 \cos 32° = 4.1$$
$$V_y = 4.8 \sin 32° = 2.5$$

This verifies our graphical result in Example 6 (Fig. 23-7).

If a vector **V** is given in polar form

$$\mathbf{V} = V \angle \theta$$

we use Eqs. 117 and 118 to find its rectangular components, V_x and V_y. The components are the coordinates of the tip of the vector **V**. We can specify the vector **V** by giving the coordinates of the tip of the vector:

$$\mathbf{V} = (V_x, V_y)$$

This is called the *rectangular form* of the vector.

■ FIGURE 23-9

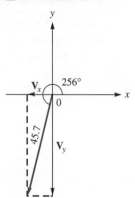

Example 8

Sketch and express the vector **V** in rectangular form.

$$\mathbf{V} = 45.7 \angle 256°$$

Solution: We sketch the vector in Fig. 23-9. Using Eqs. 117 and 118 above, with $V = 45.7$ and $\theta = 256°$, we have

$$V_x = 45.7 \cos 256° = -11.1$$

and

$$V_y = 45.7 \sin 256° = -44.3$$

Thus, the rectangular form of vector **V** is $(-11.1, -44.3)$.

Determine whether each of the following describes a scalar quantity or a vector quantity.

1. The area of a page.

2. The ship traveled 8 km north.

3. The airplane flew 15° west of south at 225 mi/h.

4. We averaged 49.6 mi/h on our trip.

5. The block weighs 75 kg.

6. The newspaper boy had a 12.4-mile route.

7. The woman pushed the block up the 5% incline with a force of 45 lb.

8. A circuit has an impedance of 975 Ω and a phase angle of 28°.

9. A free-falling object accelerates at approximately 81 m/s².

10. The pipe held 45.7 gallons of water.

11. A temperature of 136° F is the world's highest recorded temperature.

12. The sled went down the 34° hill at 15.6 m/s.

Resolving a Vector Graphically

Sketch these displacement vectors, and graphically approximate the magnitudes of the rectangular components to two significant digits.

13. 25 km at 32° north of east

14. 5.6 km at 35° north of west

Sketch these velocity vectors, and graphically approximate the magnitudes of the rectangular components.

15. 6.8 km/h at 43° south of west

16. 83 mi/h at 52° south of east

Sketch these force vectors on a plane inclined at the given angle, and graphically approximate the magnitudes of the normal and tangential components.

17. 54 kg at 15°

18. 15 pounds at 28°

Resolving a Vector Algebraically

Given the magnitude V of each vector and the angle θ that it makes with the positive x axis, find the x and y components, rounded to three significant digits.

19. $V = 785$, $\theta = 35.0°$ 20. $V = 346$, $\theta = 67.2°$

21. $V = 135$, $\theta = 225°$ 22. $V = 758$, $\theta = 145°$

Write the vector **V** in rectangular form, rounded to three significant digits.

23. $\mathbf{V} = 4.77 \angle 115°$ 24. $\mathbf{V} = 6.85 \angle 315°$

25. $\mathbf{V} = 0.125 \angle 285°$ 26. $\mathbf{V} = 0.375 \angle 255°$

23-2   APPLICATIONS OF VECTOR COMPONENTS

In this section, we consider applications of vectors, such as displacement, velocity, and force vectors. When solving vector applications, it's always helpful to draw a sketch. Often a sketch on the coordinate plane is useful.

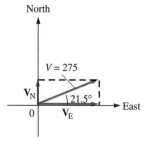

FIGURE 23-10

━━━ **Example 9** ━━━

A pilot is flying a plane headed 21.5° north of east with a speed of 275 mi/h. What are the east and north components of her velocity?

Solution: First we sketch the velocity vector **V** (Fig. 23-10), noting that the angle of 21.5° is drawn from the *east* direction, toward the north. Then we sketch the east and north components, **V**$_E$ and **V**$_N$. It is clear from the sketch that we are asked to find the rectangular components of the velocity vector **V**. Using Eqs. 117 and 118 with $V = 275$ and $\theta = 21.5°$,

$$V_E = 275 \cos 21.5° = 256$$
$$V_N = 275 \sin 21.5° = 101$$

So, the east component of the velocity is 256 mi/h, and the north component is 101 mi/h.

The sketch for Example 9 (Fig. 23-10) was made on the coordinate axes. That was convenient, since the angle of 21.5° was in standard position. In some applications, the given angle may not be in standard position, and we have to calculate the angle in standard position.

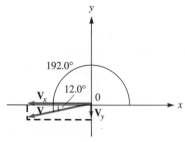

FIGURE 23-11

━━━ **Example 10** ━━━

A ship pilot heads the ship due west at a velocity of 21.0 km/h. However, he finds that the ship is pushed off course by water current, so that his actual direction of travel is 12.0° south of west. If the current is heading due south, find the ship's actual velocity in the direction of travel.

Solution: We make a sketch (Fig. 23-11), letting

V = actual velocity
V_x = west component of the actual velocity = -21.0 km/h
V_y = velocity of the current

The angle θ, in standard position, is

$$\theta = 180° + 12.0° = 192.0°$$

Using Eq. 117

$$V_x = V \cos \theta$$
$$-21.0 = V \cos 192.0°$$

Solving for V,

$$V = \frac{-21.0}{\cos 192.0°}$$
$$= 21.5 \text{ km/h}$$

So the ship's actual velocity is 21.5 km/h in the direction of 12.0° south of west.

Alternate Solution: It's not always necessary to use Eqs. 117 and 118 in which the angle is in standard position. Sometimes it's more reasonable to sketch a right triangle and use right triangle trigonometry.

We sketch the vectors (Fig. 23-12), letting

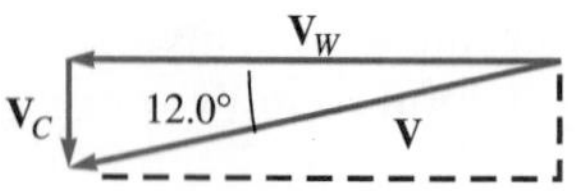

V = actual velocity
V_W = west component of the actual velocity
V_C = velocity of the current

FIGURE 23-12

We are given V_w,

$$V_w = 21.0 \text{ km/h}$$

and we need to find V. Using right triangle trigonometry,

$$\cos 12.0° = \frac{V_w}{V}$$

Or,

$$V = \frac{V_w}{\cos 12.0°} = \frac{21.0}{\cos 12.0°} = 21.5 \text{ km/h}$$

This confirms our first solution.

In some applications, the direction of the given vector has nothing to do with the positive x axis, and we just sketch a right triangle and apply simple right triangle trigonometry to the problem.

Example 11

A force of 15.8 pounds is being applied in the direction of the handle to push a lawn mower. The handle is at a 36.0° angle to the horizontal force. Find the vertical component which is pushing the mower into the ground.

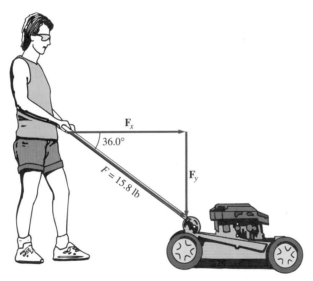

■ **FIGURE 23-13**

Solution: First we make a sketch (Fig. 23-13), letting F be the 15.8-pound force exerted on the handle of the lawn mower. We want to find F_y, the vertical component of the force. We're given that $F = 15.8$ and $\theta = 36.0°$. From Eq. 118 we get

$$F_y = F \sin \theta$$

So,

$$F_y = 15.8 \sin 36.0°$$
$$F_y = 9.29$$

So, the vertical component (the force pushing the mower into the ground) is 9.29 pounds.

In some applications, the components asked for are not x and y components, and we must use right triangle trigonometry to solve them.

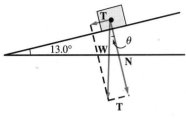

FIGURE 23-14

——— **Example 12** ———

A 225-kg block is placed on the inclined plane in Fig. 23-14. Find the normal force vector (holding the block on the inclined plane) and the tangential force vector (causing the block to slide down the inclined plane) acting on the block. Round results to three significant digits.

Solution: It can be shown with geometry that the angle θ between the force vector **W** and the normal vector **N** is the same as the angle of the inclined plane. So

$$\theta = 13.0°$$

and we're given that

$$W = 225 \text{ kg}$$

Using right triangle trigonometry,

$$N = W \cos \theta = 225 \cos 13.0°$$
$$N = 219 \text{ kg}$$

and

$$T = W \sin \theta = 225 \sin 13.0°$$
$$T = 50.6 \text{ kg}$$

So, the normal force is 219 kg and the tangential force is 50.6 kg.

SECTION 23-2 **Exercises** **Applications of Vector Components**

Solve these applications. Round results to three significant digits.

1. A ship travels 425 miles in a direction 36.3° north of east. What are the north and east components of the displacement?

2. An airplane is 126 km in a direction 64.4° south of west of an airport. What are the components of the plane's displacement from the airport?

3. A helicopter pilot, flying at an altitude of 751 m, sights a landing position at an angle of depression of 21.2°. What is the horizontal component of her displacement from the landing, and what is her distance from the landing?

4. An escalator has a angle of inclination of 32.7° between two floors that are 15.4 ft apart. When going from the bottom of the escalator to the top, what is the horizontal component of the displacement, and how long is the escalator?

5. A plane is flying in a direction 35.5° west of north at 325 mi/h. Find the north and west components of its velocity.

FIGURE 23-15

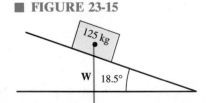

6. The wind is blowing at 56.6 mi/h from the southeast (toward 45.0° west of north). Find its velocity components directed toward the north and the west.

7. Find the normal and tangential force vectors of this 125-kg block on the inclined plane (Fig. 23-15).

8. What force, neglecting friction, must be exerted to push a 75.0-lb weight up a slope inclined 14.7° from the horizontal?

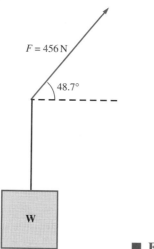

■ **FIGURE 23-16**

9. A cable exerts a force of 456 N at an angle of 48.7° with the horizontal (Fig. 23-16). Resolve this force into vertical and horizontal components. What is the maximum weight (in Newtons) that can be lifted by this cable?

10. A woman is pulling a sled with a rope with a force of 40.7 lb. If the rope makes an angle of 35.8° with the ground, find the component of the pulling force that pulls the sled along the ground and the component which is lifting the sled off the ground.

11. At what speed with respect to the water should a ship head south in order to follow a course 12.3° west of south if a current is flowing west at the rate of 8.75 mi/h?

12. At what speed should an airplane head east in order to follow a course 75.3° east of north if a 18.5 km/h wind is blowing from due south?

23-3   GRAPHICAL ADDITION OF VECTORS

In Sec. 23-1 we saw that any vector can be replaced by its component vectors which, acting together, exactly duplicate the effect of the original vector. Conversely, any two vectors can be combined into a single vector, which has the same effect as the original two vectors. This vector is called the *resultant* of the two vectors. In this section, we estimate graphically the magnitude and direction of the resultant of two vectors.

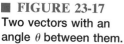

■ **FIGURE 23-17**
Two vectors with an angle θ between them.

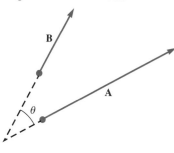

Addition of Vectors

The process of combining vectors into a resultant is called *vector addition*. The resultant is also called the *vector sum*. If the vector sum of vectors **A** and **B** is the resultant **R**, we can write

$$\mathbf{A} + \mathbf{B} = \mathbf{R}$$

Note that this is *not* ordinary addition, but *vector* addition.

In general, if the vectors **A** and **B** are two vectors, with an angle θ between them (Fig. 23-17), we can find the resultant graphically using either of two methods: the *parallelogram* method or the *triangle* method.

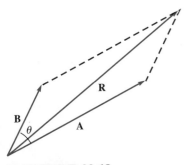

FIGURE 23-18
Vector addition:
parallelogram method.

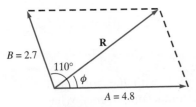

FIGURE 23-19

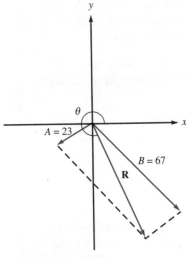

FIGURE 23-20

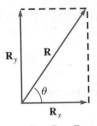

$R_x + R_y = R$

FIGURE 23-21
Vector sum.

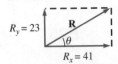

FIGURE 23-22

Parallelogram Method

The method of adding two vectors *tail to tail* is called the *parallelogram* method. First we draw the vectors with their *tails* at a common endpoint and with the given angle between them. Then we complete a parallelogram by drawing line segments from their tips parallel to the given vectors (Fig. 23-18). The resultant **R** is a vector represented by the diagonal of the parallelogram, drawn from the common endpoint of the original vectors. Since the resultant is a vector, it has both magnitude and direction. If the sketch is drawn to scale, both the magnitude and direction (the angle it makes with one of the vectors) can be estimated using a ruler and protractor.

Example 13

Two vectors, **A** and **B** make an angle of 110° with each other and have magnitudes $A = 4.8$ and $B = 2.7$. Sketch the vectors and show their resultant **R**. Estimate the magnitude of **R** and the angle that it makes with vector **A**.

Solution: The vectors **A** and **B** can be drawn in any direction, as long as we maintain the 110° angle between them. Using a ruler, we draw the vectors using a convenient scale, say 1 unit = 1 cm. So we make vector **A** 4.8 cm long and vector **B** 2.7 cm long. We sketch the other two sides of the parallelogram, draw the resultant (Fig. 23-19), and measure the length of the resultant. We find it is about 4.6 cm long, so

$$R \approx 4.6$$

Using a protractor, we measure the angle θ that **R** makes with **A**, and find

$$\phi \approx 33°$$

Example 14

Find the resultant (magnitude and direction) of two force vectors, **A** and **B**, with $A = 23$ N at 210°, and $B = 67$ N at 315°.

Solution: We sketch these vectors, to scale, in standard position with the given angles: $\theta_A = 210°$ and $\theta_B = 315°$ (Fig. 23-20). Using ruler and protractor we find,

$$R \approx 65 \text{ N} \quad \text{and} \quad \theta \approx 296°$$

If the vectors to be added are perpendicular to each other, the parallelogram is a rectangle, and the process is the same as above. A good example of perpendicular vectors are horizontal and vertical components (Fig. 23-21). *The vector sum of* $\mathbf{R}_x$ *and* $\mathbf{R}_y$ *is the resultant vector* **R**:

$$\mathbf{R}_x + \mathbf{R}_y = \mathbf{R}$$

Example 15

Vector **R** has horizontal and vertical components

$$R_x = 41 \quad \text{and} \quad R_y = 23$$

Estimate the magnitude R and the angle θ that **R** makes with $\mathbf{R}_x$.

Solution: We sketch vectors $\mathbf{R}_x$ and $\mathbf{R}_y$ (Fig. 23-22), and sketch the rectangle and the diagonal from the common tail of the two vectors. From the sketch we estimate that

$$R \approx 47 \quad \text{and} \quad \theta \approx 29°$$

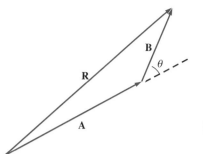

■ FIGURE 23-23
Vector addition: vector triangle method.

Triangle Method

The method of adding two vectors *tip to tail* is called the vector *triangle* method. We will again find the magnitude and direction of the resultant **R** of the two vectors in Fig. 23-17. First we draw one vector, say **A** (it doesn't matter which vector you start with). Then we draw vector **B** by connecting the tail of **B** to the tip (arrowhead) of **A**, keeping the angle θ between vector **A** and vector **B** (Fig. 23-23). Note that we need to extend the line in the direction of vector **A** before we sketch angle θ. The resultant **R** is drawn by connecting the tail of **A** to the tip of **B**. To find the resultant **R**, we need to measure the magnitude (length) of **R** and the angle ϕ that **R** makes with **A**. Notice that this triangle is identical to the lower triangle of Fig. 23-18. Thus the two methods give the same results.

— **Example 16** —

Two vectors **A** and **B** make an angle of 65° with each other. If their magnitudes are

$$A = 39 \quad \text{and} \quad B = 24$$

use the vector triangle method to estimate the magnitude of **R** and the angle that **R** makes with **B**.

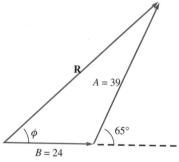

■ FIGURE 23-24

Solution: Since we need to find the angle that **R** makes with **B**, we begin by drawing vector **B**. We chose to draw **B** horizontal and to the right, since we're accustomed to sketching angles in standard position. We choose a convenient scale, say 1 unit = 0.1 cm, and draw **B** to scale. Vector **A** is drawn to scale from the arrowhead (tip) of **B** at a 65° angle with the extension of **B**. Then the resultant **R** is drawn from the tail of **B** to the tip of **A** (Fig. 23-24). With a ruler, we find that

$$R \approx 54$$

and with a protractor, we find that the angle ϕ that **R** makes with **B** is

$$\phi \approx 41°$$

Addition of Three or More Vectors

The vector triangle method (tip to tail) is especially useful when three or more vectors are to be added. However, with three or more vectors, the sketch is not a triangle, but a *polygon*.

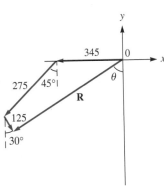

FIGURE 23-25

┌─ **Example 17** ─────

A pilot flies due west for 345 km, turns and flies southwest (45° west of south) for 275 km, turns again and flies south-southeast (30° east of south) for 125 km. What is his displacement (distance and direction) from the starting point?

Solution: Using ruler and protractor, we make a sketch using the tip to tail method (Fig. 23-25). We measure the resultant vector and estimate that

$$R \approx 565 \text{ km}$$

Using a protractor, we estimate that

$$\theta \approx 58°$$

So the pilot is about 565 km 58° west of south of his starting point.

──────────

SECTION 23-3 Exercises Graphical Addition of Vectors

Parallelogram Method

In Exercises 1–8, use the parallelogram method to sketch the resultant of the vectors.

The angle θ between the vectors is given. Estimate the magnitude of the resultant **R** and the angle ϕ that **R** makes with **A**.

1. $A = 73$, $B = 89$, $\theta = 56°$

2. $A = 8.3$, $B = 6.1$, $\theta = 152°$

The angle θ between the vectors is given. Estimate the magnitude of the resultant **R** and the angle ϕ that **R** makes with **B**.

3. $A = 0.75$, $B = 0.25$, $\theta = 130°$

4. $A = 38$, $B = 78$, $\theta = 35°$

Estimate the magnitude and direction of the resultant of these vectors.

5. $A = 4.5 \angle 80°$, $B = 3.3 \angle 160°$

6. $A = 60 \angle 340°$, $B = 85 \angle 200°$

The rectangular components of a vector **R** are given. Estimate the magnitude of **R** and the positive angle that **R** makes with the positive x axis.

7. $R_x = 85$, $R_y = -51$

8. $R_x = -1.2$, $R_y = 2.1$

Triangle Method

In Exercises 9–17, use the vector triangle method to sketch the resultant of the vectors.

The angle θ between the vectors is given. Estimate the magnitude of the resultant **R** and the angle ϕ that **R** makes with **A**.

9. $A = 83$, $B = 98$, $\theta = 100°$

10. $A = 4.2$, $B = 8.6$, $\theta = 42°$

The angle θ between the vectors is given. Estimate the magnitude of the resultant **R** and the angle ϕ that **R** makes with **B**.

11. $A = 80$, $B = 50$, $\theta = 23°$

12. $A = 350$, $B = 410$, $\theta = 160°$

Addition of Three or More Vectors

Estimate the magnitude and direction of the resultant of the three given vectors.

13. $A = 6$ at $50°$, $B = 9$ at $130°$, $C = 4$ at $245°$

14. $A = 23$ at $315°$, $B = 42$ at $195°$, $C = 65$ at $75°$

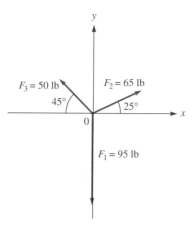

■ FIGURE 23-26

Sketch and estimate.

Applications

15. A boat sails 57° west of north at 15 km/h. There is a current from the south at 3.5 km/h. Find the boat's actual velocity, (speed and direction).

16. A plane flies 130 miles due north, turns and flies 190 miles northeast (45° east of north), turns again and flies 250 miles southeast-east (60° east of south). Find the displacement from the starting point (magnitude and direction).

17. Three forces are acting on a body as shown (Fig. 23-26). If $F_1 = 95$ lb, $F_2 = 65$ lb, and $F_3 = 50$ lb, find the magnitude and direction of the resultant force.

23-4

ADDITION OF PERPENDICULAR VECTORS

In the last section we saw how the resultant of two or more vectors could be represented graphically and the magnitude and direction estimated from the graph. In this section we show how to calculate the magnitude and direction of resultants using trigonometry. We begin with perpendicular vectors and then consider applications. Non-perpendicular vectors are treated in the next two sections.

In general if **A** and **B** are two perpendicular vectors (Fig. 23-27), we may use the vector triangle method (or the parallelogram method) to sketch the resultant. If we are asked to find the angle between **R** and **A**, we choose to draw **A** first, and then connect the tail of **B** to the tip of **A** (Fig. 23-28).

Since the triangle is a right triangle, we use the Pythagorean Theorem to determine the magnitude of **R**:

$$R = \sqrt{A^2 + B^2}$$

The direction of **R** can be described by giving the angle ϕ that **R** makes with one of the vectors, say **A**.

Using right triangle trigonometry,

$$\tan \phi = \frac{B}{A}$$

or,

$$\phi = \tan^{-1} \frac{B}{A}$$

where ϕ is the angle between the resultant **R** and the vector **A**.

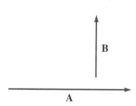

■ FIGURE 23-27
Perpendicular vectors.

■ FIGURE 23-28
Addition of perpendicular vectors.

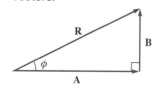

	Magnitude	$R = \sqrt{A^2 + B^2}$	119
RESULTANT OF TWO PERPENDICULAR VECTORS $R = A + B$	Angle between **R** and **A**	$\phi = \tan^{-1}\dfrac{B}{A}$	120

Example 18

Vectors **A** and **B** are perpendicular vectors, with

$$A = 481 \quad \text{and} \quad B = 762$$

Find the magnitude of the resultant and the angle that the resultant makes with vector **A**.

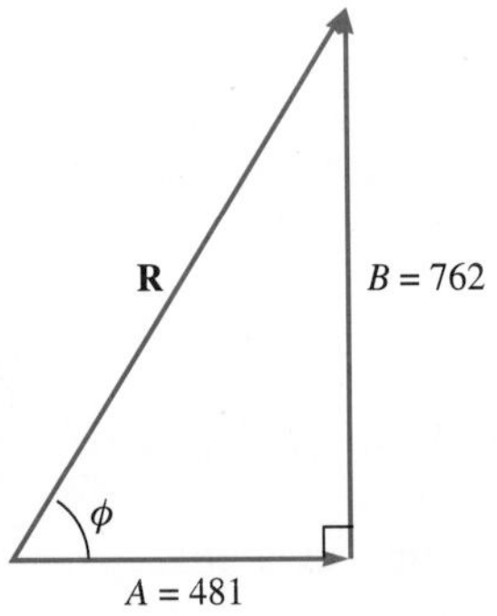

Solution: Using the vector triangle method, we sketch the vectors (Fig. 23-29). From Eqs. 119 and 120, we have

$$R = \sqrt{481^2 + 762^2} = 901$$

and

$$\phi = \tan^{-1}\frac{762}{481} = 57.7°$$

where $R = 901$ and $\phi = 57.7°$ is the angle between **R** and **A**.

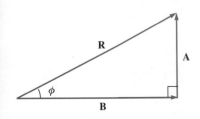

■ **FIGURE 23-29**

Take special note that in Eq. 120, ϕ is the angle between the resultant and vector **A**. If we need to find the angle between **R** and vector **B** (Fig. 23-30), then using right triangle trigonometry, we see that

$$\tan \phi = \frac{A}{B}$$

or,

$$\phi = \tan^{-1}\frac{A}{B}$$

So, if ϕ is the angle between **R** and **B**, then B must be in the *denominator*.

■ **FIGURE 23-30**

Example 19

Vectors **A** and **B** are perpendicular vectors, with

$$A = 16.2 \quad \text{and} \quad B = 28.5$$

Find the magnitude of the resultant and the angle that the resultant makes with the longer vector.

Solution: Using the vector triangle method, we sketch the vectors, with the vertex at the origin (Fig. 23-31). From Eq. 119 we have

$$R = \sqrt{16.2^2 + 28.5^2} = 32.8$$

Since **B** is the longer vector, we use the equation above to find the angle ϕ, taking special care that $B = 28.5$ is in the denominator,

$$\phi = \tan^{-1}\frac{16.2}{28.5} = 29.6°$$

■ **FIGURE 23-31**

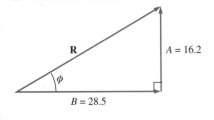

Applications

Many vector problems involve finding the resultant of two perpendicular vectors.

Example 20

A river flows with a current, at a velocity of 5.75 km/h. A rower, who can row 8.25 km/h in still water, heads directly across the current but is pushed downstream by the current. Find the actual velocity (rate and direction of travel) of the boat.

Solution: Let

$\mathbf{V}_c$ = the current's velocity vector,
$\mathbf{V}_r$ = the rower's velocity vector (in still water), and
$\mathbf{V}_b$ = the boat's actual velocity vector with respect to the shore.

Now we sketch the vectors (Fig. 23-32). We can see that $\mathbf{V}_b$ is the resultant vector of $\mathbf{V}_c$ and $\mathbf{V}_r$.

Applying Eqs. 119 and 120, with $V_c = 5.75$ and $V_r = 8.25$, we have

$$V_b = \sqrt{V_c^2 + V_r^2}$$
$$= \sqrt{5.75^2 + 8.25^2}$$
$$= 10.1$$

And,

$$\phi = \tan^{-1}\frac{V_r}{V_c}$$
$$= \tan^{-1}\frac{8.25}{5.75} = 55.1°$$

where ϕ is the angle between the boat and the current.

So, the boat's actual rate with respect to the shore is 10.1 km/h, and the direction of travel is 55.1° from the shoreline (the current's direction).

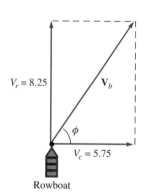

$V_r = 8.25$ $\mathbf{V}_b$

ϕ

$V_c = 5.75$

Rowboat

■ FIGURE 23-32

SECTION 23-4   **Exercises**   **Addition of Perpendicular Vectors**

The magnitudes of two perpendicular vectors are given. Sketch the vectors and the resultant. Find the magnitude of the resultant **R** and the positive angle that **R** makes with vector **A**, rounded to three significant digits.

1. $A = 585, B = 358$ **2.** $A = 846, B = 297$

3. $A = 7.53, B = 8.96$ **4.** $A = 239, B = 382$

The magnitude of two perpendicular vectors are given. Sketch the vectors and the resultant. Find the magnitude of the resultant **R** and the positive angle that **R** makes with vector **B**.

5. $A = 16.5, B = 50.1$ **6.** $A = 396, B = 677$

7. $A = 5.13, B = 8.95$ **8.** $A = 9.46, B = 4.57$

Vector **V** has the given x and y components. Find the magnitude of **V** and the positive angle it makes with the positive x axis.

9. $V_x = 12.4, V_y = 15.6$ **10.** $V_x = -6.35, V_y = 4.77$

11. $V_x = -7.42, V_y = -5.16$ **12.** $V_x = 876, V_y = -124$

Sketch, solve, and round to three significant digits.

13. A cyclist bikes 14.0 miles north, makes a left turn, and bikes 23.4 miles west. Find her displacement from her starting point (both distance and direction).

14. An airplane flies due east for 185 km, makes a 90.0° turn and flies due south for 275 km. Find its displacement from its starting point.

15. Given two displacements with magnitudes of 12.0 miles and 15.0 miles.
    (a) Find the magnitude of the maximum resultant displacement possible.
    (b) Find the magnitude of the minimum resultant displacement possible.
    (c) Find the magnitude of the resultant if the given displacements are perpendicular.

16. A driver goes 27.8 miles north, turns and drives 18.2 miles east, turns and drives 32.1 miles south. Find the total distance traveled and the displacement (both magnitude and direction) from his starting point.

17. A plane is headed due west with an airspeed of 235 km/h. It is driven from its course by a wind *from* due south blowing at 35.5 km/h. Find the groundspeed of the plane and the actual direction of travel.

18. A canoeist is paddling straight across a river with a current of 6.50 km/h. Find the rate and the direction of travel, with respect to the shore, of the canoe, if the canoeist can paddle 7.45 km/h in still water.

19. A plane is traveling 325 km/h in still air.
    (a) Find its groundspeed if it had an 85.0 km/h head wind.
    (b) Find its groundspeed if it had an 85.0 km/h tail wind.
    (c) Find its groundspeed if it were flying perpendicular to an 85.0 km/h wind. Find the angle the flight path makes with the direction of the wind.

20. A boater wants to go straight across a river, whose current rate is 9.25 mi/h. If the boat travels at 35.7 mi/h in still water, in what direction, with respect to the shore, should the boater head to go straight across the river?

21. A pilot wants to fly due north at a groundspeed of 475 km/h. The wind is 45.0 km/h *from* the east. At what heading (direction) and speed must the pilot fly the plane to compensate for the wind?

22. Two people are pushing a rectangular box on adjacent sides. One pushes with a force of 55.3 lb and the other with a force of 75.8 lb. What is the magnitude and direction, with respect to the larger force, of the resultant force?

23. The lift of an airplane wing is 852 lb and the drag is 325 lb. Find the magnitude F and direction ϕ of the resultant force (Fig. 23-33).

24. Two forces of 67.6 N and 78.9 N act on a body at right angles to each other. Find the magnitude of the resultant force and the angle it makes with the 78.9 N force.

25. A sign is being acted on by a horizontal force of 25.6 lb acting to the left and a vertical force of 15.7 lb acting down. Find the magnitude and direction of the resultant force.

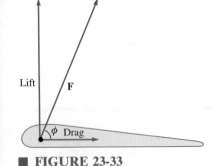

Lift   F

ϕ Drag

■ FIGURE 23-33

23-5 VECTOR ADDITION BY COMPONENTS

In the last section we added perpendicular vectors using the Pythagorean Theorem and right triangle trigonometry. In this section we add non-perpendicular vectors algebraically, by components. We resolve the vectors into horizontal and vertical components and combine like components. Then we use the Pythagorean Theorem and right triangle trigonometry to find the resultant.

Vector Addition by Components

1. Sketch the vectors and their resultant **R** on the coordinate plane and determine the quadrant of the resultant.
2. Find the x and y components, algebraically, of each given vector.
3. Add the x components to get the x component R_x.
4. Add the y components to get the y component R_y.
5. Find the magnitude R using the Pythagorean Theorem:

$$R = \sqrt{R_x^2 + R_y^2}$$

6. Find the reference angle of the resultant:

$$\theta_R = \tan^{-1}\left|\frac{R_y}{R_x}\right|$$

7. Find the angle θ that **R** makes with the positive x axis, from the quadrant of the resultant.

The absolute value sign in Step 6 above is necessary because a reference angle is never negative. If the vectors being added are in quadrants II, III, or IV, some components will be negative. By taking the absolute value, we disregard the signs of R_x and R_y, so that the reference angle is not negative.

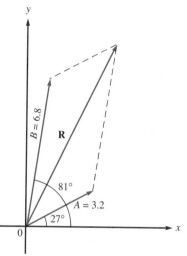

■ **FIGURE 23-34**

Example 21

Find the resultant of vector **A** = 3.2 at 27° and vector **B** = 6.8 at 81°.

Solution: We sketch the vectors and their resultant (Fig. 23-34) and note that the resultant is in quadrant I. Next we resolve vectors **A** and **B** into components and place the results in a table. We carry a third significant digit in the calculations, so that we can round in two significant digits at the end.

Vector	x component	y component
A	$3.2 \cos 27° = 2.85$	$3.2 \sin 27° = 1.45$
B	$6.8 \cos 81° = \underline{1.06}$	$6.8 \sin 81° = \underline{6.72}$
Sums	$R_x = \overline{3.91}$	$R_y = \overline{8.17}$

To find R:

$$R = \sqrt{R_x^2 + R_y^2}$$
$$= \sqrt{3.91^2 + 8.17^2}$$
$$= 9.1$$

Finding the reference angle θ_R:

$$\theta_R = \tan^{-1}\left|\frac{R_y}{R_x}\right|$$

$$\theta_R = \tan^{-1}\left|\frac{8.17}{3.91}\right|$$

$$\theta_R = 64°$$

Since θ is a quadrant I angle,

$$\theta = \theta_R = 64°$$

So the resultant is the vector

$$\mathbf{R} = 9.1 \angle 64°$$

We do a visual check by estimating the magnitude and direction of the resultant in Fig. 23-34. Indeed, our sketched resultant agrees with our numerical results.

Example 22

Find the resultant of vector $\mathbf{A} = 85.1$ at $15.5°$ and vector $\mathbf{B} = 39.2$ at $254°$.

Solution: We sketch the vectors and their resultant (Fig. 23-35) and note that the resultant is in quadrant IV. Next we resolve vectors $\mathbf{A}$ and $\mathbf{B}$ into components and place the results in a table:

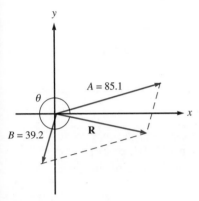

■ FIGURE 23-35

Vector	x component		y component	
A	$85.1 \cos 15.5° =$	82.00	$85.1 \sin 15.5° =$	22.74
B	$39.2 \cos 254° =$	-10.80	$39.2 \sin 254° =$	-37.68
Sums	$R_x =$	71.20	$R_y =$	-14.94

To find R:

$$R = \sqrt{R_x{}^2 + R_y{}^2}$$
$$= \sqrt{71.20^2 + (-14.94)^2}$$
$$= 72.8 \text{ (rounding to three significant digits)}$$

Finding the reference angle θ_R:

$$\theta_R = \tan^{-1} \left| \frac{R_y}{R_x} \right|$$

$$\theta_R = \tan^{-1} \left| \frac{-14.94}{71.20} \right|$$

$$\theta_R = 11.9°$$

Since the resultant is in quadrant IV, the angle

$$\theta = 360° - 11.9° = 348.1°$$

So the resultant is the vector

$$\mathbf{R} = 72.8 \angle 348°$$

■ FIGURE 23-36

We do a visual check by estimating the magnitude and direction of the resultant in Fig. 23-35. Indeed, our sketched resultant agrees with our numerical results.

Example 23

Two vectors of magnitudes $A = 25.2$ and $B = 72.8$ have an angle of $126°$ between them. Find the magnitude of the resultant $\mathbf{R}$ and the angle between $\mathbf{R}$ and $\mathbf{A}$.

Solution: Since the directions of $\mathbf{A}$ and $\mathbf{B}$ are not specifically given, we choose to draw $\mathbf{A}$ along the positive x axis for simplicity (Fig. 23-36). Vector $\mathbf{A}$ is then represented by

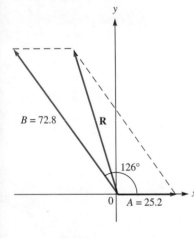

$$\mathbf{A} = 25.2 \angle 0°$$

We represent vector $\mathbf{B}$ as

$$\mathbf{B} = 72.8 \angle 126°$$

We sketch the resultant and see that it is in quadrant II. We calculate the resultant as we did in Example 21.

Vector	x component		y component	
A	25.2 cos 0°	= 25.2	25.2 sin 0°	= 0
B	72.8 cos 126°	= −42.79	72.8 sin 126°	= 58.90
Sums		$R_x = -17.59$		$R_y = 58.90$

To find R:

$$R = \sqrt{R_x{}^2 + R_y{}^2}$$
$$= \sqrt{(-17.59)^2 + 58.90^2}$$
$$= 61.5$$

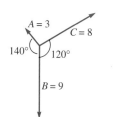

A = 3   C = 8
140°   120°
B = 9

■ **FIGURE 23-37**

To find θ_R:

$$\theta_R = \tan^{-1}\left|\frac{R_y}{R_x}\right|$$

$$\theta_R = \tan^{-1}\left|\frac{58.90}{-17.59}\right|$$

$$\theta_R = 73.4°$$

Since θ is quadrant II angle,

$$\theta = 180° - 73.4° = 106.6°$$

So, the resultant **R** is a vector whose magnitude is 61.5 and makes an angle of 107° with vector **A**.

We do a visual check by estimating the magnitude and direction of the resultant in Fig. 23-36. Indeed, our sketched resultant agrees with our numerical results.

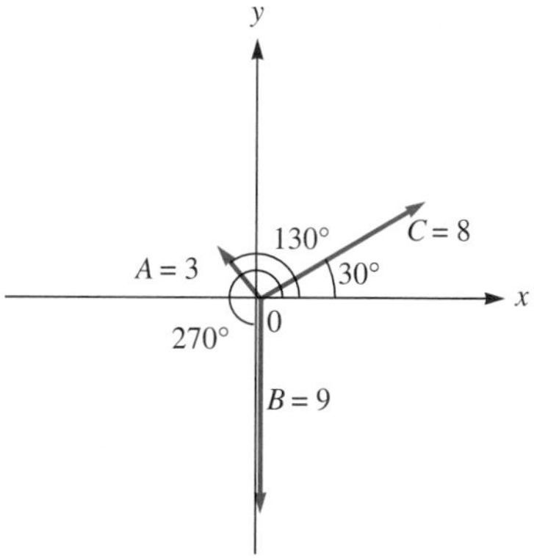

Resultant of Three or More Vectors

The component method is especially useful in finding the resultant of three or more vectors.

■ **FIGURE 23-38**

■ **FIGURE 23-39**

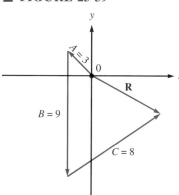

— **Example 24** —

Find the resultant of the three vectors in Fig. 23-37. Round results to three significant digits.

Solution: First we determine the angles the given vectors make with the positive x axis (Fig. 23-38) and find that

$$\mathbf{A} = 3 \angle 130°$$
$$\mathbf{B} = 9 \angle 270°$$

and
$$\mathbf{C} = 8 \angle 30°$$

Next we sketch the vector sum, on the coordinate plane with the vertex at the origin, using the vector triangle (tip to tail) method (Fig. 23-39), and see that the resultant **R** is in quadrant IV.

To find the magnitude and direction of **R**, we make the table of components:

Vector	x component	y component
A	$3 \cos 130° = -1.928$	$3 \sin 130° = 2.298$
B	$9 \cos 270° = 0$	$9 \sin 270° = -9$
C	$8 \cos 30° = 6.928$	$8 \sin 30° = 4$
Sums	$R_x = 5.000$	$R_y = -2.702$

Next we find R:

$$R = \sqrt{(5.000)^2 + (-2.702)^2} = 5.68$$

Now we find the reference angle θ_R to help determine θ:

$$\theta_R = \tan^{-1} \left| \frac{-2.702}{5.000} \right| = 28.4°$$

Since θ is a fourth quadrant angle, then

$$\theta = 360° - 28.4° = 331.6°$$

So,

$$\mathbf{R} = 5.68 \angle 332°$$

We do a visual check by estimating the magnitude and direction of the resultant in Fig. 23-39. Indeed, our sketched resultant agrees with our numerical results.

SECTION 23-5 Exercises Vector Addition by Components

Sketch the given vectors and their resultant. Find the resultant of the given vectors by components. Round to three significant digits.

1. $\mathbf{A} = 57 \angle 20°$, $\mathbf{B} = 87 \angle 75°$

2. $\mathbf{A} = 3.5 \angle 36°$, $\mathbf{B} = 2.6 \angle 80°$

3. $\mathbf{A} = 243 \angle 301°$, $\mathbf{B} = 726 \angle 155°$

4. $\mathbf{A} = 1.37$ at $248°$, $\mathbf{B} = 1.73$ at $65.8°$

The magnitudes of vectors **A** and **B** and the angle between them are given. Sketch the vectors and their resultant. Use components to find the magnitude of the resultant **R** and the angle that **R** makes with **A**.

5. $A = 6.88$, $B = 5.25$, $\theta = 112°$

6. $A = 737$, $B = 811$, $\theta = 85.0°$

The magnitudes of vectors **A** and **B** and the angle between them are given. Sketch the vectors and their resultant. Use components to find the magnitude of the resultant **R** and the angle that **R** makes with **B**.

7. $A = 0.151$, $B = 0.752$, $\theta = 165°$

8. $A = 739$, $B = 837$, $\theta = 65.4°$

Resultant of Three or More Vectors

Solve, and round to three significant digits.

9. Find the resultant of the three force vectors in Fig. 23-40, given that **B** is a horizontal force.

10. Find the resultant of the three force vectors in Fig. 23-41.

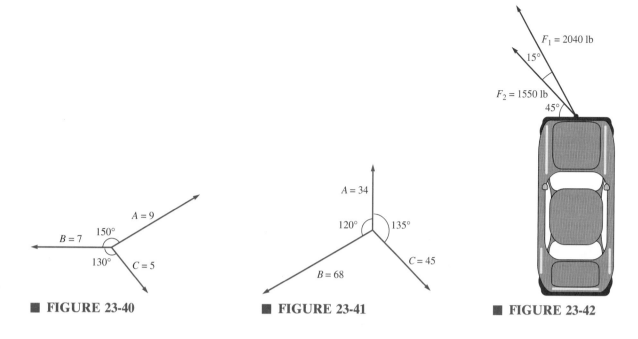

FIGURE 23-40

FIGURE 23-41

FIGURE 23-42

Solve, and round to three significant digits.

11. A ship leaves port and sails 345 km due west, then turns and sails 275 km southwest (45° south of west). Find its displacement from the port (both distance and direction).

12. A pilot is heading his plane due south with an airspeed of 325 mi/h. The wind is from the southeast (45° east of south) at 58.7 mi/h. Find the actual direction of travel and the plane's groundspeed.

13. A tractor and a truck are trying to pull a car out of a snowbank, with respective forces $F_1 = 2040$ pounds and $F_2 = 1550$ pounds (Fig. 23-42). Find the magnitude and direction (from the front of the car) of the resultant force.

14. An airplane takes off from an airport, flies 78.8 miles 35.7° east of north, turns and flies 95.0 miles 58.0° east of north, and again turns and flies 46.5 miles 75.8° east of north. Find the plane's displacement from the airport (both distance and direction).

23-6

FIGURE 23-43
Vector triangle method.

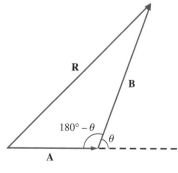

VECTOR ADDITION USING THE LAWS OF SINES AND COSINES

In the last section we added non-perpendicular vectors by the component method. In this section we add vectors using the law of sines and the law of cosines.

Given two non-perpendicular vectors and the angle between them, we can sketch the resultant using the vector triangle method (Fig. 23-43) on the coordinate plane. The resultant is one side of an *oblique* triangle. We are given two sides of the oblique triangle and can easily calculate the included angle, since it is supplementary to the angle between the vectors. To find the length of the resultant and the angle it makes with one of the original vectors, we simply solve the oblique triangle. To find the magnitude R of the resultant, we use the law of cosines. To find the angle the resultant makes with one of the vectors, we use the law of sines.

Vector Addition Using the Laws of Sines and Cosines

1. Use the vector triangle method to sketch the two vectors and their resultant.
2. Find the angle between the two given sides of the triangle.
3. Use the law of cosines to calculate R.
4. Use the law of sines to calculate the angle (direction) of $\mathbf{R}$.

Example 25

Two vectors, $\mathbf{A}$ and $\mathbf{B}$, make an angle of $41.5°$ with each other. If their magnitudes are $A = 56.7$ and $B = 87.6$, find the magnitude of the resultant $\mathbf{R}$ and the angle it makes with vector $\mathbf{B}$.

Solution: First we sketch the vectors and the resultant, using the vector triangle method. We start with vector $\mathbf{B}$ (Fig. 23-44), since we need to find the angle that $\mathbf{R}$ makes with $\mathbf{B}$. To use the law of cosines to find R, we need the included angle ϕ between the given sides A and B. Since ϕ is supplementary to the $41.5°$ angle,

$$\phi = 180° - 41.5° = 138.5°$$

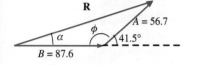

FIGURE 23-44

Using the law of cosines,

$$R^2 = 87.6^2 + 56.7^2 - 2(87.6)\,(56.7)\,(\cos 138.5°) = 18329$$
$$R = 135.4$$

Now we use the law of sines to find the angle α between $\mathbf{R}$ and $\mathbf{B}$:

$$\frac{135.4}{\sin 138.5°} = \frac{56.7}{\sin \alpha}$$

$$\sin \alpha = \frac{(\sin 138.5°)\,(56.7)}{135.4} = 0.2775$$

$$\alpha = 16.1°$$

As usual, we round our answers to the same number of significant digits as in the given data, even though we have kept more digits for intermediate steps. So, the resultant $\mathbf{R}$ has a magnitude of 135 and makes a $16.1°$ angle with vector $\mathbf{B}$.

Example 26

Find the resultant of vectors $\mathbf{A}$ and $\mathbf{B}$:

$$\mathbf{A} = 3.25 \angle 76.5° \quad \text{and} \quad \mathbf{B} = 6.75 \angle 11.2°$$

Solution: First we sketch the vectors (Fig. 23-45). To use the vector triangle method, we need to find the angle between the vectors:

$$76.5° - 11.2° = 65.3°$$

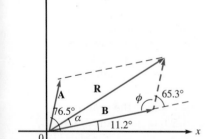

FIGURE 23-45

In the oblique triangle, the angle ϕ between the two given sides is

$$\phi = 180° - 65.3° = 114.7°$$

Now we use the law of cosines to find R:

$$R^2 = 6.75^2 + 3.25^2 - 2(6.75)\,(3.25)\,(\cos 114.7°) = 75.46$$

Thus,

$$R = 8.629$$

We use the law of sines to find the angle α between **R** and **B**.

$$\frac{8.629}{\sin 114.7°} = \frac{3.25}{\sin \alpha}$$

$$\sin \alpha = 0.3421$$

$$\alpha = 20.0°$$

So the resultant is at an angle of

$$11.2° + 20.0° = 31.2°$$

or,

$$\mathbf{R} = 8.63 \angle 31.2°$$

In Exercises 1–8, the magnitudes of vectors **A** and **B** and the angle θ between them are given. Round to three significant digits.

Find the magnitude of the resultant and the angle it makes with vector **A**.

1. $A = 345$, $B = 652$, $\theta = 75.2°$

2. $A = 5.32$, $B = 1.61$, $\theta = 158°$

3. $A = 8.22$, $B = 3.45$, $\theta = 164°$

4. $A = 735$, $B = 858$, $\theta = 63.4°$

Find the magnitude of the resultant and the angle it makes with vector **B**.

5. $A = 378$, $B = 294$, $\theta = 102°$

6. $A = 2.78$, $B = 1.05$, $\theta = 89.0°$

7. $A = 74.5$, $B = 27.7$, $\theta = 15.6°$

8. $A = 83.7$, $B = 27.4$, $\theta = 178°$

Find the resultant of vectors **A** and **B**, rounded to three significant digits.

9. $\mathbf{A} = 343 \angle 29.4°$, $\mathbf{B} = 734 \angle 85.2°$

10. $\mathbf{A} = 5.61 \angle 175°$, $\mathbf{B} = 2.26 \angle 319°$

11. $\mathbf{A} = 6.39 \angle 152°$, $\mathbf{B} = 4.23 \angle 295°$

12. $\mathbf{A} = 29.3 \angle 258°$, $\mathbf{B} = 25.4 \angle 48.3°$

Solve, and round to three significant digits.

Applications

■ FIGURE 23-46

13. A motorist drives 75.0° west of north for 35.2 miles and then turns and drives 60.0° west of south for 25.3 miles. Find the distance and direction of her displacement.

14. A surveyor locates a marker 56.3 m southwest (45.0° west of south) of him. He knows that the marker is 32.3 m south of a telephone pole. Find the displacement (distance and direction) of the telephone pole from the surveyor.

15. Two tugboats are pulling a barge toward the west as shown (Fig. 23-46), with pulling forces

$$F_1 = 1850 \text{ pounds} \quad \text{and} \quad F_2 = 1260 \text{ pounds}$$

Find the magnitude and direction of the resultant force.

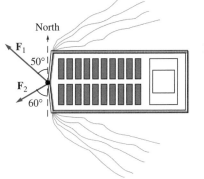

23-6 Vector Addition Using the Laws of Sines and Cosines **539**

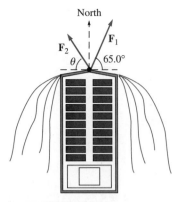

North

F_2 F_1

θ 65.0°

■ **FIGURE 23-47**

16. Two tugboats are trying to pull a barge due north. One tugboat, with a pulling force of $F_1 = 2070$ pounds, pulls at a 65.0° angle as shown (Fig. 23-47). If the smaller tugboat pulls with a force of $F_2 = 1580$ pounds, what must the angle θ of the smaller tugboat be?

17. A pilot wishes to fly 30.0° east of north with a groundspeed of 285 mi/h. A wind is blowing from the west at 65.0 mi/h. What should the pilot's heading and airspeed be?

18. A ship needs to sail directly to a port at 35.0° west of south. The current is flowing due north at 18.0 km/h. If the ship's speed in still water is 38.5 km/h, find the heading the ship must take and its actual velocity with respect to land.

■ **Chapter 23 REVIEW EXERCISES**

1. Sketch the vector **V** whose magnitude is 75.0 and makes an angle of 125° with the positive x axis. From the sketch, estimate the horizontal and vertical components, V_x and V_y. Then calculate the components.

2. Find the x and y components of the vector

$$\mathbf{V} = 4.96 \angle 265°$$

3. A ship travels 275 km in a direction 64.7° south of west. What are the south and west components of the displacement?

4. An airplane pilot, flying at an altitude of 655 ft, sights a runway at an angle of depression of 18.0°. What is the horizontal component of his displacement from the runway, and what is his distance from the runway?

5. An escalator travels at a rate of 15.4 m/min and its angle of inclination is 34.6°. What is the vertical component of the velocity? How long will it take a passenger to travel 8.05 m vertically?

6. A projectile is launched at an angle of 49.0° to the horizontal with a speed of 6750 m/min. Find the vertical and horizontal components of this velocity.

7. A truck weighing 17.0 tons stands on a hill inclined 16.0° from the horizontal. How large must the braking force be to prevent the truck from rolling downhill?

8. Use the vector triangle method to sketch the resultant **R** of two perpendicular vectors that have magnitudes of 278 and 847. Estimate the magnitude R and the angle **R** makes with the larger vector.

9. Use the parallelogram method to sketch the resultant **R** of these two vectors. Estimate the magnitude R and direction of θ.

$$\mathbf{A} = 6.85 \angle 48.8° \quad \text{and} \quad \mathbf{B} = 5.65 \angle 197.5°$$

10. Sketch the resultant **R** of these three vectors. Estimate the magnitude and direction of **R**. Round to three significant digits.

$$\mathbf{A} = 5 \angle 150°, \mathbf{B} = 9 \angle 65°, \quad \text{and} \quad \mathbf{C} = 7 \angle 245°$$

11. Two perpendicular vectors have magnitudes of 278 and 847. Calculate the magnitude of the resultant **R** and the angle **R** makes with the larger vector. Check your results with the answer in Exercise 8.

12. The components of vector **V** are $V_x = -0.376$ and $V_y = 0.644$. Find the magnitude and direction of **V**.

13. A jogger jogs 13.0 km north, then turns and jogs 11.0 km west. Find her displacement (both distance and direction) from her starting point.

14. A driver, traveling at 65.5 mi/h, throws an apple core out his window perpendcular to his direction at a speed of 3.00 mi/h. What is the velocity (magnitude and direction) of the apple core as it leaves his hand?

15. A swimmer wants to swim straight across a river, whose current rate is 1.00 km/h. If her swimming rate is 2.00 km/h in still water, in what direction should she head to go straight across the river? How long will it take her (in minutes) to cross the river, if it is 0.40 km wide?

16. Two forces of 45.5 N and 73.8 N act on a body at right angles to each other. Find the magnitude of the resultant force and the angle it makes with the smaller force.

17. Use the method of components to find the resultant of the vectors

$$\mathbf{A} = 6.85 \angle 48.8° \quad \text{and} \quad \mathbf{B} = 5.65 \angle 197.5°$$

Check your results with the answer in Exercise 9.

18. Calculate the resultant of the vectors

$$\mathbf{A} = 5.00 \angle 150.0°, \mathbf{B} = 9.00 \angle 65.0°, \mathbf{C} = 7.00 \angle 245.0°$$

Check your results with the answer in Exercise 10.

19. Two vectors of magnitude 76.0 and 32.0 have a 84.0° angle between them. Using the laws of cosines and sines, find the magnitude of the resultant and the angle it makes with the smaller vector.

20. Use the laws of cosines and sines to find the magnitude and direction of the resultant of these two vectors:

$$\mathbf{A} = 3.57 \angle 175° \quad \text{and} \quad \mathbf{B} = 1.46 \angle 287°$$

21. Two forces of 754 pounds and 986 pounds act on a point of a bridge girder at an angle of 84.8° to each other. Find the resultant force and the angle it makes with the larger force.

22. As an airplane heads due north with an airspeed of 375 mi/h, a wind with a speed of 45.0 mi/h causes the plane to travel slightly west of north with a groundspeed of 355 mi/h. What is the wind direction and in what direction does the plane actually fly?

23. We have learned three methods of adding vectors. Describe each method, including its advantages and disadvantages. Which do you prefer and why? Make up two non-perpendicular vectors and use these to illustrate all three methods.

24. A displacement vector is a distance (magnitude) traveled in a specific direction. In this problem your team will create displacement vectors and find their vector sum.

Begin in an open area of your campus. Mark a beginning point A. Measure 12 m along a compass heading of due east; mark that point B. This is your first displacement vector **AB**. Now measure 19 m along the heading 45° north of east, and mark that point C, to create vector **BC**. From point C, walk 15 m in a direction 35° west of north, and mark that point D to create vector **CD**.

The vector from A to D is the vector sum of the three displacement vectors you have just walked. With a tape or other measuring device, determine the distance (magnitude) from A to D. Use your compass to determine the direction from A to D.

Draw a scaled vector diagram of your displacement vectors. Verify the vector sum using one of the analytical methods learned in this chapter. What is the percent error? How do you account for this error?

Applications of Radian Measure

We introduced radian measure in Chapter 13 and learned that angles are measured in *radians* for some applications. In this chapter we review radian measure and consider applications of radian measure. Although we often work with angles in degrees, radians are very useful for topics such as arc length, area of a sector, and angular velocity.

24-1   RADIAN MEASURE

In this section, we learn to express radian measure in terms of π. Recall that one radian (1 rad) is the measure of a central angle that intercepts an arc whose length is equal to the radius (Fig. 24-1). (A central angle is an angle with its vertex at the center of a circle and has radii for sides.) In a circle, one revolution (360°) is equal to 2π radians.

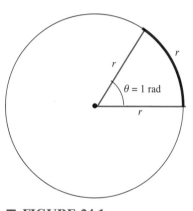

■ **FIGURE 24-1**
Angle of 1 radian.

CONVERSIONS	1 rev = 360° = 2π radians	79

We also learned to convert an angle from degrees to radians by multiplying the degree measurement by the conversion factor $\left(\dfrac{2\pi \text{ rad}}{360°}\right)$. This gives us the radian measure of an angle *in decimal form*.

--- **Example 1** ---
Convert 65.8° to radians.

Solution:

$$65.8° = (65.8°)\left(\frac{2\pi \text{ rad}}{360°}\right) = 1.15 \text{ rad}$$

Radian Measure in Terms of π

It is quite common to express the radian measure of an angle in terms of π. Eq. 79 provides the basic conversion:

$$360° = 2\pi \text{ rad}$$

543

Dividing, we get

$$180° = \pi \text{ rad}$$

$$90° = \frac{\pi}{2} \text{ rad}$$

$$60° = \frac{\pi}{3} \text{ rad}$$

and so forth. Generally, radian measure in terms of π is expressed *in fraction form*, rather than in decimal form. This table shows the common conversions of angles in degrees to radians in terms of π.

Angle measure	
Degrees	Radians
360°	2π
180°	π
90°	$\frac{\pi}{2}$
60°	$\frac{\pi}{3}$
45°	$\frac{\pi}{4}$
30°	$\frac{\pi}{6}$

In Example 1, we converted the angle from degrees to radians when we multiplied by the conversion factor $\dfrac{2\pi \text{ rad}}{360°}$, but the conversion factor may be reduced:

$$\frac{2\pi \text{ rad}}{360°} = \frac{\pi \text{ rad}}{180°}$$

To convert angles in degrees to radian measure in terms of π, we multiply the angle in degrees by $\dfrac{\pi \text{ rad}}{180°}$ *in fraction form,* and then reduce the fraction to lowest terms.

Example 2

Convert 120° to radians in terms of π.

Solution:

$$120° = 120° \left(\frac{\pi \text{ rad}}{180°} \right)$$

Reducing,

$$120° = \frac{2\pi}{3} \text{ rad}$$

To convert an angle in radians to degrees, we multiply the angle in degrees by the conversion factor $\dfrac{180°}{\pi \text{ rad}}$ and express the result in decimal form.

Example 3

Convert 1 rad to degrees, rounded to tenths.

Solution:

$$1 \text{ rad} = (1 \text{ rad})\left(\frac{180°}{\pi \text{ rad}}\right) = 57.3°$$

So, an angle of one radian is approximately 57.3° (Fig. 24-1).

$$1 \text{ rad} \approx 57.3°$$

Example 4

Convert $\frac{5\pi}{6}$ rad to degrees.

Solution:

$$\frac{5\pi}{6} \text{ rad} = \left(\frac{5\pi \text{ rad}}{6}\right)\left(\frac{180°}{\pi \text{ rad}}\right)$$
$$= 150°$$

In Chapter 13, we noted that when the unit of the angle measure is not marked, it is assumed that the angle is in radians.

Example 5

An angle of $\frac{5\pi}{6}$ rad can be expressed as just $\frac{5\pi}{6}$ (without the "rad"). So

$$\frac{5\pi}{6} \text{ rad} = \frac{5\pi}{6}$$

COMMON ERROR Students sometimes confuse the decimal value of π with the degree equivalent of π radians.

What is π? 180 or 3.1416?

A decimal approximation of the number π is

$$\pi \approx 3.1416$$

But an angle measurement of π is π radians,

$$\pi = \pi \text{ rad} = 180°$$

Trigonometric Functions of Angles in Radians

In Chapters 15 and 22, we found the trigonometric functions of angles in radians in decimal form, but not in terms of π. The process is the same when the angle is given in terms of π. When using a calculator, put the calculator in radian mode, and be careful to use parentheses properly when you enter the function.

Example 6

Use a calculator to evaluate to four significant digits:

$$\cos \frac{5\pi}{6}$$

Solution: Since the angle $\frac{5\pi}{6}$ has no written units, it is assumed to be in radians. With the calculator in radian mode, we enter

$$\cos \left(\frac{5\pi}{6} \right)$$

Note the use of parentheses when we enter the function into the calculator! We find that

$$\cos \frac{5\pi}{6} = -0.8660$$

COMMON ERROR Remember, if an angle measure has no written units, it is assumed to be in *radians*!

SECTION 24-1 | Exercises | Radian Measure

Radian Measure in Terms of π

Convert each angle from degrees to radian measure in terms of π (in fraction form).

1. 360°	**2.** 180°	**3.** 90°	**4.** 45°
5. 60°	**6.** 30°	**7.** 15°	**8.** 75°
9. 135°	**10.** 225°	**11.** 72°	**12.** 40°
13. 340°	**14.** 36°	**15.** 84°	**16.** 210°
17. 270°	**18.** 144°	**19.** 200°	**20.** 324°

Convert each angle from radians to degrees.

21. $\dfrac{\pi}{6}$	**22.** $\dfrac{\pi}{3}$	**23.** $\dfrac{\pi}{4}$	**24.** $\dfrac{\pi}{2}$
25. $\dfrac{2\pi}{3}$	**26.** $\dfrac{3\pi}{4}$	**27.** $\dfrac{4\pi}{5}$	**28.** $\dfrac{4\pi}{3}$
29. $\dfrac{17\pi}{10}$	**30.** $\dfrac{23\pi}{20}$	**31.** $\dfrac{11\pi}{15}$	**32.** $\dfrac{13\pi}{12}$

Trigonometric Functions of Angles in Radians

Use a calculator to evaluate to four significant digits.

33. $\sin \dfrac{\pi}{3}$	**34.** $\cos \dfrac{\pi}{6}$	**35.** $\tan \dfrac{\pi}{6}$	**36.** $\sin \dfrac{\pi}{4}$
37. $\cos \dfrac{5\pi}{8}$	**38.** $\tan \dfrac{4\pi}{3}$	**39.** $\sin \dfrac{6\pi}{5}$	**40.** $\cos \dfrac{7\pi}{9}$
41. $\tan \left(-\dfrac{7\pi}{4} \right)$	**42.** $\sin \dfrac{3\pi}{10}$	**43.** $\cos \dfrac{19\pi}{12}$	**44.** $\tan \left(-\dfrac{7\pi}{18} \right)$

ARC LENGTH AND AREA OF A SECTOR

In this section, we discuss arc length and the area of a sector. These are useful for topics such as latitude and longitude, magnetic tapes and disks, circular highway curves, and pendulums.

Radian Measure and Arc Length

In a circle of radius r, a central angle of 1 radian intercepts an arc of length r and the measure of any central angle, in radians, is the number of "radius units" contained in the intercepted arc (Fig. 24-2).

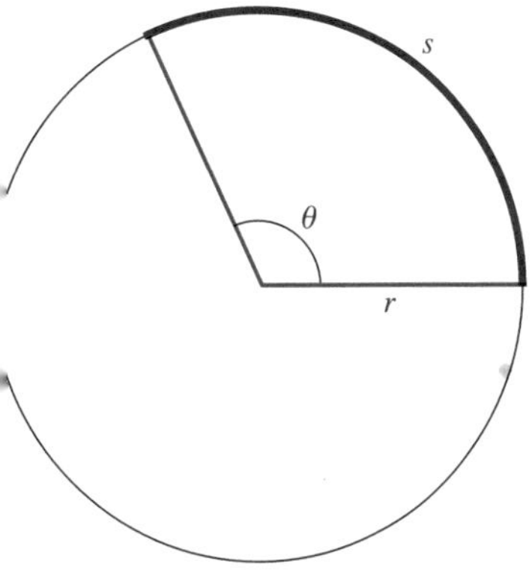

$\theta = \frac{s}{r}$

FIGURE 24-2
Radian measure of an angle.

───── **Example 7** ─────

In Fig. 24-2, it looks like the arc length s is about 2 "radius units," so the angle θ is about 2 radians.

We define the radian measure of an angle as the length of the intercepted arc s divided by the radius r (Fig. 24-2).

RADIAN MEASURE OF AN ANGLE	$\theta = \frac{s}{r}$	77

───── **Example 8** ─────

Find the angle that intercepts an arc of 8.52 cm in a circle of radius 7.00 cm (Fig. 24-3).

Solution: We use Eq. 77 with $s = 8.52$ cm and $r = 7.00$ cm.

$$\theta = \frac{s}{r} = \frac{8.52 \text{ cm}}{7.00 \text{ cm}} = 1.22$$

Notice that the units (cm) cancel out and the angle measure is a pure number (*dimensionless*). This feature makes it more useful than degrees for some calculations. We call this measure "radians" as an indication that it is an angle measurement. Thus, the angle measurement is 1.22 radians, and we write

$$\theta = 1.22 \text{ rad} \quad \text{or} \quad \theta = 1.22$$

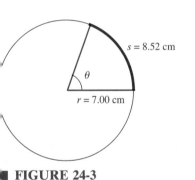

FIGURE 24-3

$s = 8.52$ cm

$r = 7.00$ cm

Arc Length

To find an expression for arc length s, we solve Eq. 77 for s.

$$s = \theta r, \text{ where } \theta \text{ is in radian measure}$$

FIGURE 24-4

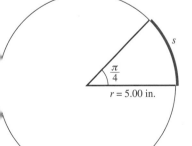

s

$\frac{\pi}{4}$

$r = 5.00$ in.

───── **Example 9** ─────

Find the length of the arc intercepted by a central angle of $\frac{\pi}{4}$ in a circle of radius 5.00 in. (Fig. 24-4).

Solution: We use the equation for arc length with $\theta = \frac{\pi}{4}$ and $r = 5.00$ in.

$$s = \left(\frac{\pi}{4}\right)(5.00) = 3.93 \text{ in.}$$

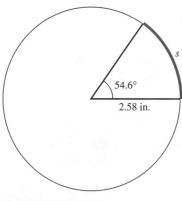

■ FIGURE 24-5

─── **Example 10** ───

Find the length of the arc intercepted by a central angle of 54.6° in a circle of radius 2.58 in. (Fig. 24-5).

Solution: Before we can use the equation for arc length, we must convert 54.6° to radians:

$$54.6° = 0.9529 \text{ rad}$$

Now, substituting $\theta = 0.9529$ and $r = 2.58$ into

$$s = \theta r$$

$$s = (0.9529)(2.58) = 2.46 \text{ in.}$$

Note that the "rad" may be put in or taken out as needed.

─── **Example 11** ───

Find the radius of a circle in which a central angle of 75.8° intercepts an arc of length 8.75 cm (Fig. 24-6).

Solution: We solve Eq. 77 for r:

$$r = \frac{s}{\theta}$$

First we convert $\theta = 75.8°$ to radians:

$$\theta = 75.8° = 1.323 \text{ rad}$$

Substituting $s = 8.75$ and $\theta = 1.323$ into the equation for r, we have

$$r = \frac{s}{\theta} = \frac{8.75}{1.323} = 6.61 \text{ cm}$$

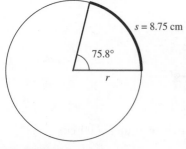

■ FIGURE 24-6

> **COMMON ERROR** Remember, when using Eq. 77 that the angle θ must be in *radians*.

Area of a Sector

■ FIGURE 24-7
Sector of a circle.

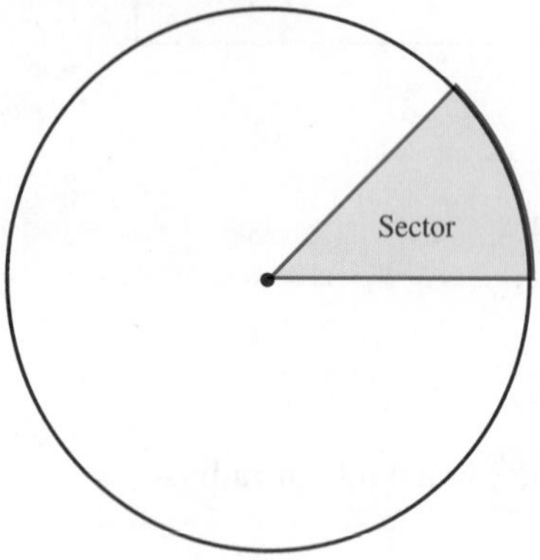

Recall from Sec. 13-5 that a *sector* of a circle is a pie-shaped region bounded by two radii and an arc (Fig. 24-7).

The area of a sector having a radius r and a central angle θ (in radians) is found using ratios that compare a sector of a circle to the whole circle. The ratio of the areas is equal to the ratio of the central angles:

$$\frac{\text{area } A \text{ of the sector}}{\text{area of the whole circle}} = \frac{\text{central angle } \theta \text{ of the sector}}{\text{central angle of the whole circle}}$$

Or,

$$\frac{A}{\pi r^2} = \frac{\theta}{2\pi}$$

where A is the area of the sector and θ is the central angle of the sector, in radians. Solving for A, we have

$$A = \left(\frac{\theta}{2\pi}\right)\pi r^2$$

or,

$$A = \frac{r^2 \theta}{2}$$

AREA OF A SECTOR	$A = \dfrac{r^2\theta}{2}$, where θ is in radians	78

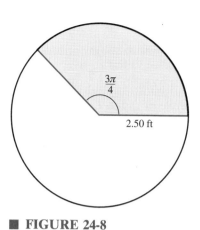

■ **FIGURE 24-8**

Example 12

Find the area of a sector having a radius 2.50 ft and a central angle of $\dfrac{3\pi}{4}$ (Fig. 24-8).

Solution: Using Eq. 78 with $r = 2.50$ and $\theta = \dfrac{3\pi}{4}$, we have

$$A = \frac{r^2\theta}{2} = \frac{(2.50)^2\left(\dfrac{3\pi}{4}\right)}{2} = 7.36 \text{ ft}^2$$

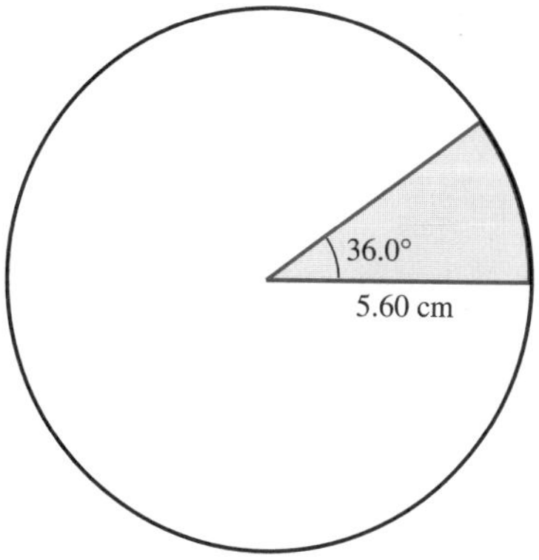

Example 13

Find the area of a sector having a radius 5.60 cm and a central angle of 36.0° (Fig. 24-9).

Solution: To use Eq. 77, we need to convert $\theta = 36.0°$ to radians:

$$\theta = 36.0° = 36.0°\left(\frac{\pi}{180°}\right) = 0.6283 \text{ rad}$$

Using Eq. 78, we have

$$A = \frac{r^2\theta}{2} = \frac{(5.60)^2(0.6283)}{2} = 9.85 \text{ cm}^2$$

■ **FIGURE 24-9**

Applications

Eqs. 77 and 78 can be used to solve many applications involving arc length and sectors.

■ **FIGURE 24-10**

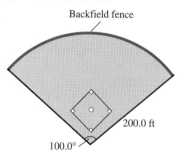

Backfield fence

200.0 ft

100.0°

Example 14

A community is building a new baseball field in the shape of a sector (Fig. 24-10). The central angle is 100.0° and the radius is 200.0 ft.

(a) How much fence is needed for the backfield?
(b) What is the area of the entire field?

Solution: To use Eq. 77, we change 100.0° to radians.

$$\theta = 100.0° = 1.7453 \text{ rad}$$

Using the equation derived from Equation 77 to find the arc length, the length of the fence, with $r = 200.0$ ft:

$$s = \theta r = (1.7453)(200.0)$$
$$s = 349.1 \text{ ft}$$

We use Eq. 78 to find the area of the sector, with $\theta = 1.7453$ rad and $r = 200.0$ feet:

$$A = \frac{r^2\theta}{2} = \frac{(200.0^2)(1.7453)}{2}$$
$$A = 34910 \text{ ft}^2$$

So, (a) the fence needs to be 349.1 ft long, and (b) the area of the field is 3491 ft².

Arc Length

Find the angle (in radians) that intercepts an arc s in a circle of radius r.

1. $s = 6.23$ cm, $r = 4.50$ cm   **2.** $s = 3.54$ in., $r = 1.27$ in.

Find the angle (in degrees) that intercepts an arc s in a circle of radius r.

3. $s = 15.0$ ft, $r = 3.43$ ft   **4.** $s = 67.4$ mm, $r = 78.2$ mm

Find the length of the arc intercepted by a central angle θ in a circle of radius r.

5. $\theta = \dfrac{\pi}{3}$ rad, $r = 4.83$ m   **6.** $\theta = \dfrac{5\pi}{6}$ rad, $r = 97.4$ in.

7. $\theta = 105°$, $r = 538$ mm   **8.** $\theta = 235°$, $r = 12.5$ cm

Find the radius of a circle in which a central angle θ intercepts an arc of length s.

9. $\theta = 1.56$ rad, $s = 9.56$ cm   **10.** $\theta = 3.17$ rad, $s = 56.7$ m

11. $\theta = 315°$, $s = 14.5$ in.   **12.** $\theta = 335°$, $s = 9.61$ ft

Area of a Sector

Find the area of a sector having a radius r and a central angle θ.

13. $r = 3.80$ cm, $\theta = \dfrac{2\pi}{3}$ rad   **14.** $r = 17.1$ ft, $\theta = 4.00$ rad

15. $r = 12.4$ mm, $\theta = 125°$ **16.** $r = 145$ m, $\theta = 80.5°$

Applications

17. Latitude is the angle, measured at the earth's center, between a point on the earth and the equator. New York City is at a latitude of 40.8° N. Find the distance in miles from New York City to the equator (Fig. 24-11). The radius of the earth is about 3960 miles.

18. One "track" on a magnetic disk used for computer data storage is located at a radius of 121 mm from the center of the disk. If 10 "bits" of data can be stored on 1.00 mm of track, how many bits can be stored in the length of track intercepted by an arc of 15.3°?

19. A circular highway curve has a radius of 351 ft and a central angle of 15.0° measured to the center line of the road. Find the length of the curve.

20. Tape passes over a circular magnetic tape head, forming a central angle of 125° (Fig. 24-12). If the radius of the tape head is 2.50 cm, how long is the section of tape that is in contact with the tape head?

21. A pendulum swings through an angle of 12.4°. If the pendulum is 72.5 cm long, find the area swept by the pendulum.

22. A cone is to be made from a circle of sheet metal by removing a 135° sector from the circle and bending the remaining sector to form the cone. If the radius of the circle is 18.0 in., find the lateral area of the cone.

■ **FIGURE 24-11**
Latitude.

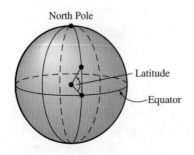

■ **FIGURE 24-12**

23. Which has the greater area: a piece of a 10-in. pizza which has been cut into 6 pieces, or a piece of a 12-in. pizza which has been cut into 8 pieces?

24. The area of a sector-shaped piece of land is 1.75 acres. Find the radius if the central angle is 142°. (1 acre = 43,560 ft²)

25. Use Eqs. 77 and 78 to find another formula for the area of a sector in terms of r and s. Use the new formula to find the area of a sector with radius 5.63 cm and arc length 14.5 cm.

24-3

UNIFORM CIRCULAR MOTION

In this section we discuss uniform circular motion, including angular velocity, angular displacement, and linear speed. Many applications involve rotations such as rotating disks, propellers, flywheels, satellites, and turbines.

Uniform Circular Motion

If an object is rotating in a circle about a point O, the *angular velocity* ω is the rate at which the object is rotating. If the angular velocity is constant, the motion is called *uniform circular motion*.

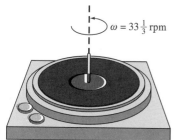

FIGURE 24-13

Example 15

The phonograph record in Fig. 24-13 is rotating about the center point with uniform circular motion. The angular velocity of the record is the number of revolutions per unit of time. This record rotates with an angular velocity ω of $33\frac{1}{3}$ revolutions per minute, or $33\frac{1}{3}$ rpm.

Angular Displacement

The angle θ through which the object rotates in time t is called the *angular displacement*. For uniform *linear* motion, we saw that

$$\text{Displacement} = (\text{velocity})(\text{time})$$

or,

$$D = v \cdot t$$

Similarly, for uniform *circular* motion,

$$\text{Angular displacement} = (\text{angular velocity})(\text{time})$$

or,

$$\theta = \omega \cdot t$$

Angular Velocity

Solving for ω in the equation above, the angular velocity (Fig. 24-14) is defined as

$$\omega = \frac{\theta}{t}$$

The units of angular velocity are degrees, radians, or revolutions, per unit time. Angular velocity is frequently expressed in revolutions per minute (rpm), or in radians per second.

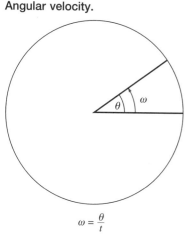

FIGURE 24-14
Angular velocity.

$$\omega = \frac{\theta}{t}$$

Example 16

Convert the angular velocity 33.3 rev/min

(a) to rad/s, and

(b) to deg/s.

Solution: We convert, as in Chapter 4, by multiplying by appropriate conversion factors:

(a) $33.3 \text{ rev/min} = \left(\dfrac{33.3 \text{ rev}}{\text{min}}\right)\left(\dfrac{2\pi \text{ rad}}{1 \text{ rev}}\right)\left(\dfrac{1 \text{ min}}{60 \text{ s}}\right) = 3.49 \text{ rad/s}$

(b) $33.3 \text{ rev/min} = \left(\dfrac{33.3 \text{ rev}}{\text{min}}\right)\left(\dfrac{360°}{1 \text{ rev}}\right)\left(\dfrac{1 \text{ min}}{60 \text{ s}}\right) = 200°/\text{s}$

Example 17

A propeller on a wind generator rotates 90.0° in 1.45 s. Find the angular velocity of the propeller in revolutions per minute.

Solution: Using the equation

$$\omega = \frac{\theta}{t}$$

$$\omega = \frac{90°}{1.45 \text{ s}} = 62.07°/\text{s}$$

We now convert to rev/min:

$$\omega = \left(\frac{62.07°}{\text{s}}\right)\left(\frac{1 \text{ rev}}{360°}\right)\left(\frac{60 \text{ s}}{1 \text{ min}}\right)$$

or

$$\omega = 10.3 \text{ rev/min}$$

Example 18

A flywheel makes 721 revolutions in a minute. How many degrees does it rotate in 0.500 s?

Solution: We need to have a common unit of time and want degrees instead of revolutions. So we change 721 rev/min to deg/s:

$$\omega = 721 \text{ rev/min} = \left(\frac{721 \text{ rev}}{\text{min}}\right)\left(\frac{360°}{1 \text{ rev}}\right)\left(\frac{1 \text{ min}}{60 \text{ s}}\right) = 4326°/\text{s}$$

We are asked for the angular displacement θ. We substitute $\omega = 4326°/\text{s}$ and $t = 0.500$ s into

$$\theta = \omega t$$
$$\theta = (4326°/\text{s})(0.500 \text{ s})$$
$$\theta = 2163°$$

Rounding to three significant digits, $\theta = 2160°$.

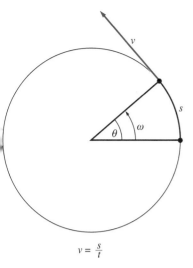

$v = \frac{s}{t}$

■ **FIGURE 24-15**
Linear velocity.

Linear Velocity

For any point on a rotating body, the linear displacement per unit time along the circular path is called the *linear velocity v* (Fig. 24-15). The linear velocity is zero for a point at the center of rotation, and it increases as the distance from the center increases.

Example 19 ─────────

Two bugs sit on a rotating phonograph record (Fig. 24-16). They are both rotating with the same angular velocity ω. But, during any specific period of time, the bug on the outside edge of the record moves a greater distance than the bug on the inside of the record. The bug on the outside moves *faster* than the bug on the inside, because it moves a greater distance in the same amount of time. So, the linear velocity of the bug on the outside is greater than the linear velocity of the bug on the inside.

■ **FIGURE 24-16**

To derive an expression for linear velocity v, we begin with the equation

$$s = \theta r, \text{ where } \theta \text{ is in radians}$$

Dividing both sides by t, we have

$$\frac{s}{t} = \frac{\theta}{t} r$$

Since $\frac{s}{t} = v$ and $\frac{\theta}{t} = \omega$, we have

$$v = \omega r$$

where ω is in *radians* per unit time.

Example 20 ─────────

A record is rotating at 33.3 rev/min. Find the linear velocity of a bug 15.2 cm from the center.

Solution: First we change 33.3 rev/min to rad/min.

$$\omega = \left(\frac{33.3 \text{ rev}}{\text{min}}\right)\left(\frac{2\pi \text{ rad}}{1 \text{ rev}}\right) = 209.2 \text{ rad/min}$$

Substituting into

$$v = \omega r$$
$$v = (209.2 \text{ rad/min})(15.2 \text{ cm})$$

we have

$$v = 1380 \text{ cm/min}$$

SECTION 24-3 Exercises Uniform Circular Motion

Angular Velocity

Find the missing values. Round to three significant digits.

	rev/min	rad/s	deg/s
1.	1520		
2.		4.50	
3.		$4\pi/5$	
4.			135

5. A propeller on a wind generator rotates 54.5° in 0.750 s. Find the angular velocity of the propeller in rev/min.

6. A satellite circles the earth 6.25 times each 24.0 hours. Find its angular velocity in deg/min.

7. The flywheel of an engine is rotating at 452 rpm. Find the angular displacement, in revolutions, of the flywheel in 5.00 s.

8. A blade on a water turbine turns 175° in 1.25 s. Find its angular velocity, and the angular displacement of the blade in 10.0 s.

Linear Velocity

9. A flywheel rotates at an angular velocity of 3.00 rad/s. If the radius of the flywheel is 14.8 in., find the linear velocity of a point on the rim, in ft/s.

10. A car wheel 24.0 inches in diameter is rotating at 775 rev/min. Find the linear velocity of the car (in mi/h).

11. The earth rotates on its axis at 1.00 revolution every 24.0 hours. Assuming that the earth is a sphere 3960 miles in radius, find the linear velocity (in mi/h) of a point on the earth's surface.

12. The earth rotates around the sun at 1 revolution every 365 days. Assuming that the earth's orbit around the sun is a circle with a radius of 92,960,000 miles, find the linear velocity of the earth around the sun, in miles per hour.

13. A car is traveling at a rate of 88.0 km/h and has tires that have a diameter of 63.2 cm. Find the angular velocity of the wheels of the car, in rev/s.

14. A race car is driven around a circular track at an average velocity of 85.0 mi/h. Find the angular velocity of the car (in rev/min) if the radius of the track is 842 ft.

Convert each angle from degrees to radian measure in terms of π (in fraction form).

1. $12°$ **2.** $80°$ **3.** $216°$ **4.** $300°$

Convert each angle from radians to degrees.

5. $\dfrac{\pi}{5}$ **6.** $\dfrac{5\pi}{4}$ **7.** $\dfrac{17\pi}{12}$ **8.** $\dfrac{14\pi}{9}$

9. Find the angle (in radians) that intercepts an arc of 4.65 cm in a circle of radius 3.15 cm.

10. Find the angle (in degrees) that intercepts an arc of 12.5 in. in a circle of radius 8.50 in.

11. Find the length of the arc intercepted by a central angle of 175° in a circle of radius 75.0 mm.

12. Find the radius of a circle in which a central angle of 105° intercepts an arc of length 7.25 m.

13. Find the area of a sector having a radius of 1.53 ft and a central angle of 75.8°.

14. A pendulum of a clock swings through an angle of 8.50°. If the pendulum is 30.0 inches long, find the length of the arc determined by the end of the pendulum.

15. Find the area of a sector-shaped window with a radius of 87.4 cm and a central angle of 152°.

16. Change 315°/s to rev/min.

17. Find the angular velocity (in rad/s) of a wheel that rotates 37.5 rev in 1.75 min.

18. The propeller of an airplane is 6.00 ft in diameter and rotates at 2250 rpm. Find the linear speed of the tip of the propeller (in ft/s).

WRITING

19. We were pretty happy measuring angles in degrees. Why do we need to study angles measured in radians? What applications use angles measured in radians, but not in degrees? Make up a problem for which you need to use radian measure. Solve your problem and revise it if necessary.

TEAM PROJECT

20. The kinetic energy of a rotating body is given by

$$E = \frac{1}{2}mk^2\omega^2 \text{ Joules}$$

where m is the mass in kilograms, ω is the angular velocity in radians/second, and k is the radius of rotation in meters. For a sphere of radius r,

$$k = \frac{\sqrt{2}}{5}r$$

Weigh and measure the radius of five balls of various sizes (such as a golf ball, a billiard ball, a baseball, a softball, and a basketball). Then compute the kinetic energy of rotation for each ball, assuming that it is rotating at 3.00 revolutions per second. What conclusions can you draw about mass, radius, and kinetic energy?

Graphs of Trigonometric Functions

25

In this chapter, we apply the graphing techniques of Chapter 12 to graph some *periodic* functions—the sine, cosine, and tangent—and we'll see how the constants in the equation effect the shape and position of the curves. We define some terms that will be applicable to the study of periodic functions in general, such as *cycle, period,* and *amplitude.* We solve applications with topics such as current, voltage, and simple harmonic motion.

25-1

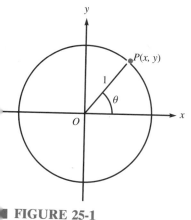

FIGURE 25-1

oint *P* on a rotating unit circle.

FIGURE 25-2

ertical displacement *y* on the
nit circle.

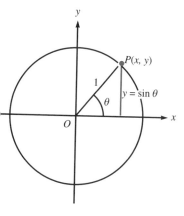

GRAPH OF $y = a \sin x$

In this section we graph the functions $y = \sin x$ and $y = a \sin x$. We discuss the concepts of period and amplitude of periodic functions and use our knowledge of period and amplitude to make a "quick sketch" of the graph.

Graph of $y = \sin \theta$

We can visualize the graph of the sine function

$$y = \sin \theta$$

Consider a point $P(x, y)$ on a rotating circle of radius 1 (a unit circle), with its center at the origin and θ the central angle as shown in Fig. 25-1. As the radius OP revolves counterclockwise through one revolution, beginning with P at $(1, 0)$, θ goes from 0° to 360°. At any instant, the y coordinate of the point P is a function of θ. In fact, since $r = 1$,

$$\sin \theta = \frac{y}{r} = \frac{y}{1} = y$$

So

$$y = \sin \theta$$

Thus, in the unit circle, $\sin \theta$ is the *vertical displacement* of the point P from the x axis (Fig. 25-2). As θ changes from 0° to 90°, the vertical displacement ($\sin \theta$) increases from 0 to 1, changing rapidly at first, and more slowly as it nears 1. As θ changes from 90° to 180°, the vertical displacement ($\sin \theta$) decreases from 1 to 0, slowly at first and more rapidly as it nears 0. As θ changes from 180° to 270°, the vertical displacement ($\sin \theta$) decreases from 0 to −1, rapidly at first and then more slowly as it nears −1. And as θ changes from 270° to 360°, the vertical displacement ($\sin \theta$) increases from −1 to 0, slowly at first and more rapidly as it nears 0 (Fig. 25-3). From this model we can see that the curve pattern *repeats itself* with each revolution, or every 360°. A curve that repeats itself is called *periodic.*

Now that we have a general idea of how the sine function behaves, let's graph the function $y = \sin \theta$ in the same way as we graphed other equations in Chapter

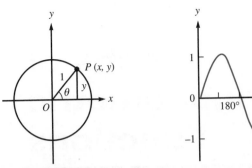

■ FIGURE 25-3
Vertical displacement *y* as a function of θ.

12. We select values for the independent variable θ and compute values for the dependent variable *y*, producing a table of values for θ and *y*. We then plot these points and connect them with a smooth curve.

Example 1

Graph $y = \sin \theta$ for θ from 0° to 360°.

Solution: We make a table of values for θ and *y*, choosing intervals for θ of 30° each. We use the calculator to find the *y* values, making sure that the calculator is in the degree mode. If *y* values are not exact, we round to tenths. (When you make a graph "by hand," you can't plot points with much more accuracy than tenths.)

Note: Many calculators have a "table" feature you can use to make a table. See your manual for specific instructions.

$y = \sin \theta$

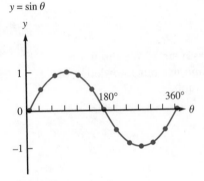

■ FIGURE 25-4
Sine curve.

θ	0°	30°	60°	90°	120°	150°	180°	210°	240°	270°	300°	330°	360°
y	0	0.5	0.9	1	0.9	0.5	0	−0.5	−0.9	−1	−0.9	−0.5	0

These points are plotted and connected with a smooth curve in Fig. 25-4.

If we had plotted points for θ from 360° to 720°, we would have gotten another curve pattern, identical to the one just sketched. The curve will *repeat* to the right of θ = 360° and also to the left of θ = 0° (Fig. 25-5).

■ FIGURE 25-5
Three cycles of the sine curve.

$y = \sin \theta$

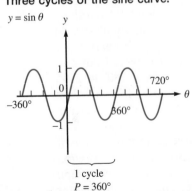

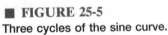
1 cycle
P = 360°

Period

Since the sine curve repeats, it is periodic. The smallest repeating portion of the graph is called a *cycle* and the length of one cycle is called the *period*. A cycle of the curve in Fig. 25-5 repeats every 360°, so the period *P* of $y = \sin \theta$ is 360°:

$$P = 360° \quad \text{in degrees}$$

Graph of $y = \sin x$

In the model above, we used θ to represent the angle, and we expressed the sine function as a function of the angle θ. However, the sine is often expressed as a function of the variable *x*, where *x* is an angle, expressed either in degrees or in radians. From this point on, we write the sine function as a function of *x*; that is, the independent variable will be *x*, not θ. The graph of $y = \sin x$ is similar to

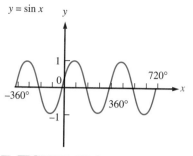

FIGURE 25-6
Sine as a function of *x*.

Fig. 25-5, except that the horizontal axis is the *x* axis, not the θ axis. Fig. 25-6 shows several cycles of the curve $y = \sin x$.

Graphing Calculator

A graphing calculator makes it easy to graph the trigonometric functions and to explore how the constants in the equation affect the graph. First we must set the calculator to the appropriate angle mode, either degree or radian. We define the window variables, which are determined by the domain and the range of the function. Then we enter the function and graph it. Check your calculator manual for specific instructions.

Example 2

Use a graphing calculator to graph $y = \sin x$ for *x* from $-180°$ to $450°$.

Solution: First we set the calculator angle mode to degree. We are given the domain (*x* is between $-180°$ and $450°$), and we have noted that *y* values for $y = \sin x$ are always between -1 and 1. Thus, we set the window so that

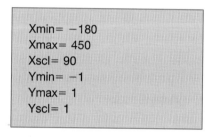

```
Xmin= -180
Xmax= 450
Xscl= 90
Ymin= -1
Ymax= 1
Yscl= 1
```

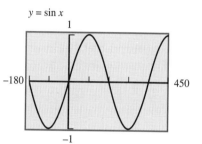

FIGURE 25-7
Sine curve on a graphing calculator.

We enter the function $y = \sin x$, and graph (Fig. 25-7). The graphing calculator does not show labels on the axes; you must check the window variables to determine what each tick mark represents on the graph. In this book, when a figure contains a graphing calculator screen, we show the values of Xmin, Xmax, Ymin, and Ymax, although they do not appear on the graphing calculator screen. Experiment with the different predefined windows on your calculator. Many calculators have a special window for graphing trigonometric functions.

Many applications of the sine function use angles measured in radians, so we must also be able to graph the sine function in radians.

Example 3

Use a graphing calculator to graph $y = \sin x$, for *x* from -2π to 2π radians.

Solution: First we set the calculator angle mode to radians. Then we set the window.

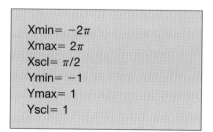

```
Xmin= -2π
Xmax= 2π
Xscl= π/2
Ymin= -1
Ymax= 1
Yscl= 1
```

25-1 Graph of $y = a \sin x$ **559**

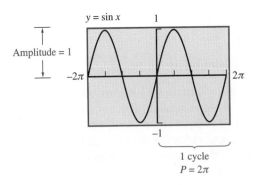

$y = \sin x$

Amplitude = 1

-2π 2π

1 cycle
$P = 2\pi$

■ **FIGURE 25-8**

Note that when we enter -2π for Xmin, the calculator immediately evaluates this as a decimal and displays it as $-6.2831853.\ldots$. Similarly, it evaluates 2π and $\pi/2$ as decimals. We enter the function

$$y = \sin x$$

and graph (Fig. 25-8). The graph shows two complete cycles of the sine curve. In radians, the period P of the function is 2π, so

$$P = 2\pi \quad \text{in radians}$$

Amplitude

We see from Fig. 25-8 that if $y = \sin \theta$, the y values vary between -1 and 1. The maximum variation of the curve $y = \sin \theta$ from zero is called its *amplitude*. So, the amplitude of the curve $y = \sin \theta$ is 1.

Graph of $y = a \sin x$

Now we'll look at the effect of a constant coefficient a on the graph of the equation.

■ **FIGURE 25-9**
A change in amplitude.

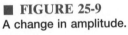

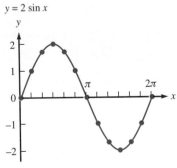

— **Example 4** —

Graph $y = 2 \sin x$, for x from 0 to 2π radians.

Solution: We make a table of values, taking x intervals of $\pi/6$. We use a calculator in radian mode to find the y values and round to tenths.

x	0	$\dfrac{\pi}{6}$	$\dfrac{\pi}{3}$	$\dfrac{\pi}{2}$	$\dfrac{2\pi}{3}$	$\dfrac{5\pi}{6}$	π	$\dfrac{7\pi}{6}$	$\dfrac{4\pi}{3}$	$\dfrac{3\pi}{2}$	$\dfrac{5\pi}{3}$	$\dfrac{11\pi}{6}$	2π
y	0	1	1.7	2	1.7	1	0	-1	-1.7	-2	-1.7	-1	0

We plot the points and draw a smooth curve (Fig. 25-9). The graph shows one complete cycle. The curve repeats every 2π units, and therefore the period P is:

$$P = 2\pi$$

We also note that for $y = 2 \sin x$, the y values vary between -2 and 2, so

$$\text{amplitude} = 2$$

The amplitude represents the maximum y value of the curve. The graph of the function $y = a \sin x$ will rise up to $y = |a|$ and down to $y = -|a|$. We use absolute values here, because *a* may be negative. In general, if $y = a \sin x$, the amplitude will be the *absolute value* of the coefficient of sin *x*. So

$$\text{amplitude} = |a|$$

The period *P* of the curve $y = a \sin x$ is always 2π or 360°, depending on whether *x* is in radians or degrees.

y = a sin x	amplitude = |a|	143
	Period = 2π or 360°	144

Example 5

Given the function $y = -3 \sin x$,

(a) Find the amplitude and the period, in degrees.

(b) Graph the function for *x* from 0° to 360°.

Solution:

(a) We use Eq. 143 to find the amplitude. Since $a = -3$:

$$\text{amplitude} = |-3| = 3$$

This means that *y* varies between −3 and 3. Since *a* is negative ($a = -3$), the curve is *inverted* (turned upside down). That is, if we start at $x = 0$, the curve goes *down* to −3, instead of *up* to 3.

We use Eq. 144 to find the period *P* of $y = -3 \sin x$. Since *x* is in degrees.

$$P = 360°$$

(b) To graph the function, we use a graphing calculator in degree mode, setting the window so that

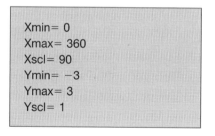

```
Xmin= 0
Xmax= 360
Xscl= 90
Ymin= -3
Ymax= 3
Yscl= 1
```

We enter the function and graph (Fig. 25-10).

■ **FIGURE 25-10**

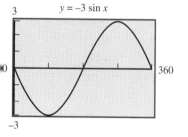

$y = -3 \sin x$

■ **FIGURE 25-11**

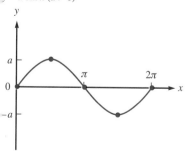

$y = a \sin x \ (a > 0)$

If $y = a \sin x$, the coefficient *a* acts as a vertical curve *stretcher* if $|a| > 1$ or as a vertical curve *compressor* if $|a| < 1$. The curves of $y = \sin x$ and $y = a \sin x$ have the same general shape; the second curve is just stretched or compressed vertically depending on the value of *a*.

All the curves $y = a \sin x$ have the same general shape and the same period. They vary only in their amplitude. To make a "complete" graph of a sine curve, we graph *at least one* complete cycle. This includes three *x* intercepts and two turning points—a maximum point and a minimum point for each cycle. We plot these five key points and connect them with a smooth curve (Fig. 25-11).

25-1 Graph of $y = a \sin x$ **561**

Example 6

Given the function

$$y = \frac{1}{2} \sin x$$

where x is in radians,

(a) Find the period and the amplitude.

(b) Graph the function.

Solution:

(a) Since the equation is of the form $y = a \sin x$, the period P is 2π (in radians). The amplitude is

$$\text{amplitude} = \left| \frac{1}{2} \right| = \frac{1}{2}$$

(b) Using $P = 2\pi$ and $a = \frac{1}{2}$, we sketch the graph in Fig. 25-12.

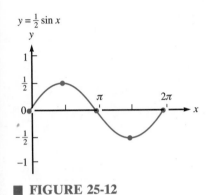

$y = \frac{1}{2} \sin x$

■ **FIGURE 25-12**

Example 7

Using a graphing calculator in degree mode, graph these three functions on the same screen, for x from $0°$ to $360°$:

(a) $y = \sin x$

(b) $y = 3 \sin x$

(c) $y = -2 \sin x$

Solution: We set the window so that

```
Xmin= 0
Xmax= 360
Xscl= 90
Ymin= -3
Ymax= 3
Yscl= 1
```

We graph the functions, as in Fig. 25-13. Now we can really see the effect of the coefficient a on the curve. The coefficient 3 in equation (b) stretches the curve vertically, to three times the original amplitude. The coefficient -2 in equation (c) inverts the curve and stretches it vertically, to two times the original amplitude.

■ **FIGURE 25-13**
Three sine curves with different amplitudes.

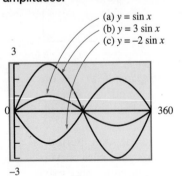

(a) $y = \sin x$
(b) $y = 3 \sin x$
(c) $y = -2 \sin x$

Graph of y = sin x

1. Complete the table for the equation

$$y = \sin x$$

rounding to tenths, and graph the equation for x from $-\pi$ to 3π.

x	$-\pi$	$-\dfrac{3\pi}{4}$	$-\dfrac{\pi}{2}$	$-\dfrac{\pi}{4}$	0	$\dfrac{\pi}{4}$	$\dfrac{\pi}{2}$	$\dfrac{3\pi}{4}$	π	$\dfrac{5\pi}{4}$	$\dfrac{3\pi}{2}$	$\dfrac{7\pi}{4}$	2π	$\dfrac{9\pi}{4}$	$\dfrac{5\pi}{2}$	$\dfrac{11\pi}{4}$	3π
y																	

2. Complete the table for the equation

$$y = 3 \sin x$$

rounding to tenths, and graph the equation for x from $-90°$ to $360°$.

x	$-90°$	$-60°$	$-30°$	$0°$	$30°$	$60°$	$90°$	$120°$	$150°$	$180°$	$210°$	$240°$	$270°$	$300°$	$330°$	$360°$
y																

Graph of y = a sin x

Write the equations of the sine curves.

3.

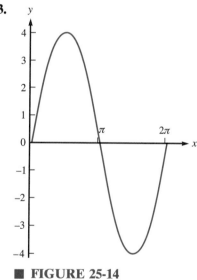

■ **FIGURE 25-14**

4.

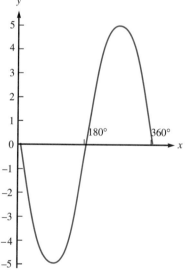

■ **FIGURE 25-15**

■ **FIGURE 25-16**

5.

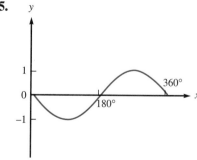

6.

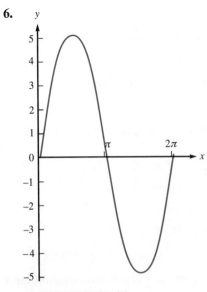

■ **FIGURE 25-17**

7. Write an equation of the sine curve that has an amplitude of 2 and a period of 360°.

8. Write an equation of the sine curve that has an amplitude of 5 and a period of 2π.

Find the amplitude and the period, in degrees. Graph each equation for x from 0° to 360°.

9. $y = 5 \sin x$ **10.** $y = 4 \sin x$

11. $y = -\sin x$ **12.** $y = -2 \sin x$

13. $y = \dfrac{1}{3} \sin x$ **14.** $y = -\dfrac{3}{2} \sin x$

Find the amplitude and the period, in radians. Graph each equation for x from 0 to 2π.

15. $y = -2.5 \sin x$ **16.** $y = 7.5 \sin x$

17. $y = 100 \sin x$ **18.** $y = -60 \sin x$

Graph each equation for integral values of x from 0 to 7 radians.

19. $y = \sin x$ **20.** $y = -\dfrac{1}{2} \sin x$

25-2   **GRAPH OF $y = a \sin bx$**

In this section we graph the function

$$y = a \sin bx$$

and see how the period of the curve is affected by the constant b.

Graph of $y = \sin bx$

First we graph the function

$$y = \sin bx$$

to see the effect of b, the coefficient of x.

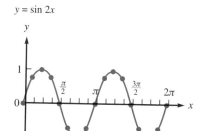

$y = \sin 2x$

FIGURE 25-18
A change in period.

Example 8

Graph the function $y = \sin 2x$, for x from 0 to 2π.

Solution: We make a table of values, taking intervals of $\pi/8$ and computing the values for y using a calculator, in radian mode, rounding to tenths.

x	0	$\frac{\pi}{8}$	$\frac{\pi}{4}$	$\frac{3\pi}{8}$	$\frac{\pi}{2}$	$\frac{5\pi}{8}$	$\frac{3\pi}{4}$	$\frac{7\pi}{8}$	π	$\frac{9\pi}{8}$	$\frac{5\pi}{4}$	$\frac{11\pi}{8}$	$\frac{3\pi}{2}$	$\frac{13\pi}{8}$	$\frac{7\pi}{4}$	$\frac{15\pi}{8}$	2π
y	0	0.7	1	0.7	0	−0.7	−1	−0.7	0	0.7	1	0.7	0	−0.7	−1	−0.7	0

These points are plotted and connected with a smooth curve in Fig. 25-18. We see from the graph that the amplitude is still 1. But the curve repeats itself every π units, instead of repeating every 2π. So the period of the graph is *not* 2π, but π. Now we can see that the effect of the 2 in the equation $y = \sin 2x$ is to divide the period (of the basic curve $y = \sin x$) in half.

Example 9

Graph $y = \sin x$ and $y = \sin 2x$ on the same axes, for x from 0 to 2π.

Solution: We graph these functions on a graphing calculator in radian mode, setting the window so that

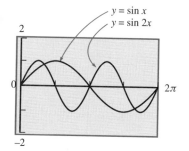

$y = \sin x$
$y = \sin 2x$

FIGURE 25-19
Two sine curves with different periods.

```
Xmin= 0
Xmax= 2π
Xscl= π/2
Ymin= −2
Ymax= 2
Yscl= 1
```

We enter the functions and graph (Fig. 25-19). Notice how the constant 2 compresses the curve horizontally, indicating a change in the period.

In general, the graph of the function

$$y = \sin bx$$

has a period P equal to 2π divided by b (in radians), or $360°$ divided by b (in degrees):

$$P = \frac{2\pi}{b} \quad \text{or} \quad \frac{360°}{b}$$

If $|b| > 1$, then the curve is horizontally *compressed*. If $|b| < 1$, then the curve is horizontally *stretched*.

Example 10

Find the period of the function $y = \sin 2x$, where x is in radians.

Solution: Since x is in radians, we use

$$P = \frac{2\pi}{b}$$

We see that $b = 2$, so

$$P = \frac{2\pi}{2} = \pi$$

This period is verified by the graph in Fig. 25-18, since the curve repeats itself every π units.

Graph of $y = a \sin bx$

The constant coefficient a of the sine function

$$y = a \sin bx$$

determines the amplitude of the curve:

$$\text{amplitude} = |a|$$

The constant a does not affect the period of the curve $y = a \sin bx$, so we still divide 2π (or $360°$) by b to find the period.

	amplitude $= |a|$	143
$y = a \sin bx$	Period $= \dfrac{2\pi}{b}$ or $\dfrac{360°}{b}$	144

Example 11

Find the amplitude and the period, in degrees, of the function

$$y = \sin 3x$$

Solution: Since

$$y = 1 \sin 3x$$

then $a = 1$ and by Eq. 143

$$\text{amplitude} = |1| = 1$$

We use Eq. 144 to find the period P, in degrees, with $b = 3$:

$$P = \frac{360°}{b} = \frac{360°}{3} = 120°$$

Each cycle of the graph of $y = \sin 3x$ repeats every $120°$.

Example 12

Find the amplitude and the period, in degrees, of the function

$$y = 2 \sin \frac{x}{4}$$

Solution: Using Eq. 143, with $a = 2$

$$\text{amplitude} = |2| = 2$$

To find the period, we note that

$$\sin \frac{x}{4} = \sin \frac{1}{4}x, \qquad \text{so} \quad b = \frac{1}{4}$$

Using Eq. 144, with $b = \frac{1}{4}$, the period P is

$$P = \frac{360°}{\frac{1}{4}} = 360° \div \frac{1}{4} = 360° \left(\frac{4}{1}\right) = 1440°$$

Each cycle of the graph repeats every 1440°.

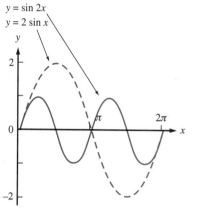

$y = \sin 2x$
$y = 2 \sin x$

FIGURE 25-20

Example 13

On the same axes, graph the functions

$$y = \sin 2x \quad \text{and} \quad y = 2 \sin x$$

for x from 0 to 2π. Compare the amplitudes, periods, and graphs of the two functions.

Solution: For the function

$$y = \sin 2x$$

the 2 is the coefficient of x. The constant 2 affects the *period* of the graph of $y = \sin 2x$. The graph of the function $y = \sin 2x$ has a period of π and an amplitude of 1 (Fig. 25-20).

For the function

$$y = 2 \sin x$$

the 2 is the coefficient of *sin x*. Here, the 2 affects the *amplitude,* but not the period, of the graph of $y = 2 \sin x$. The graph of the function $y = 2 \sin x$ has a period of 2π and an amplitude of 2.

So we see in Fig. 25-20 that the graphs of these two functions have different periods and different amplitudes.

Quick Sketch of the Sine Function

A fast way to graph one complete cycle of a sine function is to make a "quick sketch." All the curves $y = a \sin bx$ have the same general shape; they vary only in their period and amplitude.

Quick Sketch of the Sine Curve

1. Find the period P and amplitude $|a|$ of the curve.
2. Draw two horizontal lines $|a|$ units above and below the x axis, and label $|a|$ on the y axis.
3. Draw two vertical lines, one on the y axis and another P units to the right of the y axis. Label P on the x axis. (This completes a rectangle, bounded by the two horizontal lines drawn in Step 2 and the two vertical lines drawn in Step 3.)
4. On the x axis, make three marks to subdivide the period P into four equal parts; draw vertical lines through the marks.
5. Use the lines as guides to sketch one complete cycle of the sine curve.

FIGURE 25-21
Quick sketch of the sine curve.
$y = a \sin bx \ (a > 0)$

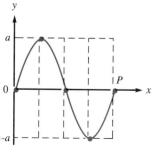

One complete cycle of the sine curve includes three x intercepts, one maximum point, and one minimum point. Plot these five key points and sketch the sine curve in the rectangle. The quick sketch is illustrated in Fig. 25-21, for $a > 0$.

COMMON ERROR Remember that the sine curve is *never straight* and *never pointy.*

Given $y = 2 \sin \dfrac{x}{3}$, where x is in radians,

(a) Find the amplitude and the period.
(b) Make a quick sketch of the graph.

Solution:

(a) We use Eq. 143 to find the amplitude:

$$\text{amplitude} = |a| = |2| = 2$$

We use Eq. 144 to find the period, with $b = \dfrac{1}{3}$

$$P = \frac{2\pi}{b} = \frac{2\pi}{\dfrac{1}{3}} = 2\pi \left(\frac{3}{1}\right) = 6\pi$$

So one cycle of the curve repeats every 6π radians.

(b) We make a quick sketch of the graph, using $a = 2$ and $P = 6\pi$. After drawing and labeling the x and y axes, we draw two horizontal lines, one at $y = 2$ and the other at $y = -2$. Then we draw two vertical lines at $x = 0$ (y axis) and $x = 6\pi$, forming a rectangle of width 6π and height 2. We subdivide the period 6π into four equal parts and draw the three inner vertical lines. We mark the five key points, and connect them with a smooth curve (Fig. 25-22).

$y = 2 \sin \frac{x}{3}$

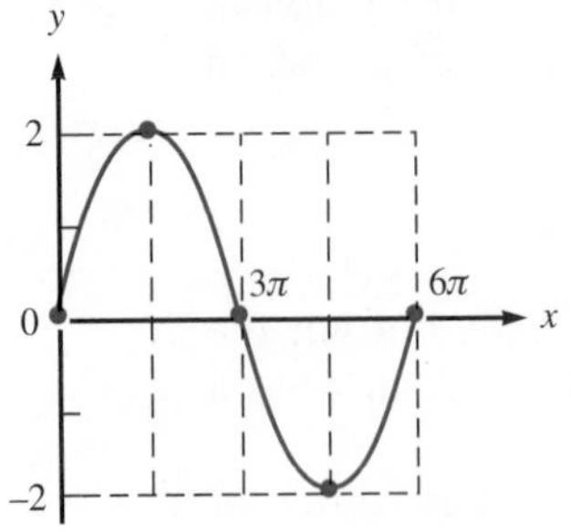

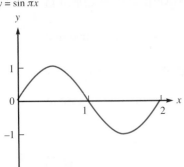

■ **FIGURE 25-22**

TIP When graphing a trigonometric function in which the domain is not specified, graph at least one complete cycle of the curve.

─ **Example 15** ─

Given the function $y = \sin \pi x$, where x is in radians,

■ **FIGURE 25-23**

$y = \sin \pi x$

(a) Find the amplitude and the period.
(b) Graph the function.

Solution:

(a) Since $y = \sin \pi x = 1 \sin \pi x$, the amplitude is 1. Since $b = \pi$, the period is

$$P = \frac{2\pi}{\pi} = 2$$

(b) To graph the function, we either make a quick stretch or use a graphing calculator (Fig. 25-23).

Graph of *y* = sin *bx*

1. Complete the table for the equation

$$y = \sin 3x$$

rounding to tenths, and graph the equation for *x* from 0 to 2π. Find the amplitude and the period. On the same axes, graph *y* = sin *x* with a dotted line.

x	0	$\frac{\pi}{12}$	$\frac{\pi}{6}$	$\frac{\pi}{4}$	$\frac{\pi}{3}$	$\frac{5\pi}{12}$	$\frac{\pi}{2}$	$\frac{7\pi}{12}$	$\frac{2\pi}{3}$	$\frac{5\pi}{6}$	π	$\frac{7\pi}{6}$	$\frac{4\pi}{3}$	$\frac{3\pi}{2}$	$\frac{5\pi}{3}$	$\frac{11\pi}{6}$	2π
y																	

2. Complete the table for the equation

$$y = \sin 5x$$

rounding to tenths, and graph the equation for *x* from 0° to 180°. Find the amplitude and the period. On the same axes, graph *y* = sin *x* with a dotted line.

x	0°	9°	18°	27°	36°	45°	54°	63°	72°	90°	108°	126°	144°	162°	180°
y															

Write the equation of the curve.

3.

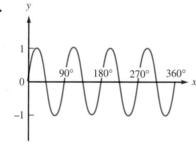

■ **FIGURE 25-24**

4.

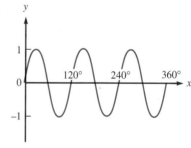

■ **FIGURE 25-25**

5. Write an equation of the sine curve that has an amplitude of 1 and a period of π.

6. Write an equation of the sine curve that has an amplitude of 1 and a period of 60°.

Find the amplitude and the period, in degrees. Then graph the equation.

7. $y = \sin \frac{x}{2}$ **8.** $y = \sin 8x$

Find the amplitude and the period, in radians. Then graph the equation.

9. $y = \sin 6x$ **10.** $y = \sin \frac{x}{4}$

Write the equation of the sine curve.

11.

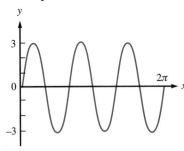

■ **FIGURE 25-26**

12.

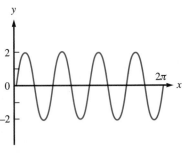

■ **FIGURE 25-27**

13. Write an equation of the sine curve that has an amplitude of 5 and a period of 90°.

14. Write an equation of the sine curve that has an amplitude of 4 and a period of $\frac{\pi}{3}$.

Find the amplitude and the period, in degrees. Then graph the equation.

15. $y = 2 \sin 3x$ **16.** $y = -3 \sin 2x$

17. $y = -3 \sin \dfrac{x}{3}$ **18.** $y = 5 \sin \dfrac{x}{5}$

Find the amplitude and the period, in radians. Then graph the equation.

19. $y = 3 \sin 2x$ **20.** $y = 2 \sin 3x$

21. $y = -\dfrac{\sin \pi x}{2}$ **22.** $y = -\sin \dfrac{\pi x}{2}$

23. The weight in Fig. 25-28 is pulled down 6.0 inches and released, causing it to oscillate with simple harmonic motion. If it takes 0.5 seconds to return to its lowest position, then the equation of motion is

$$y = 6 \sin 4\pi t$$

where y is the vertical displacement from its rest position (in inches) and t is the elapsed time (in seconds). Find the amplitude and the period of the oscillation, and graph the equation for t from 0 to 3.0 seconds.

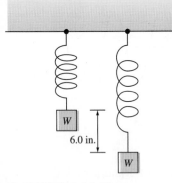

■ **FIGURE 25-28**

25-3   GRAPH OF $y = a \sin (bx + c)$

In the previous two sections we considered examples in which the amplitude and the period of the sine curve were changed. In this section we consider how a certain change in the equation causes a *shift* of the sine curve to the right or left.

Phase Shift

If a constant c is added to the x term of a sine function, it causes the graph to be shifted in the x direction (to the right or to the left). The amount of shift is called the *phase shift*.

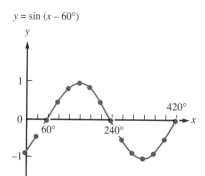

$y = \sin(x - 60°)$

FIGURE 25-29
Sine curve with phase shift.

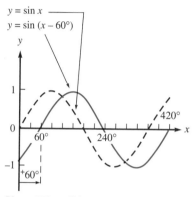

$y = \sin x$
$y = \sin(x - 60°)$

Phase shift = +60°
FIGURE 25-30

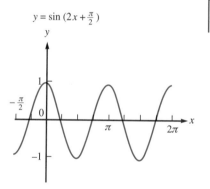

$y = \sin\left(2x + \frac{\pi}{2}\right)$

FIGURE 25-31

FIGURE 25-32

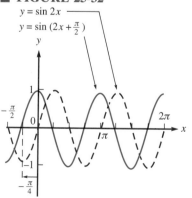

$y = \sin 2x$
$y = \sin\left(2x + \frac{\pi}{2}\right)$

Phase shift = $-\frac{\pi}{4}$

Example 16

Graph the function $y = \sin(x - 60°)$, for x from $0°$ to $420°$.

Solution: We make a table of values, taking intervals of $30°$. We use a calculator, in degree mode, to find the values for y, rounded to tenths.

x	0°	30°	60°	90°	120°	150°	180°	210°	240°	270°	300°	330°	360°	390°	420°
y	−0.9	−0.5	0	0.5	0.9	1	0.9	0.5	0	−0.5	−0.9	−1	−0.9	−0.5	0

These points are plotted and connected with a smooth curve in Fig. 25-29.

We see from the graph that the curve does not pass through the origin, but has been shifted in the x direction by an amount $60°$ to the *right*. For example, $\sin x = 0$ at $x = 60°$ instead of at $x = 0$. We call this a phase shift of $+60°$. The phase shift is the amount of horizontal shift in the curve.

To compare the graphs of $y = \sin x$ and $y = \sin(x - 60°)$, we graph both functions on the same axes (Fig. 25-30), using a dotted line for $y = \sin x$ and a solid line for $y = \sin(x - 60°)$. Note that the amplitude of both curves is 1 and that the period of both curves is $360°$. The $60°$ causes the graph (of the basic curve $y = \sin x$) to be shifted $60°$ to the right.

Graph of $y = \sin(bx + c)$

Now we'll see what happens if the coefficient of x is not 1.

Example 17

Graph the equation $y = \sin\left(2x + \frac{\pi}{2}\right)$, for x from $-\frac{\pi}{2}$ to 2π.

Solution: Once again, we graph this equation by plotting points. We take intervals of $\frac{\pi}{4}$ compute the values for y, using the calculator, in radian mode.

x	$-\frac{\pi}{2}$	$-\frac{\pi}{4}$	0	$\frac{\pi}{4}$	$\frac{\pi}{2}$	$\frac{3\pi}{4}$	π	$\frac{5\pi}{4}$	$\frac{3\pi}{2}$	$\frac{7\pi}{4}$	2π
y	−1	0	1	0	−1	0	1	0	−1	0	1

We plot these points and connect them with a smooth curve in Fig. 25-31.

To help us visualize the phase shift we graph both $y = \sin 2x$ and $y = \sin\left(2x + \frac{\pi}{2}\right)$ on the same axes (Fig. 25-32). We see from the graph that the curve $y = \sin\left(2x + \frac{\pi}{2}\right)$ is shifted $\frac{\pi}{4}$ to the left of the curve $y = \sin 2x$. We call this a phase shift of $-\frac{\pi}{4}$. You will note that the amplitude of both curves is 1 and the period of both curves is π. In the equation $y = \sin\left(2x + \frac{\pi}{2}\right)$, the only effect of the $\frac{\pi}{2}$ is to cause a phase shift of $-\frac{\pi}{4}$ (to the left). This phase shift is found by dividing $-\frac{\pi}{2}$ by the coefficient 2.

Graph of $y = a \sin (bx + c)$

In general, if
$$y = a \sin (bx + c)$$
then the constant c affects the phase shift, and the phase shift is found by dividing $-c$ by b, so
$$\text{Phase shift} = -\frac{c}{b}$$

When the phase shift is *positive,* the curve will be shifted to the *right.* When the phase shift is *negative,* the curve will be shifted to the *left.* The constant c does not affect the amplitude or the period.

The general sine curve
$$y = a \sin (bx + c)$$
contains three constants: a, b, and c. Their effects are summarized as follows:

	amplitude $= \lvert a \rvert$	143
$y = a \sin (bx + c)$	Period $= \dfrac{2\pi}{b}$ or $\dfrac{360°}{b}$	144
	Phase shift $= -\dfrac{c}{b}$	145

Quick Sketch of $y = a \sin (bx + c)$

We can use the quick sketch method of Sec. 25-2 to graph a sine function with a phase shift. First we make a *dotted* quick sketch of the graph of $y = a \sin bx$ (without the phase shift). Then calculate the phase shift and shift the curve by the phase shift.

─── **Example 18** ───

Given $y = 2 \sin \left(3x + \dfrac{\pi}{2} \right)$,

(a) Find the amplitude, the period, and the phase shift.
(b) Graph the function.

Solution:

(a) For this function, $a = 2$, $b = 3$, and $c = \dfrac{\pi}{2}$. Thus,
$$\text{amplitude} = 2$$
$$\text{Period} = \frac{2\pi}{3}$$
$$\text{Phase shift} = \frac{-\dfrac{\pi}{2}}{3} = -\frac{\pi}{6}$$

(b) First we make a quick sketch of $y = 2 \sin 3x$, shown with a *dotted* line. Since the phase shift is negative, we shift the curve $\dfrac{\pi}{6}$ units to the left as shown in Fig. 25-33.

■ **FIGURE 25-33**

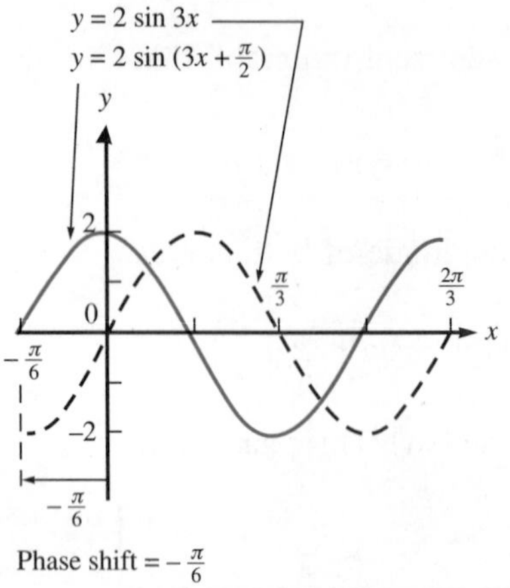

$y = 2 \sin 3x$
$y = 2 \sin (3x + \frac{\pi}{2})$

Phase shift $= -\frac{\pi}{6}$

Check: You can verify that this graph is correct by graphing the function on a graphing calculator.

SECTION 25-3   Exercises   Graph of *y* = *a* sin (*bx* + *c*)

Phase Shift

Complete the table for the equation for *x* from −90° to 360°, rounding to tenths. Sketch the graph and give the period, amplitude, and phase shift. On the same axes, graph *y* = *a* sin *bx*, for the given *a* and *b*, with a dotted line, and indicate the phase shift on the graph.

x	−90°	−60°	−30°	0°	30°	60°	90°	120°	150°	180°	210°	240°	270°	300°	330°	360°
y																

1. $y = \sin(x - 90°)$ **2.** $y = \sin(3x + 180°)$

Graph of
y = *a* sin (*bx* + *c*)

Write the equation of the curve.

3.

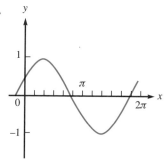

■ **FIGURE 25-34**

4.

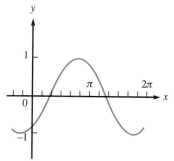

■ **FIGURE 25-35**

5.

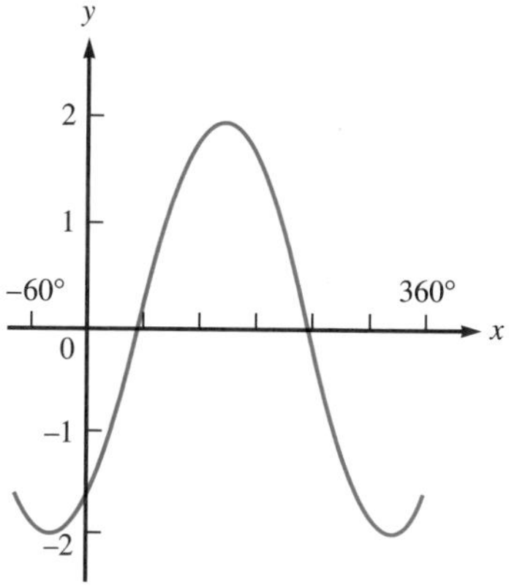

■ **FIGURE 25-36**

6.

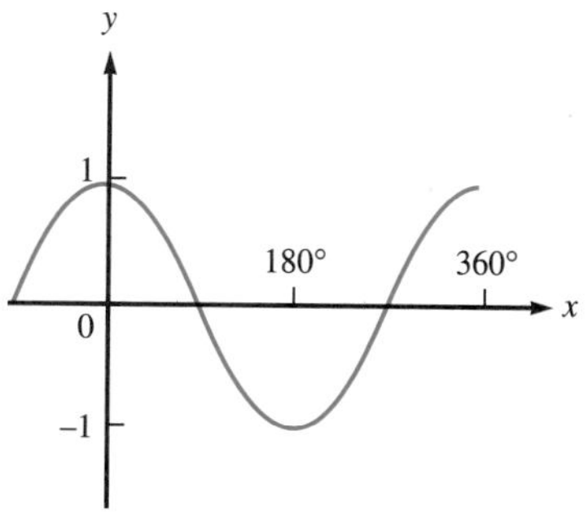

■ **FIGURE 25-37**

7. Write an equation of the sine curve that has an amplitude of 2, a period of π, and a phase shift of $\dfrac{\pi}{2}$.

8. Write an equation of the sine curve that has an amplitude of 3, a period of 90°, and a phase shift of 45°.

For each function, find the amplitude, period, and phase shift. Then sketch at least one complete cycle of the graph.

9. $y = \sin(x + 45°)$ **10.** $y = \sin\left(x - \dfrac{\pi}{3}\right)$

11. $y = \sin\left(2x - \dfrac{\pi}{3}\right)$ **12.** $y = \sin(2x + 180°)$

13. $y = -2\sin\left(x - \dfrac{\pi}{6}\right)$ **14.** $y = 3\sin(x + 90°)$

15. $y = 0.5\sin(3x + 180°)$ **16.** $y = -2\sin(3x - 2\pi)$

17. $y = \sin(x + 1)$ **18.** $y = -2\sin(3x - 1)$

19. In a certain ac circuit, the current i (in amps) is given by the equation

$$i = 60\sin\left(120\,\pi t + \dfrac{\pi}{2}\right)$$

where t is the time (in seconds). Find the amplitude, period, and phase shift, and graph two complete cycles of the curve.

25-4   GRAPH OF $y = a\cos(bx + c)$

In this section we see that the graph of the cosine function is just a sine curve with a phase shift. We find the period, amplitude, and phase shift of the graph of $y = a\cos(bx + c)$ and sketch its graph.

Graph of $y = \cos x$

Now we graph the function

$$y = \cos x$$

by plotting points from a table of values.

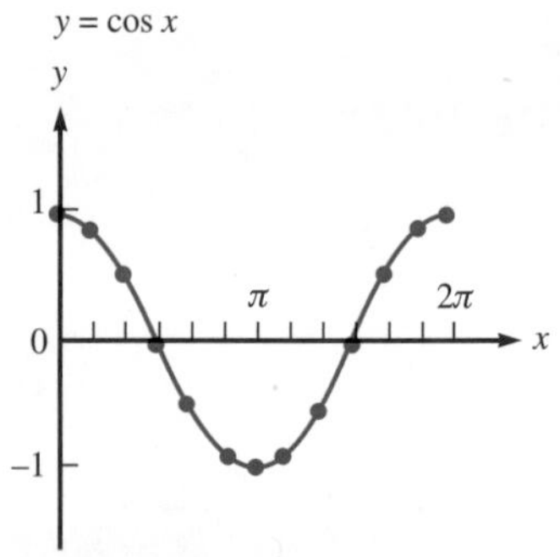

┌─── **Example 19** ───────────

Graph $y = \cos x$ for x from 0 to 2π.

Solution: We make a table of values using intervals of $\dfrac{\pi}{6}$ each, rounding y values to tenths.

x	0	$\dfrac{\pi}{6}$	$\dfrac{\pi}{3}$	$\dfrac{\pi}{2}$	$\dfrac{2\pi}{3}$	$\dfrac{5\pi}{6}$	π	$\dfrac{7\pi}{6}$	$\dfrac{4\pi}{3}$	$\dfrac{3\pi}{2}$	$\dfrac{5\pi}{3}$	$\dfrac{11\pi}{6}$	2π
y	1	0.9	0.5	0	−0.5	−0.9	−1	−0.9	−0.5	0	0.5	0.9	1

These points are plotted and connected with a smooth curve in Fig. 25-38. It looks like the *shape* of the cosine curve is identical to that of the sine curve.

$y = \cos x$

■ **FIGURE 25-38**
Cosine curve.

┌─── **Example 20** ───────────

Graph $y = \cos x$ and $y = \sin x$ on the same axes, for x from $-90°$ to $360°$.

Solution: We graph the functions using a graphing calculator, in degree mode, and the window set so that

| Xmin= −90 |
| Xmax= 360 |
| Xscl= 90 |
| Ymin= −1 |
| Ymax= 1 |
| Yscl= 1 |

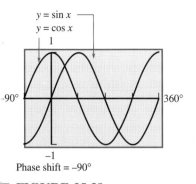

$y = \sin x$
$y = \cos x$
1

−90°   360°

−1
Phase shift = −90°

■ FIGURE 25-39
Sine and cosine curves.

and sketch the curve in Fig. 25-39. We see that the *shapes* of the curves are identical (the period of $y = \cos x$ is 2π and the amplitude is 1), and that the cosine curve is just a sine curve with a phase shift of $\dfrac{\pi}{2}$ to the left. Thus,

$$\cos x = \sin\left(x + \frac{\pi}{2}\right)$$

Graph of $y = a \cos (bx + c)$

For the function

$$y = a \cos (bx + c)$$

the constants a, b, and c have the same effect as for the sine function.

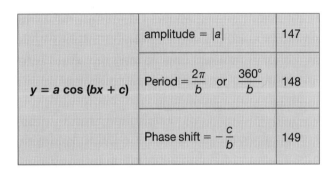

	amplitude = $\lvert a \rvert$	147
$y = a \cos (bx + c)$	Period $= \dfrac{2\pi}{b}$ or $\dfrac{360°}{b}$	148
	Phase shift $= -\dfrac{c}{b}$	149

Quick Sketch of $y = a \cos (bx + c)$

We can use the quick sketch method to graph a cosine function. We first graph the unshifted curve

$$y = a \cos bx$$

as we did for the sine function. Then we shift the curve as before.

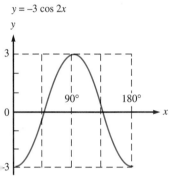

$y = -3 \cos 2x$
y

3

90°   180°

0   x

−3

■ FIGURE 25-40

■ FIGURE 25-41

$y = -3 \cos 2x$
$y = -3 \cos (2x - 180°)$
y

3

0   90°   180°   270° x

−3

+90°

Phase shift = +90°

┌─── **Example 21** ───
Given $y = -3 \cos (2x - 180°)$,

(a) Find the amplitude, period, and phase shift.

(b) Graph the function.

Solution:

(a) For this equation, $a = -3$, $b = 2$, and $c = -180°$. Thus, from Eqs. 147–149, we have

$$\text{amplitude} = \lvert -3 \rvert = 3$$

$$\text{Period} = \frac{360°}{2} = 180°$$

$$\text{Phase shift} = -\frac{-180°}{2} = +90°$$

(b) First we make a dotted quick sketch of $y = -3 \cos 2x$ (Fig. 25-40). Since the phase shift is positive, we shift the curve 90° units to the right, as shown in Fig. 25-41.

25-4 Graph of $y = a \cos (bx + c)$ **575**

Check: We verify that the graph is correct by graphing the function on the graphing calculator, setting the window so that

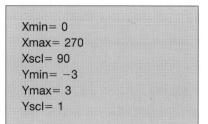

Xmin= 0
Xmax= 270
Xscl= 90
Ymin= −3
Ymax= 3
Yscl= 1

Example 22

Use a graphing calculator to graph one complete cycle of the function

$$y = 2.75 \cos \left(4.85x + \frac{\pi}{3} \right)$$

Solution: For this equation, $a = 2.75$, $b = 4.85$, and $c = \frac{\pi}{3}$. Thus, from Eqs. 147–149, we have

$$\text{amplitude} = |2.75| = 2.75$$

$$\text{Period} = \frac{2\pi}{4.85} = 1.30$$

$$\text{Phase shift} = -\frac{\frac{\pi}{3}}{4.85} = -0.216$$

On a graphing calculator, in radian mode, we set the window so that

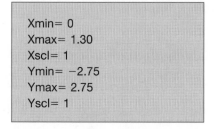

Xmin= 0
Xmax= 1.30
Xscl= 1
Ymin= −2.75
Ymax= 2.75
Yscl= 1

We enter the function and graph (Fig. 25-42).

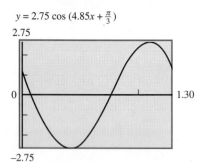

$y = 2.75 \cos (4.85x + \frac{\pi}{3})$

■ **FIGURE 25-42**

SECTION 25-4   Exercises   Graph of $y = a \cos (bx + c)$

Graph of $y = \cos x$

Complete the table for the given equation and plot the curve for x from 0° to 360°, rounded to tenths.

x	0°	30°	60°	90°	120°	150°	180°	210°	240°	270°	300°	330°	360°
y													

1. $y = \cos x$ **2.** $y = \cos \dfrac{x}{2}$

Write the equation of the cosine curve.

3.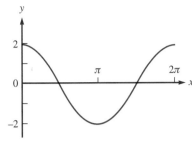

■ **FIGURE 25-43**

4.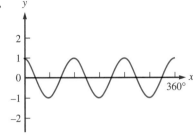

■ **FIGURE 25-44**

5.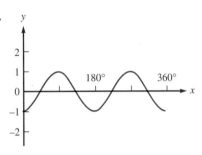

■ **FIGURE 25-45**

6.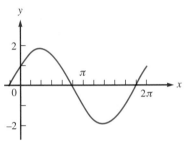

■ **FIGURE 25-46**

7. Write an equation of a cosine curve that has an amplitude of 2 and a period of $\dfrac{2\pi}{3}$.

8. Write an equation of a cosine curve that has an amplitude of 5 and a period of 90°.

9. Write an equation of a cosine curve that has an amplitude of 3, a period of 720°, and a phase shift of π.

10. Write an equation of a cosine curve that has an amplitude of 10, a period of 6π, and a phase shift of -3π.

For each function, find the amplitude and the period (in degrees). Then graph the function for x from 0° to 360°.

11. $y = 3 \cos x$ **12.** $y = -2 \cos x$

13. $y = \cos 3x$ **14.** $y = \cos 4x$

For each function, find the amplitude, the period (in radians), and the phase shift. Then graph at least one complete cycle of the function.

15. $y = -5 \cos 4x$ **16.** $y = 3 \cos 2x$

17. $y = \cos (x + \pi)$ **18.** $y = \cos \left(x - \dfrac{\pi}{2} \right)$

For each function, find the amplitude, period, and phase shift. Then graph at least one complete cycle of the function.

19. $y = \cos (2x - 180°)$ **20.** $y = \cos (3x + 360°)$

21. $y = 2 \cos (3x + \pi)$ **22.** $y = -4 \cos \left(2x + \dfrac{\pi}{2} \right)$

Use a graphing calculator to graph one complete cycle of the function.

23. $y = 4.25 \cos (2.56x - 328°)$ **24.** $y = -2.48 \cos (3.87x + 225°)$

25. The equation for the current i (in amps) in a certain alternating current circuit with a frequency of 60 hertz (cycles/s) is

$$i = 1.5 \cos 120\pi t$$

where t is time in seconds. Find the amplitude and the period, and graph the current i as a function of t for two complete cycles.

26. The equation for the voltage V in a certain alternating current circuit is given by

$$V = 120 \cos \left(120\pi t + \frac{\pi}{2} \right)$$

where t is time in seconds. Find the amplitude, period, and phase shift, and graph the voltage V as a function of t for two complete cycles.

25-5 GRAPH OF $y = a \tan (bx + c)$

In this section we graph the tangent function $y = \tan x$ and find its period. Then we examine the graph of $y = a \tan (bc + c)$ and see how the coefficients a, b, and c affect the graph. The function $y = \tan x$ will be graphed by plotting points from a table of values.

Example 23

Graph the function $y = \tan x$ for x from $-90°$ to $180°$.

Solution: With the aid of a calculator, in degree mode, we make a table, rounding to tenths.

x	$-90°$	$-60°$	$-45°$	$-30°$	$0°$	$30°$	$45°$	$60°$	$90°$	$120°$	$135°$	$150°$	$180°$
y	none	-1.7	-1	-0.6	0	0.6	1	1.7	none	-1.7	-1	-0.6	0

Note that the calculator gives no value for $\tan -90°$ and $\tan 90°$; it gives an *ERROR* message. In Sec. 22-1, we saw that some trigonometric functions are *undefined* for some angles. Indeed, the tangent function $y = \tan x$ is undefined for $x = -90°$ and $x = 90°$. On the graph (Fig. 25-47), at the points where the tangent is undefined, we draw dotted vertical lines. These lines are used as guidelines in sketching the curve. We plot the remaining points and connect them with a smooth curve, taking care *not* to cross these vertical lines.

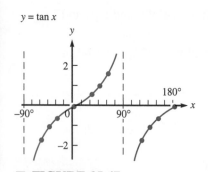

$y = \tan x$

■ FIGURE 25-47
Tangent curve.

Asymptotes

Now we look at the behavior of the function near the vertical line at $x = 90°$. For angles slightly *smaller* than $90°$ (just to the left of $90°$), say $x = 89°$, the tangent is large and positive:

$$\tan 89° = 57.3$$
$$\tan 89.9° = 573.0$$
$$\tan 89.99° = 5729.6$$

As the angle x approaches $90°$ from the left, $\tan x$ gets larger and larger.

For angles slightly *larger* than 90° (just to the right of 90°), say $x = 91°$, the tangent is negative:

$$\tan 91° \quad = -57.3$$
$$\tan 90.1° \quad = -573.0$$
$$\tan 90.01° = -5729.6$$

As the angle x approaches 90° from the right, tan x gets more and more negative. We see that the vertical line at $x = 90°$ is a line which the curve *approaches but never reaches*. Such a line is called an *asymptote*. We complete the graph of the tangent function by showing the curve approaching the asymptote but not reaching it (Fig. 25-47).

Example 24

Use a graphing calculator to graph $y = \tan x$ for x between $-\dfrac{3\pi}{2}$ and $\dfrac{3\pi}{2}$. Find the period of the curve.

Solution: On a graphing calculator, in radian mode, we set the window so that

```
Xmin= -3π/2
Xmax= 3π/2
Xscl= π/2
Ymin= -3
Ymax= 3
Yscl= 1
```

We enter the function and graph (Fig. 25-48). The tangent function is a series of "wavy curves" between asymptotes at $x = \ldots -\dfrac{\pi}{2}, \dfrac{\pi}{2}, \dfrac{3\pi}{2}, \ldots$. The period of the curve $y = \tan x$ is the distance between two consecutive asymptotes, say $x = -\dfrac{\pi}{2}$ and $x = \dfrac{\pi}{2}$. So, the period P of the tangent function is

$$P = \frac{\pi}{2} - \left(-\frac{\pi}{2}\right) = \pi$$

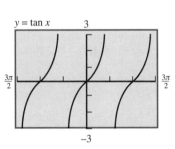

FIGURE 25-48

TIP If your graphing calculator is in "connected" graph mode, then it may show vertical lines near the position of the asymptotes. The calculator has really just connected a point high on one wave with a point low on the next wave. You can avoid this "error" by graphing in the "dot" mode, instead of "connected."

As with sine and cosine functions, the period of function $y = \tan x$ depends on whether x is in degrees or radians. Example 24 shows that, if x is in radians, the period P is

$$P = \pi \quad \text{in radians}$$

If x is in degrees, as in Example 23 (Fig. 25-47), then the period P is

$$P = 90° - (-90°)$$

or

$$P = 180° \quad \text{in degrees}$$

Graph of $y = a \tan x$

The constant a has the effect of stretching (or compressing) the curve vertically.

Example 25

Graph $y = \tan x$ and $y = 2 \tan x$ on the same axes, for x from $-\dfrac{\pi}{2}$ to $\dfrac{3\pi}{2}$.

Solution: We use a graphing calculator in radian mode, setting the window so that

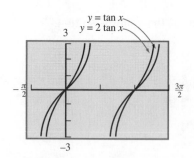

■ **FIGURE 25-49**

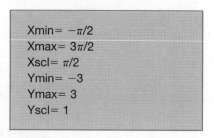

Xmin= $-\pi/2$
Xmax= $3\pi/2$
Xscl= $\pi/2$
Ymin= -3
Ymax= 3
Yscl= 1

We enter the two functions and graph (Fig. 25-49) in the "connected" mode. Note that the constant 2 has the effect of stretching the curve vertically.

Graph of $y = a \tan bx$

As with the sine and cosine curves, the coefficient b in $y = a \tan bx$ affects the period of the curve.

Example 26

Graph $y = \tan 2x$ for x from $-90°$ to $90°$. Find the period of the function.

Solution: We use a graphing calculator in degree mode, and set the window so that

■ **FIGURE 25-50**
Calculator graph in dot mode.

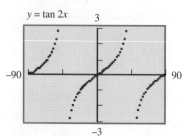

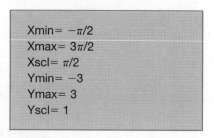

Xmin= -90
Xmax= 90
Xscl= 45
Ymin= -3
Ymax= 3
Yscl= 1

We enter the function and graph the function in the "dot" graph mode (Fig. 25-50). Note that there are asymptotes at $x = -45°$ and at $x = 45°$, but they are not shown on the graph.

The period of this curve is

$$P = 45° - (-45°) = 90°$$

This period is exactly half the period of the curve $y = \tan x$.

In general, the period P of the function $y = a \tan bx$ is

$$P = \frac{\pi}{b} \quad \text{or} \quad \frac{180°}{b}$$

depending on whether x is in radians or degrees.

The graph of $y = a \tan bx$ always passes through the origin $(0, 0)$. One asymptote is at $x = P/2$ and another asymptote is at $x = -P/2$. One complete cycle of the function is between this pair of asymptotes.

Example 27

Find the period of the function $y = 2 \tan 3x$ (x in radians), and graph two complete cycles of the curve.

Solution: We use the equation for the period of the tangent function to calculate the period:

$$P = \frac{\pi}{b} = \frac{\pi}{3}$$

To find a pair of asymptotes, we take half of $\frac{\pi}{3}$

$$\frac{1}{2}\left(\frac{\pi}{3}\right) = \frac{\pi}{6}$$

So there will be asymptotes at $x = -\frac{\pi}{6}$ and $x = \frac{\pi}{6}$. To get two complete cycles, we set

$$\text{Xmin} = -\frac{\pi}{6} \quad \text{and} \quad \text{Xmax} = \frac{\pi}{6} + \frac{\pi}{3} = \frac{\pi}{2}$$

Using a graphing calculator in radian mode, we set the window to

```
Xmin= -π/6
Xmax= π/2
Xscl= π/6
Ymin= -3
Ymax= 3
Yscl= 1
```

We enter the function and graph in dot mode (Fig. 25-51).

■ **FIGURE 25-51**

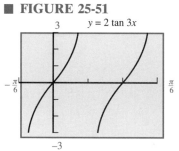

$y = 2 \tan 3x$

25-5 Graph of $y = a \tan (bx + c)$ **581**

Graph of $y = a \tan (bx + c)$

Similarly, the constant c effects a phase shift of $-c/b$ units, just as with the sine and cosine curves. In general,

$y = a \tan (bx + c)$	Period $= \dfrac{\pi}{b}$ or $\dfrac{180°}{b}$	151
	Phase shift $= -\dfrac{c}{b}$	152

─────── **Example 28** ───────

Given $y = \tan \left(x + \dfrac{\pi}{3} \right)$,

(a) Find the period and the phase shift.
(b) Graph the function.

Solution:

(a) In this equation $a = 1$, $b = 1$, and $c = \dfrac{\pi}{3}$. Using Eq. 151, with $b = 1$, the period is

$$P = \pi$$

$$\text{Phase shift} = -\frac{\pi}{3}$$

which indicates a shift of $\dfrac{\pi}{3}$ units to the left.

(b) We graph the function $y = \tan \left(x + \dfrac{\pi}{3} \right)$ for x between $-\dfrac{5\pi}{6}$ and $\dfrac{7\pi}{6}$ (Fig. 25-52).

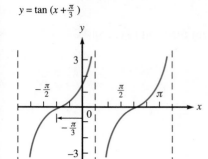

$y = \tan (x + \frac{\pi}{3})$

Phase shift $= -\frac{\pi}{3}$

■ **FIGURE 25-52**
Tangent curve with phase shift.

SECTION 25-5 Exercises Graph of $y = a \tan (bx + c)$

Graph of $y = a \tan bx$

Complete the table for the given equation, and graph the equation for x from $-90°$ to $360°$, rounding to tenths.

x	$-90°$	$-60°$	$-45°$	$-30°$	$0°$	$30°$	$45°$	$60°$	$90°$	$135°$	$180°$	$225°$	$270°$	$315°$	$360°$
y															

1. $y = \tan x$ **2.** $y = -3 \tan x$

Find the period, in degrees, and graph two complete cycles of each equation, indicating asymptotes by dotted lines.

3. $y = 3 \tan x$ **4.** $y = 4 \tan x$

5. $y = \tan 3x$ **6.** $y = \tan 4x$

Find the period, in radians, and graph two complete cycles of each equation, indicating asymptotes by dotted lines.

7. $y = -2 \tan x$

8. $y = -5 \tan x$

9. $y = \tan \dfrac{x}{2}$

10. $y = \tan \dfrac{x}{3}$

Graph of
$y = a \tan (bx + c)$

For each equation, (a) find the period and the phase shift; and (b) graph two complete cycles of the equation, indicating asymptotes by dotted lines.

11. $y = \tan \left(x - \dfrac{\pi}{2} \right)$

12. $y = \tan (x - 45°)$

13. $y = \tan (x + 45°)$

14. $y = \tan (x + \pi)$

15. $y = 2 \tan (2x + \pi)$

16. $y = -3 \tan (4x - 360°)$

■ **CHAPTER 25** | **REVIEW EXERCISES**

For each equation, find the amplitude, the period (in degrees), and the phase shift. Graph at least one complete cycle of the function.

1. $y = -4 \cos x$

2. $y = 3 \sin 2x$

3. $y = \sin (x - 45°)$

4. $y = -\dfrac{2}{3} \cos x$

5. $y = \cos 2x$

6. $y = 0.5 \sin (2x + 60°)$

7. $y = \cos (2x - 90°)$

8. $y = \cos (x + 180°)$

9. $y = 5 \sin 3x$

10. $y = 75 \sin x$

For each equation, find the period (in radians) and the phase shift. Graph two complete cycles of the function, indicating asymptotes by dotted lines.

11. $y = \tan \dfrac{x}{4}$

12. $y = \tan \left(x + \dfrac{3\pi}{2} \right)$

For each function, find the amplitude, the period (in radians), and the phase shift. Graph at least one complete cycle of the function.

13. $y = \sin 3x$

14. $y = \sin \dfrac{x}{4}$

15. $y = 5 \cos x$

16. $y = -2 \sin \dfrac{\pi x}{3}$

17. $y = -3 \sin \left(x + \dfrac{\pi}{8} \right)$

18. $y = \sin (3x - \pi)$

19. $y = 2 \cos 3x$

20. $y = -3 \cos (x + \pi)$

21. Write an equation of a sine curve that has an amplitude of 2, a period of 180°, and a phase shift of 90°.

22. Write an equation of a cosine curve that has an amplitude of 3, a period of 4π, and a phase shift of -2π.

For each function, find the period (in radians) and the phase shift. Graph two complete cycles of the function, indicating asymptotes by dotted lines.

23. $y = \tan 5x$

24. $y = 2 \tan \left(x - \dfrac{\pi}{2} \right)$

25. Given the general sine function, $y = a \sin (bx + c)$, explain in your own words how a change in each of the constants (a, b, or c) will affect the graph of the function. Show three examples (one for each constant) which illustrate the effect of each constant.

TEAM PROJECT

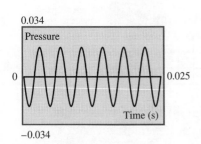

0.034

Pressure

0

−0.034

Time (s)

0.025

■ **FIGURE 25-53**

26. When a sound is made, it disturbs nearby air molecules, creating regions of higher-than-normal pressure and regions of lower-than-normal pressure. If we graph pressure y versus time t, we get a sine wave which represents the sound. Important characteristics of the sound wave are its period, frequency, and amplitude. Using a tuning fork and a microphone, pressure readings were taken every 0.0002 second for about 0.025 second, and plotted on a pressure versus time graph to get the graphing calculator graph in Fig. 25-53.

Note that the y axis is the left border of the graph. The coordinates of the first and last maximum points are

(0.0031030, 0.022389) and (0.0222112, 0.019033)

Use this information to estimate the amplitude a of the sound wave. Round all calculations to three significant digits. How many cycles of the sine curve are there between the first and last maximum points? Estimate the period T of the sound wave, in seconds per cycle. The frequency f is found using

$$f = \frac{1}{T}$$

Frequency is measured in cycles per second, or hertz (Hz). This tuning fork was labeled 256 Hz. Compare this frequency to your estimated frequency and calculate the percent error. The equation for a sound wave is

$$y = a \sin 2\pi f t$$

Write the equation of a sound wave using your estimated values for a and f. Graph the equation. How does your graph compare to the graph in Fig. 25-53? The sine curve in Fig. 25-53 has been shifted to the right or left, so you must find the phase shift. (Note that the phase shift is not a property of the sound, but that the phase shift depends on when the first data point was taken.) Estimate the phase shift d for this data. The equation for this sound wave is

$$y = a \sin (2\pi f(t + d))$$

Write the equation of this sound wave using your estimated values for a, f, and d. Graph the equation. Your graph should match the graph in Fig. 25-53 pretty well.

Ratio, Proportion, Variation

26

In this chapter, we deal with ratio, proportion, the power function, and four types of variation: direct, joint, inverse, and combined. Many applications use the concepts of ratio, proportion, and variation, such as maps and scale drawings.

26-1 RATIO

The ratio concept is an extremely important one for math students. Maps are printed to scale where the scale is the ratio of a map distance to an actual distance. Auto mechanics are concerned with compression ratios, machinists with gear ratios, and carpenters with the pitch ratio of a roof. In this section, we define a ratio, consider two forms of a ratio, learn to simplify a ratio, and do calculations with similar figures and rates.

We are often concerned with comparing one quantity to another. For example, we may need to compare the weights of two objects, the first 6 kg and the second 3 kg. We could say that the first weighs 3 kg *more* than the second; this is a comparison by *subtraction* (the *difference*). Or we could say that the first weighs *twice as much as* the second; this is a comparison by *division*. The comparison of two quantities by division is called a *ratio*.

Notation

The ratio of quantity a to quantity b may be written with a colon or as a fraction.

RATIO OF a TO b	$a:b$ or $\dfrac{a}{b}$	45

The quantities a and b are called the *terms* of the ratio.

Example 1

The ratio of 3 to 2 may be written

with a colon as

$$3:2$$

or as a fraction

$$\frac{3}{2}$$

585

Simplification of Ratios

To *simplify* a ratio, we write the ratio as a fraction, and reduce that fraction.

—————— **Example 2** ——————

Simplify:

(a) the ratio of 12 cm to 9 cm

(b) the ratio of 6 pounds to 3 pounds

Solution:

(a) The ratio of 12 cm to 9 cm may be written as a fraction:

$$\frac{12\ cm}{9\ cm}$$

We simplify the ratio by reducing the fraction:

$$\frac{12\ cm}{9\ cm} = \frac{4}{3}$$

Thus the ratio of 12 cm to 9 cm is the same as the ratio of 4 to 3, or 4:3. Notice that since the quantities were expressed in the same units, the ratio is *dimensionless*.

(b) The ratio of 6 pounds to 3 pounds may be written in fractional form and reduced:

$$\frac{6\ pounds}{3\ pounds} = \frac{2}{1}$$

Notice that the denominator 1 is retained. Since a ratio is a comparison of *two* quantities, one of them cannot disappear by being canceled. Thus the ratio of 6 pounds to 3 pounds is the same as the ratio of 2 to 1, or 2:1.

Dimensionless Ratios

If the quantities being compared can be expressed in the same units, it is customary to do so, so that the units will cancel and leave the ratio dimensionless.

■ **FIGURE 26-1**

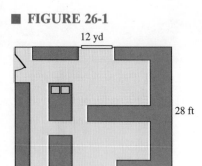

—————— **Example 3** ——————

A laboratory is 28 ft wide and 12 yd long (Fig. 26-1). Find the ratio of the length to width and simplify.

Solution: It is important to note that we need to find the ratio of length to width. Since length is the first term of the ratio, the *length* must be the *numerator* of the fraction. We change 12 yd to 36 ft and proceed to simplify the ratio.

$$\frac{12\ yd}{28\ ft} = \frac{36\ ft}{28\ ft} = \frac{9}{7}$$

So the ratio of length to width is 9 to 7, or 9:7.

Dimensionless ratios are often used in mathematics.

Example 4

(a) The number *pi* (π) is the ratio of circumference C to diameter D.

$$\pi = \frac{C}{D}$$

(b) The radian measure θ of an angle is the ratio of arc length s to radius r.

$$\theta = \frac{s}{r}$$

(c) Trigonometric ratios are ratios of two sides of a right triangle.

$$\tan \theta = \frac{\text{opposite leg}}{\text{adjacent leg}}$$

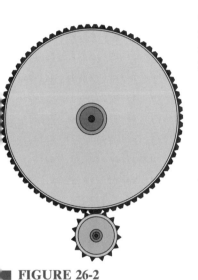

FIGURE 26-2

Other examples of ratios in technical fields are gear ratio, turn ratio, load ratio, fuel-air ratio, efficiency, Mach number, mechanical advantage, and the pitch of a roof. The word *specific* is often used to denote a ratio, such as in specific heat (the ratio of two thermal capacities) and specific gravity (the ratio of two densities).

Example 5

A flywheel has 75 teeth and a starter drive gear has 15 teeth (Fig. 26-2). Find the ratio of flywheel teeth to starter drive teeth.

Solution:

$$\frac{\text{flywheel teeth}}{\text{starter drive teeth}} = \frac{75}{15} = \frac{5}{1}$$

So the ratio is 5 to 1 or 5:1. Note that we do *not* change $\frac{5}{1}$ to 5, since this is a *ratio,* not a fraction.

Scale Drawings

An important application of ratio is in the use of scale drawings, such as maps, engineering drawings, surveying layouts, and so on. The ratio of a distance on the scale drawing to the corresponding distance on the actual object is called the *scale* of the drawing. A scale is often given using a colon.

$$\text{Scale} = \text{distance on the drawing} : \text{actual distance}$$

A scale on a drawing is often given as a ratio of two distances with different units (dimensions). To express the scale as a dimensionless ratio, convert the distances to the same unit. This may be done before writing the ratio, or afterward, as in the following example.

Example 6

On a certain map, 1 inch represents 10 miles. Express the scale of the map as a dimensionless ratio.

Solution: We set up a ratio for the scale and express it in fraction form:

$$\text{Scale} = 1\,\text{in.} : 10\,\text{mi} = \frac{1\,\text{in.}}{10\,\text{mi}}$$

Convert 10 mi to in.:

$$10 \text{ mi} = (10 \text{ mi}) \left(\frac{5280 \text{ ft}}{1 \text{ mi}} \right) \left(\frac{12 \text{ in.}}{1 \text{ ft}} \right) = 633{,}600 \text{ in.}$$

Expressing the scale as a dimensionless ratio, we have

$$\frac{1 \text{ in.}}{10 \text{ mi}} = \frac{1 \text{ in.}}{633{,}600 \text{ in.}} = \frac{1}{633{,}600}$$

So the scale of the map, as a dimensionless ratio, is $1 : 633{,}600$.

SECTION 26-1 Exercises Ratio

Notation

Write the ratio in both colon and fractional form.

1. 5 to 4

2. 3 to 5

Simplification of Ratios

Simplify the ratio in fraction form.

3. 6 cm to 15 cm

4. 4 lb to 6 lb

5. 8 ft² to 4 ft²

6. 15 kg to 5 kg

7. 7.1 mℓ to 28.4 mℓ

8. 9.3 in. to 18.6 in.

Dimensionless Ratios

Simplify the ratio in fraction form.

9. 4 inches to 1 foot

10. 4 feet to 18 inches

11. 1 meter to 25 cm

12. 400 grams to 2 kg

13. 3 quarts to 1 gallon

14. 1 pound to 4 ounces

15. 2 mℓ to 2 liters

16. 45 minutes to 1 hour

17. 75 cents to 2 dollars

18. 1500 m to 1 km

19. A corridor is 9 ft wide and 15 yd long. Find the ratio of length to width.

Applications

20. A bronze alloy is made by mixing 1.40 kg of tin with 11.2 kg of copper. Find the ratio of copper to tin.

21. Estimate the value of π if a circle with a diameter of 2.50 cm has a circumference of 7.85 cm.

■ **FIGURE 26-3**

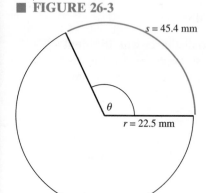

22. Find the radian measure of an angle θ that intercepts an arc of 45.4 mm in a circle of radius 22.5 mm (Fig. 26-3).

23. The *mechanical advantage* of a simple machine is the ratio of load force to effort force. Find the mechanical advantage of a pulley system in which a force of 85.6 N (effort) is needed to lift a 3470-N load. Express the ratio as a decimal.

24. A mixture of gasohol contains 8.60 gallons of alcohol and 90.3 gallons of gasoline. Find the ratio of gasoline to alcohol.

25. The power output of any machine or device is always less than the power input, because of inevitable power losses within the device. The *efficiency* of the device is a measure of those losses, and is defined as the ratio of output

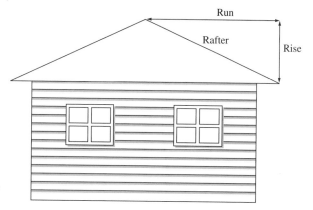

■ **FIGURE 26-4**

to input. Find the efficiency of a motor which consumes 795 W and has an output of 1 horsepower. Express the efficiency as a percent. (1 hp = 746 W)

26. Carpenters define the pitch of a roof as the ratio of the rise to the run of the rafter (Fig. 26-4). What is the pitch of a roof with a rise of 9.0 feet and a run of 12.0 feet?

27. The *Mach number* of a moving object is defined as the ratio of its velocity to the velocity of sound (340.2 m/s). A Mach number is generally expressed as a decimal. Find the Mach number of a jet traveling at 275 m/s.

28. The *specific gravity* (also called *relative density*) of a substance is the ratio of its density to the density of water. Specific gravity is generally expressed as a decimal. Taking the density of water as 62.4 lb/ft^3, find the specific gravity of copper if its density is 554.7 lb/ft^3.

26-2

PROPORTION

In this section, we define a proportion and discuss the terms, means, and extremes of a proportion. We learn how to find a missing term in a proportion, state the fundamental principle of proportions, and solve some application problems.

The Fraction Form of a Proportion

A *proportion* is a statement that two ratios are equal. A proportion is an equation, which in fraction form looks like this:

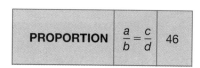

| PROPORTION | $\dfrac{a}{b} = \dfrac{c}{d}$ | 46 |

The proportion in Eq. 46 is read

"the ratio of *a* to *b* equals the ratio of *c* to *d*"

or

"*a* is to *b* as *c* is to *d*"

Example 7

Write as a proportion in fraction form.

(a) The ratio of x to y equals the ratio of 9 to 12.

(b) x is to 5 as y is to 15.

Solution: **(a)** $\dfrac{x}{y} = \dfrac{9}{2}$ **(b)** $\dfrac{x}{5} = \dfrac{y}{15}$

In the proportion in Eq. 46, the quantities a, b, c, and d are called the *terms* of the proportion. The first and fourth terms, a and d, are called the *extremes* and the second and third terms, b and c, are called the *means* of the proportion.

$$\frac{a}{b} = \frac{c}{d}$$

extremes — a, d

means — b, c

Fundamental Principle of Proportions

A very important principle of proportions is found by multiplying both sides of the general proportion

$$\frac{a}{b} = \frac{c}{d}$$

by the LCD, bd:

$$\frac{a}{b}(bd) = \frac{c}{d}(bd)$$

$$ad = bc$$

Thus we have the *fundamental principle of proportions:*

FUNDAMENTAL PRINCIPLE OF PROPORTIONS	In a proportion, the product of the extremes equals the product of the means. If $\quad \dfrac{a}{b} = \dfrac{c}{d}$ then $\quad ad = bc$	47

Note: This procedure is often referred to as "cross-multiplication," and ad and bc are called cross-products.

Example 8

Use the fundamental principle of proportions to show that this is a true proportion:

7 is to 4 as 63 is to 36

Solution: First we write the proportion in fraction form:

$$\frac{7}{4} = \frac{63}{36}$$

Calculating cross-products, we have

$$(7)(36) = 252 \quad \text{and} \quad (4)(63) = 252$$

Since the cross-products are equal, the proportion is true.

Finding a Missing Term in a Proportion

To find a missing term in a proportion, cross-multiply and solve the resulting equation.

Example 9

Find x in the proportion:

$$\frac{3}{5} = \frac{2}{x}$$

Solution: Cross-multiplying, we have

$$3x = 10$$

Solving for x:

$$x = \frac{10}{3}$$

Applications

Many applications use proportions. Set these up as you set up the word problems in Chapter 11. Be careful not to reverse the terms in the proportion. Match the first item in the verbal statement with the first number in the ratio, and the second with the second.

Example 10

Separate \$180 into two parts so that one part is to the other part as 7 is to 5.

Solution: We follow the process of Chapter 11, starting with defining the variables.

Let $\quad x \quad$ = one part
and $\quad 180 - x$ = other part

Writing the proportion, we have

$$\frac{\text{one part}}{\text{other part}}: \quad \frac{x}{180 - x} = \frac{7}{5}$$

Cross-multiplying,

$$5x = 7(180 - x)$$

Solving for x,

$$5x = 1260 - 7x$$

$$12x = 1260$$

$$x = 105$$

Finding the other part:

$$180 - x = 180 - 105 = 75$$

So the two parts of $180 are $105 and $75.

Check: Verify that these two parts have a ratio of 7 to 5:

$$\frac{105}{75} \stackrel{?}{=} \frac{7}{5}$$

Calculating cross-products:

$$(105)(5) = 525 \text{ and } (75)(7) = 525 \quad \text{checks}$$

Rate Problems

Many applications involving rates may be solved using the method of proportions.

─── **Example 11** ───

If a car can go 131 miles on 5.25 gallons of gasoline, how far can it go on 12.5 gallons of gasoline?

Solution: Set up a proportion with dimensionless ratios, using like quantities in each ratio (that is, miles with miles and gallons with gallons).

$$\overset{\text{miles}}{\underset{x}{\frac{131 \text{ mi}}{}}} = \overset{\text{gallons}}{\frac{5.25 \text{ gal}}{12.5 \text{ gal}}}$$

We cancel the units on the right:

$$\frac{131 \text{ mi}}{x} = \frac{5.25}{12.5}$$

Cross-multiply, and solve:

$$5.25x = (131 \text{ mi})(12.5)$$

$$x = 312 \text{ mi}$$

So the car will go 312 miles on 12.5 gallons of gasoline.

■ **FIGURE 26-5**
Similar triangles.

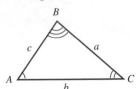

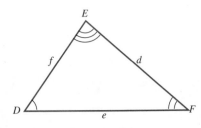

Similar Figures

Two geometric figures are said to be similar if they have the *same shape,* but not necessarily the same size (Fig. 26-5). This means that the angles of the two figures have the same measure, but the sides are not necessarily equal. Figure 26-5 shows two similar triangles. There are three pairs of equal angles:

$$A = D \qquad B = E \qquad C = F$$

Each pair of equal angles are called *corresponding* angles. So, we say, angle A and angle D are corresponding angles, and so forth.

Sides opposite corresponding angles are called *corresponding sides*. In Fig. 26-5, corresponding pairs of sides are a and d, b and e, and c and f. In similar figures, the ratios of these corresponding sides are equal, so

$$\frac{a}{d} = \frac{b}{e} = \frac{c}{f}$$

We can use this to set up a proportion to solve similar triangles.

Example 12

A right triangle has sides 3.00 cm, 4.00 cm, and 5.00 cm. The hypotenuse of a similar triangle is 12.7 cm.

(a) What is the ratio of corresponding sides?

(b) Find the other two sides of the similar triangle.

Solution: We sketch the two similar triangles in Fig. 26-6.

(a) Since the triangles are similar, the ratio of corresponding sides must be the same for all pairs of sides. The hypotenuse of the first triangle (5.00 cm) corresponds to the hypotenuse of the similar triangle (12.7 cm). We form the ratio and note that we can cancel the units:

$$\frac{5.00 \text{ cm}}{12.7 \text{ cm}} = \frac{5.00}{12.7}$$

(b) To find the other two sides, x and y, we need to identify their corresponding sides in the first triangle. Side x is the shorter leg in the second triangle and corresponds to the shorter leg in the first triangle (the 3.00-cm side). We set up a proportion:

$$\frac{3.00}{x} = \frac{5.00}{12.7}$$

Cross-multiply and solve:

$$5.00x = (3.00)(12.7)$$

$$x = 7.62 \text{ cm}$$

Side y is the longer leg in the second triangle and corresponds to the longer leg in the first triangle (the 4.00-cm side).

We set up a proportion:

$$\frac{4.00}{x} = \frac{5.00}{12.7}$$

Cross-multiply and solve:

$$5.00x = (4.00)(12.7)$$

$$x = 10.2 \text{ cm}$$

So the other two sides of the similar triangle are 7.62 cm and 10.2 cm long.

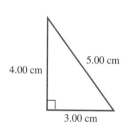

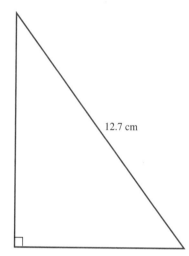

■ **FIGURE 26-6**

SECTION 26-2 Exercises Proportion

The Fraction Form of a Proportion

Write as a proportion in fraction form.

1. The ratio of x to y equals the ratio of 6 to 9.
2. The ratio of 15 to 18 equals the ratio of x to y.

3. p is to 20 as q is to 5.

4. q is to 25 as p is to 100.

5. The ratio of 12 cm to 30 cm equals the ratio of y to 5.

6. 6 kg is to 33 kg as x is to 11.

Fundamental Principle of Proportions

Use the fundamental principle of proportions to show that these proportions are true proportions.

7. The ratio of 2 to 3 equals the ratio of 6 to 9.

8. The ratio of 15 to 18 equals the ratio of 5 to 6.

9. 12 is to 20 as 3 is to 5.

10. 13 is to 25 as 52 is to 100.

11. The ratio of 12 cm to 30 cm equals the ratio of 2 to 5.

12. 6 kg is to 33 kg as 2 is to 11.

Finding a Missing Term in a Proportion

Solve the proportion.

13. $\dfrac{x}{2} = \dfrac{15}{6}$ **14.** $\dfrac{5}{x} = \dfrac{25}{30}$

15. $\dfrac{12}{6} = \dfrac{6}{x}$ **16.** $\dfrac{3}{7} = \dfrac{x}{3}$

17. $\dfrac{4}{x} = \dfrac{x}{9}$ **18.** $\dfrac{x}{3} = \dfrac{12}{x}$

19. $\dfrac{x}{5} = \dfrac{6}{7}$ **20.** $\dfrac{7}{4} = \dfrac{x}{9}$

21. $\dfrac{x}{8} = \dfrac{18}{x}$ **22.** $\dfrac{8}{x} = \dfrac{x}{11}$

23. $\dfrac{x}{4} = \dfrac{x+3}{5}$ **24.** $\dfrac{x-2}{3} = \dfrac{x}{5}$

25. $\dfrac{x}{30-x} = \dfrac{7}{8}$ **26.** $\dfrac{75-x}{x} = \dfrac{8}{7}$

Applications

27. A 16.0-ft board must be cut into two pieces so that the lengths of the pieces are in a ratio of 3 to 5. Find the lengths of the two pieces.

28. An electrician and an apprentice worked together on a job and received $585 in wages. Separate the wages so that the parts are in a ratio of 8 to 5.

29. Separate $650 into two parts that are in a ratio of 4 to 9.

30. Two women jointly bought some stock. The first put in $600 more than the second, and the stock of the first was to that of the second as 8 is to 7. How much money did each invest?

31. A certain automobile cooling system contains a coolant that is a mixture of antifreeze and water in a ratio of 2 to 11. How much antifreeze is contained in 9.00 liters of coolant?

32. A certain solder contains tin and lead in a ratio of 2 to 3. How much tin is contained in 65.5 kg of the solder?

In the following problems, assume that the rates remain constant.

33. If a 5.00-lb bag of sand weighs 2.27 kg in metric units, how much would an 8.00-lb bag of sand weigh in kg?

34. The property taxes on a property valued at $95,500 are $1710. What are the property taxes on a similar property valued at $115,000?

FIGURE 26-7

35. A certain map uses the scale 1.00 inch represents 25.0 miles. If New York City and Philadelphia are 4.04 inches apart on the map (Fig. 26-7), find the actual distance between the cities.

36. A map of the United States uses the scale 1.00 cm represents 85.0 km. If Chicago and Los Angeles are 31.6 cm apart on the map, find the actual distance between the cities.

37. A carpenter finds that an 8.00-ft length of a two by four board weighs 21.6 lb. What is the weight of a 9.00-ft length of a similar board?

38. A technician can assemble 3.5 instruments in a 7.5-hour day. How many hours would it take her to assemble 25 instruments? How many days?

39. An airplane travels 585 km in 2.75 hours. At the same rate, how long would it take the airplane to travel 825 km?

40. If a box of 250 screws costs $3.49, how much would a box of 400 screws cost?

41. A truck can carry 5.05 yd^3 of sand weighing 6.57 tons. A larger truck can carry 8.25 yd^3 of sand. What is the weight of the sand?

42. A contractor can build an 1850 ft^2 house for $81,900. How much would it cost to build a similar house of 2250 ft^2?

43. A right triangle has sides 5.00 cm, 12.0 cm, and 13.0 cm. The hypotenuse of a similar triangle is 18.5 cm. Find the other two sides of the similar triangle.

44. A triangle has sides 24.0 in., 28.0 in., and 32.0 in. The shortest side of a similar triangle is 15.5 in. Find the other two sides of the similar triangle.

45. A tree casts a shadow 36.0 ft long at the same time a boy 4.00 feet tall casts a shadow 6.25 ft long (Fig. 26-8). How tall is the tree?

46. Use similar triangles to find the width of the river in Fig. 26-9.

FIGURE 26-8

36.0 ft

4.0 ft

6.25 ft

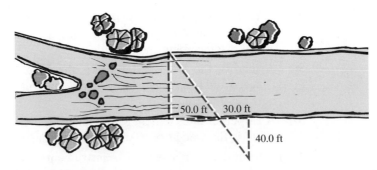

50.0 ft   30.0 ft

40.0 ft

■ **FIGURE 26-9**

26-3   DIRECT VARIATION

We saw in Chapter 12 that two quantities, x and y, may be related in such a way, that if the value of one quantity changes, then the value of the other quantity also changes. If a change in x causes a specific and predictable change in y, then we say that *y varies as x varies.*

     In this section we will look at one type of variation: direct variation. We will see that direct variation is closely associated with the concepts of ratio and proportion. We consider many applications of direct variation and solve problems involving direct variation.

Direct Variation

If two variables, x and y, are related in such a way that the ratio $\dfrac{y}{x}$ is a constant, then we say *y varies directly as x*, and we may write

$$y \propto x$$

The symbol $\propto$ means *is directly proportional to.* We may also say that *y is proportional to x.*

Example 13 ───────

It is obvious that the circumference C of a circle changes as its diameter D changes. Consider this table of pairs of values of D (in cm) and C (in cm):

D	1.50	2.00	2.50	3.00	3.50
C	4.71	6.28	7.85	9.42	11.0

If we calculate the ratio $\dfrac{C}{D}$ for each pair of values, we find that the ratio $\dfrac{C}{D}$, rounded to three significant digits, is a constant:

$$\frac{4.71}{1.50} = \frac{6.28}{2.00} = \frac{7.85}{2.50} = \frac{9.42}{3.00} = \frac{11.0}{3.50} = 3.14$$

Thus, we say that the circumference of a circle *varies directly as* the diameter, or the circumference *is directly proportional to* the diameter, and we write $C \propto D$. (In this case, the ratios are equal to the familiar mathematical constant π.)

One important property of direct variation is that if one variable *increases,* then the other *increases,* and if one variable *decreases,* then the other also *decreases.* In Example 13, we see that the circumference of a circle varies directly as the diameter. If the diameter *increases,* then the circumference *increases.* Similarly, if the diameter *decreases,* then the circumference *decreases.*

Constant of Proportionality

If the ratio of two variables, y and x, is equal to a constant k, then we express the relationship by the equation

$$\frac{y}{x} = k$$

where k is called the *constant of proportionality.* If we multiply both sides of the equation by x, we have a familiar form of the equation for direct variation.

DIRECT VARIATION	$y = kx$ where k is the constant of proportionality	48

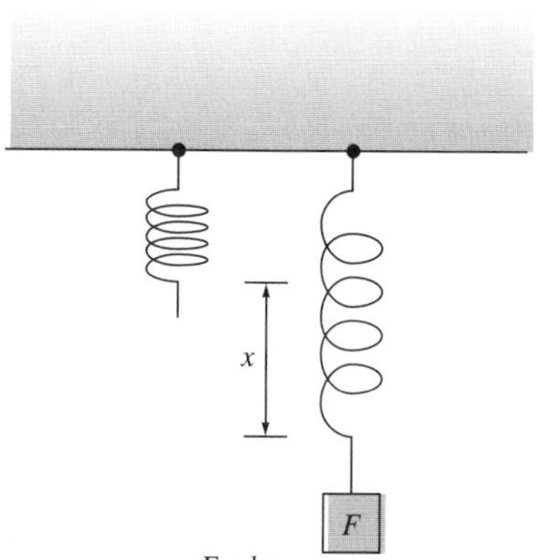

$F = kx$

■ FIGURE 26-10
Direct variation.

Example 14

Write an equation for the force needed to stretch a spring, if the force F varies directly as the distance x the spring is stretched (Fig. 26-10).

Solution: We use Eq. 48 to write the equation. Substituting the force F for y, we get

$$F = kx$$

(This is called Hooke's Law and k is called the spring constant.)

Example 15

Write an equation for the electrical resistance R of a wire, if the resistance is proportional to the length L of the wire.

Solution: We use Eq. 48 to write the equation. Substituting the resistance R for y and the length L for x, we get

$$R = kL$$

where k is related to the coefficient of resistivity of the wire.

Example 16

Write an equation for the weight w of an object if it is directly proportional to its mass m.

Solution: Again we use Eq. 48 to write the equation.

$$w = km$$

In this equation, k is the constant rate of acceleration due to gravity, g.

In most direct variation problems involving physical relationships, the constants of proportionality have *units* associated with them. In Example 16, if the weight w is in grams, then the constant k is the rate of acceleration due to gravity, so that

$$k = 981 \text{ cm/s}^2$$

The units are an *essential* part of the constant of proportionality.

Solution of Variation Problems

To solve a variation problem in which a value is unknown, at least one pair of corresponding values of the related variables must be known.

Example 17

If s varies directly as t, and $s = 118$ km when $t = 1.25$ hours, find s when $t = 2.75$ hours.

Solution: Since s varies directly as t, we use Eq. 48 to write the equation

$$s = kt$$

To find the constant of proportionality, k, we substitute the given pair of values for s and t into the equation.

$$118 \text{ km} = k(1.25 \text{ h})$$

Solving for k:

$$k = \frac{118 \text{ km}}{1.25 \text{ h}} = 94.4 \text{ km/h}$$

Substituting $k = 94.4$ (km/h) into the equation $s = kt$, we have

$$s = 94.4t$$

where s is in kilometers and t is in hours.

Although the units of the constant of proportionality are of the utmost importance, we usually don't write the units of k in the equation. We do, however, state the units used for the other variables. The constant is 94.4 (km/h) *only if* s is in kilometers and t is in hours.

To find the missing s, we substitute $t = 2.75$ into the equation $s = 94.4t$ to get

$$s = (94.4)(2.75) = 260 \text{ km}$$

The example above suggests a method for solving variation problems in which a value is missing.

The Solution of Variation Problems

1. Using Eq. 48, write an equation that describes the expressed relationship.
2. Substitute the two given values in the equation and solve for the constant of proportionality, k.
3. Rewrite the equation by substituting the value for the constant into the equation from Step 1.
4. To find the missing value, substitute the corresponding given value into the equation and solve.

Sometimes we find that one quantity varies directly as a *power* of another quantity.

Example 18

The area A of a circle varies directly as the square of its radius r (Eq. 76):

$$A = \pi r^2$$

where π is the constant of proportionality.

If y varies as a power of x, then we write $y = kx^n$, where k is the constant of proportionality.

Example 19

(a) If y *varies directly as the square of* x, then

$$y = kx^2$$

(b) The period T of a pendulum *is proportional to the square root of* its length L, so

$$T = k\sqrt{L}$$

Example 20

The volume V of a sphere *is directly proportional to the cube of* the radius r, so

$$V = kr^3$$

This reminds us of the familiar formula for the volume of a sphere, Eq. 95:

$$V = \frac{4}{3}\pi r^3$$

We see that the constant of proportionality is

$$k = \frac{4}{3}\pi$$

Applications

Many practical problems can be solved using the idea of direct variation.

Example 21

The electric current i through a resistor is directly proportional to the voltage drop E across it. If the current through a resistor is 18.6 amps when the voltage drop across it is 45.6 volts, find the current when the voltage drop is 60.0 volts.

Solution: Use Eq. 48 to find the equation of proportionality:

$$I = kE$$

We substitute the given values ($I = 18.6$ A when $E = 45.6$ V) and solve for k:

$$18.6 \text{ A} = k(45.6 \text{ V})$$

$$k = \frac{18.6 \text{ A}}{45.6 \text{ V}} = 0.4079 \frac{\text{A}}{\text{V}}$$

(Note that for accuracy, we're carrying one extra digit for k.)
Substituting $k = 0.4079$ into the equation:

$$I = 0.4079E \quad \text{where } I \text{ is in amps and } E \text{ is in volts}$$

Substitute $E = 60.0$ to find the current I:

$$I = (0.4079)(60.0) = 24.5 \text{ A}$$

Example 22

The flow rate Q through a pipe varies directly as the square of the diameter D of the pipe. If water flows through a pipe 3.75 inches in diameter at the rate of 162 gal/min, what is the rate of flow, under similar conditions, through a pipe 2.25 inches in diameter?

Solution: We write an equation for Q:

$$Q = kD^2$$

We substitute $Q = 162$ gal/min and $D = 3.75$ inches into the equation.

$$162 \text{ gal/min} = k(3.75 \text{ in.})^2$$

We solve for k:

$$k = \frac{162 \text{ gal/min}}{(3.75 \text{ in.})^2} = \frac{162 \text{ gal/min}}{3.75^2 \text{ in.}^2} = 11.52 \frac{\text{gal/min}}{\text{in.}^2}$$

We substitute the value for k into the equation $Q = kD^2$ to get

$$Q = 11.52D^2 \quad \text{where } Q \text{ is in gal/min and } D \text{ is in inches}$$

To find the rate of flow through the 2.25-in. pipe, we substitute $D = 2.25$ into the equation.

$$Q = (11.52)(2.25)^2 = 58.3 \text{ gal/min}$$

SECTION 26-3 Exercises   Direct Variation

Show that the following tables of values represent direct variation by calculating the ratio $\frac{y}{x}$ for each pair of values.

1.

x	1	2	3	4	5
y	3	6	9	12	15

2.

x	1	3	5	7	9
y	4	12	20	28	36

3. Given the following experimental data, show that the velocity v of a freely falling object varies directly as the time of the fall.

t (in s)	1.50	2.00	2.50	3.00	3.50
v (in m/s)	14.7	19.6	24.5	29.4	34.3

4. Given the following experimental data, show that the force F applied to a spring is proportional to the distance x the spring is stretched.

x (in cm)	1.25	2.50	3.75	5.00	6.25
F (in kg)	0.50	1.00	1.50	2.00	2.50

Constant of Proportionality

Write an equation that describes the relationship.

5. y varies directly as x.

6. p is directly proportional to q.

7. The weight w of steel balls is proportional to the number n of balls.

8. The number of parts n that a machine can make varies directly as the time t of operation.

9. The power P generated by a hydroelectric plant is directly proportional to the flow rate r through the turbines.

10. The volume V of a cylinder varies directly as its height h.

11. y varies as the cube of x.

12. p varies as the square root of r.

13. The distance s varies as the square of the time t.

14. The volume V of a cylinder is proportional to the square of the radius.

Solution of Variation Problems

Find the constant of proportionality and write the equation that describes the relationship.

15. y varies directly as x, and $y = 20$ when $x = 5$.

16. w is proportional to z, and $w = 6.4$ when $z = 4$.

17. w is directly proportional to L, and $w = 3.0$ lb when $L = 15$ in.

18. E varies directly as I, and $E = 12$ volts when $I = 5$ amps.

Write the equation and find the missing value.

19. If y is proportional to x, and $y = 50.4$ when $x = 3.20$, find y when $x = 2.50$.

20. If p varies directly as q, and $p = 5.7$ when $q = 15$, find p when $q = 35$.

Find the constant of proportionality and write the equation.

Applications

21. The area A of a circle is directly proportional to the square of the radius, and $A = 63.6$ cm^2 when $r = 4.50$ cm.

22. The power P of an engine varies directly with the square of the radius r of its piston, and a 90.0-hp engine has a piston of radius 3.00 inches.

23. The power P dissipated in a resistor varies directly as the square of the current I in the resistor, and the power dissipated in a resistor is 502 W when the current is 2.80 A.

24. The kinetic energy E_K of a moving object varies as the square of its velocity v and the kinetic energy is 9360 ft-lb at 10.0 ft/s.

Find the missing value.

25. According to Charles' Law, the volume V of a gas at constant pressure is directly proportional to it absolute temperature T (in Kelvin units). If the volume of a gas is 1220 cm³ at 295 K, what is the volume of the gas, at the same pressure, at 273 K?

26. The thermal resistance R of a wall (the R value) is proportional to the thickness T of the wall. If the R value of an 8.00-in. wall is 2.40, find the R value of a 9.00-in. wall.

27. The amount of heat Q necessary to melt ice is directly proportional to the mass m of ice melted. If it takes 717 calories of heat to melt 9.00 grams of ice, how much heat is required to melt 5.00 grams of ice?

28. A certain automobile engine has a displacement (the total volume swept out by the pistons) of 3.85 liters and delivers 67.8 horsepower. If the power is proportional to the displacement, what horsepower would you expect from a similar engine that has a displacement of 3.00 liters?

29. The exposure time t for a photograph is directly proportional to the square of the f stop. A certain photograph will be correctly exposed at a shutter speed of $\dfrac{1}{50}$ s with a lens opening of $f8(f = 8)$. What shutter speed, is required if the lens opening is changed to $f\,5.6$? (Hint: Round the shutter speed, as a decimal, to one significant digit and express as a fraction.)

30. The surface area A of a sphere varies directly as the square of its radius r. If the surface area of a sphere of radius 1.50 cm is 28.3 cm², find the surface area of a sphere of radius 2.50 cm.

31. The period T of a pendulum is proportional to the square root of its length L. If the period of a 1.50-m pendulum is 1.75 s, what is the period of a 1.25-m pendulum?

32. The rate of flow of liquid from a hole in the bottom of a tank is proportional to the square root of the liquid depth. If the flow rate is 175 liters/min when the depth is 2.75 m, find the flow rate when the depth is 2.00 m.

26-4   JOINT, INVERSE, AND COMBINED VARIATION

In this section we study quantities that vary directly as *two or more* quantities. We also study quantities that vary *inversely* as other quantities, and quantities that vary directly as some quantities and inversely as other quantities. In mathematics, science, and technology there are many such quantities.

Joint Variation

In the last section, we studied examples of a quantity that varied directly as *one* other quantity. That is, if y varied directly as x, then y was a function of the *single* variable x. But in the real world, many quantities are functions of *two or more* variables. For example, the area A of a triangle varies directly as both its base b and its height h (Fig. 26-11). We call this *joint variation*.

In general, if y varies directly as x, and also if y varies directly as w, we say that

$$y \text{ varies } \textit{jointly} \text{ as } x \text{ and } w.$$

■ **FIGURE 26-11**
Joint variation.

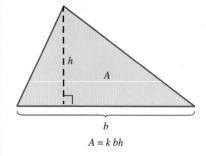

$A = k\,bh$

Or, we may say that

y is proportional to x and w.

We may write

$$y \propto xw$$

This means that y varies directly as the product of x times w. Therefore, the three variables (y, x, and w) and the constant of proportionality, k, can be related by the equation:

JOINT VARIATION	$y = k\,xw$ where k is the constant of proportionality	49

Example 23

(a) If p varies jointly as q and r, then

$$p = k\,qr$$

(b) The area A of a triangle varies jointly as the base b and the height h (Fig. 26-11), so

$$A = k\,bh$$

■ **FIGURE 26-12**
Joint variation.

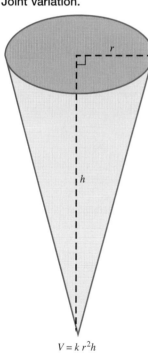

$V = k\,r^2h$

Example 24

The volume V of a conical pendulum weight is jointly proportional to the height h and the square of the radius r (Fig. 26-12), so

$$V = k\,r^2h$$

This reminds us of the familiar formula for the volume of a cone, Eq. 92:

$$V = \frac{1}{3}\pi r^2 h$$

We see that the constant of proportionality is

$$k = \frac{1}{3}\pi$$

Inverse Variation

We recall that when two quantities vary directly, and one quantity increases, then the other increases. Sometimes, however, quantities are so related that if one *increases,* the other *decreases.*

Example 25

A woman regularly drives between two towns, 100 miles apart. The rate of travel (speed) and the time of travel are related. If we *increase* the rate, then the travel time *decreases.* Similarly, if we decrease the rate, then the travel time increases. We say that, for a car traveling a constant distance with uniform motion, the travel time t varies inversely as the rate r.

In general, if two variables, x and y, are related in such a way that $y = \dfrac{k}{x}$, then we say that y *varies inversely as x,* or y *is inversely proportional to x.* We may write

$$y \propto \frac{1}{x}$$

INVERSE VARIATION	$y = \dfrac{k}{x}$ where k is the constant of proportionality	50

Example 26

(a) If s varies inversely as w, then

$$s = \frac{k}{w}$$

(b) The current I in a resistor is inversely proportional to the resistance R, so

$$I = \frac{k}{R}$$

(c) The force F of attraction between two bodies varies inversely as the square of the distance d between them (Fig. 26-13).

$$F = \frac{k}{d^2}$$

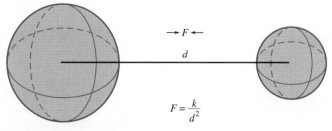

$$F = \frac{k}{d^2}$$

■ **FIGURE 26-13**
Inverse variation.

If y varies inversely as x, then $y = \dfrac{k}{x}$, where k is a constant. If we multiply both sides of this equation by x, we find that

$$k = xy$$

or that k is the product of x times y. In other words, in inverse variation the *product* of the quantities is a constant.

Example 27 ─────────

Show from the following data that the current I in a resistor is inversely proportional to the resistance R.

R (in ohms)	10	25	50	100	250
I (in amps)	25	10	5	2.5	1

Solution: To show that I is inversely proportional to R, we calculate the product RI for each pair of numbers. If the product is a constant, then I varies inversely as R.

$$10(25) = 25(10) = 50(5) = 100(2.5) = 250(1) = 250$$

Since

$$RI = 250$$

for all R and I, then

$$k = 250$$

and the current I varies inversely as the resistance R, so that

$$I = \frac{250}{R}$$

───────────

Combined Variation

If one quantity varies with two or more quantities in a way that combines direct and inverse variation, we have what is called *combined variation*. In general, if y varies directly as x and inversely as w, we may write

$$y \propto \frac{x}{w}.$$

We write an equation by placing the variables which vary directly in the numerator and the variables which vary inversely in the denominator. Of course, we must include the constant of proportionality in the numerator.

COMBINED VARIATION	If y varies directly as x and inversely as w then $\qquad y = \dfrac{kx}{w}$	51

───────────

Example 28 ─────────

(a) If p varies inversely as q and directly as r, then

$$p = \frac{kr}{q}$$

(b) The resistance R in a wire is directly proportional to the length L and inversely proportional to the square of the diameter d.

$$R = \frac{kL}{d^2}$$

───────────

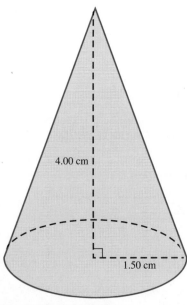

■ FIGURE 26-14

Solution of Variation Problems

To solve a variation problem for an unknown quantity, we use the method described in the last section.

Example 29

The volume V of a cone varies as the square of the radius r of the base, and directly as the height h. When the radius is 1.50 cm and the height is 4.00 cm, the volume is 9.42 cm³ (Fig. 26-14). Find the constant of proportionality and write the equation for V.

Solution: We use Eq. 49 for joint variation to write the equation:

$$V = k\, r^2 h$$

We substitute the given values ($V = 9.42$ cm³, $r = 1.50$ cm, and $h = 4.00$ cm) into the equation above:

$$9.42 \text{ cm}^3 = k\, (1.50 \text{ cm})^2 (4.00 \text{ cm})$$

Solving for k:

$$k = 1.057$$

Substituting k into the equation $V = k\, r^2 h$, we get

$$V = 1.057\, r^2 h$$

Applications

Many applications involve variation.

Example 30

Boyle's Law states that for a confined gas at a constant temperature, the pressure p is inversely proportional to the volume V. If the air in a cylinder is at a pressure of 14.7 lb/in.² and occupies a volume of 225 in.³, find the pressure when it is compressed to 125 inches³.

Solution: We use Eq. 50 for inverse variation to write the equation:

$$p = \frac{k}{V}$$

where k is the constant of proportionality. To find k, we substitute the given values ($p = 14.7$ lb/in.² and $V = 225$ in.³) into the equation and solve for k.

$$14.7 \text{ lb/in.}^2 = \frac{k}{225 \text{ in.}^3}$$

$$k = \left(14.7 \frac{\text{lb}}{\text{in.}^2}\right)(225 \text{ in.}^3) = 3307.5 \text{ lb-in.}$$

To find the equation, we substitute the value for k in the equation $p = \dfrac{k}{V}$.

$$p = \frac{3307.5}{V}$$

where p is in lb/in.² and V is in inches³.

To find the unknown pressure, we substitute the volume associated with it ($V = 125$ in.^{3}) into the equation and solve.

$$p = \frac{3307.5}{125} = 26.5 \text{ lb/in.}^2$$

Joint Variation

Show that the following table of values represent that z varies jointly as x and y by calculating the ratio $\dfrac{z}{xy}$ for each set of values.

1.

x	1	2	3	4	5
y	2	4	6	8	10
z	4	16	36	64	100

2.

x	2	4	6	8	10
y	1	3	5	7	9
z	6	36	90	168	270

3. Given the following data, show that the interest i earned on a savings account varies jointly as the interest rate r and the time t.

r (in %)	6.28	6.54	6.86	6.98
t (in yr)	0.25	0.50	1.25	1.50
i (in $)	3.14	6.54	17.15	20.94

4. Given the following experimental data, show that the expansion E of a metal rod varies jointly as its original length L and the temperature change T.

L (in cm)	25	50	50	75
T (in °C)	10	10	20	20
E (in cm)	0.005	0.010	0.020	0.030

Write an equation.

5. p varies jointly as q and r.

6. x is proportional to y and to z.

7. The expansion E of a metal rod varies jointly as its original length L and the temperature change T.

8. The kinetic energy E_K of a moving object is proportional to its mass m and the square of its velocity v.

Inverse Variation

Show that the following table of values represent that y varies inversely as x by calculating the product xy for each pair of values.

9.

x	1	2	3	4	5
y	60	30	20	15	12

10.

x	2	4	6	8	10
y	30	15	10	7.5	6.0

Write an equation.

11. z varies inversely as y.

12. p is inversely proportional to q.

13. y is inversely proportional to the square of z.

14. a varies inversely as the cube of b.

15. The capacitive reactance X_c of a circuit varies inversely as the capacitance C of the circuit.

16. The amount of light I that falls on an object is inversely proportional to the square of the distance d from the source of light.

Combined Variation

Write an equation.

17. y is inversely proportional to x and directly proportional to z.

18. p varies directly as q and inversely as r.

19. Electric current I flowing in a circuit varies directly as the voltage V and inversely as the resistance R.

20. The intensity of illumination I at a given point is directly proportional to the intensity of the light source L and inversely proportional to the square of the distance d from the light source.

Solution of Variation Problems

Write the equation and find the unknown value.

21. x varies jointly as y and z, and $x = 36$ when $y = 3$ and $z = 4$. Find x when $y = 4$ and $z = 5$.

22. a is directly proportional to b^2 and to c, and $a = 4.5$ when $b = 1.5$ and $c = 4$. Find a when $b = 2.5$ and $c = 6$.

23. If y varies directly as x and inversely as z, and $y = 12$ when $x = 3$ and $z = 4$, find y when $x = 5$ and $z = 8$.

24. a is inversely proportional to the square of b and directly proportional to c. If $a = 7.5$ when $b = 12$ and $c = 10$, find a when $b = 10$ and $c = 12$.

Applications

Solve.

25. The amount A paid to a work crew varies jointly as the number n of persons working and the length of time t worked. If a crew of 4 workers earns \$150.00 in 5.00 hours, how much will a crew of 7 workers earn in 6.00 hours?

26. The amount W of work done varies jointly as the number n of persons working and the length of time t worked. If 6 stonemasons can build 60.0 ft of a stone wall in 4.00 hours, how long a wall can 8 stonemasons build in 3.00 hours?

27. The time t needed to empty a vertical cylindrical tank varies as the square root of the height h of the tank and the square of the radius r. If it takes 81.0 minutes to empty a tank whose height is 6.00 ft and whose radius is 2.25 ft, how long would it take to empty a tank 5.00 ft high with radius 1.44 ft?

28. The weight W of a solid metal cylinder varies jointly as the height h and the square of the radius r. If a solid lead cylinder of height 3.50 cm and radius 1.20 cm weighs 179 g, find the weight of a lead cylinder of height 2.50 cm and radius 1.10 cm.

29. When an object travels a constant distance in uniform motion, the travel time t is inversely proportional to the velocity v. If a driver usually makes a trip in 1.5 hours at an average speed of 88 km/h, find the traveling time if the speed were reduced to 66 km/h.

30. The wavelength w of a radio wave varies inversely as its frequency f. If the wavelength is 70.0 m when the frequency is 4250 kHz, find the wavelength when the frequency is 4560 kHz.

31. Newton's law of gravitation states that any two bodies attract each other with a force F that is inversely proportional to the square of the distance d between them. The force of attraction between two certain spheres is 2.25×10^{-5} dynes when the spheres are 24.0 cm apart. What is the force of attraction when the spheres are 15.0 cm apart?

32. The intensity I of sound is inversely proportional to the square of the distance d from the source. If the intensity is 3.6×10^{-7} W/m^2 at a distance of 6.0 m from a speaker, find the intensity at 9.0 m.

33. When an electric current flows through a wire, the resistance R varies directly as the length L and inversely as the cross-sectional area A of the wire. If the length of the wire is tripled and the cross-sectional area is doubled, by what factor will the resistance change?

34. The volume V of a given weight of gas varies directly as the absolute temperature T and inversely as the pressure p. If the volume is 3.76 m³ when the pressure is 275 kPa and the absolute temperature is 315 K, find the volume when the pressure is 285 kPa and the absolute temperature is 295 K.

35. The number n of vibrations per second made when a stretched wire is plucked varies directly as the square root of the tension T in the wire, and inversely as the length L. If a 1.50-m wire vibrates 226 times a second when the tension is 125 N, find the number of vibrations per second if the wire is shortened to 1.25 m and the tension increased to 175 N.

36. Newton's law of gravitation states that every body in the universe attracts every other body with a force that varies directly as the product of their masses and inversely as the square of the distance between them. By what factor will the force change when the distance is tripled and each mass is doubled?

■ Chapter 26 REVIEW EXERCISES

Solve the proportion.

1. $\dfrac{3}{7} = \dfrac{x}{42}$ **2.** $\dfrac{x}{21} = \dfrac{5}{6}$ **3.** $\dfrac{9}{x} = \dfrac{x}{16}$ **4.** $\dfrac{8}{9} = \dfrac{x}{136 - x}$

Simplify the ratio in fraction form.

5. 9 kg to 15 kg **6.** 35 ft to 7 ft **7.** 600 g to 3 kg

Write an equation.

8. y varies jointly as x and z.

9. p varies inversely as the square of r.

10. t varies directly as the square root of s and inversely as r.

Write an equation that describes the relationship.

11. The thermal resistance R of a wall is proportional to the thickness T of the wall.

12. According to Newton's second law, the acceleration a of a body is directly proportional to the force F applied to it.

Solve.

13. A 48-ft beam must be separated into two parts so that the parts are in a ratio of 5 to 3. How long must the parts be?

14. A settling tank contains 6250 gallons of water and can be emptied in 16.0 hours. How much water was in the tank if it was emptied in 9.75 hours? (Assume the volume is directly proportional to time.)

15. Given the following experimental data, show that the resistance R of a wire is directly proportional to the length L of the wire.

L (m)	0.5000	1.500	2.500	3.500	4.500
R (ohms)	57.50	172.5	287.5	402.5	517.5

16. Steel is made by mixing nickel and iron. If 10 tons of nickel is mixed with 95 tons of iron, find the ratio of iron to nickel.

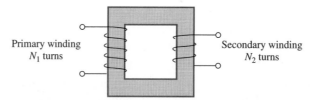

■ FIGURE 26-15
Transformer.

17. The turn ratio of a transformer (Fig. 26-15) is

$$a = \frac{\text{number of turns in primary winding}}{\text{number of turns in secondary winding}} = \frac{N_1}{N_2}$$

Find the turn ratio of a transformer that has 250 turns in the secondary winding and 10,000 turns in the primary winding.

18. Write the following as a proportion in fraction form and then use the fundamental principle of proportions to show that the proportion is true.

63 is to 81 as 7 is to 9

Give the general equation and find the missing value.

19. The interest i earned on a savings account varies jointly as the interest rate r and the time t. A certain savings account earned \$176.40 interest at a rate of 6.72% for $\frac{3}{4}$ of a year. How much interest would the account earn at 6.68% for $1\frac{1}{4}$ year?

20. The rate r of heat loss through insulation is inversely proportional to the thickness t of the insulation. If 2200 Btu/h are lost through a 3.5-in. thickness of insulation, what would the rate of heat loss be through 5.0 in. of insulation?

21. The time required to fill a tank varies inversely as the square of the diameter of the pipe used. If it takes 3.5 hours to fill a tank using a 2.5-in. pipe, how long would it take to fill the tank using a 4.5-in. pipe?

22. When an electric current flows through a wire, the resistance R varies directly as the length and inversely as the square of the diameter of the wire. If 958 m of wire 4.00 mm in diameter has a resistance of 19.7 ohms, find the resistance in a wire 825 m long with a diameter of 4.50 mm.

Write the equation and find the missing value.

23. If E varies directly as I, and $E = 24.8$ volts when $I = 15.2$ amps, find E when $I = 20.5$ amps.

24. The amount of sulfuric acid in a solution varies directly as the concentration. If 3.65 liters of sulfuric acid is used to make a quantity of a 12.0% solution, how many liters is needed to make the same quantity of a 15.0% solution?

25. If w is proportional to r^3, and $w = 2.13$ lb when $r = 1.25$ in., find w when $r = 1.50$ in.

26. The horizontal distance s traveled by a projectile is directly proportional to the square of its initial velocity v. If a projectile travels 728 m horizontally when its initial velocity is 157 m/s, how far will it travel when its initial velocity is 218 m/s?

27. The velocity v of a ship varies directly as the cube root of the power P used to propel the ship. If 6570 hp can propel the ship at 15.0 knots, what will be the velocity under 4950 hp?

28. Find the constant of proportionality and give the general equation of variation that describes this relationship: y varies directly as x, and $y = 9.6$ when $x = 1.6$.

WRITING

29. Suppose your company makes plastic trays and is planning new ones with dimensions double those now being made. Your company president is convinced that the new trays will need only twice as much plastic as the older version. "Twice the size, twice the plastic," he proclaims, and no one is willing to challenge him. Write a presentation to the president in which you tactfully point out that he is wrong and you explain that the new trays will require eight times as much plastic.

TEAM PROJECT

30. We saw that the volumes (and hence the weights) of solids are proportional to the cube of corresponding dimensions. Applying that idea, suppose that a sporting goods company has designed a new type of ski based on the statement:

    The weights of people of similar build are proportional to the cube of their heights.

    The goal of this project is to prove or disprove the given statement. Gather data from students on your campus to reach a conclusion. In addition to a person's height and weight, it may be useful to note a person's build, such as thin, average, heavy. Should you note gender and/or age? Make a conclusion and prove it.

27 Exponential Functions

In this chapter we define the exponential function, examine its graph, and review the laws of exponents. We also study two very important and special examples of exponential functions: exponential growth and exponential decay. We look at applications of the exponential function, such as biological growth, compound interest, population growth, radioactive decay, electromotive force, heat loss, inflation, electric current, light intensity, atmospheric pressure, and capacitance.

27-1 INTRODUCTION TO EXPONENTIAL FUNCTIONS

In this section we define the exponential function and see how the laws of exponents are applied with exponential functions.

Definitions

An *exponential function* is a function in which the independent variable is in the exponent. The quantity that is raised to the power is called the *base*.

Example 1

Some exponential functions are

$$y = 4^x, \; y = 2 \cdot 3^x, \; y = b^x, \text{ and } y = 3e^{2x}$$

where b and e represent positive constants, not equal to 1.

Note that the exponential function is different from the *power function* of Chapter 12. In the power function $y = x^n$, the exponent n is a constant and the base x is a variable. In the exponential function $y = b^x$, the exponent x is a variable and the base b is a constant.

Example 2

The function $y = 3^x$ is an *exponential* function, since the variable is in the exponent. However, the function $y = x^3$ is a *power* function, since the variable is in the base.

Laws of Exponents

All of the laws of exponents previously studied in Chapter 20 are also true here. We rewrite the laws below for expressions in which the base is a positive constant and the exponent is a variable.

LAWS OF EXPONENTS	Product	$b^x \cdot b^y = b^{x+y}$	19
	Quotient	$\dfrac{b^x}{b^y} = b^{x-y}$	20
	Power	$(b^x)^y = b^{xy}$	21
	Product Raised to a Power	$(ab)^x = a^x b^x$	22
	Quotient Raised to a Power	$\left(\dfrac{a}{b}\right)^x = \dfrac{a^x}{b^x}$	23
	Negative Exponent	$b^{-x} = \dfrac{1}{b^x}$	25

The following examples illustrate the laws of exponents for exponents that are variables.

Example 3

$$2^x 2^y = 2^{x+y}$$

Example 4

$$\frac{3^z}{3^y} = 3^{z-y}$$

Example 5

$$(4^x)^y = 4^{xy}$$

Example 6

$$(5a)^x = 5^x a^x$$

Example 7

$$\left(\frac{2}{3}\right)^x = \frac{2^x}{3^x}$$

Example 8

$$3^{-x} = \frac{1}{3^x}$$

The next two examples follow from the laws of exponents above and the fact that

$$\frac{1}{2} = 2^{-1}$$

Example 9 ─────

$$\left(\frac{1}{2}\right)^x = (2^{-1})^x = 2^{-x}$$

Example 10 ─────

$$\left(\frac{1}{2}\right)^{-x} = (2^{-1})^{-x} = 2^x$$

SECTION 27-1   Exercises   Introduction to Exponential Functions

Definitions

1. Define an exponential function.
2. What is the difference between an exponential function and a power function? Give an example of each.

Laws of Exponents

Use the laws of exponents to complete these equations.

3. $5^x 5^y = ?$

4. $\dfrac{6^y}{6^z} = ?$

5. $(8^y)^z = ?$

6. $(9m)^x = ?$

7. $\left(\dfrac{5}{3}\right)^x = ?$

8. $\left(\dfrac{1}{3}\right)^x = ?$

9. $2^{-x} = ?$

10. $8^{-x} = ?$

11. $\left(\dfrac{1}{4}\right)^{-x} = ?$

12. $\left(\dfrac{1}{5}\right)^{-x} = ?$

27-2 ───── GRAPH OF THE EXPONENTIAL FUNCTION

In this section we graph exponential functions and solve some applications involving exponential functions.

We make a graph of an exponential function in the same way we graphed other functions in Chapter 12. We choose values of the independent variable x and compute the corresponding value of the dependent variable y. We then plot the resulting ordered pairs and connect the points with a smooth curve.

Example 11 ─────

Graph the exponential function $y = 3^x$ for x from -2 to $+2$.

Solution: We compute the values of y and obtain the following table of values:

x	-2	-1	0	1	2
y	$\dfrac{1}{9}$	$\dfrac{1}{3}$	1	3	9

These pairs are plotted and connected with a smooth curve in Fig. 27-1. Note that the y intercept is 1 and there is no x intercept. In fact, the x axis is an asymptote. (Recall that an asymptote is a line that a curve approaches but never reaches.) Also as x increases, y increases, and the curve rises and gets steeper. We say that y is an *increasing* function.

■ **FIGURE 27-1**
An exponential function.

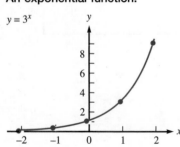

$y = 3^x$

Example 12

Use the graphing calculator to graph $y = 3^x$ for x from -2 to $+2$.

Solution: We are given the domain of the function (x is between -2 and 2). In the table of values above, y is between $\frac{1}{9}$ and 9. We set the window so that

```
Xmin=-2
Xmax=2
Xscl=1
Ymin=0
Ymax=9
Yscl=1
```

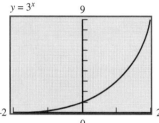

$y = 3^x$

FIGURE 27-2

We enter the function $y = 3^x$ and graph (Fig. 27-2).

Example 13

Graph $y = \left(\dfrac{1}{2}\right)^x$ for x from -3 to 3.

Solution: We compute the values of y and make the table of values:

x	-3	-2	-1	0	1	2	3
y	8	4	2	1	$\frac{1}{2}$	$\frac{1}{4}$	$\frac{1}{8}$

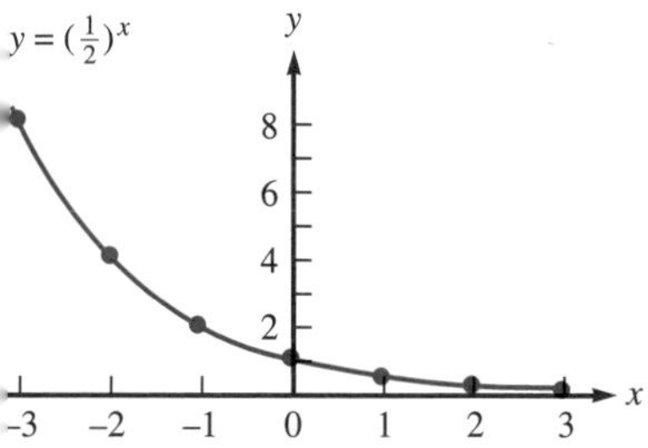

$y = (\frac{1}{2})^x$

FIGURE 27-3

We plot these points and connect them with a smooth curve in Fig. 27-3. Notice that the y intercept is 1 again, there again is no x intercept, and the x axis is an asymptote. Also, as *x increases, y decreases* and the curve "falls" and gets "flatter." We say that y is a *decreasing* function.

Example 14

Use a graphing calculator to graph $y = 2^{-x}$ for x from -3 to 3.

Solution: We set the window so that

```
Xmin=-3
Xmax=3
Xscl=1
Ymin=0
Ymax=8
Yscl=1
```

FIGURE 27-4

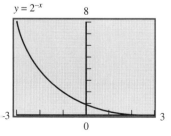

$y = 2^{-x}$

We enter the function and graph (Fig. 27-4). Note that this graph of $y = 2^{-x}$ is identical to the graph of $y = \left(\dfrac{1}{2}\right)^x$ in Fig. 27-3. This is because the functions are the same function, since

$$\left(\frac{1}{2}\right)^x = \frac{1^x}{2^x} = \frac{1}{2^x} = 2^{-x}$$

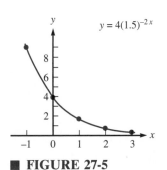

$y = 4(1.5)^{-2x}$

■ FIGURE 27-5

—— Example 15 ——

Graph $y = 4(1.5)^{-2x}$ for x from -1 to 3.

Solution: We make a table of values, using a calculator to compute the y values, and rounding to the nearest tenth.

x	-1	0	1	2	3
y	9	4	1.8	0.8	0.4

The points are plotted and connected with a smooth curve in Fig. 27-5.

Applications

Applications of exponential functions include biological growth of organisms such as bacteria and viruses.

—— Example 16 ——

A biologist has a culture consisting of about 1500 bacteria, which doubles its number every hour.

(a) Write an equation for the number of bacteria n after t hours.
(b) Find the number of bacteria in the culture after 8.0 hours.
(c) Graph the equation.
(d) Graphically find the number of bacteria after 6.0 hours.

Solution:

(a) To find an equation for the number of bacteria n as a function of time t, we make a table to see if there is a pattern:

t	0.0	1.0	2.0	3.0	...	t
n	1500	1500(2)	1500(2)(2)	1500(2)(2)(2)	...	$1500(2^t)$

Indeed there is a pattern and we see that an equation for n as a function of t is

$$n = 1500(2^t)$$

(b) To find how many bacteria there are in the culture after 8.0 hours, we substitute $t = 8.0$ into the equation above:

$$n = 1500(2^{8.0}) = 380{,}000 \text{ bacteria, after 8.0 hours}$$

(c) We graph this function on the graphing calculator. Using y as a function of x instead of n as a function of t, we enter the function

$$y = 1500(2^x)$$

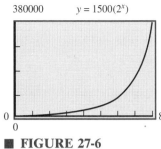

FIGURE 27-6

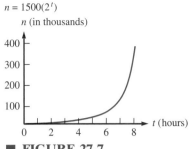

FIGURE 27-7

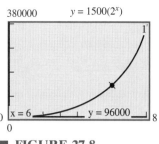

FIGURE 27-8

and set the window so that

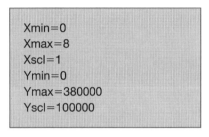

```
Xmin=0
Xmax=8
Xscl=1
Ymin=0
Ymax=380000
Yscl=100000
```

and graph (Fig. 27-6). To graph the function

$$n = 1500(2^t)$$

on paper, we label the axes with n and t, and label the scales (Fig. 27-7).

(d) We find how many bacteria there are after 6.0 hours graphically. We find the point where $x = 6.0$ and read the y value. Many graphing calculators will find this value for you (Fig. 27-8). So, after 6.0 hours, there are about 96,000 bacteria in the culture.

SECTION 27-2 Exercises   Graph of the Exponential Function

Graph each exponential function for x from -3 to 3.

1. $y = 2^x$ **2.** $y = 4^x$

3. $y = -3^x$ **4.** $y = -2^x$

5. $y = 3(2^x)$ **6.** $y = 2(4^x)$

7. $y = 3^{-x}$ **8.** $y = 4^{-x}$

9. $y = 2^{3x}$ **10.** $y = 2^{-2x}$

11. $y = \left(\dfrac{2}{3}\right)^x$ **12.** $y = \left(\dfrac{3}{2}\right)^{-x}$

13. $y = (2.3)^{-x}$ **14.** $y = (1.4)^x$

15. $y = 4(1.5)^{3x}$ **16.** $y = 0.3(2.1)^{-2x}$

Solve. If answers are not exact, round to two significant digits.

17. A biologist has a culture of 2500 bacteria, which triples its number every hour.
    (a) Write an equation for the number of bacteria n after t hours.
    (b) Find the number of bacteria in the culture after 5.0 hours.
    (c) Graph the equation for t from 0 to 5 hours.
    (d) Graphically find the number of bacteria after 3.5 hours.

18. A biologist has a culture of 200 bacteria, which quadruples its number every day.
    (a) Write an equation for the number of bacteria n after t days.
    (b) Find the number of bacteria in the culture after 4.0 days.
    (c) Graph the equation for t from 0 to 4 days.
    (d) Graphically find the number of bacteria after 2.5 days.

19. The radioactive isotope technetium-99m is used to identify brain tumors. Half of it disintegrates in 6.0 hours. If a physician injects 4.8 mg, then the amount present y (in mg) after t hours is found using the equation

$$ y = 4.8 \left(\frac{1}{2} \right)^{t/6} $$

    (a) How much technetium-99m will be present after 12.0 hours?
    (b) Graph y as a function of t, for t from 0 to 12 hours.
    (c) Graphically find how many hours must elapse for the amount present to be 2.0 mg.

20. A census in Ethiopia found that the population was about 55 million people. It was expected to double every 22 years, so that the expected population P, in millions of people, was found using

$$ P = 55(2^{t/22}) $$

    where t is the number of years after the census.
    (a) Find the expected population in 50.0 years.
    (b) Graph the expected population as a function of time, for t from 0 to 50 years.
    (c) After how many years was the population expected to reach 100 million people?

21. A yeast manufacturer finds that the yeast will double in weight every hour.
    (a) Find the weight of yeast the manufacturer must start with to obtain 650 kilograms in 4.0 hours.
    (b) Using the result in part (a), write an equation for the weight w of yeast as a function of time t.
    (c) What is the weight of yeast after 2.5 hours?
    (d) Graph the equation for t from 0 to 4 hours.
    (e) Graphically find how many hours it takes to produce 500 kg of yeast.

22. A pharmaceutical company is growing an organism to be used in a vaccine. The organism grows at a ratè so it triples every day.
    (a) Find the number of organisms the company must start with to have 2800 units after 6.0 days.
    (b) Using the result in part (a), write an equation for the number n of organisms as a function of time t.
    (c) How many organisms will there be after 10 days?
    (d) Graph n as a function of t, for t from 0 to 10 days.
    (e) Graphically find how many days it takes to grow 100,000 organisms.

EXPONENTIAL GROWTH

In the last section, we looked at growth that occurred in *steps* (by doubling or tripling periodically). In this section we look at a type of growth that is *continuous;* it is called *exponential growth.* The irrational number *e* is introduced, as an important constant in the equation for exponential growth. We study the graphs of exponential functions which use *e* as the base. We also look at applications, including compound interest and inflation.

Compound Interest

When an amount of money (the *principal*) is invested for a period of time, it will earn *interest* at a specified rate. If this interest is then *added* to the principal and both continue to earn interest, we have what is called *compound interest,* the type of interest commonly given on bank accounts. The bank determines how often the interest is compounded; it may be compounded every year, month, day, hour, minute, second, or even continuously. If the interest is compounded *m* times per year at an annual interest rate *r*, then the original amount *a* will accumulate to an amount *y* in *t* years, according to the formula

$$y = a\left(1 + \frac{r}{m}\right)^{tm}$$

The interest rate *r* is generally given as a percent, but it must be expressed as a decimal in the equation. Also, the units for time *t* and rate *r* must agree; for example, if the rate is an annual rate, then the time must be in years.

Example 17
Find the amount to which $700 will accumulate in 4 years at an annual interest rate of 6.5%, if the interest is compounded annually. Round to the nearest cent.

Solution: Converting the rate *r* to a decimal,

$$r = 6.5\% = 0.065$$

Interest is compounded annually (once a year), so $m = 1$. Using the formula for compound interest with $a = 700$, $r = 0.065$, $m = 1$, and $t = 4$,

$$y = 700\left(1 + \frac{0.065}{1}\right)^4$$

$$= 900.53$$

So if the interest is compounded annually, $700 will accumulate to $900.53 after 4 years.

Example 18
Find the amount to which $700 will accumulate in 4 years at an annual interest rate of 6.5%, if the interest is compounded monthly. Round to the nearest cent.

Solution: Interest is compounded monthly, so it is compounded 12 times each year, making $m = 12$. Using the formula for compound interest with $a = 700$, $r = 0.065$, $m = 12$, and $t = 4$,

$$y = 700\left(1 + \frac{0.065}{12}\right)^{4(12)}$$

$$= 907.21$$

So, if the interest is compounded monthly, $700 will accumulate to $907.21 after 4 years. Compare that to the $900.53 we found in Example 17 with annual compounding!

But interest can even be compounded *continuously!* When interest is compounded continuously, then the accumulated amount *y* grows *exponentially.*

Exponential Growth

If a quantity grows continuously in this manner, we use the equation for *exponential growth:*

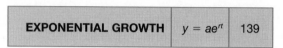

| EXPONENTIAL GROWTH | $y = ae^{rt}$ | 139 |

where *a* is the original amount of a quantity which is growing continuously at a rate *r* and *y* is the amount present after time *t*. The base *e* is a constant, described below. We say that *y* *increases* (or *grows*) *exponentially* with respect to *t*.

The number *e*

So what is this number *e*, which is the base in the exponential equation for continuous growth? The number *e* is an irrational number and has been calculated to millions of digits. The first 16 digits of *e* are

$$e = 2.718281828459045$$

Evaluating e^x

With a calculator, we can evaluate *e* or any power of *e*.

Example 19

Use a calculator to find the value of the number *e*.

Solution: Since $e = e^1$, we use the $\boxed{e^x}$ key on a calculator to find

$$e = e^1 = 2.718281828 \qquad \text{to 10 significant digits}$$

Example 20

Use a calculator to approximate $e^{2.75}$.

Solution: Using the $\boxed{e^x}$ key and rounding to three significant digits,

$$e^{2.75} = 15.6$$

Example 21

Find the amount to which $700 will accumulate in 4 years at an annual interest rate of 6.5%, if the interest is compounded continuously. Round to the nearest cent.

Solution: Since the interest is being compounded continuously, the accumulated amount grows exponentially and is given by Eq. 139, with $a = 700$, $r = 0.065$, and $t = 4$:

$$y = 700e^{0.065(4)} = 907.85$$

So, $700 will accumulate to $907.85 in 4 years. Compare that to the $907.21 we found with monthly compounding. We earn *a little more* with continuous compounding!

Graph of Exponential Growth

We saw in Sec. 27-2 that an exponential function $y = b^x$ is an increasing function if $b > 1$. The function $y = e^x$ is a special exponential function in which the base is the positive number e. Since $e > 1$, then $y = e^x$ is an increasing function. The graph of $y = e^x$ rises to the right and gets steep, and the x axis is an asymptote.

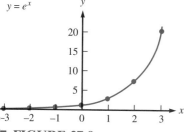

$y = e^x$

■ FIGURE 27-9
Exponential growth.

Example 22

Graph the function $y = e^x$ for x from -3 to 3.

Solution: We make a table of values and use the calculator to find the y values, rounding to tenths.

x	-3	-2	-1	0	1	2	3
y	0.0	0.1	0.4	1	2.7	7.4	20.1

We plot the points and connect them with a smooth curve (Fig. 27-9). Note that the function is increasing, the y intercept is 1, the x axis is an asymptote, and the values of y are all positive.

Example 23

Graph $y = 3e^{2x}$ for x from -2 to 2.

Solution: We use a graphing calculator to graph this function. But first we determine the range of values for y. The values of y will be positive, and the largest y value will be when $x = 2$, since this is an increasing function. So, we calculate the y value when $x = 2$:

$$3e^{2(2)} = 164$$

We set the window so that

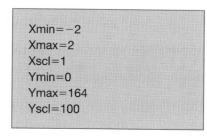

```
Xmin=-2
Xmax=2
Xscl=1
Ymin=0
Ymax=164
Yscl=100
```

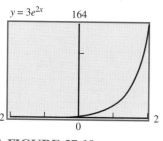

$y = 3e^{2x}$ 164

■ FIGURE 27-10

and graph (Fig. 27-10).

Example 24

The prices of some goods rise exponentially due to inflation. The price of a certain new radio is $125. Assume that the rate of inflation remains constant at 4.50% per year.

(a) Find the price of the radio 5.00 years from now.

(b) Graph the price p as a function of time t from 0 to 25 years.

(c) Graphically, find the price of the radio 20.0 years from now.

Solution:

(a) Since price is increasing exponentially, we use Eq. 139, with $y = p$, $a = 125$, $r = 4.50\% = 0.0450$, and $t = 5.00$

$$p = 125e^{0.0450(5.00)}$$
$$= \$157$$

(b) We graph the function on the graphing calculator. The price p as a function of time t is

$$p = 125e^{0.0450t}$$

We enter the function as

$$y = 125e^{0.0450x}$$

To find the range of the function, we evaluate y for $x = 25$ to find the maximum y value:

$$y = 125e^{0.0450(25)}$$
$$= \$385$$

So we set the window so that

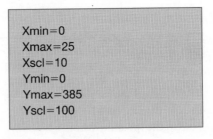

```
Xmin=0
Xmax=25
Xscl=10
Ymin=0
Ymax=385
Yscl=100
```

and graph (Fig. 27-11).

(c) To find the price of the radio 20.0 years from now, we find the point on the graph where x is 20.0 (Fig. 27-12), and read the y value. The price of this radio will be \$307 in 20.0 years.

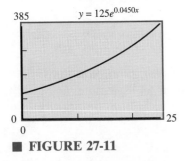

■ **FIGURE 27-11**

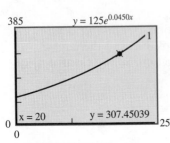

■ **FIGURE 27-12**

Note: Many graphing calculators will evaluate a function at a specified value. Check your calculator manual for instructions.

SECTION 27-3 | Exercises | Exponential Growth

Evaluating e^x

Use a calculator to evaluate these expressions, and round to three significant digits.

1. e^2 **2.** e^3 **3.** $e^{1.36}$ **4.** $e^{2.75}$ **5.** $e^{-0.234}$

6. $e^{-1.75}$ **7.** $3e^5$ **8.** $7e^{-2}$ **9.** $4.56e^{0.765}$ **10.** $78.3e^{-8.56}$

Compound Interest

Round to the nearest cent.

11. Find the amount to which \$650 will accumulate in 4 years at an annual interest rate of 6.5%, if the interest is compounded (a) annually, (b) monthly, and (c) continuously.

12. Find the amount to which \$2800 will accumulate in 5 years at an annual interest rate of 7.5%, if the interest is compounded (a) annually, (b) daily, and (c) continuously.

13. Find the amount to which $1.00 will accumulate in 10 years at an annual interest rate of 10%, if the interest is compounded (a) annually, (b) monthly, (c) daily, and (d) continuously.

14. Find the amount to which $1.00 will accumulate in 10 years at an annual interest rate of 12%, if the interest is compounded (a) annually, (b) monthly, (c) daily, and (d) continuously.

15. A savings bank is offering an annual interest rate of 9.5%, compounded monthly. How much do you need to invest to have an accumulated amount of $5250 at the end of 5 years?

16. A savings bank is offering an annual interest rate of 8.5%, compounded daily. How much do you need to invest to have an accumulated amount of $1550 at the end of 3 years?

Graph of Exponential Growth

17. Graph on the same set of axes, for x from -2 to 2:
    (a) $y = e^x$ (b) $y = e^{2x}$ (c) $y = 2e^x$

18. Graph on the same set of axes, for x from -2 to 2:
    (a) $y = e^x$ (b) $y = 3e^x$ (c) $y = e^{3x}$

19. Graph $y = 2e^{3x}$, for x from -2 to 2.

20. Graph $y = 4e^{2x}$, for x from -2 to 2.

Round to three significant digits.

Applications

21. The population of the United States on July 1, 1995, was estimated to be 264 million people. The growth rate was estimated to be 1.02% per year. If the population grows exponentially, what will the population be on July 1, 2000?

22. If the rate of inflation is 3.90% per year, how much would a stereo system that now costs $575 be expected to cost 4.00 years from now, assuming that the rate of inflation remains constant and that cost increases exponentially?

23. A yeast manufacturer finds that a certain kind of yeast will grow exponentially at a rate of 13.0% per hour.

    (a) How many pounds of yeast will the manufacturer have at the end of 10.0 hours, if there is 45.0 pounds originally?

    (b) Graph the weight of yeast as a function of t, for t from 0 to 10 hours.

    (c) Graphically find the weight of yeast after 7.25 hours.

    (d) Graphically find how many hours it takes to grow 100 pounds of yeast.

24. A utility company claims that the demand for electricity in its region is growing exponentially at an annual rate of 11.5%. At present, the company's customers use 275 billion kWh of electricity.

    (a) Assuming that the customers make no effort to conserve electricity, how much electrical power will be needed in 10 years?

    (b) Graph the demand for electricity as a function of time, for t from 0 to 10 years.

    (c) Graphically find the demand for electricity after 7.50 years.

    (d) Graphically find how many years it would take for the demand to be 500 billion kWh.

27-4   EXPONENTIAL DECAY

There are many quantities in nature that *decrease* continuously; we say that the quantity *decays exponentially,* usually with respect to time. In this section we solve applications involving quantities including radioactive materials and temperature.

Exponential Decay

The equation for exponential decay is the same as for exponential growth, except that the rate is expressed as $-r$ (a *negative* rate, since the quantity is *decaying*):

EXPONENTIAL DECAY	$y = ae^{-rt}$	140

where a is the original amount of a quantity that is decaying continuously at a rate r and y is the amount present after time t. We say that y *decreases* (or *decays*) *exponentially* with respect to t.

Example 25

Radioactive materials disintegrate (decay) exponentially. Strontium-90, a material found in nuclear power plants, decays at the rate of 3.05% per year. A certain power plant had 125 kg of strontium-90.

(a) Write the equation for the amount of strontium-90 remaining t years later.
(b) Find the amount remaining 1.00 year later.

Solution:

(a) First we change the rate to a decimal,

$$r = 3.05\% = 0.0305$$

We use Eq. 140, with $a = 125$ and $r = 0.0305$,

$$y = 125e^{-0.0305t}$$

(b) To find the amount remaining one year later, we let $t = 1.00$ in the equation above:

$$y = 125e^{-0.0305(1)} = 121 \text{ kg}$$

Graph of Exponential Decay

We saw in Sec. 27-1 that an exponential function $y = b^{-x}$ is a decreasing function if $b > 1$. So $y = e^{-x}$ is a decreasing function, since $e > 1$.

■ FIGURE 27-13
Exponential decay.

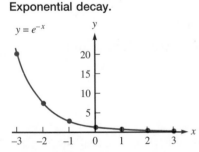

Example 26

Graph $y = e^{-x}$ for x from -3 to 3.

Solution: We make a table of values, rounding to the nearest tenth.

x	-3	-2	-1	0	1	2	3
y	20.1	7.4	2.7	1	0.4	0.1	0.0

We sketch the graph in Fig. 27-13. We see that the function is decreasing and that the graph "falls" to the right and gets "flatter." The x axis is an asymptote.

Example 27

In Example 25, the amount y of strontium-90 remaining after t years was found to be

$$y = 125e^{-0.0305t}$$

(a) Graph this function for t from 0 to 100 years.

(b) Graphically find the amount of strontium-90 remaining after 50.0 years.

Solution:

(a) We use a graphing calculator to graph the function, letting $t = x$:

$$y = 125e^{-0.0305x}$$

The domain is given (x is between 0 and 100) and we must find the range of y values. Since the amount of strontium-90 is decaying, the maximum amount present is 125 kg (at $x = 0$). Certainly the amount will never be negative. So y must be between 0 and 125. We set the window so that

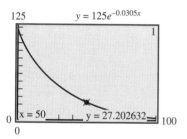

FIGURE 27-14

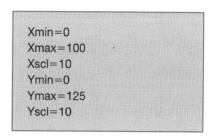

```
Xmin=0
Xmax=100
Xscl=10
Ymin=0
Ymax=125
Yscl=10
```

We enter the function and graph (Fig. 27-14). This curve is a typical exponential decay curve. Note that the x axis is an asymptote, indicating that the mass remaining *approaches* (but never reaches) zero. In other words, there will *always* be *some* radioactive material left.

(b) To graphically find how much remains after 50.0 years, we find the point on the graph where $x = 50.0$ (Fig. 27-15), and read that its y value is 27.2. So, after 50.0 years, there will be 27.2 kg of strontium-90 remaining.

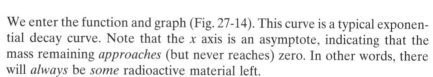

FIGURE 27-15

Other quantities that decay exponentially include temperature, light intensity, and current through a capacitor. We examine temperature in the next example.

Example 28

If an object is placed in a freezer at 0°C, the temperature of the object decreases exponentially according to the formula

$$T = T_0 e^{-rt}$$

where T_0 is the original temperature of the object, r is the rate of temperature change, and T is the temperature of the object t time units later.

A man puts a hot cup of coffee, at 100.0°C in a freezer at 0°C. If it cools at the rate of 5.15% per minute, what is the temperature of the coffee after 1.00 hour in the freezer?

27-4 Exponential Decay   **625**

Solution: First we note that the given units of the rate and the time do not agree (yet).

$$r = 5.15\% \text{ per minute} \quad \text{and} \quad t = 1.00 \text{ hour}$$

We change the time to minutes, to agree with the rate units:

$$t = 1.00 \text{ hour} = 60.0 \text{ minutes}$$

Changing the rate to a decimal,

$$r = 5.15\% = 0.0515$$

Using Eq. 140 with $a = 100$, $r = 0.0515$, and $t = 60.0$,

$$T = 100e^{-0.0515(60.0)}$$
$$= 4.55°C$$

COMMON ERROR When using Eq. 140 for exponential decay, the time units of rate r and time t must agree.

SECTION 27-4 Exercises Exponential Decay

Graph of Exponential Decay

1. Graph $y = 2e^{-x}$ for x from -3 to 3.
2. Graph $y = 3e^{-0.5x}$ for x from -3 to 3.
3. Graph $y = 0.8e^{-0.3t}$ for t from 0 to 5.
4. Graph $y = 0.5e^{-2t}$ for t from 0 to 3.

Solve. Round to three significant digits.

Applications

5. Tritium, a radioactive isotope of hydrogen, decays at the rate of 6.02% per year. A power plant has 175 kg of tritium.
   (a) Write the decay equation.
   (b) Graph the equation for 0 to 25 years.
   (c) Find the amount remaining after 7.50 years.

6. In a certain chemical reaction, the amount of salt in a solution decreases exponentially at the rate of 35.3% per hour. A solution has 7.85 grams of salt when the chemical reaction begins.
   (a) Write the equation for the amount of salt after time t.
   (b) Find the amount of salt in the solution 2.00 hours later.
   (c) Graph the amount of salt as a function of time for 0 to 5 hours.

7. A flywheel is rotating at a speed of 2400 rev/min. When the power is disconnected, the speed decreases exponentially at the rate of 28.2% per minute. Find the speed 25.0 seconds after the power has been disconnected.

8. A steel forging is 2500°F above room temperature. If it cools exponentially at the rate of 2.38% per minute, how many degrees above room temperature will it be in 2.00 hours?

9. A certain isotope of uranium decays at the rate of 2.85% per day. What percent of the original amount of radioactive uranium will remain after 31 days?

10. In a certain fabric mill, cloth is removed from a dye bath and is then observed to dry exponentially at the rate of 22.5% per hour. What percent of the original moisture will still be present after 40.0 minutes?

11. When a capacitor with capacitance C (in farads), charged to a voltage v_0 (in volts), is discharged through a resistor, the current i (in amperes) will decay exponentially with respect to time (in seconds) according to the formula

$$i = \frac{v_0}{R} e^{-t/RC} \quad \text{amperes}$$

Find the current, in milliamps, after 37.0 milliseconds in a circuit where $v_0 = 220$ volts, $C = 125$ microfarads, and $R = 2500$ ohms.

12. Graph $y = 3(1 - e^{-t})$ for t from 0 to 8.

13. When an object is placed in surroundings that are at a higher temperature T_0, it will increase in temperature exponentially according to the equation

$$T = T_0(1 - e^{-rt})$$

A steel casting initially at 0°C is placed in a furnace at 1000°C.

(a) If it increases in temperature at the rate of 5.00% per minute, find its temperature after 120 minutes.

(b) Graph T as a function of t, for t from 0 to 120 minutes.

■ Chapter 27 REVIEW EXERCISES

Graph each function for x between -3 and 3.

1. $y = 4^x$
2. $y = 2 \cdot 3^x$
3. $y = 4e^x$
4. $y = e^{4x}$
5. $y = 3e^{-x}$
6. $y = 2e^{-2x}$
7. $y = 5^{-x}$
8. $y = 3(2.5)^{2x}$

Use the laws of exponents to complete these equations.

9. $\left(\frac{1}{4}\right)^x = ?$
10. $3^0 = ?$

11. $2^{-x} = ?$
12. $\left(\frac{1}{8}\right)^{-x} = ?$

13. Find the amount to which $500 will accumulate in 5.00 years at an annual interest rate of 7.25%, if the interest is compounded (a) annually, (b) monthly, (c) daily, (d) continuously. Round to the nearest cent.

14. A certain virus will increase according to the equation

$$N = N_0 2^{0.2t}$$

where N_0 is the original number of viruses and N is the number after time t (in hours). How many viruses will there be after 12.0 hours if there were 1250 to begin with?

15. The demand y for a certain mineral grows exponentially at the rate of 15.7% per year.

(a) If the demand was for 12.3 tons in 1997, what will the demand be in 2007?

(b) Graph the demand as a function of time t for 0 to 20 years.

16. The atmospheric pressure P (in kPa) at an altitude h (in miles) is given by

$$P = 100e^{-0.325h}$$

assuming constant air temperature. Find the atmospheric pressure at the summit of Alaska's Mount McKinley, which is about 3.85 miles high.

17. A certain yeast will triple every 6.00 hours.
    (a) If you start with 15.0 pounds, how many pounds of yeast will there be after 18.0 hours?
    (b) How many pounds do you need to start with to have 1500 pounds after 1.00 day?

18. A savings bank is offering an annual interest rate of 7.5%, compounded continuously. How much do you need to invest to have an accumulated amount of $3000 at the end of 4 years? Round to the nearest cent.

19. Radioactive carbon-14 (C-14) is present at a constant level in all living things, but decays exponentially after death at the rate of $r = 1.24 \times 10^{-4}$ per year. Scientists use the formula

$$m = m_0 e^{-rt}$$

where m_0 is the original amount of C-14 in the material and m is the amount of C-14 present after t years, to determine the age of organic material.
    (a) What percent of the original C-14 (m_0) would remain in the body of a person who died 100 years ago?
    (b) Graph the percent of m_0 as a function of time t, for 0 to 10,000 years.

WRITING

20. You are applying for a job as the business manager of a new biotech company. On the application you are asked how exponential functions are used in business and in biotechnology. Write a response.

TEAM PROJECT

21. A kettle of boiling water is taken off the stove, and the temperature of the water is measured every minute, as the water cools from 212.0°F at $t = 0$. The room temperature is 70.0°F. Data is recorded as follows:

212.0 196.5 188.4 175.6 161.2 155.3 146.4 141.4 130.0 127.6 120.1

116.3 109.1 104.6 103.1 102.5 101.8 95.9 93.1 91.0 89.0 87.0

Make a graph of temperature versus time. Plot the data points and write an exponential equation for the temperature as a function of time, with all constants evaluated.

Logarithms  28

In this chapter we define a logarithm, evaluate common and natural logarithms, graph the logarithmic function, learn the properties of logarithms, and solve exponential and logarithmic equations.

Logarithms have many applications, including the measurement of sound intensity in decibels, pH levels in chemistry, earthquake intensity by the Richter scale, and applications in electronics and mechanics.

28-1   INTRODUCTION TO LOGARITHMS

We introduce logarithms in this section. We look at the exponential and logarithmic forms of an equation and solve some simple logarithmic equations.

Recall from Sec. 27-2 the 1500 bacteria that double their number every hour. How long would it take to have 900,000? The equation for the number of bacteria after t hours was

$$n = 1500 \cdot 2^t$$

So, letting $n = 900{,}000$,

$$900{,}000 = 1500 \cdot 2^t$$

Dividing both sides by 1500, we have

$$600 = 2^t$$

We need to solve this equation. Try it. What is the value of t?

We *can't* solve it with the operations of adding, subtracting, multiplying, or dividing both sides of the equation. We need a new operation. The base, 2, is being raised to the tth power. We need a new operation—we need to be able to "find the exponent"; that is,

$$\text{if } 2^t = 600, \quad \text{then } t = ?$$

We call the exponent t a *logarithm* and the new operation is called "taking the logarithm." We use the symbol *log* for logarithm, and we write

$$\text{if } 2^t = 600, \quad \text{then } t = \log_2 600$$

This is read "t is the logarithm of 600 to the base 2."

Definition of a Logarithm

A logarithm is defined from the exponential function.

$$\text{If } y = b^x \quad \text{then } x = \log_b y$$

where x is called the *logarithm* of y to the base b.

The two equations above are equivalent equations. That is, they are two different forms of the same equation.

EXPONENTIAL AND LOGARITHMIC FORMS	$y = b^x$	is called the *exponential form*	121
	$x = \log_b y$	is called the *logarithmic form*	

So, we see that a logarithm is really an *exponent*. The logarithm $\log_b y$ is the exponent (the power) that you raise the base b to, to get y.

---- **Example 1** ----

Consider the equation $4^3 = 64$, in which 4 is the base and 3 the exponent. The exponent 3 is a logarithm, and we can write

$$3 = \log_4 64$$

We read this equation "3 is the logarithm of 64 to the base 4."

In working with logarithmic and exponential expressions, we sometimes need to change an equation from exponential form to logarithmic form, or vice versa.

---- **Example 2** ----

Change $2^{-3} = \dfrac{1}{8}$ to logarithmic form.

Solution: Using Eq. 121, we get

$$-3 = \log_2 \left(\frac{1}{8}\right)$$

---- **Example 3** ----

Change $7^{1.50} = 18.5$ to logarithmic form.

Solution: Using Eq. 121, we get

$$1.50 = \log_7 18.5$$

---- **Example 4** ----

Change $\log_5 125 = 3$ to exponential form.

Solution: Using Eq. 121, we get

$$5^3 = 125$$

The constant e is often used in logarithmic and exponential equations. Recall from Sec. 27-3, that e is an irrational number, approximately equal to 2.718.

Example 5

Find x, rounded to three significant digits, and write the equation in logarithmic form.

$$x = e^{1.62}$$

Solution: Using the calculator with the $\boxed{e^x}$ key, we find that

$$x = e^{1.62} = 5.05$$

In logarithmic form, $e^{1.62} = 5.05$ becomes

$$1.62 = \log_e 5.05$$

Logarithmic Equation

A *logarithmic equation* is an equation that contains a logarithm of a variable.

Example 6

The equation

$$\log_{10} x = 3$$

is a logarithmic equation, but the equation

$$x \log_{10} 2 = 5$$

would not usually be called a logarithmic equation, since it does not contain a logarithm of a variable.

Solution of Simple Logarithmic Equations

We know that the two equations

$$y = b^x \quad \text{and} \quad x = \log_b y$$

are equivalent equations; that is, they represent the same relationship between x and y. To solve a simple logarithmic equation, write the equation in logarithmic form, change it to the exponential form, and solve.

Example 7

Solve for y: $\log_7 y = 2$.

Solution: Changing to exponential form, we have

$$7^2 = y$$

or

$$y = 49$$

Example 8

Solve for b: $\log_b 8 = 3$.

Solution: Changing to exponential form, we have

$$b^3 = 8$$

To solve for b, we take the cube root of both sides,

$$\sqrt[3]{b^3} = \sqrt[3]{8}$$
$$b = 2$$

Example 9

Solve for x: $\log_2 32 = x$.

Solution: Changing to exponential form, we have

$$2^x = 32$$

Now we express 32 as a power of 2,

$$32 = 2^5$$

So,

$$2^x = 2^5$$

This can only be true if

$$x = 5$$

Therefore,

$$\log_2 32 = 5$$

SECTION 28-1 Exercises Introduction to Logarithms

Change to logarithmic form.

1. $2^3 = 8$ **2.** $4^2 = 16$

3. $625 = 5^4$ **4.** $16 = 2^4$

5. $10^6 = 1000000$ **6.** $e^2 = 7.4$

7. $\dfrac{1}{81} = \left(\dfrac{1}{3}\right)^4$ **8.** $\dfrac{1}{9} = 3^{-2}$

9. $1 = 78^0$ **10.** $2 = 8^{1/3}$

11. $6^{1.27} = 9.73$ **12.** $10^{2.48} = 200$

Without using the calculator, find the indicated power and write the equation in logarithmic form.

13. $4^3 = ?$ **14.** $2^5 = ?$

15. $10^5 = ?$ **16.** $10^{-3} = ?$

17. $9^0 = ?$ **18.** $\left(\dfrac{1}{4}\right)^2 = ?$

19. $3^{-3} = ?$ **20.** $16^{1/4} = ?$

Use the calculator to find the indicated power, rounded to three significant digits. Then write the equation in logarithmic form.

21. $4^{2.37} = ?$ **22.** $1.5^{5.9} = ?$

23. $2.4^{-3.29} = ?$ **24.** $e^{3.21} = ?$

Express each of the following in logarithmic form.

25. $5^x = 120$ **26.** $y^8 = 65$

27. $a = 9^5$ **28.** $10 = 7^p$

29. $4^{-3} = q$ **30.** $10^x = 4.5$

31. $e^x = 45$ **32.** $a^5 = b$

Change to exponential form.

33. $\log_7 343 = 3$ **34.** $\log_2 64 = 6$

35. $3 = \log_{10} 1000$ **36.** $\log_e 5 = 1.609$

37. $\log_9 3 = \dfrac{1}{2}$ **38.** $\log_5 1 = 0$

39. $1 = \log_7 7$ **40.** $-3 = \log_{1/2} 8$

41. $\log_5 \dfrac{1}{25} = -2$ **42.** $\log_x 7 = 34$

43. $6.59 = \log_4 y$ **44.** $a = \log_7 9$

45. $\log_b x = y$ **46.** $\log_c xy = p$

47. $\log_e 5 = q$ **48.** $\log_{10} 0.01 = -2$

Solution of Simple Logarithmic Equations

Change to exponential form and solve for x. (Do *not* use your calculator.)

49. $x = \log_3 27$ **50.** $x = \log_2 16$

51. $\log_5 125 = x$ **52.** $\log_{10} 1000 = x$

53. $x = \log_2 1$ **54.** $\log_6 1 = x$

55. $\log_e e^4 = x$ **56.** $x = \log_e 1$

57. $\log_x 64 = 3$ **58.** $\log_x 81 = 4$

59. $\log_3 x = 0$ **60.** $\log_x 4 = \dfrac{1}{2}$

61. $2 = \log_x 100$ **62.** $1 = \log_x e$

63. $\log_5 x = 3$ **64.** $\log_{10} x = 4$

65. $-3 = \log_6 x$ **66.** $-4 = \log_{1/2} x$

67. $\dfrac{2}{3} = \log_8 x$ **68.** $-\dfrac{2}{3} = \log_8 x$

69. $\log_x 3 = \dfrac{1}{3}$ **70.** $\log_4 x = -2$

71. $\log_x e^2 = 2$ **72.** $-3 = \log_x 0.001$

28-2   COMMON AND NATURAL LOGARITHMS

In the last section we worked with logarithms to many different bases: b, 2, e, 3, 10, and so on. In practice however, we usually use only two bases, the base 10 and the base e. In this section we examine common and natural logarithms, and use the calculator to find values of logarithms.

Common Logarithms

A logarithm to the base 10 is called a *common* logarithm. If a logarithm is written without naming the base, then it is understood to be a common logarithm—that is, the base is 10.

$$\log N = \log_{10} N$$

Example 10

$$\log 12.5 = \log_{10} 12.5$$

Exponential and Logarithmic Forms

If x is a common logarithm, we have

$$\log N = x \quad \text{the } \textit{logarithmic form}$$
$$10^x = N \quad \text{the } \textit{exponential form}$$

Example 11

Change $10^3 = 1000$ to logarithmic form.

Solution: The logarithmic form is

$$\log 1000 = 3$$

Example 12

Complete the equation and then write in logarithmic form:

$$10^{-2} = ?$$

Solution: Completing the equation (without using a calculator),

$$10^{-2} = \frac{1}{10^2} = \frac{1}{100}$$

Or, as a decimal,

$$10^{-2} = 0.01$$

The logarithmic form is

$$\log 0.01 = -2$$

Example 13

Write the exponential form of the equation

$$\log 1.74 = 0.241$$

Solution: The exponential form is

$$10^{0.241} = 1.74$$

Common Logarithms by Calculator

Common logarithms may be found using the $\boxed{\log}$ key on a calculator. Express the result using the same number of significant digits as in the given number. From now on, the term *logarithm* will refer to the *common* logarithm.

Example 14

Find $\log 25.9$ by calculator. Then write the equation in exponential form.

Solution: Using the $\boxed{\log}$ key, we get

$$\log 25.9 = 1.41$$

The exponential form is

$$10^{1.41} = 25.9$$

To find a number whose logarithm is given, use the $\boxed{10^x}$ key on the calculator. (On some calculators, the 10^x may be called the "inverse log" key.)

────── **Example 15** ──────

Find the number whose logarithm is 1.37.

Solution: Let x be the number whose logarithm is 1.37. Then

$$\log x = 1.37$$

Changing to exponential form, we have

$$10^{1.37} = x$$

Using the $\boxed{10^x}$ key to evaluate $10^{1.37}$, we get

$$10^{1.37} = 23.4$$

So, the number whose logarithm is 1.37 is about 23.4.

────── **Example 16** ──────

Solve $10^x = 7.34$.

Solution: Changing from exponential form to logarithmic form, we get

$$x = \log 7.34$$
$$x = 0.866$$

────── **Example 17** ──────

Solve $1.45(10^x) = 38.2$.

Solution: We divide both sides of the equation by 1.45 to write the equation in logarithmic form,

$$\frac{1.45(10^x)}{1.45} = \frac{38.2}{1.45}$$
$$10^x = 26.34$$

Changing to logarithmic form,

$$x = \log 26.34$$
$$x = 1.42$$

Natural Logarithms

A *natural* logarithm is a logarithm to the base e. Recall that we discussed the irrational number e in Sec. 27-3, and found that it is approximately equal to 2.718, rounded to four significant digits. The natural logarithm of N is written $\ln N$

$$\ln N = \log_e N$$

If x is a natural logarithm, we have

$$\ln N = x \qquad \text{the } logarithmic\ form$$
$$e^x = N \qquad \text{the } exponential\ form$$

Example 18

Change $e^2 = 7.39$ to logarithmic form.

Solution: The logarithmic form is

$$\ln 7.39 = 2$$

Natural Logarithms by Calculator

To find the natural logarithm of a number by calculator, use the $\boxed{\ln}$ key.

Example 19

Find $\ln 78.5$ and write the exponential form.

Solution: Use the $\boxed{\ln}$ key to get

$$\ln 78.5 = 4.36$$

Changing to exponential form, we have

$$e^{4.36} = 78.5$$

To find a number whose natural logarithm is given, use the $\boxed{e^x}$ key on a calculator. (On some calculators, the e^x key may be called the "inverse ln" key.)

Example 20

Use a calculator to find the number whose natural logarithm is 5.43.

Solution: Let x be the number whose natural logarithm is 5.43. Then we need to solve the equation

$$\ln x = 5.43$$

Changing to exponential form, we have

$$e^{5.43} = x$$

Use the $\boxed{e^x}$ key to find

$$e^{5.43} = 228$$

So the number whose natural logarithm is 5.43 is about 228.

Graph of the Logarithmic Function

The function $y = \log_b x$ is called the *logarithmic function.* The base b is a positive number other than 1.

Example 21

Make a table of values for the common logarithmic function

$$y = \log x$$

for x from $\frac{1}{10}$ to 10. Graph $y = \log x$ for x from $\frac{1}{10}$ to 10.

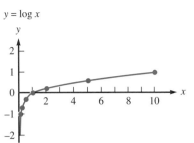

$y = \log x$

FIGURE 28-1
Common logarithmic function.

Solution: We make a table of values for x from $\frac{1}{10}$ to 10, and use the calculator to determine the y values, rounding to tenths.

x	$\frac{1}{10}$	$\frac{1}{5}$	$\frac{1}{2}$	1	2	5	10
y	-1	-0.7	-0.3	0	0.3	0.7	1

We plot these points and connect them with a smooth curve in Fig. 28-1. We see that the graph of $y = \log x$ is asymptotic to the y axis and that the x intercept is 1. The equation is defined for only positive values of x. The curve is not asymptotic to any horizontal line, and indeed, the y values increase without bound. This will be true of all graphs of the equation $y = \log_b x$, where $b > 0$ and $b \neq 1$. The domain of logarithmic function $y = \log_b x$ is all positive numbers (x), and the range is all real numbers (y).

Example 22

Use a graphing calculator to graph the natural logarithmic function

$$y = \ln x$$

for x from 0.1 to 10.

Solution: To determine the range of the function, we find its smallest and its largest values. The logarithmic function is an increasing function, so its smallest value (in the given domain) is

$$y = \ln 0.1 = -2.30$$

and its largest value is

$$y = \ln 10 = 2.30$$

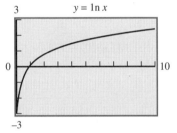

$y = \ln x$

FIGURE 28-2
Natural logarithmic function.

We set the window so that $0 \leq x \leq 10$ and $-3 \leq y \leq 3$. We enter the function $y = \ln x$ and graph (Fig. 28-2). Again we see that the graph of $y = \ln x$ is asymptotic to the y axis and that the x intercept is 1. The equation is defined for only positive values of x. The curve is not asymptotic to any horizontal line, and indeed, the y values increase without bound.

SECTION 28-2 Exercises Common and Natural Logarithms

Common Logarithms

Write the logarithmic form of the equation.

1. $10^2 = 100$ **2.** $10^4 = 10000$ **3.** $10^{-3} = 0.001$ **4.** $10^{-1} = 0.1$

Without using the calculator, complete the equation, and then write the logarithmic form.

5. $10^5 = ?$ **6.** $10^0 = ?$ **7.** $10^{-4} = ?$ **8.** $10^1 = ?$

Write the equation in exponential form.

9. $\log 15.5 = 1.19$ **10.** $\log 155 = 2.19$

11. $\log 1.55 = 0.19$ **12.** $\log 1550 = 3.19$

Use a calculator to find the value of the following logarithms, rounding to three significant digits. Then write the equation in exponential form.

13. log 5.37   **14.** log 34.5   **15.** log 74.5

16. log 6.97   **17.** log 0.456   **18.** log 0.624

Write the equation in exponential form and evaluate x to three significant digits.

19. $\log x = 3$ **20.** $\log x = 5$ **21.** $\log x = 1.56$

22. $\log x = 2.48$ **23.** $\log x = -1.62$ **24.** $\log x = -0.521$

Natural Logarithms

Write the logarithmic form of the equation.

25. $e^3 = 20.1$ **26.** $e^{-1} = 0.368$

Complete the equation, and write the logarithmic form.

27. $e^{2.42} = ?$ **28.** $e^{1.45} = ?$ **29.** $e^0 = ?$ **30.** $e^{-2} = ?$

Find the value of the natural logarithm and then write the equation in exponential form.

31. $\ln 56.7 = ?$ **32.** $\ln 1.05 = ?$ **33.** $\ln 0.0238 = ?$ **34.** $\ln e = ?$

Write the equation in exponential form and evaluate x to three significant digits.

35. $\ln x = 3.54$ **36.** $\ln x = 8.64$

37. $\ln x = -0.00950$ **38.** $\ln x = -2.75$

Graph of the Logarithmic Function

39. Complete this table of values for the logarithmic function $y = \ln x$, rounding to tenths.

x	$\frac{1}{10}$	$\frac{1}{5}$	$\frac{1}{2}$	1	2	5	10
y							

Graph the function $y = \ln x$ by plotting these points and connecting them with a smooth curve.

40. Use a graphing calculator to graph the function $y = \log x$.

28-3   PROPERTIES OF LOGARITHMS

In this section we study the properties of logarithms, including the logarithm of a product, quotient, power, and root. We will use these properties in the next section to solve both logarithmic and exponential equations using logarithms.

Before discussing any specific properties of logarithms, we need to consider an important fact concerning powers of numbers: *Any positive number can be expressed as a power of any other positive number, other than 1.*

--- **Example 23** ---

We know that 8 can be expressed as a power of 2. Indeed,

$$8 = 2^3 \quad \text{and in logarithmic form,} \quad \log_2 8 = 3$$

But 8 can also be written as a power of any other positive number, other than 1:

$$8 = 10^{0.9031} \quad \text{and in logarithmic form,} \quad \log 8 = 0.9031$$

and

$$8 = 7^{1.0686} \quad \text{and in logarithmic form,} \quad \log_7 8 = 1.0686$$

In general, if M is any positive number, then M can be expressed as a power of a base b, where $b \neq 1$:

$$M = b^c \quad \text{and in logarithmic form,} \quad \log_b M = c$$

Similarly, if N is a positive number, it also can be expressed as

$$N = b^d \quad \text{and in logarithmic form,} \quad \log_b N = d$$

We use these equations to derive the property for the logarithm of a product.

Logarithm of a Product

We consider the product of two positive numbers M and N. Substituting $M = b^c$ and $N = b^d$ in the product MN, we have

$$MN = b^c \cdot b^d$$

By the laws of exponents,

$$MN = b^{c+d}$$

Changing to logarithmic form,

$$\log_b MN = c + d$$

Substituting $c = \log_b M$ and $d = \log_b N$, we have

LOGARITHM OF A PRODUCT	$\log_b MN = \log_b M + \log_b N$	122

In other words, *the logarithm of a product equals the sum of the logarithms of the factors.* This is reasonable, since logarithms are *exponents,* and when two quantities with the same base are multiplied, we add the exponents!

Example 24

Using Eq. 122,

$$\log_b 2x = \log_b 2 + \log_b x$$

Example 25

Express $\log_4 3 + \log_4 y$ as a single logarithm.

Solution:

$$\log_4 3 + \log_4 y = \log_4 3y$$

COMMON ERRORS The logarithm of a product is *not* equal to the product of the logarithms.

$$\log_b MN \neq (\log_b M)(\log_b N)$$

The logarithm of a sum is *not* equal to the sum of the logarithms.

$$\log_b (M + N) \neq \log_b M + \log_b N$$

The product of two logs is *not* equal to the sum of the logarithms.

$$(\log_b M)(\log_b N) \neq \log_b M + \log_b N$$

Logarithm of a Quotient

Similarly, the property for the logarithm of the quotient of two numbers is

LOGARITHM OF A QUOTIENT	$\log_b \dfrac{M}{N} = \log_b M - \log_b N$	123

In other words, *the logarithm of a quotient equals the logarithm of the numerator minus the logarithm of the denominator.* This is reasonable, since logarithms are exponents, and when two quantities with the same base are divided, we subtract the exponents!

Example 26

Using Eq. 123,

$$\log_5 \frac{x}{3} = \log_5 x - \log_5 3$$

Example 27

Using Eq. 123,

$$\log \frac{2}{3x} = \log 2 - \log 3x$$
$$= \log 2 - (\log 3 + \log x)$$
$$= \log 2 - \log 3 - \log x$$

Similar errors are made with quotients as with products.

COMMON ERRORS

$$\log_b \frac{M}{N} \neq \frac{\log_b M}{\log_b N}$$

$$\log_b (M - N) \neq \log_b M - \log_b N$$

$$\frac{\log_b M}{\log_b N} \neq \log_b M - \log_b N$$

Logarithm of a Power

Now we write an expression for the logarithm of a power, say M^p, where M is a positive number. Again we let $M = b^c$, so that $c = \log_b M$. Substituting $M = b^c$, we get

$$M^p = (b^c)^p$$

By the laws of exponents,

$$M^p = b^{cp}$$

Changing to logarithmic form,

$$\log_b M^p = cp$$
$$\log_b M^p = pc$$

Substituting $c = \log_b M$, we get

| LOGARITHM OF A POWER | $\log_b M^p = p \cdot \log_b M$ | 124 |

In other words, *the logarithm of a power of a number equals the exponent times the logarithm of the number.*

```
┌──── Example 28 ────────
│ Using Eq. 124,
```
$$\log_b x^2 = 2 \cdot \log_b x$$

```
┌──── Example 29 ──────────
│ Express as a single logarithm, with a coefficient of 1.
```
$$3 \ln 2 + 2 \ln x - \ln 3$$

Solution:
$$3 \ln 2 + 2 \ln x - \ln 3 = \ln 2^3 + \ln x^2 - \ln 3$$
$$= \ln \left(\frac{2^3 \cdot x^2}{3} \right)$$
$$= \ln \left(\frac{8x^2}{3} \right)$$

Logarithm of a Root

Similarly, the property for the logarithm of the root of a number is

| LOGARITHM OF A ROOT | $\log_b \sqrt[r]{M} = \dfrac{1}{r} \log_b M$ | 125 |

In other words, *the logarithm of the rth root of a number equals $\dfrac{1}{r}$ times the logarithm of the number.*

```
┌──── Example 30 ────────
│ Using Eq. 125,
```
$$\log_2 \sqrt[3]{x} = \frac{1}{3} \log_2 x$$

```
┌──── Example 31 ────────
│ Using Eq. 125,
```
$$\log_b \sqrt{y} = \log_b \sqrt[2]{y} = \frac{1}{2} \log_b y$$

Example 32 ──────────

Using Eq. 125,

$$\log_3 \sqrt{x^5} = \frac{1}{2}\log_3 x^5 = \frac{5}{2}\log_3 x$$

Let's consider some special logarithms now.

Logarithm of 1

First we look at $\log_b 1$. We know that

$$b^0 = 1$$

Changing $b^0 = 1$ to logarithmic form, we get

$$\log_b 1 = 0$$

LOGARITHM OF 1	$\log_b 1 = 0$	126

In other words, *the logarithm of 1 to any base is always 0.*

Logarithm of the Base

Now we look at $\log_b b$. We know that

$$b^1 = b$$

Changing $b^1 = b$ to logarithmic form, we get

$$\log_b b = 1$$

LOGARITHM OF THE BASE	$\log_b b = 1$	127

Logarithm of the Base Raised to a Power

The logarithm of a base b raised to a power n is $\log_b b^n$. By Eq. 127,

$$\log_b b^n = n \cdot \log_b b$$

Substituting $\log_b b = 1$, we get

$$\log_b b^n = n \cdot 1$$

LOGARITHM OF THE BASE RAISED TO A POWER	$\log_b b^n = n$	128

Example 33 ──────────

By Eq. 128, $\log_2 2^3 = 3$.

The properties in Eqs. 127 and 128 are especially important for common and natural logarithms:

$$\log 10 = 1 \qquad \log 10^n = n$$
$$\ln e = 1 \qquad \ln e^n = n$$

---- **Example 34** ----

By Eq. 128, $\ln e^3 = 3$.

---- **Example 35** ----

By Eq. 128, $\log 0.01 = \log(10^{-2}) = -2$.

SECTION 28-3 Exercises Properties of Logarithms

Logarithms of Products and Quotients

Use the properties of logarithms to expand the expression.

1. $\log_3 7x$
2. $\ln 3y$
3. $\log_5 \dfrac{x}{4}$
4. $\log_2 \dfrac{5}{y}$
5. $\ln 3xy$
6. $\log abc$
7. $\log_2 \dfrac{5x}{3}$
8. $\log_3 \dfrac{xy}{7}$
9. $\log_b \dfrac{3}{4a}$
10. $\log_b \dfrac{2}{5x}$

Express as a single logarithm.

11. $\log_4 5 + \log_4 a$
12. $\log_7 8 - \log_7 c$
13. $\log_b 2 - \log_b y$
14. $\log_3 z + \log_3 4$
15. $\log_2 x + \log_2 y + \log_2 z$
16. $\log_b x - \log_b y + \log_b 3$
17. $\log_3 a - \log_3 5 - \log_3 c$
18. $\log_4 3 + \log_4 a - \log_4 7$

Logarithms of Powers and Roots

Use the properties of logarithms to expand the expression.

19. $\log_b x^3$
20. $\log a^x$
21. $\log_3 2^x$
22. $\ln x^{1.76}$
23. $\log_b \sqrt{x}$
24. $\ln \sqrt[5]{2}$
25. $\log_2 \sqrt[3]{x}$
26. $\log \sqrt[4]{x}$
27. $\log_b x^2 y^3$
28. $\log_2 5\sqrt{3}$

Express as a single logarithm with a coefficient of 1.

29. $5 \log_2 x$
30. $x \log_2 5$
31. $3 \ln x + 4 \ln y$
32. $2 \log a - 3 \log c - 4 \log d$
33. $\dfrac{1}{2} \log_b x$
34. $\dfrac{1}{3} \log_2 7$
35. $\dfrac{\ln 2}{4}$
36. $\dfrac{\log 6}{3}$
37. $\dfrac{1}{2} \log_2 x + \dfrac{1}{3} \log_2 y$
38. $\dfrac{1}{3} \ln 4 - 2 \ln a$
39. $\dfrac{2}{3} \log_5 x$
40. $\dfrac{3}{4} \log_2 y$

Simplify.

41. $\log_2 1$
42. $\log_7 1$
43. $\log_3 3$
44. $\log_b b$
45. $\log_5 5^9$
46. $\log_2 2^7$
47. $\ln e^{5.3}$
48. $\log_b b^2$
49. $\ln \sqrt[5]{e}$
50. $\log_b \sqrt[3]{b^4}$

SOLUTION OF EXPONENTIAL EQUATIONS USING LOGARITHMS

In this section we use logarithms to solve exponential equations. We solve applications, including the growth of bacteria, compound interest, population growth, and radioactive decay.

To solve an exponential equation we generally just *take the logarithm of both sides of the equation.* We can use *either* common logarithms *or* natural logarithms. If the exponential equation uses the base e, then we take the natural logarithm of both sides. If the exponential equation uses the base 10, then we take the common logarithm of both sides. If it uses a base other than 10 or e, we may use either natural or common logarithms.

Example 36

Solve to three significant digits:

$$10^x = 7.34$$

Solution: In this equation, the base is 10, so take the common log of both sides:

$$\log 10^x = \log 7.34$$

Using Eq. 128,

$$x = \log 7.34$$

Evaluating $\log 7.34$,

$$x = 0.866$$

Example 37

Solve $75.3 = 91.2e^{-0.417t}$ to three significant digits.

Solution: First we divide both sides by the constant 91.2:

$$\frac{75.3}{91.2} = \frac{91.2e^{-0.417t}}{91.2}$$

Carrying one extra digit, we have

$$0.8257 = e^{-0.417t}$$

Taking the natural log of both sides,

$$\ln 0.8257 = \ln e^{-0.417t}$$

By Eq. 128

$$\ln 0.8257 = -0.417t$$

Dividing both sides by -0.417,

$$t = \frac{\ln 0.8257}{-0.417}$$

$$t = 0.459$$

Check: Substitute $t = 0.459$ into the original equation,

$$75.3 \stackrel{?}{=} 91.2e^{-0.417(0.459)}$$

Evaluate the right side by calculator,

$$91.2e^{-0.417(0.459)} = 75.3 \quad \text{checks}$$

Example 38

Solve $588 = 2^t$ to three significant digits.

Solution: In this equation the base is 2 (neither 10 nor e), so we may use either common or natural logarithms. Let us take the common logarithm of both sides:

$$\log 588 = \log 2^t$$

By Eq. 124,

$$\log 588 = t \cdot \log 2$$

Dividing both sides by $\log 2$,

$$t = \frac{\log 588}{\log 2}$$

Evaluating the right side,

$$t = 9.20$$

Check: We substitute $t = 9.20$ into the right side of the equation and evaluate:

$$588 \stackrel{?}{=} 2^{9.20}$$

$$588 = 588 \quad \text{checks}$$

Applications

In many applications we find it necessary to solve exponential equations. A number of these applications were introduced in Chapter 27 on exponential functions. These include biological growth, compound interest, and exponential growth and decay.

Example 39

How long would it take for *any* amount of money, compounded continuously at 5.75% per year, to triple?

Solution: Use Eq. 139 for continuous compounding,

$$y = ae^{rt}$$

where a is the original amount invested.

Since the money is being tripled, the amount a accumulates to $3a$. So,

$$y = 3a \quad \text{and} \quad r = 5.75\% = 0.0575$$

Substituting, we get

$$3a = ae^{0.0575t}$$

Dividing both sides by a,

$$3 = e^{0.0575t}$$

Taking the natural log of both sides,

$$\ln 3 = \ln e^{0.0575t}$$

By Eq. 128,

$$\ln 3 = 0.0575t$$

Dividing by 0.0575,

$$t = \frac{\ln 3}{0.0575}$$

$$t = 19.1$$

So, it would take about 19.1 years for any amount of money to triple, at 5.75% per year, compounded continuously.

── **Example 40** ──

Radium D decays exponentially at the rate of 3.22% per year. Find the time necessary for 5.00 kg of Radium D to disintegrate to 4.00 kg.

Solution: Since radioactive decay is an example of exponential decay, we use Eq. 140:

$$y = ae^{-rt}$$

Substituting,

$$4.00 = 5.00e^{-0.0322t}$$

Solving,

$$0.800 = e^{-0.0322t}$$

$$\ln 0.800 = \ln e^{-0.0322t}$$

$$\ln 0.800 = -0.0322t$$

$$t = \frac{\ln 0.800}{-0.0322}$$

$$t = 6.93$$

So, 5.00 kg of Radium D will disintegrate to 4.00 kg in about 6.93 years.

SECTION 28-4 Exercises Solution of Exponential Equations Using Logarithms

Solve each exponential equation, and round to three significant digits. Check each solution.

1. $10^x = 25.4$ **2.** $8.97 = 10^x$

3. $e^x = 0.789$ **4.** $33.8 = e^x$

5. $4.89 = 1.06^x$ **6.** $1.76^x = 525$

7. $0.476 = 10^{-x}$ **8.** $10^{-x} = 87.9$

9. $e^{-x} = 1.25$ **10.** $0.376 = e^{-x}$

11. $1.29 = 2.30^{-x}$ **12.** $5.45^{-x} = 4.98$

13. $6.45 = 4.64(10^x)$ **14.** $5.25 \cdot 10^x = 9.69$

15. $2e^x = 3$ **16.** $7 = 3e^x$

17. $12 = 24e^{-x}$ **18.** $15e^{-x} = 5$

19. $700 = 400(1.01)^t$ **20.** $225(1.025)^{3t} = 500$

21. $6.79 = 10^{2x}$ **22.** $1200 = 300e^{0.04x}$

23. $8.5e^{55t} = 145$ **24.** $0.023 = 0.01e^{5x}$

25. $7.6 = 9.2e^{-75x}$ **26.** $3 = 6.25e^{-150x}$

27. $6 = 18e^{-0.4t}$ **28.** $25 = 100e^{-0.075t}$

Solve for t.

29. $N = 100 \cdot 2^t$ **30.** $S = 500(1.12^t)$

31. $1000 = P \cdot e^{0.12t}$ **32.** $m = m_0 e^{-kt}$

Solve. If answers are not exact, round to three significant digits.

Applications

33. A certain bacterium triples its number every hour. If we begin with 1000 bacteria, how long would it take to have a million?

34. A yeast manufacturer finds that the yeast will quadruple every hour. If the manufacturer begins with 5 pounds of yeast, how long will it take to grow 100 pounds of yeast?

35. How long will it take for $100 to grow to $450, if it is compounded continuously at 11.5% per year?

36. How long would it take for an amount of money to quadruple, if it were compounded continuously at 12.5% per year?

37. The population of a city is growing exponentially at the rate of 1.85% per year. How many years will it take for the population to double?

38. If the rate of inflation is 3.45% per year, how long will it take for a $10.00 bag of groceries to cost $15.00? Assume the cost of goods rises exponentially.

39. A certain radioactive isotope of Uranium, Uranium Z, decays at the rate of 10.3% per hour. Find the time necessary for 75.0 grams of Uranium Z to decay to 30.0 grams.

40. A radioactive isotope called Radium A decays at the rate of 22.7% per minute. Find the time necessary for 20.0 kg of Radium A to decay to 10.0 kg.

41. The current i in a certain circuit is given by

$$i = 4.75e^{-125t} \text{ amperes}$$

where t is the time in seconds. At what time will the current be 2.00 amperes?

42. The discharge current i of a capacitor is given by

$$i = 0.0338e^{-50.2t}$$

where t is the time in seconds. At what time will the discharge current be 0.00500 amperes?

43. A cup of coffee will cool exponentially in a certain freezer at the rate of 5.06% per minute. How long will it take for a cup of coffee at 99.5°C to cool to 37.0°C?

44. A flywheel is rotating at a speed of 2450 rev/min. When the power is disconnected, the speed decreases exponentially at the rate of 31.4% per minute. How long would it take the flywheel to slow to a speed of 1.00 rev/min (virtually stop)?

45. In 1966, the world population was 5.8 billion people and expected to double in 45 years.

(a) Assuming that the growth is exponential, find the rate of growth (as a percent).

(b) In what year would we expect there to be only 1 square yard of land per person? (The earth contains approximately 1.68×10^{14} square yards of land.)

46. If a leukemic cell is injected into a healthy mouse, that cell will divide into two cells in half a day. The mouse will die when the number of leukemic cells reaches one billion.

(a) Write an equation for the number of leukemic cells n present after t days.

(b) After how many days will the mouse die?

47. When an object at 0°C is placed in surroundings that are at a higher temperature c, its temperature T will increase exponentially according to the equation

$$T = c(1 - e^{-rt})$$

where r is the rate of increase and t is the time. A steel casting initially at 0°C is placed in a furnace at 1000°C. If it increases in temperature at the rate of 5.16% per minute, how long will it take to reach 700°C? Round to three significant digits.

48. Solve for t:

$$T = c(1 - e^{-rt})$$

28-5 _____ APPLICATIONS OF LOGARITHMIC EQUATIONS

In this section we solve applications with logarithmic equations. Logarithmic equations are used to define many important quantities such as the loudness of sound, the pH value of a solution, and the magnitude of an earthquake.

Example 41 ————

The loudness of sound β (in decibels) is given by the formula

$$\beta = 10 \log \frac{I}{I_0}$$

where I is the intensity of the sound and I_0 is the faintest sound detectable, about 10^{-16} Watts/cm² (W/cm²). Normal hearing ranges from 0 to 120 decibels (dB).

The intensity of sound for street traffic is about 3.25×10^{-9} W/cm². How loud is the street traffic in decibels?

Solution: Substituting $I = 3.25 \times 10^{-9}$ and $I_0 = 10^{-16}$ into the given equation:

$$\beta = 10 \log \frac{3.25 \times 10^{-9}}{10^{-16}}$$

$$\beta = 10 \log (3.25 \times 10^7)$$

$$\beta = 75.1 \text{ dB}$$

Example 42 ————

The pH value of a solution having a concentration C of hydrogen ions (in moles per liter) is

$$pH = - \log C$$

Acids have a pH value less than 7.00 and bases have a pH value greater than 7.00.

Vinegar has a concentration of 0.006 moles per liter. Find its pH value to three significant digits.

Solution: Using the given equation for the pH value, we have

$$pH = - \log 0.006$$

$$pH = 2.22$$

Example 43

The difference in elevation h (in feet) between two locations having barometer readings of B_1 and B_2 inches of mercury is given by the equation

$$h = 60{,}470 \log \frac{B_2}{B_1}$$

where B_1 is the pressure at the upper station.

Find the difference in elevation between two stations having barometer readings of 28.4 in. at the lower station and 25.8 in. at the upper station.

Solution: Using the equation above for the difference in elevation, we have

$$h = 60470 \log \frac{28.4}{25.8}$$

$$h = 2520 \text{ ft} \quad \text{(rounding to three significant digits)}$$

SECTION 28-5 Exercises Applications of Logarithmic Equations

Solve the applications, and round the answer to three significant digits.

1. The intensity of sound for a whisper is about 10^{-14} W/cm². How many decibels is a whisper?

2. The intensity of sound for normal conversation is about 10^{-10} W/cm². What is the loudness in decibels?

3. The loudness of sound for the threshold of pain is 120 dB. What is the intensity of the sound?

4. The loudness of sound for soft music is 30.0 dB. What is the intensity of the sound?

5. Wine has a hydrogen ion concentration of 0.0004 moles per liter. Find its pH value.

6. Pure water has a hydrogen ion concentration of 10^{-7} moles per liter. Find its pH value.

7. The pH of ammonia is 10.8. Find its hydrogen ion concentration.

8. The pH of beer is 6.34. Find its hydrogen ion concentration.

9. Find the difference in elevation between two stations having barometer readings of 29.6 in. at the lower station and 24.5 in. at the upper station.

10. Find the barometer reading 927 ft above a station having a barometer reading of 27.5 in.

11. If interest is compounded continuously, the time necessary for an account to double is given by

$$t = \frac{\ln 2}{r}$$

where r is the rate of interest. How long will it take for an account to double if the interest is 6.25% per year compounded continuously?

12. What interest rate would be necessary for an account to double in 4 years, if the interest were compounded continuously?

13. The magnitude R of an earthquake of intensity I is defined on the Richter scale, as

$$R = \log\left(\frac{I}{I_0}\right)$$

where I_0 is the minimum level for comparison. Since the Richter scale is a logarithmic base 10 scale, a magnitude 7 quake is 10 times more intense than a magnitude 6 quake; a magnitude 7 quake is 100 times more intense than a magnitude 5 quake; and a magnitude 7 quake is 1000 times more intense than a magnitude 4 quake. How many times more intense was the 1964 Alaska earthquake whose magnitude was 9.2 on the Richter scale than the 1906 San Francisco earthquake whose magnitude was 7.9?

■ **Chapter 28   REVIEW EXERCISES**

If answers are not exact, round to three significant digits.
Express each of the following in logarithmic form.

1. $4^3 = 64$ **2.** $\dfrac{1}{8} = 2^{-3}$

3. $4 = 10^x$ **4.** $y^5 = 2.45$

Use the properties of logarithms to expand the expression.

5. $\log_2 ab$ **6.** $\log_3 \dfrac{6}{7}$

7. $\log_7 \dfrac{5a}{b}$ **8.** $\log_5 y^3$

9. $\ln \sqrt[3]{x}$ **10.** $\log \sqrt{x^3}$

11. $\log_3 a^2 b^3$ **12.** $\ln \dfrac{\sqrt{3}}{x}$

Complete the equation, and write the logarithmic form.

13. $10^4 = ?$ **14.** $10^{-5} = ?$

15. $e^{-1.23} = ?$ **16.** $e^{0.6} = ?$

Change to exponential form and find x, without using a calculator.

17. $x = \log_2 8$ **18.** $\log_5 1 = x$

19. $\ln e^3 = x$ **20.** $\log 0.01 = x$

21. $\log_x 9 = \dfrac{1}{2}$ **22.** $\log_8 x = 2$

23. $\log 100000 = x$ **24.** $\log_x \sqrt[3]{b} = \dfrac{1}{3}$

25. $x = \log_4 \dfrac{1}{16}$ **26.** $x = \log_7 \sqrt{7}$

27. $x = \log_b 1$ **28.** $x = \log_8 8^7$

Using a calculator, determine the following logarithms. Then write the equations in exponential form.

29. $\log 1.70$ **30.** $\log 35.7$

31. $\ln 4.50$ **32.** $\ln 0.625$

Solve each equation. Check your solution.

33. $2.50^x = 7.20$ **34.** $e^x = 6.15$

35. $0.090 = e^{-x}$ **36.** $8.34 = 3(1.23)^x$

37. $4e^x = 7$ **38.** $8.32 = 3e^{-x}$

39. $5.25 = 6.50(1.02)^{2x}$ **40.** $92.6 = 6.18e^{-0.3x}$

Use a calculator to find the indicated power, rounded to three significant digits. Then write the equation in logarithmic form.

41. $6^{2.30} = ?$ **42.** $e^{-1.52} = ?$

43. $10^{-3.65} = ?$ **44.** $2.5^{3.18} = ?$

Solve for x and evaluate.

45. $\log x = 7$ **46.** $\log x = -2.35$

47. $\ln x = -1.52$ **48.** $\ln x = 0.327$

Without using a calculator, find the indicated power and write the equation in logarithmic form.

49. $5^3 = ?$ **50.** $10^{-2} = ?$

51. $\left(\dfrac{1}{3}\right)^{-1} = ?$ **52.** $8^{1/3} = ?$

Express as a single logarithm with a coefficient of 1.

53. $\log_3 5 + \log_3 x$ **54.** $\log_2 5 + \log_2 x - \log_2 3 - \log_2 y$

55. $4 \log 3x$ **56.** $x \log_b 5$

57. $\dfrac{1}{2} \log_4 a$ **58.** $\dfrac{\log_5 x}{3}$

59. $3 \ln x - 2 \ln y$ **60.** $\dfrac{1}{4} \ln 3 + \dfrac{2}{3} \ln y$

Change to exponential form.

61. $\log_3 81 = 4$ **62.** $2 = \log 100$

63. $\ln 7 = 1.95$ **64.** $\log_4 \left(\dfrac{1}{16}\right) = -2$

Evaluate.

65. $\log_3 3^7$ **66.** $\log_2 4^3$

67. $\ln \sqrt[3]{e}$ **68.** $\log_b \sqrt{b^9}$

Solve.

69. A certain bacterium quadruples its number every day. If we began with 1000 bacteria, how long would it take to have a million?

70. How long would it take $250 to grow to $800, if it were invested at 6.25% per year, compounded continuously?

71. A radioactive isotope called Radium A decays at the rate of 22.7% per minute. Find the time necessary for 16.0 kg of Radium A to decay to 12.0 kg.

72. The time necessary for half the quantity of radioactive tritium to decay is 11.6 years (its half-life). Find the rate of decay of tritium.

73. The discharge current i of a capacitor is given by

$$i = 0.050e^{-65.0t}$$

where t is the time in seconds. At what time will the discharge current be 0.001 amperes?

74. The intensity of sound for loud music is about 10^{-6} W/cm². Use the equation

$$\beta = 10 \log \frac{I}{I_0}$$

where $I_0 = 10^{-16}$ W/cm², to find the loudness in decibels of loud music.

75. Use the equation

$$pH = -\log C$$

to find the hydrogen ion concentration C (in moles/liter) for eggs, which have a pH of 8.

76. Find the time necessary for a casting to cool from 1520°C to 825°C, if it cools at 25.1% per minute.

WRITING

77. What is a logarithm? Explain the relationship between the logarithmic form and the exponential form of an equation. Give three examples of how logarithms are used in everyday life.

TEAM PROJECT

78. The students in Ms. Smith's math class were asked to record the number of hours spent studying for their math test. For each student, Ms. Smith wrote an ordered pair (x, y). The value of x represented the number of hours of study and the value of y was the student's grade on the test. Jenni's ordered pair was (3.0, 82) because she spent 3.0 hours studying and received a grade of 82 on her math test. The set of ordered pairs which Ms. Smith recorded are shown here:

(3.0, 82), (5.5, 78), (1.0, 60), (4.9, 93), (5.1, 86), (2.5, 71), (4.2, 90), (0.5, 40), (3.5, 88), (7.0, 96), (1.5, 73), (2.4, 82), (2.0, 53), (6.2, 87), (8.4, 100), (2.6, 75), (3.7, 85), (5.4, 70), (9.3, 89), (7.6, 87)

Make a graph of test grades versus the number of hours studied. Is there is a relationship between the number of hours of study and the test grade? Find an equation of a line which seems to "fit" this data. Graph the line. Might a logarithmic curve fit this data? Find a logarithmic equation that fits the data and graph the curve. Which curve do you think yields the better fit? Use that curve to estimate the grade Becky would achieve with 5.2 hours of study.

Complex Numbers 29

Thus far we have dealt entirely with real numbers; in this chapter we introduce another set of numbers that are not real; they are called "imaginary" numbers. We also learn what complex numbers are, how to express them in several different forms, and how to perform the basic operations on them.

So why do we need to study imaginary and complex numbers? There are many reasons. The solutions of many quadratic and higher degree equations are imaginary or complex numbers; complex numbers are used in the study of electricity; and many manipulations of vectors are simplified if the vectors are represented by complex numbers.

29-1   IMAGINARY NUMBERS

In this section, we define the basic imaginary unit and imaginary number, and learn to simplify, add, subtract, multiply, and divide imaginary numbers.

Imaginary Unit

In Chapter 6 we learned that, in the real number system, we cannot take the square root of a negative number. For example, $\sqrt{-1}$ is *not* a *real* number. But $\sqrt{-1}$ *is* a number; we call it an *imaginary* number. The choice of the words *real* and *imaginary* is unfortunate. *Imaginary* was chosen as the opposite of *real*. But *imaginary* numbers are actual solutions of equations and are very *real* to people using them.

We define the *imaginary unit* as the square root of -1, and represent it by the symbol j. Some textbooks use i as the imaginary unit. Since i frequently represents electrical current, it has become common to represent the imaginary unit by j instead of i. This symbol is sometimes called the *j-operator.*

IMAGINARY UNIT	$j = \sqrt{-1}$	153

Imaginary Numbers

An *imaginary number* is the product of the imaginary unit j times a real number b; it is generally expressed as

$$jb$$

where $j = \sqrt{-1}$ and b is a real number.

Example 1

An example of an imaginary number is $j2$. It is the product of the imaginary unit j and the real number 2.

Example 2

The imaginary number $j(-3)$ is usually written as $-j3$:

$$j(-3) = -j3$$

because if the real coefficient b is negative, the negative sign is usually written in *front* of j.

Imaginary Numbers in *jb* Form

The square root of a negative number is an imaginary number. It can be expressed in *jb* form by factoring the -1 from the radicand, and expressing $\sqrt{-1}$ as j.

Example 3

Express the imaginary number $\sqrt{-7}$ in *jb* form.

Solution: First we factor -1 from the radicand -7:

$$\sqrt{-7} = \sqrt{(-1)(7)}$$

and take the square root of each factor

$$= \sqrt{-1} \cdot \sqrt{7}$$

Expressing $\sqrt{-1}$ as j, we get

$$= j\sqrt{7}$$

In Example 3, the square root of the positive factor $\sqrt{7}$ was already simplified. Often, however, the square root of the positive factor can be simplified further using the rules of radicals from Chapter 20.

Example 4

Express the imaginary numbers in *jb* form.

(a) $\sqrt{-4}$ **(b)** $\sqrt{-45}$

Solution:

(a) Following the same process as in Example 3, we get

$$\sqrt{-4} = \sqrt{(-1)(4)}$$
$$= \sqrt{-1} \cdot \sqrt{4}$$
$$= j\sqrt{4}$$

Now we simplify $\sqrt{4}$ to get

$$= j2$$

(b) Similarly,

$$\sqrt{-45} = \sqrt{-1 \cdot 45}$$
$$= \sqrt{-1} \cdot \sqrt{45}$$

Simplifying $\sqrt{45}$,

$$= j \cdot \sqrt{9 \cdot 5}$$
$$= j \cdot \sqrt{9} \cdot \sqrt{5}$$
$$= j3\sqrt{5}$$

Example 5

Express $-2\sqrt{-18}$ in jb form.

Solution: First we express $\sqrt{-18}$ in jb form, and then multiply by the factor -2.

$$-2\sqrt{-18} = -2 \cdot \sqrt{(-1)(18)}$$
$$= -2 \cdot \sqrt{-1} \cdot \sqrt{18}$$
$$= -2 \cdot j\sqrt{9 \cdot 2}$$
$$= -2 \cdot j3\sqrt{2}$$

Now multiply by the -2 to get

$$= -j6\sqrt{2}$$

Adding and Subtracting Imaginary Numbers

Imaginary numbers may be added and subtracted in their jb form. They are added or subtracted by combining the coefficients of j (as we combine other like algebraic terms).

Example 6

Combine $j2 + j5 - j3$.

Solution:

$$j2 + j5 - j3 = j(2 + 5 - 3)$$
$$= j4$$

Example 7

Combine $\sqrt{-25} - \sqrt{-16} + 2\sqrt{-9}$.

Solution:

$$\sqrt{-25} - \sqrt{-16} + 2\sqrt{-9} = j5 - j4 + 2j3$$
$$= j5 - j4 + j6$$
$$= j7$$

Multiplying Imaginary Numbers

Imaginary numbers may be multiplied by real numbers or by other imaginary numbers. First we look at the product of an imaginary number and a real number. Just treat j as a variable, multiply the real factors together, and simplify.

—— **Example 8** ——

Multiply and simplify.

(a) $5 \cdot j3$ **(b)** $-\sqrt{-49} \cdot -3$

Solution:

(a) Multiplying the real factors, we get

$$5 \cdot j3 = j15$$

(b) Writing in jb form, we get

$$-\sqrt{-49} \cdot -3 = -j7 \cdot -3$$
$$= j21$$

In Example 8, we see that the product of an imaginary number and a real number is an *imaginary* number.

Now we learn to multiply two imaginary numbers. But first we must define j^2, using the fact that $j = \sqrt{-1}$:

$$j^2 = -1$$

—— **Example 9** ——

Multiply and simplify:

$$j2 \cdot -j\sqrt{6}$$

Solution: First, multiplying the j's and the real coefficients,

$$j2 \cdot -j\sqrt{6} = j^2 \cdot -2\sqrt{6}$$

Changing j^2 to -1,

$$= (-1)(-2\sqrt{6})$$

So we get,

$$= 2\sqrt{6}$$

Recall from the product rule in Chapter 20 that

$$\sqrt{a} \cdot \sqrt{b} = \sqrt{ab}$$

is true for all *real* numbers. This product rule is true only if $\sqrt{a}$ and $\sqrt{b}$ are real numbers. Since the square root of a negative number is not real, then it must be true that

$$a \geq 0 \text{ and } b \geq 0$$

We can use the product rule when we multiply imaginary numbers, but only after expressing the imaginary numbers in jb form.

To multiply two imaginary numbers, convert them to jb form, multiply the j's, multiply the real coefficients, and then simplify.

Example 10

Multiply and simplify:

$$\sqrt{-4} \cdot \sqrt{-9}$$

Solution: Changing to jb form,

$$\sqrt{-4} \cdot \sqrt{-9} = j2 \cdot j3$$
$$= j^2 \cdot 6$$
$$= -1 \cdot 6$$
$$= -6$$

Now watch what happens if we don't change to jb form first. We are tempted to use the product rule and multiply the radicands:

$$\sqrt{-4} \cdot \sqrt{-9} \stackrel{?}{=} \sqrt{-4 \cdot -9}$$
$$\stackrel{?}{=} \sqrt{+36}$$
$$\stackrel{?}{=} 6$$

We get the wrong answer when we use the product rule with numbers that are not real; that is, when the radicand is negative, and we don't change to jb form first!

COMMON ERROR Never use the product rule with negative radicands. Always change square roots of negative numbers to jb form before operating on them.

So we see in Examples 9 and 10 that the product of two imaginary numbers is a *real* number.

Dividing Imaginary Numbers

To divide with imaginary numbers, express the division problem as a fraction and divide by canceling like factors. Then express the result in jb form.

Example 11

Divide and simplify.

(a) $j6 \div j3$ **(b)** $-j12 \div 4$

Solution:

(a) $j6 \div j3 = \dfrac{\overset{2}{\cancel{j6}}}{\underset{1}{\cancel{j3}}} = 2$

(b) $-j12 \div 4 = \dfrac{\overset{3}{-j\cancel{12}}}{\underset{1}{\cancel{4}}} = -j3$

If a factor of j remains in the denominator after you have canceled, multiply the fraction by 1 in the form of $\dfrac{j}{j}$.

—— **Example 12** ——

Divide 8 by $j2$ and simplify.

Solution:

$$8 \div j2 = \frac{\overset{4}{\cancel{8}}}{\underset{1}{\cancel{j2}}} = \frac{4}{j}$$

Since j is in the denominator, we multiply by $\dfrac{j}{j}$

$$= \frac{4 \cdot j}{j \cdot j} = \frac{j4}{-1}$$
$$= -j4$$

SECTION 29-1 Exercises Imaginary Numbers

Imaginary Numbers in *jb* Form

Express the imaginary number in *jb* form and simplify.

1. $\sqrt{-5}$ 2. $\sqrt{-2}$ 3. $\sqrt{-900}$ 4. $\sqrt{-16}$

5. $\sqrt{-8}$ 6. $\sqrt{-27}$ 7. $-\sqrt{-12}$ 8. $-\sqrt{-200}$

9. $2\sqrt{-7}$ 10. $-3\sqrt{-3}$ 11. $-5\sqrt{-50}$ 12. $7\sqrt{-48}$

13. $\sqrt{-\dfrac{1}{9}}$ 14. $\sqrt{-\dfrac{4}{25}}$ 15. $-\sqrt{-0.16}$ 16. $\sqrt{-0.0001}$

Adding and Subtracting Imaginary Numbers

Combine and simplify.

17. $j3 - j7 + j5$ 18. $-j6 + j4 - j2$

19. $\sqrt{-25} + \sqrt{-16} - \sqrt{-100}$ 20. $\sqrt{-36} - \sqrt{-64} + \sqrt{-81}$

21. $2\sqrt{-49} - 3\sqrt{-900} + 5\sqrt{-121}$ 22. $4\sqrt{-400} + 2\sqrt{-144} - 3\sqrt{-169}$

Multiplying Imaginary Numbers

Multiply and simplify.

23. $6 \cdot j3$ 24. $-3 \cdot j8$ 25. $-j7 \cdot 4$

26. $j9 \cdot 5$ 27. $j\sqrt{2} \cdot 5$ 28. $-j3 \cdot \sqrt{7}$

29. $-\sqrt{5} \cdot j\sqrt{6}$ 30. $\sqrt{6} \cdot j\sqrt{3}$ 31. $5 \cdot \sqrt{-64}$

32. $-7 \cdot \sqrt{-81}$ 33. $-\sqrt{-36} \cdot -5$ 34. $\sqrt{-49} \cdot -6$

35. $j8 \cdot j2$ 36. $j5 \cdot j4$ 37. $j\sqrt{2} \cdot -j\sqrt{7}$

38. $j\sqrt{3} \cdot j\sqrt{5}$ 39. $j\sqrt{3} \cdot j\sqrt{3}$ 40. $j\sqrt{5} \cdot j\sqrt{5}$

41. $j3\sqrt{2} \cdot -j2\sqrt{3}$ 42. $-j5\sqrt{2} \cdot j3\sqrt{2}$ 43. $\sqrt{-4} \cdot \sqrt{-9}$

44. $-\sqrt{-25} \cdot \sqrt{-36}$ 45. $\sqrt{-7} \cdot \sqrt{-7}$ 46. $\sqrt{-3} \cdot \sqrt{-3}$

47. $-\sqrt{-3} \cdot \sqrt{-12}$ 48. $\sqrt{-5} \cdot \sqrt{-10}$ 49. $-4\sqrt{-2} \cdot -6\sqrt{-2}$

50. $-5\sqrt{-7} \cdot -3\sqrt{-7}$

Dividing Imaginary Numbers

Divide and simplify.

51. $j14 \div j2$ 52. $j15 \div j3$ 53. $j12 \div -j8$ 54. $-j20 \div j12$

55. $-j16 \div 4$ 56. $j18 \div -6$ 57. $-j20 \div -8$ 58. $j16 \div -6$

59. $5 \div j10$ 60. $-12 \div j4$ 61. $-16 \div j20$ 62. $-15 \div -j25$

29-2

COMPLEX NUMBERS

In this section, we define a complex number and its conjugate and learn that the solutions of quadratic equations are complex numbers.

Complex Number

A *complex number* is a number that can be expressed in the form

$$a + jb$$

where a and b are real numbers, and j is the imaginary unit ($j = \sqrt{-1}$).

Example 13

The number $2 + j3$ is a complex number, in which $a = 2$, $b = 3$, and $j = \sqrt{-1}$.

Rectangular Form of a Complex Number

A complex number

$$a + jb$$

is said to be in *rectangular form*. The number a is called the *real part* and the number jb is called the *imaginary part*.

RECTANGULAR FORM OF A COMPLEX NUMBER	$a + jb$ ↗ ↖ real part imaginary part	154

If the coefficient b is negative, the negative sign is usually written in front of the j.

Example 14

The complex number $2 + j(-3)$ is usually written as $2 - j3$.

Example 15

Express the complex number in rectangular form:

$$-3 - \sqrt{-8}$$

Solution: Factoring $\sqrt{-1}$ from $\sqrt{-8}$, we have

$$-3 - \sqrt{-8} = -3 - \sqrt{-1}\sqrt{8}$$
$$= -3 - j\sqrt{8}$$

Simplifying $\sqrt{8}$, we have

$$= -3 - j2\sqrt{2}$$

If the real part a of a complex number is zero ($a = 0$), then the number, in rectangular form, is

$$a + jb = 0 + jb = jb$$

The number jb is called a *pure imaginary number.* Every pure imaginary number is also a complex number, since it can be written as $a + jb$.

If the coefficient b of a complex number is zero ($b = 0$), then the number, in rectangular form, is

$$a + jb = a + j0 = a$$

The number a is a *real* number, but it is also a *complex* number, since it can be written as $a + jb$. Every real number is a complex number.

Example 16

Express each number as a complex number in rectangular form.

(a) 5   **(b)** $\sqrt{-50}$

Solution:

(a) $5 = 5 + j0$

So the number 5 is both a real number and a complex number.

(b) $\sqrt{-50} = j\sqrt{50} = 0 + j5\sqrt{2}$

So the number $\sqrt{-50}$ is both a pure imaginary number and a complex number.

Conjugate of a Complex Number

The *conjugate* of a complex number $a + jb$ is the complex number $a - jb$. A complex number and its conjugate differ only in the *sign* of the imaginary part. Conjugates are used in division with complex numbers.

Example 17

Write the conjugate of the complex numbers:

(a) $-4 + j3$ **(b)** $5 - j2$

Solution:

(a) The conjugate of $-4 + j3$ is $-4 - j3$.
(b) The conjugate of $5 - j2$ is $5 + j2$.

Since real numbers and pure imaginary numbers are complex numbers, they also have conjugates.

Example 18

Write the conjugate of each number.

(a) -8 **(b)** $j7$

Solution:

(a) First we express the real number -8 as a complex number in rectangular form:

$$-8 = -8 + j0$$

The conjugate of $-8 + j0$ is $-8 - j0$. But

$$-8 - j0 = -8$$

So the conjugate of -8 is -8. In fact, *the conjugate of any real number is the real number itself.*

(b) We write the pure imaginary number $j7$ as a complex number in rectangular form:

$$j7 = 0 + j7$$

The conjugate of $0 + j7$ is $0 - j7$. But

$$0 - j7 = -j7$$

So the conjugate of $j7$ is $-j7$. In fact, *the conjugate of any pure imaginary number is its opposite.*

Complex Solutions of a Quadratic Equation

We saw in Chapter 21 that every quadratic equation has two solutions. In that chapter, we solved quadratic equations with solutions that were real numbers. Now we can solve quadratic equations with solutions that are not real numbers—they are non-real complex numbers.

Example 19

Solve the quadratic equation:

$$x^2 + 27 = 0$$

Solution: We use the method of Sec. 21-1 to solve the pure quadratic $x^2 + 27 = 0$.

$$x^2 = -27$$
$$x = \pm\sqrt{-27}$$
$$x = \pm j3\sqrt{3}$$

The solutions of this quadratic equation are two non-real pure imaginary complex numbers.

Solution by Quadratic Formula

To solve the quadratic equation $ax^2 + bx + c = 0$, we may use the quadratic formula. If the discriminant, $b^2 - 4ac$, is negative, then the two solutions of the quadratic equation are non-real complex numbers.

Example 20

Solve the quadratic equation $x^2 - 6x + 13 = 0$.

Solution: Applying the quadratic formula to the equation $x^2 - 6x + 13 = 0$, we have

$$x = \frac{-(-6) \pm \sqrt{(-6)^2 - 4(1)(13)}}{2(1)} = \frac{6 \pm \sqrt{-16}}{2}$$

The discriminant is negative, so we know the solutions are complex. Simplifying $\sqrt{-16}$, we get

$$x = \frac{6 \pm j4}{2}$$

Simplifying the fraction, we get

$$x = \frac{2(3 \pm j2)}{2} = 3 \pm j2$$

So, the two solutions of the equation $x^2 - 6x + 13 = 0$ are $3 + j2$ and $3 - j2$. The two solutions of this quadratic equation are conjugates of each other. In fact, *the complex solutions of a quadratic equation are always conjugates.*

SECTION 29-2 Exercises Complex Numbers

Rectangular Form of a Complex Number

Express the complex number in rectangular form.

1. $3 + \sqrt{-2}$ 2. $-2 - \sqrt{-5}$ 3. $\sqrt{-12} - 5$ 4. $-\sqrt{-18} - 4$
5. -7 6. 9 7. $\sqrt{-24}$ 8. $-\sqrt{-27}$

Conjugate of a Complex Number

Write the conjugate of the complex number.

9. $3 + j8$ 10. $-4 - j7$ 11. $-5 - j\sqrt{2}$ 12. $6 + j\sqrt{5}$
13. -2 14. 5 15. $j4$ 16. $-j5$

Complex Solutions of a Quadratic Equation

Solve the quadratic equation. Write the complex solutions in rectangular form.

17. $x^2 + 25 = 0$ 18. $x^2 + 49 = 0$
19. $x^2 + 48 = 0$ 20. $x^2 + 75 = 0$
21. $x^2 + 4x + 13 = 0$ 22. $x^2 - 10x + 34 = 0$
23. $x^2 - 8x + 20 = 0$ 24. $x^2 + 2x + 2 = 0$
25. $2x^2 + 2x + 3 = 0$ 26. $3x^2 - 4x + 16 = 0$

29-3   OPERATIONS WITH COMPLEX NUMBERS

In this section we learn to perform the basic operations of addition, subtraction, multiplication, and division with complex numbers. Before performing any operations, the complex numbers should be simplified and expressed in rectangular form.

Adding Complex Numbers

The addition of complex numbers is similar to the addition of binomial expressions. To add complex numbers, first we remove parentheses, and then combine the real parts and the imaginary parts separately. We simplify if possible and express the result in rectangular form.

┌─── **Example 21** ───
│ Add $(2 - j5) + (-4 + j8)$.

Solution: Removing parentheses:

$$(2 - j5) + (-4 + j8) = 2 - j5 - 4 + j8$$

Combining parts separately,

$$= (2 - 4) + (-j5 + j8)$$

Simplifying,

$$= -2 + j3$$

Alternate Solution: Complex numbers can also be added vertically.

$$\text{Add:} \quad \begin{array}{r} 2 - j5 \\ -4 + j8 \\ \hline -2 + j3 \end{array}$$

Subtracting Complex Numbers

To subtract complex numbers, we first remove the parentheses by changing signs, and then combine the real parts and the imaginary parts separately.

Example 22

Subtract $(-3 + j7) - (5 - j2)$.

Solution: Removing parentheses:

$$(-3 + j7) - (5 - j2) = -3 + j7 - 5 + j2$$

Combining parts separately,

$$= (-3 - 5) + (j7 + j2)$$

Simplifying,

$$= -8 + j9$$

Alternate Solution: To subtract vertically, we change the signs of the second (lower) number and add.

$$\text{To subtract:} \quad \begin{array}{r} -3 + j7 \\ 5 - j2 \end{array} \qquad \text{add:} \quad \begin{array}{r} -3 + j7 \\ -5 + j2 \\ \hline -8 + j9 \end{array}$$

Multiplying Complex Numbers

Use the distributive property to multiply complex numbers just as you multiplied algebraic expressions. Whenever j^2 occurs, replace j^2 with -1. Express the result in rectangular form.

Example 23

Multiply:

(a) $3(-2 + j5)$ **(b)** $j4(3 - j7)$

Solution: We use the familiar distributive property to multiply.

(a)
$$3(-2 + j5) = 3 \cdot -2 + 3 \cdot j5$$
$$= -6 + j15$$

(b)
$$j4(3 - j7) = j4 \cdot 3 - j4 \cdot j7$$
$$= j12 - j^2 \cdot 28$$

Replacing j^2 with -1,

$$= j12 - (-1)(28)$$
$$= j12 + 28$$

Writing in rectangular form,

$$= 28 + j12$$

Example 24

Multiply $(2 - j3)(-1 + j4)$.

Solution: We multiply as with binomials, using FOIL:

$$\begin{aligned}
(2 - j3)(-1 + j4) &= 2(-1) + 2(j4) + (-j3)(-1) + (-j3)(j4) \\
&= -2 + j8 + j3 + j^2(-12) \\
&= -2 + j8 + j3 + (-1)(-12) \\
&= -2 + j8 + j3 + 12 \\
&= 10 + j11
\end{aligned}$$

Example 25

Multiply $(3 + j2)(3 - j2)$.

Solution: Again we multiply as with binomials:

$$\begin{aligned}
(3 + j2)(3 - j2) &= 3(3) + 3(-j2) + (j2)(3) + (j2)(-j2) \\
&= 9 - j6 + j6 - j^2(4) \\
&= 9 - j6 + j6 - (-4) \\
&= 13
\end{aligned}$$

Note that the product in Example 25 is a real number, and that we were multiplying conjugates. In fact, *the product of complex conjugates is always a real number*. These factors have the same form as the difference of two squares, and the middle terms (the j terms) combine to make zero, leaving only the real part.

Dividing Complex Numbers

In the division of complex numbers, we generally want to express the quotient in rectangular form. Division by a single term, real or complex, is done term by term in fraction form.

Example 26

Divide $(6 - j4) \div 2$.

Solution: First we change the division problem to fraction form,

$$(6 - j4) \div 2 = \frac{6 - j4}{2}$$

Divide term by term and cancel,

$$= \frac{\cancelto{3}{6}}{\cancelto{1}{2}} - \frac{j\cancelto{2}{4}}{\cancelto{1}{2}}$$

and write in rectangular form,

$$= 3 - j2$$

We want to express a quotient in rectangular form, so whenever j appears in a denominator of a fraction, it is common practice to eliminate it. The j can be eliminated from the denominator of a fraction by multiplying the fraction by 1, in the form of $\dfrac{j}{j}$.

664 CHAPTER 29 / Complex Numbers

Example 27

Divide $(-12 + j8) \div j4$.

Solution:

$$(-12 + j8) \div j4 = \frac{-12 + j8}{j4}$$

$$= \frac{-\overset{3}{\cancel{12}}}{\underset{1}{\cancel{j4}}} + \frac{\overset{2}{j\cancel{8}}}{\underset{1}{\cancel{j4}}}$$

$$= \frac{-3}{j} + 2$$

In this case, j appears in the denominator of the fraction $\dfrac{-3}{j}$, and we eliminate it from the denominator by multiplying the fraction by $\dfrac{j}{j}$, so

$$(-12 + j8) \div j4 = \frac{-3}{j} \cdot \frac{j}{j} + 2$$

$$= \frac{-j3}{-1} + 2$$

$$= j3 + 2$$

$$= 2 + j3$$

If the divisor in a division problem is a complex number, $a + jb$, where neither a nor b is 0, then eliminating the j from the denominator is more complicated. Recall from Example 25 that multiplying a complex number by its conjugate gives you a product that is a real number. So we can eliminate the j from the denominator by multiplying both the numerator and the denominator by the conjugate of the denominator. Then we simplify and express the result in rectangular form. We observe that this process is similar to *rationalizing the denominator* of a fraction containing radicals.

Example 28

Divide $(1 + j4) \div (3 + j2)$.

Solution: We write the division problem in fraction form:

$$(1 + j4) \div (3 + j2) = \frac{1 + j4}{3 + j2}$$

Multiplying the numerator and denominator by the conjugate of the denominator $(3 - j2)$,

$$(1 + j4) \div (3 + j2) = \frac{(1 + j4)}{(3 + j2)} \cdot \frac{(3 - j2)}{(3 - j2)}$$

Simplifying,

$$= \frac{3 - j2 + j12 - j^2 \cdot 8}{9 - j^2 \cdot 4}$$

$$= \frac{3 + j10 + 8}{9 + 4}$$

$$= \frac{11 + j10}{13}$$

Expressing in rectangular form:

$$= \frac{11}{13} + j\frac{10}{13}$$

Adding and Subtracting Complex Numbers

Combine and simplify.

1. $2 - j4 + j7 - 5$

2. $j5 - 7 + j2 - 4$

3. $(-3 + j8) + (2 - j5)$

4. $(5 - j3) + (-3 - j4)$

5. $(4 - j3) + (4 + j3)$

6. $(-5 + j4) + (-5 - j4)$

7. $(-7 + j7) - (3 - j4)$

8. $(5 + j4) - (8 - j3)$

9. $(-1 + j2) - (-1 - j2)$

10. $(6 - j) - (6 + j)$

Add.

11. $-8 + j6$
$\underline{ 5 - j3}$

12. $9 - j8$
$\underline{-7 + j7}$

13. $6 - j5$
$-8 + j$
$-5 - j4$
$\underline{ 4 - j2}$

14. $-1 + j6$
$7 - j$
$-5 - j5$
$\underline{ 8 - j4}$

Subtract.

15. $9 - j7$
$\underline{5 + j4}$

16. $-4 + j3$
$\underline{-5 - j8}$

17. $-5 - j$
$\underline{-6 - j3}$

18. $7 + j4$
$\underline{-3 + j}$

Multiplying Complex Numbers

Multiply and simplify.

19. $5(-3 + j2)$

20. $-6(4 - j5)$

21. $-7(-4 - j6)$

22. $8(-5 + j)$

23. $j(2 - j3)$

24. $-j3(-3 + j4)$

25. $-j5(-6 + j4)$

26. $j7(8 - j)$

27. $(3 - j)(5 - j2)$

28. $(5 + j4)(-3 + j)$

29. $(-4 + j3)(1 - j5)$

30. $(-7 - j4)(-6 - j5)$

31. $(9 + j5)(-4 + j8)$

32. $(-8 - j3)(7 + j7)$

33. $(7 + j4)(7 - j4)$

34. $(-5 - j3)(-5 + j3)$

35. $(-6 - j5)(-6 + j5)$

36. $(8 + j3)(8 - j3)$

37. $(4 - j3)^2$

38. $(5 + j6)^2$

39. $(-7 + j2)^2$

40. $(-3 - j5)^2$

Dividing Complex Numbers

Divide and express in rectangular form.

41. $(4 - j8) \div 2$

42. $(6 - j9) \div 3$

43. $(-9 + j12) \div -6$

44. $(-10 - j25) \div -15$

45. $(15 - j20) \div j5$

46. $(-12 + j15) \div -j3$

47. $(-18 - j24) \div -j12$

48. $(8 - j6) \div j4$

49. $3 \div (1 - j)$

50. $-5 \div (2 + j3)$

51. $-j4 \div (3 + j2)$

52. $j2 \div (2 + j)$

53. $(2 + j5) \div (1 + j)$

54. $(1 + j) \div (-4 - j)$

55. $(-3 - j4) \div (2 - j)$

56. $(-3 + j) \div (-2 - j)$

57. $(6 + j5) \div (-2 - j3)$

58. $(-1 - j) \div (3 + j5)$

29-4   GRAPH OF A COMPLEX NUMBER

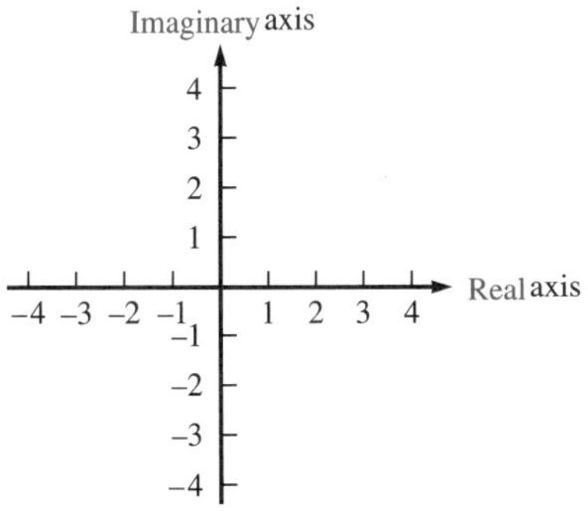

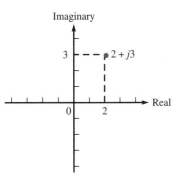

■ FIGURE 29-1
The complex plane.

In this section we define the complex plane and show that every complex number can be represented by a point on the complex plane. We show that a vector can be represented as a complex number and vice versa, and that vector addition is greatly simplified using the complex number representation.

Complex Plane

A complex number, $a + jb$, consists of two parts: a real part a and an imaginary part jb. To graph a complex number we use two axes, one for the real part and one for the imaginary part. We use a form of the rectangular coordinate system (Fig. 29-1). The horizontal axis is called the *real axis*. The vertical axis is called the *imaginary axis*. Such a coordinate system is called the *complex plane*.

Graph of a Complex Number

Every complex number can be represented by a *point* on the complex plane. To plot a complex number $a + jb$ on the complex plane, locate a point with a horizontal coordinate a and a vertical coordinate b. (Start at the origin. Go a units to the right or left of the origin, and b units up or down from the horizontal axis.)

― **Example 29** ―――――――

Graph the complex number $2 + j3$.

Solution: Draw the two perpendicular axes, label the axes *real* and *imaginary,* and label the units on each axis. To graph the complex number $2 + j3$, plot the point which is 2 units to the right of the origin on the real (horizontal) axis and 3 units up (Fig. 29-2). This point on the complex plane is labeled by writing the number $2 + j3$, and do not enclose the complex number with parentheses.

■ FIGURE 29-2

This process is similar to plotting the point (a, b) on the standard rectangular coordinate system. However, a point on the complex plane represents a single complex number, *not* an ordered pair of real numbers.

■ FIGURE 29-3

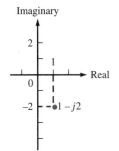

― **Example 30** ―――――――

Graph the complex number $1 - j2$.

Solution: To graph the complex number $1 - j2$, plot the point that is 1 unit to the right of the origin and two units *down*, since the imaginery coefficient b is -2 (Fig. 29-3).

Graphs of Real and Pure Imaginary Numbers

Real and pure imaginary numbers can also be graphed on the complex plane. Real numbers are graphed on the real axis and pure imaginary numbers are graphed on the imaginary axis.

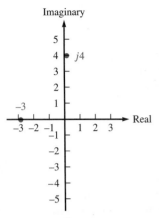

FIGURE 29-4

Example 31

Graph the numbers on the same complex plane:
(a) -3 **(b)** $j4$

Solution:

(a) First we write -3 as a complex number in rectangular form:

$$-3 = -3 + j0$$

To graph the complex number $-3 + j0$, we plot the point which is 3 units to the left of the origin and 0 units up (Fig. 29-4). Note that the point is *on* the real axis. This is reasonable, since -3 is a real number.

(b) First we write $j4$ as a complex number in rectangular form:

$$j4 = 0 + j4$$

To graph the number $0 + j4$ on the complex plane, plot the point which is 0 units to the right of the origin and 4 units up. You will note that the point is *on* the imaginary axis. This is reasonable, since $j4$ is a pure imaginary number (Fig. 29-4).

Vectors and Complex Numbers

One of the most important applications of complex numbers is that they can represent vectors, and vice versa.

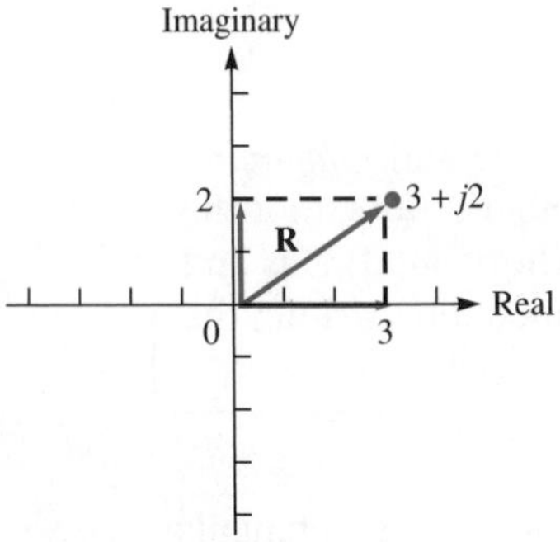

Example 32

Consider the complex number $3 + j2$, which is plotted as a point in Fig. 29-5. If we draw a line segment from the origin to the point, then we have a *vector*, which we call **R**. The vector **R** has a horizontal component of 3 units and a vertical component of 2 units.

Vector Addition

The sum of two complex numbers may be found graphically. In Sec. 23-3 we learned the parallelogram method of adding two vectors. When we consider complex numbers as vectors, we may use the parallelogram method to add them graphically.

FIGURE 29-5
Vector representation of a complex number.

FIGURE 29-6
Sum of two complex numbers.

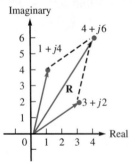

Example 33

Add $(3 + j2) + (1 + j4)$ graphically on the complex plane.

Solution: We plot the two points on the complex plane and draw the vectors associated with those points. We complete the other two sides of the parallelogram and draw the diagonal (Fig. 29-6). The fourth vertex is the point which represents the sum of the two complex numbers, $4 + j6$. Of course, if we had added the complex numbers algebraically we would have found the same result:

$$(3 + j2) + (1 + j4) = 4 + j6$$

So we have a very important result here: *The resultant of two vectors representing complex numbers is a vector that represents the sum of the complex numbers.*

Complex Plane

Graph each complex number.

1. $1 + j2$	**2.** $3 - j4$	**3.** $4 - j5$	**4.** $5 + j2$
5. $-3 + j3$	**6.** $-4 - j$	**7.** $-2 - j$	**8.** $-1 + j2$
9. 4	**10.** -2	**11.** $j2$	**12.** $-j3$
13. -5	**14.** 1	**15.** $-j4$	**16.** j

17. Write the complex numbers represented by the points in Fig. 29-7.

18. Write the complex numbers represented by the points in Fig. 29-8.

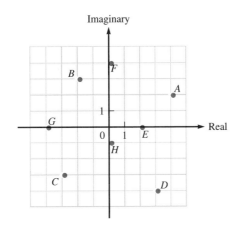

■ **FIGURE 29-7**

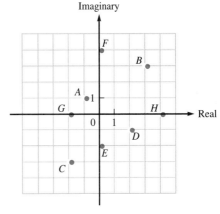

■ **FIGURE 29-8**

Vector Addition

Add the complex numbers graphically, and check the result algebraically.

19. $(2 + j) + (3 + j4)$ **20.** $(4 + j3) + (2 - j)$

21. $(3 - j2) + (-6 - j3)$ **22.** $(-5 - j4) + (3 - j2)$

23. $-1 + j2$ **24.** $3 - j5$ **25.** $2 + j$ **26.** $-4 - j3$
     $\underline{-1 - j2}$ $\underline{3 + j5}$ $-3 - j4$ $-5 + j$
                                                     $\underline{5 - j2}$ $\underline{7 - j2}$

29-5

POLAR AND TRIGONOMETRIC FORMS

Thus far we have expressed complex numbers in only one form, the rectangular form, $a + jb$. In this section we learn to express complex numbers in two other forms, the polar form and the trigonometric form.

Polar Form of a Complex Number

■ **FIGURE 29-9**
Polar form of a complex number.

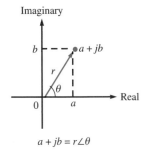

$a + jb = r\angle\theta$

Figure 29-9 shows a complex number $a + jb$, and its vector representation. We let the length of the vector be r. The vector is at an angle of θ with the horizontal axis (in standard position), where $0 \leq \theta < 360°$. The radius r is called the *absolute value* of the complex number and can be found using the Pythagorean Theorem:

ABSOLUTE VALUE r OF THE COMPLEX NUMBER $a + jb$	$r = \sqrt{a^2 + b^2}$	158

The angle θ is called the *argument* of the complex number. Note that in solving for θ, we get

ARGUMENT θ OF THE COMPLEX NUMBER $a + jb$	$\tan \theta = \dfrac{b}{a}$	159

where $0 \le \theta < 360°$.

Any complex number can be expressed in terms of r and θ, by writing the absolute value r, the angle symbol $\angle$, and then the argument θ. This is called the *polar form* of the complex number.

POLAR FORM OF A COMPLEX NUMBER $a + jb$	$a + jb = r\underline{/\theta}$	160

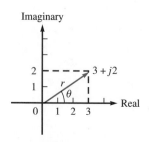

■ **FIGURE 29-10**

─── **Example 34** ───

Given the complex number $3 + j2$.

(a) Graph the number as a vector.

(b) Find the absolute value r and the argument θ, in degrees, rounding to three significant digits.

(c) Express the number in polar form.

Solution:

(a) We graph the number $3 + j2$ in Fig. 29-10.

(b) Using Eq. 158, we find the absolute value r:

$$r = \sqrt{3^2 + 2^2} = \sqrt{13} = 3.61$$

Using Eq. 159,

$$\tan \theta = \frac{2}{3}$$

Solving for θ, in quadrant I,

$$\theta = \tan^{-1} \frac{2}{3} = 33.7°$$

(c) Using Eq. 160, $3 + j2$ can be expressed in polar form, as

$$3 + j2 = 3.61\underline{/33.7°}$$

If the graph of a complex number is in quadrant II, III, or IV, then we find the reference angle of the argument, θ_R, using

$$\theta_R = \tan^{-1} \left| \frac{b}{a} \right|$$

Absolute value is used, since the reference angle is always positive. We use the quadrant and the reference angle θ_R to find θ, as we did in Chapter 22.

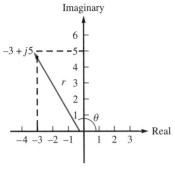

Imaginary

FIGURE 29-11

Example 35

Given the complex number $-3 + j5$,

(a) Graph the number as a vector

(b) Express it in polar form.

Solution:

(a) We graph the number $-3 + j5$ in Fig. 29-11, and see that its graph is in quadrant II.

(b) Using Eq. 158 to find r,

$$r = \sqrt{(-3)^2 + 5^2} = \sqrt{34} = 5.83$$

To find θ_R, we use the equation

$$\theta_R = \tan^{-1} \left| \frac{b}{a} \right|$$

or,

$$\theta_R = \tan^{-1} \left| \frac{5}{-3} \right| = 59.0°$$

In quadrant II,

$$\theta = 180° - \theta_R$$

Substituting,

$$\theta = 180° - 59.0°$$

Or,

$$\theta = 121°$$

So,

$$-3 + j5 = 5.83 \underline{/121°} \quad \text{in polar form}$$

Suppose we know the polar form of a complex number, $r \underline{/\theta}$, and we wish to express it in rectangular form, $a + jb$. That is, we know the absolute value r and the argument θ of the complex number, and we want to find a and b to express it in rectangular form. Referring to Fig. 29-9, we see that both a and b can be found by trigonometry:

$$\cos \theta = \frac{a}{r} \quad \text{and} \quad \sin \theta = \frac{b}{r}$$

Solving for a and b, respectively, we get

POLAR TO RECTANGULAR FORM CONVERSIONS	$a = r \cos \theta$	156
	$b = r \sin \theta$	157

Example 36

The polar form of a complex number is $9.00 \underline{/105°}$. Graph the number as a vector and express it in rectangular form.

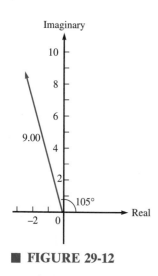

Imaginary

10

8

6

9.00

4

2

105°

-2 0 → Real

■ FIGURE 29-12

Solution: We graph the number in Fig. 29-12 and use Eqs. 156 and 157 with $r = 9.00$ and $\theta = 105°$ to get

$$a = 9.00 \cos 105° = -2.33$$
$$b = 9.00 \sin 105° = 8.69$$

So, rounding to three significant digits, and expressing the number in rectangular form, we have

$$9.00 \underline{/105°} = -2.33 + j8.69$$

Trigonometric Form of a Complex Number

If we use Eqs. 156 and 157 and substitute the values for a and b into the rectangular form, we have

$$a + jb = r \cos \theta + j \cdot r \sin \theta$$

Factoring out the r from the right side, we have the *trigonometric form* of a complex number.

TRIGONOMETRIC FORM OF A COMPLEX NUMBER, $a + jb$	$a + jb = r(\cos \theta + j \sin \theta)$	158

Example 37

Write the complex number $3 + j2$ in trigonometric form, accurate to three significant digits.

Solution: In Example 34, we expressed $3 + j2$ in polar form, finding that

$$r = 3.61 \quad \text{and} \quad \theta = 33.7°$$

Using Eq. 155, we have

$$3 + j2 = 3.61(\cos 33.7° + j \sin 33.7°)$$

Example 38

Verify that the right side of the answer in Example 37 equals the left side.

Solution: We distribute the 3.61 on the right side and use the calculator to find the products:

$$3.61(\cos 33.7° + j \sin 33.7°) = 3.61 \cos 33.7° + j(3.61 \sin 33.7°)$$
$$= 3.00 + j2.00 \quad \text{(accurate to three significant digits)}$$

Example 39

Write the complex number

$$5.28(\cos 68.1° + j \sin 68.1°)$$

(a) in polar form, and **(b)** in rectangular form, rounded to three significant digits.

Solution:

(a) We are given the trigonometric form, and can see from Eq. 155 that

$$r = 5.28 \quad \text{and} \quad \theta = 68.1°$$

Using Eq. 160, we write the complex number in polar form:

$$5.28\underline{/68.1°}$$

(b) To find the rectangular form we use the trigonometric form, distribute the 5.28, and use a calculator to find the products:

$$5.28(\cos 68.1° + j \sin 68.1°) = 5.28 \cos 68.1° + j(5.28 \sin 68.1°)$$
$$= 1.97 + j4.90$$

SECTION 29-5 Exercises Polar and Trigonometric Forms

Polar Form of a Complex Number

Graph each complex number and express it in polar form, rounding to three significant digits.

1. $2 + j3$ **2.** $4 - j$ **3.** $-4 - j2$ **4.** $-3 + j4$

Graph each complex number and express it in rectangular form, rounding to three significant digits.

5. $8.00 \underline{/56.5°}$ **6.** $43.7 \underline{/235°}$

7. $549 \underline{/235°}$ **8.** $0.702 \underline{/345°}$

Trigonometric Form of a Complex Number

Graph each complex number and express it in trigonometric form.

9. $1 + j$ **10.** $3 + j4$ **11.** $3 - j2$ **12.** $-6 - j$

13. $-4 + j7$ **14.** $-5 + j3$ **15.** $-6 - j3$ **16.** $1 - j7$

Express each complex number in both polar and rectangular forms (rounded to three significant digits).

17. $2(\cos 30° + j \sin 30°)$ **18.** $4(\cos 45° + j \sin 45°)$

19. $3.56(\cos 135° + j \sin 135°)$ **20.** $5.74(\cos 121° + j \sin 121°)$

21. $8.45(\cos 297° + j \sin 297°)$ **22.** $78.9(\cos 247° + j \sin 247°)$

23. Verify the following rule for the product of two complex numbers in polar form

$$\left(r_1\underline{/\theta_1}\right)\left(r_2\underline{/\theta_2}\right) = r_1 r_2 \underline{/\theta_1 + \theta_2}$$

by showing that

$$(2\underline{/45°})(5\underline{/30°}) = 10\underline{/75°}.$$

(Hint: Change to rectangular form, multiply, and change back to polar form.)

24. Verify the following rule for the quotient of two complex numbers in polar form

$$\frac{r_1\underline{/\theta_1}}{r_2\underline{/\theta_2}} = \frac{r_1}{r_2}\underline{/\theta_1 - \theta_2}$$

by showing that

$$\frac{2\underline{/45°}}{5\underline{/30°}} = \frac{2}{5}\underline{/15°}$$

Graph and express each complex number in polar form, rounding to three significant digits.

1. $-2 + j5$ **2.** $5 - j7$ **3.** $7 + j$ **4.** $-4 - j4$

Perform the indicated operations and express in rectangular form.

5. $2\sqrt{-64} - 5\sqrt{-25} + 4\sqrt{-36}$ **6.** $-j\sqrt{3} \cdot -\sqrt{6}$

7. $(4 - j3)(1 - j2)$ **8.** $(7 - j5)(7 + j5)$

9. $2\sqrt{-3} \cdot -3\sqrt{-3}$ **10.** $-j18 \div j8$

11. $(-6 + j8) \div 4$ **12.** $(12 - j9) \div -j6$

13. $(3 - j5) + (-2 - j7)$ **14.** $(-3 - j4) - (-6 - j5)$

15. $-4(6 - j3)$ **16.** $-j3(2 - j4)$

17. $(-3 + j4)^2$ **18.** $(-2 - j)^2$

19. $-5 \div (-1 + j3)$ **20.** $j4 \div (2 + j3)$

21. $-2 \cdot \sqrt{-8}$ **22.** $-j7 \cdot -j6$

23. $j15 \div -3$ **24.** $-26 \div -j8$

25. $j\sqrt{5} \cdot j\sqrt{5}$ **26.** $\sqrt{-16} \cdot \sqrt{-64}$

27. $(5 - j3) \div (7 + j2)$ **28.** $(3 - j4) \div (3 + j4)$

Add graphically on the complex plane and check the result algebraically.

29. $(3 - j5) + (-4 - j)$ **30.** $-2 - j5$
                                          $\underline{4 + j3}$

31. $-2 + j5$ **32.** $-4 - j2$
          $5 - j$ $\underline{-3 + j5}$
          $\underline{-5 - j4}$

Graph and express each complex number in trigonometric form, rounding to three significant digits.

33. $2 + j2$ **34.** $-4 + j6$ **35.** $-5 - j3$ **36.** $5 - j6$

Solve the quadratic equations. Write the complex solutions in rectangular form.

37. $x^2 + 72 = 0$ **38.** $x^2 + 6x + 10 = 0$

Express each complex number in both polar and rectangular forms, rounding to three significant digits.

39. $4.25(\cos 123° + j \sin 123°)$ **40.** $35.6(\cos 347° + j \sin 347°)$

WRITING

41. Some complex numbers are real, some are pure imaginary, some are neither. Explain the relationship among real, pure imaginary, and complex numbers. Give an example of each type and graph each number on the complex plane. Explain how the graph of the number relates to the type of number.

42. We know that $\sqrt[3]{1} = 1$. So, in the real number system, the solution of the equation

$$x^3 = 1$$

is

$$x = 1$$

But, in the complex number system, the equation

$$x^3 = 1$$

has three solutions. These three solutions can be found by factoring the equation

$$x^3 - 1 = 0$$

and setting the factors equal to zero. Find the three complex cube roots of 1. Check, by cubing each number.

Summary of Facts and Formulas

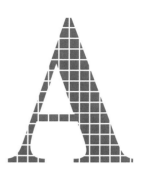

	No.				Page
ALGEBRAIC LAWS	1	Commutative Law	Addition	$a + b = b + a$	7
	2		Multiplication	$ab = ba$	11, 140
	3	Associative Law	Addition	$a + (b + c) = (a + b) + c = (a + c) + b$	7
	4		Multiplication	$a(bc) = (ab)c = (ac)b = abc$	11
	5	Distributive Law		$a(b + c) = ab + ac$	12, 142, 319
RULES OF SIGNS	6	Addition and Subtraction		$a + (-b) = a - (+b) = a - b$	105
	7			$a + (+b) = a - (-b) = a + b$	106
	8			$-(a + b) = -a - b \qquad -(a - b) = b - a$	346
	9	Multiplication		$(+a)(+b) = (-a)(-b) = +ab$	108, 139
	10			$(+a)(-b) = (-a)(+b) = -(+a)(+b) = -ab$	108, 139
	11	Division		$\dfrac{+a}{+b} = \dfrac{-a}{-b} = -\dfrac{-a}{+b} = -\dfrac{+a}{-b} = \dfrac{a}{b}$	154, 346
	12			$\dfrac{+a}{-b} = \dfrac{-a}{+b} = -\dfrac{-a}{-b} = -\dfrac{a}{b}$	154, 346
PERCENTAGE	13			Amount = base × rate, $\quad A = BP$	90
	14			Percent change $= \dfrac{\text{new value} - \text{original value}}{\text{original value}} \times 100$	95
	15			Percent error $= \dfrac{\text{measured value} - \text{known value}}{\text{known value}} \times 100$	97
	16			Percent concentration of ingredient $A = \dfrac{\text{amount of } A}{\text{amount of mixture}} \times 100$	98, 190
	17			Percent efficiency $= \dfrac{\text{output}}{\text{input}} \times 100$	98

Source: Adapted from Paul Calter, *Technical Mathematics, Third Edition,* Prentice-Hall, 1995.

	No.				Page
EXPONENTS	18	Definition		$x^n = x \cdot x \cdot x \cdot \ldots \cdot x$ $\underbrace{\hphantom{xxxxxx}}_{n \text{ factors}}$	132
	19	Laws of Exponents	Product	$x^m \cdot x^n = x^{m+n}$	132, 422, 613
	20		Quotient	$\dfrac{x^m}{x^n} = x^{m-n} \quad (x \neq 0)$	133, 154, 422, 613
	21		Power	$(x^m)^n = x^{mn} = (x^n)^m$	134, 422, 613
	22		Product Raised to a Power	$(xy)^n = x^n \cdot y^n$	134, 422, 613
	23		Quotient Raised to a Power	$\left(\dfrac{x}{y}\right)^n = \dfrac{x^n}{y^n} \quad (y \neq 0)$	135, 422, 613
	24		Zero Exponent	$x^0 = 1 \quad (x \neq 0)$	135, 422
	25		Negative Exponent	$x^{-n} = \dfrac{1}{x^n} \quad (x \neq 0)$	136, 422, 613
	26	Fractional Exponents		$\sqrt[r]{x} = x^{\frac{1}{r}}$	423
	27			$x^{\frac{p}{r}} = \left(\sqrt[r]{x}\right)^p = \sqrt[r]{x^p}$	424
RADICALS	28	Rules of Radicals		$\sqrt[r]{x^r} = x \ \text{ and } \ \left(\sqrt[r]{x}\right)^r = x$	424
	29		Root of a Product	$\sqrt[r]{xy} = \sqrt[r]{x}\sqrt[r]{y}$	427
	30		Root of a Quotient	$\sqrt[r]{\dfrac{x}{y}} = \dfrac{\sqrt[r]{x}}{\sqrt[r]{y}}$	430
SPECIAL PRODUCTS AND FACTORING	31	Binomials	Difference of Two Squares	$x^2 - y^2 = (x - y)(x + y)$	144, 325
	32		Sum of Two Cubes	$x^3 + y^3 = (x + y)(x^2 - xy + y^2)$	341
	33		Difference of Two Cubes	$x^3 - y^3 = (x - y)(x^2 + xy + y^2)$	341
	34	Trinomials	Test for Factorability	$ax^2 + bx + c$ is factorable if $b^2 - 4ac$ is a perfect square	335
	35		Leading Coefficient = 1	$x^2 + (a + b)x + ab = (x + a)(x + b)$	145, 329
	36		General Quadratic Trinomial	$acx^2 + (ad + bc)x + bd = (ax + b)(cx + d)$	145, 333
	37		Perfect Square Trinomials	$x^2 + 2xy + y^2 = (x + y)^2$	148, 339
	38			$x^2 - 2xy + y^2 = (x - y)^2$	148, 339

Category	No.	Label	Sub-label	Formula	Page
FRACTIONS	39	Simplifying		$\dfrac{ac}{bc} = \dfrac{a}{b}$	347
	40			$\dfrac{a \div c}{b \div c} = \dfrac{a}{b}$	347
	41	Multiplication		$\dfrac{a}{b} \cdot \dfrac{c}{d} = \dfrac{ac}{bd}$	352
	42	Division		$\dfrac{a}{b} \div \dfrac{c}{d} = \dfrac{a}{b} \cdot \dfrac{d}{c} = \dfrac{ad}{bc}$	353
	43	Addition and Subtraction	Same Denominators	$\dfrac{a}{b} \pm \dfrac{c}{b} = \dfrac{a \pm c}{b}$	357
	44		Different Denominators	$\dfrac{a}{b} \pm \dfrac{c}{d} = \dfrac{ad}{bd} \pm \dfrac{bc}{bd} = \dfrac{ad \pm bc}{bd}$	360
RATIO AND PROPORTION	45	Ratio of a to b		$a : b$ or $\dfrac{a}{b}$	585
	46	Proportion		$\dfrac{a}{b} = \dfrac{c}{d}$	589
	47	Fundamental Principle of Proportions $\dfrac{a}{b} = \dfrac{c}{d}$		The product of the extremes equals the product of the means $\qquad ad = bc$	590
VARIATION	48	$k =$ Constant of Proportionality	Direct	$y \propto x$ or $y = kx$	597
	49		Joint	$y \propto xw$ or $y = kxw$	603
	50		Inverse	$y \propto \dfrac{1}{x}$ or $y = \dfrac{k}{x}$	604
	51		Combined	$y \propto \dfrac{x}{w}$ or $y = \dfrac{kx}{w}$	605
SYSTEMS OF LINEAR EQUATIONS	52	Algebraic Solution	$a_1x + b_1y = c_1$ $a_2x + b_2y = c_2$	$x = \dfrac{b_2c_1 - b_1c_2}{a_1b_2 - a_2b_1}$, and $y = \dfrac{a_1c_2 - a_2c_1}{a_1b_2 - a_2b_1}$ where $a_1b_2 - a_2b_1 \neq 0$	410
	53		$a_1x + b_1y + c_1z = k_1$ $a_2x + b_2y + c_2z = k_2$ $a_3x + b_3y + c_3z = k_3$	$x = \dfrac{b_2c_3k_1 + b_1c_2k_3 + b_3c_1k_2 - b_2c_1k_3 - b_3c_2k_1 - b_1c_3k_2}{a_1b_2c_3 + a_3b_1c_2 + a_2b_3c_1 - a_3b_2c_1 - a_1b_3c_2 - a_2b_1c_3}$ $y = \dfrac{a_1c_3k_2 + a_3c_2k_1 + a_2c_1k_3 - a_3c_1k_2 - a_1c_2k_3 - a_2c_3k_1}{a_1b_2c_3 + a_3b_1c_2 + a_2b_3c_1 - a_3b_2c_1 - a_1b_3c_2 - a_2b_1c_3}$ $z = \dfrac{a_1b_2k_3 + a_3b_1k_2 + a_2b_3k_1 - a_3b_2k_1 - a_1b_3k_2 - a_2b_1k_3}{a_1b_2c_3 + a_3b_1c_2 + a_2b_3c_1 - a_3b_2c_1 - a_1b_3c_2 - a_2b_1c_3}$	415
	54	Value of a Determinant	Second Order	$\begin{vmatrix} a_1 & b_1 \\ a_2 & b_2 \end{vmatrix} = a_1b_2 - a_2b_1$	411
	55		Third Order	$\begin{vmatrix} a_1 & b_1 & c_1 \\ a_2 & b_2 & c_2 \\ a_3 & b_3 & c_3 \end{vmatrix} = a_1b_2c_3 + a_3b_1c_2 + a_2b_3c_1 - (a_3b_2c_1 + a_1b_3c_2 + a_2b_1c_3)$	416
	56	Determinants — Cramer's Rule	Two Equations	$x = \dfrac{\begin{vmatrix} c_1 & b_1 \\ c_2 & b_2 \end{vmatrix}}{\begin{vmatrix} a_1 & b_1 \\ a_2 & b_2 \end{vmatrix}}$ and $y = \dfrac{\begin{vmatrix} a_1 & c_1 \\ a_2 & c_2 \end{vmatrix}}{\begin{vmatrix} a_1 & b_1 \\ a_2 & b_2 \end{vmatrix}}$ where $\begin{vmatrix} a_1 & b_1 \\ a_2 & b_2 \end{vmatrix} \neq 0$	412
	57		Three Equations	$x = \dfrac{\begin{vmatrix} k_1 & b_1 & c_1 \\ k_2 & b_2 & c_2 \\ k_3 & b_3 & c_3 \end{vmatrix}}{\begin{vmatrix} a_1 & b_1 & c_1 \\ a_2 & b_2 & c_2 \\ a_3 & b_3 & c_3 \end{vmatrix}}$, $y = \dfrac{\begin{vmatrix} a_1 & k_1 & c_1 \\ a_2 & k_2 & c_2 \\ a_3 & k_3 & c_3 \end{vmatrix}}{\begin{vmatrix} a_1 & b_1 & c_1 \\ a_2 & b_2 & c_2 \\ a_3 & b_3 & c_3 \end{vmatrix}}$, $z = \dfrac{\begin{vmatrix} a_1 & b_1 & k_1 \\ a_2 & b_2 & k_2 \\ a_3 & b_3 & k_3 \end{vmatrix}}{\begin{vmatrix} a_1 & b_1 & c_1 \\ a_2 & b_2 & c_2 \\ a_3 & b_3 & c_3 \end{vmatrix}}$ where $\begin{vmatrix} a_1 & b_1 & c_1 \\ a_2 & b_2 & c_2 \\ a_3 & b_3 & c_3 \end{vmatrix} \neq 0$	417

	No.				Page
QUADRATICS	58	General Form	$ax^2 + bx + c = 0$		451
	59	Quadratic Formula	$x = \dfrac{-b \pm \sqrt{b^2 - 4ac}}{2a}$		465
INTERSECTING LINES	60		Vertical angles are equal		239
	61		When two lines intersect, the adjacent angles are supplementary, $A + B = 180°$		240
	62		If two parallel lines are intersected by a transversal, corresponding angles are equal and alternate interior angles are equal		241
QUADRILATERALS	63		Quadrilateral	$A + B + C + D = 360°$	257
	64			Perimeter $= a + b + c + d$	258
	65		Trapezoid	Area $= \dfrac{(a + b)h}{2}$	258
	66		Parallelogram	Perimeter $= 2(a + b)$	259
	67			Area $= bh$	259
	68		Rectangle	Perimeter $= 2(a + b)$	259
	69			Area $= ab$	259
	70		Rhombus	Perimeter $= 4a$	260
	71			Area $= ah$	260
	72		Square	Perimeter $= 4a$	261
	73			Area $= a^2$	261
CIRCLES	74		Diameter $= d = 2r$		264
	75		Circumference $= 2\pi r = \pi d$		265
	76		Area $= \pi r^2$		265
	77		$\theta = \dfrac{s}{r}$ (in radians)		547
	78		Area of sector $= \dfrac{r^2\theta}{2}$		549
	79		1 revolution $= 2\pi$ radians $= 360°$ $1° = 60'$ $1' = 60''$		236, 543

	No.				Page
SOLIDS	80	Base	Prism	Volume = (area of base)(altitude) = Bh	270
	81			Lateral area = (perimeter of base)(altitude) = ph	272
	82			Surface area = 2(area of base) + lateral area = $2B + L$	272
	83		Rectangular Solid	Volume = lwh	273
	84			Surface area = $2(lw + hw + lh)$	273
	85		Cube	Volume = a^3	274
	86			Surface area = $6a^2$	274
	87		Cylinder	Volume = $\pi r^2 h$	277
	88			Surface area = $2\pi r^2 + 2\pi rh$	278
	89	Base	Pyramid	Volume = $\frac{1}{3}$ (area of base)(altitude) = $\frac{1}{3} Bh$	281
	90			Lateral area = $\frac{1}{2}$ (perimeter of base) $\times$ (slant height) = $\frac{1}{2} ps$	282
	91			Surface area = area of base + lateral area = $B + L$	282
	92		Cone	Volume = $\frac{1}{3} \pi r^2 h$	283
	93			Lateral area = πrs	283
	94			Surface area = $\pi r^2 + \pi rs$	284
	95		Sphere	Volume = $\frac{4}{3} \pi r^3$	286
	96			Surface area = $4\pi r^2$	287

No.				Page
ANY TRIANGLE	97	Perimeter	$p = a + b + c$	247
	98	Area	$Area = \frac{1}{2} bh$	247
	99	Hero's Formula	$Area = \sqrt{s(s-a)(s-b)(s-c)}$ where $s = \frac{1}{2}(a+b+c)$	247
	100	Sum of the Angles	$A + B + C = 180°$	245
	101	Law of Sines	$\dfrac{a}{\sin A} = \dfrac{b}{\sin B} = \dfrac{c}{\sin C}$	501
	102	Law of Cosines	$a^2 = b^2 + c^2 - 2bc \cos A$ $b^2 = a^2 + c^2 - 2ac \cos B$ $c^2 = a^2 + b^2 - 2ab \cos C$	510
RIGHT TRIANGLES	103		The acute angles of a right triangle are complementary, $A + B = 90°$	251, 298
	104	Pythagorean Theorem	$c^2 = a^2 + b^2$	252, 298
	105	Hypotenuse of a Right Triangle	$c = \sqrt{a^2 + b^2}$	252, 298
	106	Leg of a Right Triangle	$a = \sqrt{c^2 - b^2}$	252, 298
	107		30-60-90 Right Triangle Relationships	253
	108		Isosceles Right Triangle Relationships	255
	109	Trigonometric Ratios — Hypotenuse	$r = \sqrt{x^2 + y^2}$	291
	110	Sine	$\sin \theta = \dfrac{y}{r} = \dfrac{\text{opposite leg}}{\text{hypotenuse}}$	292, 298, 487
	111	Cosine	$\cos \theta = \dfrac{x}{r} = \dfrac{\text{adjacent leg}}{\text{hypotenuse}}$	292, 298, 487
	112	Tangent	$\tan \theta = \dfrac{y}{x} = \dfrac{\text{opposite leg}}{\text{adjacent leg}}$	292, 298, 487
	113	Cosecant	$\csc \theta = \dfrac{r}{y} = \dfrac{\text{hypotenuse}}{\text{opposite leg}}$	311, 487
	114	Secant	$\sec \theta = \dfrac{r}{x} = \dfrac{\text{hypotenuse}}{\text{adjacent leg}}$	311, 487
	115	Cotangent	$\cot \theta = \dfrac{x}{y} = \dfrac{\text{adjacent leg}}{\text{opposite leg}}$	311, 487
	116	Reciprocal Relations	$\csc \theta = \dfrac{1}{\sin \theta}$ $\quad$ $\sec \theta = \dfrac{1}{\cos \theta}$ $\quad$ $\cot \theta = \dfrac{1}{\tan \theta}$	312
VECTORS	117	Components of a Vector **V**	$V_x = V \cos \theta$	520
	118		$V_y = V \sin \theta$	520
	119	Resultant of Two Perpendicular Vectors	$R = \sqrt{A^2 + B^2}$	530
	120		$\tan \theta = \dfrac{B}{A}$	530

LOGARITHMS	121	Exponential and Logarithmic Forms	If $b^x = y$ then $x = \log_b y$ $(y > 0, b > 0, b \neq 1)$	630
	122	Laws of Logarithms	Product	$\log_b MN = \log_b M + \log_b N$	639
	123		Quotient	$\log_b \dfrac{M}{N} = \log_b M - \log_b N$	640
	124		Power	$\log_b M^p = p \log_b M$	641
	125		Root	$\log_b \sqrt[r]{M} = \dfrac{1}{r} \log_b M$	641
	126		Log of 1	$\log_b 1 = 0$	642
	127		Log of the Base	$\log_b b = 1$	642
	128		Log of the Base Raised to a Power	$\log_b b^n = n$	642

	No.						
SOME USEFUL FUNCTIONS	129	Straight Line		Slope	$m = \dfrac{y_2 - y_1}{x_2 - x_1}$		211
	130			$m = \tan(\text{angle of inclination}) = \tan\theta$ $0 \le \theta < 180°$			211
	131			Linear Equation	$Ax + By = C$		384
	132			Horizontal Line	$y = b$		213
	133			Vertical Line	$x = a$		214
	134			Slope-Intercept Form	$y = mx + b$		211
	135	Quadratic Function		$y = ax^2 + bx + c$			216
	136			Vertex V at $x = \dfrac{-b}{2a}$			217
	137	Power Function		$y = ax^n$			599
	138	Exponential Function		$y = a \cdot b^x$ where $b > 0$ and $b \ne 1$			614
	139	Exponential Growth		$y = ae^{rt}$			620
	140	Exponential Decay		$y = ae^{-rt}$			624
	141	Logarithmic Function		$y = \log_b x$ $(x > 0, b > 0, b \ne 1)$			636

TRIGONOMETRIC FUNCTIONS	142	Sine Function		$y = a\sin(bx + c)$	572
	143			amplitude $= |a|$	561, 566, 572
	144			Period $= \dfrac{2\pi}{b} = \dfrac{360°}{b}$	566, 572
	145			Phase shift $= -\dfrac{c}{b}$	572
	146	Cosine Function		$y = a\cos(bx + c)$	575
	147			amplitude $= |a|$	575
	148			Period $= \dfrac{2\pi}{b} = \dfrac{360°}{b}$	575
	149			Phase shift $= -\dfrac{c}{b}$	575
	150	Tangent Function		$y = a\tan(bx + c)$	582
	151			Period $= \dfrac{\pi}{b} = \dfrac{180°}{b}$	582
	152			Phase shift $= -\dfrac{c}{b}$	582

COMPLEX NUMBERS	153	Imaginary Unit	$j = \sqrt{-1}$	653

	154		Rectangular Form	$a + jb$	659
	155		Trigonometric Form	$a + jb = r(\cos\theta + j\sin\theta)$	672
	156			where $a = r\cos\theta$	671
	157			$b = r\sin\theta$	671
	158			$r = \sqrt{a^2 + b^2}$	669
	159			$\tan\theta = \dfrac{b}{a}$	670
	160		Polar Form	$a + jb = r\angle\theta$	670

APPENDIX A / Summary of Facts and Formulas

Conversion Factors

Unit	Equals
Length	
1 angstrom	1×10^{-10} meter
	1×10^{-4} micrometer (micron)
1 centimeter	10^{-2} meter
	0.3937 inch
1 foot	12 inches
	0.3048 meter
1 inch	25.4 millimeters
	2.54 centimeters
1 kilometer	3281 feet
	0.5400 nautical mile
	0.6214 statute mile
	1094 yards
1 light-year	9.461×10^{12} kilometers
	5.879×10^{12} statute miles
1 meter	10^{10} angstroms
	3.281 feet
	39.37 inches
	1.094 yards
1 micron	10^{4} angstroms
	10^{-4} centimeter
	10^{-6} meter
1 nautical mile (International)	8.439 cables
	6076 feet
	1852 meters
	1.151 statute miles
1 statute mile	5280 feet
	8 furlongs
	1.609 kilometers
	0.8690 nautical mile
1 yard	3 feet
	0.9144 meter
Angles	
1 degree	60 minutes
	0.01745 radian
	3600 seconds
	2.778×10^{-3} revolution
1 minute of arc	0.01667 degree
	2.909×10^{-4} radian
	60 seconds
1 radian	0.1592 revolution
	57.296 degrees
	3438 minutes
1 second of arc	2.778×10^{-4} degree
	0.01667 minute

Unit	Equals
Area	
1 acre	4047 square meters
	43 560 square feet
1 are	0.024 71 acre
	1 square dekameter
	100 square meters
1 hectare	2.471 acres
	100 ares
	10 000 square meters
1 square foot	144 square inches
	0.092 90 square meter
1 square inch	6.452 square centimeters
1 square kilometer	247.1 acres
1 square meter	10.76 square feet
1 square mile	640 acres
	2.788×10^7 square feet
	2.590 square kilometers
Volume	
1 board-foot	144 cubic inches
1 bushel (U.S.)	1.244 cubic feet
	35.24 liters
1 cord	128 cubic feet
	3.625 cubic meters
1 cubic foot	7.481 gallons (U.S. liquid)
	28.32 liters
1 cubic inch	0.01639 liter
	16.39 milliliters
1 cubic meter	35.31 cubic feet
	10^6 cubic centimeter
1 cubic millimeter	6.102×10^{-5} cubic inch
1 cubic yard	27 cubic feet
	0.7646 cubic meter
1 gallon (imperial)	277.4 cubic inches
	4.546 liters
1 gallon (U.S. liquid)	231 cubic inches
	3.785 liters
1 kiloliter	35.31 cubic feet
	1.308 cubic yards
	220 imperial gallons
1 liter	10^3 cubic centimeters
	10^6 cubic millimeters
	10^{-3} cubic meter
	61.02 cubic inches

Unit	Equals
Mass	
1 gram	10^{-3} kilogram
	6.854×10^{-5} slug
1 kilogram	1000 grams
	0.06854 slug
1 slug	14.59 kilograms
	14,590 grams
1 metric ton	1000 kilograms
Force	
1 dyne	10^{-5} newton
1 newton	10^5 dynes
	0.2248 pound
	3.597 ounces
1 pound	4.448 newtons
	16 ounces
1 ton	2000 pounds
At Sea Level	
1 kilogram	2.205 pounds
Velocity	
1 foot/minute	0.3048 meter/minute
	0.011 364 mile/hour
1 foot/second	1097 kilometers/hour
	18.29 meters/minute
	0.6818 mile/hour
1 kilometer/hour	3281 feet/hour
	54.68 feet/minute
	0.6214 mile/hour
1 kilometer/minute	3281 feet/minute
	37.28 miles/hour
1 knot	6076 feet/hour
	101.3 feet/minute
	1.852 kilometers/hour
	30.87 meters/minute
	1.151 miles/hour
1 meter/hour	3.281 feet/hour
1 mile/hour	1.467 feet/second
	1.609 kilometers/hour
Power	
1 British thermal unit/hour	0.2929 watt
1 Btu/pound	2.324 joules/gram
1 Btu-second	1.414 horsepower
	1.054 kilowatts
	1054 watts

Unit	Equals
Power (*continued*)	
1 horsepower	42.44 Btu/minute
	550 footpounds/second
	746 watts
1 kilowatt	3414 Btu/hour
	737.6 footpounds/second
	1.341 horsepower
	10^3 joules/second
	999.8 international watt
1 watt	44.25 footpounds/minute
	1 joule/second
Pressure	
1 atmosphere	1.013 bars
	14.70 pounds/square inch
	760 torrs
	101 kilopascals
1 bar	10^6 baryes
	14.50 pounds-force/square inch
1 barye	10^{-6} bar
1 inch of mercury	0.033 86 bar
	70.73 pounds/square foot
1 pascal	1 newton/square meter
1 pound/square inch	0.068 03 atmosphere
Energy	
1 British thermal unit	1054 joules
	1054 wattseconds
1 foot-pound	1.356 joules
	1.356 newtonmeters
1 joule	0.7376 foot-pound
	1 wattsecond
	0.2391 calories
1 kilowatthour	3410 British thermal units
	1.341 horsepowerhours
1 newtonmeter	0.7376 footpounds
1 watthour	3.414 British thermal units
	2655 footpounds
	3600 joules

Source: Reprinted from P. Calter, *Technical Mathematics, Third Edition*, Upper Saddle River, NJ: Prentice Hall, 1995.

Answers to Selected Odd-Numbered Exercises

CHAPTER 1 Arithmetic with Whole Numbers

EXERCISE 1-1
The Real Numbers

1. integer, whole, natural, a digit, real, rational, positive

3. real, rational, positive

5. real, decimal, negative

7. real, rational, negative

9. real, decimal, irrational, negative

11. ten thousands, tens

13. exact

15. approximate

17. to 27.

19. 25.   27.   21. 17.
-6 -4 -2 0 2 4 6 8
23.

29. $8 < 11$

31. $-2 < 6$

33. $\dfrac{1}{4} = 0.25$

EXERCISE 1-2
Adding and Subtracting Whole Numbers

1. 1521 **3.** 1731 **5.** 16,786 **7.** 16,057
9. 164,303 **11.** 87,934 **13.** 2466 **15.** 1649
17. 340 **19.** 131 **21.** 463 **23.** 489
25. 3836 **27.** 62 **29.** 62 **31.** $42,129
33. 99

EXERCISE 1-3
Multiplying Whole Numbers

1. 2,081,937 **3.** 24,064 **5.** 410,688 **7.** 36,204
9. 7,056 **11.** 21,868 **13.** $299 **15.** $330
17. 3240 degrees

EXERCISE 1-4
Dividing Whole Numbers

1. 3 **3.** 12 **5.** 5 **7.** 0
9. undefined **11.** 3 ft³ **13.** 495 miles **15.** 4 A

EXERCISE 1-5
Powers and Roots of Whole Numbers

1. 9 **3.** 4 **5.** 121 **7.** 529 **9.** 64
11. 125 **13.** 625 **15.** 128 **17.** 256 **19.** 512
21. 4 **23.** 45 **25.** 3 **27.** 2 **29.** 2
31. 2645 W **33.** 14 cm

EXERCISE 1-6
Scientific Notation

1. 100,000 **3.** 10 **5.** 10,000 **7.** 10,000,000
9. 10^3 **11.** 10^4 **13.** 3×10^4 **15.** 2×10^6
17. 7×10^3 **19.** 3000 **21.** 200 **23.** 7,000,000

EXERCISE 1-7 Combined Operations	**1.** 3340	**3.** 5940	**5.** 5	**7.** 3	**9.** 121	**11.** 27
	13. 27	**15.** 24	**17.** 12	**19.** 2	**21.** 30	

CHAPTER 1
Review Exercises

1. 441 **3.** 451 **5.** 63 **7.** 7
9. 0 **11.** 812 **13.** 32 **15.** 625
17. 7 **19.** 5 **21.** 559 **23.** 20,988
25. 11 **27.** undefined **29.** 343 **31.** 12
33. 21 **35.** 4 **37.** 10,000
39. The 5 is in the hundreds place, the 6 is in the ten-thousands place.
41. 400,000
43. and 45.

47. 3×10^5 **49.** $2 < 3$ **51.** $\dfrac{6}{3} = 2$ **53.** 10^5
55. 6 ft³ **57.** 188 miles **59.** 3 A **61.** 44
63. \$752 **65.** 3960° **67.** 6 ft

CHAPTER 2   Common Fractions

EXERCISE 2-1
Terminology

1. common proper **3.** common improper **5.** algebraic
7. common improper **9.** algebraic complex **11.** mixed number
13. similar **15.** neither

EXERCISE 2-2
Changing the Form of a Fraction

1. $\dfrac{1}{3}$ **3.** 2 **5.** $\dfrac{4}{9}$ **7.** $\dfrac{4}{3}$ **9.** $\dfrac{9}{12}$

11. $\dfrac{56}{21}$ **13.** $2\dfrac{2}{5}$ **15.** $2\dfrac{1}{2}$ **17.** $11\dfrac{5}{7}$

EXERCISE 2-3
Multiplying Fractions

1. $\dfrac{2}{7}$ **3.** $\dfrac{5}{12}$ **5.** $\dfrac{3}{10}$ **7.** $\dfrac{7}{27}$

9. $\dfrac{1}{8}$ **11.** $\dfrac{1}{10}$ **13.** $\dfrac{7}{2}$ **15.** $\dfrac{34}{27}$

17. $\dfrac{15}{2}$ **19.** $\dfrac{1813}{135}$ **21.** $\dfrac{17}{8}$ **23.** $\dfrac{11}{24}$

25. $\dfrac{77}{24}$ **27.** $\dfrac{1225}{192}$ **29.** $9\dfrac{17}{40}$ cm **31.** $147\dfrac{29}{30}$ lb

EXERCISE 2-4
Dividing Fractions

1. $\dfrac{1}{2}$ **3.** $\dfrac{3}{2}$ **5.** $\dfrac{13}{12}$ **7.** $\dfrac{14}{15}$

9. $\dfrac{9}{16}$ **11.** $\dfrac{4}{15}$ **13.** $\dfrac{3}{25}$ **15.** 49

17. $\dfrac{27}{2}$ **19.** $\dfrac{1}{60}$ **21.** $\dfrac{2}{315}$ **23.** $\dfrac{20}{9}$

25. $\dfrac{21}{5}$ **27.** $\dfrac{4}{9}$ **29.** $\dfrac{237}{140}$ **31.** $\dfrac{49}{6}$

33. $\dfrac{3}{2}$ **35.** $\dfrac{17}{80}$ **37.** $\dfrac{13}{90}$ **39.** $\dfrac{416}{81}$

41. $\dfrac{147}{295}$ **43.** $47\dfrac{11}{15}$ acres   **45.** $21\dfrac{19}{21}$ lb/in.

EXERCISE 2-5
Adding and Subtracting
Fractions

1. $\dfrac{5}{7}$ **3.** $\dfrac{3}{4}$ **5.** $\dfrac{3}{13}$ **7.** $\dfrac{1}{21}$

9. $\dfrac{4}{5}$ **11.** $\dfrac{5}{13}$ **13.** 16   **15.** 12

17. 35   **19.** $\dfrac{5}{6}$ **21.** $\dfrac{7}{16}$ **23.** $\dfrac{13}{22}$

25. $\dfrac{5}{24}$ **27.** $\dfrac{17}{30}$ **29.** $\dfrac{53}{156}$ **31.** $\dfrac{20}{3}$

33. $\dfrac{99}{16}$ **35.** $\dfrac{19}{5}$ **37.** $\dfrac{7}{4}$ **39.** $5\dfrac{1}{2}$

41. $\dfrac{121}{16}$ **43.** $\dfrac{60}{11}$ **45.** $\dfrac{35}{24}$ **47.** $\dfrac{139}{30}$

49. $\dfrac{799}{156}$ **51.** $36\dfrac{7}{32}$ in.   **53.** $5\dfrac{43}{60}$ lb   **55.** $84\dfrac{19}{20}$ ft

EXERCISE 2-6
Complex Fractions

1. $\dfrac{1}{20}$ **3.** $\dfrac{24}{5}$ **5.** $\dfrac{15}{14}$ **7.** $\dfrac{10}{3}$ **9.** $\dfrac{1}{2}$

11. $\dfrac{7}{13}$ **13.** $\dfrac{1}{2}$

CHAPTER 2
Review Exercises

1. $\dfrac{8}{15}$ **3.** $\dfrac{17}{16}$ **5.** $\dfrac{5}{48}$ **7.** $\dfrac{104}{135}$

9. $\dfrac{7}{16}$ **11.** $\dfrac{71}{16}$ **13.** $\dfrac{11}{48}$ **15.** $\dfrac{47}{6}$

17. $\dfrac{9}{35}$ **19.** $\dfrac{3}{110}$ **21.** $\dfrac{27}{44}$ **23.** $\dfrac{1}{6}$

25. $\dfrac{37}{8}$ **27.** $\dfrac{210}{37}$ **29.** $\dfrac{61}{105}$ **31.** $\dfrac{7}{24}$

33. 1   **35.** $\dfrac{1}{27}$ **37.** $\dfrac{38}{5}$ **39.** $\dfrac{15}{8}$

41. $\dfrac{1}{7}$ **43.** $\dfrac{16}{3}$ **45.** $3\dfrac{1}{7}$ **47.** $17\dfrac{4}{5}$ acres

49. $327\dfrac{1}{4}$ ft   **51.** $359\dfrac{4}{5}$ lb   **53.** $89\dfrac{41}{60}$ lb   **55.** 44 lb/in.

CHAPTER 3   Arithmetic with Decimal Numbers

EXERCISE 3-1
Decimal Numbers

1. 1, 3   **3.** 3, 4   **5.** 0, 1   **7.** 5, 5   **9.** 58.31
11. 86.04   **13.** 689.05   **15.** 9876.76   **17.** 46.0   **19.** 9.8
21. 3984.7   **23.** 4358.4   **25.** 58,000   **27.** 8,475,300   **29.** 7350
31. 0.0476   **33.** 231,000   **35.** 43.6

EXERCISE 3-2
Adding and Subtracting
Decimal Numbers

1. 1361.6   **3.** 5.02   **5.** 37.3   **7.** 7334   **9.** 22.14
11. 0.00109   **13.** 35.92   **15.** $20,258   **17.** $70,504   **19.** 1101 lb

EXERCISE 3-3 Multiplying Decimal Numbers	**1.** 4.32	**3.** 1881	**5.** 997.78	**7.** 62.8
	9. 21.22	**11.** 0.220	**13.** 81.7	**15.** 2673.0
	17. 88.5	**19.** 252	**21.** 65.6	**23.** 866
	25. 42.2	**27.** 4386 ft, $71.50	**29.** 293 W	**31.** $430.75
	33. 38.824 cm			

EXERCISE 3-4 Dividing Decimal Numbers	**1.** 5.07	**3.** 1.808	**5.** 77.81	**7.** 0.301	**9.** 0.2168
	11. 1.82	**13.** 17.35	**15.** 80.34	**17.** 2183	**19.** 0.01066
	21. 0.001066	**23.** 0.104	**25.** 0.00178	**27.** 0.105	**29.** 1.38 ft^3
	31. 6618 km	**33.** 8.607 ft	**35.** 0.578	**37.** 0.190	

EXERCISE 3-5 Powers and Roots of Decimal Numbers	**1.** 4.54	**3.** 346	**5.** 0.0348	**7.** 1.018	**9.** 9460
	11. 572	**13.** 5.559	**15.** 4.28	**17.** 22.2	**19.** 4.08
	21. 4.56	**23.** 21.00	**25.** 9.18	**27.** 3.76	**29.** 91.79
	31. 3.85	**33.** 4.44	**35.** 559 in.^{3}	**37.** 441 ft	**39.** 1.52 s
	41. 2918 W				

EXERCISE 3-6 Scientific Notation	**1.** 2.373×10^3	**3.** 7.26×10^3	**5.** 1.92×10^6	**7.** 5.228×10^3
	9. 7.01×10^3	**11.** 902.3	**13.** 4010	**15.** 952,000
	17. 8.74×10^4	**19.** 5.49×10^3	**21.** 1.36×10^3	**23.** 10^7
	25. 10^9	**27.** 3.10×10^{12}	**29.** 4.94×10^9	**31.** 2.51×10^9
	33. 10^3	**35.** 7.81×10^4	**37.** 2×10^4	**39.** 30
	41. 5.65×10^7	**43.** 2.09 A	**45.** 24.6 lb/in.^{2}	

EXERCISE 3-7 Combined Operations	**1.** 105	**3.** 97,800	**5.** 2.28	**7.** 0.160	**9.** 189
	11. 37$\overline{0}$0	**13.** 3.59	**15.** 7.17	**17.** 3.23	**19.** 0.871
	21. 7.93				

CHAPTER 3 Review Exercises	**1.** 0.00305	**3.** 0.4241	**5.** 0.0188	**7.** 2.3×10^3
	9. 8.39×10^5	**11.** 3.02×10^3	**13.** 302.3	**15.** 8010
	17. 41.0	**19.** 25.4	**21.** 2270	**23.** 1.878
	25. 29.9	**27.** 41.0	**29.** 89.09	**31.** 14.83
	33. 20.31	**35.** 28.9	**37.** 30.9	**39.** 0.212
	41. 3.52	**43.** 3.32	**45.** 1.36	**47.** 34.6
	49. 1.5×10^6	**51.** 3.27×10^{11}	**53.** 5.07	**55.** 0.153
	57. 4.52×10^{15}	**59.** 6.44×10^4	**61.** 3.41×10^4	**63.** 8.89×10^7 bits
	65. 4.9 h			

CHAPTER 4 Units of Measurement

EXERCISE 4-1 Converting Units	**1.** 12.7 ft	**3.** 9144 in.	**5.** 58,000 lb
	7. 44.8 tons	**9.** 1.09 nautical miles	**11.** 0.217 cubic ft
	13. 45.0 board feet	**15.** 665 cubic inches	**17.** 0.377 gal
	19. 0.2028 acres	**21.** 2.154 acres	**23.** 117 gallons

EXERCISE 4-2 The Metric System	**1.** 364 km	**3.** 735.9 kg	**5.** 6.2×10^3 megaohms
	7. 0.009348 microfarad	**9.** 2.75×10^3 g	**11.** 1.19×10^6 ft
	13. 3.337×10^5 kg	**15.** 17.6 liters	**17.** 3.60 m
	19. 125 g	**21.** 2.88×10^6 ns	**23.** 5.34 kV

EXERCISE 4-3	**1.** 59.6 in.	**3.** 34.6 acres	**5.** 317 in.-lb	**7.** 30.6 cm^3
Arithmetic with Denominate Numbers	**9.** 29.3 mi/h	**11.** 1.65 m/s	**13.** 25.2	**15.** 1.66
	17. 27.4 cm^2	**19.** 86,900 mm^3	**21.** 105 in.^{3}	**23.** 5.08 mm
	25. 6.34 cm	**27.** 5250 m^2	**29.** 14.7 in.^{3}	

EXERCISE 4-4	**1.** 0.587 acre	**3.** 2.30 m^2	**5.** 0.863 in.^{2}
Converting Areas and Volumes	**7.** 243,000 acres	**9.** 982.8 yd^2	**11.** 1.27×10^7 in.^{3}
	13. 56.4 m^3	**15.** 0.0163 in.^{3}	**17.** 29.6 gal
	19. 97.8 yd^3	**21.** 6160 m^2	**23.** 1.44×10^7 in.^{3}

EXERCISE 4-5	**1.** 4.00 mi/h	**3.** 107 km/h	**5.** 18.3 births/week
Converting Rates	**7.** 1070 acres/yr	**9.** 19.8¢/m^2	**11.** 236¢/lb
	13. 0.468 liters/are		

CHAPTER 4	**1.** 47,260 ft-lb/min	**3.** 2.28×10^3 megaohms	**5.** 7.25×10^{-3} μF
Review Exercises	**7.** 969.2 in.^{2}	**9.** 112 acre/year	**11.** 15.6 slug
	13. 0.0832 km	**15.** 283.2 kg	**17.** 24.6¢/m^2
	19. 2820 N	**21.** 19.2 mi/h	**23.** 89.3 km/h
	25. 229 ft^2	**27.** 268 ft^3	

CHAPTER 5 Percentage

EXERCISE 5-1	**1.** 629%	**3.** 0.46%	**5.** 65.3%	**7.** 395.9%	**9.** 60%
Converting to and from Percent	**11.** 90%	**13.** 28.6%	**15.** 4%	**17.** 33.3%	**19.** 46.7%
	21. 0.45	**23.** 3.655	**25.** 0.074	**27.** 0.52	**29.** $\frac{1}{8}$
	31. $2\frac{1}{4}$	**33.** $\frac{5}{8}$	**35.** $\frac{9}{20}$		

EXERCISE 5-2	**1.** 75 tons	**3.** 980 kg		**5.** 1.5 bushels	**7.** 126 km
Solving Percentage Problems	**9.** 529 acres	**11.** 5000		**13.** 300	**15.** 72
	17. 2.57	**19.** 50%		**21.** 20%	**23.** 30%
	25. 10%	**27.** 6.67%		**29.** 2120 Ω	**31.** $1050
	33. 2.31 cm	**35.** 119 tons		**37.** 107 mi	**39.** $3668
	41. $3000	**43.** 18.2 million barrels		**45.** 77.8%	**47.** 56.3%
	49. 64.4%	**51.** $182,000		**53.** $520	**55.** $5567.02

EXERCISE 5-3	**1.** +97.7%	**3.** +37.2%	**5.** $28\overline{0}$
Percent Change	**7.** 8306	**9.** 360	**11.** 400
	13. 20	**15.** 775	**17.** 9.52% increase
	19. $33,000 and $33,990	**21.** 3.72%	**23.** $89.88
	25. 29.7% increase	**27.** 10.8%	**29.** 38.3 million barrels

EXERCISE 5-4	**1.** 1.66% low	**3.** 120V and 225V	**5.** 0.392% high
Percent Error, Percent Concentration and Percent Efficiency	**7.** 75 ft^3	**9.** 1110 liters	**11.** 62.5%
	13. 24.2%	**15.** 1.92 hp	

CHAPTER 5	**1.** **(a)** 61.5% increase	**(b)** 18.2% decrease
Review Exercises	**3.** **(a)** 841	**(b)** 22.8
	5. **(a)** 309	**(b)** 646

7. (a) $\dfrac{21}{40}$ (b) $\dfrac{19}{40}$

     (c) $3\dfrac{1}{4}$ (d) $\dfrac{9}{100}$

9. (a) 14.3%   (b) 46.7%
     (c) 12.0%   (d) 71.4%

11. (a) 824%   (b) 62.7%
     (c) 0.27%   (d) 35.8%

13. (a) 171   (b) 73.6

15. (a) 1358 kg   (b) 95.3 gal
     (c) 331 km   (d) 134 liters

17. (a) 7030   (b) 388
     (c) 368

19. (a) 0.15   (b) 0.0764
     (c) 0.655   (d) 0.283

21. (a) 35.7%   (b) 38.1%
     (c) 52%   (d) 24.3%

CHAPTER 6   Arithmetic with Signed Numbers

EXERCISE 6-1
Signed Numbers

1. and **3.**

5. $5 > 1$ **7.** $-5 < 1$ **9.** $-12.2 > -12.3$

11. 6   **13.** 3   **15.** 1

EXERCISE 6-2
Adding and Subtracting Signed Numbers

1. 2   **3.** 53   **5.** 2   **7.** -12 **9.** -89
11. -7.56 **13.** -5.56 **15.** 59.5   **17.** -46.16 **19.** 10
21. 126   **23.** -71 **25.** -6 **27.** 42   **29.** 26.30
31. -12.10 **33.** -127.9 **35.** 86.48   **37.** $-\$5218.60$ **39.** -415 ft

EXERCISE 6-3
Multiplying Signed Numbers

1. -32 **3.** -77 **5.** -27 **7.** 15   **9.** 46
11. -1250 **13.** $-10\overline{0}0$ **15.** 20.8   **17.** 3250   **19.** 42
21. 240   **23.** 144   **25.** -9.20 **27.** 22.6   **29.** 2140
31. -4.45 ft/s   **33.** 10,300 ft-lb

EXERCISE 6-4
Dividing Signed Numbers

1. -4 **3.** -0.3333 **5.** -2 **7.** 1.6000
9. 0.3478   **11.** -0.493 **13.** -61.1 **15.** 0.816
17. 0.0531   **19.** -11.5V/min

EXERCISE 6-5
Powers and Roots of Signed Numbers

1. -32 **3.** 25   **5.** 36   **7.** 64   **9.** -2.93
11. 5.52   **13.** 5.57   **15.** 88.2   **17.** 0.0123   **19.** 0.125
21. 0.0204   **23.** 0.00137   **25.** 0.00328   **27.** 0.0325   **29.** 0.0195
31. 1.74×10^{-5} **33.** -2 **35.** -2 **37.** -3 **39.** -2.03
41. -4.01 **43.** -2.85 **45.** -2.12 **47.** -2.21 **49.** -1.39

EXERCISE 6-6
Scientific Notation

1. 3.73×10^{-1} **3.** 2.60×10^{-2} **5.** 2.0000×10^{-3}
7. 2.28×10^{-2} **9.** 0.02323   **11.** 0.00701
13. 0.00003370   **15.** 2.02×10^{-2} **17.** 0.436

19. 2.24×10^{-10}	**21.** 5640	**23.** 2.82×10^5	
25. 3.40×10^{-5}	**27.** 3.45 A	**29.** 2.27×10^4 lb/in.²	
31. 5.15×10^8 lb/in.²			

EXERCISE 6-7
Combined Operations

1. -10.4 **3.** -1.06 **5.** 0.0651   **7.** -1.34 **9.** 2.55×10^{-5}
11. 0.214   **13.** 1.77   **15.** 0.629   **17.** 8.48

CHAPTER 6
Review Exercises

1. $-3 < -1$ **3.** $-13.2 > -18.6$ **5.** 8.54×10^{-3} **7.** 3
9. 8   **11.** 0.054704   **13.** 0.197   **15.** 2.91×10^{-4}
17. -22.5 **19.** $-91\bar{0}$ **21.** -1230 **23.** -3520
25. 8.96×10^{-5} **27.** 0.0234   **29.** -2.41 **31.** -2.698
33. -2.23 **35.** -2.18 **37.** 2840   **39.** -13.5
41. -1.84 **43.** -0.281 **45.** -13.0 **47.** -18.2
49. -0.0213 **51.** 0.00707   **53.** 15.79   **55.** -12.10
57. -2.26 **59.** 30.5   **61.** 25.8   **63.** 2480
65. 1.10×10^{-8} **67.** $25,635   **69.** $-$7218.64   **71.** 26.8°F
73. -67.1 V   **75.** 2.48×10^{-10} W

CHAPTER 7   Introduction to Algebra

EXERCISE 7-1
Algebraic Expressions

1. yes   **3.** no   **5.** no   **7.** yes   **9.** 2   **11.** 1
13. $3, a, x$ **15.** $7, x, y$ **17.** 6   **19.** -1 **21.** $\dfrac{a}{2}$ **23.** $3(c - 2)$
25. 1st   **27.** 2nd   **29.** 1st   **31.** 2nd

EXERCISE 7-2
Adding and Subtracting
Polynomials

1. $10y$ **3.** $21.0a$
5. $-6ac$ **7.** $2.71y$
9. $-2x$ **11.** $52x + 91y + 22z$
13. $2.83a - 7.32x - 9.59z$ **15.** $5z - 3$
17. $-2c - 11$ **19.** $10.48z + 0.97$
21. $0.553c - 8.13$ **23.** $7a^4 - 8a^3 - 2$
25. $2bc - 2c + a$ **27.** $7.74a - 7.56b + 10.58c$
29. $8ab + 3x^2 - 7z + 2y^2$ **31.** $y - 2y^2 + 6b - 5$
33. $a + 14b - 3c + 3d$ **35.** $-5ab + 12cd$
37. $8b - 17p - 14r + 9s$ **39.** $6x^3 - 11x^2 + 5x - 2a^3 - 5a^2 + 2a - 15$
41. $2b$ **43.** $12b - 2a - 16c - 8d$
45. $5by^5 - 72bx^5 - 8bx^4 + 23by^4$ **47.** $26a^2 + 5bd - 14c$
49. $5(b + c)$

EXERCISE 7-3
Laws of Exponents

1. 27   **3.** 16   **5.** 1×10^{-9} **7.** x^6
9. a^9 **11.** 10^8 **13.** a^2 **15.** y^3
17. 10^{b-a+2} **19.** b **21.** a^8 **23.** z^{ac}
25. a^{3x-3} **27.** $9a^2$ **29.** $32x^5y^5z^5$ **31.** $-8/125$
33. $\dfrac{16x^4}{9y^4}$ **35.** 1   **37.** 82   **39.** 1
41. $\dfrac{1}{x}$ **43.** $\dfrac{x^4}{16}$ **45.** $\dfrac{27y^6}{64x^9}$ **47.** $\dfrac{4}{w^4} - \dfrac{3}{z^2}$
49. a^{-1} **51.** c^2d^{-3} **53.** x^4y^2

1. 81

3. -64

5. $35w^2 - 2y + 4xz$

7. $\dfrac{1}{b}$

9. x

11. 1

13. b^3

15. b^{x+3}

17. 10^{14}

19. $256x^{12}y^8$

21. $\dfrac{1}{27}$

23. $\dfrac{27a^3x^9}{64y^6z^3}$

25. $-2a$

27. $2w + 3y - 2z$

29. w^{-1}

31. $x^2y^2z^{-4}$

33. $108w^3$

35. 7

37. 2

39. $\dfrac{1}{a}$

41. $\dfrac{w}{x^5y^2}$

43. $\dfrac{27b^6}{64a^9}$

45. $5z - 3$

47. $11b^2 - 15ab + w$

49. $5w - 5z - 2y - 4x$

51. $2w + 12a - w^3 - 9a^2 + 18w^2 + 4 - 6a^3$

CHAPTER 8 Multiplication of Algebraic Expressions

EXERCISE 8-1
Product of Two Monomials

1. $-xy$
3. $-abc$
5. $9.19xy$
7. $16x^2y$
9. a^7
11. $-15.8a^4x^3$
13. $-30x^5y^5z^3$
15. $-40x^4y^5z^2$
17. $9.14a^{m+3}x^2$
19. $3x^{m+3}y^{n+2}$

EXERCISE 8-2
Product of a Multinomial and a Monomial

1. $-6 - 3x$
3. $2a + 6b$
5. $3.83b^3 + 4.86b$
7. $-30x^2 - 21x$
9. $2a^3b + a^2b^2 - a^3b^2$
11. $18a^3b + 12a^2b^2 - 6ab^3$
13. $-22.0x^3y^3 - 6.35x^3y^2 + 30.1x^2y^3 + 15.2x^2y^2$

EXERCISE 8-3
Product of Two Binomials

1. $x^2 + xy + xz + yz$
3. $8m^3 + 2m^2n - 4mn - n^2$
5. $2x^2 + xy - y^2$
7. $12x^2y^4 + 7a^3bxy^2 - 12a^6b^2$
9. $2a^2 - 21x^2 - 11ax$
11. $a^2x^2 - 25b^2$
13. $2.93x^2 + 1.82xy - 1.11y^2$
15. $13.1y^4 + 6.28a^3by^2 - 18.5a^6b^2$
17. $LW - 3L + 2W - 6$

EXERCISE 8-4
Product of Two Multinomials

1. $x^2 - xy + x + 3y - 12$
3. $4w^3 + 2w^2 - 22w + 10$
5. $b^8 + 2.93b^7 - 2.82b^6 - 8.26b^5 + 4.27b^4 + 12.5b^3$
7. $4c^4 + 7c^3 + c^2 + 7c - 3$
9. $x^3 + b^3$
11. $b^3 - x^3$
13. $a^3 - xyz - axy + a^2z + 5az + 5a^2$
15. $x^5 - bx^4 - x^3y^2 + abx^3y - x^2y^3 - ax^2y^2 + bx^2y^2 - abxy^3 + y^5 + ay^4$
17. $25dz^2 - 40dwz + 16dw^2$
19. $d^3 - 6.78d^2 - 0.168d + 45.8$
21. $c^3 + c^2m + c^2n - c^2p + cmn - cmp - cnp - mnp$
23. $y^5 - 2y^4 - 16y + 32$
25. $25x^5 - 10x^4 + 10x^3y^2 - 10x^3 - 4x^2y^2 + 4x^2$
27. $x^2 - 2xz - y^2 + z^2$
29. $m^5 - 3.37m^4 + 34.7m^3 - 114m^2 + 89.5m - 10.9$

EXERCISE 8-5
Powers of Multinomials

1. $x^2 + 2xy + y^2$ **3.** $a^2 - 2ad + d^2$

5. $B^2 + 2BD + D^2$ **7.** $24.2y^2 + 30.7yz + 9.73z^2$

9. $36x^2 + 60nx + 25n^2$ **11.** $w^2 - 2w + 1$

13. $b^6 - 26b^3 + 169$ **15.** $x^2 + 2xy + 2xz + y^2 + 2yz + z^2$

17. $a^2 + 2ab - 2a + b^2 - 2b + 1$ **19.** $c^4 - 2c^3d + 3c^2d^2 - 2cd^3 + d^4$

21. $x^3 - 3x^2y + 3xy^2 - y^3$ **23.** $27.5m^3 + 59.1m^2n + 42.3mn^2 + 10.1n^3$

25. $c^3 + 3c^2d + 3cd^2 + d^3$ **27.** $x^2 + 4x + 4$

29. $4.19r^3 - 25.1r^2 + 50.3r - 33.5$

EXERCISE 8-6
Removing Symbols of Grouping

1. $2a + b$ **3.** $2x - y$ **5.** $9.03x + 34.0y$ **7.** $-5x - 11$

9. $14p + 2q + 2$

CHAPTER 8
Review Exercises

1. $-12a^4y^3$

3. $a^2 - ab + a + 3b - 12$

5. $x^2 - 3.45xy + 5.83x + 2.45y^2 - 5.83y$

7. $x^4 + 2x^2y - 15y^2$

9. $8a^3 + 4a^2b - 4ab - 2b^2$

11. $c^3 + c^2 - 21c - 45$

13. $x^3 + p^3$

15. $25a^5 - 10a^4 + 10a^3b^2 - 10a^3 - 4a^2b^2 + 4a^2$

17. $p^3 + p^2q - p^2r - pq^2 - 2pqr - pr^2 - q^3 - q^2r + qr^2 + r^3$

19. $3m^5 - 14m^4 + 2m^3 + 32m^2 - 35m + 12$

21. $a^4 - a^3b + ab^3 - b^4$

23. $b^2m^2 - b^2y^2 - bnyz + bmnz$

25. $-720x^3y^6z^5$

27. $4p^2 + 12pq + 9q^2$

29. $x^6 + 2x^5 + x^4$

31. $x^4 - 2x^3y + 3x^2y^2 - 2xy^3 + y^4$

33. $7x + 4y$

35. $6x + 2y + 2$

37. $64x^3 + 144x^2y + 108xy^2 + 27y^3$

CHAPTER 9 Division of Algebraic Expressions

EXERCISE 9-1
Quotient of Two Monomials

1. x^3 **3.** $5z$ **5.** $-2a$ **7.** $31b$ **9.** $-33d$

11. $6p^2q^3r$ **13.** $-8mn$ **15.** $-4a^3bc$ **17.** $-6n^3$ **19.** $5a^2y$

21. $-4ac$ **23.** $-3by$ **25.** $8xy$ **27.** y^6 **29.** a^{x-y}

31. $-5a^2z$ **33.** $\dfrac{19}{ab^2}$

EXERCISE 9-2
Dividing a Polynomial by a Monomial

1. $15x^2 + 3x$ **3.** $12d^4 - 2d$

5. $4c^2 + 3c^3$ **7.** $-3 - 4p$

9. $-5x^2 - 3x$ **11.** $bmn + 3b$

13. $2a^2 + ab$ **15.** $y^2z - xy$

17. $mn^2 - mn - m^2n$ **19.** $x^3 - x^2y^2 - y^3$

21. $\dfrac{c^2}{d^2} - 4c + \dfrac{d}{c}$ **23.** $\dfrac{a^2}{b^2} + 2 - \dfrac{b^2}{a^2}$

25. $\dfrac{3z^3}{x} - 4x^2 - 2z$ **27.** $r^3 - \dfrac{pq^3}{r} - \dfrac{q^2r^2}{p}$

29. $d^3 - \dfrac{4c^3}{d} - 3cd$ **31.** $12c - \dfrac{8c^2}{b} - \dfrac{4}{bc}$

EXERCISE 9-3
Quotient of Two Polynomials

1. $a + 8$ **3.** $a + 8$ **5.** $x + 5$ **7.** $a - 8$
9. $-3x - 2$ **11.** $a^4 - 2$ **13.** $a^2 + 3a + 1$

CHAPTER 9 Review Exercises

1. $q^2r - pq$ **3.** $m - 8$ **5.** $p - q$

7. $(u - v)^3$ **9.** $x^3 - xy - y^3$ **11.** $-\dfrac{11b}{a}$

13. $p + 2q$ **15.** $x - 3y + \dfrac{4y^2}{x + 2y}$ **17.** $a^2 + 2a - 1$

19. $z^4 + \dfrac{1}{z^4} + 1$ **21.** $y - x^2y^2 - x^2y$

CHAPTER 10 Simple Equations

EXERCISE 10-1
Substituting into Equations

1. 15 **3.** -23 **5.** 28 **7.** 19.78
9. 515 **11.** 4.86 **13.** \$3050 **15.** 118°F
17. 13 ft-lb **19.** \$9756 **21.** 958 Ω

EXERCISE 10-2
Solving Simple Equations

1. 7 **3.** 2 **5.** $\dfrac{9}{4}$ **7.** -4 **9.** 2

11. $\dfrac{9}{4}$ **13.** 11 **15.** 2 **17.** 4 **19.** $-\dfrac{1}{2}$

21. 5 **23.** $\dfrac{1}{21}$ **25.** 3 **27.** 3 **29.** 7

31. $-\dfrac{66}{5}$ **33.** 13 **35.** $\dfrac{1}{7}$ **37.** 16 **39.** 6

41. 0 **43.** 5 **45.** $\dfrac{2}{3}$ **47.** 22 **49.** 5

51. $\dfrac{59}{3}$ **53.** -6 **55.** $-\dfrac{12}{7}$ **57.** 2 **59.** 26

61. 56 **63.** 4 **65.** $\dfrac{15}{11}$

EXERCISE 10-3
Simple Equations with Approximate Coefficients and Literals

1. 0.583 **3.** 28.5 **5.** -6.19 **7.** -7.16 **9.** -10.3

11. 3.89 **13.** $\dfrac{3}{d}$ **15.** $\dfrac{(c - b + d)}{a}$ **17.** $-\dfrac{a + 2b}{a - b}$

CHAPTER 10 Review Exercises

1. 12 **3.** $\dfrac{5}{14}$ **5.** 1 **7.** 49

9. 4 **11.** $\dfrac{6}{5}$ **13.** 63 **15.** -8

17. 11 **19.** $-5/11$ **21.** -4 **23.** 25/33

25. $\dfrac{24}{23}$ **27.** 84 **29.** 54 **31.** -142

33. 14.7 **35.** 105 **37.** -630 ft **39.** 2.88 ft lb

41. 2.91×10^7 N/cm² **43.** 15.6 ft **45.** -0.361 **47.** 31.4

49. -0.504 **51.** $\dfrac{(d - 5)}{c}$ **53.** $\dfrac{a}{m}$

EXERCISE 11-1
Solving Word Problems

1. x and $3x + 10$ **3.** x and $x + 42$ or x and $x - 42$

5. $\dfrac{x}{(6x + 4)}$ **7.** $0.11x$ gal   **9.** **(a)** $32 - x$ **(b)** $\dfrac{320}{x}$

11. 8   **13.** 9   **15.** 5

EXERCISE 11-2
Uniform Motion Problems

1. 3.38 h   **3.** $13\overline{0}$ km, 1.52 h   **5.** 4.95 days

7. $27\overline{0}$ mi   **9.** 3:17 P.M., 902 km

EXERCISE 11-3
Financial Problems

1. $63,644   **3.** 6 masons

5. $34,050 at 6.75%   **7.** $398 for computer
$139,874 at 8.24% $597 for printer

EXERCISE 11-4
Mixture Problems

1. 11.0 gal   **3.** 89.1%   **5.** 5.71 tons of 1.15% chromium steel
2.29 tons of 1.50% chromium steel

7. 0.180 tons   **9.** 27.8 lb

CHAPTER 11
Review Exercises

1. $165,420 at 5.94%   **3.** 6.44 liters   **5.** $68,493   **7.** 7 technicians
$58,392 at 8.56%

9. 10   **11.** 5.40 gal   **13.** 1130 lb   **15.** 3060 kg

CHAPTER 12   Functions and Graphs

EXERCISE 12-1
Functions

1. 1   **3.** -64

5. explicit, x-independent, y-dependent

7. implicit   **9.** $y = 5x^2$

11. $y = x + \dfrac{1}{x}$ **13.** $A = 4\pi r^2$

15. $v = at$ **17.** $d = 55t$

19. $C = 0.75 + 0.53(t - 1)$ **21.** $A = 5h$

23. $A = \dfrac{p^2}{16}$ **25.** $A = \dfrac{C^2}{4\pi}$

27. $(-2, -5)$, $(-1, -2)$, $(0, 1)$, $(1, 4)$, $(2, 7)$

EXERCISE 12-2
Functional Notation

1. $f(x) = 2x - 1$ **3.** $f(C) = \dfrac{9}{5}C + 32$

5. domain: all real numbers x; range: all real numbers y

7. domain: $x \neq 0$; range: $y \neq 0$ **9.** domain: $d \geq 0$; range: $W \geq 0$

11. $1, -8, 3a - 2, 14$ **13.** $\dfrac{7}{4}, 1, 3c^2 + 1, -9$

15. 169 ft/s, 330 ft/s, 652 ft/s

EXERCISE 12-3
Rectangular Coordinates

1. $(4, 2)$ **3.** $(-4, 3)$ **5.** $(0, -4)$ **7.** $(-2.5, -2.5)$

9. to **25.**

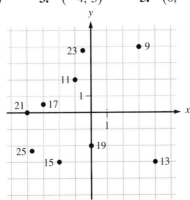

27. IV   **29.** II, III   **31.** 0   **33.** II, IV   **35.** parallelogram

37.

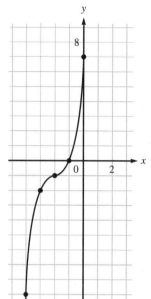

EXERCISE 12-4
Linear Functions

1. (a) $-2, -1, 0, 1, 2$ **(b)**

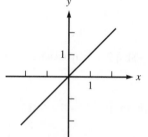

(c) $x = 0, y = 0$ **(d)** $m = 1$

3. (a) $-1, 1, 3, 5, 7$ **(c)** $x = -\dfrac{3}{2}, y = 3$

(d) $m = 2$

5. (a) $7, 4, 1, -2, -5$ **(b)**

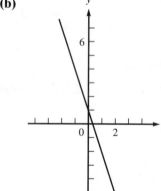

(c) $x = \dfrac{1}{3}, y = 1$ **(d)** $m = -3$

7. (a) $-2, -1.5, -1, -0.5, 0$ **(c)** $x = 2, y = -1$

(d) $m = \dfrac{1}{2}$

9. **(a)** 3, 3, 3, 3, 3

(b)

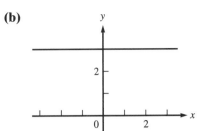

(c) no x intercept, $y = 3$ **(d)** $m = 0$

11. **(b)** $x = -1$, no y intercept   **(c)** m is undefined

13.

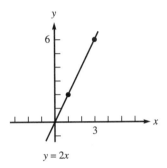

$y = 2x$

15. $y = -\dfrac{1}{5}x - \dfrac{6}{5}$

17.

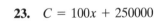

19. $V = 4.5I$

21. $W = 7.87\,V$

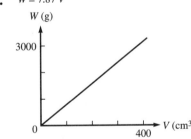

23. $C = 100x + 250000$

25.

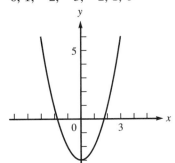

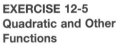
EXERCISE 12-5
Quadratic and Other
Functions

1. 6, 1, −2, −3, −2, 1, 6

3. $V(0, 0)$

5. $V(-1, 2)$

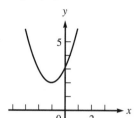

7. $V(0, 0)$

9. $-8, -1, -\dfrac{1}{8}, 0, \dfrac{1}{8}, 1, 8$

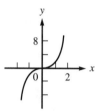

11.

x	−3	−2	−1	0	1	2	3
y	9	2.7	0.3	0	−0.3	−2.7	−9

13.

x	−3	−2	−1	0	1	2	3
y	0	4	2	0	4	20	54

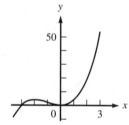

15.

x	−3	−2	−1	0	1	2	3
y	0.3	0.5	1	undef	−1	−0.5	−0.3

17.

x	−3	−2	−1	0	1	2	3
y	−0.5	−1	undef	1	0.5	0.3	0.2

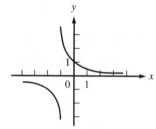

19.

x	0	1	4	9
y	0	−1	−2	−3

21.

x	−8	−3	0	1
y	3	2	1	0

25.

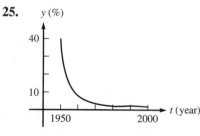

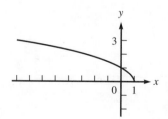

1. **(a)** $s = 160t - 16t^2$
(c) 300 ft

(b)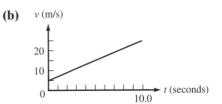

3. **(a)** $A = 12\pi h + 72\pi$
(d) 380 cm² **(c)** 7.3 cm

5. **(a)** $v = 5.0 + 2.0t$

(b)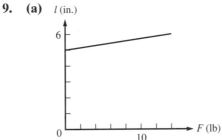

(c) 16 m/s **(d)** 7.5 s

7. **(b)** 2.5 s

9. **(a)** 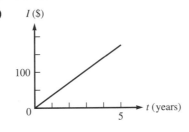 **(b)** 14 lb

11. **(b)** 96 ft/s **(c)** 3.0 s

13. **(a)** *I* ($) **(b)** $140

15. $-0.6, 4.6$ **17.** $-5.2, -0.8$ **19.** 2.4

1.

5.

9.

11. $x = -2.4, 0.4; y = 1; \max(-1, 2)$

13.

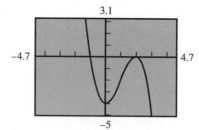

3.1

−4.7   4.7

−5

$x = -1, 2$
$y = -4$
min $(0, -4)$
max $(2, 0)$

15. $x = -1.45, 3.45; y = 5; \text{max}(1, 6)$

17.

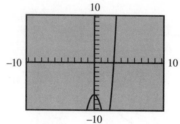

6.2

−9.4   9.4

−6.2

$x = -0.62, 1, 1.62$
$y = -1$
min $(-1, 0)$
max $(1.33, 0.19)$

19. $y = 1.32x^2 + 1.95x - 4.02;$
$x = -2.63, 1.16$

21. $y = 2.56x^3 - 6.75 - 7.25x^2$

10

−10   10

−10

$x = 3.11$

CHAPTER 12
Review Exercises

1. $y = 3x^2$

3. domain: $x \geq 3$; range: $y \geq 0$

5. $0, 6, b - b^3, \dfrac{3}{4}$

7. $-8, -5, -2, 1, 4$

9. implicit

11.

x	−3	−2	−1	0	1	2	3
y	0	−4	−6	−6	−4	0	6

13.

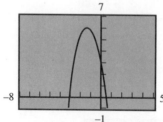

7

−8   5

−1

$x = -3, 0.5$
$y = 3.00$
max $(-1.25, 6.12)$

15. $C = 6.65 + 0.16(x - 200)$

17. $y = 300(1 + r)^2$
$340, $347, $364

21. $R = 3.35(10^{-4})L$

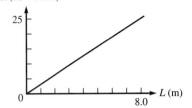

23. $x = 3.81$; $y = -30$; max $(0.33, -29.85)$; min $(1.00, -30)$

CHAPTER 13 Plane Geometry

EXERCISE 13-1
Lines and Angles

1. R —— 3 cm —— S

3. θ, $\angle Y$, $\angle XYZ$, $\angle ZYX$

5. 40°, 150°, −120°

7. A radian is the measure of a central angle that cuts off an arc of length r (the radius of the circle). A radian is about 57.3°.

9. 32.30°

11. 93.7914°

13. 235.3750°

15. 56°15′

17. 12°20′42″

19. 263°52′30″

21. 15.7 rad

23. 1.05 rad

25. 2.1962 rad

27. 0.557 rev

29. 0.836 rev

31. 0.209487 rev

33. 1620°

35. 293°33′29″

37.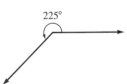

39.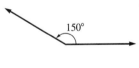

41.

43.

45.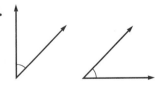

47. 64.7°, 127.4°, 123.7°

49. $A = C = D = F = 117.9°$, $B = E = G = 62.1°$

51. $A = C = 145.2°$, $B = 34.8°$ **53.** due north

EXERCISE 13-2
Triangles

1. 62.1°, isosceles, acute, oblique

3. 61.7°, scalene, right

5. 27.9°, isosceles, obtuse, oblique

7. sum = 180°

9. 41.4°

11. 65°5′

13.	yes	15.	yes
17.	yes, yes, yes	19.	4.19 in., 0.33 in.²
21.	11.44 cm, 2.98 cm²	23.	78.6°
25.	73.9°	27.	$728
29.	$1146	31.	$184

EXERCISE 13-3
Right Triangles

1. The hypotenuse is the longest side of a *right* triangle.
3. **(a)** 9.96 cm **(b)** 6.18 cm
5. **(a)** $x = 7.79$ in., $y = 4.50$ in. **(b)** $x = 4.60$ cm, $y = 3.25$ cm
7. 4.11 in., 9.77 in.² 9. 79.2 cm
11. 51.62 in. 13. 188.2 mm, 2160 mm²
15. 57.0 ft 17. 13.1 m
19. 29.4 ft 21. 22.4 ft
23. 0.866 in.

EXERCISE 13-4
Quadrilaterals

1. 360°

3. 4 properties of a parallelogram: opposite sides are parallel, opposite sides are equal, opposite angles are equal, diagonals bisect each other

5. 6 properties of a rhombus: opposite sides are parallel, all sides are equal, opposite angles are equal, diagonals bisect each other, diagonals are perpendicular to each other, diagonals bisect the angles

7. **(a)** 27.50 cm, 41.6 cm² **(b)** 13.50 in., 7.88 in.² **(c)** 209.6 mm, 2750 mm²
9. 50.0 in., 110 in.² 11. $1926 13. 1944 tiles
15. $315000 17. 0.258 acre 19. $557

EXERCISE 13-5
Circles

1. C, $\overline{AC}$ or $\overline{BC}$ or $\overline{CD}$, $\overline{BD}$, $\overline{AD}$ or $\overline{BD}$, $\widehat{AB}$ or $\widehat{AD}$ or $\widehat{BD}$, $\widehat{BD}$

3. 22.6 cm, 40.7 cm² 5. 12.9 in.²

7. The radius is perpendicular to the tangent.
9. 78.5 m 11. 584,000,000 mi 13. 11.0 in.² 15. 20.3 cm
17. $44\overline{0}$ yd 19. 37.7 mi 21. 6.61 in. 23. 661 mm²

CHAPTER 13
Review Exercises

1. 68.4°, 56.9° 3. 80.8° 5. 51.3°
7. 28° 9. 5.66 cm, 18.5 cm² 11. 36.3 cm, 50.9 cm²
13. 4 properties of a parallelogram: opposite sides are parallel, opposite sides are equal, opposite angles are equal, diagonals bisect each other

6 properties of a rectangle: opposite sides are parallel, opposite sides are equal, opposite angles are equal, all angles are right angles, diagonals bisect each other, diagonals are equal

6 properties of a rhombus: opposite sides are parallel, all sides are equal, opposite angles are equal, diagonals bisect each other, diagonals are perpendicular to each other, diagonals bisect the angles

7 properties of a square: opposite sides are parallel, all sides are equal, all angles are right angles, diagonals bisect each other, diagonals are equal, diagonals are perpendicular to each other, diagonals bisect the angles

15. 193.8 mm, 1170 mm² 17. 21.8 acres 19. 3610 ft, 235 acres
21. 14.0 in. 23. 96 ft

EXERCISE 14-1
Prisms

1. 268 cm³, 330 cm²   **3.** 0.928 in.³, 7.37 in.²   **5.** 66.7 cm³, 105 cm²
7. 61.6 in.³, 98.4 in.²   **9.** 43.6 ft³, 74.3 ft²   **11.** 0.165 m³, 1.81 m²
13. 1070 cm³   **15.** 15.3 cm³   **17.** 29$\overline{0}$ ft²
19. 3.0 ft³   **21.** 16.3 loads   **23.** 396 cm³, 3.10 kg
25. 561 in.²

EXERCISE 14-2
Cylinders

1. 358 ft³   **3.** 205 cm²   **5.** 4570 gal   **7.** 37200 ft³
9. 152 cm³   **11.** 2.81 ft³   **13.** 5760 ft²   **15.** 346 cc
17. 15600 gal   **19.** 2.94 cm³   **21.** 1,350,000 ℓ   **23.** 6.76 cm

EXERCISE 14-3
Pyramids and Cones

1. 94.6 cm³   **3.** 114 cm³   **5.** 110 in.²   **7.** 377 in.²
9. 471 in.²   **11.** 1830 g   **13.** 12$\overline{0}$0 g   **15.** 10800 lb
17. 17.6 kg   **19.** 114 ft²   **21.** 186 in.², 3.47 lb   **23.** 9.84 m
25. 5.66 gal, 47.2 lb

EXERCISE 14-4
Spheres

1. 153 cm³, 139 cm²   **3.** 2.60 × 10¹¹ mi³, 1.97 × 10⁸ mi²
5. 8.18 m³   **7.** 0.999 gal
9. 630 ℓ   **11.** $63,700
13. 125 lb   **15.** 18.6 lb
17. 1.24 ft, 4.84 ft²   **19.** 0.0940 m³

CHAPTER 14
Review Exercises

1. 179 cm³, 194 cm²   **3.** 88.5 cm³, 190 cm²
5. 8.52 gal   **7.** 113 ft²
9. 1880 kℓ   **11.** 3.77 cm³
13. 9.02 × 10⁷ ft³, 1.48 × 10⁶ ft²   **15.** 288 yd², 51.9 boxes
17. 38.5 in.², 22.4 in.³, 6.38 lb   **19.** 67.7 g
21. a million $1 bills

CHAPTER 15   Right Triangle Trigonometry

EXERCISE 15-1
Sine, Cosine, and
Tangent Functions

1.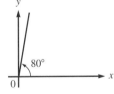

3.

5.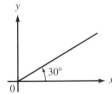

7.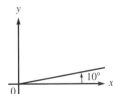

9. sin θ = 0.923, cos θ = 0.385, tan θ = 2.4
11. sin θ = 0.471, cos θ = 0.882, tan θ = 0.533
13. sin θ = 0.832, cos θ = 0.555, tan θ = 1.5

15. $\sin\theta = 0.5$, $\cos\theta = 0.866$, $\tan\theta = 0.577$

17. $\sin\theta = 0.916$, $\cos\theta = 0.401$, $\tan\theta = 2.29$

19. $\sin 60° = 0.866$, $\cos 60° = 0.5$, $\tan 60° = 1.73$

21. $\sin 45° = 0.707$, $\cos 45° = 0.707$, $\tan 45° = 1$

23. $\tan\theta = 0.75$ **25.** $\tan\theta = 6$

27. $\sin 45° = 0.707$, $\cos 45° = 0.707$, $\tan 45° = 1$

29. 0.789 **31.** 0.222 **33.** 0.801 **35.** 0.980 **37.** 2.38

39. 0.723 **41.** 32.5 **43.** 1.00

EXERCISE 15-2
Inverse Trigonometric
Functions

1. θ, angle, sine, 0.636 **3.** α, angle, tangent, 3.45

5. 30°, angle, cosine, a **7.** $\frac{\pi}{2}$, angle, tangent, x

9. B, angle, sine, a **11.** 1, angle, sine, 0.8415

13. 66.5°, angle, cosine, 0.399 **15.** $\alpha = \sin^{-1} 0.634$, $\alpha = \arcsin 0.634$

17. $\theta = \tan^{-1} 4.67$, $\theta = \arctan 4.67$ **19.** $b = \cos 60°$, $60° = \arccos b$

21. $\sin\alpha = 0.268$, $\alpha = \sin^{-1} 0.268$ **23.** $x = \cos 0.785$, $\cos^{-1} x = 0.785$

25. $89.0° = \tan^{-1} 57.3$, $89.0° = \arctan 57.3$

27. 45.0° **29.** 78.4° **31.** 30° **33.** 45° **35.** 30.0°

37. 0.39 **39.** 1.55 **41.** 80.8° **43.** 45.0° **45.** 60.0°

47. 1.47 **49.** 0.52 **51.** $\sin^{-1} 0.738 = 0.830$, $(\sin 0.738)^{-1} = 1.49$

EXERCISE 15-3
Solution of Right
Triangles

1. $c = 25.0$, $\tan A = 3.43$, $A = 73.7°$

3. $a = 40.0$, $\tan A = 4.44$, $A = 77.3°$

5. $b = 24.8$, $\tan A = 0.524$, $A = 27.7°$

7. $c = 11.7$, $\tan A = 0.854$, $A = 40.5°$

9. $b = 1.25$, $\tan A = 0.248$, $A = 13.9°$

11. $A = 42.0°$, $a = 23.4$ in., $b = 26.0$ in.

13. $B = 21.0°$, $b = 1.66$ m, $c = 4.63$ m

15. $A = 64.8°$, $b = 26.7$ cm, $c = 62.8$ cm

17. $B = 47°45'$, $b = 1.82$ m, $c = 2.45$ m

19. $A = 29.7°$, $B = 60.3°$, $c = 8.06$ in.

21. $A = 41.8°$, $B = 48.2°$, $b = 5.14$ cm

23. $A = 30.18°$, $B = 59.82°$, $a = 20.77$ m

EXERCISE 15-4
Applications

1. 155 in. **3.** 69.1° **5.** 53.8°

7. 115 ft **9.** 33,800 ft **11.** 18.4 ft

13. 32.5 ft **15.** 63.5 ft **17.** 1280 ft

19. 96.5 ft **21.** 23.8 km **23.** $19\overline{0}$ mi north, 292 mi west

25. 39.3° **27.** 21.8°, 26.9 ft **29.** 58.8°

31. 2.95 cm, 53.6° **33.** 2.782 in. **35.** 4.85 cm

EXERCISE 15-5
Cosecant, Secant,
and Cotangent
Functions

1. $\csc\theta = 1.25$, $\sec\theta = 1.67$, $\cot\theta = 0.75$

3. $\csc\theta = 3.57$, $\sec\theta = 1.04$, $\cot\theta = 3.43$

5. $\csc\theta = 1.05$, $\sec\theta = 3.16$, $\cot\theta = 0.333$

7. $\csc\theta = 2.35$, $\sec\theta = 1.11$, $\cot\theta = 2.12$

9. $\csc\theta = 1.05$, $\sec\theta = 3.22$, $\cot\theta = 0.326$

11. $\csc A = 1.13$, $\sec A = 2.12$, $\cot A = 0.533$, $A = 61.9°$

13. $\csc A = 1.02$, $\sec A = 4.56$, $\cot A = 0.225$, $A = 77.3°$

15. $\csc A = 1.33$, $\sec A = 1.51$, $\cot A = 0.882$, $A = 48.6°$

17. 0.503   **19.** 1.13   **21.** 24.0   **23.** 0.760   **25.** 3.17

27. 29.6°   **29.** 4.5°   **31.** 0.224   **33.** 55.2°   **35.** 0.190

CHAPTER 15
Review Exercises

1. $\sin \theta = 0.980$, $\cos \theta = 0.198$, $\tan \theta = 4.95$,
$\csc \theta = 1.02$, $\sec \theta = 5.05$, $\cot \theta = 0.202$, $\theta = 78.6°$

3. 0.974   **5.** 0.332   **7.** 0.224   **9.** 30.0°

11. 78.1°   **13.** 0.603   **15.** 34.1°

17. $\sin A = 0.421$, $\cos A = 0.907$, $\tan A = 0.464$,
$\csc A = 2.37$, $\sec A = 1.10$, $\cot A = 2.15$, $A = 24.9°$

19. $B = 52.4°$, $a = 40.1$ cm, $b = 52.1$ cm

21. $A = 0.746$, $b = 3.52$ cm, $c = 4.79$ cm

23. 68.5 cm   **25.** 1660 cm^2 **27.** 3.88 cm   **29.** 8210 ft

31. 19.5 ft, 77.0°

CHAPTER 16   Factoring

EXERCISE 16-1
Common Factors

1. $x(y + z)$ **3.** $3(a - b)$

5. $x(x - y)$ **7.** $6(x + 3)$

9. $3(8x^2 - 1)$ **11.** $a(c + 1)$

13. $x^2(4x - 3)$ **15.** $xy(24x^2 + 15xy - 40)$

17. $2x(2xy + 3x^2 - 1)$ **19.** $2a^2b^3(3 + 2a - 5b)$

21. $(2 + x)(x + y)$ **23.** $(p - 1)(q - p)$

25. $(c - d)(d + 1)$ **27.** $(x - y)(x - y + 1)$

29. $(3 + a)^2(2 + a)$ **31.** $(x + 3)(y + z)$

33. $(x + 3)(y - 2)$ **35.** $(x - 2)(x + y)$

37. $(m^2 + 3)(m - 5)$ **39.** $2\pi r(r + h)$

41. $\pi r(r + s)$ **43.** $16(4 + 2t - t^2)$

45. $\dfrac{1}{2}wx(L - x)$

EXERCISE 16-2
Difference of Two Squares

1. $(m + n)(m - n)$ **3.** $(x + 4)(x - 4)$

5. $(5 + p)(5 - p)$ **7.** $(3y + 1)(3y - 1)$

9. $(2a + 3b)(2a - 3b)$ **11.** $(p^2 + 2)(p^2 - 2)$

13. $3(x + 2y)(x - 2y)$ **15.** $4(ab + c)(ab - c)$

17. $(a^2 + b^2)(a + b)(a - b)$ **19.** $(p^3 + q^4)(p^3 - q^4)$

21. $(10x^2 + 9)(10x^2 - 9)$ **23.** $(x^4 + 1)(x^2 + 1)(x + 1)(x - 1)$

25. $(a + b + c)(a + b - c)$ **27.** $(p + q + r)(p - q - r)$

29. $(1 + a - b)(1 - a + b)$ **31.** $(s + d)(s - d)$

33. $\pi(R + r)(R - r)$ **35.** $\pi LD(R + r)(R - r)$

EXERCISE 16-3
Simple Trinomials

1. $(x + 7)(x + 1)$ **3.** $(a - 3)(a + 2)$

5. $(u + 9)(u - 2)$ **7.** $(y - 6)(y - 1)$

9. $(7 + k)(2 + k)$ **11.** $(x - 5y)(x - 3y)$

13. $(6 + x)(3 - x)$ **15.** $(a - 4b)(a + 2b)$

17. $(w + 4)(w - 15)$ **19.** $(x - 162)(x - 200)$

21. $(x - 10)(x - 1)$ **23.** $(x + y - 4)(x + y - 1)$

25. $(t + 1)^2(t - 1)^2$

EXERCISE 16-4		
General Trinomials	**1.** $(2x + 1)(x + 1)$	**3.** $(5y - 2)(y - 1)$
	5. $(3x + 1)(2x - 5)$	**7.** $(3a - 2)(a + 3)$
	9. $(3 + 4x)(5 - 3x)$	**11.** $(3t + 2s)(2t + 7s)$
	13. $3(b - 6)(b + 1)$	**15.** $2ab(b - 4)(b - 3)$
	17. $3(5k - 7)(k + 1)$	**19.** $16(3 - t)(2 + t)$
	21. $6(x - 15)(x + 6)$	**23.** $(2y - 5)(3y - 4)$
	25. $(3 - x + y)(5 - x + y)$	**27.** $(7m + 5n)(2m)$

EXERCISE 16-5			
Perfect Square Trinomials	**1.** $(x - 2)^2$	**3.** $(y + 3)^2$	**5.** $(6 - s)^2$
	7. $(3a + 2b)^2$	**9.** $(1 - 11y)^2$	**11.** $(10 + 3x)^2$
	13. $(9y + 4)^2$	**15.** $(4t - 3uv)^2$	**17.** $7y(y + 1)^2$
	19. $5x(2y - 3)^2$	**21.** $(x + y + 1)^2$	**23.** $(2 - a + b)^2$
	25. $(9m - n)^2$		

EXERCISE 16-6		
Sum or Difference of Two Cubes	**1.** $(a - 1)(a^2 + a + 1)$	**3.** $(x + 3)(x^2 - 3x + 9)$
	5. $(b - 4)(b^2 + 4b + 16)$	**7.** $(1 - 2a)(1 + 2a + 4a^2)$
	9. $(10 + 3x)(100 - 30x + 9x^2)$	**11.** $(4a - 5b)(16a^2 + 20ab + 25b^2)$
	13. $3(y - 2)(y^2 + 2y + 4)$	**15.** $2p(3p + 1)(9p^2 - 3p + 1)$

17. $(a + b - x)[(a + b)^2 + (a + b)x + x^2]$

19. $(2x - a + b)[4x^2 + 2x(a - b) + (a - b)^2]$

21. $(a - b + x + y)[(a - b)^2 - (a - b)(x + y) + (x + y)^2]$

23. $(x^2 + 1)(x^4 - x^2 + 1)$

25. $(a - 1)(a^2 + a + 1)(a^6 + a^3 + 1)$

27. $\dfrac{4}{3}\pi(R - r)(R^2 + Rr + r^2)$

29. **(a)** $(a + b)(a - b)(a^4 + a^2b^2 + b^4)$
(b) $(a + b)(a^2 - ab + b^2)(a - b)(a^2 + ab + b^2)$
difference of two squares

CHAPTER 16		
Review Exercises	**1.** $3y(5y - 9)$	**3.** $3a(5a - 7b + c)$
	5. $(3 + y)(x - y)$	**7.** $(2x + 5)(2x - 5)$
	9. $3(3 + 2y)(3 - 2y)$	**11.** $(m + n - 1)(m - n + 1)$
	13. $4\pi(R + r)(R - r)$	**15.** $(m + 10)(m - 9)$
	17. $(b - 7)(b - 4)$	**19.** $(b - 18)(b - 2)$
	21. $(8y - 5)(3y + 4)$	**23.** $(5x - 8)(3x + 5)$
	25. $(y + 8)^2$	**27.** $(4x + 9)^2$
	29. $(1 - y)(1 + y + y^2)$	**31.** $(3x + 2y)(9x^2 - 6xy + 4y^2)$

33. $(a - b + 2)[a^2 + a(b - 2) + (b - 2)^2]$

CHAPTER 17 Algebraic Fractions

EXERCISE 17-1			
Equivalent Fractions	**1.** $\dfrac{r}{s}$	**3.** $\dfrac{x}{2}$	**5.** $\dfrac{b - a}{1 - b}$
	7. $\dfrac{s - r}{r - 1}$	**9.** $\dfrac{12x^2}{20xy}$	**11.** $\dfrac{a^2 - b^2}{a^2 - 2ab + b^2}$
	13. $\dfrac{3x}{2y}$	**15.** $\dfrac{6x^2}{5y^2}$	**17.** $\dfrac{9p^5}{5q^4}$

19. $\dfrac{4}{x + 2}$ **21.** $\dfrac{4(x - 2)}{3(x + 2)}$ **23.** $\dfrac{5}{3}$

25. $\dfrac{x + 4}{x - 3}$ **27.** $\dfrac{x + 7}{x - 7}$ **29.** $\dfrac{x - 2}{x + 2}$

31. $\dfrac{1 + 2x}{3x - 2}$ **33.** $\dfrac{x^2 - xy + y^2}{x - y}$ **35.** $\dfrac{(x - y)^2}{x^2 + xy + y^2}$

37. -1 **39.** 1 **41.** $b - a$

43. $\dfrac{1 - 2x}{2 + x}$ **45.** $\dfrac{x - 2}{x + 2}$ **47.** $-\dfrac{3x}{3 + x}$

49. $-\dfrac{y}{y + 2x}$ **51.** $-\dfrac{x^2 + 2x + 4}{2 + x}$ **53.** $-\dfrac{3x^2}{x^2 + 2xy + 4y^2}$

55. $\dfrac{m + M}{m}$ **57.** $\dfrac{R^2 + Rr + r^2}{R + r}$

EXERCISE 17-2
Multiplying and Dividing Algebraic Fractions

1. $\dfrac{2ax}{3by}$ **3.** $\dfrac{20}{x}$ **5.** $\dfrac{2a}{c}$

7. $\dfrac{ab}{a^2 - b^2}$ **9.** $\dfrac{2}{a}$ **11.** $\dfrac{3a + 2b}{2(2a + 3b)}$

13. $\dfrac{x + 2}{x - 2}$ **15.** $\dfrac{35ay}{18bx}$ **17.** $\dfrac{3x^2}{2y}$

19. $2x^2$ **21.** $\dfrac{(x + 2)(x - 1)}{6}$ **23.** $\dfrac{a + 5}{3(6 - a)}$

25. $\dfrac{4}{3}$ **27.** $\dfrac{3(x + y)}{x - y}$ **29.** $\dfrac{2c}{b}$

31. $\dfrac{2}{3}$ **33.** $\dfrac{2(2x - 1)}{x^2 - 2x + 4}$ **35.** $\dfrac{(x + 1)^2}{(x - 2)^2}$

37. $\dfrac{3x + 1}{4(x - 3)}$ **39.** $\dfrac{5x^2y^2}{a^4}$ **41.** $\dfrac{5xy}{12}$

43. $\dfrac{3x - 1}{x - 1}$ **45.** $\dfrac{2(x - 1)(x - 2)}{x^2 + x + 1}$ **47.** $\dfrac{2(x + 1)}{2x + 1}$

49. $\dfrac{3t + 1}{2t^3}$ **51.** $\dfrac{3m}{\pi r^2 h}$

EXERCISE 17-3
Adding and Subtracting Algebraic Fractions

1. $\dfrac{7}{y}$ **3.** $\dfrac{7}{ab}$ **5.** $\dfrac{8x - 5y}{a^2b^3}$

7. $\dfrac{4a}{x - y}$ **9.** $\dfrac{3}{a - 2}$ **11.** $c + 1$

13. $x - 3$ **15.** $\dfrac{1}{n - m}$ **17.** $\dfrac{2(a + b)}{a - b}$

19. $\dfrac{x + y}{xy}$ **21.** $\dfrac{7a - 5}{a^2}$ **23.** $\dfrac{7a}{6x}$

25. $\dfrac{3x - 18y + 8z}{12}$ **27.** $\dfrac{yz - xz + xy}{xyz}$ **29.** $\dfrac{12a^2b^2 - 20a + 9b}{24a^2b^2}$

31. $-\dfrac{x + 5y}{(x + y)(x - y)}$ **33.** $\dfrac{2p^2 + 7}{(p + 2)(p - 3)}$ **35.** $-\dfrac{13b + 19}{(b + 3)(b - 2)(b + 1)}$

37. $\dfrac{1}{x - 1}$ **39.** $\dfrac{6 + 5x}{2x(x - 1)}$ **41.** $\dfrac{a^2 + 1}{a}$

43. $\dfrac{c^2 - 2c + 3}{c - 2}$ **45.** $\dfrac{-3x^3 + 2x^2 + 1}{x^3}$ **47.** $\dfrac{-x^2 - 7x + 4}{x(3x - 2)}$

49. $\dfrac{3\pi\ell^3 - 12c\ell^2 + \pi c^3}{3\pi\ell^3}$ **51.** $\dfrac{R_1R_2 + R_1R_3 + R_2R_3}{R_2 + R_3}$

EXERCISE 17-4
Complex Fractions

1. $\dfrac{2a}{b}$ **3.** $\dfrac{xz}{y} \neq \dfrac{x}{yz}$ **5.** $\dfrac{ad}{bc}$

7. $\dfrac{x}{2}$ **9.** $\dfrac{b}{4}$ **11.** $\dfrac{1}{3y-2}$

13. $\dfrac{3a}{4(3a+2)}$ **15.** $\dfrac{b}{b-a}$ **17.** $\dfrac{2a+1}{2a(a+4)}$

19. $\dfrac{x+2y}{x-2y}$ **21.** $\dfrac{R_1 R_2}{R_1 + R_2}$ **23.** $\dfrac{C_1 C_2 C_3}{C_2 C_3 + C_1 C_3 + C_1 C_2}$

CHAPTER 17
Review Exercises

1. $\dfrac{x+6}{x-6}$ **3.** $-a-2b$ **5.** $\dfrac{y^2+xy+x^2}{y-x}$

7. $\dfrac{6(x+5)}{x(x+3)}$ **9.** $-\dfrac{4}{n^2+2n+4}$ **11.** $1+a$

13. $\dfrac{2y}{y-x}$ **15.** $\dfrac{17x}{24}$ **17.** $\dfrac{17-x^2}{(x+4)(x-4)}$

19. $\dfrac{a-15}{5}$ **21.** $\dfrac{-6p^2+17p+15}{p(p-3)}$ **23.** $-\dfrac{x}{3}$

25. $\dfrac{x}{x^2+x+1}$ **27.** $\dfrac{t\lambda - xT}{T\lambda}$

CHAPTER 18   Fractional Equations

EXERCISE 18-1
Solution of Fractional Equations

1. 18   **3.** 5   **5.** 3   **7.** 60

9. 18   **11.** $-\dfrac{29}{8}$ **13.** -1 **15.** 9

17. 2   **19.** 14   **21.** 8   **23.** 3

25. 2   **27.** -4 **29.** no solution   **31.** no solution

33. $\dfrac{5}{2}$ **35.** no solution   **37.** no solution   **39.** 5.55

41. 0.386

EXERCISE 18-2
Literal Equations and Formulas

1. $x = ac - ab$ **3.** $x = \dfrac{bc}{a}$ **5.** $x = \dfrac{ab}{bc+ac}$

7. $x = ac - b$ **9.** $x = \dfrac{bc}{a-b}$ **11.** $a = \dfrac{2A}{b}$

13. $g = \dfrac{4\pi^2 L}{T^2}$ **15.** $e = IR - E$ **17.** $R_1 = \dfrac{RR_2 R_3}{R_2 R_3 - RR_3 - RR_2}$

19. $r = \dfrac{eR}{E-e}$ **21.** $T_1 = \dfrac{P_1 V_1 T_2}{P_2 V_2}$ **23.** $g = \dfrac{2vt - 2s}{t^2}$

25. $a = \dfrac{2A - bh}{h}$ **27.** 33 mg   **29.** 17.5 ohms

31. 59.9 cm   **33.** 7.58 μF

EXERCISE 18-3
Applications

1. 93.0 cm $\times$ 124 cm   **3.** 96.0 m, 48.0 m, 40.0 m   **5.** \$324

7. 2.49 h   **9.** 17.9 days   **11.** 6.95 h

13. 23.7 mi/h   **15.** 45.3 mi/h, 4.08 h   **17.** no, all or no hits

CHAPTER 18
Review Exercises

1. 4.5   **3.** −12   **5.** 10

7. no solution   **9.** $M_1 = \dfrac{Fd^2}{kM_2}$ **11.** $C = \dfrac{5F - 160}{9}$

13. 18 in. × 27 in.   **15.** 10.9 days   **17.** 3.36 h

19. $T = 2, W = 3, O = 7, S = 5, E = 6, V = 1, N = 9$

CHAPTER 19   Systems of Linear Equations

EXERCISE 19-1
Graphical Method

1. yes   **3.** no   **5.** yes

7. $(-4, 4)$ **9.** $(-1, -2)$ **11.** $(9, 3)$

13. $(-2, 1)$ **15.** $(-1, 3)$ **17.** $(-5.3, 5.7)$

19. all (x, y) such that $x - 2y = 6$ **21.** $(12, 7)$

23. $(0.3, 0.2)$ **25.** no solution   **27.** $(0.8, -0.5)$

29. $(3, -1)$ **31.** all (x, y) such that $2x - 5y = -2$

33. $(2.1, -0.2)$ **35.** no solution   **37.** $(2.5, 1.3)$

39. $(0, 1)$ **41.** $(3, 2)$ **43.** no solution

45. $(1.75, 0.39)$ **47.** $(-0.12, 1.31)$

EXERCISE 19-2
Addition Method

1. $(5, 2)$ **3.** $\left(2, -\dfrac{1}{3}\right)$ **5.** $(-1, 3)$

7. $\left(\dfrac{5}{2}, \dfrac{1}{2}\right)$ **9.** $(-2, 3)$ **11.** $(0.90, -0.38)$

13. no solution   **15.** all (x, y) such that $2x - y = 3$

17. $(2, 3)$ **19.** $\left(2, -\dfrac{1}{2}\right)$ **21.** no solution

23. $(5.04, 2.17)$ **25.** $(0.427, 1.45)$ **27.** $(2, 3)$

29. $\left(\dfrac{7}{5a}, \dfrac{1}{5b}\right)$ **31.** $\left(\dfrac{ce - bf}{ae - bd}, \dfrac{af - cd}{ae - bd}\right)$

EXERCISE 19-3
Substitution Method

1. $(5, 4)$ **3.** $(3, -1)$ **5.** $(4, 1)$

7. $\left(\dfrac{2}{11}, -\dfrac{7}{11}\right)$ **9.** $(0, 3)$ **11.** $\left(-\dfrac{14}{5}, -\dfrac{16}{5}\right)$

13. $(2, -1)$ **15.** all (x, y) such that $3x + 2y = -1$

17. no solution   **19.** $\left(\dfrac{2}{3}, \dfrac{7}{3}\right)$ **21.** $\left(\dfrac{6a + 2b}{3}, \dfrac{3a + 4b}{3}\right)$

23. $\left(\dfrac{1}{3}, \dfrac{1}{4}\right)$ **25.** $(12, 8)$ **27.** $(1, 1)$

29. $(-2, 1)$ **31.** $(3, -2)$ **33.** $(4, 3)$

EXERCISE 19-4
Applications of Systems
of Two Equations

1. 9.00 ft, 12.0 ft   **3.** TV: $350, stereo: $370

5. 72.0 mi by car, 288 mi by train   **7.** 2.00 min, 2160 ft

9. 0.471 mi/h   **11.** airplane: 837 mi/h, wind: 93.0 mi/h

13. 5.50 ft/s, 3.50 ft/s²   **15.** 200 kg of 63%, 800 kg of 78%

17. 40.0 cc of 75%, 110 cc of water   **19.** 180 ℓ

21. $3600 at 6.0%, $4400 at 10.5%   **23.** $2400 at 9.0%, $1800 at 12.0%

25. carpenter 18 days, helper 36 days   **27.** faster 4, slower 3

1. $(4, 2, 0)$ **3.** $(3, 2, 1)$ **5.** no solution **7.** $(2, 2, -1)$

9. $(1, 0, -1)$ **11.** $(2, -1, 3)$ **13.** $\left(\dfrac{1}{2}, \dfrac{1}{2}, \dfrac{1}{4}\right)$

15. 70 mℓ of 20%, 10 mℓ of 30%, 20 mℓ of 40%

17. 105 V, 65 V, 170 V

19. $1000 at 8%, $2000 at 9%, $5000 at 12%

21. 3 oz of A, 6 oz of B, 1 oz of C

1. 5 **3.** -1 **5.** 23.21

7. -8.1162 **9.** $(4, -1)$ **11.** $\left(\dfrac{1}{14}, -\dfrac{31}{14}\right)$

13. $\left(\dfrac{16}{11}, \dfrac{67}{11}\right)$ **15.** $(4, 0)$ **17.** $(4, -1)$

19. $(4.83, 1.09)$ **21.** $(0.824, 1.22)$ **23.** 34 ohms, 53 ohms

25. 24 lb at 95¢, 8 lb at 75¢ **27.** 4.5 oz of A, 5 oz of B

1. 19 **3.** -3 **5.** -29.6

7. $(3, 2, -1)$ **9.** $(-5.4, -4.6, -3.0)$ **11.** $(5, 6, 7)$

13. $(10, 15, 20)$ **15.** $(4, -3, 3)$ **17.** 1440 ohms, 340 ohms, 820 ohms

1. no **3.** $(-1, 0)$ **5.** no solution **7.** $(3, -1)$

9. $(34, 25)$ **11.** all (x, y) such that $y = 5 - 2x$

13. $(4, 5)$ **15.** no solution **17.** $(0, 2)$ **19.** 14

21. $(-2, 2)$ **23.** all (x, y) such that $5x - 2y = 4$

25. no solution **27.** $(6, 0)$ **29.** 50 V, 36 V

31. 30 acres at $550, 50 acres at $650 **33.** 40 lb, 8 lb

35. 8 lb at 70¢, 24 lb at $1.50 **37.** $(3, -4, 2)$

39. 33.951 **41.** 150 cc of 10%, 450 cc of 30%, 400 cc of 60%

43. 86.0 lb of zinc, 0.512 lb of tin, 4.62 lb of lead, 2.25 lb of nickel, 9.05 lb of manganese

CHAPTER 20 Exponents and Radicals

1. $\sqrt{x}$ **3.** $\sqrt[3]{y}$ **5.** $\dfrac{1}{\sqrt{a}}$

7. $\sqrt{x - y}$ **9.** $\left(\sqrt[3]{-c}\right)^2 = \sqrt[3]{(-c)^2}$ **11.** $-\left(\sqrt[3]{c}\right)^2 = -\sqrt[3]{c^2}$

13. $x^{1/2}$ **15.** $y^{1/4}$ **17.** $b^{2/3}$

19. $y^{-1/2}$ **21.** $(x + y)^{1/2}$ **23.** 4

25. 4 **27.** -3 **29.** 1

31. 0.1 **33.** 125 **35.** $\dfrac{1}{9}$

37. $\dfrac{1}{16}$ **39.** 16 **41.** $-\dfrac{1}{27}$

43. 27 **45.** 4 **47.** 0.111

49. -64 **51.** 2.24 **53.** -1.43

55. 4.33	**57.** 0.397	**59.** undefined	
61. $a^{7/12}$	**63.** $b^{3/7}$	**65.** $2a$	
67. $x^{1/2}$	**69.** $\dfrac{x^{3/2}}{y^3}$	**71.** $\dfrac{1}{a^{1/3}}$	
73. $x^{3/2}$	**75.** $a^{2/3}$	**77.** $a^{10/3}$	
79. $\sqrt[3]{x^2} + \sqrt[3]{y^2} = \sqrt[3]{a^2}$		**81.** 660 bacteria	

EXERCISE 20-2
Simplification of Radicals

1. $2\sqrt{3}$ **3.** $6\sqrt{2}$ **5.** $2\sqrt[3]{3}$ **7.** $2\sqrt[5]{2}$

9. $3x^2$ **11.** $2b^2c\sqrt{6b}$ **13.** $3x\sqrt[3]{x}$ **15.** $2a^2\sqrt[3]{2a}$

17. $3x\sqrt[4]{x}$ **19.** $2x\sqrt[6]{x}$ **21.** $\dfrac{\sqrt{5}}{3}$ **23.** $\dfrac{2\sqrt{3}}{3}$

25. $\dfrac{2\sqrt{5}}{5}$ **27.** $\dfrac{\sqrt[3]{18}}{3}$ **29.** $\dfrac{\sqrt{7}}{2x}$ **31.** $\dfrac{\sqrt[3]{2y}}{3x}$

33. $\dfrac{-3\sqrt{5x}}{5x}$ **35.** $\dfrac{\sqrt{3x}}{3x}$ **37.** $\dfrac{\sqrt{30y}}{6}$ **39.** $\dfrac{\sqrt[3]{10x^2}}{5x}$

41. $\dfrac{\sqrt{7ab}}{14b^2}$ **43.** $\dfrac{x\sqrt{2xy}}{10y}$ **45.** $\dfrac{y^2\sqrt{6x}}{3x^2}$ **47.** $\dfrac{2ac\sqrt{3bc}}{3b}$

49. $\dfrac{\sqrt{\pi A}}{\pi}$ **51.** $\dfrac{\sqrt{kgW}}{2\pi W}$ **53.** (a) $x\sqrt{x^2+1}$ (b) $a+b$

EXERCISE 20-3
Adding and Subtracting Radicals

1. $4\sqrt{3}$ **3.** $4\sqrt[3]{2}$ **5.** $3\sqrt{2}$

7. $18\sqrt{2} - \sqrt{3}$ **9.** $15\sqrt{6} - 38\sqrt{5}$ **11.** $10 + 2\sqrt{2} + \sqrt[4]{8}$

13. $\dfrac{3\sqrt{2}}{4}$ **15.** $\sqrt{x}$ **17.** $3\sqrt{2x} - 3\sqrt{3x}$

19. $3\sqrt{2x} + 12\sqrt{5x}$ **21.** $11\sqrt{10x} + 110\sqrt{x}$ **23.** 0

25. $\dfrac{3\sqrt{10x}}{2}$ **27.** $\dfrac{\sqrt[3]{4y^2}}{4y}$ **29.** $\dfrac{26}{5}$

EXERCISE 20-4
Multiplying Radicals

1. $\sqrt{15}$ **3.** $-\sqrt{2x}$ **5.** $\sqrt[3]{6x^2}$

7. $3y$ **9.** $12\sqrt{6}$ **11.** $30x$

13. $2\sqrt[3]{3}$ **15.** $45a$ **17.** $10a\sqrt{2}$

19. $18\sqrt{10x}$ **21.** $6x^2y\sqrt{10}$ **23.** $24x\sqrt{3}$

25. $-120\sqrt{3}$ **27.** $5\sqrt{3} - 2\sqrt{6}$ **29.** $24\sqrt{x} + 18x\sqrt{2}$

31. $8\sqrt{3} - 12 + 6\sqrt{2}$ **33.** $\sqrt{2ab} - a\sqrt{b}$ **35.** $6x^2 + x\sqrt{3} - 3$

37. $8 - \sqrt{10}$ **39.** $2x - 3\sqrt{3x} + 3$ **41.** $5 - 4a^2$

43. $x - y$ **45.** $2y + 2\sqrt{6y} + 3$

47. $\sqrt{6} + 3\sqrt{3} - 2\sqrt{10} - 6\sqrt{5}$

49. (a) $\sqrt[6]{5400}$ (b) $\sqrt[12]{5000}$

EXERCISE 20-5
Dividing Radicals

1. $\sqrt{3}$ **3.** $\sqrt{2x}$ **5.** $\sqrt{7b}$

7. $\dfrac{5\sqrt{3}}{3}$ **9.** $\dfrac{x\sqrt{14}}{14}$ **11.** $\sqrt[3]{7y}$

13. $1 + 2\sqrt{6}$ **15.** $\dfrac{2\sqrt{x}}{x} - 1$ **17.** $\dfrac{\sqrt{2}}{6} + \dfrac{2}{3}$

19. $6 + 3\sqrt{3}$ **21.** $\dfrac{4\sqrt{6} - 3}{29}$ **23.** $\dfrac{3 + 4\sqrt{x} + x}{1 - x}$

25. $6 - \sqrt{35}$ **27.** $\dfrac{2a + 5\sqrt{2a} - 6}{a - 18}$ **29.** $\sqrt{3} - \sqrt{6} + \sqrt{2} - 2$

31. $\dfrac{6 - 5\sqrt{x} + x}{4 - x}$ **33.** $\dfrac{4 - 3\sqrt{2} - \sqrt{10}}{6}$ **35.** $\dfrac{\sqrt{mv} - v}{m - v}$

37. $\dfrac{(\sqrt{d_1} - \sqrt{d_2})^2}{d_1 - d_2}$

EXERCISE 20-6
Radical Equations

1. 4	**3.** 16	**5.** no real solution	**7.** 27
9. -2	**11.** 9	**13.** no real solution	**15.** 16
17. 11	**19.** no real solution	**21.** 4	**23.** 9
25. 33	**27.** 3	**29.** -2	**31.** -8
33. 4	**35.** 4	**37.** no real solution	**39.** 1

41. 5   **43.** 5.0   **45.** $k = \dfrac{\omega^2 W}{g}$ **47.** 2.23

CHAPTER 20
Review Exercises

1. $\sqrt[5]{x}$ **3.** $\dfrac{1}{\sqrt[3]{b}}$ **5.** $(3y)^{1/2}$

7. $5^{-1/3}a$ **9.** 2   **11.** $\dfrac{1}{7}$

13. 16   **15.** 1.57   **17.** $x^{1/8}$

19. $x^{3/5}$ **21.** $a^{5/12}$ **23.** 449 units

25. $20\sqrt{2y}$ **27.** $2\sqrt[4]{2}$ **29.** $3y^2\sqrt{2xy}$

31. $4x^2\sqrt[4]{x}$ **33.** $\dfrac{\sqrt{11}}{5a^3}$ **35.** $\dfrac{\sqrt{6}}{2}$

37. $\dfrac{\sqrt{30x}}{6x}$ **39.** $\dfrac{y\sqrt{6xy}}{24x}$ **41.** $-4\sqrt{2}$

43. $-\sqrt[3]{2}$ **45.** $93 + 3\sqrt[3]{3}$ **47.** 18

49. $4\sqrt{6} - 5\sqrt{2}$ **51.** $30 - 12\sqrt{6}$ **53.** $\sqrt{3}$

55. $\dfrac{6\sqrt{10} - 5}{5}$ **57.** $\dfrac{16 - 7\sqrt{10}}{18}$ **59.** $(4x - 3y)\sqrt[3]{2}$

61. $2\sqrt{7x} - \sqrt{6x}$ **63.** $7\sqrt{3x} - 3\sqrt{7x}$ **65.** $\dfrac{\sqrt{15x}}{10}$

67. $\sqrt[3]{12x^2}$ **69.** $8x^2y\sqrt{15}$ **71.** $8x\sqrt{5} - 60\sqrt{x}$

73. $3x + 7\sqrt{xy} + 2y$ **75.** $\sqrt{2y}$ **77.** $\sqrt[3]{11y}$

79. $\dfrac{6 + 5\sqrt{a} + a}{4 - a}$ **81.** 4   **83.** no real solution

85. 17   **87.** 3   **89.** 64

91. $\dfrac{2\sqrt{\pi A}}{\pi}$ **93.** $W = \dfrac{kg}{4\pi^2 f^2}$ **95.** 21.8 cm, 41.2 cm, 32°, 58°

CHAPTER 21   Quadratic Equations

EXERCISE 21-1
Introduction to
Quadratic Equations

1. a, c, f **3.** $2x^2 - 3x - 4 = 0$ **5.** $4x^2 - 7x = 0$

7. $5x^2 - 2x + 18 = 0$ **9.** $3x^2 - 5x - 3 = 0$ **11.** 2, 3

13. $-\dfrac{1}{3}, 1$ **15.** ±5 **17.** $\pm2\sqrt{3}$

19. 0   **21.** $\pm\dfrac{\sqrt{10}}{2}$ **23.** no real solution

25. $\dfrac{\sqrt{em}}{m}$ **27.** $\dfrac{\sqrt{3\pi hV}}{\pi h}$ **29.** $\dfrac{\sqrt{kI}}{I}$

31. $\dfrac{\sqrt{KFM_1M_2}}{F}$ **33.** 0.994 cm **35.** 115 V

37. 189 cm/s

EXERCISE 21-2
Solution by Factoring

1. $-3, 3$ **3.** $-2, 2$ **5.** $-\dfrac{8}{3}, \dfrac{8}{3}$ **7.** 0, 2

9. $-2, 0$ **11.** 0, 3 **13.** $-2, 3$ **15.** $-3, 1$

17. -1 **19.** $-\dfrac{5}{2}, -2$ **21.** $-3, 5$ **23.** 3

25. 1 m **27.** **(a)** 4 s, 8 s **(b)** 1 s, 11 s

EXERCISE 21-3
Quadratic Factoring

1. $-3 \pm \sqrt{11} = -6.32, 0.317$ **3.** $\dfrac{2 \pm 2\sqrt{7}}{3} = -1.10, 2.43$

5. $\dfrac{-5 \pm \sqrt{29}}{2} = -5.19, 0.193$ **7.** no real solution

9. $\dfrac{7 \pm \sqrt{17}}{4} = 0.719, 2.78$ **11.** no real solution

13. $-1 \pm \sqrt{2} = -2.41, 0.414$ **15.** $\pm\sqrt{3} = -1.73, 1.73$

17. no real solution **19.** $\dfrac{3 \pm \sqrt{6}}{3} = 0.184, 1.82$

21. $-2, 5$ **23.** $-4 \pm 2\sqrt{2} = -6.83, -1.17$

25. $\dfrac{-3 \pm \sqrt{15}}{2} = -3.44, 0.436$ **27.** $-2, \dfrac{1}{3}$

29. $\dfrac{2 \pm \sqrt{13}}{3} = -0.535, 1.87$ **31.** $\dfrac{4 \pm \sqrt{10}}{3} = 0.279, 2.39$

33. $-2.08, 0.907$ **35.** $-5.19, 1.87$

37. 0.226 ms, 1.11 ms **39.** 1.79, 4.04

41. $\dfrac{-\pi h \pm \sqrt{\pi^2 h^2 + 2\pi A}}{2\pi}$ **43.** $\dfrac{-1 \pm \sqrt{1 + 8S}}{2}$

EXERCISE 21-4
Applications of
Quadratic Equations

1. 22.0 ft $\times$ 66.0 ft **3.** 3.82 m $\times$ 7.32 m **5.** 3.00 in.

7. 2.07 m **9.** 2.54 s **11.** 7.25 s

13. 5.38 s **15.** 2.23 s **17.** 30.0 s

19. **(a)** 20.0 s **(b)** 7.95 s, 32.0 s **(c)** 40.0 s

EXERCISE 21-5
Applications Involving
Rates

1. 600 mi/h **3.** 15 mi/h

5. 662 mi/h **7.** 57.0 mi/h, 46.6 mi/h

9. 56.1 mi/h, 46.4 mi/h **11.** 80 members

13. 9 m **15.** 362 ohms, 138 ohms

EXERCISE 21-6
Quadratic Functions

1. **(a)** **(b)** 19.6 cm²

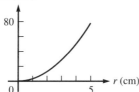

3. **(a)** **(b)** 27.0 m

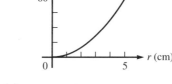

5. **(a)**

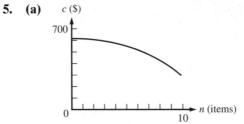

(b) $511

7. **(a)**

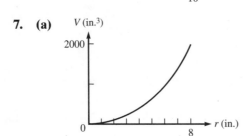

(b) 6.31 in.³

9. **(a)**

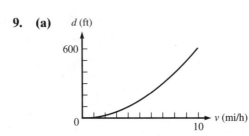

(b) 50 mi/h

11. 0.5, 0.6   **13.** −0.535, 1.87

CHAPTER 21
Review Exercises

1. (b), (d), (e)

3. $2x^2 - 3 = 0$

5. −2

7. $x = \pm 8$

9. 0

11. $\pm \sqrt{5}$

13. no real solution

15. $\pm \dfrac{\sqrt{2gs}}{g}$

17. −2, 0

19. $-\dfrac{1}{2}, 1$

21. −2

23. $-\dfrac{2}{3}, \dfrac{1}{2}$

25. $\pm \dfrac{3}{4}$

27. $-\dfrac{2}{3}, \dfrac{3}{2}$

29. −1, 12

31. $\dfrac{7 \pm \sqrt{41}}{2} = 0.298, 6.70$

33. $2 \pm \sqrt{3} = 0.268, 3.73$

35. no real solution

37. $\dfrac{2}{3}, 5$

39. −8, 2

41. 0.240, 2.32

43. $\pi r^2 + \pi sr - A = 0, r = \dfrac{-\pi s \pm \sqrt{\pi^2 s^2 + 4\pi A}}{2\pi}$

45. 3.29 cm

47. 4.00 s

49. **(a)** 1.30 s, 4.69 s

(b) 5.99 s

51. 60¢

EXERCISE 22-1
Trigonometric Functions
of Any Angle

1. $\sin \theta = -0.8$, $\cos \theta = 0.6$, $\tan \theta = -1.333$, $\csc \theta = -1.25$, $\sec \theta = 1.667$, $\cot \theta = -0.75$

3. $\sin \theta = 0.8321$, $\cos \theta = -0.5547$, $\tan \theta = -1.5$, $\csc \theta = 1.202$, $\sec \theta = -1.803$, $\cot \theta = -0.6667$

5. $\sin \theta = -1$, $\cos \theta = 0$, $\tan \theta = $ undefined, $\csc \theta = -1$, $\sec \theta = $ undefined, $\cot \theta = 0$

7. $\sin \theta = 0$, $\cos \theta = 1$, $\tan \theta = 0$, $\csc \theta = $ undefined, $\sec \theta = 1$, $\cot \theta = $ undefined

9. III, $\sin \theta = -0.2949$, $\cos \theta = -0.9555$, $\tan \theta = 0.3086$, $\csc \theta = -3.391$, $\sec \theta = -1.047$, $\cot \theta = 3.24$

11. IV, $\sin \theta = -0.4625$, $\cos \theta = 0.8866$, $\tan \theta = -0.5216$, $\csc \theta = -2.162$, $\sec \theta = 1.128$, $\cot \theta = -1.917$

13. III, $\sin \theta = -0.6922$, $\cos \theta = -0.7217$, $\tan \theta = 0.9592$, $\csc \theta = -1.445$, $\sec \theta = -1.386$, $\cot \theta = 1.043$

15. sine positive, cosine negative, tangent negative

17. sine negative, cosine positive, tangent negative

19. I, II   21. II, III   23. II, IV   25. IV   27. III

29. sine positive, cosine negative, tangent negative

31. sine negative, cosine positive, tangent negative

33. sine negative, cosine negative, tangent positive

35. sine positive, cosine negative, tangent negative

37. 0   39. undefined   41. 0   43. -1

45. -1 47. undefined   49. -0.9063 51. -10.02

53. 0.08194   55. -1.221 57. -1.428 59. -0.08716

61. 0.6841   63. 0.4833   65. -0.2748

67. $\sin \theta = -\dfrac{4}{5}$, $\cos \theta = \dfrac{3}{5}$, $\tan \theta = -\dfrac{4}{3}$

69. $\sin \theta = -\dfrac{15}{17}$, $\cos \theta = -\dfrac{8}{17}$, $\tan \theta = \dfrac{15}{8}$

EXERCISE 22-2
Finding Angles from
Trigonometric Functions

1. $524°$, $-196°$ 3. $688°$, $-32°$ 5. $67°$ 7. $75°$

9. $39°$, $\sin 219° = -\sin 39°$, $\cos 219° = -\cos 39°$, $\tan 219° = \tan 39°$

11. $53°$, $\sin 307° = -\sin 53°$, $\cos 307° = \cos 53°$, $\tan 307° = -\tan 53°$

13. $60°$, $300°$ 15. $7.1°$, $172.9°$ 17. $45°$, $225°$

19. $210°$, $330°$ 21. $113.1°$, $293.1°$ 23. $68.0°$, $292.0°$

25. $29.7°$, $209.7°$ 27. $197.9°$, $342.1°$ 29. $90°$, $270°$

31. $0°$, $180°$

EXERCISE 22-3
Law of Sines

1. 89.6 in.   3. 12.2 cm   5. 107 in.   7. 47.1 cm

9. $a = 10.6$ cm, $b = 11.0$ cm, $C = 24.0°$

11. $a = 94.7$ cm, $c = 147$ cm, $B = 28.9°$

13. 172 ft to B, 168 ft to A

15. 25.9 cm   17. 4140 ft   19. 250 ft   21. 805 mi

EXERCISE 22-4
Ambiguous Case

1. $c = 6.53$ cm, $B = 41.9°$, $C = 12.1°$

3. $a = 216$ in., $A = 92.5°$, $C = 51.5°$; or $a = 57.6$ in., $A = 15.5°$, $C = 128.5°$

5. $b = 16.0$ cm, $A = 23.2°$, $B = 4.8°$

7. $c = 81.6$ cm, $A = 37.0°$, $C = 68.0°$

9. $b = 30.1$ in., $B = 122.8°$, $C = 40.5°$; or $b = 14.5$ in., $B = 23.8°$, $C = 139.5°$

11. $a = 4.08$ m, $A = 54.4°$, $B = 67.8°$; or $a = 0.871$ m, $A = 10.0°$, $B = 112.2°$

13. $a = 9.64$ in., $A = 64.1°$, $B = 50.4°$

15. $58.5°$ **17.** 52.7 km

EXERCISE 22-5
Law of Cosines

1. $a = 562$ cm, $B = 49.1°$, $C = 67.7°$

3. $c = 103$ in., $A = 82.2°$, $B = 62.0°$

5. $b = 80.0$ cm, $A = 24.4°$, $C = 102.9°$

7. $c = 175$ mm, $A = 18.3°$, $B = 19.7°$

9. $A = 57.2°$, $B = 62.7°$, $C = 60.1°$

11. $A = 27.4°$, $B = 114.4°$, $C = 38.2°$

13. $A = 170.7°$, $B = 3.9°$, $C = 5.5°$

15. 38.8 m   **17.** $44.2°$ **19.** 9.39 km

21. 104 km   **23.** $97.2°$ **25.** 171 ft, 235 ft

CHAPTER 22
Review Exercises

1. $\sin \theta = -0.7809$, $\cos \theta = -0.6247$, $\tan \theta = 1.25$, $\csc \theta = -1.281$, $\sec \theta = -1.601$, $\cot \theta = 0.8$

3. sine positive, cosine negative, tangent negative

5. **(a)** 0   **(b)** 0   **(c)** 0   **(d)** undefined

7. $\theta_R = 68°$, $\sin 248° = -\sin 68°$, $\cos 248° = -\cos 68°$, $\tan 248° = \tan 68°$

9. $a = 5.16$ cm, $b = 3.30$ cm, $B = 36.5°$

11. $b = 82.8$ mm, $A = 34.2°$, $C = 63.8°$

13. $A = 54.5°$, $B = 75.8°$, $C = 49.7°$

15. $PR = 40.1$ ft, $QR = 23.0$ ft

17. 7.73 mi, $76.9°$ west of south

19. $\sin \theta = \dfrac{7}{25}$, $\cos \theta = -\dfrac{24}{25}$

21. 11.21 cm

CHAPTER 23   Vectors

EXERCISE 23-1
Vectors and
Components

1. scalar   **3.** vector   **5.** vector

7. vector   **9.** vector   **11.** scalar

13. 21 km east, 13 km north   **15.** 5.0 km/h west, 4.6 km/h south

17. $N = 52$ kg, $T = 14$ kg   **19.** $V_x = 643$, $V_y = 450$

21. $V_x = -95.5$, $V_y = -95.5$ **23.** $\mathbf{V} = (-2.02, 4.32)$

25. $\mathbf{V} = (0.0324, -0.121)$

EXERCISE 23-2
Applications of Vector
Components

1. 343 mi east, 252 mi north

3. 1940 m horizontal, distance $= 2080$ m

5. 265 mi/h north, 189 mi/h west

7. $N = 119$ kg, $T = 39.7$ kg

9. $F_x = 301$ N, $F_y = 343$ N, 343 N maximum

11. 40.1 mi/h

EXERCISE 23-3	**1.** $\mathbf{R} = 143 \angle 31°$	**3.** $\mathbf{R} = 0.62 \angle 112°$	**5.** $\mathbf{R} = 6.0 \angle 113°$
Graphical Addition of Vectors	**7.** $\mathbf{R} = 99 \angle 329°$	**9.** $\mathbf{R} = 117 \angle 56°$	**11.** $\mathbf{R} = 128 \angle 14°$
	13. $\mathbf{R} = 8.7 \angle 115°$	**15.** 17.2 km, 47° west of north	**17.** 40 lb, 306°

EXERCISE 23-4
Addition of Perpendicular Vectors

1. $686 \angle 31.5°$ **3.** $11.7 \angle 50.0°$
5. $52.7 \angle 18.2°$ **7.** $10.3 \angle 29.8°$
9. $19.9 \angle 51.5°$ **11.** $9.04 \angle 215°$
13. 27.3 mi, 59.1° west of north
15. (a) 27.0 mi   (b) 3.0 mi   (c) 19.2 mi
17. 238 km/h, 81.4° west of north
19. (a) 240 km/h   (b) 410 km/h   (c) 336 km/h, 75.3°
21. 477 km/h, 5.41° east of north   **23.** 912 lb, 69.1°
25. 30.0 lb, 31.5° down from left

EXERCISE 23-5
Vector Addition by Components

1. $128 \angle 53.7°$ **3.** $542 \angle 170°$ **5.** $6.92 \angle 44.7°$
7. $0.607 \angle 3.69°$ **9.** $4.06 \angle 9.49°$ **11.** 573 km, 70.2° west of south
13. 3560 lb, 53.5°

EXERCISE 23-6
Vector Addition Using the Law of Sines and Cosines

1. $812 \angle 50.9°$ **3.** $5.00 \angle 11.0°$ **5.** $428 \angle 59.8°$
7. $101 \angle 11.4°$ **9.** $969 \angle 68.2°$ **11.** $3.94 \angle 192°$
13. 56.0 mi, 176° west of south
15. 2570 lb, 77.4° west of north
17. 259 mi/h, 17.4° east of north

CHAPTER 23
Review Exercises

1. $V_x = -43.0, V_y = 61.4$ **3.** 118 km west, 249 km south
5. 8.74 m/min, 0.921 min   **7.** 4.69 tons
9. $3.56 \angle 104.2°$ **11.** $891 \angle 18.2°$
13. 17.0 km, 40.2° west of north   **15.** 26.6° upstream, 13.9 min
17. $3.56 \angle 104.2°$ **19.** $85.5 \angle 62.1°$
21. $1290 \angle 35.5°$

CHAPTER 24   Applications of Radian Measure

EXERCISE 24-1
Radian Measure

1. 2π **3.** $\dfrac{\pi}{2}$ **5.** $\dfrac{\pi}{3}$ **7.** $\dfrac{\pi}{12}$
9. $\dfrac{3\pi}{4}$ **11.** $\dfrac{2\pi}{5}$ **13.** $\dfrac{17\pi}{9}$ **15.** $\dfrac{7\pi}{15}$
17. $\dfrac{3\pi}{2}$ **19.** $\dfrac{10\pi}{9}$ **21.** 30°   **23.** 45°
25. 120°   **27.** 144°   **29.** 306°   **31.** 132°
33. 0.8660   **35.** 0.5774   **37.** −0.3827   **39.** −0.5878
41. 1   **43.** 0.2588

EXERCISE 24-2
Arc Length and Area of a Sector

1. 1.38 rad   **3.** 251°   **5.** 5.06 m
7. 986 mm   **9.** 6.13 cm   **11.** 2.64 in.
13. 15.1 cm²   **15.** 168 mm²   **17.** 2820 mi
19. 91.9 ft   **21.** 569 cm²   **23.** a piece of 12-in. pizza
25. $A = \dfrac{rs}{2}$, 40.8 cm²

CHAPTER 25   Graphs of Trigonometric Functions

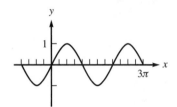
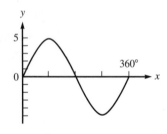
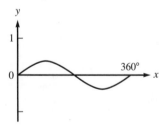
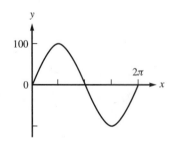

1. 0, 0.7, 1, 0.7, 0, −0.7, −1, −0.7, 0, 1, 0, −1, 0, 1, 0, −1, 0

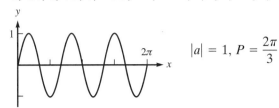

$|a| = 1, P = \dfrac{2\pi}{3}$

3. $y = \sin 4x$ **5.** $y = \sin 2x$ **7.** $|a| = 1, P = 720°$

9. $|a| = 1, P = \dfrac{\pi}{3}$

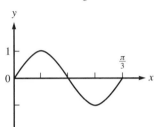

11. $y = 3 \sin 3x$ **13.** $y = 5 \sin 4x$ **15.** $|a| = 2, P = 120°$
17. $|a| = 3, P = 1080°$ **19.** $|a| = 3, P = \pi$

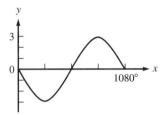

21. $|a| = \dfrac{1}{2}, P = 2$ **23.** $|a| = 6$ in., $P = 0.5$ s

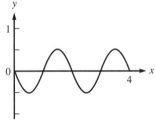

1. 0, −0.5, −0.9, −1, −0.9, −0.5, 0, 0.5, 0.9, 1, 0.9, 0.5, 0, −0.5, −0.9, −1

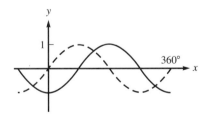

$P = 360°, |a| = 1, -\dfrac{c}{b} = 90°$

3. $y = \sin \left(x + \dfrac{\pi}{6} \right)$ **5.** $y = 2 \sin (x - 60°)$ **7.** $y = 2 \sin (2x - \pi)$

9. $|a| = 1, P = 360°, -\dfrac{c}{b} = -45°$ **11.** $|a| = 1, P = \pi, -\dfrac{c}{b} = \dfrac{\pi}{6}$

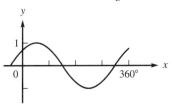

725

13. $|a| = 2, P = 2\pi, -\dfrac{c}{b} = \dfrac{\pi}{6}$ **15.** $|a| = 0.5, P = 120°, -\dfrac{c}{b} = -60°$

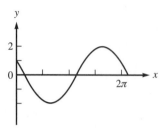

17. $|a| = 1, P = 2\pi, -\dfrac{c}{b} = -1$ **19.** $|a| = 60$ amps, $P = \dfrac{1}{60}$ s, $-\dfrac{c}{b} = -\dfrac{1}{240}$ s

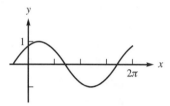

EXERCISE 25-4
Graph of
$y = a \cos (bx + c)$

1. $1, 0.9, 0.5, 0, -0.5, -0.9, -1, -0.9, -0.5, 0, 0.5, 0.9, 1$

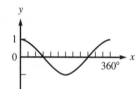

3. $y = 2 \cos x$ **5.** $y = -\cos 2x$

7. $y = 2 \cos 3x$ **9.** $y = 3 \cos \left(\dfrac{x}{2} - \dfrac{\pi}{2}\right)$ **11.** $|a| = 3, P = 360°$

13. $|a| = 1, P = 120°$ **15.** $|a| = 5, P = \dfrac{\pi}{2}, -\dfrac{c}{b} = 0$

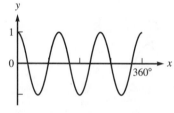

17. $|a| = 1, P = 2\pi, -\dfrac{c}{b} = -\pi$ **19.** $|a| = 1, P = 180°, -\dfrac{c}{b} = 90°$

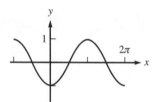

21. $|a| = 2, P = \dfrac{2\pi}{3}, -\dfrac{c}{b} = -\dfrac{\pi}{3}$ **23.** $|a| = 4.25, P = 141°, -\dfrac{c}{b} = 128°$

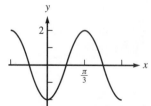

25. $|a| = 1.5$ amp, $P = \dfrac{1}{60}$ s

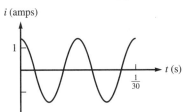

EXERCISE 25-5
Graph of
$y = a \tan (bx + c)$

1. error, -1.7, -1, -0.6, 0, 0.6, 1, 1.7, error, -1, 0, 1, error, -1, 0

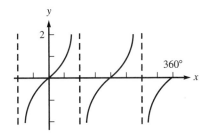

3. $P = 180°$

5. $P = 60°$ **7.** $P = \pi$

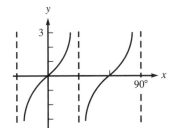

9. $P = 2\pi$ **11.** $P = \pi$, $-\dfrac{c}{b} = \dfrac{\pi}{2}$

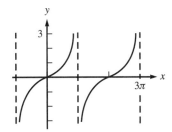

13. $P = 180°$, $-\dfrac{c}{b} = -45°$ **15.** $P = \dfrac{\pi}{2}$, $-\dfrac{c}{b} = -\dfrac{\pi}{2}$

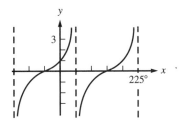

1. $|a| = 4, P = 360°, -\dfrac{c}{b} = 0°$

3. $|a| = 1, P = 360°, -\dfrac{c}{b} = 45°$

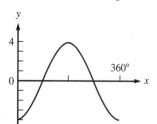

5. $|a| = 1, P = 180°, -\dfrac{c}{b} = 0°$

7. $|a| = 1, P = 180°, -\dfrac{c}{b} = 45°$

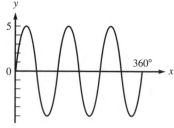

9. $|a| = 5, P = 120°, -\dfrac{c}{b} = 0°$

11. $P = 4\pi, -\dfrac{c}{b} = 0$

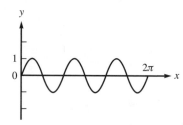

13. $|a| = 1, P = \dfrac{2\pi}{3}, -\dfrac{c}{b} = 0$

15. $|a| = 5, P = 2\pi, -\dfrac{c}{b} = 0$

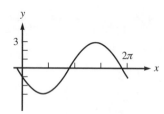

17. $|a| = 3, P = 2\pi, -\dfrac{c}{b} = -\dfrac{\pi}{8}$

19. $|a| = 2, P = \dfrac{2\pi}{3}, -\dfrac{c}{b} = 0$

21. $y = 2\sin(2x - 180°)$

23. $P = \dfrac{\pi}{5}$

EXERCISE 26-1
Ratio

1. $5:4, \dfrac{5}{4}$ **3.** $\dfrac{2}{5}$ **5.** $\dfrac{2}{1}$ **7.** $\dfrac{1}{4}$

9. $\dfrac{1}{3}$ **11.** $\dfrac{4}{1}$ **13.** $\dfrac{3}{4}$ **15.** $\dfrac{1}{1000}$

17. $\dfrac{3}{8}$ **19.** $\dfrac{5}{1}$ **21.** 3.14 **23.** 40.5

25. 93.8% **27.** 0.808

EXERCISE 26-2
Proportion

1. $\dfrac{x}{y} = \dfrac{6}{9}$ **3.** $\dfrac{p}{20} = \dfrac{q}{5}$ **5.** $\dfrac{12}{30} = \dfrac{y}{5}$

7. $2 \cdot 9 = 3 \cdot 6$ **9.** $12 \cdot 5 = 20 \cdot 3$ **11.** $12 \cdot 5 = 30 \cdot 2$

13. 5 **15.** 3 **17.** $-6, 6$

19. $\dfrac{30}{7}$ **21.** $-12, 12$ **23.** 12

25. 14 **27.** 6 ft, 10 ft **29.** $200, $450

31. 1.38 gal **33.** 3.63 kg **35.** 101 mi

37. 24.3 lb **39.** 3.88 h **41.** 10.7 tons

43. 7.12 cm, 17.1 cm **45.** 23.0 ft

EXERCISE 26-3
Direct Variation

1. 3 **3.** 9.80

5. $y = kx$ **7.** $w = kn$

9. $P = kr$ **11.** $y = kx^3$

13. $s = kt^2$ **15.** $k = 4, y = 4x$

17. $k = 0.2$ lb/in., $w = 0.2L$ **19.** $y = 15.75x, y = 39.4$

21. $k = 3.14, A = 3.14r^2$ **23.** $k = 64.0 \dfrac{\text{W}}{\text{A}^2}, P = 64.0I^2$

25. 1130 cm^3 **27.** 398 cal

29. $\dfrac{1}{100}$ s **31.** 1.60 s

EXERCISE 26-4
Joint, Inverse, and Combined Variation

1. 2 **3.** 2 **5.** $p = kqr$

7. $E = kLT$ **9.** 60 **11.** $z = \dfrac{k}{y}$

13. $y = \dfrac{k}{z^2}$ **15.** $X_c = \dfrac{k}{C}$ **17.** $y = \dfrac{kz}{x}$

19. $I = \dfrac{kV}{R}$ **21.** $x = 3yz, x = 60$ **23.** $y = \dfrac{16x}{z}, y = 10$

25. $315 **27.** 30.3 min **29.** 2.0 h

31. 5.76×10^{-5} dynes **33.** 1.5 **35.** 321 vibrations/s

CHAPTER 26
Review Exercises

1. 18 **3.** $-12, 12$

5. $\dfrac{3}{5}$ **7.** $\dfrac{1}{5}$

9. $p = \dfrac{k}{r^2}$ **11.** $R = kT$

13. 30 ft, 18 ft **15.** 115

17. $\dfrac{40}{1}$

19. $292.25

21. 1.1 h

23. $E = 1.632I$, $E = 33.4$ V

25. $w = 1.091r^3$, $w = 3.68$ lb

27. $v = 0.801\sqrt[3]{P}$, $v = 13.6$ knots

CHAPTER 27 Exponential Functions

**EXERCISE 27-1
Introduction to
Exponential Functions**

1. An exponential function is a function in which the independent variable is in the exponent.

3. 5^{x+y} **5.** 8^{yz} **7.** $\dfrac{5^x}{3^x}$ **9.** $\dfrac{1}{2^x}$ **11.** 4^x

**EXERCISE 27-2
Graph of the
Exponential Function**

1.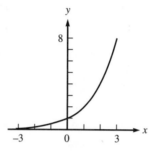

3.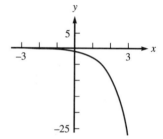

5.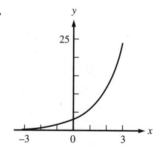

7.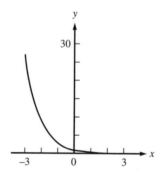

9.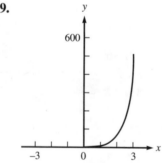

11.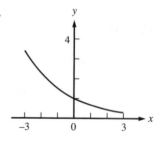

13.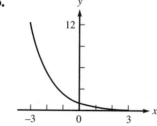

15.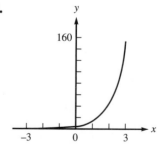

17. **(a)** $n = 2500 \cdot 3^t$ **(b)** 610,000 bacteria
(c) *n* (bacteria)   **(d)** 120,000 bacteria

n (bacteria)

600,000

0   5   *t* (h)

19. **(a)** 1.2 mg   **(b)** *y* (mg)
(c) 7.6 h

y (mg)

5.0

0   12   *t* (h)

21. **(a)** 40.6 kg   **(b)** $w = 40.6(2^t)$
(c) 230 kg   **(d)** *w* (kg)
(e) 3.6 h

w (kg)

600

0   4   *t* (h)

EXERCISE 27-3
Exponential Growth

1. 7.39   **3.** 3.90   **5.** 0.791   **7.** 445   **9.** 9.80
11. **(a)** \$836.20   **(b)** \$842.41   **(c)** \$843.00
13. **(a)** \$2.59   **(b)** \$2.71   **(c)** \$2.72   **(d)** \$2.72
15. \$3271.01

17. 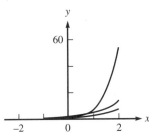   **19.**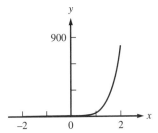

21. 278 million
23. **(a)** 165 lb   **(b)** $y = 45.0e^{0.130t}$

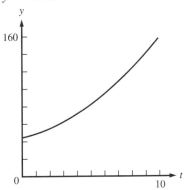

(c) 115 lb   **(d)** 6.14 h

1.

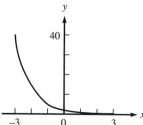

3.

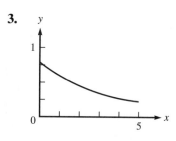

5. **(a)** $y = 175e^{-0.0602t}$ **(b)**
   (c) 111 kg

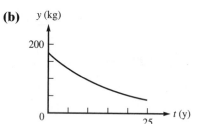

7. 2130 rev/min   **9.** 41.3%   **11.** 78.2 mA

13. **(a)** 997°C   **(b)**

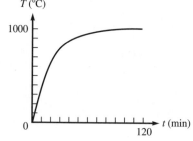

1.

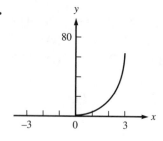

3.

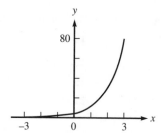

5.

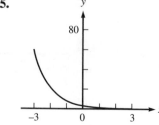

7.

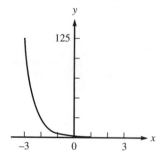

9. $\dfrac{1}{4^x}$ or 4^{-x} **11.** $\dfrac{1}{2^x}$

13. **(a)** $709.51   **(b)** $717.68   **(c)** $718.43   **(d)** $718.46

15. **(a)** 59.1 tons   **(b)** $y = 12.3e^{0.157t}$

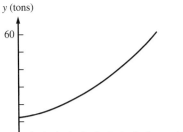

17. **(a)** 405 lb   **(b)** 18.5 lb
19. **(a)** 98.8%   **(b)** $y = 100e^{(-1.24\times10^{-4})t}$

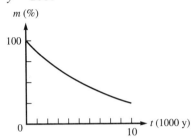

21.

T (°F)

$$T = 142e^{-0.1t} + 70$$

CHAPTER 28   Logarithms

EXERCISE 28-1
Introduction
to Logarithms

1. $\log_2 8 = 3$

3. $\log_5 625 = 4$

5. $\log_{10} 1000000 = 6$

7. $\log_{1/3}\left(\dfrac{1}{81}\right) = 4$

9. $\log_{78} 1 = 0$

11. $\log_6 9.73 = 1.27$

13. $64, \log_4 64 = 3$

15. $100000, \log_{10} 100000 = 5$

17. $1, \log_9 1 = 0$

19. $\dfrac{1}{27}, \log_3\left(\dfrac{1}{27}\right) = -3$

21. $26.7, \log_4 26.7 = 2.37$

23. $0.0561, \log_{2.4} 0.0561 = -3.29$

25. $\log_5 120 = x$

27. $\log_9 a = 5$

29. $\log_4 q = -3$

31. $\log_e 45 = x$

33. $7^3 = 343$ **35.** $10^3 = 1000$ **37.** $9^{1/2} = 3$ **39.** $7^1 = 7$

41. $5^{-2} = \dfrac{1}{25}$ **43.** $4^{6.59} = y$ **45.** $b^y = x$ **47.** $e^q = 5$

49. 3   **51.** 3   **53.** 0   **55.** 4

57. 4	**59.** 1	**61.** 10	**63.** 125

65. $\dfrac{1}{216}$ **67.** 4   **69.** 27   **71.** e

EXERCISE 28-2
Common and Natural Logarithms

1. $\log 100 = 2$ **3.** $\log 0.001 = -3$

5. $100000,\ \log 100000 = 5$ **7.** $0.0001,\ \log 0.0001 = -4$

9. $10^{1.19} = 15.5$ **11.** $10^{0.19} = 1.55$

13. $0.730,\ 10^{0.730} = 5.37$ **15.** $1.87,\ 10^{1.87} = 74.5$

17. $-0.341,\ 10^{-0.341} = 0.456$ **19.** $x = 10^3 = 1000$

21. $x = 10^{1.56} = 36.3$ **23.** $x = 10^{-1.62} = 0.0240$

25. $\ln 20.1 = 3$ **27.** $11.2,\ \ln 11.2 = 2.42$

29. $1,\ \ln 1 = 0$ **31.** $4.04,\ e^{4.04} = 56.7$

33. $-3.74,\ e^{-3.74} = 0.0238$ **35.** $x = e^{3.54} = 34.5$

37. $e^{-0.00950} = 0.991$ **39.** $-2.3, -1.6, -0.7, 0, 0.7, 1.6, 2.3$

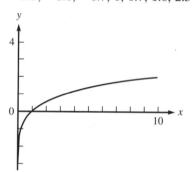

EXERCISE 28-3
Properties of Logarithms

1. $\log_3 7 + \log_3 x$ **3.** $\log_5 x - \log_5 4$

5. $\ln 3 + \ln x + \ln y$ **7.** $\log_2 5 + \log_2 x - \log_2 3$

9. $\log_b 3 - \log_b 4 - \log_b a$ **11.** $\log_4 5a$

13. $\log_b \dfrac{2}{y}$ **15.** $\log_2 xyz$ **17.** $\log_3 \dfrac{a}{5c}$

19. $3 \log_b x$ **21.** $x \log_3 2$ **23.** $\dfrac{1}{2} \log_b x$

25. $\dfrac{1}{3} \log_2 x$ **27.** $2 \log_b x + 3 \log_b y$ **29.** $\log_2 x^5$

31. $\ln x^3 y^4$ **33.** $\log_b \sqrt{x}$ **35.** $\ln \sqrt[4]{2}$

37. $\log_2 \sqrt{x}\ \sqrt[3]{y}$ **39.** $\log_5 \sqrt[3]{x^2}$ **41.** 0

43. 1   **45.** 9   **47.** 5.3

49. $\dfrac{1}{5}$

EXERCISE 28-4
Solution of Exponential Equations Using Logarithms

1. 1.40   **3.** -0.237 **5.** 27.2

7. -0.322 **9.** -0.223 **11.** -0.306

13. 0.143   **15.** 0.405   **17.** 0.693

19. 56.2   **21.** 0.416   **23.** 0.0516

25. 0.00255   **27.** 2.75   **29.** $t = \dfrac{\ln \dfrac{N}{100}}{\ln 2}$

31. $t = \dfrac{\ln \dfrac{1000}{P}}{0.12}$ **33.** 6.29 h   **35.** 13.1 y

37. 37.5 y	**39.** 8.90 h	**41.** 0.00692 s	
43. 19.6 min	**45.** **(a)** 1.54%	**(b)** 2663	**47.** 23.3 min

EXERCISE 28-5
Applications of
Logarithmic Equations

1. 20 dB   **3.** 10^{-4} W/cm^2 **5.** 3.40   **7.** 1.58×10^{-11} moles/ℓ
9. 4970 ft   **11.** 11.1 y   **13.** 20.0

CHAPTER 28
Review Exercises

1. $\log_4 64 = 3$ **3.** $\log 4 = x$
5. $\log_2 a + \log_2 b$ **7.** $\log_7 5 + \log_7 a - \log_7 b$
9. $\dfrac{1}{3} \ln x$ **11.** $2 \log_3 a + 3 \log_3 b$
13. 10000, $\log 10000 = 4$ **15.** 0.292, $\ln 0.292 = -1.23$
17. 3   **19.** 3
21. 81   **23.** 5
25. -2 **27.** 0
29. 0.230, $10^{0.230} = 1.70$ **31.** 1.50, $e^{1.50} = 4.50$
33. 2.15   **35.** 2.41
37. 0.560   **39.** -5.39
41. 61.6, $\log_6 61.6 = 2.30$ **43.** 0.000224, $\log 0.000224 = -3.65$
45. $x = 10^7 = 10000000$ **47.** $x = e^{-1.52} = 0.219$
49. 125, $\log_5 125 = 3$ **51.** 3, $\log_{1/3} 3 = -1$
53. $\log_3 5x$ **55.** $\log (3x)^4$
57. $\log_4 \sqrt{a}$ **59.** $\ln \dfrac{x^3}{y^2}$
61. $3^4 = 81$ **63.** $e^{1.95} = 7$
65. 7   **67.** $\dfrac{1}{3}$
69. 4.98 days   **71.** 1.27 min
73. 0.0602 s   **75.** 10^{-8} moles/ℓ

CHAPTER 29 Complex Numbers

EXERCISE 29-1
Imaginary Numbers

1. $j\sqrt{5}$ **3.** $j30$ **5.** $j2\sqrt{2}$ **7.** $-j2\sqrt{3}$
9. $j2\sqrt{7}$ **11.** $-j25\sqrt{2}$ **13.** $j\dfrac{1}{3}$ **15.** $-j0.4$

17. j **19.** $-j$ **21.** $-j21$ **23.** $j18$
25. $-j28$ **27.** $j5\sqrt{2}$ **29.** $-j\sqrt{30}$ **31.** $j40$
33. $j30$ **35.** -16 **37.** $\sqrt{14}$ **39.** -3
41. $6\sqrt{6}$ **43.** -6 **45.** -7 **47.** 6

49. -48 **51.** 7   **53.** $-\dfrac{3}{2}$ **55.** $-j4$

57. $j\dfrac{5}{2}$ **59.** $-j\dfrac{1}{2}$ **61.** $j\dfrac{4}{5}$

EXERCISE 29-2
Complex Numbers

1. $3 + j\sqrt{2}$ **3.** $-5 + j2\sqrt{3}$
5. $-7 + j0$ **7.** $0 + j2\sqrt{6}$
9. $3 - j8$ **11.** $-5 + j\sqrt{2}$

13. -2	**15.** $-j4$
17. $-j5, j5$	**19.** $-j4\sqrt{3}, j4\sqrt{3}$
21. $-2 - j3, -2 + j3$	**23.** $4 - j2, 4 + j2$
25. $-\dfrac{1}{2} - j\dfrac{\sqrt{5}}{2}, -\dfrac{1}{2} + j\dfrac{\sqrt{5}}{2}$	

EXERCISE 29-3
Operations with
Complex Numbers

1. $-3 + j3$	**3.** $-1 + j3$	**5.** 8
7. $-10 + j11$	**9.** $j4$	**11.** $-3 + j3$
13. $-3 - j10$	**15.** $4 - j11$	**17.** $1 + j2$
19. $-15 + j10$	**21.** $28 + j42$	**23.** $3 + j2$
25. $20 + j30$	**27.** $13 - j11$	**29.** $11 + j23$
31. $-76 + j52$	**33.** 65	**35.** 61
37. $7 - j24$	**39.** $45 - j28$	**41.** $2 - j4$
43. $\dfrac{3}{2} - j2$	**45.** $-4 - j3$	**47.** $2 - j\dfrac{3}{2}$
49. $\dfrac{3}{2} + j\dfrac{3}{2}$	**51.** $-\dfrac{8}{13} - j\dfrac{12}{13}$	**53.** $\dfrac{7}{2} + j\dfrac{3}{2}$
55. $-\dfrac{2}{5} - j\dfrac{11}{5}$	**57.** $-\dfrac{27}{13} + j\dfrac{8}{13}$	

EXERCISE 29-4
Graph of a
Complex Number

1. to 15.

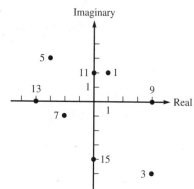

17. $A: 4 + j2$ $\quad$ $B: -2 + j3$ $\quad$ $C: -3 - j3$ $\quad$ $D: 3 - j4$	
$\quad\quad$ $E: 2$ $\quad\quad\quad\quad$ $F: j4$ $\quad\quad\quad\quad$ $G: -4$ $\quad\quad\quad\quad$ $H: -j$	
19. $5 + j5$ $\quad$ **21.** $-3 - j5$ $\quad$ **23.** -2 $\quad$ **25.** $4 - j5$	

EXERCISE 29-5
Polar and Trigonometric
Forms

1. $3.61 \; \underline{/56.3°}$	**3.** $4.47 \; \underline{/207°}$
5. $4.42 + j6.67$	**7.** $-315 - j450$
9. $1.41 (\cos 45° + j \sin 45°)$	**11.** $3.61 (\cos 326° + j \sin 326°)$
13. $8.06 (\cos 120° + j \sin 120°)$	**15.** $6.71 (\cos 207° + j \sin 207°)$
17. $2 \; \underline{/30°}, 1.73 + j$	**19.** $3.56 \; \underline{/135°}, -2.52 + j2.52$
21. $8.45 \; \underline{/297°}, 3.84 - j7.53$	**23.** $2.588 + j9.658$

CHAPTER 29
Review Exercises

1. $5.39 \; \underline{/112°}$	**3.** $7.07 \; \underline{/8.13°}$
5. $j15$	**7.** $-2 - j11$
9. 18	**11.** $-\dfrac{3}{2} + j2$
13. $1 - j12$	**15.** $-24 + j12$
17. $-7 - j24$	**19.** $\dfrac{1}{2} + j\dfrac{3}{2}$

21. $-j4\sqrt{2}$ **23.** $-j5$

25. -5 **27.** $\dfrac{29}{53} - j\dfrac{31}{53}$

29. $-1 - j6$ **31.** -2

33. $2.83\,(\cos 45° + j \sin 45°)$ **35.** $5.83\,(\cos 211° + j \sin 211°)$

37. $-j6\sqrt{2},\ j6\sqrt{2}$ **39.** $4.25\ \underline{/123°},\ -2.31 + j3.56$

Index to Applications

Index to Writing Questions

Index to Team Projects

General Index

Right angle, 238
Right cone, 283
Right cylinder, 276
Right prism, 270
Right triangle(s), 245, 251, 290–317
    solution of, 297–304
Roots:
    by calculator, 16, 63, 114
    complex, 661
    of decimal numbers, 63
    of denominate numbers, 82
    of equations, 230
    logarithm of, 641
    principal, 114
    of a product, 427
    of quadratics (see Quadratic equation)
    of a quotient, 430
    of signed numbers, 114
    of whole numbers, 16
Rotation, 551
Rounding, 54
Row of a determinant, 410
Rules of signs:
    for addition and subtraction, 105, 106
    for division, 111, 153
    for fractions, 346
    for multiplication, 108, 139

Scalar, 517
Scale drawings, 587
Scalene triangle, 245
Scientific notation, 18, 65, 116
    on the calculator, 66, 117
Secant of an angle, 311
Second of arc, 236
Secondary diagonal, 410, 415
Second order determinants, 410
Sector, 266
    area of, 548
Segment, line, 234
Segment of a circle, 266
Semicircle, 265
Sets of equations (see Systems of linear equations)
Side of an equation, 163
Signed number, 102
Significant digits, 53
Signs of fractions (see Rules of signs)
Signs of the trigonometric functions, 488
Signs, rules of (see Rules of signs)
Similar figures, 592
Similar fractions, 28, 39, 356
Similar radicals, 433
Simultaneous equations (see Systems of linear equations)
Sine of an angle, 292, 298, 487
Sine curve, 557
Sines, law of, 500
SI prefixes, 78
    table of, 78
SI units, 75, 78
Slant height of cone, pyramid, 281, 283
Slope, 211
Slope-intercept form, 211
Solid geometry, 270–289
Special products, 144, 318
Specific gravity, 589
Sphere, 286
Square, 261
Square root (see Roots)
Squares, difference of two, 144, 324
Standard position for angles, 290, 487
Straight line, 208, 234, 385
Substitution:
    into equations, 164
    into formulas, 165, 375
    into functions, 202

Substitution method, 396
Subtraction (see Addition and subtraction)
Sum or difference of two cubes, 341
Supplementary angles, 239
Symbols of grouping (see Grouping)
Systems of linear equations, 384–421
    by determinants, 410–18

Tables:
    of conversion factors, 686
    of facts and formulas, 677–85
    of metric prefixes, 78
Tangent:
    of an angle, 292, 298, 487
    to a circle, 266
Tangential component, 519
Terminal position, 235, 290, 487
Term of an expression, 124
Terms, like, 127
Third-order determinants, 415
Trace, 228
Transversal, 240
Trapezoid, 258
Triangle(s), 245–57
    area of, 247
    oblique (see Oblique triangles)
    right, 245, 251
    types of, 245
Trigonometric form of a complex number, 672
Trigonometric function(s), 291, 487
    by calculator, 293
    graphs of, 557–84
Trigonometric ratio (see Trigonometric function)
Trinomial, 142, 329
    factoring of, 329–40
    perfect square, 147, 339
    quadratic, 145, 329
    test for factorability, 335

Uniform motion, 379, 475
    circular, 551
Unit, imaginary, 653
Unit circle, 557
Units of measure, 75–87
    angular, 235, 543
    conversion of, 75
    table of, 686
Unknown, 179

Value:
    absolute, 27, 103
    of a determinant, 411, 415
Variables, 123
    dependent and independent, 197
Variation:
    combined, 605
    direct, 596
    functional, 596
    inverse, 603
    joint, 602
Vector(s), 517–42
    represented by complex numbers, 668
Vector sum (see Addition and subtraction, of vectors)
Velocity, angular, 551
    linear, 553
Vertex:
    of an angle, 235
    of a cone, 283
    of a parabola, 217, 479
    of a polygon, 244, 245
Vertical angles, 239